AF386611

Coastal Research Library

41

Series Editors

Charles W. Finkl, *Department of Geosciences, Florida Atlantic University, Boca Raton, USA*

Chris Makowski, *Coastal Education & Research Foundation, Inc. (CERF), Charlotte, NC, USA*

The aim of the Coastal Research Library (CRL) is to disseminate accurate information to both the public and research specialists around the world on all aspects of coastal issues in an effort to maintain or improve the quality of our planet's shoreline resources. By covering the entire field of coastal research [e.g., geology, biology, ecology, geomorphology, physical geography, climate change, coastal evolution and shoreline change, littoral oceanography, hydrodynamics and AI modeling, sea-level change, beaches and dunes, coastal hydraulics, erosion, waves, flooding, environmental (resource) management (law), policy and planning, coastal and marine engineering, GIS and remote sensing, socio-economic topics, coastal hazards and extreme storms, anthropogenic change, etc.], the scope of the CRL is broad with a unique cross-disciplinary nature that encompasses all subjects relevant to natural and engineered environments (freshwater, brackish, or marine) and the protection/management of their resources in the vicinity of coastlines around the world. This book series offers current cutting-edge coastal research from the world's leading scientists and is relevant to professionals, academics, and stakeholders, as well as the general public.

The CRL tends to focus on topics linked to complex multidisciplinary problems and contentious issues that surround the coastal environment and occur within the coastal zone. Welcome additions to the series can include: Contributed Volumes, Handbooks, Monographs, Proceedings, Professional Books, Textbooks, and others.

We invite you to submit your new proposals to CRL Series Editor: Chris Makowski, Ph.D. Email: cmakowski@cerf-jcr.com

Carlos Coelho · Caroline Hallin ·
Francisco Sancho · Paulo A. Silva

Editors

Coastal Dynamics 2025

Volume 1

Springer

Editors
Carlos Coelho 
Civil Engineering Department
University of Aveiro
Aveiro, Portugal

Caroline Hallin
Water Resources Engineering
Lund University
Lund, Sweden

Francisco Sancho
Hydraulics and Environment Department
National Laboratory for Civil Engineering
Lisbon, Portugal

Paulo A. Silva
Physics Department
University of Aveiro
Aveiro, Portugal

ISSN 2211-0577   ISSN 2211-0585 (electronic)
Coastal Research Library
ISBN 978-3-032-15472-9   ISBN 978-3-032-15473-6 (eBook)
https://doi.org/10.1007/978-3-032-15473-6

This work was supported by the Coastal Dynamics 2025 conference.

This Springer imprint is published by the registered company Springer Nature Switzerland AG
The registered company address is: Gewerbestrasse 11, 6330 Cham, Switzerland

If disposing of this product, please recycle the paper.

Foreword

The international *Coastal Dynamics* conference series is one of the most important meetings for scientists in the fields of nearshore dynamics and coastal evolution. The 1st edition was held in 1994 in Barcelona and envisioned to complement the *Coastal Sediments* conference series (since 1977) and the *International Conferences on Coastal Engineering* (since 1950). It emphasized the use of large-scale field experiments and laboratory facilities for understanding the underlying physics of the coastal processes and improving numerical modelling capabilities. Since then, embracing and expanding the initial aim and themes, following conferences were held in Gdansk, Poland (1995); Plymouth, UK (1997); Lund, Sweden (2001); Barcelona (2005); Tokyo, Japan (2009); Arcachon, France (2013); Helsingor, Denmark (2017); and Delft, Netherlands (2021).

The 2025 Coastal Dynamics conference, hosted by the University of Aveiro, Portugal, in April 7–11, 2025, was developed under the theme "Living with a Dynamic Coast". 379 abstracts were submitted and critically reviewed by the Scientific Committee. Scientists and experts from all over the world conducting research on the dynamics and changes of the coastal systems contributed with their insights and research results. Approaches included field (remote and *in situ*) observations, laboratory experiments, theoretical formulations, numerical simulations, and data-based learning methods. Recent research and applications concerning hydrodynamics of coastal waves and currents, interactions between wind, water, sediments and eco-systems, and morphology changes in different environments (with and without structures) such as sandy, rocky, and muddy coasts, inlets, and estuaries were presented. Emerging contributions of nature-based solutions and sediment-biota/plastics interactions were also presented.

The theme of the conference was highlighted by two Keynotes, delivered by Bregje van Wesenbeeck, a scientific director of Deltares (the Netherlands) and a senior expert in nature-based solutions, and Alec Torres-Freyermuth, a researcher in the Engineering Institute at the National Autonomous University of Mexico. Van Wesenbeeck presentation "Including nature and natural dynamics into management of coastal zones" addressed science and valorisation of working with nature and natural processes. Torres-Freyermuth presented "Linking beach morphodynamics and coastal resilience", highlighting that coastal resilience has become a key element for coastal management and for climate change adaptation plans. The two Keynotes were complemented by nearly 280 professional presentations in four concurrent oral sessions and poster sessions for three days. These sessions followed an initial day where three short courses were provided, on the topics of "Coastal Resilience Assessment – Transferring the principles of ecological resilience to coastal geomorphic system", "Incorporating Data Science and Climate in Coastal Engineering" and "Coastal Restoration Under Sea Level Rise". The last day of the Conference included three technical visits, to allow participants to contact with the dynamics of the coastal region of central Portugal, which is characterized by sandy beaches with extensive erosion problems, a Natura 2000 protected area ("Dunas de S. Jacinto Nature Reserve), important coastal protection structures and recent beach nourishment projects.

During the conference, the Coastal Award 2025 was attributed to Andrew Short, a coastal scientist specializing in beach morphodynamics and coastal processes and evolution. He has degrees from the University of Sydney, University of Hawaii and Louisiana State University and has worked on the coasts of North and South America, including north Alaska and Hawaii, Europe, Korea, New Zealand and the entire Australian coast. He is presently Honorary Professor at the University of Sydney and University of Wollongong, Visiting Scientist in the Coastal Oceanography Lab at the Universidade Santa Catarina, Brazil, Senior Coastal Scientist (part-time) with CoastalCOMS.com, and Chair of National Surfing Reserves (Australia) and serves on the NSW Coastal Panel. He has written 12 books and over 200 scientific publications, may dealing with rip currents and beach hazards. In collaboration with Surf Lifesaving Australia, he used his knowledge of surf zone processes to develop the Australian Beach Safety Program. The Organizing Committee of Coastal Dynamics 2025 joins conference participants and the extended research community in celebrating Andrew Short contributions towards advancing science and engineering for resilient coastal systems.

The Proceedings of *Coastal Dynamics 2025* contain the findings of the world's leading scientists and engineers in the area of coastal dynamics. The contents are a valuable source of up-to-date information on coastal processes and morphodynamics, ranging from basic research to case studies and lessons learned.

We would like to acknowledge all the sponsors and partners who have turned this conference possible: Rohde Nielsen; Port of Aveiro; Dravo S.A.; Teixeira Duarte—Engenharia e Construções, S.A.; Inersel; Dragus—Engenharia Hidráulica; Atlantic Land Consulting; Consulmar Projectistas e Consultores, Lda.; CPTS Underwater Services; R5 marine solutions; Geosurveys—geophysical consultants; APRH—Associação Portuguesa dos Recursos Hídricos; PIANC—The World Association for Waterborne Transport Infrastructure; OE—Ordem dos Engenheiros; EurOcean—The European Centre for Information in Marine Science and Technology; Applied Sciences—an Open Access by MDPI; Journal of Marine Science and Engineering—an open Access by MDPI.

Finally, we would like to thank the Local Organizing Committee, the International Steering Board of the Coastal Dynamics Conference series and the Technical Abstract Review Committee for supporting us in organizing the CD2025 conference session.

Carlos Coelho
Caroline Hallin
Francisco Sancho
Paulo Silva

About This Book

This open-access proceedings of Coastal Dynamics 2025 compile 231 manuscripts of the communications presented at the University of Aveiro, Portugal, between 7 and 11 April 2025. The works are divided in 15 chapters (2 volumes), in accordance with the topic of the sessions where they were presented. The conference theme was "Living with a Dynamic Coast". Most updated research on the dynamics and changes of the coastal systems include field (remote and in situ) observations, laboratory experiments, theoretical formulations, and numerical simulations. Recent research and applications concerning coastal waves and currents, interactions between wind, water, sediments and eco-systems, and morphology changes in different morphological environments (with and without structures) such as sandy, rocky, and muddy coasts, inlets, and estuaries are described in this compilation.

The extended list of highlights coming out of the conference included the following keywords: *machine learning algorithms; automatic techniques; neural networks, clustering, random forests, multivariate analysis; extensive time/space datasets; stochastic modelling/simulators; model-data hybrid approaches; climate versus seasonal drivers; dune evolution modelling; long-term trending/forecasting; high-resolution; young researchers; satellite/UAV/LIDAR imagery/data; sediment-vegetation-flow interaction under/above water; multi-disciplinary, across-scales processes; nature-based interventions; new field experiments; digital twins; adaptation pathways; climate change impacts; role of infragravity waves; process-based models; small/remote coastal environments; sea-level rise.*

Sponsors

Special Financial Support

Organization

Sponsors

Diamond

Platinum

Gold

Contents

x   Contents

Coastal Management and Risk Assessment

Coasts and Climate Change Effects and Responses

Coral and Nature-Based Reefs: Hydrodynamics and Morphodynamics

Dynamic Islands

Emerging Technology for Surfzone Observations

Estuarine and Coastal Lagoons' Processes

Beach and Dune Morphodynamics

Observations of Multidecadal Gravel Beach Dynamics from Space

Aikaterini Konstantinou[1]([envelope]), Tim Scott[1], Gerd Masselink[1], Christopher Stokes[2], and Bruno Castelle[3]

[1] University of Plymouth, Drake Circus, Plymouth 4 8AA, UK
`a.konstantinou@plymouth.ac.uk`
[2] Met Office, Fitzroy Road, Exeter EX1 3PB, UK
[3] UMR EPOC, CNRS, Université de Bordeaux, 33615 Pessac Cedex, France

Abstract. Gravel barrier systems are ubiquitous on mid- and high-latitude coasts and provide vital protection from coastal flooding and coastal erosion. They are highly dynamic systems that exhibit complex responses to hydrodynamic forcing over a range of timescales (hourly-monthly-decadal-centennial). Their dynamics differ greatly from those of sandy beaches yet have received considerably less attention in the literature, particularly at interannual to decadal scales. We use over four decades of satellite-derived shoreline (SDS) data to explore the long-term dynamic of 45 selected gravel systems around the United Kingdom and Ireland. Unlike most large-scale studies, we apply an SDS extraction methodology specifically tailored to gravel beaches to derive the long-term shoreline trends along 1554 shore-normal transects across our sites. Our findings indicate a great variability in shoreline trends, ranging from -4.73 m/year to + 10.5 m/year with the majority of transects (62%) remaining stable over the study period (1984 – 2023). Overall, 22% of transects showed statistically significant positive trends and 14% negative trends, resulting in an overall + 0.36 m/year mean shoreline trend. Large scale climatic forcing seems to have a stronger control on the evolution of gravel nesses with 29% and 26% of transects having an average correlation of -0.36 and -0.34 with North Atlantic Oscillation and Atlantic Oscillation respectively, while open beaches showed the weakest relationship with climate forcing. Importantly, climate forcing did not seem to be a key driver of change in the regions with the highest rates of progradation.

Keywords: satellite-derived shoreline · gravel beach · climate index · shoreline trend · interannual variability

1 Introduction

Gravel barrier systems are ubiquitous on mid- and high-latitude coasts. They are vital landforms that support unique coastal habitats and internationally important biodiversity, provide protection from coastal flooding and erosion, and enable long-term carbon storage. Gravel beaches possess inherent morphosedimentary characteristics arising from

© The Author(s) 2026
C. Coelho et al. (Eds.): CD 2025, CRL 41, pp. 3–9, 2026.
https://doi.org/10.1007/978-3-032-15473-6_1

their sedimentology (e.g., higher angle of repose, hydraulic conductivity, and bed roughness) that make them naturally resilient to wave action and sea level rise (SLR), effectively acting as natural wave barriers conferring coastal risks [1, 2]. As such, gravel systems represent critical natural capital [3] that is already under threat due to climate change impacts, particularly sea level rise and potential increases in storminess.

Gravel beaches occur as diverse landforms including cuspate forelands (or nesses), spits, embayed beaches, and open coastlines [4]. They encompass the full spectrum of coarse clastic environments (2mm to 256mm) and can comprise of pure gravels or mixed sand and gravel. They typically feature near-linear cross-shore beach profiles though some beaches feature composite profiles with distinct breaks in the beach profile reflecting significant changes in the cross-shore sediment distribution [5]. Gravel beaches are more dynamic than sandy beaches and tend to exhibit high alongshore and cross-shore variability controlled by a complex interplay between hydrodynamic processes, morphological changes, and the spatial heterogeneity of sediment characteristics [6, 7]. Importantly, the study of gravel beach dynamics has received considerably less attention than sandy beaches, particularly at interannual to decadal scales resulting in a knowledge gap in their morpho-sedimentary dynamics on these timescales. Most existing studies on gravel beach dynamics concentrate on specific areas, and their findings point to different key drivers of change including alongshore drift [8], storm action [3], sediment delivery [9], as well as geomorphological controls and anthropogenic influences [10]. Evidence suggests that at regional scales, climate indices associated with the regional wave climate, may be important drivers of shoreline evolution [10–13].

Climate patterns such as the Atlantic Oscillation (AO), influence wide regions and demonstrate pronounced decadal-scale cycles. Meanwhile, the morphological evolution of gravel beach systems displays significant alongshore morphological variability that often masks their longer-term signals. Consequently, the dynamics of these systems requires investigations at appropriate temporal and spatial scales. Advances in Earth Observation (EO) and image analysis technologies now offer the opportunity to study long-term shoreline change and its relationship with atmospheric patterns at the appropriate time and spatial scales, allowing the exploration of interannual shoreline variability and trends even in regions with limited satellite usability, such as the UK [14].

This work aims to evaluate the multi-annual to multi-decadal gravel beach dynamics of 45 sites around the UK and Ireland using satellite imagery. We investigate their long-term (decadal scale) evolution over the 40-year satellite data record. We further investigate the relationships between the observed shoreline variability and climate forcing indices to identify key drivers of change.

2   Methodology

We used CoastSat [15] to extract the instantaneous satellite-derived waterlines (SDW) from all available satellite images between 1984 and 2023 at 45 gravel beaches around Great Britain, Ireland and the Isle of Man (BIS) (**Fig. 1a**). Only sites where SDS could be derived with relative confidence were included and selected to provide a good representative distribution of the different system types around the coastline (embayed beaches, unconstrained (open) beaches, cuspate forelands (nesses) and spits), as well as cover areas with different climate controls on the inshore wave climate. Unlike previous large-scale studies [e.g., [12], we applied a site-specific approach for gravel beaches in meso- to macro-tidal environments. We considered the type of beach profile each beach typically exhibits (linear or composite) and applied a minimum water level threshold such that only waterlines on the reflective portion of the beach were included. We extracted instantaneous satellite-derive waterlines (SDW) along 1550 shore-normal transects across all sites, and applied a tidal correction following [14] to obtain the satellite-derived shorelines (SDS) using linear translation. The beach slope value used was statistically defined as the median slope for gravel beaches (median tan $\beta = 0.15$) as presented in [16]. A sensitivity analysis was performed to derive an estimate of the uncertainty associated with the selection of the beach slope and was combined with the measurement uncertainty reported by [14] using linear error theory. The overall uncertainty of the derived trends was conservatively estimated at ± 0.52 m/year. Long-term trends were derived at each transect to assess the long-term dynamics of each beach type and the influence of climate forcing on interannual shoreline variability.

Due to a lack of consistent winter SDS data, we address the relationship between shoreline change and climate forcing using the z-score normalised annual shoreline change along each transect. Comparisons were made with the winter-averaged (NDJFM) values of the three leading climatic indices controlling inshore wave conditions in the region as identified by [17]: the Atlantic Oscillation (AO), the North Atlantic Oscillation (NAO), and the Western Europe Pressure Anomaly (WEPA).

3 Results

3.1 Long-Term Shoreline Trends

The long-term (40 years) transect shoreline trends (t_T) across the study sites showed great variability ranging from -4.73 m/year to + 10.50 m/year with an overall average (statistically significant; $p \leq 0.05$) trend of + 0.36 m/year (Fig. 1b).

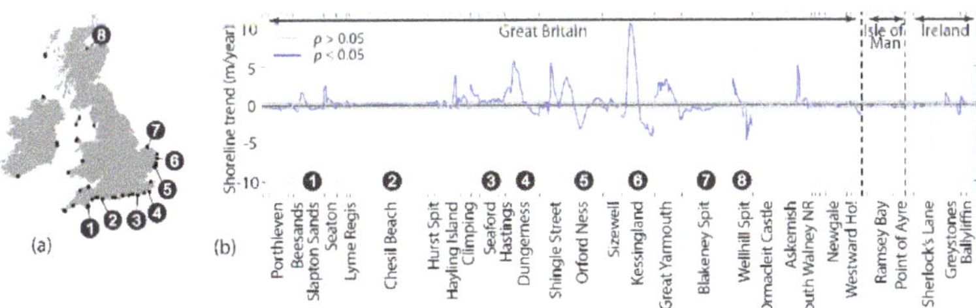

Fig. 1. (a): Map of the British Isles showing the location of the study sites (black dots) (http://www.eea.europa.eu/legal/copyright). The names and locations of example sites in (a) are indicated in (b). (b): Shoreline trends derived for all study sites. X-axis tick marks indicate the first transect for each site. Numbers in circles are included to show the location of selected sites in (a) and (b). Statistically significant trends are shown in darker colour and those with no statistical significance in fainter colour. The grey shaded area represents the uncertainty limit of 0.52 m/year.

The majority transects (961 out of 1554 transects or 62%) remained stable over the study period as they showed trends that where either not statistically significant at the 95% confidence level or were of absolute magnitude smaller than the uncertainty level of 0.52 m/year (Fig. 1b). Of the 574 transects that did exhibit significant trends, 40% showed negative (regressive) trends (mean: -1.42 m/year), and 60% showed positive (prograding) trends (mean: + 2.08 m/year). The mean trend across all transects was + 0.36 m/year ($p \leq 0.05$), below the uncertainty limit of $\pm$ 0.52 m/year. Nesses and spits exhibited the highest and lowest t_T (maximum: 10.5 m/year and 5.6 m/year; minimum: -4.73 m/year, -4.15 m/year respectively) (Fig. 2a-c), suggesting they are the most dynamic gravel features in the BIS, over the long term. Embayed and open beaches showed similar trends between them but much lower than those exhibited by nesses and spits (maximum: 2.69 m/year and 3.89 m/year; minimum: -0.79 m/year, -1.97 m/year respectively) (Fig. 2a-c).

3.2 Interannual Shoreline Variability and Climate Forcing

To address possible links of shoreline variability with climate forcing we compare (normalised, annual) shoreline change with the winter-averaged values of AO, NAO, and WEPA. Overall, 18% of transects showed statistically significant correlations at the 95% confidence level with leading teleconnection indices (Fig. 2). The AO and NAO seem to exert the strongest (but week to moderate) influence on shoreline change around the BIS (Table 1). Climate forcing seems to have stronger control on the evolution of nesses with 29% and 26% of transects having an average correlation of -0.36 and 0.34 with NAO and AO respectively. Open beaches were the least influenced with less than 15% of transects showing Statistically significant, but weak, correlations with AO, NAO, and WEPA (Table 1).

Table 1. Results summary table showing the number of transects for each beach type, the mean, 5th and 95th percentile transect trends (in brackets) where statistically significant at the 95% level. The mean correlation coefficient ($\overline{R_s}$) of shoreline change along transects with statistically significant correlations to the 95% level are included for AO, NAO, and WEPA with the percentage of transects with a statistically significant correlation (η_T) given in brackets.

Beach type	number of transects	Trend (m/year)	AO $\overline{R_s}$ (η_T)	NAO $\overline{R_s}$ (η_T)	WEPA $\overline{R_s}$ (η_T)
Embayed	341	0.36 (−0.79–2.69)	−0.33 (21)	−0.24 (17)	−0.10 (9)
Open	821	0.14 (−1.97–3.89)	−0.20 (14)	−0.15 (13)	−0.07 (10)
Nesses	266	0.74 (−4.15–10.5)	−0.34 (26)	−0.36 (29)	0.04 (2)
Spits	126	0.58 (−4.73–5.56)	−0.45 (14)	−0.24 (12)	0.30 (11)

(*continued*)

Table 1. (*continued*)

Beach type	number of transects	Trend (m/year)	AO $\overline{R_s}$ (η_T)	NAO $\overline{R_s}$ (η_T)	WEPA $\overline{R_s}$ (η_T)
Overall	1554	0.36 (−4.73–10.5)	−0.29 (18)	−0.24 (17)	−0.03 (8)

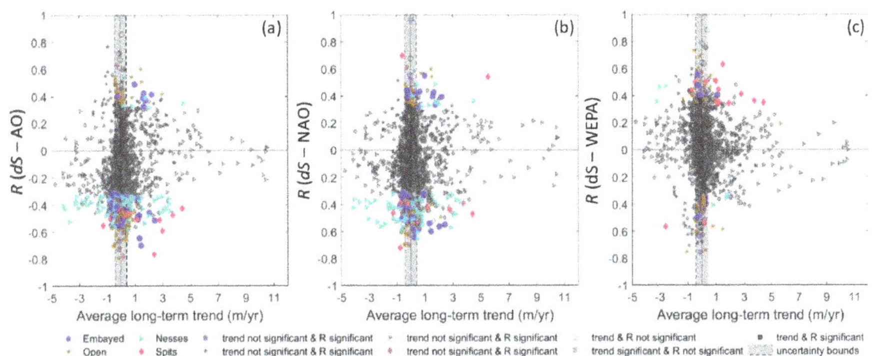

Fig. 2. Comparison of long-term transect trends (x-axis) with the correlation coefficient between shoreline change and (a) AO; (b) NAO; and (c) WEPA (y axis). The shaded areas correspond to the uncertainty limit of 0.52 m/year. Beach types are represented in different colour and shape. Points corresponding to both statistically significant trend and R are shown in solid colour.

4 Discussion

Our findings indicate that the majority of gravel beach transects in the BIS have remained stable over the last 40 years. Nesses (teal triangle in Fig. 2) and spits (red diamond in Fig. 2) are the most dynamic landforms associated with largest accretionary and transgressive trends. Figure 1b shows that in all cases, nesses and spits that showed the largest positive trends, also included transects with large negative trends (e.g., Dungeness, Orford Ness, Kessingland, South Walney Spit, Wellhill Spit). This highlights the distinctive migratory behaviour of these systems, characterized not by simple regression or transgression, but by a dynamic transformation in planform shape (e.g., Orford Ness, Kessingland, and Wellhill Spit) and/or realignment of the shoreline (Slapton, Climping, and Budleigh Salterton). As previously demonstrated [e.g., [18], these dramatic changes often result in insignificant net beach changes as observed at Slapton, Orford Ness, Climping, and Budleigh Salterton. Our results show an overall positive trend of + 0.36 m/year across all transects ($p \leq 0.05$) and suggest that chronic erosion, is a highly localized issue often associated with nesses, and was concentrated in a limited number of hotspots, mostly located in the southeast and east coasts of England (Fig. 1b).

Correlation of shoreline change with dominant climate indices was generally poor across all beach types with less than 20% of all transects showing any statistically significant correlation (Table 1). Interestingly, we found that transects with the largest positive trends showed no correlation with atmospheric forcing though that was not necessarily the case for transects showing the largest negative trends (Fig. 2a-c). Notably, all transects showing substantially transgressive behaviour and statistically significant correlations with climate forcing belonged to either ness or spit systems. The absence of strong links with climate forcing in areas of considerable progradation highlight the importance of other key factors affecting the sediment budget in these systems, such as geomorphological controls and anthropogenic interference.

References

1. Masselink G, van Heteren S (2014) Response of wave-dominated and mixed-energy barriers to storms. Mar Geol 352:321–347
2. Stéphan P (2019) et al Long, Medium, and Short-term Shoreline Dynamics of the Brittany Coast (Western France). Journal of Coastal Research: p 89–109
3. Stéphan P et al (2018) Monitoring the medium-term retreat of a gravel spit barrier and management strategies, Sillon de Talbert (North Brittany, France). Ocean Coast Manag 158:64–82
4. Pethick JS (1984) An introduction to coastal geomorphology. Edward Arnold
5. Jennings R, Shulmeister J (2002) A field based classification scheme for gravel beaches. Mar Geol 186(3–4):211–228
6. Buscombe D, Masselink G (2006) Concepts in gravel beach dynamics. Earth-Sci Rev 79(1):33–52
7. de Alegria-Arzaburu AR, Masselink G (2010) Storm response and beach rotation on a gravel beach, Slapton Sands. *UK*. Marine Geology. 278(1–4):77–99
8. Olson D et al (2012) Decadal-scale gravel beach evolution on a tectonically-uplifting coast: Wellington, New Zealand. Earth Surf Proc Land 37(11):1133–1141
9. St-Hilaire-Gravel D, Forbes DL, Bell T (2012) Multitemporal Analysis of a Gravel-Dominated Coastline in the Central Canadian Arctic Archipelago. J Coastal Res 28(2):421–441
10. Burningham H, French J (2017) Understanding coastal change using shoreline trend analysis supported by cluster-based segmentation. Geomorphology 282:131–149
11. Castelle B et al (2024) Satellite-derived sandy shoreline trends and interannual variability along the Atlantic coast of Europe. Sci Rep 14(1):13002
12. Vos K et al (2023) Pacific shoreline erosion and accretion patterns controlled by El Nino/Southern Oscillation. Nature Geosci 16(2):140-+ (2023)
13. Barnard PL, et al (2015) Coastal vulnerability across the Pacific dominated by El Nino/Southern Oscillation. Nature Geosci 8(10):801-+ (2015)
14. Konstantinou A et al (2023) Satellite-based shoreline detection along high-energy macrotidal coasts and influence of beach state. Mar Geol 462:107082
15. Vos K et al (2019) CoastSat: A Google Earth Engine-enabled Python toolkit to extract shorelines from publicly available satellite imagery. Environ Model Softw 122:7
16. Bujan N, Cox R, Masselink G (2019) From fine sand to boulders: Examining the relationship between beach-face slope and sediment size. Mar Geol 417:106012
17. Scott T, et al (2021) Role of Atmospheric Indices in Describing Inshore Directional Wave Climate in the United Kingdom and Ireland. Earths Future. **9**(5) (2021)
18. McCarroll RJ et al (2023) Coastal survey data for Perranporth Beach and Start Bay in southwest England (2006–2021). Scientific Data 10(1):17

From Fixed to Transgressive Dunes, the Conditions and Timing of the Transition Along the Aquitaine Coast, France

Alexandre Nicolae Lerma[1]([✉]), Olivier Burvingt[2], Bruce Ayache[1], Nicolas Robin[3], David Rosebery[4], and Bruno Castelle[2]

[1] BRGM Direction Nouvelle-Aquitaine, Pessac, France
a.nicolaelermal@brgm.fr
[2] UMR 5805, Univ. Bordeaux, CNRS, INP, EPOC, Pessac, Bordeaux, France
[3] Laboratoire CEFREM, UMR 5110, Université de Perpignan, Perpignan, France
[4] Office National Des Forêts, ONF, Bruges, France

Abstract. Today most of the coastal dunes in temperate latitudes, especially in the northern hemisphere, are relatively stable. However, over the last decade, the Gironde coast, southwest France, has experienced substantial natural dune remobilization following a major marine erosion event. Annual, large-scale and high-resolution, airborne LiDAR data and Satellite imagery (Sentinel-2) are combined to address the coastal dune morphological changes and establish relations with forcing and controlling factors (vegetation cover, geomorphological descriptors). Between 2014 and 2023, about 10 out of 85 km of the Gironde dunes have switched from fixed to transgressive state. The analysis showed that in the vast majority of the cases the dominant process involved was dune front cannibalism. However, there is considerable spatial and temporal variability along the coast, depending on the vegetation cover evolution, the amount of sediment remobilized and the morphological characteristics of the dunes (steepness of the front slope, width).

Keywords: Dune remobilization · transgressive dunes · LiDAR · Gironde coast

1 Introduction

Coastal dunes are complex environments among sandy coastal systems that combine multiple critical roles and services, described as 'ecosystemic', for coastal communities and biodiversity. Their morphodynamics is controlled by multiple, non-linear processes interacting with each other [1] under external forcing (e.g. meteorological and oceanic forcing, management practices) and affected by internal parameters (e.g. specific local biotic and abiotic characteristics). In many areas, particularly in temperate climates, coastal dune systems are fixed and relatively stable [2]. This stability is either considered as "natural" due to favourable conditions (greening) [3] in the Northern hemisphere or associated with management actions that limit sand blowing out of the system to protect the hinterland [4]. However, under certain conditions, coastal dunes can remobilize

C. Coelho et al. (Eds.): CD 2025, CRL 41, pp. 10–15, 2026.
https://doi.org/10.1007/978-3-032-15473-6_2

[5]. Some recent studies have presented examples of spontaneous remobilization [6, 7], but they have focused on dunes at a very local scale (a few hundreds meters) and without a detailed quantitative analysis of the factors driving the evolution from a fixed to a transgressive system. Using a dataset of Digital Terrain Models and synchronous orthophotos derived from airborne LiDAR surveys over a decade (2014 to 2023) and vegetation coverage inferred from bi-annual (2017–2023) Sentinel-2 optical satellite images, we address the conditions and timescales associated with the transition from a fixed to a transgressive dune along 85 km of coastal dunes at temperate latitudes.

2   Site and Methods

2.1   Study Site

The study site is located on the Gironde coast, the southwest France. The calibrated foredunes of the Gironde coast are part of the highest and largest coastal dunes in Europe, peaking at 30 m above mean sea level (MSL) and easily extending hundreds of meters landward [8]. The coastal system exhibits considerable morphological variability, mostly induced by the large alongshore gradient in sediment availability [9]. As a result of the chronic marine erosion affecting the Gironde coast [10], the dune belt has narrowed considerably in the northern half part of the coast and the general shape has been substantially modified [8].

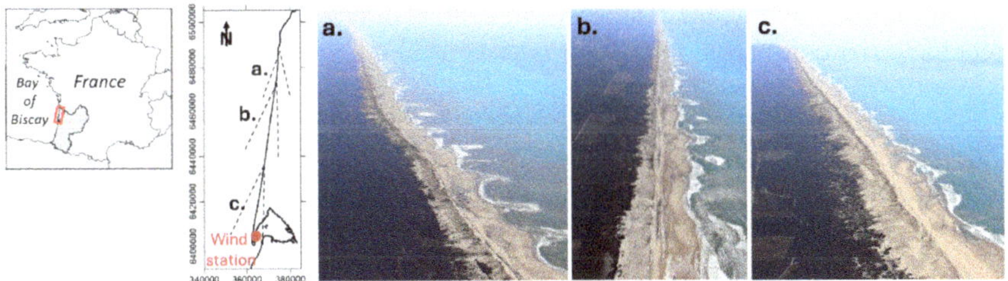

Fig. 1. Study site location map and photographs of the dune belt along the Gironde coast. Photos a. b. c. illustrate the variability of the dune morphology along the coast (photos OCNA, April 2020).

2.2   Data and Method

2.3   Wind Forcing

Hourly 10-m wind data were collected by Météo France at weather station Cap-Ferret (Fig. 1). These data were processed to estimate the potential wind-driven sand flux, following the steps outlined in [11]. In the following, Q is defined as the cumulated wind-driven sand flux between two topographic surveys for comparison with morphological changes.

2.4 Geomorphology

Several airborne LiDAR campaigns from the Observatoire de la Côte de Nouvelle-Aquitaine (OCNA) database acquired in collaboration with the French National Institute of Geography (IGN) were used. Data were acquired in the autumn of 2014, and annually at the same period since 2016. Data were extracted from LiDAR Digital Terrain Models (1-m gridded spatial resolution DTMs) along 10 m-spaced cross-shore profiles from the MSL contour of the most seaward position of all surveys and the inland dune foot location of the last survey (2023). Multiple morphological and volume descriptors were computed for each profile (Fig. 2). We focus here on the inter-survey gross mobility (M), the Dune volume (Dv) and the inland displacement of the back dune front (Tr).

2.5 Vegetation Cover

Vegetation cover characterization was derived from Sentinel-2 optical satellite imagery acquired between 2017 and 2023. Yearly digital maps of dune vegetation cover along the Gironde coast were created using Sentinel-2 satellites images adapting the methodology proposed by [12]. A total of 10 images acquired from June to October were used for this analysis. For each image, an endmember reflectance spectrum was calculated on the bands of the Sentinel-2 images [12]. Information is then used for the linear unspectral mixing model [13], used to quantify the fractional abundance of sand and vegetation of each pixel. The fractional abundance of sand and vegetation of each dune pixel was computed and constrained such that each pixel values range from 0 (bare sand) to 1 (fully vegetated). Finally, the temporal and spatial variability of the cross-shore mean vegetation cover concentration ($Covmean$) was used in the following.

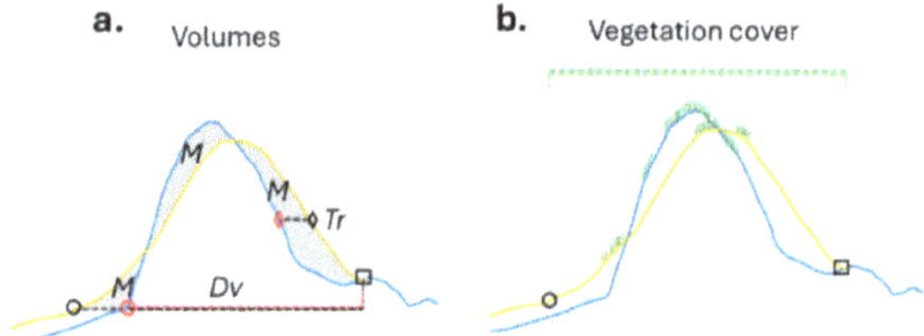

Fig. 2. Descriptors beach-dune volume and vegetation cover

3 Results

3.1 Dune Mobility

The inter-survey gross mobility (M) along the coast depicts massive changes, reaching 5% of the total dune volume (Dv) in certain years (not show). Between 2014 and 2023, a cumulative mobility of more than 20% of Dv is often observed. In the most mobile sectors around 40% of Dv has been remobilized.

We identify a clear relation between time-varying Q and M indicating that at the regional scale, over 50% of the variance in M can be explained by Q (r = 0.74; R2 =

0.55; pvalue $= 0.04$). This substantial mobility is observed with a large temporal and spatial variability.

The mobile sectors (covering 10 km long) show two distinct behaviours: (i) the dunes have gradually evolved as transgressive since at least 2019 due to continuous mobility since 2014 (e.g. $x = 6.482.10^6$ m, around $x = 6.46.10^6$ m or $x = 6.44.10^6$ m, Fig. 3.); (ii) after a period of relatively low mobility, a massive evolution during the winter 2019–2020 and a fairly continuous mobility in the following years. In these sectors, the mobility of the dunes leads to a massive increase of transgressive dunes in 2023.

3.2 Formation of Transgressive Dunes

Along the coast, the proportion of transgressive ($Tr > 2$ m/year) dunes increases from 5.7% to 17.3% between 2014 and 2023. Significant differences are observed between sectors, with the northern half of the study area already exhibiting a much higher percentage of transgressive dunes in 2014, but even more so in 2023 (Fig. 2). A notable exception is the U4 sector, where the percentage is low and stable at 1.6% in 2023. The other sectors show a large increase of transgressive dunes, by a factor of 2.1, 4.2 and 3.7 for U1, U2 and U3 respectively. In sector the most mobile sector (U2), 23% of the dune belt became transgressive during the decade.

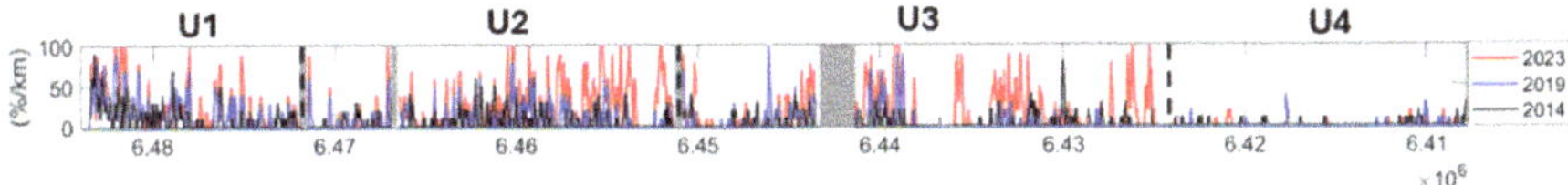

Fig. 3. Initiation of transgressive dune along the coast from North to South

Transgressive dunes are observed in areas where M reaches its highest values (Fig. 3). Notably, most of the dunes with M greater than 200 m^3/m have shifted from stable to transgressive between 2014 and 2023. Newly transgressive dunes are also logically associated with the lowest vegetation cover, with a median value of 40% of coverage in 2023. In contrast, 75% of the non-transgressive dunes have more than 40% of vegetation cover. Figure 3c shows that transgressive dunes can be also associated with contrasting sediment supply context. The proposed classification considers the ratio between ΔDv and the M both between 2014 and 2023. It can be observed:

-$\Delta Dv/M = $ -1, erosion explains the mobility;

-$\Delta Dv/M = 0$, dune mobility is exclusively due to the remobilization of pre-existing sediment (cannibalism).

-$\Delta Dv/M = 1$, dune mobility is exclusively due to new sediment supply.

For non-transgressive dunes, M is equally due to cannibalism and new sediment supply.

The classification results for transgressive dunes indicate that in over 75% of the cases, most of the mobility (more than 60%) is attibuted to remobilisation of pre-existing sediment. In only about 15% of cases the mobility is dominated by new sediment supply.

In rarer situations, transgressive dunes can be observed in the context of marine erosion or large sediment supply (-1 and 1 respectively). These cases may be due to

14   A. N. Lerma et al.

transgressive conditions prior to the study period (e.g. within U1, Fig. 5, profile A) or they may have been rapidly established after the strong erosion of 2013–2014 (e.g. in U3 y = 6.44 .10^6 m), with large sediment supply occurring posterior to the transition to transgressive state (Fig. 4).

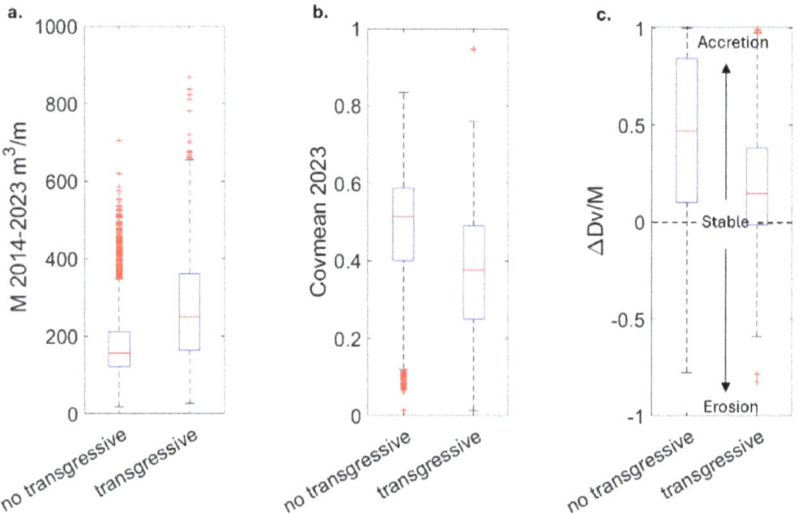

Fig. 4. Box plot of the distribution of transgressive or no-transgressive dunes depending on: a. the mobility (*M*) period 2014–2023, b. the vegetation cover (*Covmean*) in 2023, c. the beach supply context (*ΔDv/M*) period 2014–2023.

4   Conclusion

At the scale of the 85 km of dunes studied here, the proportion of transgressive dunes increased dramatically from 4.7 km in 2014 and 7.1 km in 2019 to 14.7 km in 2023. The vast majority of dunes that have transitioned from stable to transgressive did go through a process of cannibalism, demonstrating that sediment supply is not required to transition from one stage to the other.

To better understand the sand exchange between beach, dune and back dune, and the non-linear eco-geomorphic relations and feedbacks that drive morphological evolution, large-scale quantitative studies over time periods of the order of a decade are needed. It is also necessary to develop appropriate and ambitious management plans, particularly in view of the challenges posed by climate change to the maintenance of dune areas and their biodiversity.

Acknowledgements. This work was carried out as part of the DUNES project "Espace des Transitions", co-funded by the Région Nouvelle-Aquitaine and the BRGM (research agreement). We would also like to thank all the co-funders of the project and the Observatoire de la Côte Nouvelle-Aquitaine for making the data available and PSGAR CORALI which support the project.

References

1. Hesp P (2013) A: Conceptual models of the evolution of transgressive dune field systems. Geomorphology 199:138–149
2. Cooper A, Jackson D (2021) Dune gardening? A critical view of the contemporary coastal dune management paradigm. Area 53(2):345–352
3. Gao J, Kennedy DM, Konlechner TM (2020) Coastal dune mobility over the past century: A global review. Progress in Physical Geography: Earth and Environment 44:814–836
4. Jackson NL, Nordstrom KF (2013) Aeolian sediment transport and morphologic change on a managed and an unmanaged foredune. Earth Surf Proc Land 38(4):413–420
5. Hesp P.A., Da Silva M., Miot Da Silva G., Bruce D., Keane R.: Review and direct evidence of transgressive aeolian sand sheet and dunefield initiation. Earth Surface Processes and Landforms, 47(11), 2660–2675 (2022)
6. Robin N, Billy J, Nicolae Lerma A, Castelle B, Hesp PA, Rosebery D, Miot Da Silva G (2003) Natural remobilization and historical evolution of a modern coastal transgressive dunefield. Earth Surface Processes and Landforms 48(5), 1064–1083 (2023)
7. DaSilva MD, Hesp PA, Bruce D, Downes J, da Silva GM (2024) Coastal transgressive dunefield evolution as a response to multi-decadal shoreline erosion. Geomorphology 455:109165
8. Bossard V, Lerma AN (2020) Geomorphologic characteristics and evolution of managed dunes on the South West Coast of France. Geomorphology 367:107312
9. Nicolae Lerma A et al (2022) Decadal beach-dune profile monitoring along a 230-km high-energy sandy coast: Aquitaine, southwest France. Appl Geogr 139:102645
10. Castelle B et al (2018) Spatial and temporal patterns of shoreline change of a 280-km high-energy disrupted sandy coast from 1950 to 2014: SW France. Estuar Coast Shelf Sci 200:212–223
11. Laporte-Fauret Q, Castelle B, Michalet R, Marieu V, Bujan S, Rosebery D (2021) Morphological and ecological responses of a managed coastal sand dune to experimental notches. Sci Total Environ 782:146813
12. van Kuik N, de Vries J, Schwarz C, Ruessink G (2022) Surface-area development of foredune trough blowouts and associated parabolic dunes quantified from time series of satellite imagery. Aeol Res 57:100812
13. Keshava N, Mustard JF (2002) Spectral unmixing. IEEE signal processing magazine 19(1):44–57

Beach-Dune System Evolution Under the Combined Effects of Cross-Shore and Longshore Sediment Transport Processes

Ana Margarida Ferreira[1]([envelope]), Carlos Coelho[1], and Paulo A. Silva[2]

[1] RISCO & Civil Engineering Department, University of Aveiro, Aveiro, Portugal
margarida.ferreira@ua.pt
[2] CESAM & Physics Department, University of Aveiro, Aveiro, Portugal

Abstract. Coastal dunes systems are important features of the coastal morphology, providing natural protection against storm impacts and serving as habitats for diverse species. Numerical models of coastal evolution are typically categorized by the temporal and spatial scales of the processes that they describe: cross-shore numerical models address sediment transport within the beach profile over short to medium timescales related to storm events (days to months); and shoreline evolution models simulate changes in coastal morphology due to longshore sediment transport processes over timescales ranging from years to decades. In this study, the numerical approach presented by Ferreira *et al.* [1, 2] was applied to evaluate the sediment dynamics and the morphological evolution of the berm-dune system in a coastal domain. This approach, developed to conduct medium to long-term time scales analyses with low computational effort, combines longshore and cross-shore sediment transport processes, by integrating the results of two numerical models (LTC and CS-Model). The results highlight the interconnections between cross-shore and longshore sediment transport processes and emphasize the importance of incorporating longshore effects into numerical modelling of beach-dune systems, from a medium to long-term perspective (years). Furthermore, the study demonstrates the potential of the model to support coastal management through its outputs (shoreline position, berm width, dune evolution and regenerative capacity due to windblown sand transport).

Keywords: Numerical Modelling · Long-Term · Sediment Dynamics · Dune Erosion · Windblown Sand Transport

1 Introduction

The formation and evolution of sandy dunes result from interactions between the beach and dunes, driven by wind and waves. Dune formation occurs through sediment transport induced by wind, when sand is trapped by an obstacle. During storm events, wave impacts on the dune system can lead to dune erosion and sediment transfer from the dune to the beach. After storm periods, sediment transport driven by windblown sand contributes to dune reconstruction, by transferring sediments from the beach back to the dune [3].

© The Author(s) 2026

C. Coelho et al. (Eds.): CD 2025, CRL 41, pp. 16–22, 2026.
https://doi.org/10.1007/978-3-032-15473-6_3

Considering the importance of dune systems related to the diverse ecosystem services that they provide, forecasting their evolution is a key element for coastal management. Elko *et al.* [3] and Palmsten *et al.* [4] highlighted the necessity of developing medium- to long-term numerical tools to model dune systems evolutions. According to a review of the state-of-the-art about numerical modelling of coastal zones evolution, the numerical modelling of dune evolution associated with short-term storm effects is well established and therefore, the current research challenge lies in modelling post-storm dune recovery. Dune recovery depends on aeolian processes, hydrodynamic conditions, and ecological factors, and typically occurs on longer timescales than storm-driven processes [3–5].

In this study, the numerical approach presented by Ferreira *et al.* [1, 2] was applied to analyze the combined effects of longshore and cross-shore processes, related to littoral drift, wave impact in the dune system and windblown sand transport. The study focuses on the morphological evolution of the berm-dune system over a long-term period (10 years), considering five generic coastal domain conditions.

2 Methodology

2.1 Numerical Approach

The numerical approach presented by Ferreira *et al.* [1, 2] integrates longshore and cross-shore sediment transport processes from a medium- to long-term perspective by combining the results of two existing numerical models: LTC [6], which simulates longshore effects and CS-Model [7], which simulates sediment transport within the beach profiles. In summary, the methodology employs a cyclic structure repeated over a number of time steps (corresponding to the number of waves to be simulated). At each time step, both models are run, and the sediment transport results calculated by each model, for each cross-shore beach profile within the numerical domain, are used to update the beach morphology of the domain (bathymetry and topography and cross-shore profiles configuration). To apply the model, the user specifies the mesh dimension of the domain (NX and NY) and the distance between mesh points in the longshore (X) and cross-shore (Y) directions, respectively, Δx and Δy. NX is the number of cross-shore profiles to be modelled in the CS-Model.

The LTC is a one-line model that applies the equation of continuity to determine the shoreline position over time [6]. Based on the one-line model theory, changes in the shoreline position are driven by gradients in longshore sediment transport occurring between consecutive segments of the coastal domain. The CS-Model is a cross-shore profile evolution numerical model that simulates the primary processes governing beach profile evolution over a medium-term timescale [7]. Each cross-shore beach profile is defined by a set of morphological parameters allowing to characterize its configuration, which is assumed to follow a simplified schematic representation (**Fig. 1a**). The CS-Model calculates sediment transport within the beach profiles, incorporating the processes related to sandbar-berm dynamics (q_B), windblown sand transport (q_{WS}) and dune erosion due to wave impact (q_D).

2.2 Numerical Setup and Assessed Scenarios

The study spanned a 10-year period, using a 3-h time step, and it was conducted for a generic beach, with regular and parallel bathymetry, under a constant wave climate (H = 3 m, T = 10.55s and 80 degrees direction, anticlockwise, from North). The domain comprises 251 points spaced 20 m in the West-East direction and 15 cross-shore profiles spaced 100 m (5000×1400 m^2, Fig. 1b). All cross-shore profiles were defined identical, with the morphology characterized as shown in Fig. 1a, and an initial berm width of 50 m (distance between Y_B and Y_S).

Five scenarios were defined, considering dune erosion due to wave impact equal to 35 m^3/m/year. The scenarios varied in the imposed longshore sediment transport gradients at the domain boundaries, controlled by the sediment volume going in and out of the domain: A) a constant longshore sediment transport along all the domain; B) and C) an updrift longshore sediments deficit of 10% and 20%, at P15, respectively; D) and E) a downdrift longshore sediments deficit or accumulation of 10%, at P1, respectively. Each scenario was simulated twice (with and without the wind effect), to evaluate dune reconstruction due to aeolian sediment transport.

Longshore sediment transport was calculated using the CERC formula and the wave impact on the dune considered friction losses across the berm width (distance between Y_S and Y_B). In the scenarios that included wind effect, a dune seaward accretion rate of 35 m^3/m/year was applied, to balance dune erosion caused by wave impact. The sandbar volume of the cross-shore profiles (V_B) was defined to be in equilibrium with the wave climate [1].

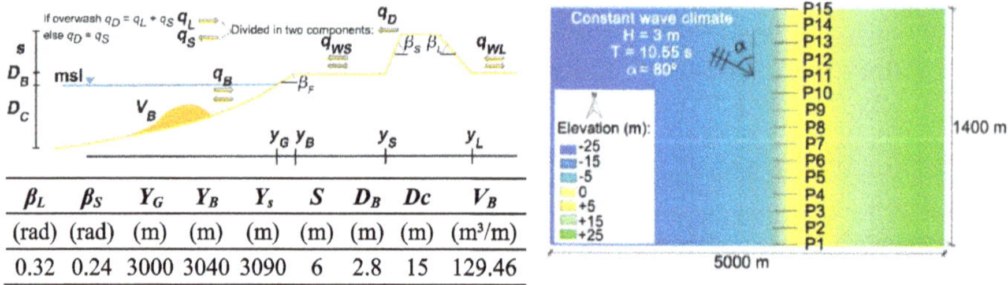

β_L	β_S	Y_G	Y_B	Y_s	S	D_B	Dc	V_B
(rad)	(rad)	(m)	(m)	(m)	(m)	(m)	(m)	(m³/m)
0.32	0.24	3000	3040	3090	6	2.8	15	129.46

Fig. 1. Study area: a) scheme of the beach profile used by the CS-Model (morphological parameters and sediment transport components, adapted from [7]) and the initial cross-shore morphology of the profiles; b) numerical domain with identification of the cross-shore beach profiles.

3 Results

Figure 2 summarizes the sediment balance within the numerical domain for the assessed scenarios. The results are presented in terms of volume variations induced by cross-shore (*ΔVCross*) and longshore (*ΔVLong*) sediment transport processes, as well as the changes in beach berm volume (*ΔVBerm*), which results from the balance between these processes. Sediment volumes that represent sediment removal from the berm are represented as negative values, while processes contributing to sediment gain are shown as positive. *ΔVcross* result from wave impacts on the dune, which transfer sediment from the dune to the berm of the cross-shore profiles (q_D, considered positive) and, in the numerical simulations that considered wind effects, windblown sand transport (q_W) moves sediment from the berm to the dune (which was considered negative). Longshore effects are driven by gradients in longshore sediment transport, which arise when consecutive shoreline segments have different orientations, and consequently, each segment have a different potential longshore sediment transport.

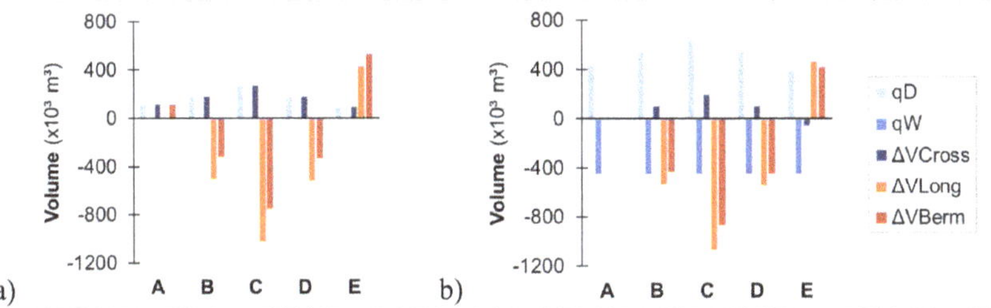

Fig. 2. Sediment volumes induced by cross-shore (*ΔVCross*) and longshore (*ΔVLong*) sediment transport processes, along with the variation of the beach berm volume (*ΔVBerm*) within the numerical domain, for scenarios A to E: a) No windblown sand transport; b) With windblown sand transport.

As shown in Fig. 2, the evolution of the coastal domain in Scenario A is only governed by cross-shore sediment transport processes. This is because no changes in potential longshore sediment transport were imposed on the boundary conditions of the numerical domain, and all profiles experienced the same cross-shore displacements, maintaining consistent shoreline orientation across profiles. In this scenario, without wind effect, the cross-shore processes driven by wave impacts on the dune results in sediment transfer from the dune to the berm. This leads to shoreline advance and dune toe retreat, gradually increasing the berm width over time. By the end of the numerical simulation, the berm width evolves to approximately 65 m (Fig. 3a).

In Scenarios B, C, D and E, both cross-shore and longshore processes influence the sediment balance (Fig. 2a). Figure 2 shows the total longshore volume changes in the profile's domain, but the magnitude of the profiles impact varies alongshore, being higher in the cross-shore profiles located near the boundary where changes in potential sediment transport were imposed, and progressively decreasing alongshore.

When the boundary conditions induce sediment deficits (scenarios B, C and D), longshore effects negatively impact the cross-shore profiles, causing sediment losses from the berm of the profiles, resulting in shoreline retreat. Additionally, the profiles experiencing the greatest negative impacts from longshore effects also exhibit the most significant dune volume losses due to wave impact, leading to dune toe recession. This result reflects the relationship between berm width and dune erosion, which indicates that under the same numerical conditions, cross-shore profiles with a narrower berm have a lower capacity to dissipate wave energy and consequently, experience greater dune erosion. Therefore, in these scenarios, without wind effect, the shoreline retreat caused by longshore sediment transport gradients initially reduces the berm width in profiles near the boundary, later balanced by the cross-shore processes (Fig. 3 - upper row). In contrast, Scenario E exhibits the opposite behavior. Here, longshore effects result in sediment gains in the berm of the cross-shore profiles, leading to shoreline advance. The increased berm width enhances wave energy dissipation, which reduces the erosive impact on the dune.

Wind effects contribute to dune reconstruction by mitigating the sediment losses caused by wave impacts, transporting sediments from the berm to the dune (Fig. 2b). Due to the opposite effects of aeolian sediment transport and dune toe recession caused by wave impact, in scenario A, the system maintains its initial morphology. In Scenarios B, C and D, the dune toe experiences less seaward retreat compared to the simulations without wind. The system evolves toward an equilibrium configuration, resulting in similar berm widths across all profiles, although these are smaller (approximately 49.4 m) than the initial berm width. The time required to reach equilibrium is influenced by the longshore sediment transport deficits updrift (see Fig. 3, lower row). In Scenario E, the dune accretion due to windblown sand transport exceeds the dune erosion caused by wave impact, leading to seaward advance of the dune toe and gain of dune volume. In this scenario, the berm width of the cross-shore profiles increases beyond its initial value, reaching approximately 50.6 m (Fig. 3e).

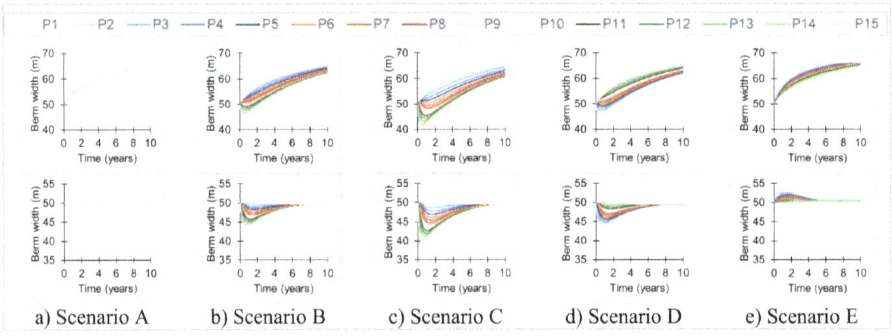

Fig. 3. Berm width in the cross-shore profiles over time: upper row - without wind effect; lower row - with wind effect.

4 Conclusion and Discussion

The results show that longshore sediment transport gradients are a key factor influencing the evolution of dune systems. Cross-shore profiles experiencing greater sediment deficits due to longshore sediment transport gradients are more susceptible to dune erosion from wave impacts, resulting in larger volumes of dune erosion. The opposite effect occurs when longshore gradients induce accretion. These numerical results align with Brodie *et al.* [5], who indicate that wave contact with dunes is a primary driver of dune erosion at both short and meso-timescales. Additionally, the numerical results presented by Hallin *et al.* [8] support this, as their comparison of CS-Model simulation results with monitoring data indicates that longshore gradients are an important parameter in numerical modeling of dune evolution.

In this study aeolian transport mitigates the eroded volume in the dune, causing the profiles to evolve toward a new equilibrium configuration, where all profiles within the domain exhibit similar berm widths. The time for the profiles to reach equilibrium depends on the balance between longshore and cross-shore processes. Once equilibrium is achieved, the positions of the shoreline and dune toe differ from their initial positions.

Overall, although the study is of a generic nature and, therefore, requires validation in future research, the results emphasize the interconnections between cross-shore and longshore sediment transport processes. According to de Vries *et al.* [9], at decadal scales, dune volume changes are governed by aeolian processes and coastal erosion, with the effects of aeolian processes being comparable in magnitude to dune erosion caused by wave storm events. The numerical approach applied, which requires low computational effort, demonstrates its ability to facilitate the interpretation of how distinct sediment transport processes contribute to the spatial and temporal evolution of the shoreline and dune system, including dune erosion and reconstruction driven by windblown sand transport. Therefore, the proposed numerical model addresses the research needs for modelling dune evolution over long-term periods, particularly in relation to dune reconstruction driven by windblown sediment transport [3–5]. As a result, the approach shows potential to support coastal management through the results provided (evolution of the shoreline position, berm width, dune evolution and its regenerative capacity due to windblown sand transport). Furthermore, the numerical approach allows the consideration of real wave climate scenarios and their seasonality, important for future real-world case studies, investigating dune recovery after storm events.

Acknowledgments. The authors thank FCT by the financial support provided to Ana Margarida Ferreira through the PhD grant 2021.07269.BD and to the research units RISCO and CESAM.

References

1. Ferreira AM, Coelho C, Silva PA (2024) Numerical evaluation of the impact of sandbars on cross-shore sediment transport and shoreline evolution. J Environ Manage 370:15
2. Ferreira AM, Coelho C, Silva PA (2024) Medium-term effects of dune erosion and longshore sediment transport on beach-dune systems evolution. J Mar Sci Eng 12 (2024)
3. Elko N et al (2016) Dune management challenges on developed coasts. Shore Beach 84:15
4. Palmsten M, Brodie K, Spore N (2017) Coastal Foredune Evolution, Part 2: Modeling Approaches for Meso-Scale Morphologic Evolution. USACE, 10 p. (2017)
5. Brodie K, Palmsten M, Spore N (2017) Coastal Foredune Evolution, Part 1: Environmental Factors & Forcing Processes affecting Morphological Evolution. USACE, 10 p.
6. Coelho C (2005) Riscos de Exposição de Frentes Urbanas para Diferentes Intervenções de Defesa Costeira, PhD Thesis, University of Aveiro, Portugal, 404 p. (2005). (in Portuguese)
7. Larson M, Palalane J, Fredriksson C, Hanson H (2016) Simulating cross-shore material exchange at decadal scale. Theory and model component validation. Coast Eng 116
8. Hallin C, Huisman B, Larson M, Walstra D, Hanson H (2019) Impact of sediment supply on decadal-scale dune evolution - Analysis and modelling of the Kennemer dunes in the Netherlands. Geomorphology 337:94–110
9. de Vries S, Southgate HN, Kanning W, Ranasinghe R (2012) Dune behavior and aeolian transport on decadal timescales. Coast Eng 67:41–53

ML Prediction of Equilibrium Shoreline Response of Beaches Protected by Natural and Human-Made Detached Structures

Arnau Garcia$^{(\boxtimes)}$, Erica Pellón, Albert Gallego, and Mauricio González

Santander, Spain
arnau.garcia@unican.es

Abstract. Detached breakwaters are critical coastal protection measures, yet conventional emerged designs often have significant environmental and aesthetic impacts. Submerged and low-crested breakwaters offer a promising alternative, but the morphodynamic responses they induce remain poorly understood, particularly in the formation of tombolos and salients—key sedimentary features. Existing definitions of these formations often rely on dimensionless ratio analysis, which can lead to contradictions and inconsistencies depending on the selected methodology. While prior studies have predominantly focused on fully emerged structures morphodynamics, this study addresses the uncertainty surrounding tombolo and salient predictions for submerged and low-crested structures. A novel framework is proposed, utilizing Machine Learning (ML) classification techniques as an alternative to traditional methodologies, aiming to improve the reliability of predictive models and reduce uncertainty in the formation limits of tombolo and salient events.

Keywords: Detached structures · Machine Learning · Tombolo-Salient · Equilibrium shoreline response

1 Introduction

Detached breakwaters are structures parallel to the coast at a certain distance from the coastline ('detached'), designed for reducing the amount of incident wave energy impacting a protected stretch of shoreline. According to their relative crest-level, the structures can be emerged (crest above) or submerged (crest below mean sea-level (MSL)). Emerged detached breakwaters have long been proven effective in promoting sediment accretion in their lee and have been widely implemented over many decades, whereas submerged breakwaters remain primarily a subject of ongoing research [1]. More specifically, detached breakwaters can be typically classified into three main types based on their relative structure-crest elevation: Emerged Detached (ED), Low-Crested (LC), and Submerged Detached (SD) structures. Each type of structure exhibits different hydrodynamical properties, influenced by its morphological dimensions.

ED breakwaters, which primarily disrupt waves through diffraction, are the most extensively studied. They have well-established design criteria and empirically documented formation limits that are widely applied in engineering practices [2]. Research

© The Author(s) 2026
C. Coelho et al. (Eds.): CD 2025, CRL 41, pp. 23–28, 2026.
https://doi.org/10.1007/978-3-032-15473-6_4

indicates that the most effective ED breakwaters for coastal protection are constructed near the shore with high crest levels, making them highly effective in generating fully developed tombolos or salients. Most ED breakwaters have been implemented along micro-tidal beaches, particularly in Japan, the USA, and the Mediterranean, with fewer examples on open, exposed meso-tidal and macro-tidal beaches [3]. Conventional emerged breakwaters, while effective, often leads to environmental and aesthetic concerns, particularly in areas where maintaining natural seascapes is paramount [1]. In contrast, SD and LC breakwaters have raised as viable surrogate solutions, offering reduced visual intrusion and lower ecological impacts. They offer several advantages over traditional emerged breakwaters, including lower wave reflection, which can lead to calmer water conditions in the protected area. Additionally, SD breakwaters have been found to be more effective in attenuating long-period waves, which often possess high energy levels and can bypass conventional barriers. However, submerged structures also have several drawbacks. Main disadvantages are the relatively high construction and maintenance costs and the inconvenience/danger to swimmers and small boats [4]. SD and LC breakwaters present significant challenges due to their non-linear processes and complex dynamics, which make it difficult to predict their long-term stability. This uncertainty extends to the shoreline response, where it becomes problematic to determine whether a tombolo or a salient will form, or worse, if the structure will inadvertently exacerbate shoreline erosion. In fact, rather than providing consistent protection, SD structures have frequently been associated with adverse effects, including accelerated beach erosion in several cases [5].

One of the critical challenges in designing DS breakwaters is accurately determining the appropriate dimensions and optimal placement of the structures to achieve the desired shoreline response, whether it is a tombolo or a salient. The efforts found in the literature consistently highlight that the main governing parameters involved in these processes, are those depicted in Fig. 1.

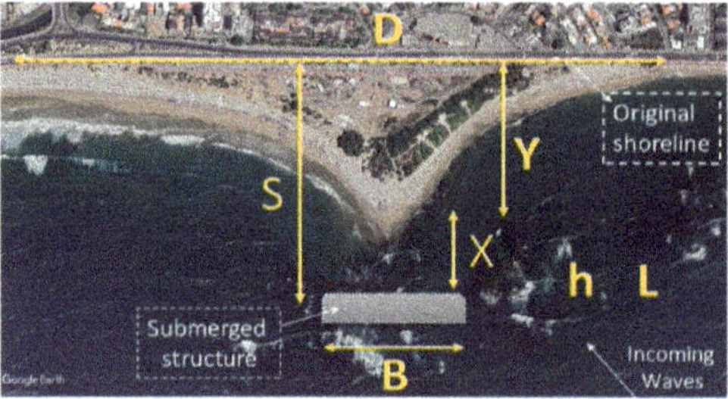

Fig. 1. Relevant parameters on a salient response. Adapted from [6].

The attributes considered here, include the following parameters: the shape and alignment of features relative to the original shoreline (D), the degree of accretion and sediment redistribution along the shore (Y), the distance from the original shoreline to the structure (S), the protrusion of sediment features towards the structure (X), the length of the structure (B), the mean water depth at the base of the structure on the lee side (h), and the characteristic wavelength (L) at this depth.

To address these challenges, two main approaches have traditionally been adopted: numerical modelling and empirical analyses. Both face significant limitations, particularly for SD and LC structures, where the non-linear dynamics and complexity of wave-current-sediment interactions lead to poor predictions of the final shoreline response. Although several limits of formation have been proposed for partially or fully submerged structures, the prior studies have relied on dimensionless analysis tailored to fully emerged configurations, frequently resulting in contradictory or inconsistent predictions under varying environmental conditions [7].

This highlights a clear lack of effective alternatives to reliably predict the behaviour of such structures. In this study, a novel alternative approach is presented, where ML-classification algorithms are implemented to address the aforementioned uncertainty problems related with the formation of tombolos and salients.

2 Data and Methods

To predict the morphodynamic response generated by detached structures using ML, important efforts for data collection were needed. Over 250 beach data were manually collected from several sources (constructive projects, papers, PhD's supplementary data), standardized and characterized in a single format. Our database includes data from Brazil, South Korea, Japan, Israel, Italy and Spain, combining all types of detached structures, natural reefs and man-made breakwaters. Following the proposed methodology in [6], the parametrization of each Salient and Tombolo event was conducted. The dataset encompasses the most relevant features, focusing on the aforementioned parameters visible in Fig. 1. For each one of the structures considered, all their parameters were carefully measured and double-checked for avoiding possible inconsistencies. A clear strength of the presented approach is that accounts for all types of detached structures, encompassing ED (85 events), LC (57 events) and especially SD (112 events) of different dimensions and orientations. Nevertheless, structures with dimensions exceeding $B > 3622.12$ m, $X > 561.25$ m and $Y > 1024.66$ m were excluded from the analysis, as they fall outside the 98th percentile of the data cloud. These extreme values (commonly referred to as "Mega salient cases") introduce a significant bias to the dataset. By skewing the statistical distribution, they could lead to inconsistencies and reduce the reliability of the predictive model. Excluding such outliers ensures the dataset better reflects the typical scenarios under evaluation, thereby improving the robustness and generalizability of the analysis.

After preprocessing, which included handling missing values, scaling the data, feature engineering by combining variables to create more informative features, and filtering irrelevant events by removing outliers or non-representative samples, the final ML model relies on five essential input parameters: B, S, B/S, S/L, and B/2L. These parameters are, in essence, different combinations of the primary variables B (length of the structure), S (shoreline distance), and L (wavelength). The combination of these directly measured features with dimensionless ratios, derived from the primary variables, was identified as particularly significant for accurate predictions.

Following the ML-classification procedure (Fig. 2) and using the processed dataset, a total of 243 beaches were analysed, with 166 events of Salient and 77 events of Tombolo distributed between the Train (65%) and Test (35%) sets.

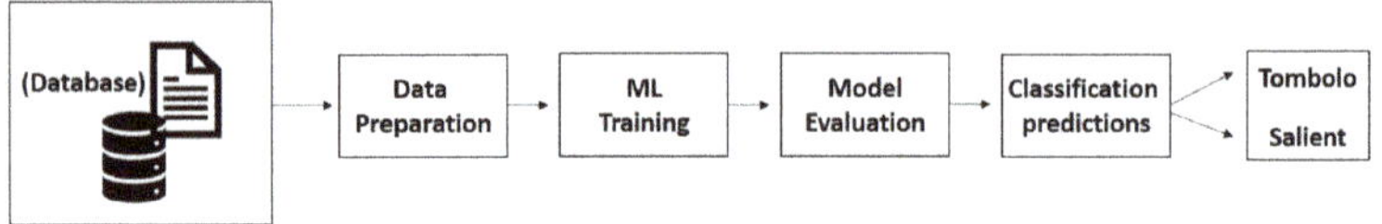

Fig. 2. Graphical abstract of the ML-classification procedure.

A combination of three base models—RandomForest, XGBoost and KNN—were used, each within a pipeline that includes their internal hyperparameters (e.g., class weights, tree depth) and evaluation metrics. To optimize each model's training, a GridSearchCV (Cross-Validation) was conducted to identify the best hyperparameters, based on the performance of the trained model. Once best hyperparameters have been identified for each base model, we combined them into a stacking model "ensemble", to integrate the strengths of each one of the calibrated base models.

3 Results

The stacking ensemble model proved to substantially enhance both the accuracy and reliability of predictions, outperforming the independent use of the base models. A good performance was obtained consistently during Train and Test sets. A total of 157 observations were used for training the model, while 86 instances were used to test and evaluate its performance. To comprehensively evaluate the model's results, four metrics—accuracy, precision, recall, and F1-score—were used. Each of these four metrics reflects a different aspect of the model's capabilities.

The confusion matrix for the test set (Fig. 3) further supports the model's reliability, offering a detailed breakdown of the classification results. It shows the correct predictions on the major diagonal (58.1% for salients and 29.1% for tombolos), with errors appearing on the minor diagonal, where 2.3% of tombolos were misclassified as salients, and 10.5% of salients were misclassified as tombolo events. This demonstrates the model's strong ability to correctly classify the major part of the predicted responses.

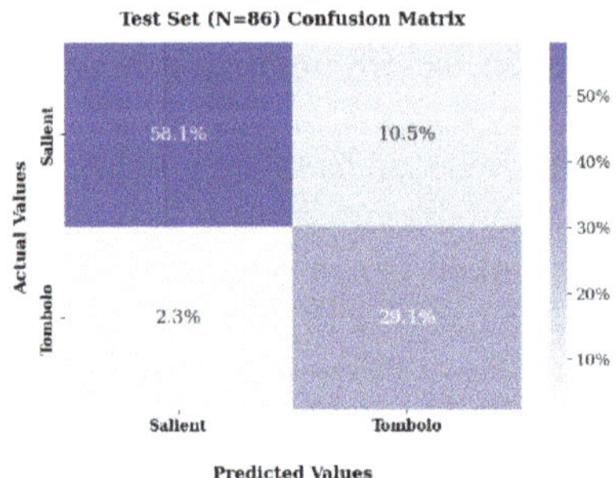

Fig. 3. Confusion matrix results obtained for the Test set.

These results suggest that the trained model consistently determines whether a structure is more likely to form a tombolo or a salient upon reaching equilibrium, highlighting a good discrimination power between classes.

This approach is applicable to beaches affected by all three types of breakwaters (ED, LC, SD), confirming the versatility and reliability of this ML-based framework, offering a significant advancement in predicting equilibrium shoreline responses for a wide range of detached structures. Figure 4 shows some examples of correctly classified shoreline responses extracted from the test set.

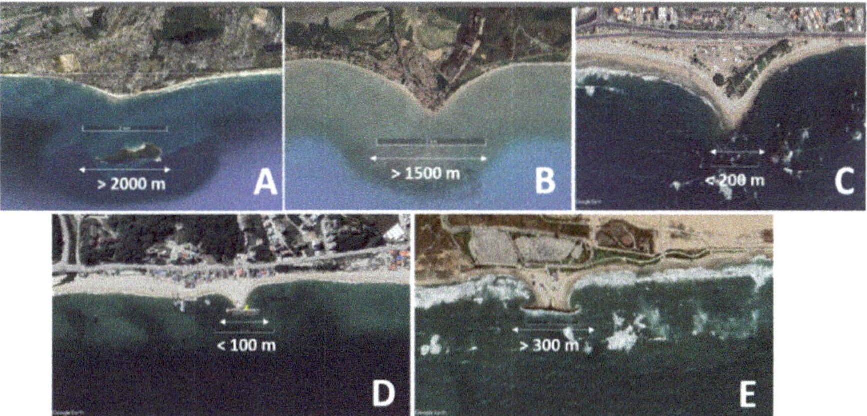

Fig. 4. Examples of correctly classified responses extracted from the Test set. Source: *Google Earth Pro: Image @2024 Maxar Technologies*

4 Conclusions

This study showcase that ML can offer a robust alternative to traditional empirical models, potentially reducing the uncertainty on the predictions of Tombolo and Salient cases of several types of detached structures. While the proposed approach seems promising in predicting tombolo and salient formations, incorporating additional hydrodynamic parameters and broader classification schemes could further improve its precision and applicability, creating more robust tools for coastal engineering.

Acknowledgements. This study is part of the ThinkInAzul programme and was supported by the Ministerio de Ciencia, Innovación y Universidades (PRTR-C17.I1) with funding from the NextGeneration EU and the Comunidad de Cantabria, as well as by the research project PID2023-153343OA-I00, funded by MICIU/AEI (https://doi.org/10.13039/501100011033) and co-financed by FEDER, EU.

References

1. Lamberti A, Archetti R, Kramer M, Paphitis D, Mosso C, Di Risio M (Nov.2005) European experience of low crested structures for coastal management. Coast Eng 52(10):841–866. https://doi.org/10.1016/j.coastaleng.2005.09.010

2. Chasten MA, Rosati JD, McCormick JW, Randall RE (1993) Engineering Design Guidance for Detached Breakwaters as Shoreline Stabilization Structures, US Army Corps of Engineers, Dec
3. Negro V, Diez Gonzalez JJ, Bricio Garberí L, López Gutiérrez JS (2009) Coastal erosion. Geometric detached breakwaters indicators for preventing the shoreline erosion. Fringe session. *Proceedings of the Coasts, Marine Structures and Breakwaters 2009*
4. Van Rijn LC (2011) Coastal erosion and control. Ocean Coastal Manage 54(12):867–887. https://doi.org/10.1016/j.ocecoaman.2011.05.004
5. Ranasinghe R, Turner I (2006) Shoreline response to submerged structures: a review. Coast Eng 53:65–79. https://doi.org/10.1016/j.coastaleng.2005.08.003
6. Macêdo R, Manso VAV, Klein AEF: The geometric relationships of salients and tombolos along a mesotidal tropical coast. Geomorphology. https://doi.org/10.1016/j.geomorph.2022.108311
7. da F Klein AH, Filho LB, Schumacher DH: Short-term beach rotation processes in distinct headland bay beach systems. J Coastal Res 18(3): 442–458

Wind and Sediment Transport Pattern Inside a Small Trough Blowout During Opposing Wind Conditions

Camille René[1]([✉]), Nicolas Robin[1], Thomas A. G. Smyth[2], Patrick A. Hesp[3], Antoine Lamy[1], and Olivier Raynal[1]

[1] CEFREM, University of Perpignan via Domitia, Perpignan, France
camille.rene@univ-perp.fr
[2] Department of Geography and Environmental Science, Liverpool Hope University, Liverpool, England
[3] BEADS Laboratory, Flinders University, Bedford Park, Australia

Abstract. In recent decades, significant anthropogenic impact has modified European coastal dunes, leading to the development of blowouts, a common wind erosional feature. This study investigates a small trough blowout altered by closed human footpath since several decades at Canet-en-Roussillon during both offshore and onshore wind events. Campaigns consisting of 26 2D ultrasonic anemometers deployed at 0.30 m across the entire surface of the blowout. Simultaneously, sediment transport runs were conducted using Hilton sand traps. The results revealed the blowout as a double-nested system with distinct wind dynamics driven by bidirectional wind regimes. Section 1 (landward) is subjected to oblique incident winds, whereas Sect. 2 (seaward) is governed by parallel flow. The interplay between the winds complexity and the morphology of the blowout results in relatively low sediment transport rates. Although offshore winds dominate the environment, it is the onshore winds that generate the highest transport rates. This dynamic raises important questions about the long-term stability and lifespan of such systems.

Keywords: Trough Blowout · Aeolian processes · Sediment transport rates

1 Introduction

Human activities, especially tourism, are heavily concentrated at the coast. This concentration of activity can lead to degradation of coastal dunes which are necessary for protection against marine flooding [1]. This degradation result in topographic weaknesses that can trigger the formation of blowouts. These erosional landforms substantially affect wind flow patterns due to their topographic complexity and the angle of the incident wind [2]. Knowledge of the aeolian processes in these systems is still limited, particularly in environments controlled by opposing wind directions (onshore/offshore) or only by offshore conditions. The aim of this research is to investigate the relationship between wind processes, aeolian sediment transport and a complex trough blowout created by trampling in a French Mediterranean coastal dune.

© The Author(s) 2026
C. Coelho et al. (Eds.): CD 2025, CRL 41, pp. 29–33, 2026.
https://doi.org/10.1007/978-3-032-15473-6_5

2   Study Site

The field site is a trough blowout located in the south of the Gulf of Lion at Canet-en-Roussillon (SE-France) (Fig. 1). The blowout is 2.5 m deep, 20 m long for 3 m wide and its morphology is disrupted by anthropogenic activity with trampling. It was naturally initiated in 2009 and rapidly extended landward along the onshore prevalent winds which led to its connection with a closed footpath. This connection before 2014 marked the cessation of further elongation and the establishment of an elbow-shaped morphology. It consists of a trough axis divided into two parts (Sect. 1 -landward, WNW- and Sect. 2 -seaward, NNW) (Fig. 1). Field site is a wave-dominated microtidal environment where tidal range is around 0.30 m [3]. Thus, dune-beach profile is scarcely affected by hydrodynamic processes. Aeolian processes are influenced by two main winds: offshore winds (NW) are the most frequent (61.6% of the time) and onshore winds (S-SE) representing 19.6% of the time (Fig. 1). The median grain size (D50) ranges from coarse to very coarse sand on the floor of the deflation basin and around medium sand on the dune.

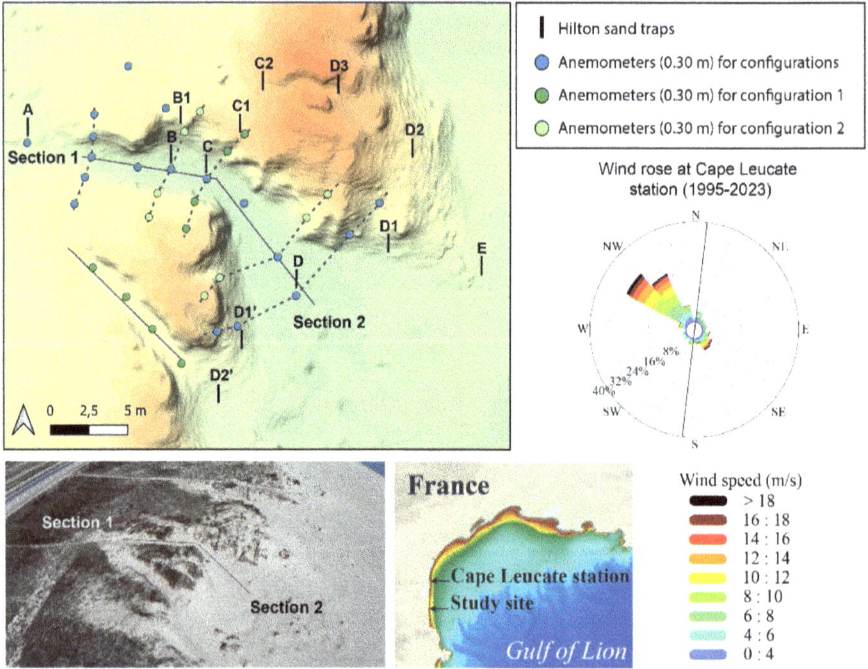

Fig. 1. Representation of the blowout dimensions and location of instruments.

3   Data and Methods

Two-days of short field campaigns were conducted during offshore (27/02/24 mean wind speed 13.7 m/s) and onshore winds (07/03/24 mean wind speed 7.0 m/s). Twenty-six 2D ultrasonic anemometers were deployed at 0.3 m above the ground, over the entire area (Fig. 1). Another sensor was placed at 4.5 m above the surface of the back dune as a reference station. Wind speed was normalized to the reference station, to reduce effects of temporal variability [4]. Five to ten sand trap stations consisting of 12 traps arranged vertically [5], were placed for 15-min runs (Fig. 1). Sand transport rates for each station were determined by weighing the sand in the traps and integrating over the vertical profile of each station.

4 Results

4.1 Spatial Wind Flow Behavior

Wind flow patterns during offshore conditions show that there are two zones of acceleration within the blowout (Fig. 2A). The first begins in Sect. 1 with maximum velocities within the constricted throat region. Then bifurcated into Sect. 2 aligned with the incident winds that allows acceleration. The second one is observed on the southern dune crest due to the compression of the flow up the slope [6]. A strong deceleration of the wind is observed in the seaward part of the blowout due to streamline expansion behind the dune crest which is also associated with an increase in turbulence (**ref**). The direction recorded within the trough of the blowout deviated parallel to the two axes of the blowout sections.

The wind flow pattern during onshore conditions was drastically different (Fig. 2B). Speed acceleration occurred in Sect. 2 while deceleration is observed in Sect. 1 due to the bifurcation. A speed up occurred also on the northern dune crest due to topographic steering on steep lateral walls inducing an ejection flow out of the blowout. In Sect. 2, which is parallel to the incoming wind, there is no change in direction. An increase in turbulence associated with deceleration is observed in Sect. 1.

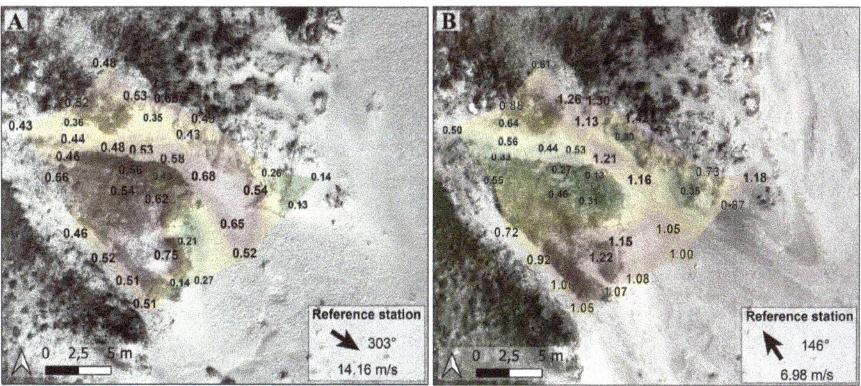

Fig. 2. Planimetric representation of direction and normalized wind speed for offshore (A) and onshore (B) campaigns.

4.2 Sediment Transport Rates

The sediment transport rate during offshore winds was very low inside Sect. 1 reaching only 1 kg.m^{-1}.h^{-1} and slightly higher (about 2 kg.m^{-1}.h^{-1}) in the enlargening entrance area of the blowout (end of Sect. 2) [7] [4] (Fig. 3A). The maximum sediment transport rate was observed on the seaward walls, reaching 14 kg.m^{-1}.h^{-1} despite lower wind speeds but in an area composed of finer sand and no vegetation cover. This transport was oriented to the SW. A different sediment transport pattern occurred during onshore conditions (Fig. 3B). In the constricted throat region, a much larger flux was recorded (up to 21.85 kg.m^{-1}.h^{-1}) than during offshore winds. The highest sediment transport rate was on the eastern flank on the dune (reaching 84 kg.m^{-1}.h^{-1}) due to streamline compression inducing transport towards the NW and the dune crest. Sediment transport rates during onshore winds were higher than during offshore winds despite the much lower mean wind speed. This may be because the upwind fetch surface consisted of a large area of bare sand, whereas the fetch surface for offshore winds was vegetated.

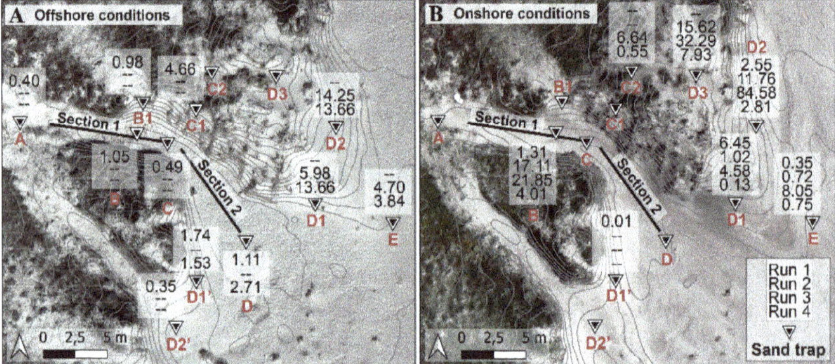

Fig. 3. Planimetric representation of sand transport rates (kg/m/h) during offshore (A), onshore (B).

5 Conclusion

This study, conducted in an environment dominated by landward winds but where seaward winds are also present, provides insights into the temporal and spatial variability of aeolian processes affecting a representative blowout along the Mediterranean coastline. Concerning the wind velocity, a similar pattern of wind flow was identified across various parts of the site, contingent on prevailing wind conditions. Acceleration zones were consistently observed within the constricted throat region and along the crests, whereas deceleration zones occurred behind the slopes, either seaward and landward, depending on offshore or onshore wind regimes, respectively. The behavior of the wind proves to be complex due to the variability in morphology, both along the axis of the blowout's development and in the perpendicular section. Specifically, offshore winds intensify flow velocities through the blowout, while onshore winds produce a deceleration effect. Moreover, the wind direction within the system evolves over time, depending on the prevailing wind regime. These findings underscore the capacity of blowouts to significantly alter or modify wind flow dynamics, with turbulence commonly associated with deceleration zones and enhanced velocities characterizing acceleration zones. The interaction of the blowout morphology with anthropogenic features, such as the bifurcation induced by the footpath modulates these wind patterns.

Furthermore, sediment transport values were obtained at several locations within the blowout under both onshore and offshore wind conditions. Once again, these results reveal a spatially and temporally complex behavior. This behavior is not necessarily correlated with the wind speeds recorded within the system. It appears that, onshore winds, despite being less dominant in the regional wind regime and characterized by lower velocities (compared to offshore regime), drive the highest sediment transport rates. This behavior arises from the fact that, under onshore conditions, the wind blows across a wide beach acting as a sediment source, whereas under offshore conditions, the vegetated contours of the blowout limit sediment transport potential to within the blowout itself. These events are pivotal for blowout evolution, as they enable the direct transport of sand from the adjacent beach system.

This interplay between wind direction, flow dynamics, and sediment supply suggests that morphogenic events associated with onshore winds, though relatively infrequent, play a critical role in sustaining blowout development. Consequently, the longevity of this geomorphic feature is likely secured by the episodic but substantial sediment flux generated during these events.

References

1. Cooper A, Jackson D (2020) Dune gardening? A critical view of the contemporary coastal dune management paradigm. Area 53(2):345–352
2. Pease P, Gares P (2013) The influence of topography and approach angles on local deflections of airflow within a coastal blowout. Earth Surf Process Landforms 38:1160–1169
3. Brunel C et al (2014) 20th century sediment budget trends on the Western Gulf of Lions shoreface (France): An application of an integrated method for the study of sediment coastal reservoirs. Geomorphology 204:625–637
4. Sun Y, Hasi E, Liu M, Du H, Guan C, Tao B (2016) Airflow and sediment movement within an inland blowout in Hulun Buir sandy grassland, Inner Mongolia. China. Aeolian Research 22:13–22
5. Hilton M et al (2017) An efficient, self-orienting, vertical-array, sand trap. Aeol Res 25:11–21
6. Davidson SG, Hesp PA, DaSilva M, Da Silva GM (2022) Flow dynamics over a high, steep, erosional coastal dune slope. Geomorphology 402:108–111
7. Smyth TAG, Jackson D, Cooper A (2014) Airflow and aeolian sediment transport patterns within a coastal trough blowout during lateral wind conditions. Earth Surf Process Landforms 39:1847–1854

The Influence of Oregon's Coastal Dunes on Tsunami Impact and Inundation

Carly A. Ringer[1]([✉]), Meagan E. Wengrove[2], Jonathan C. Allan[3], Erin K. Peck[4], Ernesto Guerrero Fernandéz[5,6], and Yong Wei[5,6]

[1] College of Earth, Ocean, and Atmospheric Sciences, Oregon State University, Corvallis, OR, USA
ringerc@oregonstate.edu
[2] School of Civil and Construction Engineering, Oregon State University, Corvallis, OR, USA
[3] Oregon Department of Geology and Mineral Industries, Newport, OR, USA
[4] Graduate School of Oceanography, University of Rhode Island, Kingston, RI, USA
[5] Ocean, and Ecosystem Studies, Cooperative Institute for Climate, University of Washington, Seattle, WA, USA
[6] NOAA Center for Tsunami Research, NOAA Pacific Marine Environmental Laboratory, Seattle, WA, USA

Abstract. Nearly half of the Oregon coast is comprised of dune-backed beaches, and these foredunes serve as protective features against coastal flood hazards. However, the ability of Oregon's coastal dunes to protect against tsunami inundation is poorly understood due to existing inundation modeling that assumes static bathymetry. With an estimated 20% chance of partial rupture of the Cascadia Subduction Zone in the next 50 years, it is imperative to investigate how coastal dunes will mitigate tsunami impacts. This study explores how including sediment transport in tsunami inundation modeling at Netarts Spit, Oregon, USA impacts maximum tsunami flow depth and inundation extent, and how a tsunami shapes coastal dunes. Using the model FUNWAVE-TVD, we simulate the M1 tsunami scenario produced by the Oregon Department of Geology and Mineral Industries (DOGAMI) and show that the hydrodynamic output is comparable to the output produced by DOGAMI. Then, with the sediment transport module in FUNWAVE-TVD, we simulate tsunami-induced morphologic change. We show that dunes with a maximum pre-tsunami elevation of less than about 12 m are eroded vertically by up to 0.42 m. Additionally, we find that allowing for sediment transport in the simulations does not greatly impact inundation extent on Netarts Spit but does change maximum flow depth in some areas. Future work will aim to validate the sediment transport simulations using a robust record of tsunami deposits from the salt marsh in Netarts Bay.

Keywords: Tsunami inundation · coastal dunes · coastal geomorphology · FUNWAVE-TVD

© The Author(s) 2026
C. Coelho et al. (Eds.): CD 2025, CRL 41, pp. 34–40, 2026.
https://doi.org/10.1007/978-3-032-15473-6_6

1 Introduction

The Oregon coast in the United States faces chronic and acute erosion hazards, including the potential of a Cascadia Subduction Zone (CSZ) earthquake and tsunami. Nearly 50% of the Oregon coast consists of dune-backed beaches [1], with communities established just landward of the dunes. While Oregon's dunes can be effective nature-based buffers against the wave climate of the U.S. Pacific Northwest [2], the role of dunes in mitigating tsunami impact and inundation is not well understood. With a ~20% chance of a partial CSZ rupture occurring in the next 50 years [3], it is crucial to explore how coastal dunes respond to tsunami inundation and how tsunami inundation is influenced by tsunami-induced dune erosion.

Current products developed for tsunami preparedness by the Oregon Department of Geology and Mineral Industries (DOGAMI) are based on model simulations that assume static bathymetry. However, field observations after tsunamis show that significant morphologic change occurs [4], often with coastal dunes eroded and sediment deposited inland [5]. Therefore, incorporating sediment transport in tsunami modeling is essential to understand feedback mechanisms resulting in greater tsunami inundation and quantify the potential protective role of dunes. In this study, we examine morphologic change and inundation during a CSZ event at Netarts Spit, Oregon, USA.

2 Site and Methods

2.1 Netarts Spit, Tillamook County, Oregon

Netarts Spit separates Netarts Bay from the Pacific Ocean in Tillamook County, Oregon, USA (Fig. 1A). The spit is approximately 8 km long and features diverse vegetation types and landforms, including a dune ridge along the ocean side. This site was selected for two reasons: first, the dune ridge allows us to model tsunami-induced dune erosion; and second, previous research has established a robust record of the 1700 C.E. CSZ tsunami deposits found in the salt marsh in the southern portion Netarts Bay [6], providing us with in-situ measurements of tsunami deposits to use in model validation.

2.2 Hydrodynamic and Morphologic Change Modeling

We used the model FUNWAVE-TVD [7] to simulate tsunami hydrodynamics and sediment transport. We simulated the "Medium" (M1) tsunami scenario developed by [8], which is the result of an ~8.9 magnitude earthquake with an average slip of about 7–9 m. Full description of the DOGAMI tsunami scenarios can be found in [8] and [9]. The bathymetry data used in this study were provided by DOGAMI as a point shapefile [10] and converted to a raster surface using inverse distance weighting interpolation in ArcGIS Pro 3.1. The model grid used in the simulations in this study was 1201×601 with $\Delta x = \Delta y = 10$ m (~12 km $\times$ 6 km), covering the entirety of Netarts Spit and Bay (Fig. 1A).

We provided FUNWAVE-TVD with a timeseries of water surface elevations (η) and flow velocities (u, v) extracted from the M1 simulation at our model grid boundaries. To ensure the tsunami was simulated accurately, we compared the FUNWAVE-TVD and existing M1 simulation η timeseries at four virtual stations (Fig. 1A). Sediment transport was not included in this validation.

After hydrodynamic validation, we then used the sediment transport module in FUNWAVE-TVD to simulate tsunami-induced morphologic change. We assumed that the bed under Netarts Bay is primarily comprised of cohesive sediment, and therefore we considered it non-erodible. The rest of model domain was assumed to be sand with a median grain size of 0.25 mm and with a maximum erodible depth of 5 m.

To investigate how incorporating sediment transport into tsunami modeling impacts inundation, we first compared the maximum flow depth and extent between simulations with and without sediment transport, run for 120 min of simulation time. Inundation extents were compared using the Intersection over Union (IoU) similarity index:

$$IoU = \frac{Numberofinundatedcellsinbothsimulations(intersection)}{Numberofinundatedcellsineithersimulation(union)} \tag{1}$$

Maximum flow depth was calculated by finding the maximum water depth for each grid cell over 120 min of simulation time. We then computed the difference between the simulations with and without sediment transport. To observe how dunes respond to tsunami impact, we examined the pre- and post-tsunami elevation of the alongshore dune crest (highest point) profile on Netarts Spit.

3   Results

3.1   FUNWAVE-TVD Hydrodynamic Validation

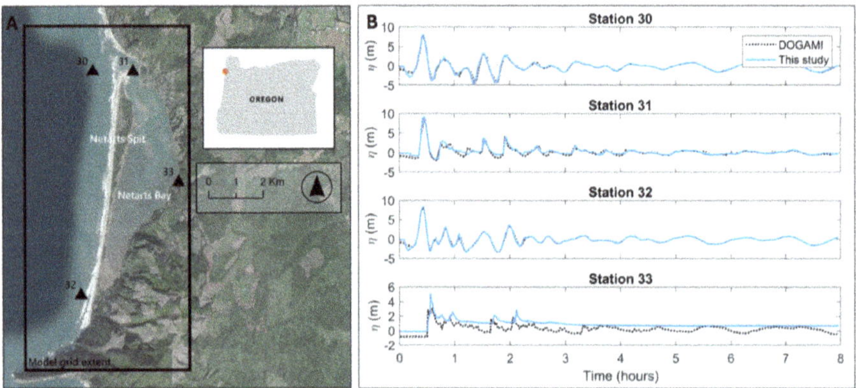

Fig. 1. (A) Map of Netarts Spit and Bay with model grid extent (black box), locations of stations used in hydrodynamic validation (black triangles), and location of study site in Oregon (inset). (B). Comparison of the CSZ M1 tsunami water surface elevation (η, in m) at four virtual stations during a full 8-h simulation from the DOGAMI simulation (dotted black line) and the FUNWAVE-TVD simulation (solid blue line).

We simulated the CSZ M1 tsunami event at Netarts Spit in FUNWAVE-TVD for 8 h of simulation time and compared the water surface elevation at four stations to time-series from the output of the DOGAMI simulation of the same event. The results from FUNWAVE-TVD show good agreement with the DOGAMI simulation with R^2 values ranging from 0.52 to 0.97 (Fig. 1B). The station 33 R^2 value of 0.52 is likely due to the spatial offset between the stations used in FUNWAVE-TVD and the DOGAMI simulation due to differences in model grids. Stations 30, 31, and 32 show strong agreement with R^2 values of 0.96, 0.85, and 0.97, respectively.

3.2   Netarts Spit Inundation and Dune Crest Elevation Change

We calculated the tsunami inundation extent and the maximum flow depth after 120 min of simulation time over the model domain. The inundation extents with and without sediment transport have a IoU score of 0.97, indicating a nearly complete overlap in inundation extent. The difference in maximum flow depth between the two simulations is close to zero in most of the model domain, with areas of increase and decrease primarily on the east (bay) side of Netarts Spit and the southern end of Netarts Spit and Bay (Fig. 2A-B). The maximum positive difference (increase in flow depth with sediment transport on) is 1.89 m, and the maximum negative difference (decrease in flow depth with sediment transport on) is -3.27 m.

Areas where the dune crest is above ~ 12 m in elevation before the tsunami were not overtopped, so there is no elevation change; however, areas where the dune crest is below ~ 12 m in elevation were overtopped, which caused both erosion and deposition on the dune crest (Fig. 2B-C). The mean change in elevation pre- and post- tsunami is − 0.02 m. The mean change at cells where elevation did change is −0.03 m. The maximum positive change (deposition) is 2.48 m. The maximum negative change (erosion) is − 0.42 m.

4   Discussion

The purpose of this study was to explore how including sediment transport in inundation modeling impacts the topography of the study area and the inundation and flow depth over the study area. First, we found that FUNWAVE-TVD simulated the M1 tsunami event comparably to the previous simulation of the same event by DOGAMI, adding confidence to our use of FUNWAVE-TVD to investigate tsunami-induced morphologic change and inundation.

Field studies after the 2011 Tohoku-Oki tsunami show that coasts with intermittent traditional (hard) protective structures experienced more severe erosion in the less protected areas [5]. If we consider Oregon's coastal dunes to be nature-based protective structures, then we see a similar pattern at Netarts Spit: where the dune crest is low (a "break" in coastal protection), dunes are overtopped by the tsunami and eroded, and where the dune crest is high, limited inundation occurs from the ocean side. While the observed crest overtopping and erosion is expected, we are not yet confident in the magnitude of the elevation changes output by the model. Further work is required to calibrate and validate the sediment transport results. We note that because Netarts Spit

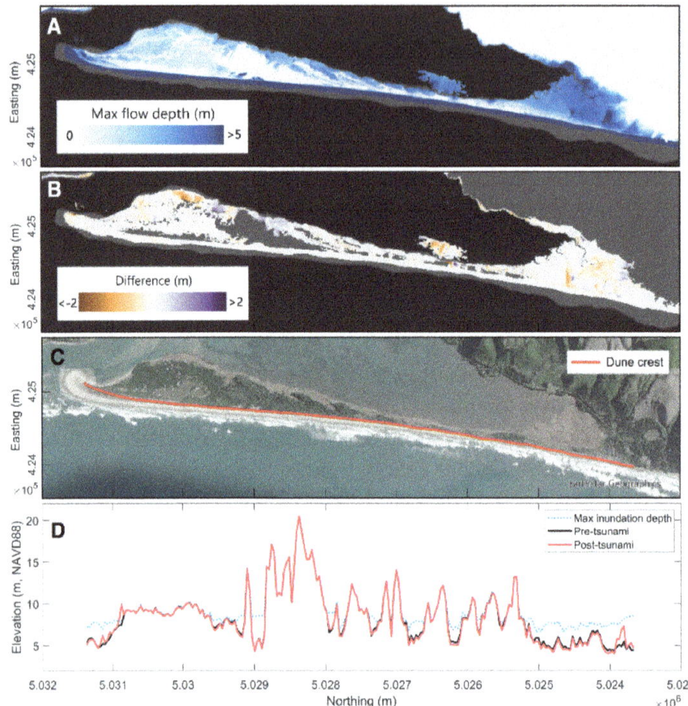

Fig. 2. (A) Maximum flow depth with sediment transport on after 120 min. (B) Difference in maximum flow depth between simulations with and without sediment transport after 120 min. Positive (negative) values indicate an increase (decrease) in flow depth when sediment transport is on. (C) Alongshore dune crest elevation pre-tsunami (solid black line) and post-tsunami (solid red line), and maximum flow depth above dune crest (dotted blue line) after 120 min. (D) Location of alongshore dune crest line on Netarts Spit.

is also inundated from the east (bay) side as the tsunami enters the bay and propagates southward, it is difficult to elucidate whether a continuous and taller dune ridge would reduce impacts on the bay side of the spit.

Tsunami flow depth can also be used to assess hazard level [11], and we see that at Netarts Spit, when including sediment transport in the simulation, the maximum flow depth changes primarily on the bay side of Netarts Spit behind the lowest elevation dunes. Over the course of the simulation, the morphology changes in response to tsunami currents, and the tsunami flow depth in turn responds to the change in bed elevation. Thus, including sediment transport in inundation modeling may provide a more accurate estimate of the flow depth as a function of time.

Future work will aim to validate the sediment transport and morphologic change observed in this study. Using the existing record of tsunami deposits from the 1700 C.E. CSZ event in Netarts Bay [6] and adding to the record using sediment cores collected in January 2025, we will compare the morphologic change simulated in FUNWAVE-TVD to the in-situ measurements of tsunami deposit thickness. The model bathymetry/topography grid will be modified using aerial imagery of Netarts Spit from 1939 to reconstruct conditions as similar to the year 1700 C.E. as possible.

5 Conclusion

Oregon's coastal dunes have the potential to protect communities from tsunami inundation and tsunami-induced morphologic change; however, previous modeling efforts in Oregon do not account for morphologic change and thus the ability of dunes to provide protection is unknown. We aim to fill that gap by using a model that simulates both the tsunami hydrodynamics and morphologic change and investigates the impacts of including morphologic change on tsunami inundation extent, depth, and dune crest elevation. This work has important implications for dune management in Oregon, as the dynamic bathymetry may provide further insight into the tsunami hazard vulnerability of communities adjacent to sandy coastlines.

References

1. Cooper WS (1958) Coastal Sand Dunes of Oregon and Washington. GSA-Memoir 72:169
2. Mull J, Ruggiero P (2014) Estimating Storm-Induced Dune Erosion and Overtopping along US West Coast Beaches. J Coastal Res 30:1173–1187
3. Goldfinger C, Galer S et al (2017) The importance of site selection, sediment supply, and hydrodynamics: A case study of submarine paleoseismology on the northern Cascadia margin, Washington USA. Mar Geol 384:4–46
4. Gelfenbaum G, Jaffe B (2003) Erosion and Sedimentation from the 17 July, 1998 Papua New Guinea Tsunami. Pure Appl Geophys 160(10):1969–1999
5. Tanaka H, Tinh N, Umeda M, Hirao R, Pradjoko E, Mano A, Udo K (2012) Coastal and Estuarine Morphology Changes Induced by the 2011 Great East Japan Earthquake Tsunami. Coastal Eng J 1525
6. Peck EK, Guilderson TP, et al (2022) Recovery Rate of a Salt Marsh From the 1700 CE Cascadia Subduction Zone Earthquake, Netarts Bay, Oregon. Geophys Res Lett 49(18):e2022GL099115
7. Shi F, Kirby J, ct al (2016) FUNWAVE-TVD Fully Nonlinear Boussinesq Wave Model with TVD Solver—Documentation and User's Manual (Version 3.0)
8. Witter R, Zhang Y, et al (2011) Simulating tsunami inundation at Bandon, Coos County, Oregon, using hypothetical Cascadia and Alaska earthquake scenarios (43; DOGAMI Special Paper)
9. Priest GR, Goldfinger C, et al (2009). Tsunami hazard assessment of the northern Oregon coast: A multi-deterministic approach tested at Cannon Beach, Clatsop County, Oregon (41; DOGAMI Special Paper)
10. Allan JC, Zhang YJ, O'Brien FE (2021) Tsunami inundation modeling update for the northern Oregon coast: Tillamook and Clatsop counties (O-21–08)
11. Zamora N, Catalán PA, Gubler A, Carvajal M (2021) Microzoning Tsunami Hazard by Combining Flow Depths and Arrival Times. Frontiers in Earth Science, 8

Morphological Development and Evolution of Barchanoid Dunes on the Cap Ferret Coastal Spit

Clélia Billieres-Kim[1,2]([✉]), Julie Billy[3], Raphaël Bourillot[4], and Hugues Fenies[5]

[1] BRGM, French Geological Survey, 33600 Pessac, France
c.billieres@brgm.fr
[2] Univ. Bordeaux, CNRS, Bordeaux INP, EPOC, UMR 5805, 33600 Pessac, Bordeaux, France
[3] BRGM, 33600 Pessac, France
[4] UMR 5805, Univ. Bordeaux, CNRS, INP, EPOC, 33600 Pessac, Bordeaux, France
[5] CV Associés Engineering, Pittsburgh, USA

Abstract. Barchans are crescent-shaped dunes with their convex side facing the wind and their horns pointing downwind, commonly observed in deserts but also in coastal environments. This study investigates the dune field of the Cap Ferret Spit (5-km wide, 23-km long; Atlantic Coast, SW France) including of series of barchanoids dunes, currently stabilized by forest cover. The aims are to classify each dune into distinct typologies and characterize their morphology in order to analyze their evolution. LiDAR-based analysis identified four distinct barchanoid dune types. Moreover, it revealed superimposed bedforms on the dunes, including draping and oscillating patterns, reflecting the influence of consistent wind regimes during dune development and migration. This study proposes an evolutionary sequence in which proto-barchans evolve into barchanoid ridges and eventually as mature isolated barchans. These findings provide insight into the geomorphological evolution of barchanoid dunes in a coastal context, and particularly their migration pattern on a coastal spit system.

Keywords: Barchanoid dunes · Spit · Morphology · Superimposed bedforms · Dune migration

1 Introduction

Barchan dunes are one of the most iconic types of dunes and can be located in desert areas, as well as along coastlines [1–4]. They are defined by a crescent shape with a convex windward side and a concave leeward side. Their geomorphology can be influenced by wind direction, sediment availability, and substrate properties. Unlike desert barchans, coastal barchanoids remain understudied, even if these dunes are prominent along many coastlines, *e.g.,* the eastern Atlantic coast (*e.g.* Cap Ferret Spit of France, Doñana Spit bar in Spain [5]), the Rio Grande do Norte coast in Brazil [4], etc. Studies referring to barchan migration or geomorphological evolution are mainly based on numerical models (*i.e.* theoretical evolutionary models based on distinct wind directions) and consider high

C. Coelho et al. (Eds.): CD 2025, CRL 41, pp. 41–47, 2026.
https://doi.org/10.1007/978-3-032-15473-6_7

dome-shaped sand deposits as starting points [6, 7]). However, these idealized cases are not always representative of barchan dune initiation, especially in coastal areas.

This study focuses on the inland domain of an emblematic coastal system on the Atlantic coast, SW France, the Cap Ferret Spit, a 23-km-long and 5-km-wide sedimentary spit (Fig. 1). Over the last 3k years, the progradation of the spit pushed the inlet of the Leyre estuary to the south, whereby shaping the modern geomorphology of the Arcachon Bay [8]. The spit hosts three distinct dune systems: proto-barchan, barchanoid ridge and barchan. Only the foredune to the west is active, all other dunes inland are fixed by maritime pine forests due to coastal management practices since the mid-19th century. The objective here is to characterize barchanoid dune systems at different stages of their formation and to highlight key elements related to their mobility.

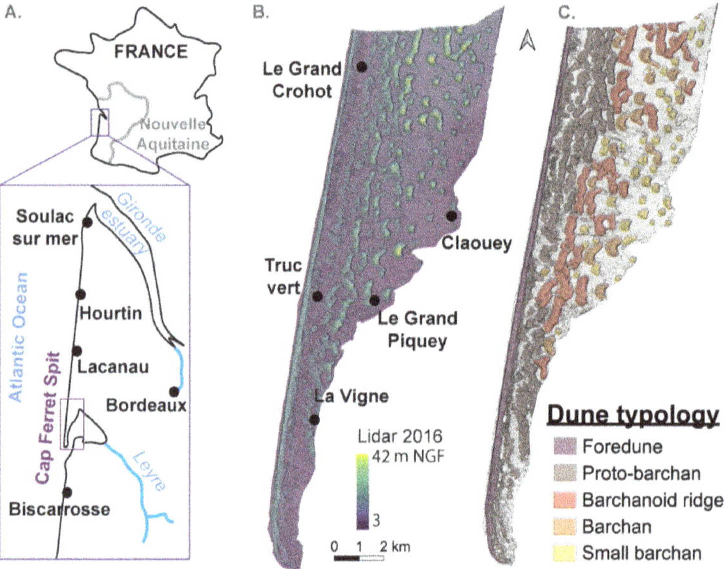

Fig. 1. A) Location map of the study area (Atlantic coast, France); B) LiDAR map (SIBA 2016, res. 1m) of the Cap Ferret Spit; C) Dune typologies identified on the spit: foredune, proto-barchan, barchanoid ridge and isolated barchan.

2   Methodology

The analysis is based on the high-resolution LiDAR data collected in 2016 and provided by the Arcachon Bay Intercommunal Syndicate (SIBA) and is supplemented by data from the Nouvelle Aquitaine Coast Observatory [9]. The 2016 LiDAR dataset (Fig. 1B) has a spatial resolution of 1 m, enabling detailed and precise dune mapping. The data were analyzed using Geographic Information System (GIS) software, specifically QGIS. Slope maps were generated from the LiDAR dataset to support a detailed analysis of dune morphology and their spatial organization. Each dune type was characterized based on key morphological parameters, including shape (*i.e.* not well-formed dune, interconnected dunes, well-defined crescent shape), maximum crest height, dune width, dune volume and surface area. These metrics were selected to enable a representative comparison among all dune types, including those with variable lengths.

3   Classification of the Cap Ferret Dune Systems

3.1   Typologies and Spatial Distribution

Prior classifications of dunes along the Aquitaine coast [10] identified four main dune types from the shore to the land: (i) foredunes are coastal dunes formed by sand deposition along the shoreline, (ii) undefined dunes are behind foredunes, (iii) barchanoid dunes composed of barchans and barchanoid ridges and (iii) parabolic dunes characterized by crescent-shaped dune with horns pointing upwind in contrast to barchans (single, digitate or climbing forms). On the Cap Ferret Spit, only foredunes and barchanoid dunes were observed (Fig. 1C, Fig. 2). Parabolic dunes, commonly observed in other regions of this coast, are absent.

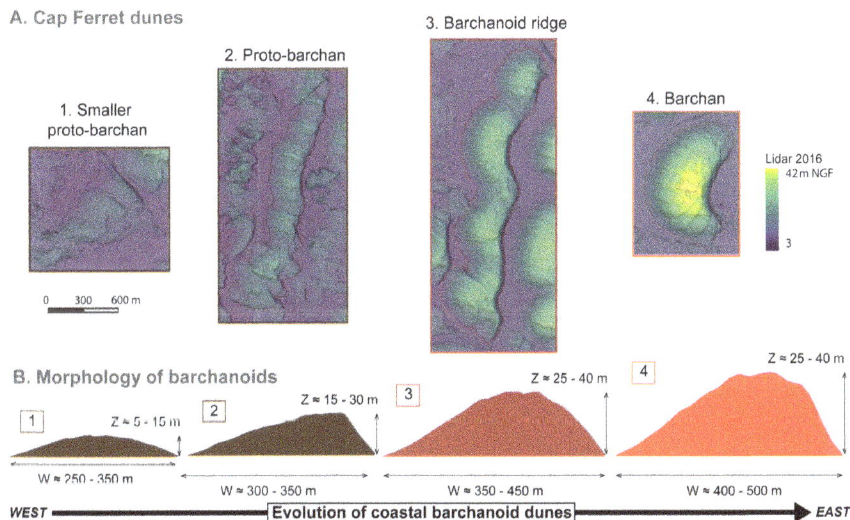

Fig. 2. (A) Dune typologies identified on the Cap Ferret Spit (LiDAR data overlaid with slope map), (B) Representative 2D sections of the main barchanoid dune morphologies (Z: elevation in m NGF; W: dune width; Vertical Exaggeration x9)

In this study, barchanoid dunes constitute the primary typology of the spit dune field, exhibiting greater complexity than previously exposed on the prior classification ([10], Fig. 2). These dunes can be categorized into three types: (i) proto-barchans (PB), which feature a low topography and form interconnected dune systems without specific shape (Fig. 2A-1,2; altitude range from 5 to 15 m and width from 250 to 350 m), (ii) barchanoid ridges (BR), consisting of interconnected dunes with crescent-shaped morphologies (Fig. 2A-3; altitude range from 25 to 40 m and width from 350 and 450 m) and (iii) isolated barchans (B) characterized by well-defined crescent-shaped morphologies (Fig. 2A-4, with altitudes of 25 to 40 m and width from 400 to 500 m). Each barchanoid dune displays distinct morphological characteristics, notably in height or shape. The distribution of these forms reveals a spatial dune progression from west to east, with the dunes becoming increasingly defined and higher in elevation toward the Arcachon Bay. For example, maximal crest heights ranged from 15 m (PB) to 40 m (B) (Fig. 2B), illustrating a positive gradient across the spit.

3.2 Barchanoid Evolution Model

Geomorphological observations reveal a pronounced west-to-east gradient across the 5-km-wide spit, characterized by variations in dune height and shape. This gradient reflects the gradual evolution pattern of the dunes (Fig. 2B). The dune evolution begins with sand accumulation near the coast, initially connected to the beach system, where wind dynamics drive their eastward migration. At this stage, embryonic dunes (PB) lack distinct forms. As this aeolian process continues, the embryonic dunes migrate inland, and crescent-shaped structures begin to emerge, remaining interconnected (BR). Over time, ongoing aeolian dynamics result in further growth and individualization, forming distinct isolated barchan dunes (B). This progression illustrates a clear dynamic sequence of barchanoid dune development, a process rarely documented in the existing literature. The transition from PB to BR represents a critical phase in the evolutionary sequence, as dunes evolve from embryonic, poorly organized sand masses (PB) into larger, more structured forms with increased internal complexity (BR). Preliminary insights into their internal architecture are presented in previous work [11]. The final stage is marked by the development of well-defined barchans, where dunes become isolated and individualized. These crescent-shaped barchans continued their eastward migration until being stabilized by forest plantations in the mid-19th century [12], completing the evolutionary sequence.

4 Superimposed Bedforms

The barchanoid dunes of the studied spit display two distinct types of superimposed bedforms with contrasting orientations: (i) oscillating patterns and (ii) draping forms (Fig. 3). Notably, similar structures have been observed on active modern dunes, such as the Pyla Dune [15] where superimposed forms migrate toward crests at rates of up to 5 m per month, driven by seasonal wind variations. Superimposed oscillating patterns (profile 1; Fig. 3) are prominent on the back (ocean facing) side of the dunes, partially smoothed or overlain by draping forms near the crest. These patterns extend 30–50 m in length and reach heights of up 8 m. Draping forms (profile 3, 4 and 5; Fig. 3) dominate the dune crests, reflecting upward sand migration driven by dominant westerly winds, which facilitate the eastward growth of the dunes. These forms span lengths of ~10 to 60 m and have elevations of 2.5 m–4.5 m.

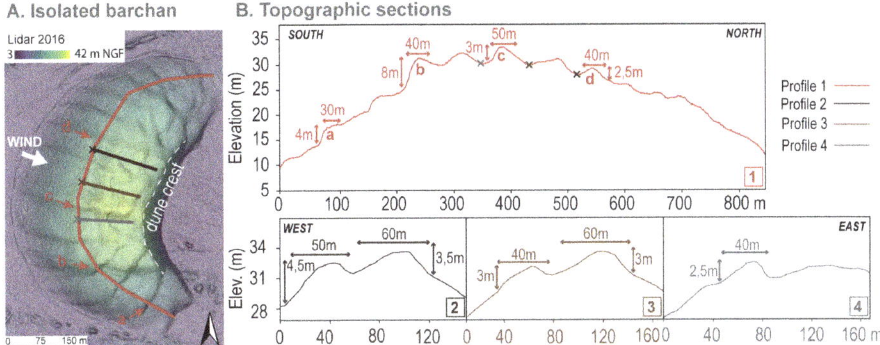

Fig. 3. Focus on (A) an isolated barchan of the Cap Ferret Spit (LiDAR data overlaid with slope map); (B) Topographic sections on superimposed bedforms (Profile 1 covering oscillating structures; Profile 2, 3 and 4 on draping forms)

These contrasting structures suggest a complex interplay of wind dynamics. While draping forms align with stable wind directions, oscillating patterns imply a more variable wind regime than a simple strictly westerly dynamic. These patterns highlight the role of winds shaping the secondary structures, adding nuance to the processes driving dune evolution. The presence of these forms often indicates specific climatic and wind conditions during dune formation [13]. Previous research [14] associates such features with stable wind patterns and low variability, enabling the progressive development of secondary structures atop primary dunes. However, the oscillating patterns observed challenge this understanding, suggesting that more variable and complex wind regimes played a role in shaping the studied dunes. Further research into wind dynamics, including directionality, strength, and variability, is crucial to understand the processes responsible for these structures. Such investigations would enhance knowledge of aeolian processes and dune evolution.

5   Conclusion

This study highlights the former (*i.e.,* pre-mid-19th century) dynamic evolution of barchanoid dunes on the Cap Ferret Spit. Barchanoids are analyzed and classified using high-resolution LiDAR data. A clear west-to-east progression pattern is observed, beginning with embryonic dunes connected to the beach system, shaped by initial sand accumulation and wind dynamics. Through aeolian processes, these structures gradually evolve into crescentic forms and ultimately into distinct isolated barchan dunes. These findings uncover the complex interactions between aeolian processes, sediment transport, and environmental factors, offering new insights into the dynamic of coastal barchanoids, which are often understudied compared to desertic barchanoid systems. The presence of superimposed bedforms adds further complexity to these systems, revealing the significant influence of wind variability and environmental constraints in shaping their evolution.

Acknowledgements. The work is a part of the DUNES research program, co-funded by the BRGM (French Geological Survey), the Nouvelle-Aquitaine Region, the ONF, the Landes Department, Vermilion REP SAS, CCGL, AEAG, and of the CAFERS project, funded by the EC2CO HYBIGE CNRS program.

References

1. Bristow C (2019) Bounding surfaces in a barchan dune: annual cycles of deposition? Seasonality or Erosion by Superimposed Bedforms? Remote sensing 11:965
2. Fu T, Wu Y, Tan L, Li D, Wen Y (2019) Imaging the structure and reconstructing the development of a barchan dune using ground-penetrating radar. Geomorphology 341:192–202
3. Gomez-Ortiz D, Martin-Crespo T, Rodriguez I, Sanchez MJ, Montoya I (2009) The internal structure of modern barchan dunes of the Ebro River Delta (Spain) from ground penetrating radar. J Appl Geophys 68:159–170
4. Oliveria J, Medeiros W, Tabosa W, Vital H (2008) From barchan to domic shape: evolution of a coastal sand dune in northeastern Brazil based on GPR survey. Brazilian J. Geophysics 26(1)
5. Goy J-L et al (2022) Holocene aeolian dunes in the National and Natural Parks of Doñana (SW Iberia): Mapping, geomorphology, genesis and chronology. Geomorphology 398(2022):108066
6. Hersen P: Morphogenese et Dynamique des Barchanes. Univ. Paris-Diderot - Paris VII
7. Yang B, Chong Y, Lin L, He N, Zhang Y, Gao X (2024) Experimental study of the stability of crescent barchan shape in subaqueous dune evolution in unsteady water flow. Europ J Mech/B Fluids 103(2024):246–258
8. Féniès H, et Lericolais (2005) Internal architecture of an incised valley-ll on a waveand tide-dominated coast (the Leyre incised valley, Bay of Biscaye, France). Comptes Rendus Geoscience 337(14):12571266
9. Observatoire de la Côte Aquitaine. https://www.observatoire-cote-aquitaine.fr/. Accessed 10 Nov 2024
10. Pontee N, Tastet J-P (1998) Morpho-sedimentary evidence of Holocene coastal changes near the mouth of the Gironde and on the Medoc Peninsula, SW France. Oceanologica Acta., 21 N°2:0399
11. Billières, C (2022) Etude du système dunaire de la flèche du Cap Ferret : Synthèse de l'état des connaissances, Analyses morphologique des dunes, SIG, Traitement, analyse et interprétation des données Géoradar. Rapport de stage de fin d'étude. BRGM: 20–23
12. Favennec J (2002) Connaissance et gestion durable des dunes de la côte atlantique. ONF Les Dossiers Forestiers n°11
13. Smith A, Jackson D, Cooper J (2017) Three-dimensional airflow and sediment transport patterns over barchan dunes. Geomorphology 278(2017):28–42
14. Whang D, Narteau C, Rozier O (2010) Morphodynamics of barchan and transverse dunes using a cellular automaton model. J Geological Res 115:F03041
15. Charpentier M (2022) Architecture et morphodynamique des dunes transgressives de la dune du Pilat. Rapport de stage de fin d'étude. Université de Perpignan, p 2–28

A Full-Size Hybrid Dune Field Experiment: Design and First Results

Daan W. Poppema[1]([✉]), Sierd de Vries[1], Ruben G. C. Rosman[1,2], Stefan Pluis[3], Myron van Damme[3], Robert T. McCall[4], and Alessandro Antonini[1]

[1] Delft University of Technology, Delft, The Netherlands
`d.w.poppema@tudelft.nl`
[2] Boskalis, Papendrecht, The Netherlands
[3] Rijkswaterstaat WVL, Utrecht, The Netherlands
[4] Deltares, Delft, The Netherlands

Abstract. Coastal defense strategies increasingly include hybrid dunes, which combine sandy dunes with hard structures. In the case of hybrid dunes, the sandy system and hard elements interact. Wave impact, reflection runup as well as dune erosion and scour are all being affected by both the hard elements and sandy parts. To better understand these interactions, we conducted a field experiment, with a hybrid dune setup to measure hydrodynamics and dune erosion during storms. The experiment setup contains four cross sections to compare the morphological development at a sandy dune (1), dike-in-dune (2), vertical wall-in-dune (3) and dike (4). The works took place in the winter of 2024/2025. The various storms that occurred caused strong erosion to the test setup. Preliminary results show that while the hard elements slowed down dune retreat behind them, the beach morphology in front remained remarkably similar. Erosion may have been governed by large-scale sediment availability in the system, beach elevation and hydrodynamics rather than by hard-soft interactions at the hybrid dune.

Keywords: Hybrid flood defense · green-gray infrastructure · hard-soft interaction · dune erosion · dike

1 Introduction

Hybrid dunes combine sandy dunes with hard structures and are increasingly being implemented for coastal flood protection. These hybrid dunes aim to combine the advantages of sandy dunes, such as space for nature and recreation, with those of structures like levees and dikes, such as predictable safety and erosion resistance. Sometimes, dunes have accreted naturally at manmade structures, similarly resulting in hybrid flood defense. Also, hybrid dunes can be formed when e.g. buildings are constructed in dunes for other means, as they unintendedly alter hydrodynamics and dune erosion during storms.

The stability of these hybrid dunes, also known as hybrid flood defense, is not only a function of the classical behavior of either individual hard or soft parts under

© The Author(s) 2026
C. Coelho et al. (Eds.): CD 2025, CRL 41, pp. 48–54, 2026.
https://doi.org/10.1007/978-3-032-15473-6_8

wave attack: complex feedback processes between dune morphology, wave action and structural stability start to play a role. In recent years, physical model experiments have been carried out to examine the hydrodynamics and sandy morphodynamics of a limited number of hybrid structures (e.g. Van Geer et al., 2009). However, these physical model experiments are often hampered by sediment scaling effects, and do not address crucial alongshore processes. Although these interactions are essential to properly design new hybrid dunes or evaluate the safety of existing ones, they remain poorly understood.

To gain knowledge and understanding on hybrid structures, we have developed a field experiment, where we measure the hydrodynamics and dune erosion during storms at an hybrid dune construction in the field. In this contribution, we will present the experimental setup, instrumentation and first results.

2 Methodology

In a large field experiment, an artificial life-size hybrid dune was constructed just above the high water line, to systematically observe its response during storms (Fig. 1). Methodologically, this resembles the RealDune experiment (Van Wiechen et al., 2023), where an artificial life-size sandy dune was studied. For the new Hybrid Dune experiment, the test dune contains four cross sections (Fig. 2), of 24 to 30 m wide: 1) a classical sand dune section as baseline; 2) a dike with a hard revetment of concrete 2 × 2m tiles; 3) a hybrid dike-in-dune section, with the same hard revetment covered by sand and 4) a dune with a vertical sea wall in the dune, made from shipping containers. The experiment took place at the Sand Motor in the Netherlands. The setup was constructed in December 2024. Measurements started on 18 December 2024 and run during the winter of 2024/2025.

Fig. 1. The test setup just after construction, with the concrete tiles of the dike section visible and the hard elements of the dike-in-dune and wall-in-dune still hidden under the sand.

The measurements include water levels, wave action, current velocities, sediment concentration, hydrodynamic loading on the hard structure in terms of pressures and morphological evolution of the dune and foreshore. Hereto an extensive array of sensors is employed (Fig. 3), including 2 offshore ADCP's; 5 ADV's with each 2 OBS sensors in front of the cross sections, 12 pressure transducers in the intertidal zone and 30 on the hard-soft interface, 4 multi-line lidar scanners and 6 GoPro's. In addition, topography is surveyed using a handheld GPS and lidar drone.

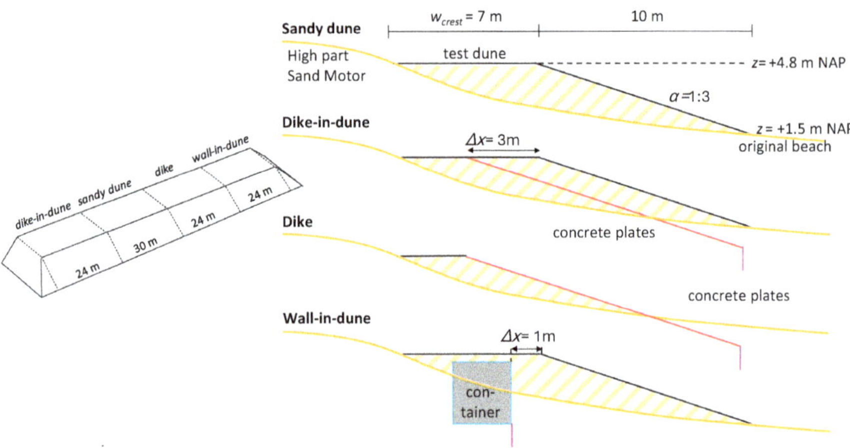

Fig. 2. The four cross sections of the experiment. The dune toe is located just above spring tide level.

Fig. 3. Overview of the instrumentation used, projected onto an aerial photo of the experiment.

3   Results and Conclusions

The experiment has resulted in a unique dataset of concurrent high-resolution observations of hydrodynamics, sediment transport, morphological evolution of the dune and foreshore, wave runup and wave load on the structures. This will enable quantitative analysis of the morphological response to hydrodynamic drivers and hybrid dune configuration. Moreover, these observations can be used to include hybrid dunes in numerical dune erosion models. Early results show that strong erosion occurred, caused by several storms during the measurement period.

Five storms occurred during the experiment to date (Fig. 4), with peak water levels ranging between 1.84 m NAP (storm 3, Jan 1 2025) and + 2.08 m NAP (storm 1, Dec 19 2024). This caused strong erosion at the setup. At the dike-in-dune, the first storm already uncovered the first of the initially-covered revetment. The subsequent storms uncovered more of the revetment (Fig. 6), allowing the revetment to interact with the waves and sand transport, and giving us to opportunity to see how this affects the morphology in the surroundings. At the wall-in-dune section, the containers of were still hidden under the sand after the first storm (Fig. 5), but they were uncovered in the second storm, with the dune cliff afterwards exactly in line with the front of the containers. During subsequent storms, erosion and dune retreat continued.

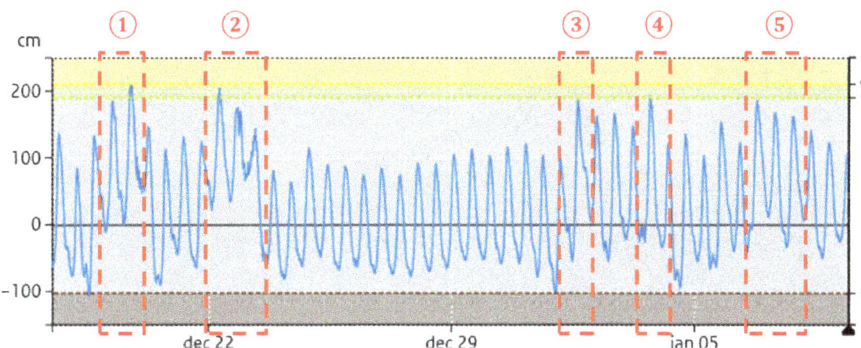

Fig. 4. Observed water levels in cm NAP during the experiments, based on a public measurement station at 8 km from the test site. Source: Rijkswaterstaat (2024).

Sand eroded from underneath the revetment, both from the edges of the dike- and dike-in-dune section and through gaps between tiles. This resulted in a gradual collapse of the revetments, starting at the edges of the dike section after the first storm (Fig. 5) and eventually resulting in the destruction of the dike- and dike-in-dune after the fourth storm. Given the early onset of revetment failure at the edges of the sections (a consequence of the loose concrete tiles used to facilitate deconstruction after the experiment), we concentrate here on the morphological development in the center of the cross sections.

Fig. 5. The setup on 20 Dec 2024, after the first storm. The tiles start to collapse at the edges of the dike section. The container (background) is still fronted by a meter of sand.

Fig. 6. The revetment of the dike-in-dune is partially uncovered after two storms. Erosion underneath the tiles causes the pates at the edges of the section to collapse.

Remarkably, morphological differences between the different cross sections appeared to be limited (Fig. 7). At the edges between the sandy dune and the concrete tiles, the beach was visibly lower, up to a couple of decimeters, leading locally to a faster retreat of the dune cliff. But no scour was visible in front of the hard revetments, nor in front of the vertical wall. The beach elevation in front of the setup, a couple of meters seaward of the position of the concrete tiles, seemed also very similar at the four cross sections. Wave runup and reflection differed by necessity between the cross sections due to the hard elements, but this had limited effect on the beach morphology. While further examination of the measurements is still needed to fully grasp the morphological differences and their drivers, it seems that the beach response in front of the

Fig. 7. The setup at 1 January 2025, the onset of high tide of storm 3. The cliff location and beach elevation are mostly uniform alongshore.

hybrid dune was primarily governed by the large-scale sediment supply, beach elevation and hydrodynamic conditions, rather than the hard-soft interactions at the hybrid dune.

Acknowledgements. This work is part of the Perspectief research programme Future flood risk management technologies for rivers and coasts with project number P21–23, financed by Domain Applied and Engineering Sciences of the Dutch Research Council (NWO). The field experiment is additionally financed by Rijkswaterstaat. The authors thank all colleagues and students whose help in the preparation and execution of the field experiment made this work possible.

References

Rijkswaterstaat. 2024. RWsOS Viewer, Waterhoogte NAP in Scheveningen. https://rwsos.rws.nl/viewer/map/noordzee/waterkwantiteit/location/SCHE?pd=-30;0 Accessed 2025/01/09

Van Geer P, Boers M, Van Gent M (2009) Measurements on the interaction between dunes and dikes during extreme storm events. Coastal Dynamics

Van Wiechen P, Rutten M, Mieras R, Anarde K, Tissier M, De Vries S (2023) Avalanching of the dune face: field observations and equilibrium theory. Coastal Sediments 2023:739–752

Species-by-Species Pattern Analysis of Coastal Dune Vegetation

Davide Demichele[⊠], Elena Belcore, Marco Piras, and Carlo Camporeale

DIATI, Department of Environment, Land and Infrastructure Engineering, Politechnic of Turin, Corso Duca degli Abruzzi, 24, 10129 Turin, Italy
davide.demichele@polito.it

Abstract. Vegetation is crucial for stabilizing and developing coastal dunes. Different plant species exhibit different spatial distributions which reflect their environmental role and adaptation strategy.

This study aims to provide a fine-scale species-by-species analysis of vegetation spatial patterns on coastal dunes within the San Rossore – Migliarino – Massacciuccoli Regional Park (Tuscany, Italy).

A comprehensive vegetation dataset generated by an Object-Based Image Analysis (OBIA) algorithm applied to high-resolution ortho-images has been utilized. A Digital Terrain Model (DTM) of the study area was created to assess the impact of dune morphology on plant distribution. Moreover, a wave runup analysis was also conducted to understand the interaction between vegetation and hydrodynamic forces.

The research highlights how the vegetation threshold distance from the coastline, L_{veg}, is superimposed by the reaching distance of wave runup during extreme events. Terrain morphology significantly affects the vegetation zonation: on taller and undisturbed dune fields, species zonation is clearer and more defined, whereas, on flatter and disturbed ones, spatial distribution is significantly fuzzier.

A positive correlation emerges between the abundance of a species and its degree of spatial clustering, indicating how less abundant species form more tightly clustered spatial patterns. Modified Ripley's L-function analysis revealed a multiscale clustered pattern for most species under examination.

The present results may provide a solid benchmark in coastal ecology research for supporting natural-based conservation plans and eco-morphodynamic modeling.

Keywords: Vegetation Pattern · Coastal eco-morphology · geostatistics · classification

1 Introduction

Coastal dunes are critical ecosystems that enhance coastal resilience, support biodiversity, and mitigate climate change impacts. They serve as natural barriers, hosting specialized flora and fauna, such as sand-tolerant species like *Ammophila arenaria* for sand stabilization and less-tolerant species like *Helichrysum stoechas* for stabilizing

© The Author(s) 2026
C. Coelho et al. (Eds.): CD 2025, CRL 41, pp. 55–61, 2026.
https://doi.org/10.1007/978-3-032-15473-6_9

back dunes [4]. Additionally, dunes dissipate wave energy, reducing flooding, while supporting recreation and local economies [7]. Despite their significance, dunes face threats from invasive species, human activities, and climate-induced erosion, impacting 70% of global sandy beaches and 42% of Italy's sandy coasts [2].

Vegetation plays a pivotal role in dune dynamics, with its distribution influenced by environmental factors like wind and salt spray [9]. Remote sensing has emerged as a powerful tool for vegetation analysis, offering high-resolution insights into spatial patterns through techniques like Object-Based Image Analysis (OBIA), enabling fine-scale classifications with UAV imagery [6].

This study leverages OBIA on UAV-acquired data for detailed vegetation mapping and examines the interplay between vegetation, topography, and hydrodynamics. Using statistical tools, the research aims to understand plant distribution, geomorphological interactions, and the impacts of human activities in one of Italy's protected coastal dune areas [1, 5]. The findings will inform conservation strategies to sustain these fragile ecosystems.

2   Material and Methods

2.1   Study Site

The study investigates the Migliarino-San Rossore-Massaciuccoli Regional Park, a UNESCO Biosphere Reserve located along 30 km of Tuscany's coastline (Fig. 1a-b). The park's dune system, spanning approximately 394 ha, exemplifies significant biological diversity but faces pressures from climate change and human activities, such as erosion and over-trampling [3]. Three study areas (A1, A2, A3) were selected to examine vegetation patterns and geomorphological features, representing varying degrees of anthropic impact and environmental conditions.

Area A1, near Lecciona Beach, experiences high anthropic activity and lacks well-developed foredunes. Instead, embryonic dunes with sparse evergreen shrubs dominate. Areas A2 and A3, located south of the Serchio and Morto Nuovo rivers respectively, are within protected zones, featuring richer psammophyte communities and taller dunes reaching 5 m. Monitoring since 2021 involved high-resolution optical data and mapping of over 900 plant samples, with vegetation patterns visualized using ground cover maps.

A Digital Elevation Model (DEM) was generated using an aerial photogrammetric survey conducted in May 2022 with a DJI Matrice 300 RTK drone, a 45 MP DJI Zenmuse P1 camera, and dual-frequency GNSS for RTK positioning. Data were processed with Agisoft Metashape following a Structure from Motion (SfM) workflow, resulting in DEMs with up to 3 cm resolution.

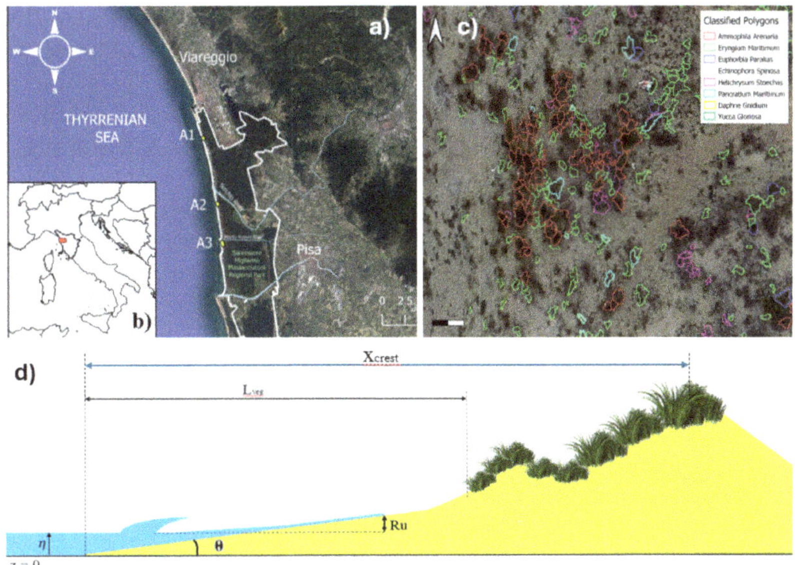

Fig. 1. a-b) Location of the three study areas inside Sanrossore-Migliarino-Massacciuccoli Regional Park (Tuscany, Italy); c) Individual plants detected and classified by the OBIA algorithm; d) Geometric configuration of the beach-dune profile. Are highlighted the wave runup R_u, vegetation threshold distance L_{veg} and crestline distance x_{crest}.

2.2 Hydrodynamics

The wave runup analysis applied the empirical model proposed by Stockdon (Eq. 1) [10], incorporating spectral wave data derived from the National wave-metric and tide-gauge network open database, run by the Italian Institute for Environmental Protection and Research (ISPRA).

$$R_u = (H_0 L_0)^{\frac{1}{2}} \left[0.385\beta_f + \left(0.17\beta_f^2 + 0.0012 \right)^{\frac{1}{2}} \right] \tag{1}$$

2.3 Vegetation Pattern Analysis

This study focuses on vegetation patterns, utilizing a field-based dataset obtained through object-based image analysis (OBIA), developed by the authors [1]. The OBIA algorithm, trained to classify 12 plant species and sand and debris, achieved an average accuracy of 75%, consistent with other vegetation classification studies in similar environments [6]. The dataset includes over 600,000 polygons representing the canopy covers of plants, with morphometric parameters such as centroid coordinates, areas, and perimeters (Fig. 1c).

The abundance of each plant species was calculated as the ratio of a species' occurrence to the total number of plants [5], with species under 1% abundance excluded from further analysis.

Geometric indices Equivalent Diameter (ED), and Mean Neighbour Distance (MND), were used to quantify size and local spatial density for each plant, reading $ED = 2\sqrt{A_i/\pi}$ and $MND_i = \frac{1}{m}\sum_{j=1}^{m} t_{i,j}$.

Point pattern analysis was performed using nearest-neighbor and distance-based methods. The mean nearest neighbor (MNN) method and Ripley's K-function [8], were used to assess spatial arrangement.

The nearest neighbor index (NNI) determines clustering or dispersion of plants, namely $NNI = \overline{d_o}/\overline{d_e}$, where $\overline{d_o}$ is the average minimum distance between observed physical centroids of the plants and $\overline{d_e}$ is the average minimum distance between points randomly distributed.

In addition, L-function $L(r) = \sqrt{K(r)/\pi} - r$ has been used to perform a distance-based point pattern analysis. It measures the average number of points that fall within a circle of a given radius r centered on each point of the population. $K(r)$ denotes Ripley's K-function. For this study, we tested the statistical significance of the clustering/dispersion patterns denoted by the L-function with a Monte Carlo simulation to build the 95% confidence envelope.

3 Key Results

3.1 Vegetation Boundaries

This study investigates the interplay between wave runup and dune morphology in defining the boundaries of coastal vegetation. The threshold distance from the coastline for vegetation growth L_{veg} is determined by the maximum reach of wave runup during extreme events, underscoring the critical role of wave dynamics in shaping vegetation distribution along coastal zones (Fig. 2a).

Furthermore, the influence of dune morphology on vegetation boundaries was analyzed. Results reveal a consistent relationship between L_{veg} and the distance to the foredune crestline x_{crest}, with a ratio of approximately 0.6 across various dune morphologies (Fig. 2d-f).

3.2 Effect on Dune Morphology on Vegetation Patterns

The size of the foredune is significantly influenced by the abundance of halophyte species, particularly *Ammophila arenaria*. This species is also highly sensitive to direct human disturbances, as in A1 the shallower dune field is reflected in the scarcity of this species (Fig. 2b-c). Furthermore, dune morphology highly impacts the zonation: the higher the foredune, the more distinct the plant zonation; on the contrary, the flatter the terrain, the fuzzier and indistinct the zonation.

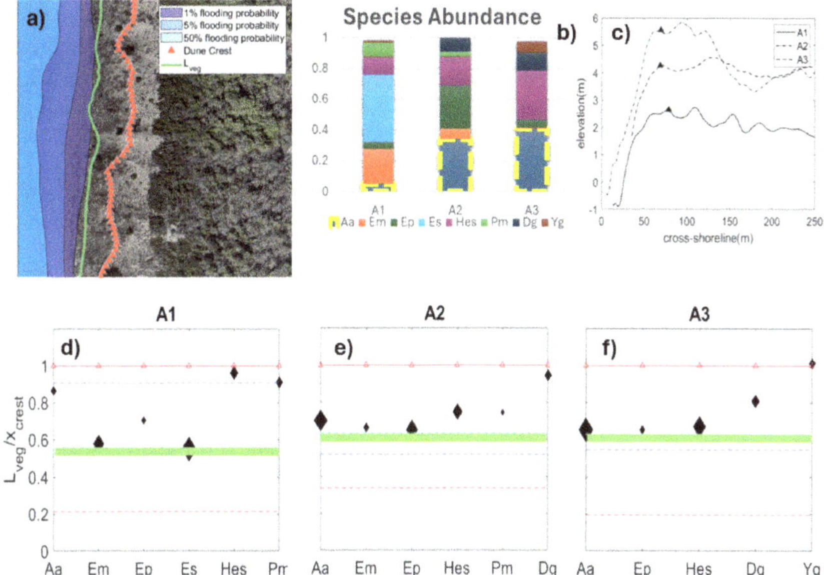

Fig. 2. a) Relationships between L_{veg}, reaching distance of wave runup and foredune crestline; b) histogram of abundance of different plant species; d) average dune profile in different study areas, highlighting the crest (black triangle); d-f) L_{veg}/x_{crest} ratio for overall vegetation (green line) and for each species (black diamonds).

3.3 The Role of Vegetation Abundance on Vegetation Patterns

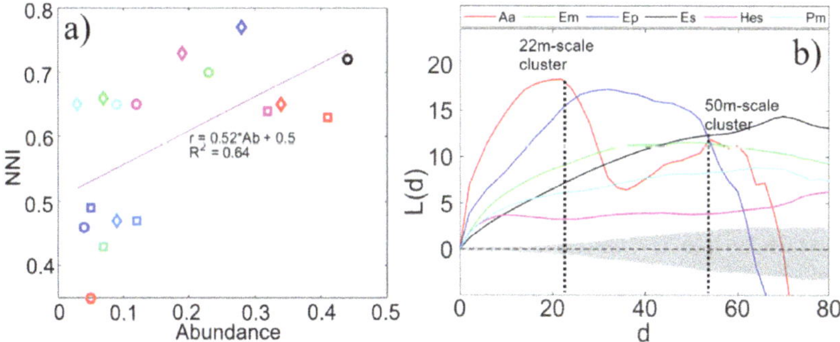

Fig. 3. *a)* Abundance vs nearest neighbor index. Linear regression of observed data is represented by the black solid line; *b)* L-function for different plant species. The 95% confidence interval for the L-function computed through Monte Carlo simulation is highlighted in grey.

Abundance, in general, stands out as one of the main eco-morphological parameters in coastal vegetation pattern as the spatial distribution of plant sizes is influenced by species abundance, with higher-abundance species exhibiting more uniform spatial distribution. The MNN point pattern analysis revealed a clustered spatial arrangement for all species taken into exam, although, the degree of clustering is correlated with the abundance, confirming that abundance is the main factor controlling the pattern type of a plant species (Fig. 3a). Notably, the L-function emerged as a valuable tool for elucidating the type and scales of vegetation clusters, revealing multi-scale clustering patterns for many coastal species (Fig. 3b).

4 Conclusions

A large vegetation dataset, derived by a machine learning classification algorithm, applied to ultra-high-resolution imagery of study areas, opened the possibility for a better insight into the complex inter-relationships between plant communities, terrain morphology, environmental forcings, and human presence, shaping sandy coastal dunes.

With this work, we provided a species-by-species extensive qualitative and quantitative description of vegetation spatial patterns using different geostatistical tools, widely used in spatial analysis and vegetation studies.

The methods proposed for studying vegetation patterns offer valuable insights into the complex interplay between biotic and abiotic processes occurring in coastal dunes, thereby enhancing environmental monitoring and conservation efforts. Additionally, the patterns identified in this study can serve as a foundation for fine-tuning and validating numerical eco-morphodynamic models in these study sites, further advancing our understanding of these dynamic ecosystems.

References

1. Belcore E et al (2021) Mapping riparian habitats of Natura 2000 network (91E0*, 3240) at individual tree level using UAV multi-temporal and multi-spectral data. Remote Sens 13(9):1756
2. Bertacchi A, Zuffi MAL, Lombardi T (2016) Foredune psammophilous communities and coastal erosion in a stretch of the Ligurian sea (Tuscany, Italy). Rendiconti Lincei 27:639–651
3. Cipriani LE et al (2010) Azioni di tutela delle dune costiere del Parco Regionale Migliarino San Rossore Massaciuccoli (Toscana settentrionale). Studi Costieri 17:165–179
4. Hesp PA (1991) Ecological processes and plant adaptations on coastal dunes. J Arid Environ 21(2):165–191
5. Kim D, Keun Bae Yu (2009) A conceptual model of coastal dune ecology synthesizing spatial gradients of vegetation, soil, and geomorphology. Plant Ecol. **202**, 135–148
6. Michez A et al (2016) Classification of riparian forest species and health condition using multi-temporal and hyperspatial imagery from unmanned aerial system. Environ. Monit. Assess. 188:1–19
7. Pranzini E (2018) Shore protection in Italy: from hard to soft engineering… and back. Ocean Coast Manag 156:43–57
8. Ripley BD (1984) 198 1. Spatial Statistics
9. Short AD, Hesp PA (1982) Wave, beach and dune interactions in southeastern Australia. Mar Geol 48(3–4):259–284
10. Stockdon HF et al (2006) Empirical parameterization of setup, swash, and runup. Coastal Eng 53(7):573–588

Decades of Change: Foredune Morphodynamics at Woonona Beach, Southeast Australia

Dylan McLaughlin[1]([✉]), Thomas B. Doyle[1,2], Colin D. Woodroffe[1], and Kerrylee Rogers[1]

[1] School of Science and Environmental Futures Research Centre, University of Wollongong, Wollongong, NSW 2522, Australia
dm068@uowmail.edu.au

[2] Water, Wetland and Coastal Science, Department of Climate Change, Energy, the Environment and Water, NSW Government, Sydney, Australia

Abstract. Coastal foredunes serve as natural protective barriers, providing defence against storm impacts, erosion, and coastal inundation. In many locations they have undergone significant modifications, including sand mining, reprofiling, urban development, and interventions to enhance recreational amenities. While foredune morphology varies by location, this variation can reveal changes associated with storm events and modifications. This study examines foredune morphodynamics at Woonona-Bellambi Beach in southeastern Australia, using high-resolution three-dimensional LiDAR data to investigate changes across the foredune. The results reveal considerable spatial variability in foredune responses to, and recovery from, large-scale storm events and soft engineering modifications over annual to decadal timescales. Using recently released nearshore wave data, we extracted key storm events over the past decade for this beach which revealed spatially diverse patterns of erosion and recovery of the foredunes. These findings highlight the importance of high-resolution coastal investigations to improve understanding of local-to-regional responses and recovery processes across varying timescales. The insights gained can inform coastal modelling, planning, and management strategies.

Keywords: foredune geomorphology · storms · human impacts

1 Introduction

Coastal dune evolution and morphology is driven by a range of factors including sand supply, vegetation type and cover, wave and wind forces, storm erosion, and increasingly by anthropogenic modification [1]. Foredune stability is influenced by storm magnitude and frequency, which drive fine-scale variations in erosion, inundation, and sediment transport [2], as well as shifts in wave climate that can disrupt longshore sediment pathways [3]. Additionally, anthropogenic modifications to foredunes and vegetation can impact shoreline stability and morphology, often necessitating ongoing maintenance and management [4]. Manipulations of foredunes have occurred globally [5, 6] typically

© The Author(s) 2026
C. Coelho et al. (Eds.): CD 2025, CRL 41, pp. 62–68, 2026.
https://doi.org/10.1007/978-3-032-15473-6_10

as a control to establish a new stable state. However, coasts are dynamic, and monitoring data is required to assess the effectiveness of the manipulation. Three-dimensional high-resolution spatial data, such as LiDAR, is crucial for characterising the morphodynamics of modified foredunes, while integrating nearshore wave data supports understanding the impact of storm events and post-event recovery and the capacity to maintain a new modified stable state. This study examines foredune morphodynamics at Woonona-Bellambi Beach in New South Wales (NSW), southeastern Australia, over a decade following intervention works [7]. The aim is to evaluate and identify foredune responses to storms and soft engineering modification over annual to decadal timescales using topographic and nearshore wave data. The findings contribute to a better understanding of storm impacts and modification effects, supporting the development of quantitative sediment budgets for coastal compartments.

1.1 Study Location

Woonona-Bellambi Beach (-34.35, 150.92, Fig. 1A) is a highly modified and managed beach and foredune system with urban development and infrastructure backing the northern foredune (i.e. Woonona Beach), a stormwater creek entrance in the middle, and Bellambi Lagoon backing Bellambi Beach at the southern end. It is a 2 km long sandy beach with an easterly aspect and a shoreface dominated by rocky reef substrate and limited sediment sharing. This region of the Australian coast is wave-dominated with a deepwater mean H_s of 1.6 m, T_p of 9.6 s with swells predominantly from the south-east [8]. It has a semi-diurnal micro-tidal regime with a range of 0.8 m neap tide to 1.2 m spring tide. Extra-tropical cyclones known as East Coast Lows (ECLs) are the dominant source of extreme wave conditions leading to coastal inundation, elevated sea levels and heavy rainfall [9].

Engineering works aimed at enhancing recreational amenity and improving coastal resilience in the 1980s included installing fencing, planting dune vegetation, and shaping frontal areas of the foredunes. At Woonona Beach, these efforts stabilised the previously sparsely vegetated northern foredunes by 1993 which then prograded and vegetation expanded seaward eventually causing community concerns regarding amenity and safety. In response, soft engineering modification works in 2014 included the removal of vegetation, re-profiling of a 250 m section of the northern dune-beach area and the establishment of a new vegetation line 30 m landward. Within 8 months, an incipient dune formed, and within 21 months, some areas had accreted more than the removed volume, and the lower beach width gained had diminished [7].

2 Materials and Methods

2.1 Foredune Geomorphic Change Analysis

Volumes for the Woonona and Bellambi foredunes were measured using triangulated irregular network (TIN) surface models derived from photogrammetric data (1958–2007), aircraft LiDAR (2013–2023), and remote piloted aircraft LiDAR data collected specifically for this research in 2024 (Fig. 1B). An area of interest (AoI) polygon was drawn to define the foredune extent, including any incipient dunes, for each year of survey data. Volumes were measured using 20 m spaced, 1 m wide transects for north–south comparison [10]. Digital elevation models were generated using ground-classified returns from point clouds derived from LiDAR surveys conducted between 2013 and 2024 (8 in total). A pixel-wise linear regression analysis was then performed to create a surface model of elevation rates of change along the foredune.

2.2 Coastal Storm Events Analysis

Wave data was sourced through the NSW Nearshore Wave Tool which provides modelled nearshore wave data using WAVEWATCH III (NOAA, USA) at 250 m spaced virtual nodes for 10 m and 30 m depth contours [11]. At the 10 m depth contour, 12 nodes covered waves arriving along the Woonona-Bellambi embayment (Fig. 1A). Coastal storm events were defined using the peak-over-threshold method [12] with the NSW threshold values [13]; wave height (H_{thresh}) $\geq$ 3.0 m; minimum storm duration (D_{storm}) $\geq$ 6 h; and the meteorological independence criterion (I) of 24 h. For storm events between 2014 and 2024, the mean and maximum H_s, T_p, and predominant wave direction were calculated using averaged 10 m depth virtual wave buoy data. Wind analysis was conducted using 1997 to 2023 data from the Bellambi AWS weather station (-34.37, 150.93) provided by the Australian Government Bureau of Meteorology. Rainfall totals for 24-h periods for each coastal storm were calculated using Bellambi AWS data. Sea level maxima were determined using tidal data from the Australian Government Bureau of Meteorology Port Kembla station (-34.47, 150.91) for each coastal storm period.

3 Results

3.1 Foredune Geomorphic Change

Linear regression analyses of elevation changes indicated spatial variability within the complex foredune system (Fig. 1A), as indicated by a relatively low overall R2 of 0.25. In the section of the Woonona Beach foredune modified in 2014, a decrease in elevation is observed along the foredune front, coinciding with a more peaked dune crest immediately landward. In the unmodified section of the Woonona Beach foredune, elevation decreases are evident along the foredune front. The most prominent elevation losses occur near the mouth of the stormwater creek near the centre of the embayment, where scarping and sand losses are evident. At the southern extent, elevation increases are observed along the foredune, and near the gully outlet channel, which shows signs of migration.

The Woonona-Bellambi Beach foredune was classified as a generally stable to slightly prograding system over multidecadal timescales (Fig. 1B). A decrease in elevation occurred along both the Woonona and Bellambi foredunes between 2013 and 2017. From 2018 onward, the entire foredune increased in volume, though recovery was interrupted in 2022, and a volume decrease occurred in the Bellambi foredune. The trends suggest continued recovery, with volumes approaching levels comparable to those estimated for 2013.

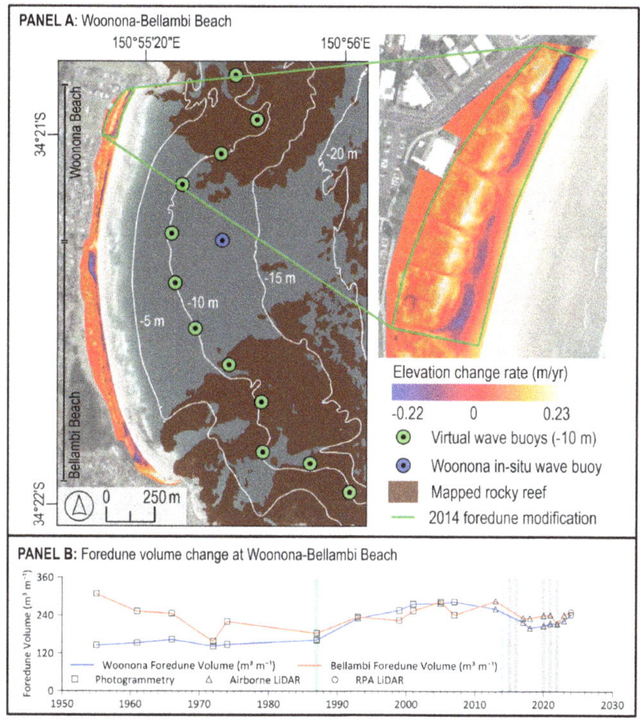

Fig. 1. Panel A: Woonona-Bellambi Beach foredune elevation change rate (m/yr) from 2013 to 2024, based on 8 years of LiDAR data. The green area indicates the section modified in 2014. White lines represent depth contours. Mapped rocky reef layer from NSW Government SEED open data portal. Panel B: Woonona Beach and Bellambi Beach foredune volumes from 1956 to 2024, with survey data types indicated. The green vertical line indicates the dune stabilisation works (c. 1987), prior to this sand mining and urban development impacted the Bellambi Beach foredune volumes. The blue vertical line marks the year of the foredune modification (2014), and the grey lines indicate years with storm event(s) as listed in Table 1.

3.2   Coastal Storm Events

A total of 11 storm events met the coastal storm thresholds at Woonona-Bellambi Beach over the focal study period (2013–2024) (Table 1). The 2016 storm (4–6 June) featured the highest significant wave height (Hs max of 4.1 m), the most rainfall (330 mm), the highest sea level (2.2 m AHD), and a long duration (60 h). Notably, the north-northeast wind direction aligned closely with the easterly wave direction and was observed from the northeast in other areas of NSW [9]. In contrast, the 1–4 March and 3–4 July storms of 2022 also exhibited close alignment between wave and wind directions but lacked a northerly wind component, which is more typical for the region. The 2022 storm series also featured considerable rainfall.

Table 1. Storm events between 2014 and 2024 for nearshore waves averaged from 12 virtual wave buoys at the 10 m depth contour along Woonona-Bellambi Beach.

	Year	2015	2016	2020			2021		2022			
	Storm Dates	20-21 Apr	4-6 Jun	8-10 Feb	14-15 Jul	26-27 Jul	20-21 Mar	6 May	1-4 Mar	1-2 Apr	3-4 Jul	10 Jul
Nearshore Waves												
H_s mean (m)		3.6	3.6	3.4	3.5	3.3	3.2	3.2	3.5	3.6	3.2	3.4
H_s max (m)		3.8	4.1	3.9	3.7	3.5	3.4	3.4	3.8	3.8	3.4	3.6
T_p mean (s)		10.9	11.8	10.1	11.8	10.4	10.0	10.4	11.4	12.5	9.6	12.4
T_p max (s)		11.3	13.2	10.8	12.1	10.7	10.3	10.5	11.9	12.8	9.7	12.6
Predominant direction		ESE	E	E	ESE	E	E	E	E	ESE	ESE	ESE
Duration H_s > 3 m (h)		30	60	32	35	17	21	11	55	21	11	22
Wind												
Wind speed mean (m/s)		16.4	8.6	8.7	9.2	8.0	5.7	9.5	7.7	9.6	15.5	6.8
Wind speed max (m/s)		20.1	17.5	17.0	12.3	16.5	10.8	15.9	13.4	13.9	18.5	12.9
Predominant direction		S	NNE	ESE	SW	SW	NE	SSE	ESE	SW	SE	SW
Rainfall (mm)		85.6	330.0	297.8	12.8	120.4	0.0	25.6	99.6	11.6	213.6	19.0
Sea level max (m AHD)		2.0	2.2	1.9	1.6	1.7	1.3	1.6	1.9	1.9	1.8	1.7

4   Discussion

The foredune system at Woonona-Bellambi Beach appears relatively stable over multidecadal timescales, with episodic storm cut and recovery. Soft engineering of the Woonona Beach foredune resulted in morphological changes over the following years and decade, including a landward increase in height and a higher dune crest, indicating sediment transfer from the front of the foredune as the re-profiled sand volume returned. This longer-term trend of increasing volume and geomorphic change appears to be a continuation of the trends observed in the 21 months following the modification works in which immediate volume increases where observed [7]. A landward shift or higher dune crest was not observed along the unmodified Woonona foredune, despite volume increases after 2018. Changes in elevation rate and volume suggest natural processes as key drivers of erosion and recovery, resulting in a morphology distinct from the modified foredune.

The timing, magnitude and extent of volume reductions observed across the entire foredune between 2013 to 2017 indicate the 2016 storm was a significant driver of erosion. This is consistent with other studies detailing the extreme erosion particularly at the southern end of NSW beaches due to large waves, anomalous wind/wave directions, and coincidence with spring high tides [9]. Along the Bellambi foredune, elevation losses at the stormwater creek mouth near the centre of the embayment were likely driven by high rainfall events. These impacts are potentially exacerbated by urban runoff directed through the creek as similar erosion is not observed at the southern extent, where the unconstrained channel disperses gully overflow.

Specific combinations of storm wave direction, wind direction, and water level are known to modify coastal erosion, cause considerable spatial variability and result in extreme impacts such as foredune overtopping and erosion [2]. This study emphasises the spatial and temporal diversity of foredune change in response to an array of potential drivers, or their combinations, including storm wave direction, wind direction, water levels, and rainfall. It also illustrates the complexity and scale of natural drivers, the dominance of natural variability over anthropogenic modifications on annual to decadal timescales and demonstrates the importance of working with nature and understanding natural patterns.

This analysis highlights the value of high-frequency and high-resolution monitoring data to improve understanding foredune responses to large-scale natural and anthropogenic changes over annual to decadal timescales. Combining this information with nearshore wave data further improves understanding of wave processes, storm impacts, and post-storm recovery. Integrating other monitoring data will continue to fill the gaps to develop comprehensive sediment budgets and create adaptable methods to improve coastal management in a changing climate.

References

1. Hesp P (2002) Foredunes and blowouts: initiation, geomorphology and dynamics. Geomorphology 48:245–268
2. Oliver TSN, Kinsela MA et al (2024) Foredune erosion, overtopping and destruction in 2022 at Bengello Beach, southeastern Australia. Cambridge Prisms: Coastal Futures 2:1–20
3. Goodwin ID, Mortlock TR et al (2016) Tropical and extratropical-origin storm wave types and their influence on the East Australian longshore sand transport system under a changing climate. J Geophys Res Oceans 121:4833–4853
4. Doyle TB, Woodroffe CD (2023) Modified foredune eco-morphology in southeast Australia. Ocean Coast Manag 240:106640
5. Bossard V, Nicolae Lerma A (2020) Geomorphologic characteristics and evolution of managed dunes on the South West Coast of France. Geomorphology 367:107312
6. Bastos AP, Taborda R et al (2024) Short-term foredune dynamics in response to invasive vegetation control actions. Remote Sensing 16:1487
7. Gangaiya P, Beardsmore A et al (2017) Morphological changes following vegetation removal and foredune re-profiling at Woonona Beach, New South Wales, Australia. Ocean Coastal Manag 146:15–25
8. Kinsela MA, Hanslow DJ et al (2022) Mapping the shoreface of coastal sediment compartments to improve shoreline change forecasts in New South Wales, Australia. Estuaries Coasts 45:1143–1169
9. Harley MD, Turner IL et al (2017) Extreme coastal erosion enhanced by anomalous extratropical storm wave direction. Sci Rep 7:6033
10. Doyle TB, Woodroffe CD (2018) The application of LiDAR to investigate foredune morphology and vegetation. Geomorphology 303:106–121
11. Baird (2024) NSW Nearshore Wave Tool – Validation Report. Prepared for NSW Department of Climate Change, the Environment and Water., Sydney, Australia
12. Harley M (2017) Coastal Storm Definition. In: Coco G, Ciavola P (eds) Coastal Storms: Processes and Impacts, pp 1–21. Wiley Blackwell, Location
13. Shand T, Goodwin I, et al (2011) NSW coastal inundation hazard study: coastal storms and extreme waves. Prepared for NSW Department of Climate Change, the Environment and Water, Sydney, Australia

Physics-Informed Machine Learning for Beach Morphodynamics Prediction

Elizabeth R. Holzenthal[1]([✉]), Dylan L. Anderson[2], Nicholas T. Cohn[2], Bradley D. Johnson[1], and Katherine L. Brodie[2]

[1] U.S. Army Corps of Engineers (USACE) Engineer Research and Development Center (ERDC) Coastal and Hydraulics Laboratory (CHL), Vicksburg 39180, USA
`Elizabeth.R.Holzenthal@usace.army.mil`
[2] USACE ERDC CHL Field Research Facility, Duck 27919, USA

Abstract. Predicting beach morphodynamics over multiple decades for coastal hazard risk analyses requires accurate and efficient computation of beach recovery following highly energetic erosive storms. However, process-based sediment transport modeling of long duration, lower energy conditions generally requires time-intensive numerical models that often exhibit limited skill unless they are calibrated with high-fidelity, site-specific measurements that are difficult, expensive, and generally unavailable. As an alternative, data-driven approaches that leverage machine learning (ML) of physical processes can be used to create a surrogate model from observations. Although data-driven methods are also vulnerable to overfitting to closely match observed data, this work demonstrates how physical principles can be embedded into data-driven emulators to predict bed elevation change as a function of cross-shore coordinate, or $\Delta z(x)$. Specifically, the methodology for building an ML model that incorporates spatial dependence between cross-shore bed elevations, temporal dependence between consecutive observations, and conservation principles is presented. Nearly a decade of beach profile elevations and offshore wave statistics collected by the U.S. Army Corps of Engineers Field Research Facility (FRF) in Duck, North Carolina are leveraged to demonstrate the proposed methodology.

Keywords: beach morphology · beach morphodynamics · machine learning

1 Introduction

As an alternative to time-intensive process-based numerical models required to estimate the multi-decadal evolution of sandy coasts facing climate change impacts, machine learning (ML) of physical processes can be used to create a data-driven surrogate model from simultaneous observations of beach profile elevations and wave hydrodynamics. A variety of ML methods have been applied to beach morphology prediction, with more recent efforts aimed at direct prediction of cross-shore beach profiles' response to storms [1, 2]. However, the feasibility of predicting changes in bed elevation Δz during quiescent conditions using ML has not been adequately tested.

© The Author(s) 2026
C. Coelho et al. (Eds.): CD 2025, CRL 41, pp. 69–75, 2026.
https://doi.org/10.1007/978-3-032-15473-6_11

Like numerical process-based models, improperly posed data-driven models are vulnerable to overfitting. However, physical intuition can be retained by selecting ML frameworks and components that promote generality and first principles [3]. For example, changes in beach elevation at a series of points in the cross-shore cannot be assumed to be entirely independent observations, a common assumption of ML models, due to the spatial relationship between nearby locations. To retain spatial dependence, beach profiles can be dimensionally reduced into spatial empirical orthogonal functions (EOFs) via principal component analysis (PCA) [1, 4]. Additionally, customizable loss functions can be developed to train ML models on not only observed data but also first principles (e.g., conservation laws). Finally, recurrent neural networks with long short-term memory (LSTM) are designed to predict time series by training on segments of sequential data, but their use has been underexplored in beach recovery applications. To demonstrate how these physics-informed ML methods can be applied to real observational datasets, hydrodynamic and topo-bathymetric data collected at the U.S. Army Corps of Engineers Field Research Facility (FRF) in Duck, North Carolina are leveraged to develop an LSTM model.

2 Methods

2.1 Observation Data

Since 2015, a lidar tower has autonomously collected cross-shore transects of surface elevations at a rate ~ 7Hz, which are then processed into hourly topographic profiles, extreme runup levels, and swash statistics [5]. Combined with the FRF's historic wave data records, these lidar datasets provide an ideal opportunity for ML model exploration. First, the entire record spanning the FRF's lidar data collection (2015–2024) was first classified as either "stormy" or "not-stormy" so that the ML model could be trained only on data during quiescent periods that promote recovery. Manually collected bathymetric survey data collected periodically between 2015–2024 were added to increase data availability at the lower shoreface and inner surfzone. Storm periods were defined using significant wave heights H_s collected at the 26 m wave buoy, and using the Wave Information Study (WIS) hindcast to calculate H_s at a depth of 26 m when buoy data was not available [6]. Times when H_s exceeded the 6% highest value for at least 12 h were designated as stormy and therefore excluded from ML test/train data, aggregating with next storm if the following storm was within 12 h.

Non-storm time periods were then split into discrete segments when sufficient hourly data was available over a continuous four-day period. Specifically, it was required that hourly observations of nearshore waves (8 m depth) and mean water surface elevation were available for at least 90% of the four-day period, and 18 h of profile elevation data were required for each of the four days. Gaps in the topo-bathymetric profiles preliminarily selected for the ML testing/training datasets were then filled using a combination of interpolating through time and through space by fitting the sub-tidal portion of the profile to an equilibrium beach profile ($h = Ax^{2/3}$). Profile extension via the equilibrium method was constrained such that the coefficient A was between 0.05 and 0.165 [7], and resulting profiles were only included in the test/train dataset if the fit root mean square error (RMSE) was less than 30 cm. Including profiles with relatively high fit error was a tradeoff to enable profile extension to a consistent cross-shore length for the purposes of performing PCA on all profiles comprising the ML test/train datasets.

2.2 ML Model Development

Before PCA was performed, the topo-bathymetric profiles selected for ML model testing/training were normalized in two key steps. First, the x-coordinate was shifted from local fixed coordinate system to a relative coordinate system x^*, where $x^* = 0$ m corresponds to the regional average elevation of the dune toe [8]. Next, the profiles were normalized by subtracting the mean profile and dividing by the standard deviation. The normalized profiles were then decomposed using PCA. Much like a random wave time series can be described as a summation of harmonic functions with varying amplitudes, beach profiles can be composed of a series of spatial functions, known as empirical orthogonal functions (EOFs), with variable amplitudes S. Each of the EOFs found by applying PCA to the full set of profiles represents a fraction of the dataset's variability.

Using the EOFs identified by PCA to reconstruct the input topo-bathymetric profiles, the volume of each profile can then be calculated by integrating over the profile discretized with a step size dx of 10 cm. Calculating each profile's volume consequently allows for the calculation of change in volume from one profile to the next. Thus, the validity of assuming mass conservation between profile observations can be tested. If mass conservation holds, a physically intuitive constraint can be added to the ML model. In other words, the ML model can be trained to minimize a two-part loss function L defined as

$$L = W_{\text{data}} \left\| y_{\text{obs}} - y_{\text{pred}} \right\|_2^2 + W_{\text{phys}} \left\| dV_{\text{obs}} - dV_{\text{pred}} \right\|_2^2 \tag{1}$$

where W_{data} is the weight assigned to the RMSE between the tested EOF amplitudes $y_{obs} = [S1_{obs}, S2_{obs}, S3_{obs}, ...]$ and the predicted EOF amplitudes $y_{pred} = [S1_{pred}, S2_{pred}, S3_{pred}, ...]$, and W_{phys} is the weight assigned to the RMSE between the observed change in beach volume between time steps dV_{obs} and the predicted change in volume dV_{pred}.

Finally, the test/train matrices were built using the non-stormy portions of the dataset with adequate data availability, segmented into four-day periods of topo-bathymetrically derived EOF amplitudes, water level, H_s, peak wave period T_p, and dominant wave direction θ_p (96 hourly observations of each). One-third of the dataset segments were set aside for training, and two-thirds remained for testing. When developing an LSTM model, the number of sub-partitions (batches) and full-data cycles (epochs) can be adjusted such that the model's short- and long-term "memory" and "forget" gates can be averaged across an optimal number of iterations. Thus, the number of epochs, batches, and length of time series segments are all hyperparameters. Tuning the number of epochs and batches can yield optimal algorithm learning, meaning the model parameters are balanced between overfitting and underfitting the solution. Tuning the length of the input time series segments, however, can yield insight into the response lag between changes in hydrodynamic forcing and profile elevation change [9].

3 Results

In total, nearly 4500 unique profiles were found in the non-stormy periods that could be extended reasonably assuming an equilibrium profile (fit RMSE < 30 cm) and showed nominal volume change between hourly time steps ($dV_{obs} < 5\,\text{m}^3/\text{m}$) (Fig. 1a). Figure 1a shows profile extents relative to the vertical datums for mean low water (MLW), mean water level (MWL), mean high water (MHW), and the regional approximate dune toe elevation [8].

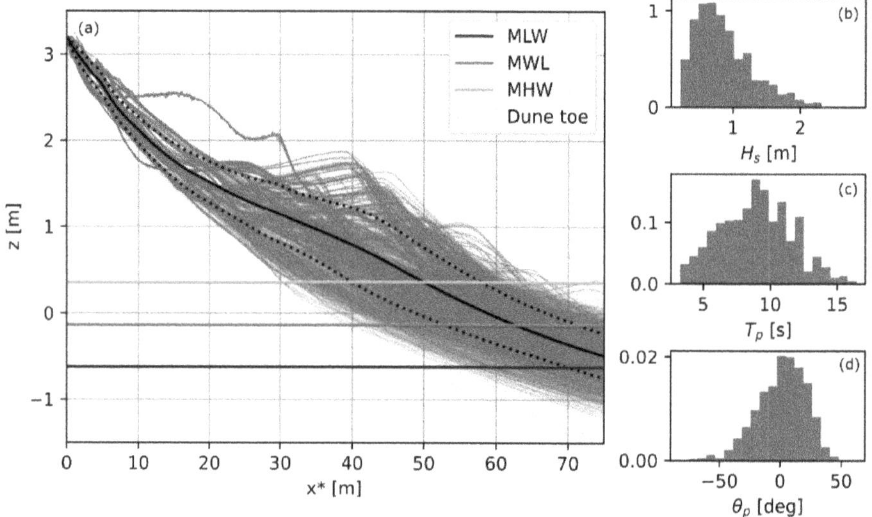

Fig. 1. (a) Profiles used in ML model development, where black is the mean profile and dashed black lines indicate one standard deviation from the mean. Normalized distributions of (b) H_s, (c) T_p, and (d) θ_p observed concurrently with profiles used for ML model.

Using PCA, the ~4500 individual profiles were found to be well represented as the summation of five spatial EOFs, which explained over 95% of their variance (Fig. 2). Mode 1 corresponds to the spatial EOF that explains the largest portion (53.1%) of profile elevation variance and likely physically corresponds to a net gain or loss in profile volume. Modes 2–5 explain 23.2%, 12.3%, 4.8%, and 2.3%, respectively, of profile elevation variance. Physically, Modes 2–5 appear to correspond with variable positions of a shoreface berm, or multiple berms.

Over 1300 continuous four-day time segments of wave statistics (H_s, T_p, and θ_p), water surface elevation (i.e., tidal stage), and EOF amplitudes were assembled from the deconstructed profiles and simultaneous hydrodynamic observations. Observations ultimately used for the ML model span between 2016 and 2023, including data from all years in between. The potential for over- or underfitting to observations is best demonstrated by the learning rate over the full-data cycles (epochs) imposed for the training and testing data (Fig. 3). As shown in Fig. 3a, the flattening of the learning curve indicates the model is generally balanced in fitness. Note that the vertical axis in Fig. 3a corresponds to the total error across five predicted EOF amplitudes that have been scaled between 0 and 1 (i.e., min-max scaling). The actual fit error is summarized in Figs. 3b-3f, which shows that the ML model is well posed to predict the first four modal amplitudes (correlation r^2 between 0.80 and 0.96), but less able to predict the fifth modal amplitude (r^2 of 0.48).

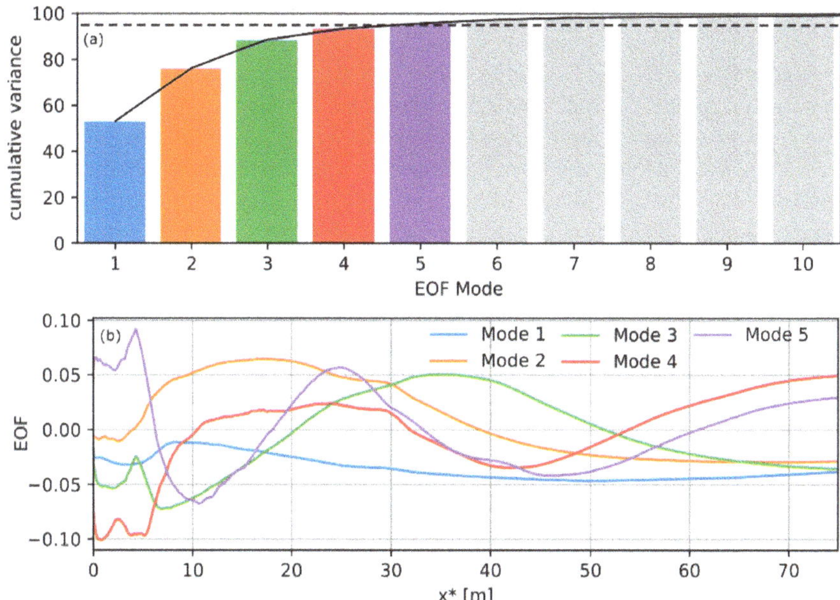

Fig. 2. (a) Cumulative variance in profile elevation explained by EOF modes. (b) Five dominant EOF modes plotted as function of distance from the dune toe x^* (see Fig. 1a).

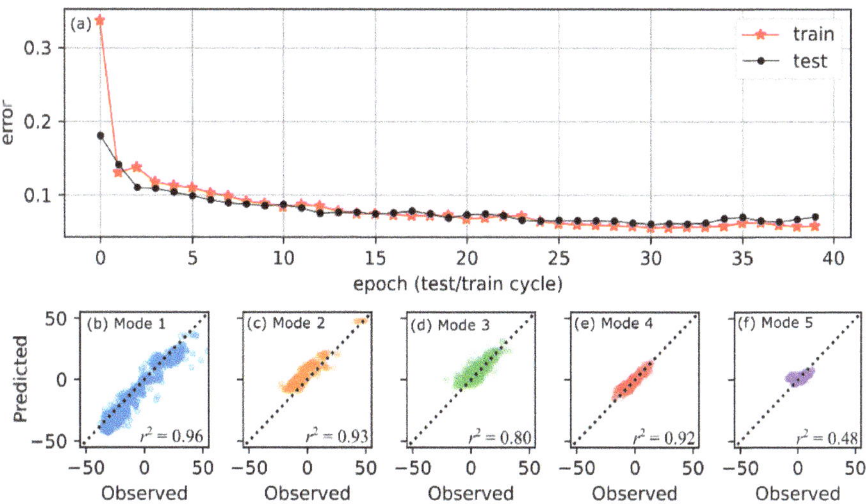

Fig. 3. (a) LSTM model learning curve with increasing test/train epochs. Observed vs. ML-predicted EOF amplitude for (b) Mode 1, (c) Mode 2, (c) Mode 3, (e) Mode 4, (f) Mode 5.

4 Conclusion

The methodology presented herein demonstrates how physical principles can be embedded into data-driven models of beach profile morphodynamics. Profile reduction into spatial EOFs with time-varying amplitudes can help retention of physical intuition and promote generality in two key ways. First, it simplifies the number of variables needed to predict $\Delta z(x)$ while retaining the spatial dependence between cross-shore observations. Secondly, excluding EOF modes that explain insignificant ($<\%1$) amounts of the cross-shore profiles' variability can eliminate the small perturbations that may otherwise confuse an ML model. Additionally, EOF spatial modes can be used to build custom loss functions that penalize the model for deviating from observations as well as from conservation laws. While further testing is needed to assess the applicability of the demonstrated LSTM model to system characteristics and dynamics not present in the training data, this preliminary result establishes a baseline upon which additional improvements can be made.

References

1. Gharagozlou A, Anderson DL, Gorski JF, Dietrich JC (2022) Emulator for eroded beach and dune profiles due to storms. J Geophys Res Earth Surf 127(8):e2022JF006620
2. Athanasiou P, Van Dongeren A, Giardino A, Vousdoukas M, Antolinez JA, Ranasinghe R (2022) Estimating dune erosion at the regional scale using a meta-model based on neural networks. Nat Hazard 22:3897–3915
3. Brunton SL, Kutz JN (2022) Data-driven science and engineering: Machine learning, dynamical systems, and control, 2nd edn. Cambridge University Press, Cambridge
4. Anderson D, Bak AS, Cohn N, Brodie KL, Johnson B, Dickhudt P (2023) The impact of inherited morphology on sandbar migration during mild wave seasons. Geophys Res Lett 50(3):e2022GL101219
5. O'Dea A, Brodie KL, Hartzell P (2019) Continuous coastal monitoring with an automated terrestrial lidar scanner. J Marin Sci Eng 7(2):37
6. Halls C, Jensen RE, Collins CO, Hesser TJ, Brown ME (2024) US Army Corps of Engineers (USACE) Wave Information Study: 2021 Annual Update. ERDC/CHL CHETN-I-xx. Vicksburg, MS: U.S. Army Engineer Research and Development Center. https://doi.org/10.21079/11681/48731
7. Dean R, Rosati J (2010) Beach nourishment. In: Handbook of coastal and ocean engineering, pp 843–866. World Scientific, Singapore
8. Cohn N, Brodie K, Conery I, Spore N (2022) Alongshore variable accretional and erosional coastal foredune dynamics at event to interannual timescales. Earth Space Sci 9(12):e2022EA002447
9. Davidson MA, Splinter KD, Turner IL (2013) A simple equilibrium model for predicting shoreline change. Coast Eng 73:191–202

Coastal Impact and Shoreline Recovery from Hurricane Irma (2017) on Barbuda Island, Caribbean

Emilia Guisado-Pintado[1]([✉]), Derek W. T. Jackson[2,3], Tony Dolphin[4], and Richard Heal[5]

[1] Department of Physical Geography and AGR, University of Sevilla, C/Maria de Padilla s/n, 41004 Sevilla, Spain
eguisado@us.es
[2] School of Geography and Environmental Sciences, Ulster University, Coleraine, Northern Ireland, UK
[3] School of Agricultural, Earth and Environmental Sciences, College of Agriculture, Engineering and Science, University of Kwazulu-Natal, Durban, South Africa
[4] CEFAS, Pakefield Road, Lowestoft, Suffolk NR33 0HT, UK
[5] CEFAS, Barrack Road, Weymouth DT4 8UB, UK

Abstract. Hurricane Irma was a rare, Category 5 hurricane that occurred September 2017 and was considered the third-strongest Atlantic hurricane ever recorded at landfall in the Caribbean. Irma made landfall in Barbuda early on September 6th with observed wind gusts reaching 260 kmh^{-1} (160 mph) and a minimum barometric pressure of 916.1 mbar (27.05 in Hg). Lasting 75 h as a Category 5 hurricane, it sustained wind speeds of 260 Kmh^{-1} (160 mph) for 37 consecutive hours. Irma resulted in widespread destruction and environmental impacts on the eastern Caribbean, particularly in Barbuda and Antigua.

Here, we assess morphological changes across-shore and offshore of Barbuda island using pre-and post-event airborne (terrestrial) LiDAR topography, and satellite-derived bathymetry. To understand nearshore changes pre-hurricane bathymetry was used to simulate Hurricane-modelled waves (using SWAN) across the nearshore, making use of *in situ* and far-field measurements of initial wave conditions. Results show significant bathymetric changes throughout the nearshore (and offshore) zone, with dramatic erosion and accumulation patterns not previously observed to such levels and a dramatic flattening of terrestrial dune ridge (and vegetation).

Keywords: Bed shear stress · SWAN · bathymetric changes · DoD · LiDAR

1 Introduction

Hurricanes, also known as tropical cyclones or typhoons, can have devastating impacts on low-lying islands, especially those in the Caribbean, the Pacific, and the Atlantic oceans. Islands are particularly vulnerable to hurricanes due to their geographical characteristics and exposure.

C. Coelho et al. (Eds.): CD 2025, CRL 41, pp. 76–81, 2026.
https://doi.org/10.1007/978-3-032-15473-6_12

Hurricane Irma made landfall on Barbuda, part of Antigua and Barbuda, in the Caribbean, 6th September 2017, at approximately 1:47 AM local time. It attained a Category 5 event level, the most powerful category on the Saffir-Simpson scale, with maximum sustained winds reaching 160 mph (260 km/h). Barbuda was impacted directly by Irma's winds and heavy rain and as a result the island suffered widespread damage and disruption [1]. Reports indicated that about 95% of the buildings on the island were damaged or destroyed with complete evacuation to nearby Antigua. The strong storm surge associated with the Irma also led to severe coastal erosion, and recorded as one of the most severe impacts of the 2017 hurricane season, highlighting the vulnerability of small island nations to extreme, high-energy weather events intensified by climate change. The objective of our work is to understand the impacts that Hurricane Irma had on the coast of Barbuda, as well as the changes in the nearshore including the topographic changes on the initial 0–10 m elevation coastal fringe of the island.

1.1 Study Site

Barbuda Island is one of the two main islands that form the nation of Antigua and Barbuda in the Caribbean Sea. Located approximately 17.63° N and 61.88° W, Barbuda is about 161 km^2 in area. The island is largely a flat and low-lying island with the highest point, the Mount Saint John, rising only 38 m above sea level.

The island is primarily composed of coral limestone and sedimentary rock, which is typical of many Caribbean islands [2]. This gives it a relatively stable base but also makes it more vulnerable to storm surges and coastal erosion. Barbuda is known for its long stretches of pristine, white sandy beaches, such as the famous 90-Mile Beach along the western side of the island backed by sand dunes. Coral reefs and a shallow lagoon, particularly visible along the western coast, surround the island. These features usually help protect its coastline from the full force of ocean storms and waves.

2 Methodology

The methodology consisted of two main steps: a) analysis of nearshore hydrodynamics during the event; b) calculation of morphological terrestrial and bathymetric changes from derived DTM models before after Irma (DEM of Difference).

Initial boundary conditions for wave simulations were forced using offshore data from the Wave Watch 3 (WW3) to develop the wave spectrum for nested (nearshore) high-resolution grids. Two grids were created, the offshore one generated through GEBCO's gridded bathymetric data sets (WGS84 UTM20) with a cell grid size of 914 m for the study area. This bathymetric grid was resampled to 100 m in order to minimise the differences between coarser offshore bathy and finer (10m resolution) nearshore bathymetry. The SWAN [3] wave model was run in non-stationary mode using the wave climate from WW3 data in its boundaries (N, S and E) and a boundary JONSWAP spectrum for developing waves. Input conditions considered from 3rd September 2017 06 am to 7th September 2017 00 am with a peak condition on the 6th September 2017 00 am (Hs: 13.87 m, T: 15.36 s and wave direction: 88 degrees). Water levels from the local tidal cycle were also taken account of by using a set of non-stationary runs and local tide gauge information with an input grid of every 6 h used to simulate tides during the passage of the storm. Results from simulations on wave orbital velocity amplitude (ms^{-1}) allowed the estimation of the limit of significant sediment transport (DoT) through the analysis of bed shear stress induced by waves. Bed shear stress (BSS) produced by waves was computed using the formula from Soulsby (1997), and is generally the main forcing control on sediment transport in shallow water (<30m depths). Further, related sediment motion threshold of the seabed was calculated using the Shields parameter (D50 of 0.3mm size fraction) [4, 5]. Through the combination of sediment density and gravitational acceleration, the critical shear stress (T_{wr}) or depth of transport (DoT) for non-cohesive and well-sorted particles was calculated.

Topographical and bathymetric changes (a DEM of Difference) was calculated for pre- and post-storm 3D models to assess topographical/bathymetric changes. For bathymetric changes, DSM were created through photogrammetry using high-resolution satellite imagery (Sentinel) for dates ranging from Jan 1, 2015–Sept 4, 2017 (pre-Irma) and Sept 6, 2017–Feb 15, 2018 (post Irma). Further, pre-and post-event airborne (terrestrial LiDAR) allowed 3D examination of coastal zones in the 0-10m range. The elevation changes on the coast after the passage of very high-energy events informs us of the response of the coastal zone in terms of sediment redistribution (erosion and build up) as a direct result of the storm event.

3 Results

3.1 Bed Shear Stress During Irma Peak Conditions for Barbuda

The bed shear stress was calculated for the peak of the storm occurring on the 6th of September 2017 at 6:00 am. For Barbuda, high shear stress values (exceeding 18–20 Nm^2) are focused around the NE, SE and some of the southern nearshore areas during Irma (Fig. 1).

Significant energy appears to extend right across the nearshore and onto the beach areas along these shorelines. The western side of the island received relatively much less energy (shear stress) due to sheltering effects of the island's coastline. Fringing reefs on the eastern side, show strong stress gradients between then and shorelines of the island. The threshold for sediment motion (DoT) (critical bed shear stress, or depth of transport following works from Valiente et al., 2019) corresponds to values $>4\,Nm^{-2}$.

3.2 Topographic and Nearshore Changes

Extracting two surfaces (before and after, DoD) enables a detailed dataset of vertical changes that have occurred as a direct result of events. Resulting surface changes were examined using only the 0–10 m elevation band around the coastal fringe. Results (Fig. 2) show that terrestrial dune ridge topography was altered with largescale lowering of relief and marked differences according to orientation (to the Hurricane) and antecedent topographic heights. Largescale deflation (reds and yellows) of the sandy dune ridges and beaches that what were previously vegetated areas (shrub/tree cover) experiencing the most lowering of surface elevation (-4 to -6 m). Overall for Barbuda, within the

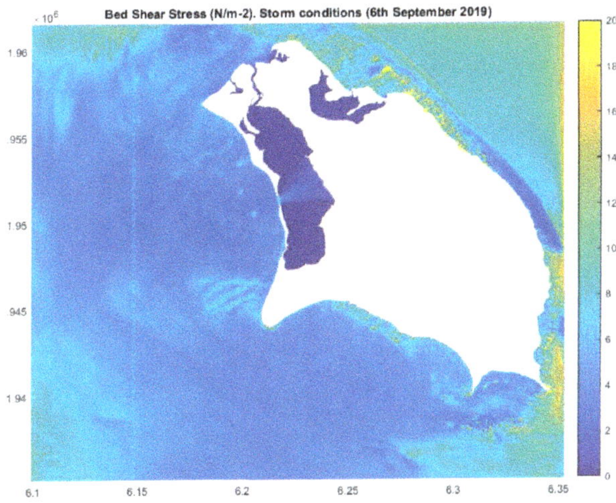

Fig. 1. Bed shear stress values for Barbuda under Hurricane Irma conditions

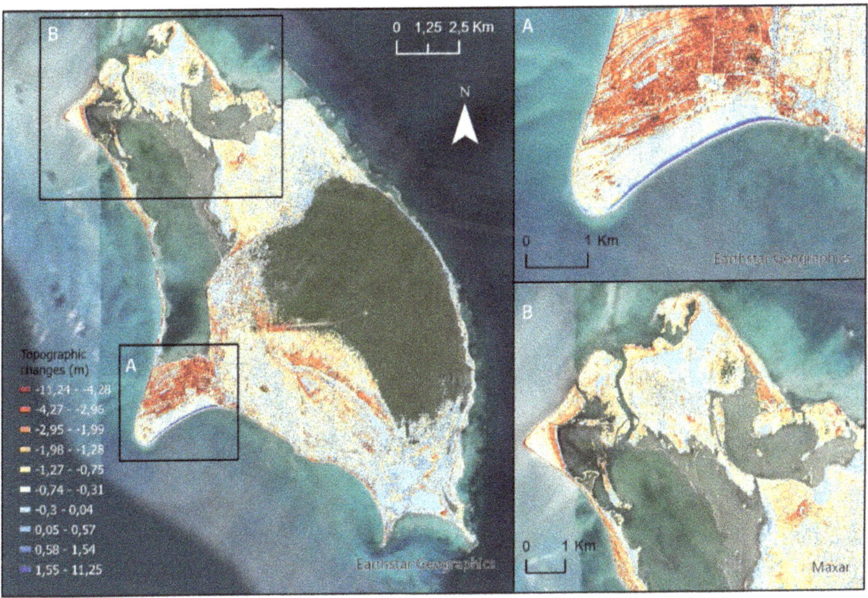

Fig. 2. Topographical difference map between post- and pre-Irma showing wide-spread surface lowering particularly in the 0–10 contour around Barbuda. A and B show different patterns of erosion deposition on the S and N respectively.

0–10 m height range of pixels interrogated 85% if the surface was lowered with 15% increasing.

Beach levels show largescale deflation (>2 m) but with what looks like to be a storm ridge (>2 m) (Fig. 2A) formed along the length of the southern and SE beaches. The northern beaches show relatively less lowering in elevation compared to eastern and southern beach systems (Fig. 2B).

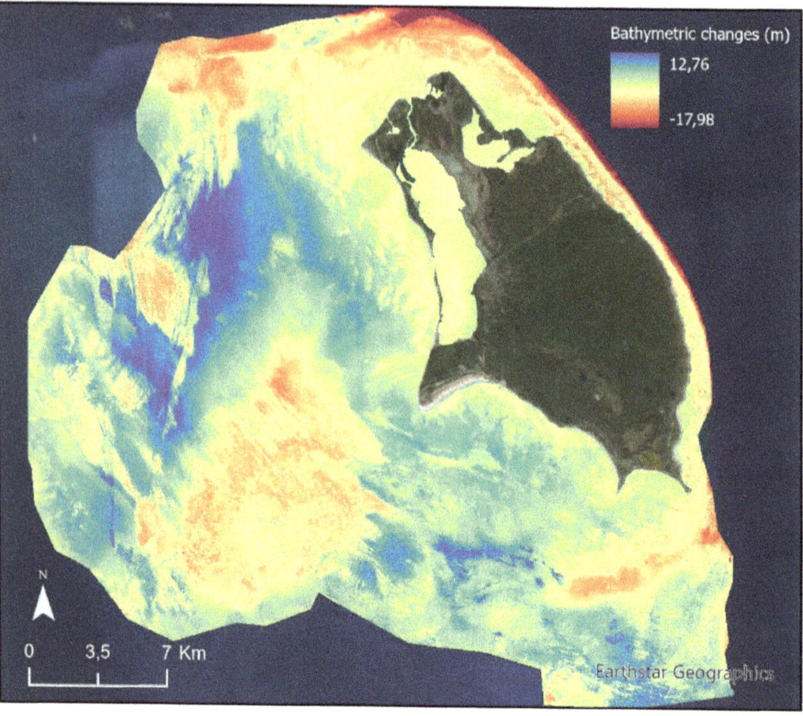

Fig. 3. Bathymetric changes map showing erosion and accumulation patterns and the turn-over sediment offshore.

Furthermore, significant bathymetric changes throughout the nearshore zone because of the Hurricane event with distinctive erosion and accumulation patterns could be observed (Fig. 3). Overall, 27.6% of the surface experienced erosion versus 72.4% of deposition. We observed seabed deflation of −7 m in places offshore (SW of Barbuda) with 12 m of accretion close by. They total volume of erosion represented around 625 million cubic metres of sand (7.42% error) whilst 1,642 (5.74% error) million cubic metres were deposited showing a new gain of sediment over the area of interest, representing a sedimentary gain from the hurricane event. The nearshore zone is notably shallower after the event, having gained sediment and therefore likely adding to potential future sediment supply for beach ridge building on land. The average depth of erosion patterns occurred around −2.51 m, whereas depositional areas are found around −3.36 m depth. Results highlight direct wave forcing (bed shear stress, Fig. 1) and its coincidence with these patterns of sediment dispersal which seem to be the main responsible agent for those changes [6]. Past phases of beach ridge building may therefore be indicative of having pre-cursor high energy events adding enormous sediment volumes into the shallower nearshore, thereby acting as a vast (regenerated) sediment source area. The supposition is that sediment locked within previously inaccessible water depths (beyond storm wave base) are re-activated and transported to inshore areas, helping supply low lying island geomorphology with sand to feed beach ridge building phases.

4 Conclusions

Hurricanes present significant risks to island nations, threatening their infrastructure, economy, and the lives of their residents. While preparedness can mitigate some damage, the increasing frequency and severity of these storms, exacerbated by climate change, highlight the need for understanding the impacts of these high-energy events.

Detailed topographical quantification of the seabed and landform response to this extraordinary event presented a unique set of forcing conditions with which to observe category 5 impact on low-lying island environments and their offshore areas. Furthermore, our study helps demonstrate the heterogeneous nature of the impact that hurricanes of this magnitude have on low-lying island environments and shows the dramatic before and after changes these can have on the local coastal landscape.

References

1. FEMA. https://www.fema.gov/fact-sheet/5-years-later-hurricane-irma-florida
2. Ruttenberg B et al (2018) Ecological assessment of the marine ecosystems of Barbuda, West Indies: using rapid scientific assessment to inform ocean zoning and fisheries management. PLoS ONE 13(1):e189355. https://doi.org/10.1371/journal.pone.0189355
3. Booij N, Ris RC, Holthuijsen LH (1999) A third-generation wave model for coastal regions: 1. Model description and validation. J Geophys Res Oceans 104(C4):7649–7666. https://doi.org/10.1029/98JC02622
4. Soulsby RL (1997) Dynamics of Marine Sands: a manual for practical applications. Thomas Telford, London, ISBN 0-7277-2584-X
5. Valiente NG, Masselink G, Scott T, Conley D, McCarroll RJ (2019) Role of waves and tides on depth of closure and potential for headland bypassing. Marine Geol 407:60–75
6. Jackson D, Guisado-Pintado E, Dolphin T, Heal R (2022) Nearshore and coastal impact of Hurricane Irma (2017) on Barbuda, eastern Caribbean. EGU General Assembly 2022, Vienna, Austria, 23–27 May 2022, EGU22-13108. https://doi.org/10.5194/egusphere-egu22-13108

Monitoring and Modelling Cross-Shore Processes: Calibration for Long-Term Simulations

Frederico Romão[1]($\boxtimes$), Márcia Lima[1,2], and Carlos Coelho[1]

[1] RISCO and Civil Engineering Department of the University of Aveiro, Aveiro, Portugal
`fredericoromao@ua.pt`
[2] Porto University Center (CUP), Lusofona University, Porto, Portugal

Abstract. The existence of extensive and updated datasets is useful for model calibration and validation. In this work, the CS-Model was applied at IJmuiden coastline, The Netherlands, evaluating the evolution of berm and dune toe positions by comparing the model results with a long series of observations (1979–2020). The model was calibrated considering the first 11 years of existing data and the model performance was evaluated every decade, considering the simulation of 2 scenarios: using the initial conditions; and updating the profiles characteristics, according to the observed positions along the decades. The results indicated better model performance when the profile characteristics are updated every decade, with the berm and seaward dune toe positions projections being nearest the observations.

Keywords: CS-Model · Beach Profiles · Accretion Rate · Validation

1 Introduction

Climate change put significant threats to coastal areas through rising sea levels, erosion, storm surges, and extreme weather events. Understanding the related risks is important to inform urban planning, risk management, and climate adaptation strategies [1]. Numerical models have become essential tools to help in decision making [2], but their accuracy is directly related to the proper calibration and validation, to better represent the observations [3]. The availability of data [4] and continuous monitoring [5] are important for model calibration and validation, to help in the development of effective strategies to mitigate coastal risks.

This work analyzes the influence of alongshore and cross-shore sediment transport processes by applying a cross-shore numerical model (CS-Model), calibrated considering approximately 40 years of wave and surveyed profile data (1979–2020), for medium-to long-term projections. The first 11 years of data were used for model calibration, while the subsequent groups of 10 years, totalizing 30 years, were used for validation. Shoreline variation rates were incorporated during calibration, to represent the effects of longshore sediment transport gradients. This analysis provides valuable insights for modelers, highlighting the impact of initial calibration assumptions and emphasizing the importance of monitoring profile evolution trends to refine modeling projections.

C. Coelho et al. (Eds.): CD 2025, CRL 41, pp. 82–88, 2026.
https://doi.org/10.1007/978-3-032-15473-6_13

2   Methodology

This section describes the applied model, and the study area selected to develop this work. Four cross-shore profiles were chosen to represent the study area (P1 to P4).

2.1   Model Description

The Cross-Shore model (CS-Model), developed by Larson *et al.* [6], was selected to analyze the berm and seaward dune toe position evolution along 41 years. The model simulates sediment transport processes accounting for dune erosion (q_S) and overwash (q_L), aeolian sediment transport from land- (q_{WL}) and seaward (q_{WS}) directions, wave-driven berm-bar sediment exchanges (q_B), and sea-level rise impacts [7]. Its simplified scheme includes land- (Y_L) and seaward (Y_S) dune toe, berm (Y_B), and shoreline (Y_G) positions, depth of closure (D_C), dune (S) and berm (D_B) heights, and a submerged bar volume (V_B), according to Fig. 1.

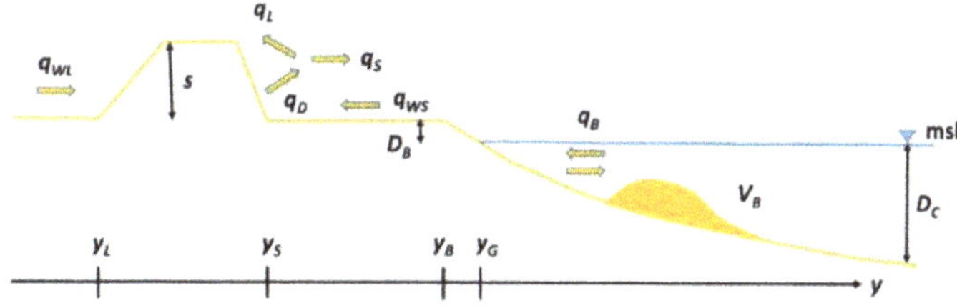

Fig. 1. Simplified model scheme and variables used to define the profile [6].

The model was previously calibrated by Romão *et al.* [4] for the same period of analysis and study area, highlighting the model capacities for simulating the cross-shore behavior at decadal scale (5–10 years). In this work, the model was calibrated considering the period between 1979–1989, changing the aeolian sediment transport parameter (QWINDS, in $m^3/s/m$), the coefficient that represents the dune eroded volumes after wave impacts (CIMPACT) and the accretion rate (AR, in m/time step), that represents the alongshore sediment transport gradients contribution (Table 1).

Table 1. Accretion Rate (AR, m/time step), Coefficient for Dune Erosion after Wave Impact (CIMPACT) and Aeolian Sediment Transport (QWINDS, $m^3/s/m$), for all the profiles.

Profiles	P1	P2	P3	P4
AR ($\times 10^{-7}$)	27.0	−8.4	86.0	0.0002
CIMPACT ($\times 10^{-4}$)	1.0	0.1	0.1	0.1
QWINDS ($\times 10^{-7}$)	0.9	5.7	1.5	2.0

The model performance was evaluated through the Mean Absolute Error (MAE). CS-Model demonstrates to be highly sensitive to the accretion rates and wind contribution.

2.2 Study Area

The CS-Model was applied to IJmuiden, located on the Holland coast, in The Netherlands (Fig. 2), a well-monitored area with extensive data, helping to calibrate and validate numerical models. At the site, the wave heights range from 0.1 m to 1.5 m (from 1979–2020), predominantly from the southwest (225°, measured clockwise from North). Tidal levels vary between 1 m and 2 m during the neap-spring cycle [4, 8]. The four profiles selected for the analysis are shown in Fig. 2.

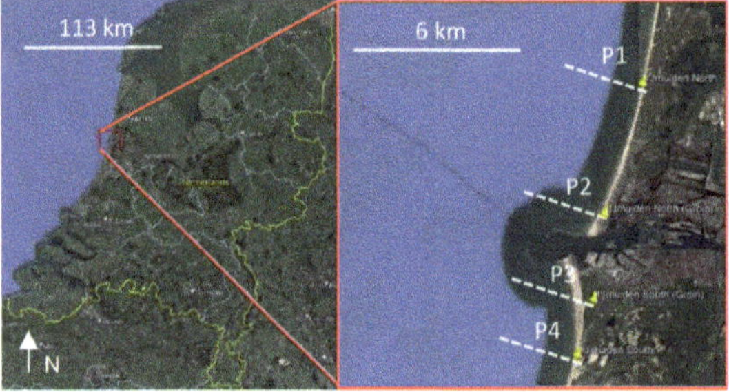

Fig. 2. IJmuiden study area, in The Netherlands (left) and profile's location (right).

Longshore sediment transport in IJmuiden southern area is estimated at 115,000–395,000 m³/year, moving from south to north. The harbor entrance disrupts this flow, causing erosion of about 40,000 m3/year in the northern area [4, 8], which is mitigated by artificial nourishments. As a result, shoreline positions exhibit varying trends. Profiles P3 and P4 show shoreline advancement due to sediment accumulation from the southern groin, while Profile P2 experiences erosion, being in the shadow zone of the northern groin. Profile P1 demonstrates accretion, driven by regular artificial nourishments. In this analysis, the first 11 years of data (1979–1989), covering the total period between 1979–2020, were selected for model calibration. For model validation, two different scenarios were created, considering ten years intervals (1990–1999; 2000–2009; 2010–2019): A) model runs without updating the cross-shore profiles characteristics over time; and B) model runs performing a correction every ten years, based on the surveyed conditions at the beginning of each decade. The calibration parameters were kept the same in all the runs of both scenarios.

3 Results

The results of berm and seaward dune toe position evolution are shown in Fig. 3 and 4, respectively. Distinct behaviors are illustrated in Fig. 3, from the initial calibration to the modeled scenarios. In scenario A (without profiles updates, blue line), a consistent deviation from observed behavior is evident across all profiles. In contrast, scenario B (with profiles updates, green line) aligns closer to the observed evolution of both the berm and seaward dune toe positions.

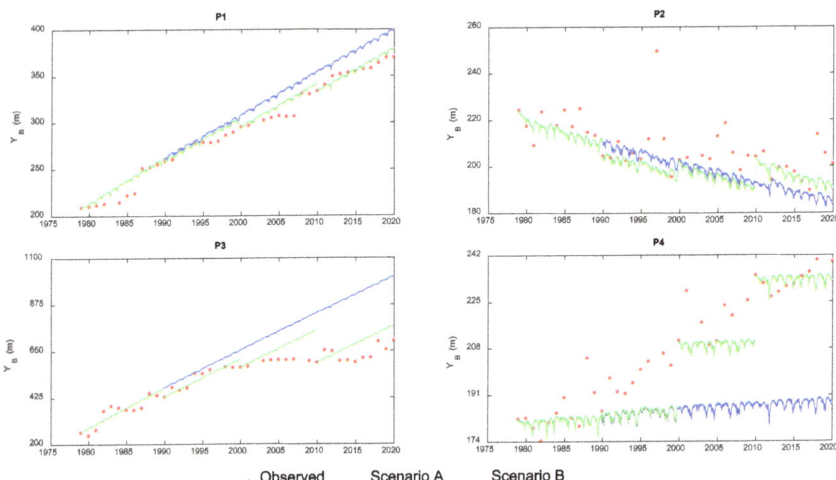

Fig. 3. Berm position evolution (Y_B) with and without updating the profiles, for all locations.

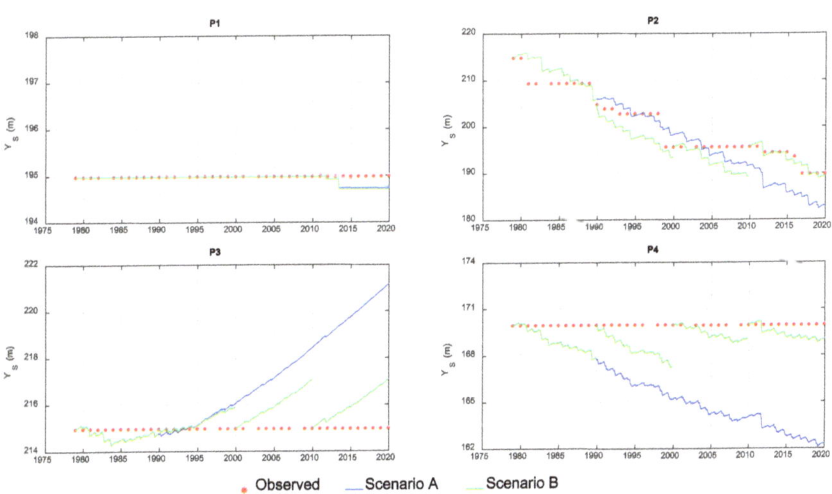

Fig. 4. Dune toe position evolution (Y_S) with and without updating the profiles, for all locations.

The seaward dune toe position (Fig. 4) demonstrates the same behavior, with major deviations observed in P3 and P4, for the scenario when the profiles are not updated. The evaluation of the modelling results is supported by the MAE calculations, summarized in Table 2 and Table 3, for berm and dune toe positions, respectively.

In general, the MAE increases in every decade, if no updates in the profile characteristics are considered over time (scenario A). All the profiles present the higher MAE for the 2010–2019 period, followed by the 2000–2009 and the best performance is observed in the first 10 years (1990–1999).

Table 2. Mean Absolute Error for calibration and validation of berm position, considering different scenarios: (A) without profile updates; and (B) with profile updates every decade.

Time Intervals	P1		P2		P3		P4	
	A	B	A	B	A	B	A	B
1979–1989	6.36		5.02		24.41		4.73	
1990–1999	5.76	4.32	8.19	9.18	33.22	19.30	14.48	13.86
2000–2009	18.60	6.55	8.45	10.19	138.48	55.46	33.59	10.29
2010–2019	19.54	3.76	11.41	5.42	278.93	52.57	46.92	3.36
1990–2019	16.95	6.38	10.29	9.14	162.52	49.80	31.71	9.64

When scenario B is considered, lower MAE are obtained for every decade, compared to the first scenario. The maximum MAE generally occurs in the 2000–2009 period, being the recent decade (2010–2019) the one that presented the best model performance. The scenario B results demonstrate the model's reliability over three decades, when corrections are applied to account for intermediate deviations, being better than scenario A. An exception occurs for the P2 profile, demonstrating worst MAE when the profiles 'are updated, compared to the scenario A, for the initial decades (1990–1999 and 2000–2009), related to the small representativeness of the first profiles of those decades, that are not aligned to the observed trends. For the total period of 30 years, the MAE obtained for validation shows better results for scenario B, when compared to scenario A.

Table 3. Mean Absolute Error for calibration and validation of dune toe position, considering different scenarios: (A) without profile updates and (B) with profile updates every decade.

Time Intervals	P1		P2		P3		P4	
	A	B	A	B	A	B	A	B
1979–1989	0.00		2.63		0.25		0.82	
1990–1999	0.00	0.00	1.78	3.00	0.33	0.26	3.41	1.28
2000–2009	0.00	0.00	2.05	2.87	2.12	0.90	5.41	0.50
2010–2019	0.17	0.19	6.20	0.64	4.68	0.88	6.86	0.55
1990–2019	0.07	0.07	3.67	2.58	2.57	0.82	5.36	0.91

The evolution of the seaward dune toe position has variations in an order of magnitude much smaller, compared to the berm position evolution. Scenario A demonstrates an increase of the MAE, from the first decade (1990–1999) to the final decade (2010–2019) of the validation period, considering all the profiles. Observing the same decades, the evolution of the MAE when the profiles are updated, presents an irregular behavior. However, the obtained results for all the validation period (1990–2019) are better when the profiles are updated every decade, compared to the scenario without updates.

4 Conclusions

The work carried out highlights that is important to continuously monitoring coastal areas. The numerical model applied was initially calibrated using 11 years of data, using the next 30 years for model validation, under two different scenarios (with and without updating the profiles every decade). When the profiles are updated, the model performance generally improves, reducing the associated errors when comparing model simulations with observations. This updated information allows a better representation of the phenomena under study. Therefore, the surveyed data should be incorporated in updated modelling, in order to improve predictive capacity over time, diminishing eventual propagation of errors.

The regular acquisition of high-quality data is essential for refining models' calibration and validation and evaluating their performance over time. For long-term simulations, the input information should be updated with observed conditions and eventual correction of calibration parameters, to improve model performance, allowing an effective coastal management. This approach is crucial to support decision-making and long-term strategies to mitigate the impacts of climate change and ensure coastal resilience. The integration of continuous monitoring, robust data acquisition, and adaptative modeling forms the foundation for effective long-term coastal management and climate change adaptation strategies.

References

1. Palanikkumar D, Maashi M, Alsamri J, Obayya M (2015) Machine learning driven multi hazard risk framework for coastal resilience. J S Am Earth Sci 152:105331
2. Yassine R, Cassan L, Roux H, Frysou O, Pérès F (2023) Numerical modelling of the evolution of a river reach with a complex morphology to help define future sustainable restoration decisions. Earth Surf Dyn 11(6):1199–1221
3. Roelvink D, Reniers A, Van Dongeren AP, De Vries JVT, McCall R, Lescinski J (2009) Modelling storm impacts on beaches, dunes and barrier islands. Coast Eng 56(11–12):1133–1152
4. Romão F, Coelho C, Lima M, Ásmundsson H, Myer EM (2024) Cross-shore modeling features: calibration and impacts of wave climate uncertainties. J. Marine Sci. Eng. 12:760
5. Khatun F, Saha OR, Matin N (2025) Morphological changes in the offshore islands of Meghna estuary: analysis of the erosion and accretion dynamics. HydroRes. 8:294–306
6. Larson M, Palalane J, Fredriksson C, Hanson H (2016) Simulating cross-shore material exchange at decadal scale. Theory and model component validation. Coast Eng 116:57–66
7. Hallin C, Huisman BJA, Larson M, Walstra D-JR, Hanson H (2019) Impact of sediment supply on decadal-scale dune evolution Analysis and modelling of the Kennemer dunes in the Netherlands. Geomorphology 337:94–110

8. Van Rijn L (2023) Improvement of coastline modelling software using field data. AX-COAST Report. (non published)

Analysis of Rip-Channel Formation at an Idealized Beach in the Framework of Coastal Sediment Connectivity

H. Shafiei[(⊠)], Stuart G. Pearson, and A. J. H. M. Reniers

Department of Hydraulic Engineering, Delft University of Technology, Delft 2600GA,
The Netherlands
h.shafiei@tudelft.nl

Abstract. Tracking coastal sediments can provide useful information about coastal dynamics, thereby helping coastal management. However, the highly dynamic conditions of the coasts makes analyzing the trajectories of a huge number of particles challenging. To solve this limitation, the framework of coastal sediment connectivity is designed. In this framework, recent advances in graph theory are used to quantify coastal systems as complex networks. In this context, sediment sinks/sources and pathways represent the graphical nodes and links, respectively. In this work, we take the first step to evaluate the ability of this newly-developed framework in quantifying the basic processes on a sandy beach. Firstly, we used Delft3D to obtain the velocity field and bed-level changes. Then, the Eulerian results were fed into SedTRAILS to simulate the sediment pathways. We show that the current version of the model can correctly calculate the basic metrics of the sediment-connectivity network (e.g., network link strength which is a proxy for sediment fluxes). More specifically, we show that this framework is capable of exploring the initiation of the rip channel formation.

Keywords: Sediment Connectivity · Rip-channeled beach · Complex Networks

1 Introduction

Sediment dynamics of sandy beaches are highly variable both temporally and spatially. Understanding gross sediment transport (which indicates the sediment fluxes) can be calculated using Lagrangian models. It can provide a wide range of understandings about the dynamics of coastal systems, e.g., the source (destination) of the eroded (deposited) sediments, the sediment-mixing dynamics, and morphological hot spots. However, the large number of sediment trajectories impedes efficient analysis of the results and drawing solid conclusions. Therefore, development of a toolbox for analyzing the sediment pathways altogether is invaluable. Considering these sediment pathways as a complex network paves the way to take advantage of recent advancements in network sciences and quantify the practical characteristics of coastal sediment dynamics.

C. Coelho et al. (Eds.): CD 2025, CRL 41, pp. 89–94, 2026.
https://doi.org/10.1007/978-3-032-15473-6_14

Coastal sediment connectivity is a recently-developed framework through which properties of sediment pathways are quantified to better understand sediment dynamics and help coastal management [1, 2]. To validate the simulated sediment pathways, the spatial distribution of their direction and strength can be compared to the short-term morphological changes. This comparison can give us insight into the interpretation of these connectivity metrics. Consequently, the main goal of this investigation is to evaluate the relevance of some basic connectivity metrics to sediment dynamics of a relatively well-studied environment, i.e., embayed rip-channeled beach [3].

2 Methodology

In the first step, an Eulerian hydro-morphodynamic model (Delft3D, here) is set up to simulate the effects of forcings on bed-level changes and obtain the intermediate bathymetries. Then, those intermediate bathymetries are used as the initial bed level in the Eulerian morpho-static model while the other settings are kept the same as in the morphodynamic model. In the third and final step, the simulations in the second step are fed into the SedTRAILS model to be used as the initial conditions as well as the particle drivers. Then, SedTRAILS tracks the seeded sediments following the approach of Soulsby et. al. [4] and calculates the connectivity metrics using the toolkit of Rubinov and Sporns [5].

SedTRAILS (Sediment TRAnsport vIsualization & Lagrangian Simulator) is a newly-developed model adapted to calculate, visualize, and analyse the sediment connectivity network [1]. In the current version, it assumes a morpho-static bed, and the particles are seeded as the initial conditions. The main outputs are the trajectories of each seeded particle until the end of the simulation and the respective 2D adjacency matrix. In this matrix, each row and column represents one of the nodes and each cell indicates whether its respective row and column are connected or not.

Once the sediment pathways have been compiled as a complex network, a suite of toolboxes and methodologies are available to quantify the properties of the resultant network [1, 2]. In this work, we focus on node strength which is a basic network metric. In the context of graph theory, the metric strength is the sum of the weights of all the links connected to a given node. Furthermore, in-strength and out-strength focus on the total weight of incoming and outgoing links, respectively. In the context of sediment transport, strength represents the amount of sediment exchanged between a given node with its neighbors. The spatial distribution of the net strength (i.e., out-strength minus in-strength) is expected to be similar to the maps of initial erosion/sedimentation. It should be noted that, in this manuscript, absolute strength is used interchangeably with strength to emphasize that the sign of the strength values is not taken into account.

3 Results

3.1 Delft3D

The Delft3D model settings are the same as Reniers et al. [3]. It simulates the morphological evolution of an idealized, embayed sandy beach with grain size of 0.22 mm under wave group forcing. Root-mean-squared wave height and directional spreading are 1 m and 12.2°, respectively. Therefore, the waves are erosive and alongshore variable while the initial bathymetry is an alongshore-uniform bar profile (see Fig. 1(a)). Besides, the hydrodynamic model simulates 8 h with morphological factor of 12. Figure 1(b) illustrates the final bed level and velocity field in which rip channels and currents are fully developed. However, in this paper, we focus only on the initial time steps during which the formation of the rip channel is initiated. Consequently, the morpho-static Delft3D model is set up using the alongshore-uniform bar and 1 h simulation period. A 30 min spin-up period is considered and therefore not included in the output. Although other aspects of the model configuration are kept as in the morphodynamic model, the outputs, which will be used in the SedTRAILS model, are written in small time steps (i.e., 4.8 s).

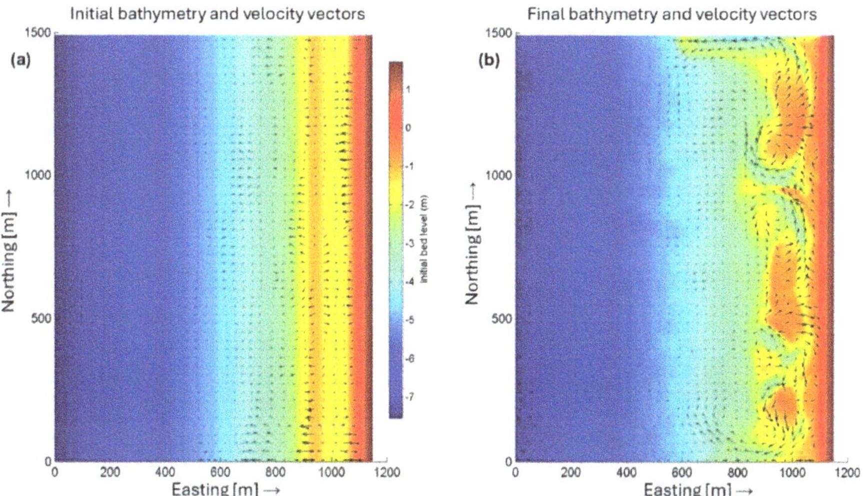

Fig. 1. Delft3D morphodynamic model results. (a) the initial bar is along-shore uniform while the velocity field is not uniform due to wave group forcing. (b) the rip currents and channels are fully developed after four days of morphological simulation.

3.2 SedTRAILS

In our SedTRAILS model, the particles are seeded uniformly within the central section of the surf zone as initial conditions. The distance between each particle is 20 m and additional particles are not introduced during the simulation. The sand properties and simulation timeframe are the same as the morpho-static Delft3D model.

Various plots can be obtained using the post-processing tool implemented into SedTRAILS. For example, particles can be illustrated at every given time step and setting their colors according to their age since their release (see Fig. 2(a)). The main observations from this figure are (I) the particles over the bar become scattered around the bar (II) the largest sediment accumulation occurs at around $(x, y) \sim (800, 900)$. This is consistent with the location of the rip-shoal system in Fig. 2(b). Furthermore, the resultant network of sediment nodes/sinks and pathways can be plotted in geographic space. As highlighted in Fig. 2(b), it reveals the main sediment pathways. We can conclude that the initial bar migrates non-uniformly in both longshore and cross-shore directions. Visually, the areas with the densest pathways can be considered as the most persistent transport patterns. Consistently, the persistent offshore- and onshore-directed sediment pathways are located around the main rip channel and shoal (compare Fig. 2(b) with Fig. 1(b)).

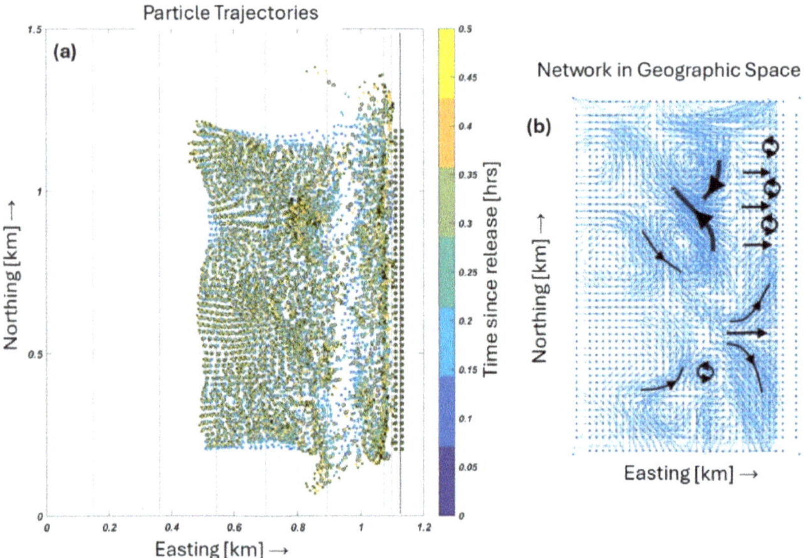

Fig. 2. Tracking the particles using SedTRAILS. (a) each particle is shown at different time steps and colored according to their age since release. (b) initial particle position and their respective pathways till the end of the simulation reveal the main sediment pathways.

3.3 Sediment Connectivity

In this manuscript, as presented in Fig. 3(a), the absolute and net values of strength of the nodes are chosen as proxies to sediment fluxes and erosion/sedimentation, respectively.

To verify the observed spatial pattens of these parameters, the initial sediment erosion/deposition during the first 30 min of the simulation is shown in Fig. 3(b). At this stage, it is important to predict the formation of the initial rip channel. This can be done by identifying the highest positive values of net strength (i.e., in-strength minus out-strength), which is consistent with Fig. 3 and is located at around (x, y) ~ (800 m, 900 m).

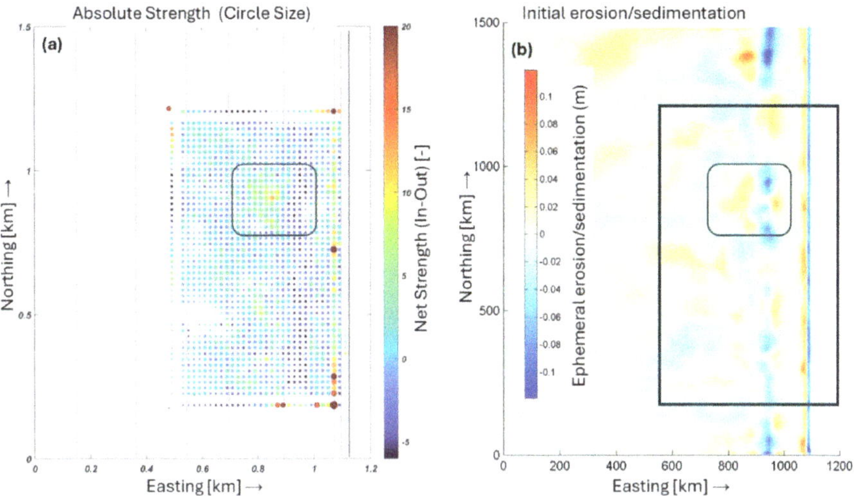

Fig. 3. Exploring the formation of a rip channel using SedTRAILS (a) Considering net and absolute values of node strength as a proxy for initial erosion and sedimentation. (b) spatial distribution of initial erosion and sedimentation in the first 30 min of the Delft3D morphodynamic simulation. The black boxes highlight the location of rip-channel formation.

4 Conclusions and Discussion

In this work, we evaluate the conceptual framework of coastal sediment connectivity to explore the formation of a rip channel at a sandy beach. SedTRAILS, as the main toolbox in this framework, simulates the sediment pathways and their respective connectivity network within the timeframe of 30 min. During this time, the wave group forcing entrains and transports all the particles on the sandbar towards various directions. The persistent patterns of sediment pathways trigger the formation of the first rip channel. Distribution of the particles at the end of the simulation and the node strength as the proxy for sediment fluxes are consistent in identifying the location where the formation of the first rip channel is initiated. Future work will focus on further development of network-based metrics for quantifying sediment pathways on evolving beaches.

References

1. Pearson SG, et al (2021) Lagrangian sediment transport modelling as a tool for investigating coastal connectivity. In: Proceedings, Coastal Dynamics Conference (2021)
2. Pearson SG, van Prooijen BC, Elias EP, Vitousek S, Wang ZB (2020) Sediment connectivity: a framework for analyzing coastal sediment transport pathways. J Geophys Res Earth Surface **125**(10), e2020JF005595 (2020)
3. Reniers AJ, Roelvink JA, Thornton EB (2004) Morphodynamic modeling of an embayed beach under wave group forcing. J Geophys Res Oceans **109**(C1) (2004)

4. Soulsby RL, Mead CT, Wild BR, Wood MJ (2011) Lagrangian model for simulating the dispersal of sand-sized particles in coastal waters. J Waterw Port Coast Ocean Eng 137(3):123–131
5. Rubinov M, Sporns O (2010) Complex network measures of brain connectivity: uses and interpretations. Neuroimage 52(3):1059–1069

Analysis of Intensification and Mitigation of Erosion Due to Sea Level Rise According to Shoreline Feature Type

Hyeong-Jun Kim[1,2], S. W. Lee[3], M. Choi[3], W. C. Cho[1], and J. L. Lee[3(✉)]

[1] Department of Civil and Environmental Engineering, Chung-Ang University, Seoul, Republic of Korea

[2] Coastal and Ocean Engineering Laboratory, DY Engineering Corporation Ltd., Seoul, Republic of Korea

[3] Graduate School of Water Resources, Sungkyunkwan University, Suwon, Republic of Korea

`jllee6359@hanmail.net`

Abstract. This study examines the impact of sea level rise on shoreline erosion at convex and concave beaches. While previous studies based on Bruun's hypothesis primarily focused on straight coastlines, this research extends the analysis to incorporate the distinctive responses of different shoreline geometries. Using Sentinel-2 satellite imagery, long-term shoreline changes over a nine-year period were analyzed and compared for Gangneung Beach (convex) and Jangho Beach (concave) in South Korea. The findings indicate that convex beaches experience intensified erosion due to reduced wave refraction and increased curvature. In contrast, concave beaches show mitigated erosion as sediment loss decreases with rising sea levels. These results deepen our understanding of erosion dynamics associated with sea level rise and provide critical insights for coastal management and adaptation strategies.

Keywords: Sea Level Rise · Shoreline Erosion · Convex and Concave Beaches · Sentinel-2 Satellite Imagery

1 Introduction

Shoreline retreat due to sea level rise is an already observed phenomenon, and there are concerns that it will soon lead to a global disaster. Various studies have been conducted on erosion caused by sea level rise, starting with Bruun's hypothesis. In this study, we extended existing research results applicable to straight coasts to convex protruding beaches or extremely concave small beaches to investigate the phenomenon of enhanced or alleviated erosion. Although sea level rise is not yet at a serious level, the results proposed in this study were verified based on long-term trend analysis of the shorelines extracted from Sentinel-2 satellite images.

C. Coelho et al. (Eds.): CD 2025, CRL 41, pp. 95–99, 2026.
https://doi.org/10.1007/978-3-032-15473-6_15

2   Background Erosion Caused by Sea Level Rise

Bruun proposed the hypothesis that shorelines recede as sea levels rise, as shown in the following equation [1].

$$W = \frac{L}{h_B + h_c} S = m_a S \tag{1}$$

where m_a is considered a constant rather than a variable, L is the horizontal limit width along which sand moves and is the horizontal distance from the berm to the closure depth, h_B is the berm height, and h_c is the closure depth.

Many studies have been conducted based on this hypothesis, and despite a critical review of Cooper and Pilkey [2], the global impact has recently been analyzed by Vousdoukas et al. [6]. Figure 1 below is a diagram illustrating the phenomenon in which erosion is strengthened at the peak of the shoreline of convex beaches and overall erosion is alleviated on concave beaches due to sea level rise.

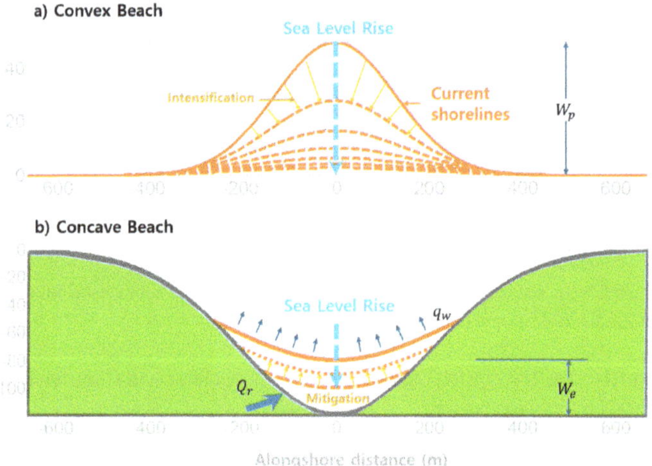

Fig. 1. Schematic diagram of changes in shoreline feature on convex and concave beaches.

3   The Concept of Wave Phase Potential (WPP)

Lee et al. [5] proposed a methodology for deriving the equilibrium beach profile under the concept of wave phase potential (WPP). The WPP concept is analogous to movable materials in a gravitational field, where materials move from regions of high potential energy to low potential energy. Similarly, the WPP describes how sediment is consistently transported from areas of high wave phase to the lowest point. WPP is defined as the integral of the wave phase, which decreases as waves propagate from deep offshore regions, and is expressed as follows:

$$\psi_p = \int k(h)dy = \int \frac{\sigma}{C} dy \tag{2}$$

where $k(h)$ is the wave number as a function of water depth, ψ_p represents the wave phase potential, σ is the angular frequency, and C is the phase celerity.

Lee et al. [5] derived the relationship between the equilibrium beach profile under shallow water conditions and the variation in WPP as follows:

$$\Delta \Psi_p = \frac{3\sigma}{2\sqrt{g}A^{3/2}} \Delta h_{eq}$$ (3)

Here, $\Delta \Psi_p$ and Δh_{eq} represent the changes in WPP and equilibrium water depth, respectively. Building on this concept, Lee et al. [5] proposed a shoreline position evolution model derived under the same framework.

By applying the WPP concept to the seabed depth variation model proposed by Laksmi et al. [3] and assuming a steady-state condition, the equilibrium beach profile equation is derived as follows:

$$y_{eq} = \frac{5v}{2\mu\sqrt{g}} \int_0^h Cdh = \frac{3}{2\sqrt{g}A^{3/2}} \int_0^h Cdh$$ (4)

According to the derived equation, under shallow water conditions (i.e., $C = \sqrt{gh}$), the equilibrium beach profile formula is consistent with the empirical equation proposed by Bruun [1]. Under deep water conditions, the equilibrium beach profile formula exhibits a constant slope proportional to the average wave period, as shown below:

$$m_o = \frac{3\sqrt{g}\overline{T}}{4\pi A^{3/2}} \cong 0.7477\frac{\overline{T}}{A^{3/2}}$$ (5)

3.1 Convex Protruding Beach

A convex beach is a beach where wave lines are concentrated on the rear coast due to the influence of the topography of the seafloor, such as a reef, where the front seafloor is not formed entirely of sand. However, as sea level rises, the refraction phenomenon is alleviated and the radius of curvature increases, resulting in more severe shoreline retreat. A more serious case of shoreline retreat occurs when both sides of the beach are open and littoral drift is free, and erosion at the convex peak point is approximately expressed by the following equation.

$$\Delta R = m_a \frac{\Delta S}{\Delta t} + \lambda \Delta W_p$$ (6)

where λ is the refraction relaxation coefficient greater than 0 and less than 1 and is estimated through a numerical model of wave deformation, and ΔW_p is the peak width of the protruding part as shown in Fig. 1a.

3.2 Concave Pocket Beach

A concave beach is a coastal feature where the incoming wave direction is restricted by the surrounding terrain. The equilibrium shoreline on such beaches is established at the boundary between the sea and land, where the amount of sediment transported from rivers is balanced by the sediment lost due to waves, as illustrated in Fig. 1b. When sea levels rise, the shoreline retreats; however, in concave beaches, the shoreline length shortens, leading to a reduction in sediment loss rate. This mitigates the overall shoreline retreat compared to that of a straight coast. If the first term in the following equation is greater than the second, background erosion due to sea level rise does not occur:

$$\Delta R = KW_e - m_a \frac{\Delta S}{\Delta t} \tag{7}$$

$$W_e = Q_r / KD_s L_b \tag{8}$$

where K is the sediment loss factor, W_e represents the equilibrium beach width where the sediment budget is balanced, Q_r is the sediment inflow rate, D_s is The vertical height of the sediment transport zone, and L_b is the beach length.

An increase in K, D_s, and L_b in the first term demonstrates that background erosion caused by sea level rise can be mitigated. This implies that as the sediment loss factor, sediment density, and beach length increase, the equilibrium beach width W_e expands, reducing the impact of erosion due to sea level rise. Therefore, the formation of concave beach shapes can serve as an alternative to slow down shoreline retreat.

4 Verification by Sentinel-2 Satellite Images

To validate the methodologies briefly presented in Sect. 2, long-term trends were obtained from Sentinel-2 satellite images for the study sites. The long-term trend of the shorelines was analyzed from the past 9 years of images of the convex beach of the Gangneung Beach and the concave pocket beach of Jangho Beach located in Korea, and the results were compared with each other.

Acknowledgements. This research was supported by the Korea Institute of Marine Science & Technology Promotion (KIMST), funded by the Ministry of Oceans and Fisheries, Korea (RS-2023-00256687).

References

1. Bruun P (1962) Sea-level rise as a cause of shore erosion. J Waterways Harbors Div 88(1):117–130
2. Cooper JAG, Pilkey OH (2004) Sea-level rise and shoreline retreat: time to abandon the Bruun Rule. Global Planet Change 43(3–4):157–171
3. Laksmi AA, Kim TK, Lee JL (2019) Numerical approaching of beach profile change by suspended sediment transport process. J Coast Res 91(SI):86–90
4. Lee S, Lee JL (2020) Estimation of background erosion rate at Janghang Beach due to the construction of Geum estuary tidal barrier in Korea. J Marine Sci Eng 8(8):551
5. Lee S, Lim C, Lee JL (2022) Prediction of long-term beach profile evolution due to episodic wave incidence under tidal environment. Front Mar Sci 9:831262

6. Vousdoukas MI et al (2020) Sandy coastlines under threat of erosion. Nat Clim Chang 10(3):260–263

Daily Sediment Transport and Intertidal Sandy Beach Response Under Calm to Moderate Waves

Isaac Rodríguez-Padilla[1(✉)], Amaia Ruiz de Alegría-Arzaburu[1],
and Ismael Mariño-Tapia[2]

[1] Instituto de Investigaciones Oceanológicas, Universidad Autónoma de Baja California, Ensenada 22860, Baja California, México
`isaac.rodriguez.padilla@uabc.edu.mx`

[2] Escuela Nacional de Estudios Superiores Unidad Mérida, Universidad Nacional Autónoma de México, Mérida 97357, Yucatán, México

Abstract. An intensive one-week field experiment was conducted on a wave-dominated sandy beach (Barra del Estero, NW Mexico) during summer, with the objective of capturing the natural beach recovery processes associated with typically milder wave conditions. The field campaign covered a 200-m stretch of beach and involved several in-situ measurements, such as daily topographic profiles, offshore wave data, near-bottom velocities, and suspended sediment concentrations at two different locations within the surf zone. Surfzone sediment transport rates were estimated using an extended energetics-based model, and were spatially and daily integrated to derive deposited volumes of sand, that were compared against measured topographic changes within the intertidal zone. The results showed contrasting patterns of sediment transport, with accretion dominating the southern zones and erosion occurring in the northern zone under higher waves. While the model captured overall trends, it underestimated accretion and overestimated erosion, emphasizing the need to include additional physical processes. These findings advance the understanding of short-term beach recovery under low-to-moderate wave conditions.

Keywords: Sediment Transport · Sandy Beach · Beach Accretion

1 Introduction

Wave-dominated sandy beaches typically experience seasonal subaerial erosion during high-energy wave events (e.g., winter storms) and accretion during low-energy conditions (e.g., summer). Within the surf zone, offshore sediment transport is primarily driven by undertow and rip currents, while skewed and asymmetric waves promote onshore transport [1]. Offshore sediment transport dominates during intense wave breaking, as the undertow becomes stronger, leading to sandbar retreat and beachface erosion. Conversely, onshore sediment transport due to non-linear waves is dominant during weakly to non-breaking wave conditions, facilitating beach accretion. This imbalance between wave-induced processes determines the magnitude and direction of cross-shore sediment transport, which, combined with the longshore current, persistently reshapes the

© The Author(s) 2026

C. Coelho et al. (Eds.): CD 2025, CRL 41, pp. 100–105, 2026.

https://doi.org/10.1007/978-3-032-15473-6_16

coastline. Thus, understanding the natural beach response requires accurately quantifying sediment transport, a task that remains challenging and essential for effective coastal management. This work aims to investigate the short-term beach evolution of a wave-dominated sandy beach by relating predicted surf zone sediment transport rates to measured intertidal bed-level changes under low-to-moderate wave conditions over a period of days.

2 Methods

2.1 Study Site

Barra del Estero is a 7 km-long natural barrier beach adjacent to the coastal lagoon of Estero de Punta Banda, located within Todos Santos Bay on the northwestern coast of the Baja California Peninsula in Mexico (Fig. 1). The beach is composed of medium to fine sand ($D_{50} \sim 0.2$ mm), typically has a single subtidal sandbar, a gentle beach face ($\tan\beta \sim 1/50$), and exhibits a semidiurnal micro/meso tidal regime (0.5–2.3 m). The beach is predominantly exposed to Pacific W-NW incoming swells and is subject to interannual wave energy variability (e.g., El Niño-Southern Oscillation). The offshore annual mean significant wave height and peak period are $H_s = 1$ m (up to 4 m during winter storms) and $T_p = 11$ s, respectively [2]. The rocky headland situated south of the bay protects the beach from Pacific southwestern waves that are typical during summer.

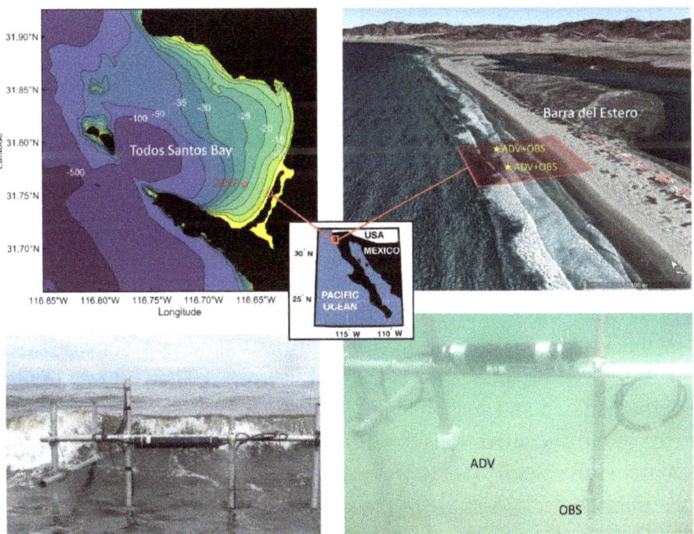

Fig. 1. Location of the field site (Barra del Estero Beach, northwestern Mexico; red rectangle) and the instrument deployment array: ADCP (Acoustic Doppler Current Profiler; red star), ADV (Acoustic Doppler Velocimeter; yellow star) and OBS (Optical Backscatter Sensor; yellow star).

2.2 Field Experiment and Data Acquisition

An intensive field campaign was conducted at Barra del Estero Beach from 10 to 16 June 2016 to examine the gradual recovery of the beach following the outstanding 2015–2016

El Niño event. The campaign included topographic surveys covering 22 cross-shore beach profiles per day, spaced 10 m apart along the shoreline, that were interpolated to produce daily topographic maps of the intertidal zone. Additionally, hourly wave and sea level data were collected using an Acoustic Doppler Current Profiler (ADCP) deployed 2.7 km offshore at a depth of 22 m. Near-bed velocity and suspended sediment concentration (SSC) data were obtained in 30-min bursts every hour from two arrays of Acoustic Doppler Velocimeters (ADVs) coupled with Optical Backscatter Sensors (OBSs). These ADVs+OBSs arrays were positioned within the surf zone at a depth of 0.5 m, separated by 90 m, and designated as the northern (N-ADV and N-OBS) and southern (S-ADV and S-OBS) arrays, respectively (see Fig. 1). A detailed description of the field conditions during the experiment is provided in [3].

2.3 Sediment Transport Model

An extended-based energetics sediment transport model was proposed by combining the energetics approach of Bailard [4] and the acceleration skewness model of Hoefel and Elgar [5], defined as:

$$q_{Tot} = q_{Bailard} + q_{acc} \tag{1}$$

where,

$$
\begin{aligned}
q_{Bailard} &= \frac{\rho_w C_f \varepsilon_b}{g(\rho_s - \rho_w)(1-p)tan\phi}\left(\langle |\boldsymbol{u}|^2 \boldsymbol{u}\rangle - \frac{tan\beta}{tan\phi}\langle |\boldsymbol{u}|^3\rangle\right) \\
&+ \frac{\rho_w C_f \varepsilon_s}{g(\rho_s - \rho_w)(1-p)W_s}\left(\langle |\boldsymbol{u}|^3 \boldsymbol{u}\rangle - \frac{\varepsilon_s}{W_s}tan\beta\langle |\boldsymbol{u}|^5\rangle\right)
\end{aligned}
\tag{2}
$$

$$
q_{acc} =
\begin{cases}
\frac{K_a}{1-p}\left(\frac{\langle \boldsymbol{a}^3\rangle}{\langle \boldsymbol{a}^2\rangle} - sign\left[\frac{\langle \boldsymbol{a}^3\rangle}{\langle \boldsymbol{a}^2\rangle}\right]a_{crit}\right), & \left|\frac{\langle \boldsymbol{a}^3\rangle}{\langle \boldsymbol{a}^2\rangle}\right| \geq a_{crit} \\
0, & \left|\frac{\langle \boldsymbol{a}^3\rangle}{\langle \boldsymbol{a}^2\rangle}\right| < a_{crit}
\end{cases}
\tag{3}
$$

From the above equations, $\boldsymbol{u} = (u, v)$ is the near-bed velocity vector (measured 0.2 m above the seabed); u and v are the cross-shore and longshore components, respectively. Vertical bars indicate the magnitude of the vector and the angular brackets the time-averaging of the vector. ρ_w and ρ_s are the fluid and sediment densities (set to $\rho_w = 1025$kg m^{-3}; $\rho_s = 2650$kg m^{-3}), C_f is the bed friction coefficient (set to $C_f = 0.003$), ε_b and ε_s are the bedload and suspended load efficiency factors (set to $\varepsilon_b = 0.135$; $\varepsilon_s = 0.015$), g is the gravity acceleration ($g = 9.81$m s^{-2}), p is the porosity (set to $p = 0.35$), W_s is the sediment fall velocity (set to $W_s = 0.026$m s^{-1}), $tan\beta$ and $tan\phi$ are the local bed slope and the internal friction angle of the grains (set to $tan\beta = 0.017$; $tan\phi = 0.63$), $\boldsymbol{a}$ is the near-bed acceleration vector, K_a is a scaling constant (set to $K_a = 1.4 \times 10^{-4}$m s), $sign$ [] is the sign of the argument and a_{crit} is a threshold that must be exceeded for initiation of sediment transport (set to $a_{crit} = 0.2$m s^{-2}).

The effects of orbital wave velocity skewness and asymmetry, which are key drivers of onshore sediment transport in the surf zone, are accounted for in q_{Tot} (Eq. 1). The model was applied to predict daily variations in sand volumes, and the results were compared with topographic changes measured in the intertidal zone.

3   Results and Discussion

The daily morphological evolution of Barra del Estero Beach during the one-week field experiment is summarized in Fig. 2. Distinct morphological patterns emerge in different localized regions within the intertidal zone, which are defined as follows: the northern high-tide inner surf zone (N-surf zone: 50 m $< x <$ 120 m; 200 m $< y <$ 300 m; 0 m $< z <$ 1 m), the southern high-tide inner surf zone (S-surf zone: 50 m $< x <$ 120 m; 100 m $< y <$ 200 m; 0 m $< z <$ 1 m), and the swash zone (30 m $< x <$ 50 m; 100 m $< y <$ 300 m; 1 m $< z <$ 2 m).

Overall, the field experiment was characterized by low-to-moderate wave conditions ($H_s <$ 1.2 m; $T_p <$ 10 s) that promoted consistent onshore sediment transport and beach accretion in both the swash zone and the S-surf zone. In contrast, the N-surf zone experienced accretion only during the initial days, when wave heights were relatively small (Hs $<$ 0.7 m; Fig. 2 a-d). By the end of the experiment, as wave heights increased ($H_s >$ 0.9 m), this region exhibited offshore sediment transport and erosion (Fig. 2 e-f). It is important to note that the bathymetry is alongshore-irregular approximately 50 m offshore of the instrument arrays at the -1.5 m isobath (not shown in Fig. 2). This morphological feature, located in the outer surf zone, can induce nonuniform wave breaking, potentially modifying sediment transport rates along the intertidal zone.

These contrasting morphological patterns between S-surf zone and N-surf zone align qualitatively with the magnitude and direction of the predicted sediment transport vectors (black arrows in Fig. 2) derived from Eq. 1. Quantitatively, however, the model underestimates intertidal beach accretion by a factor of three and overestimates erosion by a factor of two. This discrepancy can be explained by the omission of additional physical processes (e.g., swash zone dynamics, bore turbulence, aeolian transport, among others), which may significantly influence onshore sediment transport, but are not accounted for in the model [6].

The relationship between SSC, tidal stage, and H_s was further examined. As shown in Fig. 3, higher SSC values were observed during low tide and were associated with larger wave heights. With reduced water depths, larger waves can reach the seabed, generating turbulence that resuspends sediments into the water column. These sediments may subsequently be mobilized onshore by non-linear oscillatory wave motions or transported offshore by the undertow.

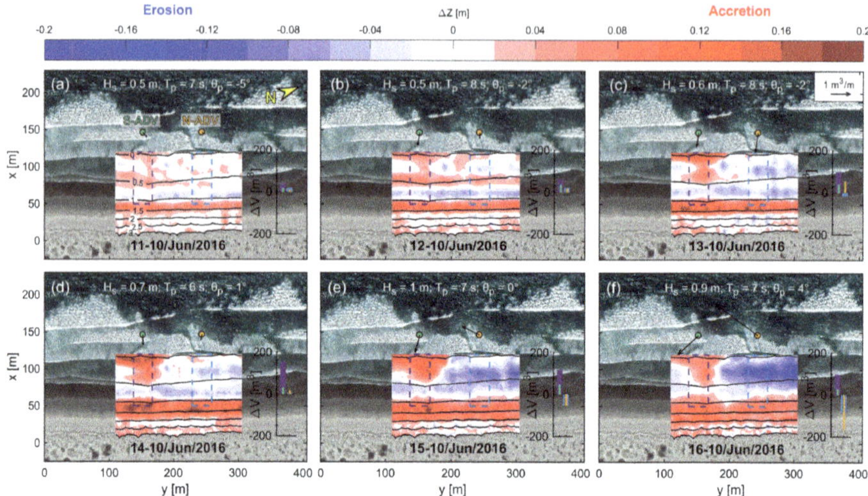

Fig. 2. Daily morphological changes at Barra del Estero Beach under low-to-moderate wave conditions. (a-f) Blue-to-red colors represent elevation changes in the intertidal zone based on topographic surveys. Elevation contours (black lines) are referenced to the local mean lower low-tide level (MLLT) and correspond exclusively to the first day of the survey. Black arrows indicate the daily integrated surfzone sediment transport variations estimated from near-bed ADV velocities. The measured (purple and light-blue bars) and estimated (green and orange bars) daily sand volume changes ΔV correspond to the dashed-line boxes. Daily averaged offshore wave conditions are provided for reference: H_s (significant wave height), T_p (peak period) and θ_p (peak wave direction relative to shore normal). The aerial image is identical in all panels and corresponds to 1/July/2016, as it was the closest available image in Google Earth to the June field experiment.

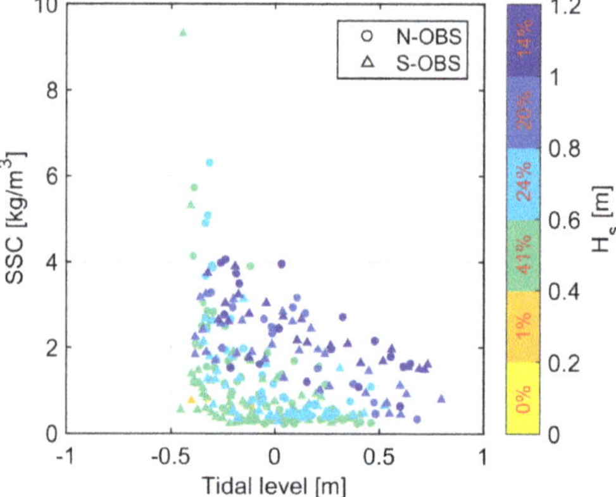

Fig. 3. Near-bed suspended sediment concentration (SSC) response under varying tidal level and offshore significant wave heights (H_s). Circles and triangles represent respectively the SSC derived from the northern and southern optical backscatter sensors (OBS) deployed in the surf zone. The colors denote the SSC associated with different ranges of H_s; percentage of occurrence is shown in red.

4 Conclusions

This study investigates the short-term morphological evolution of Barra del Estero Beach under low-to-moderate wave conditions following the 2015–2016 El Niño event. Distinct patterns of sediment transport and beach response were observed between the southern and northern regions of the surf zone, with consistent onshore sediment transport and accretion occurring in the S-surf zone and swash zone, while the N-surf zone exhibited accretion during calm conditions and erosion as wave energy increased. The energetics-based sediment transport model qualitatively captured the observed patterns, but its quantitative predictions underestimated accretion and overestimated erosion. The relationship between suspended sediment concentration, tidal stage, and wave height further underscores the complex interactions driving sediment mobilization, emphasizing the need for more comprehensive models that integrate additional physical processes to improve predictive accuracy.

Acknowledgements. We acknowledge financial and institutional support of UABC, UC-Alianza-MX, DGAPA-UNAM and CONAHCyT CB-2014-238765 project.

References

1. Castelle B, Masselink G (2023) Morphodynamics of wave-dominated beaches. Cambridge Prisms: Coastal Futures 1:e1
2. Ruiz de Alegría-Arzaburu A, Vidal-Ruiz JA (2018) Beach recovery capabilities after El Niño 2015–2016 at Ensenada Beach, northern Baja California. Ocean Dynamics 68: 749–759 (2018)
3. Rodríguez-Padilla I, Mariño-Tapia I, Ruiz de Alegría-Arzaburu A (2024) Daily timescale analysis of sediment transport and topographic changes on a mesotidal sandy beach under low to moderate wave conditions. Marine Geol. 474, 107323 (2024)
4. Bailard JA (1981) An energetics total load sediment transport model for a plane sloping beach. J. Geophys. Res. Oceans 86:10938–10954
5. Hoefel F, Elgar S (2003) Wave-induced sediment transport and sandbar migration. Science 299:1885–1887
6. Masselink G, Austin M, Tinker J, O'Hare T, Russell P (2008) Cross-shore sediment transport and morphological response on a macrotidal beach with intertidal bar morphology, Truc Vert, France. Marine Geol 251:141–155

When Dune Erosion Release Remains: Hidden Shipwreck and Fortification of the Atlantic Wall Revealed by UAV Magnetic Survey and Georadar (SW France)

Julie Billy[1]([✉]), Pauline Le Maire[2], and Nicolas Bernon[3]

[1] BRGM, French Geological Survey, 33600 Pessac, France
`j.billy@brgm.fr`
[2] CARDEM PYRO, 57280 Hauconcourt, France
[3] BRGM, 33600 Pessac, France

Abstract. This study investigates the potential of geophysical methods to identify a recently uncovered wooden shipwreck on the foredune of the Cap Ferret Spit (SW France), briefly exposed during winter 2020–2021. The objectives were to characterize the internal architecture and development of the dune system, and to precise the shipwreck's location and size. A multi-geophysical approach was deployed, combining ground-penetrating radar and magnetic surveys with UAV. The GPR survey revealed that the dune system is composed of a natural former dune overlain by a modern managed dune, influenced by coastal management practices since the mid-19th century (including wooden fences). Magnetic surveys identified a significant anomaly in the foredune area, in the same location as hyperbola-shaped anomalies observed in GPR profiles. This concordance suggests the presence of a single wreck (12–15 m in length), rather than fragmented remains. Further magnetic mapping at higher altitudes (7 m) identified additional dipolar anomalies, with one attributed to a bunker at the top of the dune, and others likely corresponding to buried WWII Atlantic Wall fortifications. These results highlight the value of combining geophysical techniques to improve knowledge of the coastal zone, for both geological and archeological issues.

Keywords: Geophysics · Dune architecture · Coastal evolution · Wreck · Bunker

1 Introduction

Coastal erosion and dune migration can occasionally uncover historical or archaeological remains. The Aquitaine Coast, in France, has a substantial coastal heritage, encompassing salt production, shipping, maritime trade with the West Indies, and Second World War fortification, such as the Atlantic Wall. While some remnants are documented (*e.g.* [1, 2]) others remain undiscovered or unsuspected. Additionally, in the mid-19th century, extensive coastal management efforts were undertaken to protect forests and villages from sand invasion, leading to the built of wooden fences to trap the sand, which has evolved into the modern foredune.

© The Author(s) 2026

C. Coelho et al. (Eds.): CD 2025, CRL 41, pp. 106–113, 2026.
https://doi.org/10.1007/978-3-032-15473-6_17

During the winter 2020–2021, coastal erosion briefly exposed a part of a wooden shipwreck of the front dune in Cap Ferret Spit (SW France), sparking interest from local authorities and scientists (Fig. 1). This study aims to gather preliminary information of this sector prior to any archaeological actions. The objectives are: i) to characterize the internal architecture of the dune system and reconstruct its past evolution, from the wreck beaching to the present day; ii) to investigate around the wreck area in order to precise its location and its dimension (a single intact ship or scattered fragments).

Geophysical methods have a great potential for coastal investigations. A multi-geophysical approach was deployed, combining ground-penetrating radar and magnetic methods surveying with UAV.

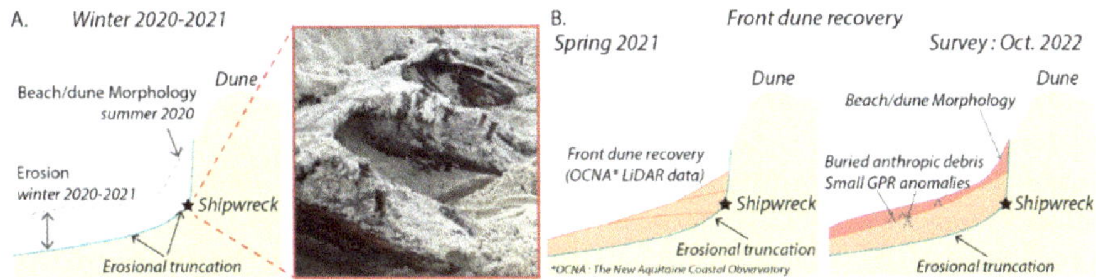

Fig. 1. Conceptual model of the beach evolution from 2020 to 2022: A) Before and after the winter 2020–2021, with a picture of the visible part of the wreck (©E. Lenain, ONF), and B) recovery of the front dune up to the survey (oct. 2022).

2   Methods: Georadar and Magnetic Survey with UAV

2.1   Ground-Penetrating Radar (GPR)

Ground-penetrating radar technique is based on the propagation and reflection of transmitted electromagnetic pulses to image sedimentary structures. Georadar survey was carried out using a Mala ProEx system, with two antennae centered at a frequency of 250 and 500 MHz, respectively. Each antenna served distinct purposes:

1- The dune system characterization: Four 250 m-long profiles, were surveyed with the 250 MHz antenna (2 profiles north and 2 south of the suspected wreck location) to characterize the entire dune system.
2- A detailed foot-dune imaging: For a high-resolution 3D imaging of the shipwreck area, both longshore and cross-shore surveys were carried out with the 500 MHz antenna, with profile spacings of 1–2 m (90 profiles in total across both directions).

The data were processed using REFLEXW V10.5 (*i.e.* time-zero drift, background removal, band-pass filtering, gain, topographic correction) and interpreted following principles of radar stratigraphy [3] using DELPH Geo Interpretation (V5). A constant radar velocity is estimated at 0.14 m/ns [4] and used for the time-depth conversion.

2.2 Magnetic Survey with UAV

The magnetic method involves measuring variations in the Earth's magnetic field, referred to as anomalies. These anomalies can identify the presence of subsurface magnetic materials and are widely applied in field such as archaeology or soil decontamination (e.g. detecting munitions or buried structured). In this study, magnetic surveys were conducted over the beach and dune system using a UAV (M300 RTK DJI) equipped with four fluxgate magnetometers (Sensys FGM3D) and carried out by CARDEM PYRO. The survey covered an 8-ha area with 20 km of profiling completed at two distinct altitudes 2 m and 7 m above ground level. The processed and interpreted magnetic data provide two magnetic maps, showing the presence of magnetic anomalies within the target area.

3 Morphological Context of the Beaching

Transverse GPR profiles reveal the internal architecture of the dune and its distinct development stages (Fig. 2). This coastal dune system comprises two superimposed dunes formed by different mechanisms: (i) a former natural dune overlain by (ii) a modern dune influenced by coastal management. Coastal zone management in this region began in the mid-19th century, and the GPR profiles capture the dune's morphology before and after this pivotal period (additional profiles in [5]). The former dune reached a maximum elevation of 15 m NGF. Wooden fences were built on the stoss slope of this dune (stages 1–2) to trap sand and prevent its encroachment into nearby villages (*e.g.* [6–8]). Over time, sand accumulation around the fences (stage 3) raised the dune to an elevation of 17–18 m NGF. Subsequently, dune development shifted from vertical to spatial expansion, occurring both landward and seaward. On the landward side, aeolian processes transported sand to cover the former dune and its fences (stage 4). On the seaward side, continued sand accumulation resulted in the dune's progradation (reaching an altitude of 25 m NGF and its current morphology; stage 5).

Preliminary historical analyses indicate that the boat dates from the 17th century [9]. The coastal landscape contemporaneous to the beaching would match with the former dune and beach morphology. The shipwreck was then completely buried by sand accumulation during the development of the modern dune.

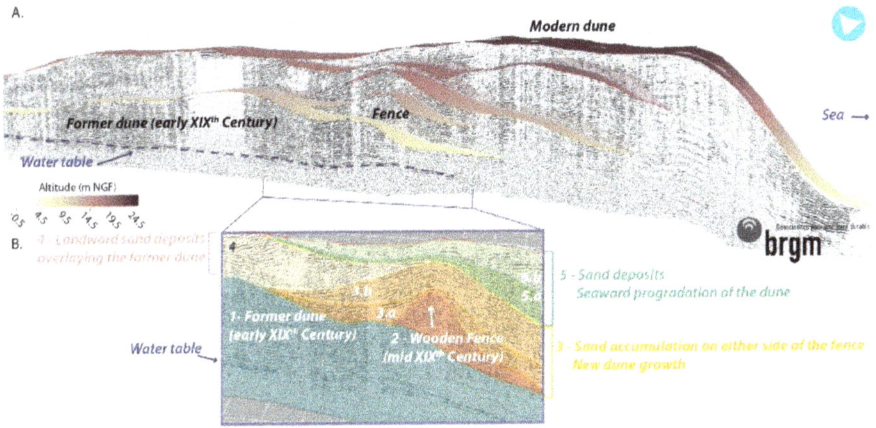

Fig. 2. Interpreted GPR profile showing A) the internal architecture of the coastal dune illustrating the former and modern dunes and B) zooming on the wooden fence and the different stage of dune development (stage 1 to 5).

4 Shipwreck Recognition

The dense georadar survey carried out on the beach/foredune system and its interpretation provide a detailed view of sand deposit patterns. First, it enables to identify the erosional truncations induced by the storms impacts during the winter 2020–2021 (Fig. 3, green dotted line). This interface was digitized across all GPR profiles to generate a DTM of the erosion scar. Additionally, the survey images the post-storm recovery of the beach and foredune from winter 2020–2021 to summer 2022 (Fig. 3, orange).

Cross-shore profiles can be classified into two categories: in the central part of the study area, numerous hyperbola-shaped anomalies are observed near the erosional truncation surface (Fig. 3A-2), whereas, in the northern and southern parts, the erosional truncation surface remains undisturbed by such anomalies. Longshore profiles perpendicular to this central zone display a concave-shape reflector. These findings suggest that the wreck is located within the central sector, where the hyperbola-shaped anomalies are concentrated (Fig. 3B, pink).

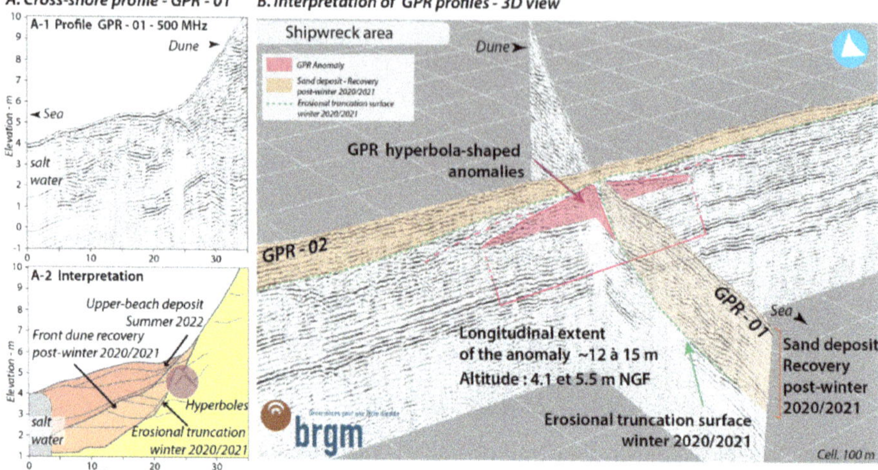

Fig. 3. A) Processed and interpreted GPR cross-shore profile at the wreck, identified by hyperbola-shaped anomaly; B) 3D-view of the GPR anomaly interpreted as the location of the wreck (pink) and post-winter 2020–2021 sand deposits (orange).

Wooden ships were constructed using large metal nails (Fig. 1A; [9]), and the main hypothesis is that the accumulation of these nails could generate magnetic anomaly around the wreck. Analysis of magnetic surveying at 2 m above ground level generated a magnetic map of the study area, revealing a significant anomaly at the front dune (Fig. 4; N#1: 4–10 nT/m;).

Magnetic and georadar anomalies are matched, both being located in the same area as appear as a single structure (Fig. 4). These concordant anomalies are therefore interpreted as the shipwreck, estimated to be 12–15 m in length, and identified as a single structure and not as fragmented parts dispersed beneath the dune. To preserve the archeological archive, the coordinates of geophysical anomalies are not given here.

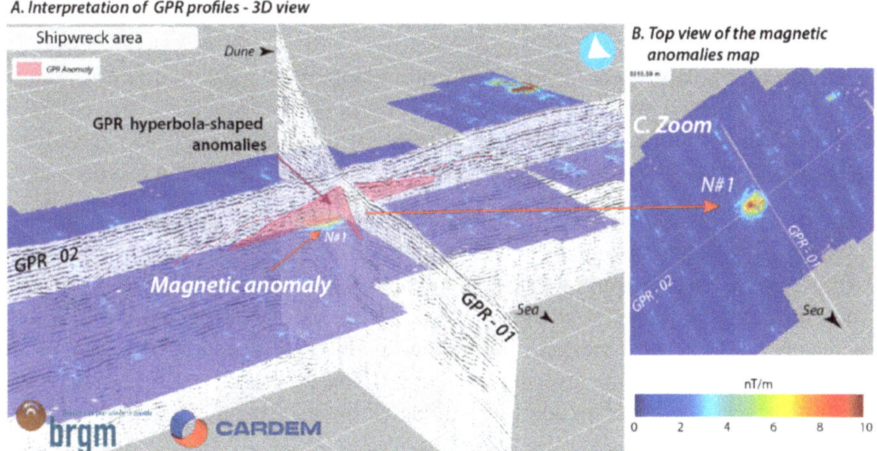

Fig. 4. A) 3D-view of the GPR and magnetic anomalies; B) top view of the absolute analytic signal map (survey at 2 m altitude above the beach/dune system).

5   Magnetic Anomalies on the Dune

Analyses of magnetic surveying at 7 m above ground level produced a magnetic map of the dune system, identifying five major dipolar anomalies (Fig. 5A). Among these, only one anomaly correlates with a structure at the top of the dune, specifically a bunker (Fig. 5B). For the remaining four anomalies, forward modelling and comparison with magnetic maps of known bunkers, suggest the presence of four additional buried bunkers, or Atlantic Wall-affiliated constructions of varying sizes. Two lower amplitude anomalies are visible to the north and a right-angle anomaly (oriented to the northeast) to the south. These structures are certainly also fortifications of the Atlantic Wall.

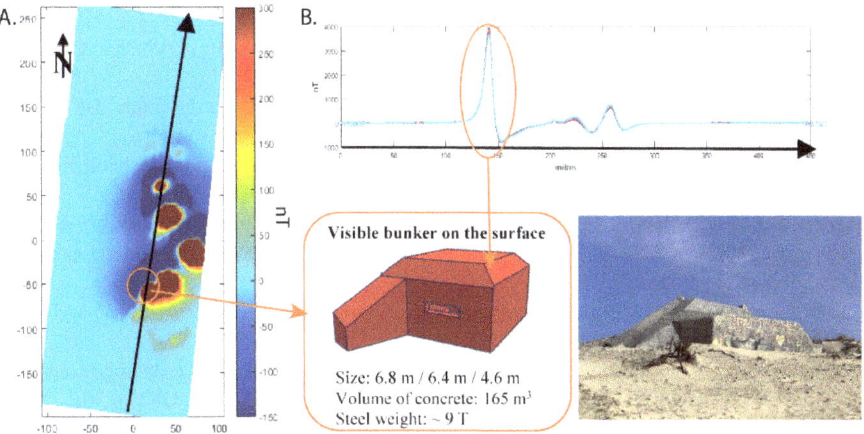

Fig. 5. A) Total magnetic intensity map, and B) profile of the only visible bunker in this area

6 Conclusion

Coastal evolution can reveal historical and archeological remains that require careful management either to preserve heritage and archives or to ensure the safety of coastal sites. This study demonstrates the effectiveness of combining GPR and UAV-based magnetic surveys to investigate coastal geosystems. Data interpretation reveals that:

1. The foredune consists of a former natural dune overlain by a modern dune influenced by coastal management practices (since mid-19th century).
2. The erosional truncations induced by winter 2020–2021 storms and subsequent beach/front dune recovery, up to summer 2022, are clearly imaged by GPR
3. Concordant GPR and magnetic anomalies precise the location and size of the shipwreck, suggesting the presence of a single body, estimated at 12–15 m in length.
4. Magnetic surveys identified additional potential buried Atlantic Wall fortifications.

These results at the Cap Ferret costal dune system highlight the value of combining geophysical techniques to improve knowledge of the coastal zone, demonstrating the utility of such investigation for both geological and archeological issues.

Acknowledgements. This work was supported by the National Forestry Office (ONF), the municipality of Lège-Cap Ferret, the *Parc Naturel Marin du Bassin d'Arcachon* and the BRGM (French geological survey). The authors would like to thank Marie BRANELLEC and Eric LENAIN for their field assistance.

References

1. Stéphan P, Verdin F, Arnaud-Fassetta G, Bertrand F, Eynaud F, García-Artola A et al (2019) Holocene coastal changes along the Gironde estuary (SW France): new insights from the North Médoc peninsula beach/dune system. Quaternaire 30(1):47–75
2. Verdin F et al (2019) Humans and their environment on the Médoc coastline from the Mesolithic to the roman period. Quaternaire 30(1):75–77
3. Neal A (2004) Ground-penetrating radar and its use in sedimentology: principles, problems and progress. Earth Sci Rev 66(3–4):261–330
4. Billy J, Nicolae-Lerma A (2021) Prospection de l'architecture interne des dunes. BRGM (NT-2021-021), p 7
5. Billy J, Bernon N (2023) Reconnaissance par la géophysique de vestiges archéologiques, au sein du cordon dunaire littoral de Lège-Cap-Ferret. BRGM (RP-72591-FR), p 32
6. Buffault P (1942) Histoire des dunes maritimes de Gascogne. Bordeaux, Delmas, p 446
7. Favennec J (2002) Connaissance et gestion durable des dunes de la côte atlantique. Les dossiers forestiers n°11, p 394
8. Robin N et al (2021) 150 yrs of foredune initiation and evolution driven by human and natural processes. Geomorphology 374(107516):1–15
9. Vinet-Poirier S (2023) Etude historique de l'Histoire maritime de la côte de Lège-Cap-Ferret de la fin du XVIIIe S au début du XIXe S. Master thesis, Univ. Poitiers, ONF, p 36

Decadal Data Set of Hourly Beach Profile Evolution in Duck, NC

Katherine L. Brodie[(✉)], Annika O'Dea, Dylan Anderson, Elizabeth R. Holzenthal, Nicholas T. Cohn, Jeremy Braun, Brad Johnson, and Pat Dickhudt

Coastal and Hydraulics Laboratory, US Army Engineer Research and Development Center, Duck, NC, USA
`katherine.l.brodie@usace.army.mil`

Abstract. This paper presents initial results of a decadal dataset of sub-aerial beach profile evolution collected at the U.S. Army Engineer Research and Development Center's Field Research Facility in Duck, NC using a stationary terrestrial lidar scanner (TLS) mounted above the dune. The data are processed using an automated work flow in near realtime and are made public for the coastal science and engineering community to use at (https://chlthredds.erdc.dren.mil/). At submission, the data set consisted of 48,911 processed cross-shore profiles that passed the automated quality assurance quality control checks (81% of the 60,6046 collected scans). A basic description of the dataset is provided, including example storm impact and beach recovery sequences, and an initial investigation is presented that relates pre-storm beach morphology, storm wave power, and profile response and highlights the importance of initial beach state.

Keywords: Beach morphology · open source · terrestrial lidar data · coastal hazards

1 Introduction

Sandy beaches continuously evolve in response to changing environmental forcings such as waves, currents, water levels, and wind. In 1963, the US Army Corps of Engineers proposed the Field Research Facility (FRF) as a coastal observatory where long-term records of waves, currents, water levels, and beach and nearshore elevation data could be collected continuously. To improve understanding of coastal change, monthly surveys of the beach and nearshore were initiated in 1980 as part of this charter, and this data record remains one of the coastal science and engineering community's hallmark data sets. While at times these morphologic data have been collected daily (during large experiments) or every few days (surrounding a storm), resource availability has prevented collection of higher temporal resolution survey data using traditional vessel or amphibious vehicle-based surveying techniques. As a result, observed morphological changes in the data record can only be interpreted as the integrated response to the cumulative environmental forcings between each subsequent survey, making it difficult to link to the explicit physical process and to develop models of integrated short and long-term coastal change.

© The Author(s) 2026
C. Coelho et al. (Eds.): CD 2025, CRL 41, pp. 114–120, 2026.
https://doi.org/10.1007/978-3-032-15473-6_18

To increase the collection frequency of coastal morphodynamic measurements near the shoreline, the FRF mounted a terrestrial lidar scanner above the dune (Fig. 1a; [9]), which has been recording observations of sub-aerial beach topography and swash and inner-surf zone hydrodynamics on hourly intervals from 2015 through present (with some downtime). This data collection effort has generated a close to decadal data set of high temporal resolution sub-aerial beach evolution. These data are processed in near-realtime and are available to the community (https://chlthredds.erdc.dren.mil/). Development of an automated processing framework for this new data type has been complex, and the approaches have evolved over the years. Recent implementation of a machine learning algorithm to aid in the separation of beach and water data points [5] has initiated the re-processing of the entire record and led to improvements in data quality. This paper presents this decadal dataset of sub-aerial beach evolution to the community, discusses data reliability and potential sources of error, and highlights how these observations can be used to better understand the integrated processes controlling short and long-term beach evolution.

2   Methods

2.1   Lidar Uses in Coastal Morphodynamics

The use of high-accuracy terrestrial lidars scanners (TLS) has increased significantly in the earth sciences in the last 25 years [11], including many applications within coastal science and engineering fields. TLS has been used to advance knowledge on topics including sand and cobble beach, dune, and cliff morphology evolution [e.g. 1, 3, 8, 10], swash-zone processes [e.g. 2], wave breaking [e.g. 7], and aeolian transport [e.g. 4], among many others. Most efforts focused on measurements of beach morphology evolution have utilized mobile data collection approaches, where TLS are mounted on a moving vehicle and used to map large regions, or short-duration static scans, where field campaigns range from hours to months. Two notable exceptions include a 3-year continuous, static data set collected at an urban beach-dune system in the Netherlands [6, 12] and a multi-year data set collected on a micro-tidal beach in Australia [10].

2.2   Duck Lidar Cross-Shore Profile Data Set

Cross-shore transects of beach topography extracted from hourly 30-min lidar linescan time series collected by the FRF's permanently deployed Riegl VZ-1000 or Riegl VZ-2000 TLS at the north end of the FRF's property (FRFy coordinate 945m; Fig. 1a). The 7 Hz linescan time-series data are processed using a fully-automated workflow [9] that spatially rectifies the scans, filters, grids the scans at a 10-cm resolution, and finally generates a range of hydrodynamic and morphologic data products [9]. Although a number of data products are produced from the lidar linescans, only the beach morphology transects and statistics are presented in this work. A notable update to the processing described in [9] is the inclusion of a machine learning algorithm to separate data points into beach and water returns following [5]. Once lidar returns are classified as "beach" or "water", the mean elevation of "beach" points is then determined at each cross-shore location to generate an elevation transect for that hour. Additional statistics, such as beach slope, are also calculated and served with the data. The overall quality of the automated data processing is determined using a series of metrics and is provided with the resulting data products as a Quality Control flag.

2.3 Analysis Approaches

To identify relationships between beach profile shape and state, observed morphology evolution (erosion vs. accretion), and the environmental forcings, storms were identified by finding wave events that exceeded the 95th percentile of observed significant wave heights (2.51 m) for a minimum of 12 h at a buoy in 26 m water depth offshore of the FRF. Cumulative wave energy flux, or storm power, was then calculated as the summation of all hours associated with a continuously elevated wave event. The high density of the lidar data provides the ability to track contour position and beach volume change, as well as details of the beach profile shape. However, since the cross-shore data coverage varies as a function of beach width, tide level, and wave conditions, in this effort a low-tide beach profile is defined which is a temporal average over the data collected between each high-tide identified in NOAA tide gauge 8651370 located in 6-m water depth at the end of the FRF pier. The cross-shore position of the MHW contour is then identified in each temporally-averaged transect and used to quantify storm impact and beach recovery for each storm during the data record.

3 Results and Discussion

3.1 Data Set Characteristics and System Performance

At submission, the decadal dataset includes over 60,600 observations of beach transect data since 2015, or roughly 75% of the total number of hours in the time-period. Ninety-eight percent of the collected data have been processed with the automated workflow, with 81% of the fully processed scans passing the automated quality assurance quality control (QAQC) procedure. The largest data gap was 137 days in 2022 (2022 Jan 21–2022 June 6), and the most complete year was 2023, with 8345 hourly scans collected and 6823 scans reaching the QAQC requirements (approximately 78% of the total hours in the year). The primary source of error leading to failure to meet QAQC protocols in the linescan-derived beach elevation data is from scan misalignment due to a poor fit in the co-registration process, described in detail in [9]. Additional sources of error in the final dataset may include returns off of aerosols, splashes, or objects on the beach that are not fully removed during the filtering process and a misclassification of returns off the water.

The observed beach elevation envelope over the 10-year data set is shown in Fig. 1b. The widest portion of the decadal sub-aerial beach envelope occurs on the mid-beach roughly 50-m offshore of the dune lidar (105 m X FRF), where beach elevations ranged approximately 3.3 m in the vertical (Fig. 1). This portion of the profile is frequented by beach cusps [8], which likely contribute to the high variance in beach shape. The narrowest portion of the sub-aerial beach envelope occurred near the seaward-most position of the dune toe (63 m X FRF), where beach elevations ranged 1.2 m in the vertical. The position of the dune toe varied 20 m in the cross-shore in response to multiple cycles of dune erosion and regrowth.

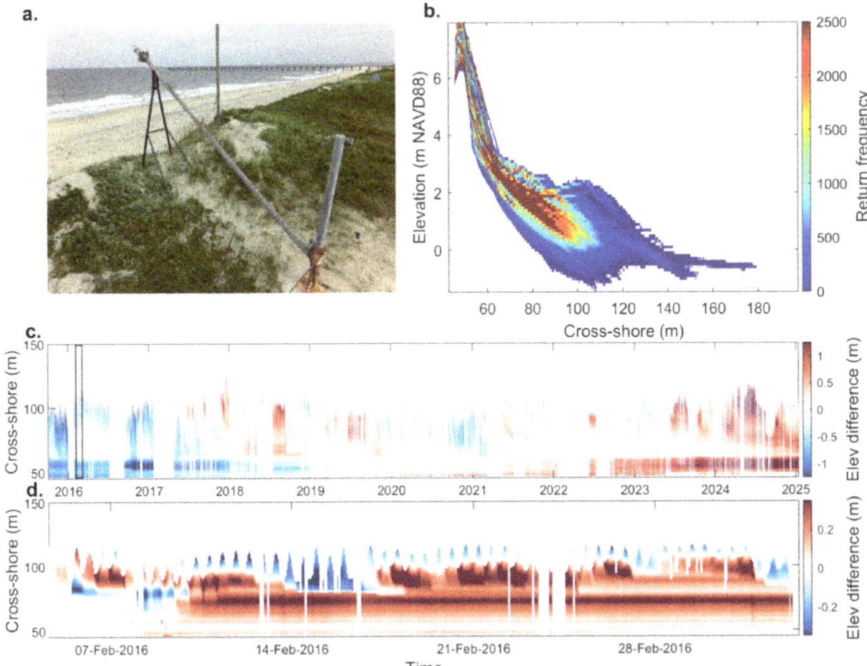

Fig. 1. (a) Image of the lidar tower on the FRF property; (b) transect position frequency as a function of elevation and cross-shore position; (c) elevation difference from a mean profile across the entire dataset as a function of cross-shore position and time; and (d) elevation difference from 2016 Feb 4, 02:00 UTC over a 1-month time period as a function of cross-shore position and time. The black box in panel (c) shows the time period of panel (d).

Data temporal density decreased from >90% of the record above the 2-m contour to only 40% at the 0-m contour (Fig. 1b). Following an initial period of erosion in the first two years of the record, the dune crest and face generally increased in elevation over the remainder of the data record (Fig. 1b). A local beach nourishment occurred just north of the FRF in Summer 2017, which led to widening of the beach in front of the lidar and likely provided a large sediment source for beach and dune growth. The beach remained wide through 2020 before returning to it's pre-nourishment width. A second nourishment occurred in Spring of 2023, which preceded a second period of beach and dune growth.

In addition to quantification of the long-term trends, the high temporal resolution of this data set also enables investigation of beach morphology evolution on the timescale of the forcing processes. Figure 1d highlights the spatio-temporal variability in beach elevations occurring over 1 month in Feb 2016; a time period of relatively little change relative to the long-term. Multiple multi-day sequences of erosion and accretion of the berm are visible, as well as elevation changes of up to 0.4 m occurring on the timescale of a tidal cycle. Not shown in Fig. 1 are the changes occurring to the beach on the timescale of individual waves and wave-groups, which are also quantifiable by this system within the 30-min linescans. No other open-source data sets exist that provides as high spatial and temporal resolution data over as long time-periods, coincident to hydrodynamic forcing conditions.

3.2 Storm Impacts

Seventy-eight storms were observed during the data collection period (Fig. 2a). The storm with the largest wave heights was Hurricane Dorian in 2019, with significant wave height reaching over 6 m. Relationships between beach profile response, pre-storm beach profile state, and wave power over the study period are explored in Fig. 2. In general, small storms impacting wide beaches (small, yellow circles) led to the most erosion of the MHW contour, consistent with equilibrium models of beach evolution. The storms with the largest wave power tended to impact already eroded beaches, which resulted in minimal change, or in some cases, even accretion of the MHW contour (large purple circles). Initial investigation suggests these relationships are sensitive to the contour chosen and the given storm definition, highlighting the spatio-temporal complexities of sub-aerial beach evolution on intermediate beaches. Future work will investigate modes of beach response through application of complex empirical orthogonal functions to these data and analysis of relationships to multi-scale forcing parameters, including the impact of storm sequencing.

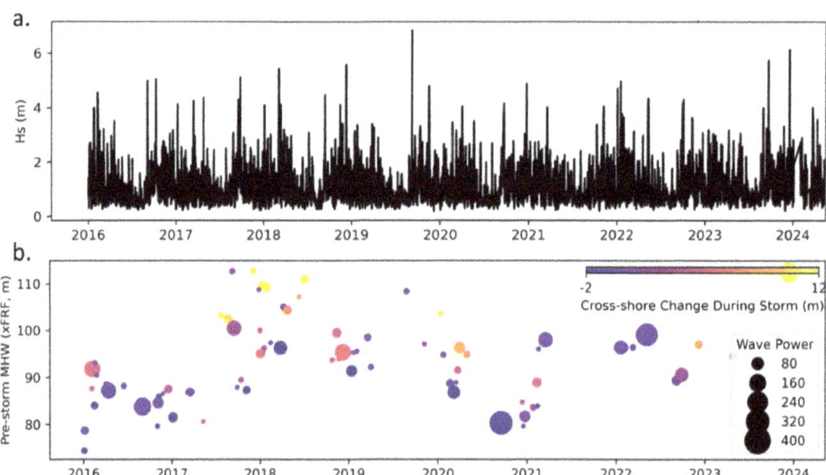

Fig. 2. (a) Significant wave height recorded in 26 m water depth offshore of the FRF between 2016 and 2024; (b) Horizontal displacement of the MHW contour (colors; positive erosion) during a storm as a function of pre-storm MHW contour position (y-axis), cumulative storm wave power (circle size), and time (x-axis).

4 Conclusions

A near 10-year (with gaps) and growing data set of hourly sub-aerial beach evolution collected at an intermediate beach in Duck, NC is presented to the community which can be freely accessed here: https://chlthredds.erdc.dren.mil/. The high temporal and spatial resolution and duration of this cross-shore beach transect data set are unprecedented in the community and we hope will be used widely to advance coastal scientific and engineering needs.

References

1. Bayle PM, Blenkinsopp CE, Martins K, Kaminsky GM, Weiner HM, Cottrell D (2023) Swash-by-swash morphology change on a dynamic cobble berm revetment: high-resolution cross-shore measurements. Coast Eng 184:104341
2. Blenkinsopp C, Mole M, Turner I, Peirson W (2010) Measurements of the time-varying free-surface profile across the swash zone obtained using an industrial lidar. Coast Eng 57:1059–1065
3. Cohn N, Brodie KL, Johnson B, Palmsten ML (2021) Hotspot dune erosion on an intermediate beach. Coast Eng 170:103998
4. Cohn N, Dickhudt P, Brodie K (2022) Remote observations of aeolian saltation. Geophys. Res Lett 49:e2022GL100066
5. Collins AM et al (2023) Automated extraction of a depth-defined wave runup time series from lidar data using deep learning. IEEE Trans Geosci Remote Sens 61:1–13
6. Kuschnerus M, de Vries S, Antolínez JA, Vos S, Lindenbergh R (2024) Identifying topographic changes at the beach using multiple years of permanent laser scanning. Coast Eng 193:104594
7. Martins K, Blenkinsopp CE, Power HE, Bruder B, Puleo JA, Bergsma EW (2017) High-resolution monitoring of wave transformation in the surf zone using a lidar scanner array. Coast Eng 128:37–43
8. O'Dea A, Brodie K (2024) Analysis of beach cusp formation and evolution using high-frequency 3d lidar scans. J Geophys Res Earth Surface 129:e2023JF007472
9. O'Dea A, Brodie KL, Hartzell P (2019) Continuous coastal monitoring with an automated terrestrial lidar scanner. J Mar Sci Eng 7:37
10. Phillips M, Blenkinsopp C, Splinter K, Harley M, Turner I (2019) Modes of berm and beach-face recovery following storm reset: observations using a continuously scanning lidar. J Geophys Res Earth Surf 124:720–736
11. Telling J, Lyda A, Hartzell P, Glennie C (2017) Review of earth science research using terrestrial laser scanning. Earth Sci Rev 169:35–68
12. Vos S et al (2022) A high-resolution 4d terrestrial laser scan dataset of the Kijkduin beach-dune system, the Netherlands. Sci Data 9:191

Numerical and Experimental Investigation of the Impact of Beach Deformation on Wind-Blown Sand

Kosuke Nobusawa[1], Akiyoshi Katano[2], Taiki Sekiguchi[1], Soma Kakizawa[1], Yota Enomoto[1], and Taro Arikawa[1(✉)]

[1] Chuo University, 1-13-27 Kasuga, Bunkyo-Ku, Tokyo, Japan
taro.arikawa.38d@g.chuo-u.ac.jp
[2] Ecoh Corporation, 2-6-4 Kitaueno, Taito-Ku, Tokyo, Japan

Abstract. This study investigates the effects of beach deformation on wind-blown sand transport through physical model experiments and numerical simulations. The experiments were conducted using a 34-m-long open channel, focusing on three scenarios: flat surfaces (Experiment A), sloped surfaces (Experiment B), and sloped surfaces with wave action (Experiment C). A piston-type wave paddle and a blower were used to simulate wind and wave conditions. Wind-blown sand transport rates were measured, and each scenario's experimental coefficients for Kawamura's equation were calculated. The results showed a gradual increase in K values, with 0.482 for Experiment A, 0.538 for Experiment B, and 0.617 for Experiment C, highlighting the influence of slope and wave action. Numerical simulations were performed using the CADMAS-STR-STM model, which integrates fluid, structure, and sediment transport. The simulation results were consistent with experimental data, demonstrating the increasing trend of wind-blown sand transport due to wave effects.

Keywords: Wind-Blown Sand · Kawamura's Equation · Numerical Simulation · Wave-Wind Interaction

1 Introduction

Coastal sandy beaches are characterized by significant sand transport caused by wind and waves. Previous studies have investigated sediment transport and wind-blown sand separately through field surveys, numerical calculations, and experiments. However, few studies have simultaneously analyzed sediment transport and wind-blown sand or assessed the influence of waves on wind-blown sand. This highlights the need for numerical evaluation methods that consider these combined effects. Experiments on flat surfaces have shown differences in wind-blown sand transport caused by variations in moisture content. Similarly, studies using inclined sand surfaces have evaluated the impact of slope gradients on sand transport. However, experiments mimicking sandy beaches remain limited, and the mechanisms governing sand transport between water and air and the interaction between waves and wind remain insufficiently understood.

© The Author(s) 2026
C. Coelho et al. (Eds.): CD 2025, CRL 41, pp. 121–127, 2026.
https://doi.org/10.1007/978-3-032-15473-6_19

Katano et al. [1] reported that approximately 50% of the wind-blown sand on a sandy beach originates from the swash zone based on field observations at Niigata Port. This finding underscores the importance of considering the combined effects of waves and wind when evaluating sand transport on sandy beaches.

This study aims to improve the accuracy of numerical simulations of wind-blown sand and establish a calculation method capable of simultaneously analyzing both sediment transport and wind-blown sand, using experiments designed to mimic sandy beach environments.

2   On Wind-Blown Sand in the Air Phase

2.1   Physical Model Experiments

Experimental Setup. The model experiments were conducted using an open channel with a total length of 34 m and a width of 0.5 m. Two experiments were performed: Experiment A, conducted on a flat surface, and Experiment B, conducted on a sloped surface with water present while only blowing air. The cross-sectional diagram of Experiment A is shown in Fig. 1(left), and that of Experiment B is shown in Fig. 1 (right).

A portable blower connected to a flexible duct delivered airflow into the experimental section. Silica sand with a median grain size of approximately 0.27 mm was used for this study. The amount of wind-blown sand was measured by placing a collection box at the rear end of the experimental section. After the experiment, the sand was air-dried for five minutes, and its weight was measured using an electronic balance. The sand was prepared for the experiments by removing the front and rear partitions just before the start, allowing the sand to settle naturally until no further deformation occurred. Each experiment was conducted for five minutes. The blower was activated at the start of the experiment and stopped after five minutes. In Experiment A, two cases were conducted: one with a single blower and the other with two blowers. In Experiment B, only the two-blower case was conducted. In Experiment A, dry sand was used, while in Experiment B, the sand was shaped into a slope, and water was allowed to permeate the sand until the moisture content stabilized before starting the experiment. During the experiments, a hot-wire anemometer was used to measure wind speed, airflow rate, and temperature data.

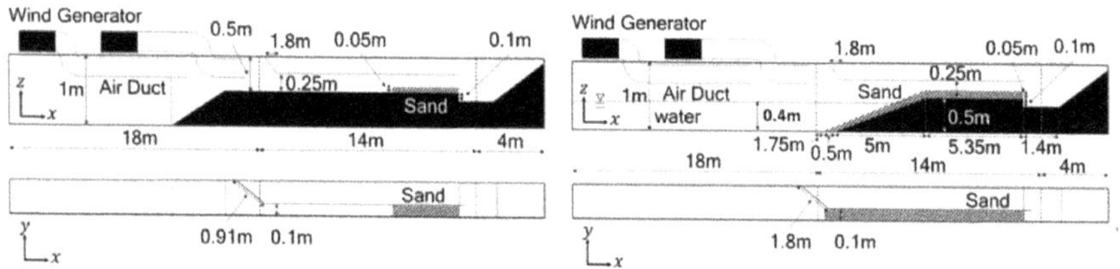

Fig. 1. Experimental cross-section (Left: Experiment A, Right: Experiment B)

Results of Experiments. Figure 2 shows the experimental results plotted with the values from Kawamura's equation and previous experimental data. The horizontal axis represents the friction velocity, and the vertical axis represents the wind-blown sand transport rate per unit width and per unit time. The data from the previous studies and Kawamura's Eq. (1) [2] are indicated in the legend.

$$q = K \frac{\rho_a}{g} (u_* + u_{*c})^2 (u_* - u_{*c}) \tag{1}$$

where ρ_a is the air density, g is the gravitational acceleration, u_* is the friction velocity, u_{*c} is the critical friction velocity, and K is the experimental coefficient.

As shown in Fig. 2, the experimental results obtained in this study generally agree with the values from previous experiments. Furthermore, the experimental coefficient K for Kawamura's equation was calculated for each experiment. The average experimental coefficient K was 0.482 for Experiment A and 0.538 for Experiment B. Based on Kawamura's equation, the wind-blown sand transport rate increased in Experiment B compared to Experiment A.

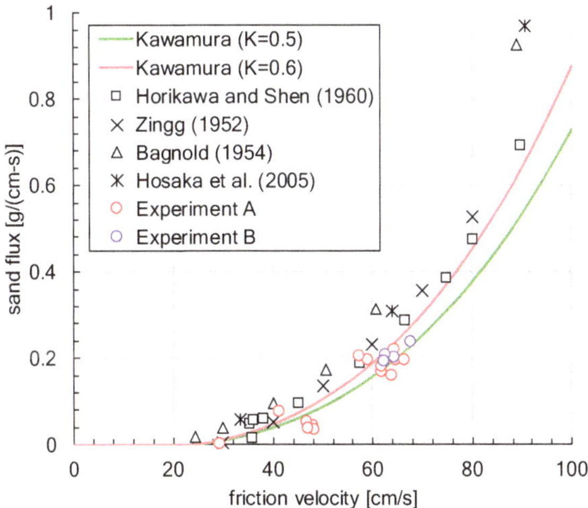

Fig. 2. Comparison of Results from Air-Phase Experiments with Existing Experimental Data and Kawamura's Equation

2.2 Numerical Simulations

Basic Equations. The calculations in this study were performed using CADMAS-STR-STM [3], a fluid-structure-sediment transport model. CADMAS-STR-STM is a numerical model that couples CADMAS-STR [4], which integrates the three-dimensional numerical wave tank CADMAS-SURF/3D-2F with the structural analysis module STR3D, with the Sediment Transport Model (STM).

The STM is based on the fundamental equations proposed in [5]. Equation (2) represents the continuity equation for the bedload transport layer in the air and liquid phases,

124   K. Nobusawa et al.

while Eq. (3) is the advection-diffusion equation for the suspended sediment layer. Additionally, the bedload transport equations used in the STM are formulated for the air and liquid phases as shown in Eqs. (4) and (5), respectively. Equation (4) is derived from Kawamura's Eq. (1). The exchange rate of sediment between the bedload and suspended sediment layers is described in Eq. (6).

$$\frac{\partial Z}{\partial t} + \frac{1}{1-\lambda}\left\{ \frac{\partial q_{Bx}}{\partial x} + \frac{\partial q_{By}}{\partial y} + w_{ex} \right\} = 0 \tag{2}$$

$$\frac{\partial (C_s h_s)}{\partial t} + \frac{\partial (M C_s)}{\partial x} + \frac{\partial (N C_s)}{\partial y} - w_{ex} = 0 \tag{3}$$

$$q_{Ba} = K \frac{\rho_a}{g \rho_s} (u_{*a} + u_{*ac})^2 (u_{*a} - u_{*ac}) \tag{4}$$

$$q_{Bw} = \beta \tau_*^{\frac{3}{2}} \sqrt{sgd^3} + e_s \left| \beta \tau_*^{\frac{3}{2}} \sqrt{sgd^3} \right| \tan \theta \tag{5}$$

$$w_{ex} = \alpha \tau_*^2 \sqrt{sgd} - w_s C \tag{6}$$

where z is height from the reference plane, t is the time, λ is void ratio of sand particles, q_{Bx}, q_{By} are sand transport rate in the x and y directions, respectively, w_{ex} is exchange sand rate between the bedload layer and suspended sand layer, C_s is average suspended sand concentration, h_s is suspended sand layer thickness, M, N are flow flux in the x and y directions, respectively, K is experimental coefficient, ρ_a is gas density, ρ_s is density of soil particles, g is gravitational acceleration, q_{Ba} is sand transport rate in the gas phase, q_{Bw} is sand transport rate in the liquid phase, u_{*a} is friction velocity in the gas phase, u_{*ac} is critical friction velocity in the gas phase, α, β are grain size-dependent coefficient, τ_* is Shields parameter, s is submerged specific gravity of sand, d is particle diameter, e_s is gradient correction coefficient, θ is slope angle, w_s is settling velocity of soil particles, C is concentration near the boundary between the bedload layer and suspended sand layer.

Calculation Conditions. The computational domain was set to match the experimental sections of Experiment A and Experiment B, and the numerical analysis cases were referred to as Calculation A and Calculation B, respectively. The grid size for CADMAS was 5 cm, and the mesh size for STR was 10 cm. The time step was automatically determined using the CFL condition feature in CADMAS. The inflow boundary wind velocity was configured to correspond with growth rate observed in the physical model experiments. 23 cases were simulated with inflow wind velocities ranging from 2.5 m/s to 10 m/s. For K in Eq. (4), a value of 0.5 was applied based on the experimental results.

Calculation Results. Figure 3 compares the numerical simulation results with the experimental results and Kawamura's equation. The wind-blown sand transport rate was calculated by dividing the amount of sand lifted from the rear edge of the sand surface by the analysis duration and the section width, following the same procedure as in the experiments. The friction velocity used in the analysis corresponds to the experiment's measurement points.

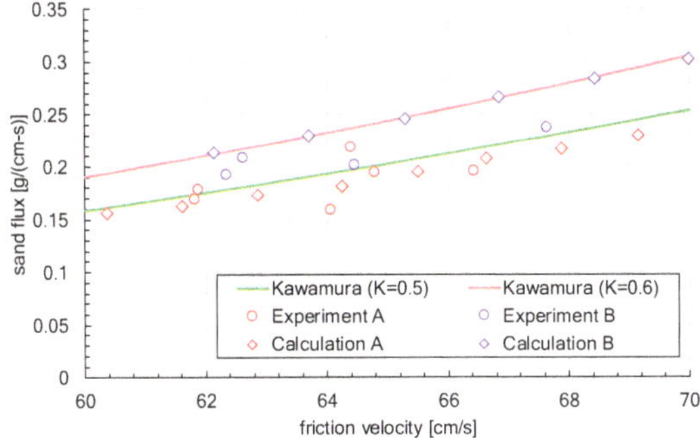

Fig. 3. Comparison of Numerical Simulation Results with Experimental Results and Kawamura's Equation

The simulation results for Calculation A are consistent with the experimental results of Experiment A and Kawamura's equation using an experimental coefficient $K = 0.5$. On the other hand, the simulation results for Calculation B show an overestimation compared to the experimental results of Experiment B. However, they are consistent with Kawamura's equation when $K = 0.6$ is applied. The experiments and numerical simulations indicate an increasing trend in wind-blown sand transport when water is on the slope. The cause of this trend requires further detailed investigation in the future.

3 Experiment in a Combined Wave and Wind Field

3.1 Experimental Conditions

The experimental topography was set to be the same as in Experiment B, with wave generation added using a piston-type wave paddle installed at the left end of a 34-m-long open channel. This case is referred to as Experiment C. The experiment duration was 5 min, the same as in Experiment B. The blower and wave paddle were activated simultaneously at the start of the experiment and stopped at the end. The wave conditions were set such that sand is deposited in swash zone. Seven cases were conducted with wave heights ranging from 0.5 cm to 2 cm.

3.2 Experimental Results and Discussions

From the model experiments conducted in this study, the experimental coefficient K of Kawamura's equation was determined to be 0.482, 0.538, and 0.617 for Experiments A, B, and C, respectively, showing a gradual increase. Figure 4 compares Kawamura's equation plotted using the average experimental coefficient K and the measured wind-blown sand transport rates.

The results indicate that the wind-blown sand transport rate increased in Experiment B compared to Experiment A due to the influence of the slope and further increased in Experiment C compared to Experiment B because of wave action. This demonstrates that, on sandy beaches, wind and wave action contribute to the increase in wind-blown sand transport. On the other hand, in the field survey results by [6], K values of 0.83 and 0.9 were observed. This highlights the need for future studies to elucidate the mechanisms driving this increase and to develop numerical calculation methods.

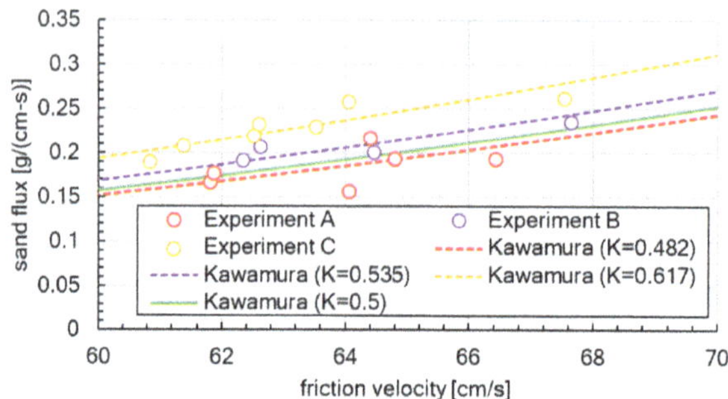

Fig. 4. Comparison of Kawamura's Equation Using the Average Experimental Coefficient Calculated from the Experimental data

Notes

This paper is an English adaptation of the Japanese paper published in the Japanese Journal of JSCE, titled "Improvement of Accuracy of Calculation of Wind-Blown Sand Volume Due To Wave Effects" by Arikawa, T., Sekiguchi, T., Enomoto, Y., Katano, A., Nobusawa, K., Kakizawa, S. and Ohara, H (Volume 80, Issue 17, Released on J-STAGE November 01, 2024, https://doi.org/10.2208/jscejj.24-17126). This paper translates the content into English and deepens the discussion with further analysis.

References

1. Katano A, Shimizu T, Senda N, Sanei R, Arikawa T (2022) The volume of wind-blown sand generating from the swash zone estimated from beach topographic changes. J Jpn Soc Civ Eng Ser B3 (Ocean Eng) 78(2):I_97–I_102. https://doi.org/10.2208/jscejoe.78.2_I_97
2. Kawamura R (1951) Study on Sand Movement by Wind, The reports of the Institute of Science and Technology, University of Tokyo, pp 95–112
3. Arikawa T, et al (2024) Improvement of accuracy of calculation of wind-blown sand volume due to wave effects. Jpn J JSCE 80(17). https://doi.org/10.2208/jscejj.24-17126. (in Japanese)
4. Arikawa T, Chida Y, Seki, K, Takagawa T, Shimosako K (2019) Development and applicability of multiscale multiphysics integrated simulator for tsunami. J Disaster Res 14(2):225–234. https://doi.org/10.20965/jdr.2019.p0225
5. Takahashi T, Shuto N, Imamura F, Asai D (1999) Development of tsunami moving bed model considering exchange sand volume between bed sand layer and suspended sand layer. J Jpn Soc Civ Eng Ser B2 (Coast Eng) 46:606–610. (in Japanese)
6. Katano A, Hayashi H, Senda N, Kabe K (2021) Sand Accumulation forms around the sand fences and estimation of sand transport rate by wind. J Jpn Soc Civ Eng Ser B2 (Coast Eng) 77(2):I_439–I_444

A Multi-tool Approach, Including Citizen-Science, to Study Beach and Dune Morphodynamics in Co. Cork, Ireland

Melanie Biausque[1,3]($\boxtimes$), Emma Verling[1], Darragh O'Suilleabháin[2], and Lee Wah-Pay[2]

[1] MaREI, the SFI Research Centre for Energy, Climate and Marine Environmental Research Institute, University College Cork, Haulbowline Road, Ringaskiddy, Cork, Ireland
`mbiau@bgs.ac.uk`
[2] Cork County Council, County Hall, Carrigrohane Road, Cork, Ireland
[3] British Geological Survey, 39-49 Adelaide Street, Belfast, Northern Ireland

Abstract. In the complex context of climate change and sea level rise, Co. Cork's coastline (Ireland) is exposed to high risks of coastal erosion and flooding. As part of the European A-AAgora project, a multi-tool approach has been developed to better understand coastal processes in action and the role of coastal protection structures at different study sites in East and West Cork. To support better decision-making, we combined methods such as DGPS surveys, a static camera video and citizen-science (e.g. CoastSnap and Maptionnaire). Preliminary topographic results show the morphological evolution of Pilmore and Inchydoney beaches under various hydrodynamic forcing, including high energy events. Data collected at Pilmore will also allow us to compare different management solutions and their efficiency in protecting the shoreline. Inchydoney beach is, in contrast, a more natural environment, where the combination of the complex local geology and hydrodynamics from the Atlantic drives significant morphological changes. Although results extracted from DGPS surveys give a good idea of sediment transport patterns within the system, further work is needed to better understand physical processes under various hydrodynamic conditions (such as low and high-water levels, storms and calm weather) using video images. To gather more data and tackle coastal challenges, we also work with citizens and communities using education and survey tools such as Maptionnaire and CoastSnap. These tools facilitate an active and participatory approach to community engagement, ultimately improving our understanding of coastal processes and ensuring a more sustainable future for coastal areas and communities.

Keywords: Coastal changes · Beach morphodynamics · Citizen-science · Video Images · Coastal management

1 Introduction

With about 1200 km of coastline, the Co. Cork shore is one of the three longest shorelines in Ireland and covers various environments such as cliffs, sandy and mixed sand and gravel beaches, and sand dunes [1]. The Co. Cork coastline is submitted to various natural

C. Coelho et al. (Eds.): CD 2025, CRL 41, pp. 128–134, 2026.
https://doi.org/10.1007/978-3-032-15473-6_20

(e.g. climate change and sea level rise) and anthropic pressures leading to significant coastal changes and further risks such as erosion and flooding [2]. In Ireland, the Horizon Europe funded A-AAgora project (https://a-aagora.eu/) aims to collect and process data to better understand those risks and how to tackle and prevent them. An approach using multiple tools, including both scientific research methods (DGPS surveys and video images) and citizen-science, was here undertaken to support more sustainable decision-making.

2 Methods

2.1 Study Site

Co. Cork is located along the south-west coast of the Republic of Ireland (Fig. 1). Three main study sites have been chosen along Co. Cork coastline representing different environments under various management strategies and exposed to different pressures. Youghal is a coastal town located at the eastern end of Co. Cork (Fig. 1) and considered as a hotspot for tourism in Ireland. The beach is classified as mixed sand and gravel, subjected to mesotidal conditions [3] and features multiple management strategies: old unmaintained wooden groynes, old seawall and rock armours, recent rock armours and a boardwalk. Youghal beach is highly managed and undergoes phases of erosion and flooding [4]. West of Youghal, Pilmore beach is a narrow beach associated with a very wide intertidal terrace, covered with sand, gravels and pebbles. While the western end of the beach is natural, the centre part is managed by the local council using a rock armour to protect a back road and the eastern end is managed by groynes which have been present for many years and undergone various levels of maintenance. In contrast, Inchydoney, located in West Cork, is a sandy beach backed by a relatively wide sand dune mainly managed with delimited paths to protect it from over-trampling and to allow easier access to the beach. The beach is delimited by the presence of a deep channel and a small lagoon to the west, and a rocky headland to the east. Inchydoney also presents an intertidal sandbar accessible at low tide (Fig. 1) and is more exposed to the Atlantic swell than Pilmore and Youghal [4].

2.2 Topographic Data

Topographic surveys are conducted once a month at both Pilmore and Inchydoney beaches, at low spring tide. The surveys consist of 17 and 13, cross-shore profiles collected at Inchydoney and Pilmore, respectively (Fig. 1), using a Trimble R580 GNSS mobile unit mounted on a pole, in RTK mode. At Inchydoney, the profiles are running from the dune foot to the low-tide waterline, while in Pilmore, they cover the area from the edge of the glacial till? to a random physical limit offshore; the low-tide terrace extends up to 900 m offshore, but covering the entire area is not necessary to understand coastal processes at Pilmore.

Since September 2024, a camera (the Tikee3 from Enlaps) has been deployed at Inchydoney, overlooking the dune, the beach and the intertidal sandbars (for more information about the camera: https://www.marei.ie/project/atlantic-arctic-agora/). The camera combines two lenses covering a large area (Fig. 2) and is powered by a solar panel for full autonomy; an additional internal battery is also included. Images are sent to the cloud every day via the 4G system. Images are primarily collected every 30 min during daylight hours; however, the frequency of recording increases to every 30 s for an hour (or more) during storms or specific events (Fig. 2). Video images provide access to morphological information such as the position of the dune foot, the position of the waterline at different tidal moments, or the location of the sandbar and channel (Fig. 1). The collection of ground control points and data for image rectification and georeferencing is ongoing.

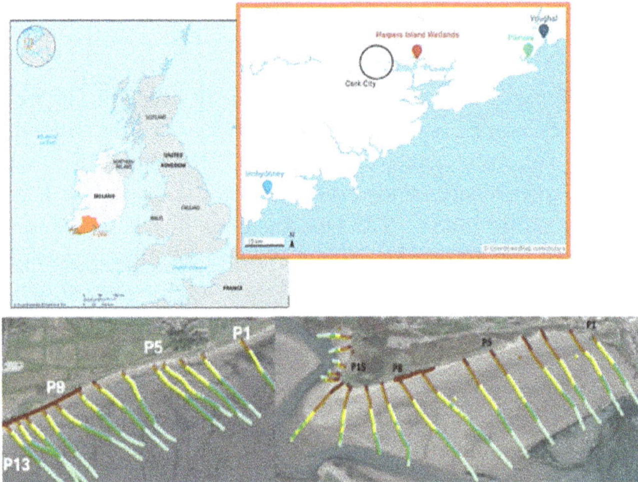

Fig. 1. Location of Co. Cork and study sites (top); Cross-shore DGPS profiles collected at Pilmore (bottom left) and Inchydoney (bottom right).

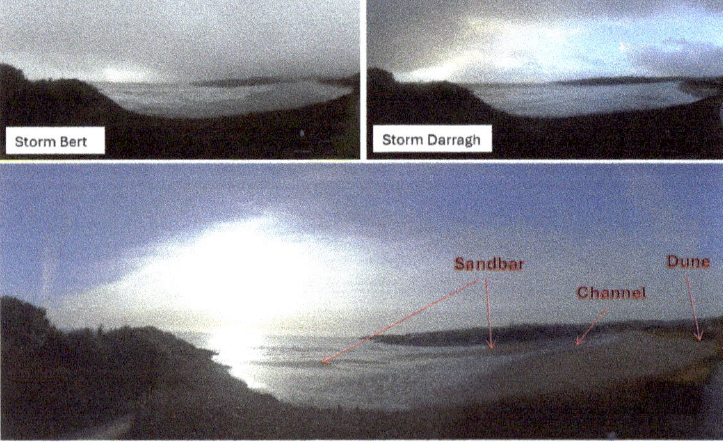

Fig. 2. Example of images extracted from the camera deployed at Inchydoney.

2.3   Citizen Science

A key aspect of the A-AAgora project is the engagement with citizens, communities and stakeholders. Two tools were here used to initiate long-term engagement with citizens and increase awareness regarding coastal changes: Maptionnaire and CoastSnap. The Maptionnaire tool can be defined as a questionnaire based on maps, designed to both collect data from participants and educate on a specific topic; our Maptionnaire focuses on the challenges related to coastal erosion. Through this tool, we aim to educate citizens about coastal erosion and how to identify signs of erosion, and coastal management strategies from hard engineering to Nature-based Solutions (NbS). We have focused our Maptionnaire on Youghal, Pilmore and Inchydoney and are currently in the process of collecting information about erosion issues in these areas.

CoastSnap (https://www.coastsnap.com/) was originally developed and co-founded by Mitchel Harley and his team from the UNSW Sydney, Australia [5]. The main purpose of CoastSnap is to provide a low-cost solution for coastal monitoring and citizen science. Beach visitors can take recurring photos from cradles located at strategic positions (for instance looking at the beach or the dune foot, easily accessible by visitor, etc....) to track coastal changes over time. Two stations were deployed along Youghal boardwalk on the 20th of December 2024 (Fig. 3).

Fig. 3. CoastSnap station deployed at Youghal.

3   Preliminary Results and Discussion

3.1   Morphological Changes: DGPS Surveys and Video Images

Figure 4 show profiles recorded at Pilmore since March 2024; each profile plotted here represents a different management strategy area. For instance, P1 is located in the area of wooden groynes which have been rebuilt, P5 is in front of the rock armour, P9 is in the transitional area from rock armour to no management, and P13 is facing the carpark. It is clear that the different areas are evolving differently over time and exhibit various features. However, more work is needed to better understand the role of different management strategies, sediment size (some areas are mostly covered by pebbles, while others are a mix of sand and gravel), and the relationship between morphological changer and waves, tides, and currents.

The DGPS profiles collected at Inchydoney are not presented here due to space constraints, but our results show a strong alongshore variability in the beach's response to seasonality and storms. Indeed, the profiles located closer to the headland (P1 to P3, Fig. 1) mainly exhibit changes in the lower intertidal area. Profiles located in the central area of the beach (P4 to P12) undergo the most morphological changes in relation to both dune foot retreat and sediment redistribution, and cross-shore intertidal bars migrations. Profiles at the western end of the beach (P13 to P17) appear to be closely linked to the evolution of the river's channel and the sediment transported within it.

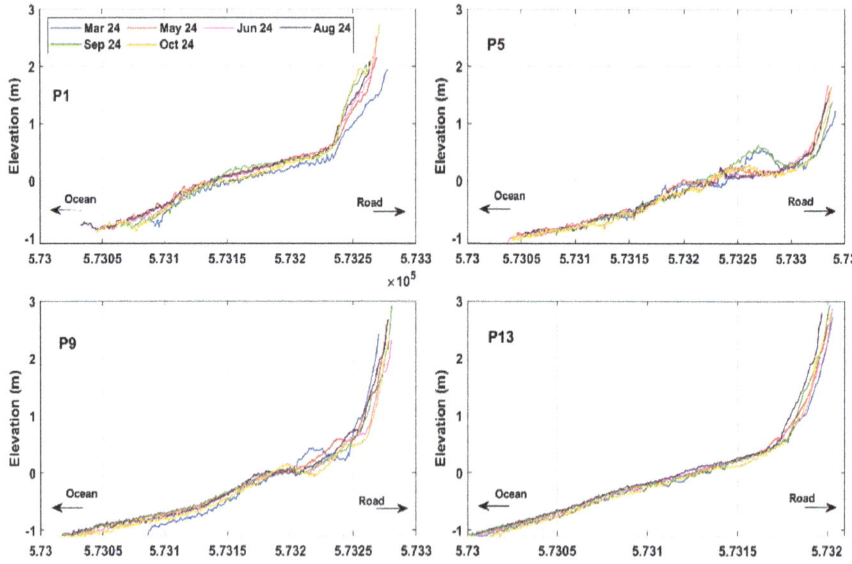

Fig. 4. Example of morphological evolution of Pilmore from DGPS profiles.

If sediment transport and morphological evolution can be extracted from DGPS surveys, there is now a need to link these observations to hydrodynamic factors. Video images collected at Inchydoney should help us make that connection, serving two main purposes: first, to extract topographic profiles at higher frequencies and assess storm impacts and post-storm recovery periods, as well as sandbar and channel cross-shore migrations; and second, to gather information regarding the waterline at low and high tides under low to high energy conditions. For instance, the storm Darragh (December 2024) had limited impacts on the dune foot and the beach despite high-energy waves, while storm Bert (November 2024) caused significant morphological changes. The images extracted from the camera (Fig. 2) clearly show that the high waterline during storm Darragh never reached the dune, whereas waves during storm Bert attacked the dune and upper beach.

3.2 Benefits of Multi-tools Approach and Engagement with Citizens

The first benefit of using a multi-tools approach is to engage a broader audience with impacts of climate change on the shore, beyond coastal scientists. Communities and users of the coast are in the front lines of the effort for sustainable coastal management plans, especially in Ireland. There is a strong interest in connecting with decision-makers and participating in events related to coastal challenges. We have been working with networks such as the Blue and Green Network in Youghal and the Dune Conservation network in Inchydoney, but Maptionnaire and CoastSnap tools are allowing us to create a direct dialogue with citizens and beach users. Moreover, results from our Maptionnaire and workshops showed a lack of knowledge regarding green management options such as NbS. There is also a demand and need for better education around coastal challenges from younger ages (e.g. CoastForKids videos).

From a scientific perspective, each site was chosen based on the challenges it faces (erosion and flooding patterns), the type of environment, and its location along the coast. However, various methods were required for practical and political reasons. For instance, a camera could only be deployed at Inchydoney due to practical considerations, while Pilmore Beach is not a touristic area and citizen science may not be efficient. Moreover, the use of different methods enables more comprehensive data collection at varying frequencies, accuracies, and timescales. The A-AAgora project also supports the Co. Cork Coastal Vulnerability project, which adds drone imagery and modelling to the list of available datasets [6], in the context of a data shortage for coastal processes research in Ireland [4].

4 Conclusions

Through the European A-AAgora project, we combined scientific methods with citizen science to better understand current and future coastal challenges along the Co. Cork coastline (Ireland). Monthly DGPS surveys were conducted at Pilmore and Inchydoney beaches, providing insights into short- and long-term sediment transport patterns and morphological changes. However, a more comprehensive study using video imagery to investigate hydrodynamic forcing is necessary to analyse morphodynamics and processes in place in these areas. In addition, citizen-science via Maptionnaire and Coastsnap not only allows for the collection of scientific data but also enables a direct dialogue with local communities and beach users. In the future, the combined use of multiple monitoring tools, citizen engagement, and interdisciplinary research will be key in developing effective and sustainable strategies for coastal management and resilience in the context of climate change.

References

1. Sinnot AM, Devoy RJN (1992) The geomorphology of Ireland's coastline: patterns, processes and future prospects. Hommes et Terres du Nord 3(1):145–153
2. Olbert AI, Comer J, Nash S, Hartnett M (2017) High-resolution multi-scale modelling of coastal flooding due to tides, storm surges and rivers inflows. A Cork City example. Coast Eng 121:278–296
3. Kandrot S (2012) Beach-dune morphological relationships at Youghal Beach, Cork. In: Gensel J, Josselin D, Vandenbroucke D (eds) AGILE 2012. Springer, Heidelberg, pp 367–390 (2012). https://doi.org/10.1007/978-3-642-29063-3_20
4. O'Sullivan J (2014) Classification of Irish beaches. Theses [online]

5. Harley MD, Kinsela M, Sánchez-García E, Vos K (2019) Shoreline change mapping using crowd-sourced smartphone images. Coast Eng 150:175–189
6. Chalençon E, Cawkwell F, O'Shea M, Murphy J.: Automating the quantification of coastal change using historical aerial photography: a case study along the coastline of county Cork, Ireland. Cambridge Prisms: Coastal Futures (2025)

Simulating Washover Deposits: Effects of the Initial Topobathymetry

Nil Carrion-Bertran[✉], Daniel Calvete, and Francesca Ribas

Universitat Politècnica de Catalunya, 08034 Barcelona, Spain
nil.carrion@upc.edu

Abstract. Low-lying beaches are highly dynamic and vulnerable coastal environments susceptible to flooding and erosion episodes, which will be exacerbated in the context of climate change. Highly complex processes take place during storms, and numerical models are useful tools to assess them. In this study, the XBeach model is applied to an idealised beach to analyse the role of the initial topobathymetry in the formation of washover deposits under an overwash episode caused by a storm event. Initial representative profiles were obtained from real measured 12-year time-lapse topobathymetries, after performing a PCA and a cluster analysis. Alongshore-periodic morphological patterns (mega-cusps, rhythmic dunes, and crescentic bars) of different realistic wavelengths were also included. After running the model with the storm, results of the different scenarios were compared. The washover volume deposited after the overwash event changed when different initial topobathymetries were used and when the morphological pattern wavelength was varied. These outcomes highlight the importance of using real, accurate and up to date topobathymetric data in numerical modeling studies. Also, it is important to consider the possible presence of morphological patterns, since not incorporating them in the initial topobathymetry or misrepresenting their wavelength, could lead to significant changes in washover deposit volumes, and in sediment transport processes in general.

Keywords: Washover deposits · XBeach · Initial topobathymetry · Morphological patterns · Beach morphodynamics

1 Introduction

Deltaic areas and low-lying beaches represent the most dynamic and vulnerable coastal environments, affected by different natural and anthropogenic impacts. Despite concentrating around 10% of the world's population [1], these regions are particularly susceptible to flooding erosion episodes resulting in significant beach width loss and shoreline retreat [2]. At the same time, these storms can create sand deposits on the emerged beach that are essential for the maintenance and development of the beach. In the context of climate change, the projected sea-level rise and the variation of weather patterns are expected to increment the frequency and intensity of these extreme weather events [3]. Consequently, the risks to these natural protection systems will increase, threatening both coastal ecosystems and human infrastructures.

© The Author(s) 2026
C. Coelho et al. (Eds.): CD 2025, CRL 41, pp. 135–140, 2026.
https://doi.org/10.1007/978-3-032-15473-6_21

Moreover, these dynamic coastal areas are extremely complex, with many different processes acting at different scales. For instance, during extreme conditions such as storm events or spring tides, the water level rises exceeding the natural beach defenses (berms, dunes, or barriers) causing a beach flooding event or overwash. These processes often leave morphological formations on the dry beach, known as washover deposits, which can reshape the morphology of the beach affecting its natural features to the incoming storms. Nevertheless, there exists a knowledge gap in the understanding of the role of the initial topobathymetric configuration on sediment transport processes and morphological evolution during extreme events, despite its known importance [4, 5].

To understand the mechanisms that drive these processes and their potential impacts, numerical models play an essential role, also enabling the design and development of effective beach protection strategies to adapt to climate change impacts. Therefore, the aim of this study is determining how the initial topobathymetric variability influences the modeling of the formation of washover deposits during extreme events.

2 Methods

2.1 Model and Set-Up

In this study, the surfbeat mode of version 1.23 of the XBeach 2DH open-source model was employed [6]. XBeach is a process-based numerical model designed to simulate morphodynamic effects of beaches during storm events, being well-suited to model the complex morphodynamic interactions between the backshore and nearshore beach regions. The model incorporates a random parameter that contributes to simulate the real effect of wave chronology groupiness on the beach morphodynamic evolution. To consider this effect and to add robustness and reliability to the outcomes obtained, a total of nine realizations were performed for each scenario. Due to the lack of sufficient data to perform a calibration procedure, the XBeach model parameters were taken from [7] where similar beach conditions were analysed. Nevertheless, a sensitivity assessment of the most sensitive parameters was previously carried out.

2.2 Representative Topobathymetries Extraction

For the purpose of this study, representative synthetic initial topobathymetries were generated by modifying real data while preserving key features, inspired in Castelldefels beach (NW Mediterranean), an open microtidal beach with ($D_{50} = 300$ μm).

From 2011 to 2022, 15 topobathymetries were measured [8]. 5-m equidistant cross-shore profiles were extracted out from a defined 500 m domain and a Principal Component Analysis (PCA) and a k-means clustering analysis were performed to obtain four representative profiles (P). Additionally, the Dean's profile was also added, conforming 5 representative profiles (Fig. 1a). These were replicated alongshore to obtain the initial synthetic representative topobathymetries for the model. The emerged part of the profiles was defined to be equal for all profiles.

To mimic observed periodically distributed morphological patterns, structures were placed in the representative topobathymetries. More specifically, the scenarios studied were the alongshore uniform (AU), without structures, mega cusps (MC), near the shoreline (Fig. 1b), crescentic bars (CB), in the inner zone, and rhythmic dune (RD), on the dry beach area. Also, several realistic wavelengths of these patterns (100–500 m) were tested.

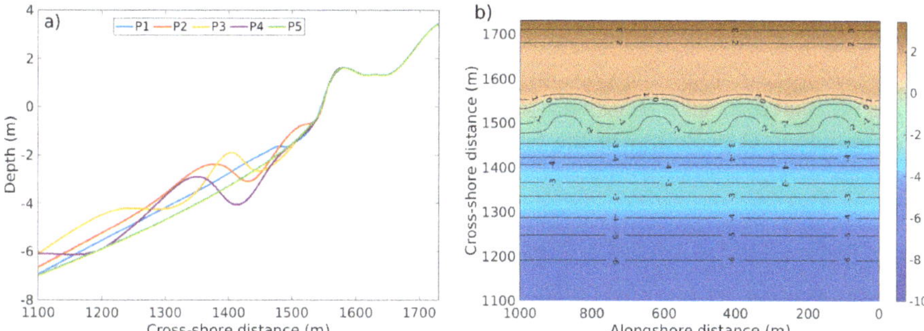

Fig. 1. Panel a) shows the representative profiles extracted after the PCA and clustering analysis of the real topobathymetric data available while panel b) represents the initial topobathymetry with mega cusps ($\lambda = 250$ m) included in the P4 as an example.

2.3 Waves and Sea-Level Conditions

The waves and sea-level conditions were extracted from CoExMed hindcast [9] and Port of Barcelona tide gauge respectively.

A 5-day storm in December 2019 was selected. It reached ~2.4 m height in Castelldefels area and overwashed the beach. The wave direction was set to be normal to the beach orientation in all scenarios.

The sea-level during that period reached around 0.55 m height. Nevertheless, as the XBeach surfbeat mode underestimates swash, runup and consequently, the intensity of flooding events, a "boost" of 0.5 m was added, following [10].

2.4 Metrics for the Analysis

To evaluate the results obtained for each scenario, the variation between the initial and final seabed was quantified. For the analysis, the beach was divided into 3 distinct zones. The first zone (Z1) was defined as the depositional zone, characterized by accumulation of washover deposits. The second zone (Z2) was the swash area, where erosional processes predominantly occur, and the third zone (Z3) extended seaward up to the breaking depth, approximately 6 m at Castelldefels beach.

To quantify, qualify and visualise the observed changes, the 2D variation was reduced to 1D computing the alongshore-average variation for each cross-shore grid cell. Subsequently, the mean bathymetric variation in the three zones was also computed.

Furthermore, to account for the influence of the XBeach random parameter within the analysis, the mean and the standard deviation of the nine realizations for each scenario was quantified, providing information on the variability and robustness of the obtained results.

3 Results and Discussion

The AU scenarios exhibited similar trends across most of the initial topobathymetries, characterized by visible washover deposits in Z1, erosion in Z2, and a combination of erosion-deposition patterns in Z3 (Fig. 2). However, when using the P2 topobathymetry, the washover deposit rate decreased ~40% compared to the other scenarios (Fig. 2c). Despite similar erosion volumes in Z2, the sediment in P2 was primary moved and accumulated to Z3 rather than Z1, as observed in the other four cases. This difference can be attributed to P2's shallower profile around 2 m depth (Fig. 1), which makes the morphological evolution of the beach to evolve differently under the same storm conditions.

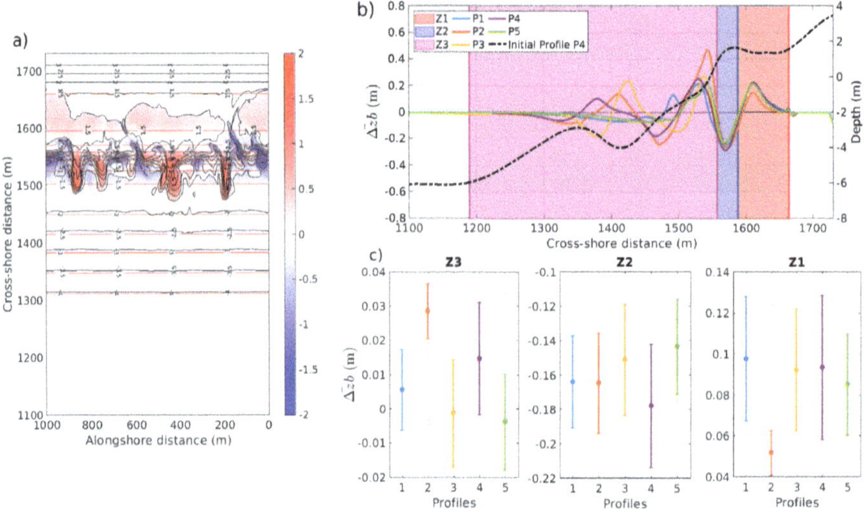

Fig. 2. Panel a) shows a 2D plot of seabed variation for one realization of the AU, P4 scenario. Panel b) shows alongshore-averaged profile variations (mean of 9 realizations) for all AU scenarios, showing the three-zone division and the initial profile (P4) as an example. Panel c) represents the mean variation per zone for each initial AU scenario, with error bars indicating standard deviation due to wave chronology effects.

When morphological patterns (MC, RD, CB) were introduced, the results deviated from those observed in the AU scenarios, with additional variability depending on the wavelength of the patterns. For instance, in the MC scenarios (Fig. 3), the washover depositional rates in Z1 generally tended to increase as the wavelength was larger, with a significant rise observed when wavelengths exceeded $\lambda = 250$ m. In contrast, for shorter wavelengths, the depositional rates in Z1, erosional rates in Z2, and sediment distributions in Z3 were similar to those obtained in the AU and the same initial P4 scenario (Fig. 2). Quite similar results were obtained for the MC scenarios on all the other profiles, with analogous deposition rates on Z1, erosion rates in Z2 and a net rate 0 variation in Z3.

This behavior could be attributed to the fact that the beach has a natural alongshore wavelength in the depositional patterns (around 300 m for P4, see Fig. 2a) inherent to each initial topobathymetry. When the wavelength of the morphological pattern aligns or exceeds this natural wavelength, a reinforcement effect on washover depositional rates can be observed. This is currently being studied in more detail by analysing the time-evolution of the pattern wavelengths within each zone and the hydrodynamic processes throughout the simulation.

The results in the RD scenario in all profiles also showed a similar behavior than the one seen in MC, with an increasing deposition rate as the wavelengths are also raised. Nevertheless, the accumulation within these scenarios is less than in MC.

In contrast, in the CB scenarios despite the accumulation is also seen in Z1, the gradually increasing rates as the wavelength also raises it is not seen in all profiles.

The results for the other patterns (CB and RD) in all profiles are also being currently analysed to detect similarities and differences. A common outcome in all the cases is the relatively large dispersion of the results of the different wave group realisations.

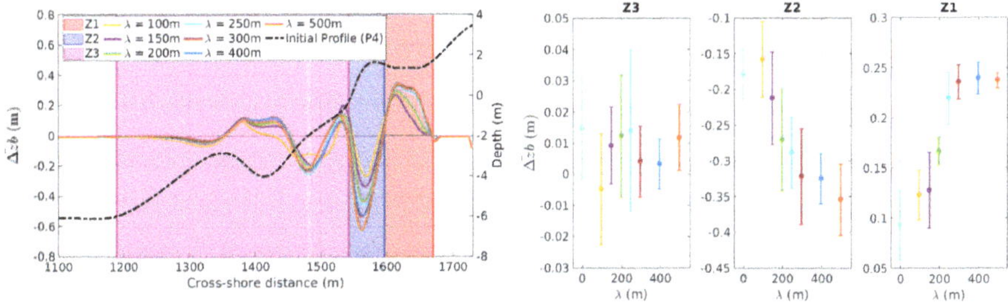

Fig. 3. Panel a) shows alongshore-averaged profile variations (mean of 9 realizations) for all lambda of the MC, P4 scenarios. Panel b) represents the mean variation per zone for each scenario, including in cyan the value for the AU scenario for the same profile, with error bars indicating standard deviation due to wave chronology effects.

4 Conclusions

The results obtained demonstrate the importance of considering the variability of topo-bathymetry in the prediction of deposit formation. For a detailed numerical modelling it seems essential to use real topo-bathymetric data. The use of inaccurate or outdated initial topobathymetric data can lead to unrealistic volumes of washover deposits. The presence of morphological patterns leads to different morphological evolution of a beach under the same storm conditions. Not including them or including them but with an inaccurate wavelength can change significantly the results obtained even if the alongshore uniform profile is up to date.

The results also highlight the importance of not just focusing on shoreline evolution, as the 3D behavior of the beach (including the dry area) can be very different depending on the initial topobathymetry and wave temporal series used. This fact introduces a high degree of uncertainty in the prediction of washover deposits and hence in the modelling of the long-term evolution of beaches. Finally, including the effect of wave chronology by performing several realizations of each scenario is crucial for giving robustness to the model outcomes and make them more comparable to what could happen in a real extreme event.

Acknowledgements. This study was supported by grants PID2021-124272OB-C22 and TED2021-130321B-I00, funded by MCIN/AEI/10.13039/501100011033/ of the Spanish government and by "ERDF A way of making Europe".

References

1. MacManus K, Balk D, Engin H, McGranahan G, Inman R (2021) Estimating population and urban areas at risk of coastal hazards, 1990–2015: how data choices matter. Earth Syst Sci Data 24:5747–5801
2. Cohn N, Ruggiero P (2016) The influence of seasonal to interannual nearshore profile variability on extreme water levels. Coast Eng 115:79–92
3. Luijendijk A, Hagenaars G, Ranasinghe R, Baart F, Donchyts G, Aarninkhof S (2018) The state of the world's beaches. Sci Rep 8:6641
4. Calvete D, Coco G, Falque A, Dodd N (2007) (Un)predictability in rip channel systems. Geophys Res Lett 34–5
5. Castelle B et al (2015) Impact of the winter 2013–2014 series of severe Western Europe storms on a double-barred sandy coast. Geomorphology 238:135–148
6. Roelvink D, Reniers A, van Dongeren A, de Vries JVT, Mc-Call R, Lescinski J (2009) Modelling storm impacts on beaches, dunes and barrier islands. Coast Eng 53:1133–1152
7. Gharagozlou A, Dietrich JC, Karanci A, Luettich RA, Overton MF (2020) Storm-driven erosion and inundation of barrier islands from dune-to region-scales. Coast Eng 158 (2020)
8. Guillén J, et al (2024) Sediment leakage on the beach and upper shoreface due to extreme storms Mar Geol 468:107207. ISSN 0025-3227
9. Toomey T, Amores A, Marcos M, Orfila A (2022) Coastal sea levels and wind-waves in the Mediterranean Sea since 1950 from a high-resolution ocean reanalysis. Front Mar Sci 9:991504
10. Kombiadou K, Costas S, Roelvink D (2021) Simulating destructive and constructive morphodynamic processes in step beaches. J Mar Sci Eng 9:86

Observations and Modelling of Coastal Dune Dynamics Along the Gironde Coast, France

Olivier Burvingt[1(✉)], Alexandre Nicolae Lerma[1], and Bruno Castelle[2]

[1] BRGM French Geological Survey, Regional Direction Nouvelle- Aquitaine, Pessac, France
olivier.burvingt@u-bordeaux.fr
[2] Univ. Bordeaux, CNRS, Bordeaux INP, EPOC, UMR 5805, 33600 Pessac, France

Abstract. Considered as reservoirs of biodiversity, coastal dunes also represent natural barrier against coastal flooding and large source of sediment to mitigate coastal erosion. Dynamics of coastal dunes are forced and controlled by marine, aeolian and biological processes. A better understanding of the interactions between all these processes based on field observations or numerical modelling is crucial to define management strategies that aim to develop the resilience of coastal dune against sea level rise. The analysis of multi-annual topographic data collected along the Gironde coast in SW France show a strong landward migration of the coastal dunes caused by strong wind events and a decrease in vegetation cover. The same data were also used to calibrate and validate a numerical model, AeoLiS, that simulated Aeolian sediment transport. This model showed good performance to reproduce the landward migration of non-vegetated dune.

Keywords: coastal dunes · Aeolian transport · airborne LiDAR · numerical modelling

1 Introduction

The south west coast of France is characterized by large (150–200 m) and relatively high (15–25 m) coastal dunes [1]. Considered as ecosystems with a rich biodiversity, these dunes also represent natural barriers against coastal flooding and large sources of sediment to mitigate coastal erosion. Coastal managers implement different strategies to protect these exposed and dynamic environments [2]. Previous strategies aimed to limit the mobility of coastal dunes so they do not migrate landward and cover existing infrastructures or pines forests located behind. However, limiting the mobility of dunes along coastal areas that have been eroding for decades increase their exposure to wave action and reduce fetch distances that control sediment transport from the beach to the dunes [3]. A good understanding of the dune morphological changes and the ability to model them are required to define new strategies that aim to increase coastal dunes resilience to sea level rise.

© The Author(s) 2026
C. Coelho et al. (Eds.): CD 2025, CRL 41, pp. 141–146, 2026.
https://doi.org/10.1007/978-3-032-15473-6_22

2 Data and Methods

Three study sites, Carcans, Lacanau and Grand Crohot, located few kilometers apart (Fig. 1), but showing different morphologies and vegetation cover, were selected.

Representative cross-shore dune profiles were extracted from digital elevations models derived from airborne LiDAR data collected from 2011 to 2023 over these study sites, and are used to analyze the morphological changes observed along dune systems.

To model morphological dune changes, airborne LiDAR data are also used to calibrate and validate the AeoLiS model [4] that includes temporal and spatial variations of sediment transport in function of wind forcing and limiting factors such as sediment supply, vegetation cover, etc. Measured significant wave height, water levels, wind speed and direction time-series are also used as inputs for the model.

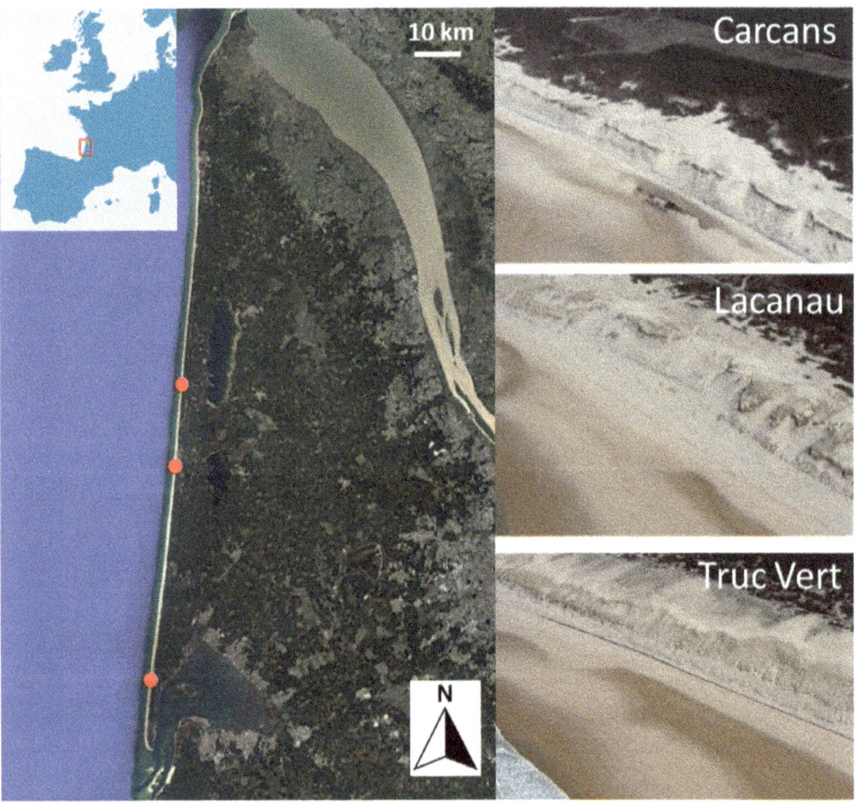

Fig. 1. Location map and aerial pictures of the three study sites in SW France: Carcans, Lacanau and Grand Crohot.

3 Results

3.1 Observations

Cross-shore dune profiles from 2011 to 2023 showed different morphological changes at the three study sites (Fig. 2, left panels). A first sequence of erosion can be observed from 2011 to 2014 where all dunes lost a large amount of sand from -114 to -614 m^3/m (Fig. 2, right panels). Over the nine following years, dunes either migrated landwards by tens of meters with a constant volume of sand (Carcans), migrated landwards while capturing sand (Lacanau), or remain relatively stable and gain sediment in comparison to the volume of 2011. Results that are not shown here, also demonstrated that the amplitude of the observed translation is not entirely governed by wind forcing, since periods of strong winds were not necessarily associated with large dune mobility. Instead, results suggest that the observed decrease of the vegetation cover at Carcans and Lacanau triggered dune translation that accelerated during subsequent windy periods. The denser vegetation cover at Grand Crohot can explain the absence of dune mobility during the same period.

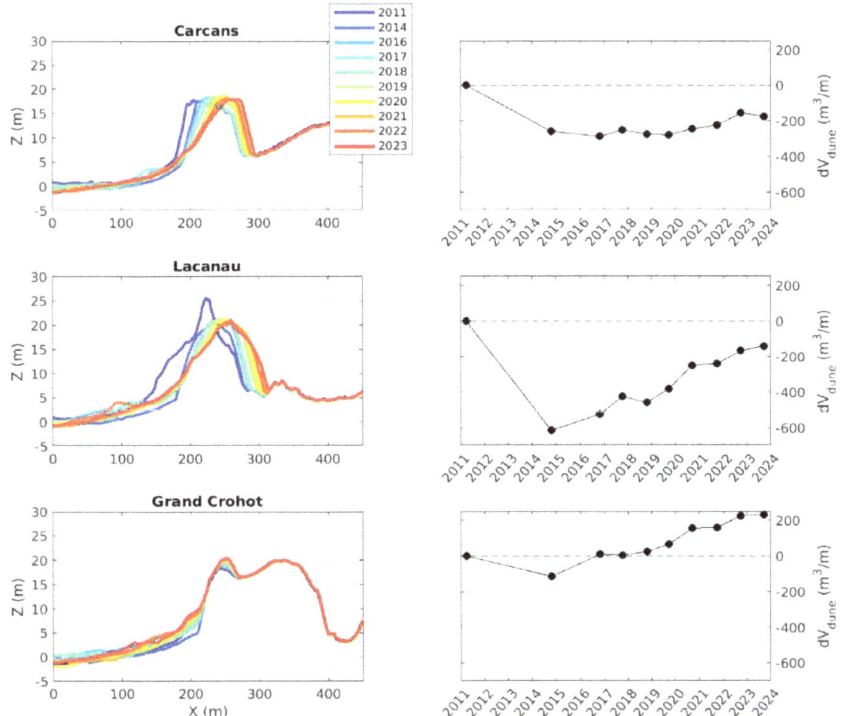

Fig. 2. Cross-shore dune profiles extracted from airborne LiDAR data from 2011 to 2023 (left panels), and the corresponding time series of dune sand volume changes, dV_{dune} in m/m^3 (right panels), at Carcans, Lacanau and Grand Crohot.

3.2 Modelling

The AeoLiS model simulates aeolian sediment transport and given that dune erosion between 2011 and 2014 was mainly caused by marine processes, only the following sequence of dune migration (2014–2023) was considered for this modelling work. Model results show that the AeoLiS model was able to well reproduce the landward dune translation at Carcans and Lacanau after two, four and six years (Fig. 3, top and middle panels). Some differences between observations and model outputs can be observed at the dune crest, and at the lee side of the dune where the avalanching slope is not properly estimated.

Large differences between observations and model results are shown at Grand Crohot (Fig. 3, bottom panel), where the model predicts dune migration while the dune remained relatively stable. The presence of a denser vegetation at this site, which was not considered here in these first model tests, highlights its strong effect in controlling dune morphological changes.

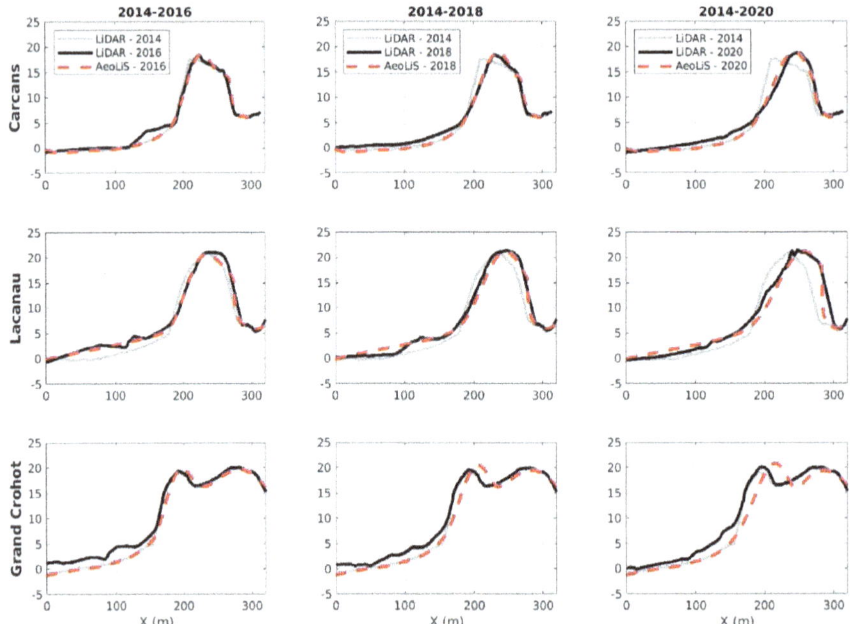

Fig. 3. Cross-shore dune profiles extracted from LiDAR data in 2014 (grey lines), in 2016/2018/2020 (black lines), and AeoLiS model outputs (dashed red lines) at Carcans, Lacanau and Grand Crohot.

4 Discussion and Conclusions

Field observations collected from 2011 to 2023 along three coastal dunes with different morphologies but exposed to similar wind and wave conditions show different morphological changes through time. A sequence of strong erosion caused by extreme storms

[5] significantly changed the dune morphology of two study sites, characterised by a steep dune face, while only the dune foot was eroded at the third study site. Landward migration was observed over the following years at the two sites with a steep dune face while the third one remained relatively stable. In the first case of dune migration, the dune translated landward with a constant volume while the dune gained sediment while migrating in the second case. Other observations that were not shown here demonstrated that the strongest migration sequences corresponded to strong wind events associated with a decrease of the vegetation cover. All these observations suggest that a steep dune face caused by marine erosion triggered dune migration, and the amplitude of this migration is controlled by vegetation dynamics and sediment supply. The temporal resolution of the topographic data does not allow the vegetation cover decrease is caused by sand burial due to landward sediment transport or cause by external factors such as drought. The complex interactions between all these processes advocate for the collection of more field data and the use of numerical models.

The first tests of modelling dune migration along the relatively high and large dunes in SW France, using the AeoLiS model, were promising. Based on previous work using this model [6], only one constant value which corresponds to the empirical constant of the Bagnold formula [7] was modified. Results showed that the model was able to fairly reproduce the landward dune migration at two study sites after two, four and six years. The largest differences between the observations and the model outputs correspond to the current limits of the model identified in other studies [8], such as complex wind field on steep topographic gradients and dune avalanching on the lee side. The large differences between the observations and the model outputs at the third study site also highlight that vegetation characteristics (cover, density) must be included to properly reproduce dune morphodynamics. Indeed, the model simulated a landward migration of the dune crest while no horizontal displacement and a gradual vertical increase were observed along this crest.

Next efforts will focus on improving the model calibration by introducing new sources of field observations and measurements (vegetation cover, wind gradients, etc.).

References

1. Tastet JP, Pontee N (1998) Morpho-chronology of coastal dunes in medoc. A new interpretation of holocene dunes in southwestern France. Geomorphology 25:93–109
2. Bossard V, Nicolae Lerma A (2020) Geomorphologic characteristics and evolution of managed dunes on the South West Coast of France. Geomorphology 367:107312
3. Bauer BO, Davidson-Arnott RG (2003) A general framework for modeling sediment supply to coastal dunes including wind angle, beach geometry, and fetch effects. Geomorphology 49:89–108
4. Hoonhoot BM, de Vries S (2016) A process-based model for aeolian sediment transport and spatiotemporal varying sediment availability. J Geophys Res Earth Surf 121(8):1555–1575
5. Burvingt O, Castelle B (2023) Storm response and multi-annual recovery of eight coastal dunes spread along the Atlantic coast of Europe. Geomorphology 435:108735
6. Cohn N et al (2019) Exploring marine and aeolian controls on coastal foredune growth using a coupled numerical model. J Mar Sci Eng 7(1):13
7. Bagnold RA (1937) The transport of sand by wind. Geogr J 89(5):409–438

8. van Westen B, de Vries S, Cohn N, van IJzendoorn C, Strypsteen G, Hallin C (2024) Aeo-LiS: numerical modelling of coastal dunes and aeolian landform development for real-world applications. Env Model Softw 179:106093

Beach Ridges and Sea Level – A Partial Critique

Patrick A. Hesp[✉]

Beach and Dune Systems (BEADS) Laboratory, College of Science and Engineering, Flinders University, Bedford Park, SA 5042, Australia
Patrick.Hesp@flinders.edu.au

Abstract. Relict wave formed berms and storm ridges, aeolian formed foredunes, and a mix of these are all termed beach ridges. This paper briefly reviews sand or mixed sand/gravel ridge formation, and then examines whether sea level at the time of formation can be determined from beach ridge height, -crest-swale elevational differences, and the foredune toe-top of backshore boundary. It further examines whether it is possible for sand beach ridges to be currently situated above present sea level yet be formed by lower sea levels. The results indicate that the foredune toe-top of backshore position can be relatively well established from both field surveys and GPR studies and relates well to sea level. Beach ridge height is a highly unreliable indicator of sea level. Claims of terrestrially prominent sand beach ridges formed by lower sea levels appear spurious.

Keywords: Foredunes · beach ridges · sea level

1 Introduction

The principal types of beach ridges including those formed as foredunes are briefly examined, and the relationships between sea level and ridge formation are discussed and critiqued for a few studies.

There are several reviews of beach ridges and their formation and terminology [e.g. 1–5]. Most so-called 'beach ridges' range from (i) sandy swash built terraces and ridges formed during either swell waves or storm waves/elevated water level conditions (and perhaps both) [6, 7], (ii) sand dunes (typically foredunes) formed by aeolian processes interacting with vegetation [e.g. 3, 8], sometimes on top of berms [9] or on linear to concave (berm-less) backshores [1], (iii) coarse sand-pebble-gravel-cobble ridges formed by storm waves [e.g. 10], or (iv) shell, gravel-cobble ramps, terraces and ridges formed by swell and storm waves overlain by aeolian foredune ridges/deposits [10, 11]. Figure 1 illustrates some examples all situated within a contiguous 15 km stretch of coast, and it is important to note that the alongshore height and morphology of the storm gravel-cobble ridges varies considerably by up to 4 m depending on the sediment availability/local supply and inshore wave energy *for the same sea level*.

C. Coelho et al. (Eds.): CD 2025, CRL 41, pp. 147–152, 2026.
https://doi.org/10.1007/978-3-032-15473-6_23

Fig. 1. A foredune, beach ridge, and mixed beach ridge-foredune ridge all formed within a contiguous 15 km length of coast near Port Macdonnell, South Australia. (A) foredune ridge (i.e. entirely aeolian) formed on a backshore-beach with no berm. Dominant pioneer species is *Atriplex cinerea*; (B) Gravel and cobble ridge, ~4 m high, formed by storm waves, (C) a mixed ridge with gravel storm ramp base and upper aeolian foredune ridge, and (D) the point (arrowed) where the mixed ridge type terminates and a foredune starts.

2   Foredunes and Sea Level

The boundary between the top of the backshore and the seaward toe of the foredune has been identified in multiple studies examining strata via GPR, and has been considered a reliable indicator of sea level [e.g. 7, 12, 13].

Foredunes form by aeolian sand deposition in pioneer vegetation species on and above the backshore [8]. On a prograding coast, a new incipient foredune may develop and leave the previous foredune relict or inactive, and a continuation of this process can result in the formation of multiple ridges forming a foredune plain (aka beach ridge plain; [8]; Fig. 2). The height that a foredune terrace or ridge may reach principally depends on the length of time the foredune is active, and sediment supply. In general, the longer a foredune is active, the larger and higher it can become. Rapid progradation produces multiple low foredunes and vice versa.

The top of the backshore-toe of the foredune is a position directly related to sea level since coastal pioneer plants cannot consistently grow below the spring high tide line due to their inability to survive one, or multiple saline emersions. Hesp [8] showed that over a 34 year period, the seaward edge of live vegetation (in essence the top of the backshore-toe of the foredune position) varied by a maximum of 60 cm, and typically within 30 cm demonstrating the utility of this position for determining a point directly related to mean sea level (Fig. 2). Typical sandy backshore stratification tends to exhibit flat to low gradients and long strata and is often a mix of primarily swash and some aeolian deposition. In contrast, foredunes exhibit shorter slighter convex strata which are initially low gradient but become higher angle cross-strata with time [cf. 14].

The *height* of a sandy foredune (or of relict foredunes termed beach ridges) is a poor indicator of a paleo-sea level, especially since a berm (which may be present under the foredune) can be eroded by aeolian processes prior to vegetation establishment, berm heights vary considerably as beaches erode or accrete [15], and once a foredune forms on top of the backshore, its height may vary by more than 20 m depending on the length of time the ridge remains in a seawardmost position, and sediment supply.

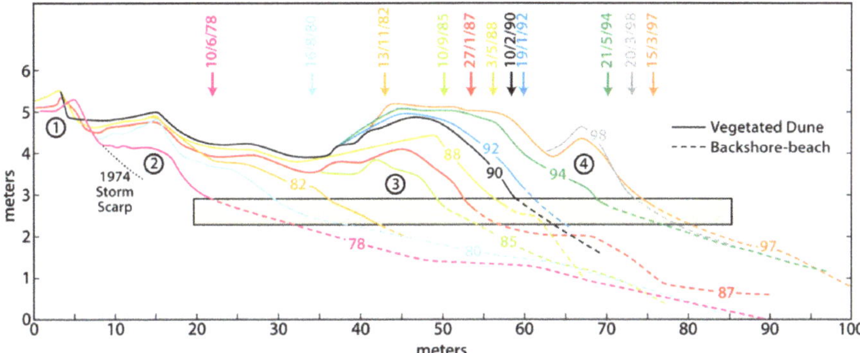

Fig. 2. Topographic surveys of foredune formation at Dark Point, NSW post-1974 to 1997. The dated arrows indicate the position of the seaward edge of vegetation. The box indicates the zone where dune vegetation ceases, and thus, the boundary between foredune and backshore deposits (modified from [8]).

3 Wave Formed Beach Ridges and Sea Levels

In the following, I examine two studies relating to 'beach ridges' to demonstrate how principally wave formed beach ridges cannot easily, if at all, in the cases examined, be related to constructional waves and/or sea levels.

3.1 Gulf Coast, USA

One of the issues driving hypotheses on the genetic origin of beach ridges in the Gulf of Mexico, USA coast relates to interpretations of Holocene sea levels, with one group arguing for slowly rising sea levels through the mid-Holocene to present [16], and another group [17, 18] indicating sea level crossed the present level around 6,500 years BP and may have been higher than present levels.

Donnelly and Giosan [19] accept that the sea level curves of Tornqvist et al. [16] and others are correct, and argue that the Holocene beach ridge systems in the Florida to Alabama region were formed when sea level was −4.0 m to −0.5 m below present sea level.

Donnelly and Giosan [19] state that "many of the modern beach ridge complexes in the northern Gulf of Mexico appear to have formed during a period of increased hurricane activity when sea level was less than 2 m below the modern level", and older ridges "may have developed as early as 5.4–4.2ka when sea level was 3–4 m below modern level"

(p. 752). Donnelly and Giosan [19] argue that increased storm activity in the Gulf would "potentially lead to an increase in the frequency of constructional swells" (p. 752) and swell frequency and wave heights would therefore be greater. Apparently, such waves would produce beach ridges "developing a few meters above their contemporaneous sea level" (p. 752).

Donnelly and Giosan [19] fail to indicate exactly how such ridges might be formed, but presumably a wave built origin is the preferred mechanism as indicated by their "constructional swells" statement, despite the evidence that beach ridges in the region have aeolian sedimentary units present in the GPR records. "Constructional swell" waves do not occur during, or even waning post-storm periods, and are most unlikely to produce beach ridges formed "a few metres" higher that their contemporary sea level. Hurricanes and large storms in the Gulf of Mexico produce largescale erosion and overwash; they do not produce sandy beach ridges well above mean sea level. Additionally, a 4 m subsequent rise in sl would likely erode many of the beach ridges, unless the sediment supply was high to very high.

Foredunes currently form on the modern backshores of Alabama and Florida beaches. This modern day observation of foredune development provides a strong key to the past formation mode of the "beach ridges" and it is highly likely that the beach ridges in the Gulf were largely formed as foredunes and even the oldest ridges (~5000–6000 years BP) are terrestrially situated well above present mean sea level. Hence, since they very likely formed as foredunes, they could not possibly be formed at a sea level lower than present, and then survive continued rising sea levels by as much as ~4 m as indicated by Donnelly and Giosan [19].

3.2 The Gulf of Almeria, Spain, Beach Ridges

Goy et al. [20] studied the beach ridges of the Gulf of Almeria, Spain coast. The region is microtidal (tidal range <0.3 m), waves range from W-SW swell of 1.5–2.5 m and E-SE wind waves of 0.5–1.0 m. The beach ridges are composed of coarse sand to fine gravel and were "formed by wave/swash" (p. 259) on reflective beaches. They appear to argue that the ridges are formed by fair-weather waves since they note that the "concave-up morphology of the prograding units is consistent with fair-weather conditions" (p. 260), although they admit to finding evidence of storm activity and erosional surfaces. Concave-up morphology is, in fact, typical of eroding to accreting reflective beaches, and common on eroded intermediate beaches (e.g. [15], figure 7.7, p. 184). The Almeria beach ridge system displays ridges of various heights (~2 m height variations) and Goy et al. [20] argue that these are the result of an oscillating sea level. In their hypothesis, higher ridges are formed by higher sea levels and lower ridges formed by lower sea levels [cf. 21]. In order to correlate sea levels with ridge heights, sea level had to change by at least 0.8 m on a minimum 100 year interval.

It has been well established that high water levels including storm surges associated with storm waves build coarse sand and gravel berms, storm ridges and beach ridges well above sea or lake level [e.g. 22]. Goy et al. [20] make no reference to storm wave formed ridges, yet many authors note that the crest height of such ridges is likely a reflection of the frequency and intensity of storms. Orford et al. [23] state that barrier crest elevation depends on wave runup and breaker height, beach slope and bed roughness. The Almeria wave climate ranges from 0.5 to 2.5 m (Goy et al. [20], a wave height range which could account for significant differences in the heights of berm crests or "beach ridges"/relict berms without any change in sea level. In addition, Orford et al. [23] state that crest elevation is also driven by sediment supply and the time the ridge remained active [cf. 24]. None of these drivers of beach ridge formation and height are considered by Goy et al. [20].

Goy et al.'s. [20] maximum beach sea level difference is +0.8 m above the top of the present active fair-weather beach crests. Evidence exists elsewhere of up to 2.6 m elevational differences in ridge heights for ridges formed at the same sea level [e.g. 25]. Tsunami would account for substantial differences in ridge heights also. Short [15] shows from medium term topographic profiles of reflective and intermediate beaches that a berm can change in elevation from 1m up to at least 3m depending on the mean grain size and wave energy. Some of Goy et al.'s. [20] ridges are only 0.2 m in elevation difference well within the vertical elevation beach changes found by Short [15]. In addition, if the Almearia ridges actually contain some aeolian strata or are possibly (even partially) foredunes, then variations of a few decimetres to several metres in ridge height can merely be due to variation in rates of beach progradation alone [cf. 8].

Conclusion

The boundary between the foredune toe and the top of the backshore is a reliable indicator of sea level as shown by topographic surveys and GPR data. Foredune and relict foredune heights, especially where foredunes are misinterpreted as berms/relict berms [e.g. 21], are inaccurate indicators of sea level height. Beach ridges formed by wave action alone are unreliable indicators of sea level. It is highly unlikely, and nigh impossible that wave formed beach ridges situated above present msl would have been formed by sea levels up to −4 m lower than present.

References

1. Hesp PA (2006) Sand beach ridges: definition and re-definition. J Coast Res SI 39:72–75
2. Otvos EG (2000) Beach ridges—definitions and significance. Geomorph 32(1–2):83–108
3. Tamura T (2012) Beach ridges and prograded beach deposits as palaeoenvironment records. Earth Sci Rev 114(3–4):279–297
4. Hein CJ, Ashton AD (2020) Long-term shoreline morphodynamics: processes and preservation of environmental signals. Sandy Beach Morphodyn 487–531
5. Isla MF, Moyano-Paz D, FitzGerald DM, Simontacchi L, Veiga GD (2023) Contrasting beach-ridge systems in different types of coastal settings. Earth Surf Proc Land 48(1):47–71
6. Anthony EJ (1991) Beach-ridge plain development: Sherbro Island Sierra Leone. Z. Geomorphol N.F. Suppl 81:85–98
7. Johnston JW, Thompson TA, Baedke SJ (2007) Systematic pattern of beach-ridge development and preservation: conceptual model and evidence from ground penetrating radar. Geol Soc Am Spec Paper 432:47–58
8. Hesp PA (2013) A 34 year record of foredune evolution, Dark Point, NSW, Australia. J Coast Res S.I. 65:1295–1300
9. Thompson TA (1992) Beach-ridge development and lake-level variation in southern Lake Michigan. Sed Geol 80(3–4):305–318
10. Johnson DW (1919) Shore processes and shoreline development. Wiley

11. Suursaar Ü, Kall T, Steffen H, Tõnisson H (2019) Cyclicity in ridge patterns on the prograding coasts of Estonia. Boreas 48(4):913–928
12. Billy J, Robin N, Hein CJ, Certain R, FitzGerald DM (2015) Insight into the late Holocene sea-level changes in the NW Atlantic from a paraglacial beach-ridge plain south of Newfoundland. Geomorphology 248:134–146
13. Dougherty AJ (2018) Prograded coastal barriers provide paleoenvironmental records of storms and sea level during late Quaternary highstands. J. Quat Sci 33(5):501–517
14. Costas S, Ferreira Ó, Plomaritis TA, Leorri E (2016) Coastal barrier stratigraphy for Holocene high-resolution sea-level reconstruction. Sci Rep 6(1):38726
15. Short AD (1999) Wave dominated beaches. Chap 7 In: Short AD (ed) Handbook of beach and shoreface morphodynamics. Wiley, pp 173–203
16. Törnqvist TE, Gonzalez JL, Newsom LA, van der Borg K, de Jong AFM, Kurnik CW (2004) Deciphering Holocene sea-level history on the U.S. Gulf coast: a high resolution record from the Mississippi Delta. Geol Soc Am Bull 116:1026–1039
17. Blum MD, Sivers AE, Zayac T, Goble RJ (2003) Middle Holocene sea-level and evolution of the Gulf of Mexico coast. Gulf Coast Assoc Geol Soc Trans 53:64–77
18. Rodrigues K, Stapor FW, Rink WJ, Dunbar JS, Doran G (2022) A 5700-year-old beach-ridge set at Cape Canaveral, Florida, and its implication for Holocene sea-level history in the southeastern USA. The Holocene 32(1–2):40–56
19. Donnelly JP, Giosan L (2008) Tempestuous highs and lows in the Gulf of Mexico. Geology 36(9):751–752
20. Goy JL, Zazo C, Dabrio CJ (2003) A beach-ridge progradation complex reflecting periodical sea-level and climate variability during the Holocene (Gulf of Almería, Western Mediterranean). Geomorphology 50(1–3):251–268
21. Tanner WF (1995) Origin of beach ridges and swales. Mar Geol 129(1–2):149–161
22. Hayes MO, Michel J, Betenbaugh DV (2010) The intermittently exposed, coarse-grained gravel beaches of Prince William Sound, Alaska: comparison with open-ocean gravel beaches. J Coastal Res 26(1):4–30
23. Orford JD, Forbes DL, Jennings SC (2002) Organisational controls, typologies and time scales of paraglacial gravel-dominated coastal systems. Geomorphology 48(1–3):51–85
24. Forbes DL, Syvitsky JPM (1994) Coastal evolution: late Quat shoreline morphodynamics
25. Adams KD, Wesnousky SG (1998) Shoreline processes and the age of the Lake Lahontan highstand in the Jessup embayment, Nevada. Geol Soc Am Bull 110(10):1318–1332

Measuring Interactions of Surface Elevation and Vegetation Dynamics on the Zandmotor Dune Landscape

Romy Hulskamp[✉], Maria Pregnolato, and Sierd de Vries

Department of Civil Engineering and Geosciences, Delft University of Technology, Delft, The Netherlands
r.l.hulskamp@tudelft.nl

Abstract. The natural functions of coastal dune systems are under threat. Rising sea level, changing wave and wind climate and increasing human activities in coastal areas lead to 'coastal squeeze'. An option to mitigate the effects of coastal squeeze can be creating seaward space through artificial beach widening. The Zandmotor mega nourishment placed in 2011 near The Hague (The Netherlands) is used as an example of creating new coastal space. Morphological changes at the Zandmotor have been studied extensively. The sedimentation-vegetation dynamics and its influence on long-term dune landscape development remain underexplored. Here we show that the presence of vegetation has contributed significantly to the increase in dune volume on the Zandmotor landscape. We used publicly available remote sensing data: elevation maps and aerial images. We found only sedimentation in the dune area on a bi-annual scale indicating that wind does not cause erosion at this timescale. Furthermore, we found that the vegetation cover is increasing over time and is often growing around the peaks of the elevation profile. Also, vegetation often stays at/around the same location and multiplies. The presence and growth of vegetation contribute to a significant increase in dune growth, reaching up to 53 m^3/m/year. Our results demonstrate that the presence of vegetation has contributed significantly to the increase in dune volume on the Zandmotor landscape. The insights about sedimentation-vegetation dynamics and its influence on dune growth, gained at the Zandmotor, may inform the development of strategies to mitigate coastal squeeze, enhance dune resilience and adapt to the challenges of climate change and human activities.

Keywords: Sand Engine · Coastal squeeze · Coastal dunes · Remote sensing

1 Introduction

Coastal dune systems face pressures from rising sea levels, changing wave and wind climates, as well as from increasing human activities in coastal areas. These pressures lead to 'coastal squeeze' [1] at many locations. Coastal squeeze threatens the natural functions of dunes, including flood protection, habitat provision and recreational space [2]. To mitigate coastal squeeze, two options to create more space for coastal dune

C. Coelho et al. (Eds.): CD 2025, CRL 41, pp. 153–159, 2026.
https://doi.org/10.1007/978-3-032-15473-6_24

landscapes are proposed: releasing landward pressure by removing coastal infrastructure, or creating seaward space through artificial beach widening. An example of creating seaward space is the Zandmotor project, which was initiated near The Hague (The Netherlands) in 2011 [3] (see Fig. 1). The Zandmotor has successfully expanded the beaches, formed new dunes, enhanced biodiversity and provided recreational space. Initial studies have focused on its morphological changes and sediment dynamics [4, and references therein], however its long-term landscape evolution and potential additional benefits remain underexplored.

The long-term landscape evolution is closely related to the interrelation between vegetation growth and sedimentation/erosion processes. This sedimentation-vegetation dynamics determine when, how and where new dune will develop [8–10]. Initially, the Zandmotor was a barren surface of sand only without any vegetation. Over time, however, patches of vegetation began to establish, acting as roughness elements that trap sediment and reduce erosion. This interaction has facilitated the growth of some new dunes to heights exceeding 5 m. The mutual feedback between sedimentation processes and vegetation patterns highlights the dynamic nature of the Zandmotor's landscape and its evolving natural features.

This study aims to measure interactions of sedimentation processes and vegetation dynamics on the Zandmotor and to explore the long-term evolution of its landscape. By using publicly available elevation maps and aerial photographs, and analysing changes in sediment distribution, vegetation cover and overall landscape dynamics, we will develop a comprehensive understanding of the Zandmotor's impact.

Fig. 1. Aerial images of the Zandmotor in a) September 2011 and b) July 2024 provided by Rijkswaterstaat.

2 Study Area and Methods

The Zandmotor is a mega nourishment of approximately 20 million m^3 of sand that was placed in an area of 2.5 km^2 at the beach [3]. It was created as a pilot project, following the'building with nature' approach. Building with nature aims to make use of natural processes in a proactive way to build, maintain and reinforce dunes and their accompanying protective functions, while providing opportunities for nature as part of the infrastructure development process [6]. Three goals were defined: 1) stimulate natural dune growth for the benefit of safety, nature and recreation; 2) generate knowledge development and innovation; 3) create attractive (temporary) recreational and natural areas.

2.1 Transect Profiles

Over the length of the whole study area cross-shore transects are drawn with a 100 m spacing. The length of the transects depend on the location of the dune foot in March 2024. The dune foot level along the Dutch coast is widely assumed to be +3 m MSL [5, 6] and is therefore used in this analyses as well. The landward end of the transect is determined by the bicycle path along the Zandmotor, i.e. a stable point in time. In total 70 cross-shore transects are created. Along every meter of each transect the values of the DTM's are extracted. A moving average window is applied with a window size of 6 (m).

2.2 Dune Volume

The area under each transect profile is calculated for every year and multiplied by 1 m^2; this area results in a total volume per year for each transect profile. We found that the trend over time of these volumes shows a (surprisingly) linear behaviour for many transects. Therefore, a linear line is fitted through the volumes over time. The fit of the line is computed with R^2. The slope of the linear line represents the volume change per alongshore meter per year of the transect.

2.3 Vegetation Detection

The vegetation presence is determined by analysing yearly RGB aerial photographs. First these RGB images are converted to greyscale images. To detect vegetation from these greyscale images, Otsu's method [7] was used. Essentially, Otsu's method identifies the optimal threshold based on the light intensity of pixels in the images. In this case sandy pixels (bright) are separated from vegetated pixels (dark) (see Fig. 2 a) and b)).

For every m^2 along each transect the percentage of vegetated pixels is determined, shown in Fig. 2 c). These percentages of vegetated pixels are plotted on top of the elevation profile, to show the vegetation presence and changes in an intuitive way and couple them to sedimentation patterns.

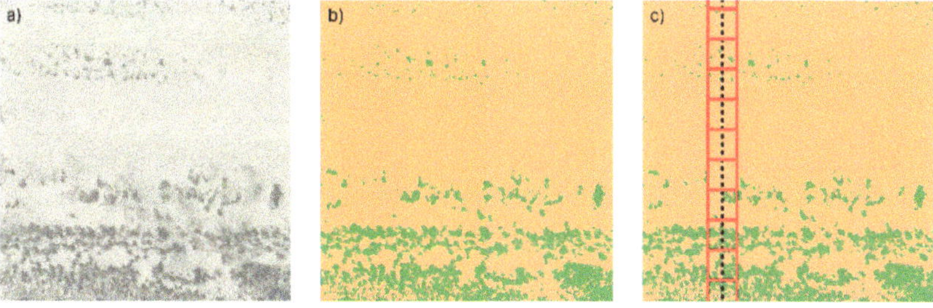

Fig. 2. Method to analyse the presence of vegetation. a) RGB image, b) classified image, sandy (yellow) and vegetated (green) pixels, derived from Otsu's method, c) example to calculate vegetation percentage per cell.

3   Results

Significant accumulation of sediment is found at many transect locations (see Fig. 3 for an example). The cross-shore transect in Fig. 3 is located just south of the dune lake on the Zandmotor. The first years (until 2016), sedimentation is mostly taking place just in front of the dune (x = 40 m). In front of the dune a new (fore)dune is developing. After 2016, there is also sedimentation between x = 150–300 m. It is notable that over the whole transect (and the other analysed transects as well), there is (almost) only accretion. This accretion implies that the eroding factor of the wind force is not present and that (most) sediment is coming from the beach and the intertidal zone more seaward from the dune toe.

Vegetation develops mainly in places where accumulation takes place. In Fig. 3 the presence of vegetation is plotted also for each year where data is available. It shows that the dune (x = 0 − x = 50) is (almost) completely vegetated after the vegetation has started to grow in 2016. It can be seen that the vegetation presence is increasing over time along this part of the transect. Between x = 160 and x = 250 the elevation profile shows a clear accretion of sediment. Along this part of the transect some vegetated patches have started to grow. Over time the vegetation cover is increasing in this area. Notably, the vegetation is often growing around the peaks of the elevation profile. Often vegetation stays at/ around the same location and multiplies, however occasionally vegetation has disappeared.

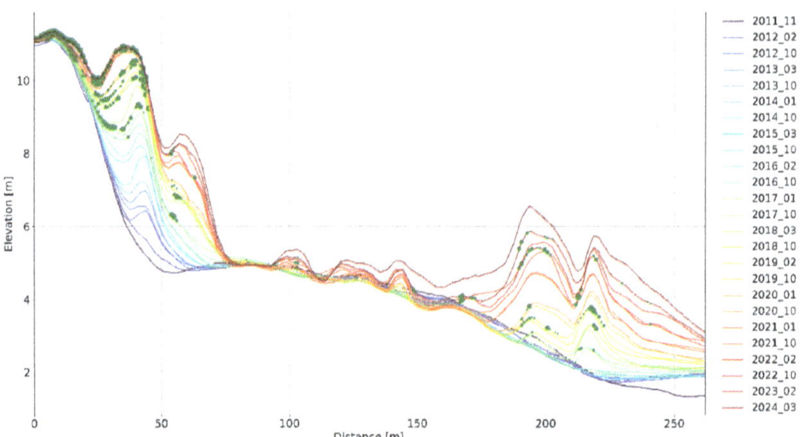

Fig. 3. Transect profile: cross-shore distance on the x-axis (0 m is bicycle lane, right is the seaward side), elevation in meters on the y-axis. Coloured lines represent profiles of different years. Green dots represent the presence of vegetation, extracted from RGB images. The larger the dot the higher percentage of vegetation per m^2. Note that vegetation data is only present for the following timestamps (year_month): 2016_02, 2017_01, 2018_03, 2019_02, 2020_01, 2022_02, 2023_02.

During the period between 2011–2024, dune growth rates seem to vary. Figure 4 shows the volume changes over time of the cross-shore transect shown in Fig. 3. It is clear that the volume of this transect is increasing over time. The linear fit (red line) shows a trend of 40 m^3/m/year, which is at the upper limit of the volume changes found by [6]. The R^2 of this fit is 0.94. However, a break in the slope in the development of the measured dune volumes in time can be seen. From 2016 onwards we found from aerial images that vegetation has started to develop. Vegetation traps sediment [8, 9 10, 11, 12] and therefore an increase in the trend of volume changes can be seen. The blue and green dashed lines show the linear fit over the two periods. The trend before the presence of vegetation is 15 m^3/m/year (R^2 = 0.90), while the trend after the vegetation has started to develop is significant larger with 53 m^3/m/year (R^2 = 0.99).

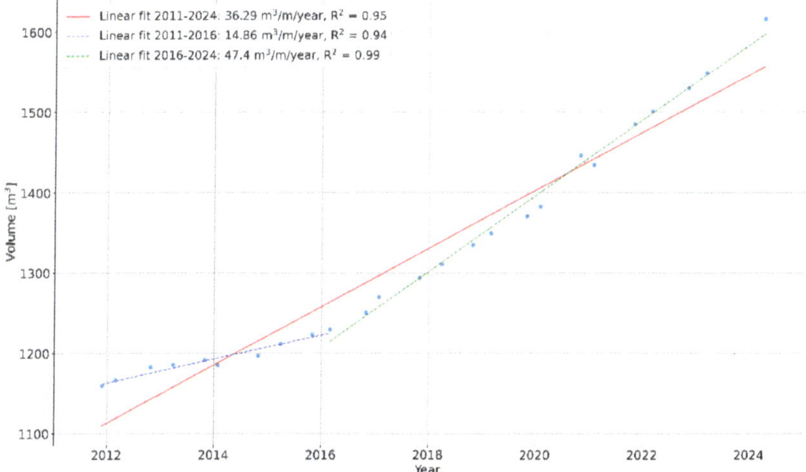

Fig. 4. Volume change within one cross-shore transect profile over time. The blue dots represent the measured volumes of the profile for each year. The red line is a linear fit through all points, the blue dashed line and the green dashed line are the linear fit through a period when vegetation was not yet present, and a period when vegetation has started to grow, respectively. The trend of the fitted lines and the R^2 value are given in the left upper corner.

4   Discussion and Conclusions

Over the past 13 years, on the Zandmotor natural processes have led to the growth of a new dune area. These dunes have formed as a result of natural sediment dynamics and vegetation development, both of which are influenced by time, location and environmental conditions. Consequently, the distribution of the new dunes is not uniform and shows spatial variation. The volume growth of the dunes on the Zandmotor are consistent with dune growth elsewhere along the Dutch coast.

In this study we found that the vegetation cover is increasing on the Zandmotor over time. We showed that the vegetation is often growing around the peaks of the elevation profile. Often vegetation stays at/ around the same location and multiplies, however occasionally vegetation has disappeared. Besides, we found that the volume change per year before the presence of vegetation is 15 m^3/m/year, while the trend after the vegetation has started to develop is significant larger with 53 m^3/m/year. Therefore we can conclude that the presence of vegetation has contributed significantly to the increase in dune volume on the Zandmotor landscape, highlighting its importance in coastal development.

These insights gained at the Zandmotor, about the interrelation between sedimentation and vegetation and its influence on dune growth, may inform the development of strategies to mitigate coastal squeeze, enhance dune resilience and adapt to the challenges of climate change and human activities.

Acknowledgements. The authors gratefully thank Zabawas for funding this research.

References

1. Silva R, Martínez M, van Tussenbroek B, Guzmán-Rodriguez L, Mendoza E, López-Portillo J (2020) A framework to manage coastal squeeze. Sustainability 12(24):1–21
2. Lansu E et al (2024) A global analysis of how human infrastructure squeezes sandy coasts. Nat Commun 15(1):432
3. Stive M et al (2013) A new alternative to saving our beaches from sea-level rise: the sand engine. J Coast Res 29(5):1001–1008
4. Roest B, de Vries S, de Schipper M, Aarninkhof S (2021) Observed changes of a mega feeder nourishment in a coastal cell: five years of sand engine morphodynamics. J Mar Sci Eng 9(1):37
5. Ruessink B, Jeuken M (2002) Dunefoot dynamics along the Dutch coast. Earth Surf Proc Land 27(10):1043–1056
6. De Vries S, Southgate H, Kanning W, Ranasinghe R (2012) Dune behaviour and aeolian transport on decadal timescales. Coast Eng 67:41–53
7. Otsu N (1979) A threshold selection method from gray-level histograms. IEEE Trans Syst Man Cybern 1:62–66
8. Maun M (1998) Adaptations of plants to burial in coastal sand dunes. Can J Bot 76(5):713–738
9. Keijsers J, de Groot A, Riksen M (2015) Vegetation and sedimentation on coastal foredunes. Geomorphology 228:723–734
10. Van Puijenbroek M, Teichmann C, Meijdam N, Oliveras I, Berendse F, Limpens J (2017) Does salt stress constrain spatial distribution of dune building grasses Ammophila arenaria and Elytrichia juncea on the beach? Ecol Evol 7(18):7290–7303
11. Nolet C, van Puijenbroek M, Suomalainen J, Limpens J, Riksen M (2018) UAV-imaging to model growth response of marram grass to sand burial: implications for coastal dune development. Aeol Res 31:50–61
12. Hesp P, Dong Y, Cheng H, Booth J (2019) Wind flow and sedimentation in artificial vegetation: field and wind tunnel experiments. Geomorphology 337:165–182

Surf Zone Bathymetry from Colour-Based Method Calibrated with Wave Breaking via Satellite Imagery

Salomé Frugier[1](✉), Rafael Almar[1], Erwin Bergsma[2], A. Spicer Bak[3], and Katherine Brodie[3]

[1] Laboratory of Space Geophysics and Oceanography Studies (LEGOS), Toulouse, France
`salome.frugier@ird.fr`
[2] National Centre for Space Studies (CNES), Toulouse, France
[3] Coastal and Hydraulics Laboratory, U.S. Army Engineer Research and Development Center, Duck, NC, USA

Abstract. While satellite-derived bathymetry (SDB) techniques have been developed to map seafloor depth over large coastal regions, they still face challenges in providing robust estimates very close to shore (less than 400 m). These existing methods struggle in the nearshore zone due to shallow water conditions and wave dynamics. Traditional bathymetric approaches, relying heavily on in situ measurements, are limited in terms of spatiotemporal resolution and coverage, particularly in these challenging coastal areas. This study demonstrates the applicability of a colour-based method for deriving nearshore bathymetry from satellite imagery, entirely independent of ground-based data. The methodology combines wave-breaking detection via the SandBar Index (SBI), empirical relationships derived from wave dynamics, and spectral analysis of optical data from Sentinel-2 images. Breaking wave zones are identified and linked to corresponding wave heights from buoy data. Using the classical breaker height-to-depth ratio, water depths are estimated along cross-shore transects. The spectral ratio $\frac{\ln(B)}{\ln(G)}$, sensitive to shallow water depths, is calibrated with these depth estimates, enabling the reconstruction of full bathymetric profiles.

Results highlight the strength and accuracy of this approach, demonstrating that reliable nearshore bathymetric estimates can be obtained directly from satellite observations, even in areas traditionally considered difficult for SDB. This method provides a cost-effective, scalable solution for nearshore bathymetry, offering significant potential for use in remote or data-scarce regions.

Keywords: Colour-based bathymetry · Sentinel-2 · Sandbars · Surf zone

1 Introduction

The submerged coastal zone represents the largest portion of the coastal area but remains invisible to the human eye and largely overlooked in coastal management studies. Yet, coastal bathymetry plays a fundamental role in coastal dynamics. It directly influences

© The Author(s) 2026
C. Coelho et al. (Eds.): CD 2025, CRL 41, pp. 160–166, 2026.
https://doi.org/10.1007/978-3-032-15473-6_25

hydrodynamic processes such as wave behaviour, tides, and currents [1] but is often absent from analyses due to a lack of comprehensive data. Indeed, available bathymetric surveys are generally limited to localized and economically strategic areas [2], leaving vast portions of global coastlines under-sampled.

Accurate and up-to-date bathymetry is essential for reliably simulating coastal dynamics and predicting morphological changes. Without this information, forecasting coastal hazards such as wave-induced surges, sea states, flooding, or storm waves, is fraught with significant uncertainties [3]. However, shallow water survey campaigns, covering depths ranging from 0 to 30 m, remain rare and often expensive. Traditional technologies such as sonar give high-resolution data but typically perform poorly in bubbly environments like the surfzone [4], while more recent tools like airborne LiDAR [5] or drones/UAVs [6] provide high-resolution data but cover only a limited fraction of coastlines and require substantial investments. Moreover, the dynamic nature of these environments raises questions about the validity duration of such point-in-time surveys [2].

In this context, Satellite-Derived Bathymetry (SDB) emerges as a promising alternative. It offers unparalleled spatiotemporal coverage and resolution, and the ability to observe coastal processes across various scales. Two main approaches stand out: the water colour method [7–9] and the wave kinematics approach [6, 10, 11]. The first relies on the interaction of light with the water column and is well-suited to clear and non-turbid environments. Wave kinematics techniques, which exploit the influence of bathymetry on waves in shallow waters, analyse wave motion modulated by depth. These methods enable access to critical data for understanding coastal processes across previously inaccessible temporal and spatial scales, thereby enhancing our ability to monitor and predict coastline evolution.

This study presents a method combining water colour analysis [7, 8] and a breaking zone identification approach [19] in the surf zone, enabling the complete reconstruction of nearshore bathymetric profiles.

2 Methods

2.1 Breaking Zones Identification

The first step involves identifying the breaking zones visible on satellite images using the SandBar Index (SBI) [19], a new methodology designed to optimise the detection of wave-breaking pixels induced by the underlying sandbar while minimising the SBI value pixels from the surrounding environment such as sand, land and water.

The transverse positions of the farthest offshore breaking points are extracted along cross shore transects, but only when wave breaking occurs at distances greater than 30 m and less than 300 m from the shore, as the focus is on the surf zone rather than the swash zone. Additionally, the analysis is currently limited to micro-tidal beaches, which justifies the 30-m and 300-m threshold.

For each breaking position ($X_{breaker}$), the wave height at the moment of breaking ($H_{breaker}$) is estimated using data from a buoy located at 8 m depth. These wave heights are temporally synchronized with the acquisition time of the satellite images, ensuring the coherence of the observations. By applying this procedure to a time series of Sentinel-2 images, we generate a spatio-temporal database associating $H_{breaker}$ with $X_{breaker}$, providing a detailed view of the distribution of wave heights at breaking in relation to their cross-shore position (see Fig. 1a).

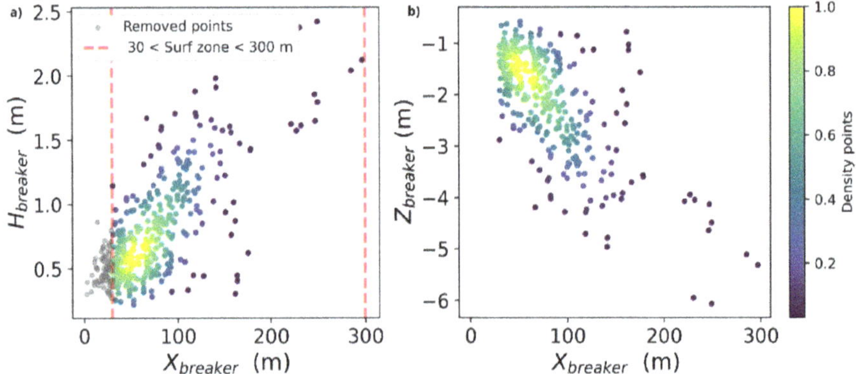

Fig. 1. By identifying the position of the outermost breaking wave and the associated breaking wave height from the buoy at the date and time of Sentinel-2 satellite acquisitions, (a) presents the scatterplot showing the distribution of wave heights and (b) the distribution of depth as a function of the most offshore breaking position ($\gamma_{breaker} = 0.4$).

2.2 Breaker Height-To-Depth Ratio

Wave heights are linked to the height-to-depth ratio [12, 13], corresponding to the classical relationship between wave height ($H_{breaker}$) and depth $Z_{breaker}$ at the breaking point, expressed as:

$$Z_{breaker} = \frac{H_{breaker}}{\gamma_{breaker}} \tag{1}$$

where $\gamma_{breaker}$ represents the breaking coefficient. The ratio was once considered constant in the surf zone, with a value of 0.78 for solitary waves on a linear beach [13]. However, existing studies showed it varies [14, 15]. [14] found it decreases from 0.8 at the breaking point to 0.5 in the inner surf zone. Recent in situ data report values from 0.2 to 1.4, increasing near the shore [16, 17]. Other researchers highlight the need for parameter adjustments in bar profiles for better numerical models [12]. This relationship allows us to estimate the depths associated with each cross-shore position $X_{breaker}$, thus generating a database linking associating $Z_{breaker}$ with $X_{breaker}$ based solely on satellite observations and buoy-derived wave height data (see Fig. 1b).

2.3 Stumpf Spectral Ratio

To leverage the spectral signatures of satellite images, we use the spectral ratio $\frac{\ln(B)}{\ln(G)}$ [7]. This ratio is particularly suited for analyzing shallow marine environments, as the blue and green spectral bands are those where light penetrates most deeply into the water before being attenuated. This property is due to the relatively low absorption of these wavelengths by water, compared to red and infrared bands, which are rapidly absorbed even in shallow waters.

[7] demonstrated that this ratio exhibits a linear relationship with depth in shallow waters, making it a robust tool for estimating bathymetry from optical data. This relationship is particularly useful because, in shallow areas, reflectance measured in the blue and green bands is strongly influenced by the substrate while remaining sufficiently sensitive to depth.

In this context, a satellite image free of artifacts (such as turbidity, clouds, or surf zones) is selected (see an example Fig. 2a). The ratio $\frac{\ln(B)}{\ln(G)}$ is calculated along cross-shore transects, producing a distribution of light intensities as a function of cross shore position. This distribution exhibits a qualitative correlation with the underlying bathymetry but is expressed in relative units and does not allow for direct depth estimation (see Fig. 2b). To convert these light intensities into real depths, we use the depths ($Z_{breaker}$) obtained from surf zones and associated with their positions ($X_{breaker}$). A linear regression is then performed between the spectral ratio $\frac{\ln(B)}{\ln(G)}$ and $Z_{breaker}$ at the corresponding positions (see Fig. 2c), according to the formula:

$$Z = a \cdot \frac{\ln(B)}{\ln(G)} + b \tag{2}$$

where a and b are coefficients determined by linear regression. Once this relationship is calibrated, it becomes possible to convert the observed Stumpf ratio intensities into depths, thereby reconstructing a complete bathymetric profile.

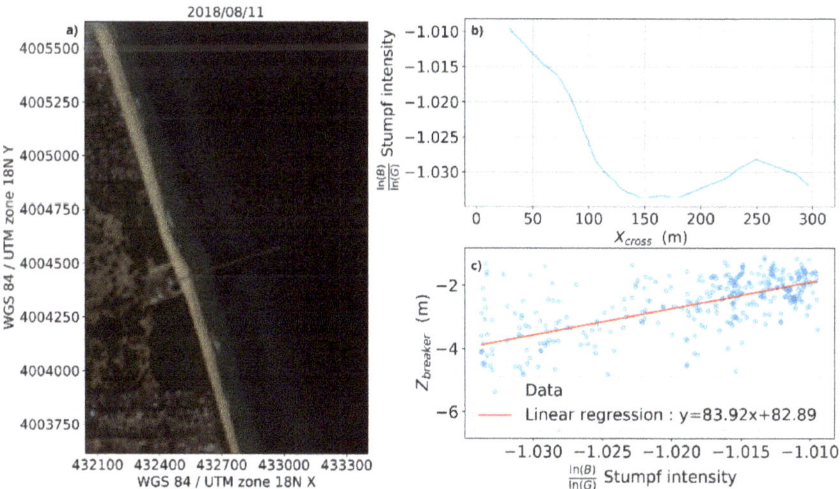

Fig. 2. (a) Sentinel-2 satellite acquisition from August 11, 2018, where the $\frac{\ln(B)}{\ln(G)}$ ratio is applicable (non-turbid and free of artefacts). (b) Median profile of this ratio along transects provided by the FRF in Duck, NC (US). (c) Linear regression between depth and this ratio.

Given that tide corrections can be impactful for coastal morphological changes, it is essential to account for tidal effects to accurately compare in situ data with satellite-derived measurements. To achieve this, tidal data from the NOAA Tides and Currents record collected at Duck, were used to apply the necessary corrections and reference all measurements to Mean Sea Level (MSL). Similarly, the bathymetric data provided by Duck are referenced to NAVD88. Therefore, a correction of -0.128 m was applied based on NOAA's tidal benchmarks to align these data with the MSL reference.

3 Results

Figure 3a illustrates the comparison between in situ bathymetric data from Duck collected on August 14, 2018 and satellite-derived bathymetry obtained on August 11, 2018. The satellite-derived method successfully captures the curvature of the sandbar and the trough between the shore and the bar. However, the upper part of the profile is less accurately represented. Overall, as shown in Fig. 3b, the results are promising, with a coefficient of determination (R^2) of 0.99, a Root Mean Square Error (RMSE) of 0.26 m, and a standard deviation (STD) of 0.84 m after applying the tidal correction.

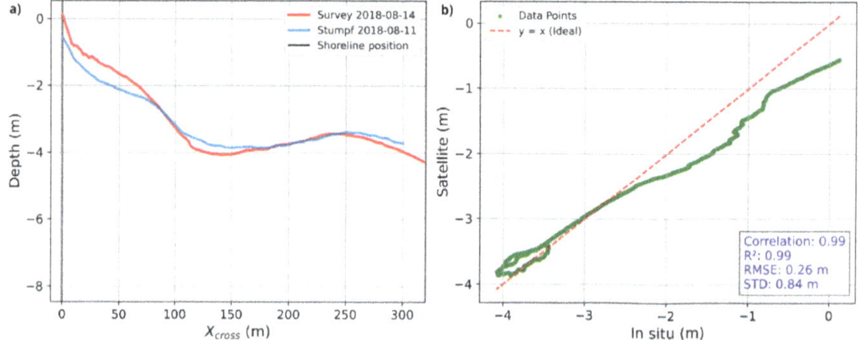

Fig. 3. (a) Comparison of in situ and satellite-derived bathymetry at Duck on August 11 and 14, 2018, respectively, showing depth along cross-shore transect. (b) Validation plot showing the correlation between in situ and satellite-derived depths, including the ideal regression line and statistical validation scores (R2, RMSE, and STD).

4 Discussion and Conclusion

This methodology provides an initial approach for deriving bathymetry of very shallow coastal areas (<300 m) from satellite images, without requiring in situ data for calibration. It still needs validation by exploring its limitations. The breaker height-to-depth ratio should not be considered constant; varying it could improve results, particularly near the upper beach and shoreline, where literature suggests the ratio increases [16, 17]. Nevertheless, this approach paves the way for large-scale temporal monitoring of bathymetry, crucial for studying coastal processes and managing coastal zones, transitioning from a "data-poor" to a "data-rich" field, as [18] stated.

References

1. Bergsma EWJ., et al (2023): S2SHORES: a python library for estimating coastal bathymetry. In: Coastal Engineering Proceedings, vol. 37, sediment.36
2. Wölfl AC, Snaith H, et al.: (2019) Seafloor mapping—the challenge of a truly global ocean bathymetry. Front. Mar. Sci.(6), 283
3. Vousdoukas MI, Mentaschi L, Voukouvalas E, Bianchi A, Dottori F, Feyen L (2018) Climatic and socioeconomic controls of future coastal flood risk in Europe. Nat Clim Chang 8:776–780
4. Mayer L, et al.: (2018) The nippon foundation—GEBCO seabed 2030 project: The quest to see the world's oceans completely mapped by 2030. Geosciences(8), 63
5. Saylam K, Hupp JR, Averett AR, Gutelius WF, Gelhar BW (2018) Airborne lidar bathymetry: Assessing quality assurance and quality control methods with Leica Chiroptera examples. Int J Remote Sens 39:2518–2542
6. Bergsma EWJ, Almar R, de Almeida L PM, Sall M (2019) On the operational use of UAVs for video-derived bathymetry. Coast. Eng.(152), 103527
7. Stumpf RP, Holderied K, Sinclair M (2003) Determination of water depth with high-resolution satellite imagery over variable bottom types. Limnol Oceanogr 48:547–556
8. Lyzenga DR, Malinas NP, Tanis FJ (2006) Multispectral bathymetry using a simple physically based algorithm. IEEE Trans Geosci Remote Sens 44:2251–2259
9. Caballero I, Stumpf RP (2019) Retrieval of nearshore bathymetry from Sentinel-2A and 2B satellites in South Florida coastal waters. Estuarine Coast. Shelf Sci.(226), 106277
10. Poupardin A, Idier D, de Michele M, Raucoules D (2016) Water depth inversion from a single SPOT-5 dataset. IEEE Trans Geosci. Remote Sens. 54:2329–2342
11. Almar R, (2021) Global Satellite-Based Coastal Bathymetry from Waves. Remote Sens.(13), 4628
12. Ruessink BG, Walstra DJR, Southgate HN (2003) Calibration and verification of a parametric wave model on barred beaches. Coastal Eng. 48), 139–149
13. McCowan J.: On the solitary wave. Philos. Mag. (36), 430–437 (1891)
14. Horikawa K, Kuo CT, A study of wave transformation inside surf zone. In: Proc. Coastal Engineering Conference, 10th, 1, pp 217–233 (1966)
15. Nakamura M, Shiraishi H, Sasaki Y (1966) Wave decaying due to breaking. In Proc. Coastal Engineering Conference, 10th, 1, 234–253
16. Stive MJF.: Energy dissipation in waves breaking on gentle slopes. Coastal Eng. (8), 99–127 (1984)
17. Raubenheimer B, Guza RT, Elgar S (1996) Wave transformation across the inner surf zone. J Geophys Res(101), 25589–25597
18. Vitousek,S, Buscombe D. Vos K, Barnard P., Ritchie, A.C, Warrick J.A.: The future of coastal monitoring through satellite remote sensing. Camb. Prisms: Coast. Futures (1), e10 (2023)
19. Frugier S. Almar R, Bergsma EWJ, Granjou A.: SBI: A sandbar extraction spectral index for multi-spectral satellite optical imagery. Submitted in Coastal Engineering

The Effect of Beach Buildings on Decadal Dune Volume Development

Sander Vos[1(✉)], Daan Hulskemper[1], Christa IJzendoorn[2], Alain de Wulf[3], Roderik Lindenbergh[1], and José A. A. Antolinez[1]

[1] Delft University of Technology, Mekelweg 5, 2628 CD Delft, The Netherlands
`s.e.vos@tudelft.nl`
[2] Utrecht university, Princetonlaan 8a, 3584 CB Utrecht, The Netherlands
[3] Ghent University, Sint-Pietersnieuwstraat 33, 9000 Gent, Belgium

Abstract. Dutch beaches are increasingly urbanized with both permanent beach pavilions and seasonal sheds and holiday houses. The effect of these buildings on long term dune development between 1999 and 2024 is studied in this paper along ~ 100 km of coast on the outer delta in the south western part of the Netherlands.

A total of ~ 7000 beach buildings have been manually identified in this period based on satellite images and the time line function of Google earth desktop. The effect of the buildings is determined and analyzed at 477 cross-shore profiles with dune volumes and properties like dune toe, top and heel based on airborne lidar datasets of 1999 and 2024. On natural beaches the dune toe position is derived from profile information, whereas on urbanized beaches near buildings the dune toe is based on the location of the buildings.

Yearly volume changes at the profile locations vary between -10 m^3/m/y and up to 40 m^3/m/y. The results indicate that smaller and standalone buildings allow for larger variations in dune volume changes and suggest that larger buildings and connected buildings impede natural dune dynamics which could impact coastal resilience in the long run.

Keywords: Sandy coast · Urbanization · LiDAR datas

1 Introduction

Sandy beach-dune systems are increasingly urbanized [1, 2] in certain countries with beach pavilions and other recreation/holiday buildings along the dune toe and in the dunes (Fig. 1). In the Netherlands a significant increase in beach sheds, buildings and pavilions has been observed in the last 20 years [3]. Although beach pavilions are part of the Dutch landscape since the 1950's, more and more local municipalities consider.

the beach-dune system as an active economic development zone. This results in an increase of year round beach pavilions and seasonal smaller buildings like beach sheds and holiday houses.

Buildings affect local wind patterns [4, 5] along the dune toe which influences both the local morphology [6] and the shoreward sand transport patterns. In the long run

C. Coelho et al. (Eds.): CD 2025, CRL 41, pp. 167–173, 2026.
https://doi.org/10.1007/978-3-032-15473-6_26

Fig. 1. An example of a Dutch urbanized beach with buildings along the dune toe (©Sander Vos).

this can lead to non-uniform dune development [7] and can impact coastal resilience. Especially when dunes cannot keep up anymore with the expected sea level rise [8] due to increasing urbanization of our coasts, these problems can pose increasing challenges for the future.

Recent research [5–7] has mainly focused on building influences on a local scale but little information is available on larger spatial and temporal scales. In order to study the effect of buildings on a larger spatio-temporal scale, the dune development of a stretch of 100 km of Dutch coast is here analyzed over a 25 year period.

2 Research Method

2.1 Research Area

The influence of buildings on local dune development is researched on the outward delta of the Dutch coast (Fig. 2A) which is located in the south-western part of the Netherlands. The study area consists of 100 km of natural and urbanized sandy beach dune systems. The coastline orientation ranges from N-W-S with dominating NW and SW orientations. A total of 477 dune cross-shore profiles are analyzed (Fig. 2B). For the profile-locations a subset (only the sandy profiles) of the Jarkus (Dutch coastal dataset) cross-shore profile dataset [9] has been utilized. The average distance between profiles is about 200–250 m.

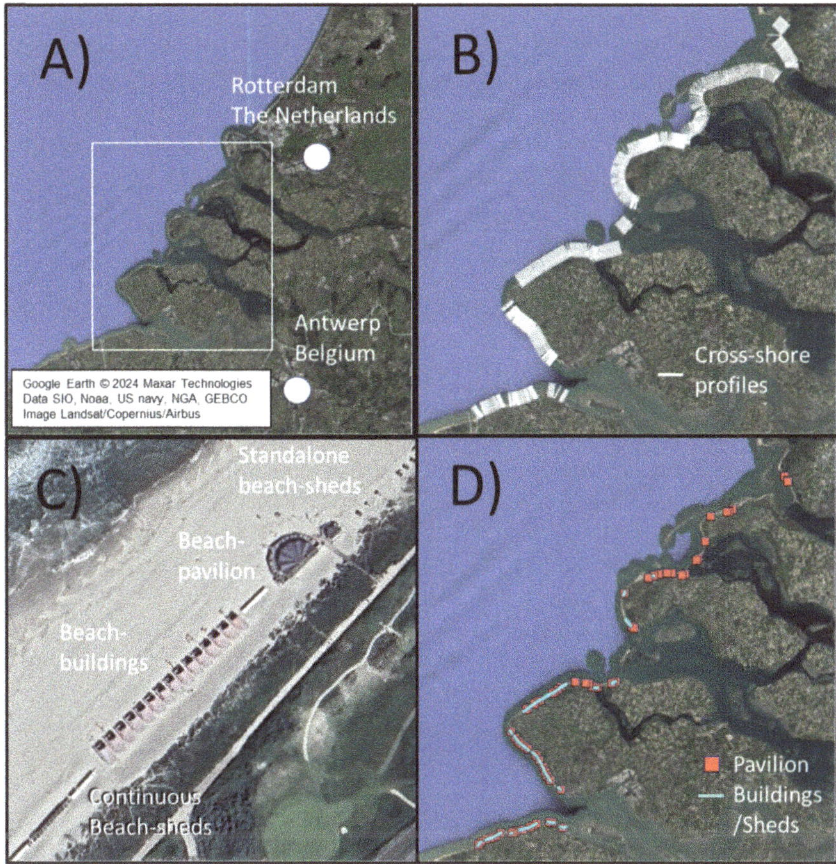

Fig. 2. A) Study area (51.632°N, 3.896°E) along the Dutch coast. B) Cross-shore profiles used in the dune analysis. C) Example of different beach buildings and configurations; Standalone and continuous beach sheds, beach houses and beach pavilions. D) Detected beach buildings (2024) along the Dutch delta.

2.2 Beach Buildings

Beach building locations have been determined manually with the Google earth desktop [GED, 10] software which provided the largest available satellite/aerial archive with submeter resolution. Buildings have been characterized based on size (Fig. 2C). The smallest beach sheds are used typically for personal storage of beach objects (like seats), while the medium beach houses are used as (overnight) holiday homes. Both buildings are seasonal and are removed during the winter storm season and can either be placed continuously (i.e., side to side) or standalone on the beach. Finally, the largest buildings are year-round beach pavilions.

Buildings have been counted either separately (standalone buildings and pavilions) or per unit length. For the unit length it was assumed that beach sheds are on average 2 m wide and beach houses on average 3.5 m wide. Figure 2D gives an overview of all beach seasonal (shed/houses) and permanent (pavilions) buildings along the Delta coast in 2024.

The history of the buildings was determined using the time-line function of GED. For each location the first available detection was based on the first available satellite image with recognizable buildings. On average first detections could be done using satellite images between 2003 and 2005, depending on location.

2.3 Dune Properties

Dune properties like dune toe, top and heel positions were based on a modified scheme of [7] and [9] and dune profiles of 2024. Dune properties were kept constant through time. On natural beaches the location of the dune toe is determined with a modified derivative profile detection method; the first location from the seaward side with a 1st and 2nd derivative less than 0.01 (m/m and m/m^2) or a maximum dune height profile of 6 m. For urbanized beaches with buildings detected within the neighborhood of the profile (distance between profile and buildings less than 20 m) the dune toe is moved 5 m towards the dune beyond the location of the buildings.

The dune top is defined as the first dune top beyond the dune toe that is larger than 6 m and with a prominence larger than 0.5 m. The prominence of the peak indicates how much the peak stands out and is defined as the vertical distance between the peak and its lowest surrounding contour line. Finally, the dune heel (end of the dune) is defined either as either 200 m landwards of the dune top or the end of the profile.

Dune volumes were calculated as the total volume of the dune between the dune toe and dune heel location. An example of detected dune properties is shown in Fig. 3.

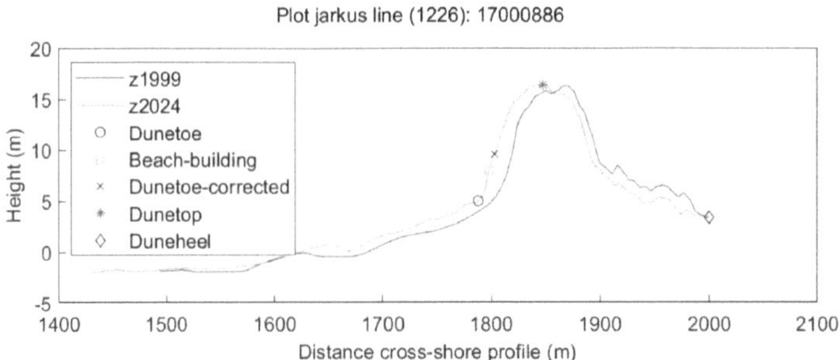

Fig. 3. Example of detected dune properties. The plots shows a cross-shore profile of 2019 and 2024 at Jarkus line 17000886. On the 2024 profile the position of the dune toe based on the derivative method, detected beach building, corrected dune toe, dune top and dune heel are indicated.

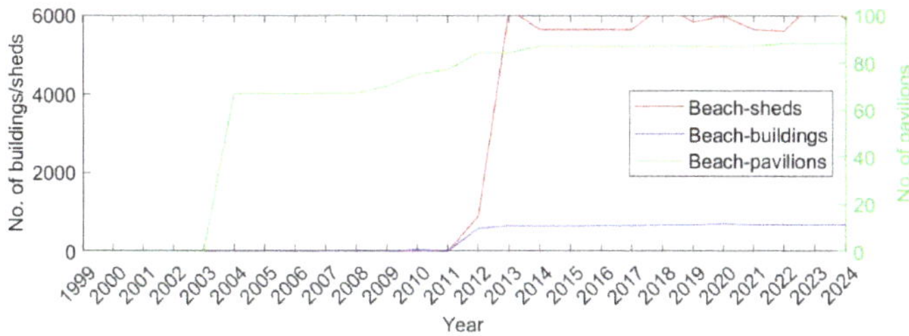

Fig. 4. Timeline of the number of detected beach sheds, buildings, and pavilions in the Dutch outer delta from 1999 till 2024.

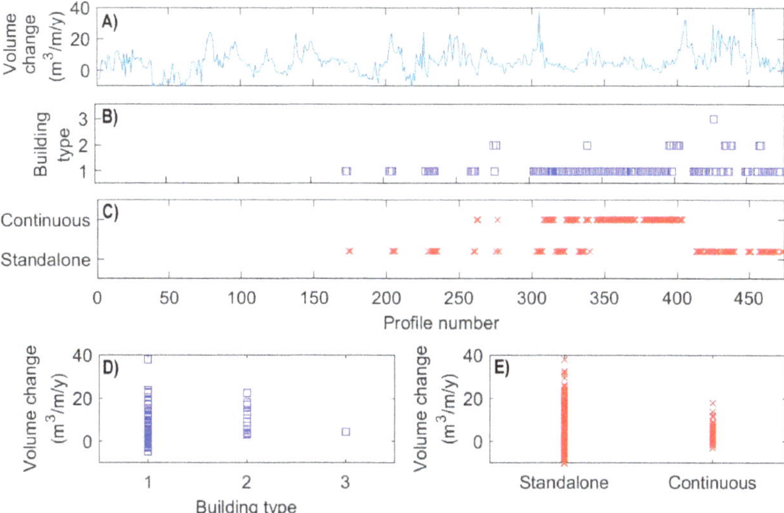

Fig. 5. Yearly volume changes (A) from North to South along the cross-shore profiles in the Dutch delta, with corresponding building type (B) [1) Shed, 2) Beach house and 3) Beach pavilion] and configuration (C) [Continuous or standalone]. Panel (D) and (E) show the relation between detected volume changes versus building type and building configuration.

3 Results

3.1 Beach Buildings

Figure 4 shows an overview of the number of detected buildings through time with a maximum of about 6000 sheds, 600 beach houses and 90 beach pavilions in the Dutch outer delta in 2024. The detection of pavilions is possible from around 2003 when the first high resolution satellite images were available. Detection of smaller buildings is hard in this period as images are only available from the winter period. High resolution summer period satellite images are available starting from 2013 after which beach sheds and buildings can clearly be identified.

Figure 5 shows the yearly volume changes (A) per cross-shore profile from north to south with the detected beach building type (B) and configuration (C). Yearly volume changes range between -10 m^3/m/y (i.e., dune loss) to 40 m^3/m/y while most beach buildings are found between profile 300 and 477.

In conclusion panel 4D and 4E show that building type and continuity have an influence on the observed range of yearly dune volume changes. Smaller and standalone buildings show a larger variation in dune volume changes than larger and continuous connected buildings. This indicates that buildings limit natural dune dynamics which could impact coastal resilience in the long run.

References

1. Hoonhout BM, Waagmeester N (2014) Influence on beach buildings on aeolian sand transport, an inventarisation of methods, measurements and models, Technical Report, Deltares, Delft
2. Malavasi M, Santoro R, Cutini M, Acosta ATR, Carranza ATR (2013) What has happened to coastal dunes in the last half century? A multitemporal coastal landscape analysis in Central Italy, Landscape urban planning 119:54–63
3. van Bergen J, Mulder J, Nijhuis S, Poppema D, Wijnberg K, Kuschnerus M (2021) Urban dunes: Towards BwN design principles for dune formation along urbanized shores. Research in Urbanism Series 7:101–128
4. Hunt J (1971) The effect of single buildings and structures. Philos Trans R Soc 269(1199):457–467
5. Pourteimouri P, Campmans GHP, Wijnberg KM, Hulscher SJMH (2022) A Numerical Study on the Impact of Building Dimensions on Airflow Patterns and Bed Morphology around Buildings at the Beach. J Marine Sci Eng 10
6. Poppema DW, Wijnberg KM, Mulder JPM, Vos SE, Hulscher SJMH (2021) The effect of building geometry on the size of aeolian deposition patterns: scale model experiments at the beach. Coastal Eng 168
7. Vos SE, van IJzendoorn C, Lindenbergh RC, de Wulf (2024) A Non-uniform dune development in the presence of standalone beach buildings. Geomorphology 466
8. Intergovernmental Panel on Climate Change. Climate Change 2021: The Physical Science Basis. Contribution of Working Group I to the Sixth Assessment Report of the Intergovernmental Panel on Climate Change [Masson-Delmotte, V, Zhai, P, Pirani, A, Connors, S, Péan, C, Berger, N. Caud, Y. Chen, L. Goldfarb, M.I. Gomis, M. Huang, K. Leitzell, E. Lonnoy, J.B.R. Matthews, T.K.Maycock, T. Waterfield, O. Yelekçi, R. Yu & B. Zhou], (2021)
9. van IJzendoorn CO, de Vries S, Hallin C, Hesp A (2021) Sea level rise outpaced by vertical dune toe translation on prograding coasts. Nat Sci Rep 11, 12792,.
10. Google earth desktop (software), https://www.google.com/earth, last accessed 3–2–2024
11. RWS (Rijkswaterstaat, Ministry of Public Works), data obtainable via https://download.rij kswaterstaat.nl (Year 1999 and 2024), last visited 15–11–2024
12. Poppema DW, Wijnberg KM, Mulder JPM, Hulscher SJMH (2022) Deposition patterns around buildings at the beach: Effects of building spacing and orientation. Geomorphology 401:108–114

Comparing Process-Based and Reduced Complexity Dune Evolution Models: A Case Study of Long Beach, WA, USA

Selwyn S. Heminway[1,2(✉)], Nicholas T. Cohn[2], Christa van IJzendoorn[3], Peter Ruggiero[3], Meagan E. Wengrove[3], Heather Weiner[4], George M. Kaminsky[4], Sally D. Hacker[3], and Danielle Whalen[3]

[1] Oak Ridge Institute of Science and Education, Corvallis, OR, USA
ssheminway@gmail.com
[2] U.S. Army Engineer Research and Development Center, Duck, NC, USA
[3] Oregon State University, Corvallis, OR, USA
[4] Department of Ecology, Lacey, WA, USA

Abstract. As populations expand in the coastal zone and climate change increases the risk of coastal flooding and erosion hazards, there is a growing need to understand how coastal dunes will evolve into the future. In recent decades, the development and extension of numerical models for dune evolution have demonstrated capabilities to successfully replicate ecomorphodynamic interactions at various spatial or temporal scales. Here, we compare the ability of two numerical models, the analytical Dune Response Tool (DRT) [1] and the process-based AeoLiS model [2], to replicate historical dune growth along a rapidly prograding section of coast in Long Beach, WA. After calibrating coefficients in the erosion and accretion modules of the analytical model, the DRT accurately reproduced historical dune evolution using minimal computational resources. AeoLiS was also able to reproduce patterns in dune growth when detailed field observations and a shoreline change rate module were added to the model. The final DRT profile does not perfectly mimic detailed topographic patterns observed in the field, but it has a lower RMSE than results from AeoLiS. Moreover, DRT's efficiency and minimal calibration requirements make it an easily implementable tool. In contrast, AeoLiS requires extensive field observations as inputs for accurate hindcasts, but the final result better matches the observed complex dune evolution. Both the DRT and AeoLiS models have the potential to offer valuable insight into predicted dune evolution and the coastal protective services dunes may provide in the future.

Keywords: Coastal dune · dune evolution · Dune Response Tool · AeoLiS

1 Introduction

Coastal foredunes often act as the first line of defense for low-lying coastal communities against coastal hazards such as storm surge and wave-driven flooding during storms. As populations expand in the coastal zone and climate change simultaneously increases

© The Author(s) 2026
C. Coelho et al. (Eds.): CD 2025, CRL 41, pp. 174–179, 2026.
https://doi.org/10.1007/978-3-032-15473-6_27

the risk of coastal hazards, there is a growing need to understand how coastal dunes will evolve into the future and if adaptations are warranted to protect infrastructure near the coast. In recent decades, the development and extension of numerical models for dune evolution have demonstrated the capability to successfully replicate physical and vegetation (ecomorphodynamic) interactions at various spatial and temporal scales. For example, process-based dune evolution models (e.g., Coastal Dune Model [3], AeoLiS [2], DUBEVEG [4]) all simulate the accretional behavior of dunes by representing wind flow dynamics, multi-fraction sediment transport, sediment supply effects, and/or the vegetation-sediment interactions that contribute to landform change of coastal dunes. While process-based models can synthesize complex interactions, these simulations are often computationally demanding and prone to compounding errors over long timescales. Comparatively, geometric or reduced complexity dune models (e.g., Dune Response Tool [DRT] [1], CS-model [5], PCR [6]) are based on empirical equations which do not represent as many physical processes as full physics models and therefore do not require as many detailed field observations as input. As such, the simplified frameworks of reduced complexity models are often easier to implement and can provide faster predictions of meso-scale dune behavior.

As process-based and reduced complexity tools address different processes and are generally appropriate for different timescales, limited model intercomparison has been completed to date to determine their relative performance over a broad range of scales. This study aims to compare the ability of AeoLiS and the DRT to hindcast seasonal- to decadal-scale dune evolution along the Long Beach Peninsula (LBP), Washington (WA), USA, which is a sandy, progradational, dune-backed barrier plain. This effort serves as a test of the ability of current ecomorphodynamic models to simulate dune evolution, including erosion and growth, along a high-energy, dissipative coastal dune system.

2 Methods

Quarterly cross-shore transects of beach-dune topography have been collected across the ~ 40 km alongshore extent of LBP since 1997 [7]. These data provide a unique long-term dune evolution dataset for comparing the two modeling tools. Here, we focus on a single profile, located on the southern end of LBP, that is representative of the general pattern of dune growth on the peninsula. Both AeoLiS and DRT were run at an hourly time step from 1997 to 2023. To calibrate the DRT, a Genetic Algorithm approach [8] was used to optimize model free parameters (see Cohn et al. [1]) based on the root mean square error (RMSE) of the upper beach and dune profile. Simulations were run with the available input datasets (as needed for each model [Fig. 1]). Model inputs include dune sand grain size data, collected April 2024, extracted dune morphometrics, and calculated shoreline change rates from long-term beach monitoring at LBP [7]. In addition, transect surveys of vegetation cover (dominated by the beachgrass *Ammophila breviligulata)* collected every 2–3 years from 2012 to 2023 [9] were used as input. Hourly tide, wave, and wind data were respectively sourced from nearby NOAA tide gauges [10], Center of Australian Weather and Climate Research (CAWCR) model output [11], and the nearest ECMWF Reanalysis v5 (ERA5) node [12].

A new shoreline change rate function was added to AeoLiS to create a linear beach below the input dune toe elevation based on the measured beach slope from 1997 and

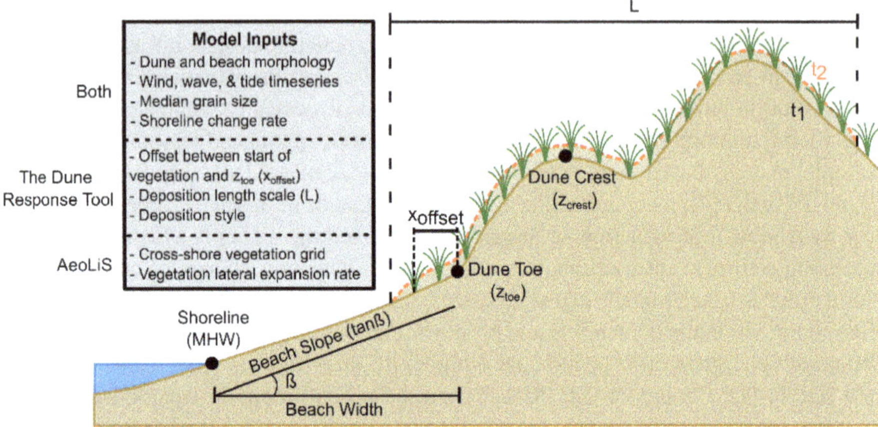

Fig. 1. Schematic of dune profile including key morphometrics and described inputs to each model.

shoreline change rate from 1997–2023. The root mean square error (RMSE) between each modeled and field dune profile is calculated above the 4 m contour (NAVD88), which is the approximate regional dune toe elevation, and the final topography and volume change above 4 m is compared.

3 Results

Both the DRT and AeoLiS models successfully replicate the observed net dune growth along the representative profile (Fig. 2a-d), a transect that has been growing on average 7.0 m³/m/yr. The DRT took minutes to run and, with calibration, was able to replicate dune growth and erosion from 1997 to 2023 with an RMSE of 0.38 m (Fig. 2b). However, DRT underpredicts foredune height (Fig. 2a-b,d). The AeoLiS simulation, which took over an hour to run, models a similar foredune height to field measurements but underpredicts the progradation of the dune toe and has an RMSE of 0.76 m, higher than the DRT (Fig. 2c). However, AeoLiS includes more detailed topographic patterns that qualitatively mimic spatially varying dune topography, including dune ridges, that result from ecological feedback effects not captured by the comparatively smoother DRT profile (Fig. 2d). Both models displayed similar trends in final elevation differences, though differences between the field profile and the AeoLiS profile were more extreme (Fig. 2e). Interannual rates of dune volume growth in DRT and AeoLiS generally matched those of the field calculations, though both models did not capture as much seasonal variation as observed in the field (Fig. 2f).

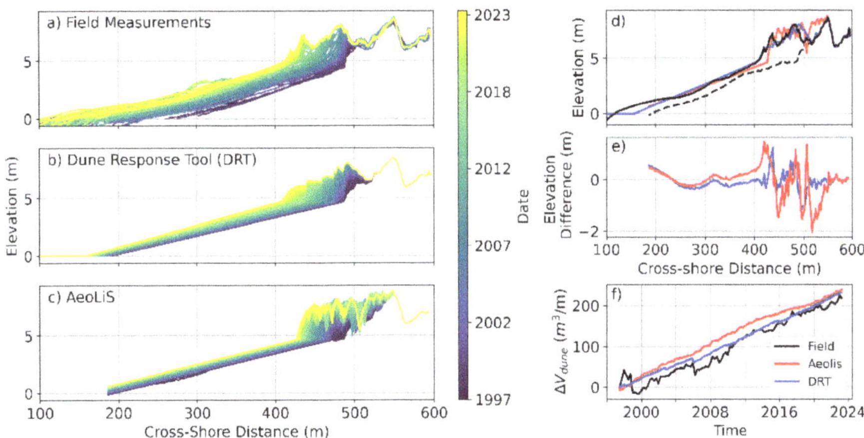

Fig. 2. The measured field dune profiles from 1997 to 2023 (a) compared to the profile evolution modeled by the Dune Response Tool (DRT) (b) and AeoLiS (c). The final measured and modeled dune profiles are plotted (d), the difference in elevation between the final measured and modeled profiles are compared (e) and the associated calculated dune volume change (ΔV_{dune}) above the 4 m contour is shown (f).

4 Discussion

Quantitative tools are necessary for risk-based planning along the coastal margin. It is only recently that numerical models have been developed that can predict coastal dune profile evolution across engineering-relevant timescales, enabling a more robust set of tools to inform coastal planning and adaptation. However, these models have not yet been widely tested across the full range of morphodynamic systems, including along prograding coastal beaches. Here, we show that both the DRT and AeoLiS models can predict dune evolution with reasonable skill. To date, the DRT has been used to simulate longer timescales (decades to centuries) (e.g., [4]), while AeoLiS has generally been used to answer short-term (hours to years), process-based questions (e.g., [13, 14]). To the best of the authors' knowledge, prior to this study, AeoLiS has not successfully reproduced multidecadal dune evolution along prograding coastal areas.

Simulating almost 30 years of dune growth with no increase in beach sediment supply causes inaccurate beach erosion in the AeoLiS model (Fig. 3). Therefore, sediment supply via a shoreline change rate module was added to the model and was successful in replicating the sediment supply that this prograding beach system naturally receives. This underscores the critical importance of sediment supply for dune evolution. Because vegetation plays a key role in capturing aeolian-transported sediment, including vegetation also proved to be a necessary input to simulate complex topographic dune features (Fig. 3). When the shoreline change rate and field cover of vegetation data were included, the model performed well and was able to predict both profile change and trends in dune volume change.

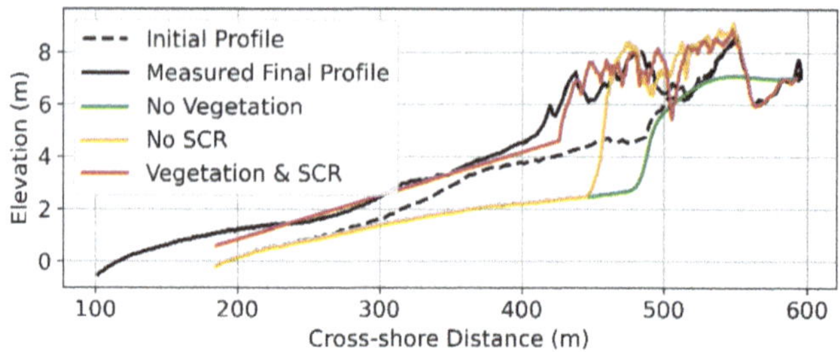

Fig. 3. The final AeoLiS modeled dune profile (red) compared to the modeled profiles without the shoreline change rate (SCR) module (yellow) or vegetation data (green). (Color figure online)

Interestingly, both DRT and AeoLiS predict relatively constant rates of interannual dune growth, whereas the field data show more seasonal and interannual variability in dune volume inputs over time (Fig. 2). Both models exclude the seasonal changes in vegetation cover (i.e., vegetation has lower cover in the winter months) and nearshore morphological variability (i.e., seasonal sand bar formation and onshore welding cycles) observed in the field, potentially explaining the mismatch.

Ultimately, hindcasts with the two calibrated tools yield similar behavior. Future work is aimed at understanding the transferability of calibrated parameter sets to other time periods and sites and further improving the implementation of ecomorphodynamic interactions in AeoLiS to enable longer term simulations in this system where vegetation processes play a predominant role in landform change [15]. Continued exploration of the capabilities and limitations of these dune evolution models will help improve model accuracy and thus their ability to predict dune behavior to inform suitable coastal management actions.

Acknowledgements. This research was supported in part by an appointment to the Department of Defense (DOD) Research Participation Program administered by the Oak Ridge Institute for Science and Education (ORISE) through an interagency agreement between the U.S. Department of Energy (DOE) and the DOD. ORISE is managed by ORAU under DOE contract number DE-SC0014664. All opinions expressed in this paper are the authors' and do not necessarily reflect the policies and views of DOD, DOE, or ORAU/ORISE. This work was also partially supported by the US Army Corps of Engineers (USACE) Grant Number W912HZ2120045 and the USACE Coastal Inlets Research Program (CIRP) Coastal Forecasting to Reduce Infrastructure Flooding (CFRIF) program. Additional support was provided by the Cascadia Coastlines and Peoples Hazards Research Hub, a National Science Foundation (NSF) Coastlines and People Large-Scale Hub (NSF award number 2103713).

References

1. Cohn N, Anderson D, Skaden, J., van Westen B (2023) Simulating tipping points in future dune state with dune response tool. COASTAL SEDIMENTS 2023, 596–608 New Orleans, Louisiana, USA

2. van Westen B, de Vries S, Cohn N, van IJzendoorn C, Strypsteen, G, Hallin, C (2024) AeoLiS: Numerical modelling of coastal dunes and aeolian landform development for real world applications. Environ. Model. Softw. 179, 106093
3. Duran O, Moore LJ (2013) Vegetation controls on the maximum size of coastal dunes. Proc Natl Acad Sci 110(43):17217–17222
4. Keijsers JGS, de Groot AV, Riksen MJPM (2016) Modeling the biogeomorphic evolution of coastal dunes in response to climate change: Modeling coastal dunes. J Geophys Res Earth Surf 121(6):1161–1181
5. Hallin C, Huisman BJA, Larson M, Walstra D-JR, Hanson H (2019) Impact of sediment supply on decadal-scale dune evolution — Analysis and modelling of the Kennemer dunes in the Netherlands. Geomorphology 337:94–110
6. Ranasinghe R, Callaghan D, Stive MJF (2012) Estimating coastal recession due to sea level rise: beyond the Bruun rule. Clim Change 110:561–574
7. Ruggiero P, Kaminsky GM, Gelfenbaum G, Cohn N (2016) Morphodynamics of prograding beaches: a synthesis of seasonal- to century-scale observations of the Columbia River littoral cell. Mar Geol 276:51–68
8. Goldstein EB, Moore LJ (2018) A calibration workflow for coastal dune models. Shore & Beach 86(3):47–51
9. Hacker SD et al (2012) Subtle differences in two non-native congeneric beach grasses significantly affect their colonization, spread and impact. Oikos 121(1):138–148
10. NOAA Tides & Currents Node 9440910 Toke Point, WA. https://tidesandcurrents.noaa.gov/waterlevels.html?id=9440910, Accessed 1 Aug 2025
11. Durrant, T, Hemer, M., Smith, G., Trenham, C., Greenslade, D.: CAWCR Wave Hindcast - Aggregated Collection. v5. CSIRO. Service Collection. (2019)
12. ECMWF Homepage, https://www.ecmwf.int/en/forecasts/dataset/ecmwf-reanalysis-v5, last accessed 2025/1/8
13. van IJzendoorn CO, Hallin C, Reniers AJHM, de Vries S (2023) Modeling multi-fraction coastal aeolian sediment transport with horizontal and vertical grain-size variability. J. Geophys. Res. Earth Surface 128 (7), e2023JF007155
14. Dickey J, Wengrove M, Cohn N, Ruggiero P, Hacker SD (2023) Observations and modeling of shear stress reduction and sediment flux within sparse dune grass canopies on managed coastal dunes. Earth Surf Proc Land 48(5):907–922
15. Zarnetske PL et al (2012) Biophysical feedback mediates effects of invasive grasses on coastal dune shape. Ecology 93(6):1439–1450

The Dynamics of Coastal Dunes Affected by Shoreface Nourishments and Storm Events

Sierd de Vries[1(✉)], Theo Meijer[1], Bas Hofland[1], and Christa van IJzendoorn[2]

[1] Delft University of Technology, Stevinweg 1, 2628CN Delft, The Netherlands
sierd.devries@tudelft.nl
[2] Oregon State University, Corvallis, OR, USA

Abstract. The dynamics of coastal dunes are often affected by a combination of environmental conditions and human interventions. During storms and calm periods, marine and aeolian flows cause sediment transport that results in sedimentation and erosion. Dunes typically grow due to aeolian transport of sediment and erode due to marine forcing by waves and currents. Shoreface nourishments can increase the sediment budget in the coastal profile which may influence the marine and aeolian sediment transport. To what extent shoreface nourishments influence sediment transport and the connected dynamics of coastal dunes is yet unknown. In this paper we investigate coastal dune volume change from yearly profile measurements and relate this to a shoreface nourishment program. The measured dune growth along the entire Dutch sandy coastline is characterized at most places by a break in slope where the dune growth increases after a specific 'breakpoint' year. The derived breakpoint years generally correspond with the start of the nourishment program with a delay of several years (1–5 years). These results provide a starting point for further elaborating on the relationship between the dynamics of coastal dunes and nourishments. The next step would be to delineate the potential influence of nourishments on either the growth of dunes through aeolian sediment transport or the erosion due to marine events.

Keywords: Coastal Dunes · Coastal Nourishments · Storm events · Data Analysis · JARKUS Analysis Toolbox

1 Introduction

The morphology of coastal dunes is dynamic on both short and longer timescales. Short term storm events in the order of days can cause coastal dunes to erode episodically. Long term sediment supply and aeolian wind forcing may cause dunes to grow gradually. Several quantitative studies have shown consistent linear trends in decadal dune growth rates at several locations (e.g. de Vries et. al., 2012; Strypsteen et. al., 2019). However, the spatially varying magnitude of these growth rates on decadal scales cannot be explained yet.

The offshore and beach sediment budget is connected to the erosion and sedimentation of dunes through mass conservation in a sediment sharing system. An increase in

C. Coelho et al. (Eds.): CD 2025, CRL 41, pp. 180–184, 2026.
https://doi.org/10.1007/978-3-032-15473-6_28

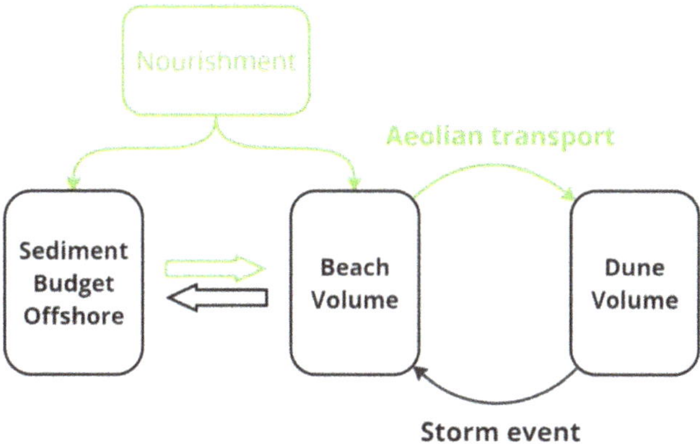

Fig. 1. The conceptual model of the links between the (nourished) offshore and beach volume and the dune volume. The dune volume may experience indirect effects of nourishments due to increased aeolian transport towards the dunes or reduced storm erosion.

sediment volume in the foreshore may therefore stimulate the growth of coastal dunes indirectly by reducing storm erosion and/or increase the sediment supply to the aeolian system (see conceptual Fig. 1). Decadal trends in dune growth rates could therefore be affected by the offshore and beach sediment budget.

Shoreface nourishments can influence offshore and beach sediment budgets (see Fig. 2). Along the Dutch coast, a sustained national nourishment program is in place since 1990 (Brand et. al., 2022). This national nourishment program aims to compensate for structural sediment erosion in the coastal profile. The program has involved the systematic nourishment of the coastal profile, in the order of 10 million cubic meters of sediment added to the coastal system on a yearly basis, for several decades.

Fig. 2. Photo shows an example of nourishments of the coastal profile feeding the offshore sediment budget (photo courtesy of Boskalis).

Measurements of the offshore, beach and dune morphology are required to analyze their interactions in the coastal profile. The extensive dataset of morphological measurements collected along the Dutch coastal zone (JARKUS; since 1965 to date) can be used to study sediment exchange between foreshore, beach and dune.

In this paper the decadal development of coastal dune volumes affected by shoreface nourishments and storms is investigated. The extensive dataset and the nourishment program in the Netherlands provide a unique opportunity to unravel the effects of shoreface nourishments and storms on decadal scale dune dynamics. The outcomes of this study may inform future nourishments and nourishment programs.

2 Methods

For this study the Dutch JARKUS dataset (JAaRlijkse KUStmeting) is employed. The JARKUS dataset consists of topographic measurements that include foreshore, beach and dune area along approximately 350km coastline collected yearly since 1965. All yearly data are interpolated to fixed transects that are spaced 200-250m alongshore. The JARKUS Analysis Toolbox is used to analyze the measured topographic profiles (see Fig. 3 for an example). Van IJzendoorn et. al. (2021) describe the JARKUS dataset and 'JARKUS Analysis Toolbox' (JAT) in more detail. Dune volumes at 1033 transect locations are derived for every available year according to the procedure described by De Vries et. al. (2012). Only transect locations that contain more than 40 years of data are considered in the analysis. This attributes to over 47000 analyzed profile measurements distributed over time and space along the Dutch coast.

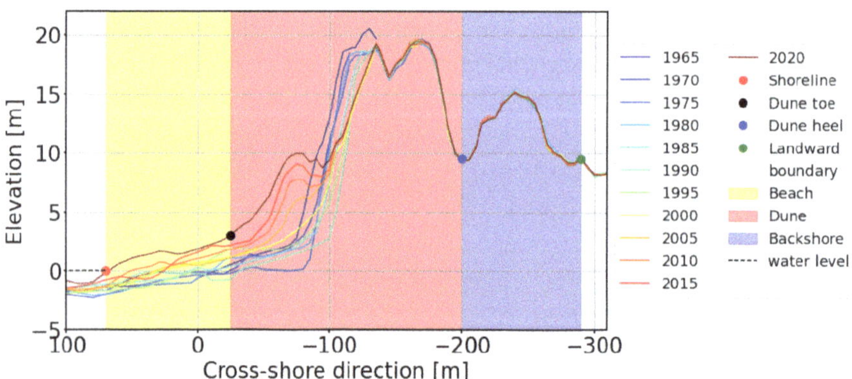

Fig. 3. Example of a JARKUS-transect showing the measured development of beach and dune topography. Elevation is normalized at 0 m NAP (about mean sea level). Dune growth is shown to be relatively large after 1990 compared with earlier years (transect ID 8009700).

For every transect location a piecewise linear relationship is fitted where it is hypothesized that after the start of the nourishment program a break in trend of dune volume development is caused. The year of the break in trend and the magnitude of the trends before and after the 'break point year' are the free fitting parameters that are optimized by minimizing the root mean square error of the piecewise fits combined.

3   Results

The extracted timeseries of the development of dune volumes indeed show a break in trend at most transect locations along the Dutch coast (see Fig. 4 for a clear example).

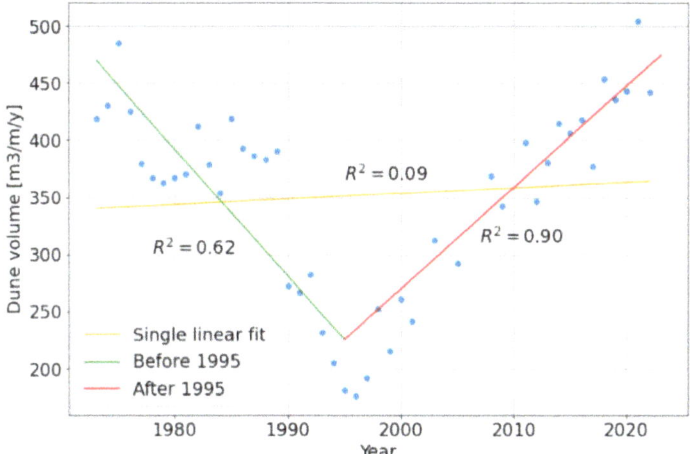

Fig. 4. The derived piecewise linear fit of dune volume development in time with a break in trend at 1995 (transect ID-7002965). Before 1995 the dune volume decreased by about 10 m^3/m/yr and after 1995 the volume increased by about 10 m^3/m/yr

Applying the piecewise linear fit to the extracted dune volume timeseries at all 1033 transect locations leads to a generally large increase in fitting quality compared to the normal linear fit. The 'break point year' seems to occur mostly around 1995 when considering all 1033 transect locations (Fig. 5).

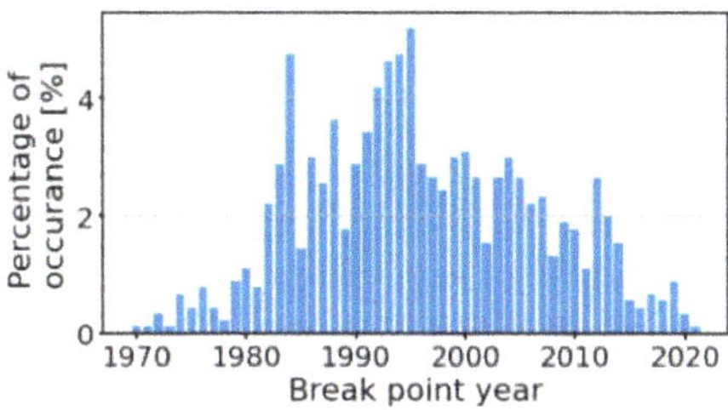

Fig. 5. The percentage of occurrence of the break point year when applying the piecewise linear fit to the 1033 considered transect locations. Most occurring break point years seem to be around 1990–1995.

4 Discussion and Conclusion

The current bulk-analysis of the dune volume development and the piecewise linear fits suggests that the nourishment program along the Dutch coast may have caused a break in trend of dune volume development. This break in trend seems to occur mostly around 1990–1995 at many locations. The break in trend may be explained due to the start of the national nourishment program in 1990. The nourishment program may have resulted in a shallower foreshore that could limit storm erosion and accretion of the beach which could have increased the sediment supply for aeolian sediment transport. The effects of individual storms and their aftermaths remain unclear. As the nourishment program may have stimulated dune growth, the results of this study have important implications for the integral management of dynamic coastal dunes as part of the sediment sharing coastal profile (foreshore-beach-dune).

References

Brand E, Ramaekers G, Lodder Q (2022) Dutch experience with sand nourishments for dynamic coastline conservation–an operational overview. Ocean Coast Manag 217:106008

De Vries, S, Southgate HN, Kanning W, Ranasinghe RWMRJB (2012) Dune behavior and aeolian transport on decadal timescales. Coast Eng 67:41–53

Strypsteen G, Houthuys R, Rauwoens P (2019) Dune volume changes at decadal timescales and its relation with potential aeolian transport. J Marine Sci Eng 7(10):357

van IJzendoorn CO, de Vries S, Hallin C, Hesp PA (2021) Sea level rise outpaced by vertical dune toe translation on prograding coasts. Sci Rep 11(1), 12792

Can Global Datasets Be Used to Predict Storm-Induced Coastal Erosion?

Valeria Fanti[1(✉)], Óscar Ferreira[1], and Carlos Loureiro[1,2]

[1] CIMA/ARNET, Faculdade de Ciências e Tecnologia, Universidade do Algarve, 8005-139 Faro, Portugal
vfanti@ualg.pt

[2] School of Agricultural, Earth and Environmental Sciences, University of KwaZulu-Natal, Durban 4000, South Africa

Abstract. The scarcity of high-resolution in situ data hampers effective storm impact assessment and the development of early warning systems for coastal barriers worldwide. This study introduces a novel methodology utilising a SWAN-XBeach modelling chain, leveraging global hydrodynamic (WAVERYS and GTSM) and topo-bathymetric (TanDEM-X and ETOPO2022) datasets to perform first-order, globally applicable storm impact assessments for barrier coasts. The approach is validated with high-resolution data from Duck, North Carolina, using pre- and post-storm topographic LiDAR, recorded wave conditions and water levels.

The results indicate that global datasets can reasonably reproduce barrier erosion patterns, despite underestimation of total water levels and dune morphology due to the coarse resolution and inaccuracies in the GTSM and TanDEM-X data. The simulations effectively capture dune retreat and erosion trends, aligning with measured data, but inconsistencies are observed in morphological changes for the upper beach face. These discrepancies are likely due to differences in intertidal elevation and beach face steepness between the LiDAR data and global models at Duck. Additionally, the approach does not account for short-scale (~100 m) longshore variability, which can contribute to deviations in modelled results.

This study demonstrates the potential of global datasets for storm impact modelling, providing a first-order assessment with relevant information about storm-induced erosion, which can be particularly useful in areas lacking high resolution data. However, it also highlights the need for further validation across diverse sites to enhance reliability and address methodological constraints.

Keywords: Coastal Erosion · Storm Impact Assessment · Barrier Islands · XBeach Modelling · Global Datasets

1 Introduction

Barrier islands account for more than 10% of the world's open coastline [1], and with coastal demographic and socio-economic pressures and climate change, their vulnerability to erosion from extreme storms and sea-level rise will increase. At a local level, the

C. Coelho et al. (Eds.): CD 2025, CRL 41, pp. 185–190, 2026.
https://doi.org/10.1007/978-3-032-15473-6_29

geomorphological impact of extreme coastal storms in barrier islands can be predicted with increasing accuracy using calibrated numerical models and high-resolution in situ oceanographic and topo-bathymetric data [2]. At a global level, the lack of high-quality in situ datasets and large spatio-temporal data gaps hinder the assessment of storm-related risks and the implementation of storm early warning systems for most coastal barriers worldwide.

Despite such limitations, readily and increasingly available global hydrodynamic and topo-bathymetric data may enable preliminary coarse-resolution storm impact assessments, which represent a low-cost and low-effort solution. However, uncertainties in the simulated results will arise both from the coarseness and inherent errors in the global datasets, as well as from the simplifications and assumptions required for a globally applicable approach.

This study presents a novel methodology based on SWAN and XBeach modelling with input data from global wave, tide and surge models and topo-bathymetric datasets, validated by comparison with high-resolution datasets for Duck, North Carolina.

2   Methodology

A simplified XBeach 1D approach was developed to assess the erosional response of storms at any barrier island worldwide (Fig. 1). This approach was implemented and tested on a natural sector of Duck, in the Outer Banks barrier islands of North Carolina (Fig. 2a), where pre- and post-storm topographic LiDAR data were available from CLARIS surveys from 2012 to 2022 [3].

Firstly, a storm de-clustering algorithm [4] was applied to significant wave heights from the WAVERYS 3-hourly wave heights available from 1993 to 2019 at around 22 km grid resolution and to the Virginia Beach buoy (NDBC station 44014), available at hourly resolution from 1990 to 2024 to select extreme events (95th percentile threshold) with available pre- and post-surveyed profiles a few days apart from the storm event itself. An additional storm wave direction criterion was used to include only north-easterly storms to avoid pier interference. The water levels from the Duck tide gauge with hourly data from 1991 to 2023 were used to validate the Global Tide and Surge Model (GTSM) that is available for the same period at around 25 km spatial resolution.

The WAVERYS wave characteristics of the 8 identified storms were calibrated using a previously developed global calibration equation [5], then propagated to 30 m depth with SWAN and used as boundary conditions with tide and surge data from GTSM for a 1D surfbeat XBeach simulation with default parameters. The topo-bathymetric profile of the barrier was derived by averaging cross-shore 20 m spaced TanDEM-X profiles over a natural homogeneous sector of the barrier island (~150 m, red profiles in Fig. 2a) and merging this with ETOPO2022 bathymetry using the Dean equilibrium profile until the depth of closure, estimated from WAVERYS significant wave height and peak period. The results from the XBeach simulations with the global models were validated against cross-shore LiDAR profiles extracted for the study area.

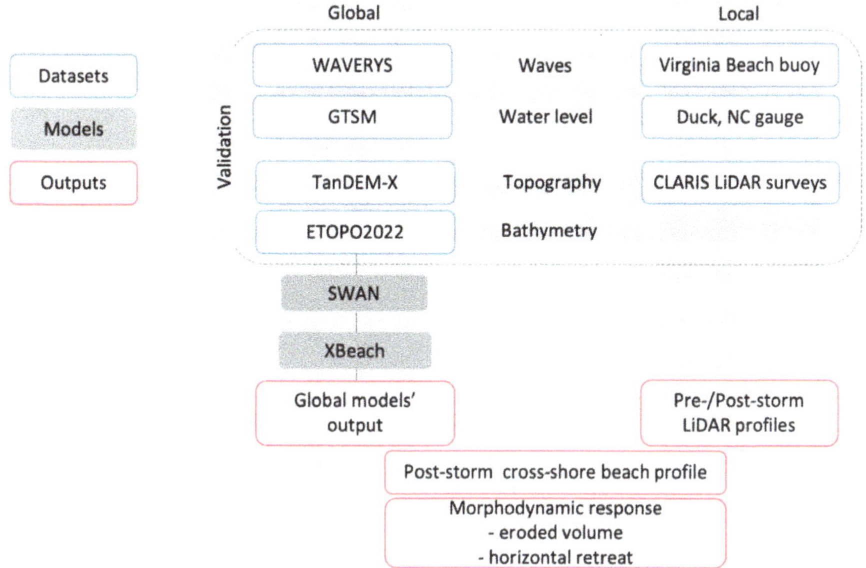

Fig. 1. Schematic representation of the methodology applied to the barrier island of Duck in North Carolina for validation of a storm impact assessment performed with global models.

3   Results

For all 8 storms, the wave conditions extracted from WAVERYS were in good agreement with the nearby Virginia buoy measurements (Fig. 2.b), while GTSM showed an underestimation of the total water level compared to the local gauge (Fig. 2b). The profile dune shape was poorly represented by TanDEM-X (Fig. 3a), with an underestimation of the dune top by over 1 m, and also of the intertidal area by up to 2 m, with a consistently shallower beach face slope compared to the LiDAR profiles. Despite this, simulated storm response for nearly all profiles evidences the occurrence of retreat (Fig. 3b) and erosion (Fig. 3c) of the dune face, in good agreement with the values obtained from the LiDAR data. The modelled profiles also show an accumulation of sediment at the upper beach face, which was not observed in the measured profiles.

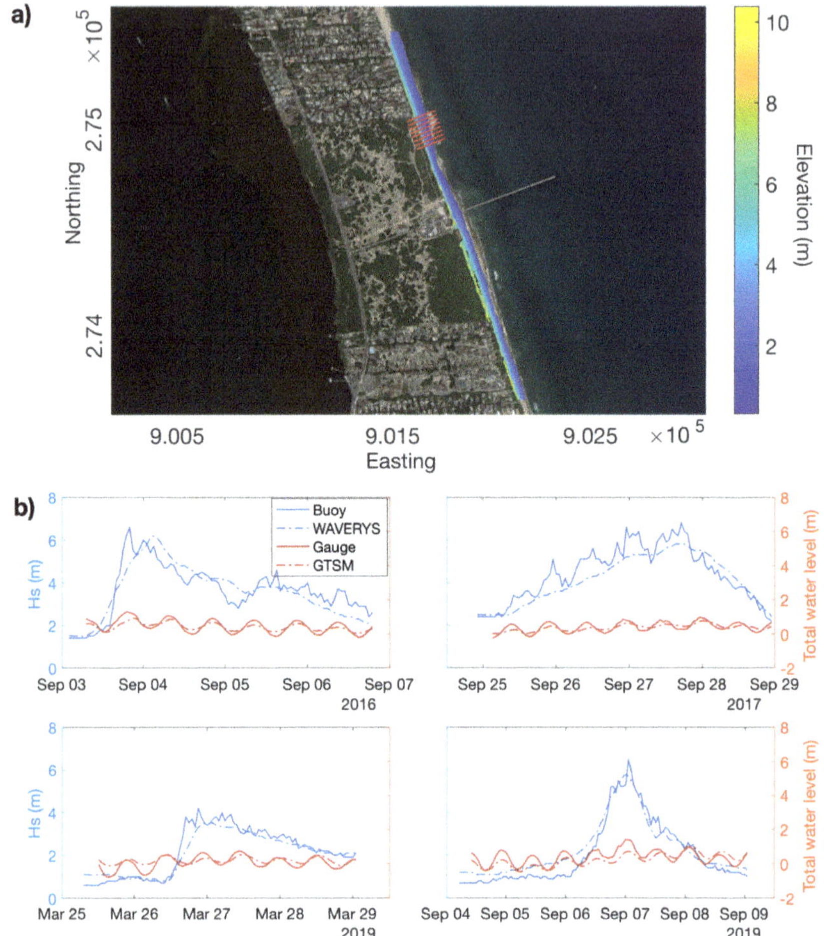

Fig. 2. a) ESRI Satellite aerial image of the barrier at Duck, North Carolina with CLARIS survey (22 September 2017) and cross-shore sampled profiles. b) Examples of storm significant wave height (blue) from Virginia buoy and WAVERYS as input for SWAN and total water level (red) from Duck tide gauge and GTSM as input for XBeach.

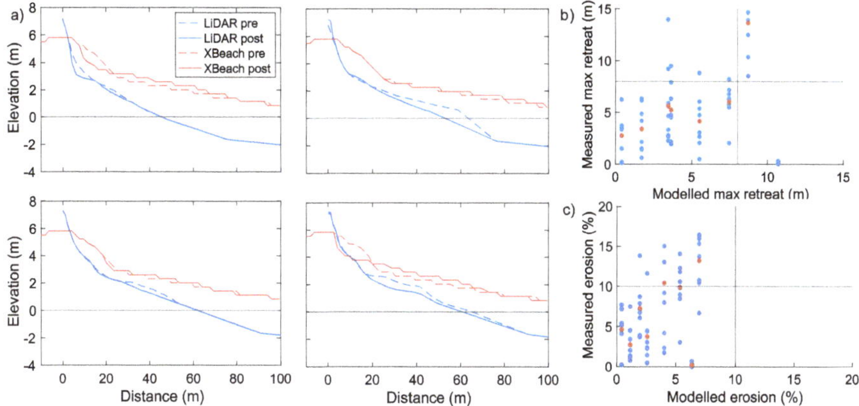

Fig. 3. a) Pre- and post-storm topo-bathymetric profile for 4 of the storms using global models as an input for XBeach and high-resolution profile data. b) Modelled versus measured maximum dune retreat with the mean for all storms in red. c) Modelled versus erosion relative to total volume (in %) with the mean for all storms in red. Modelled results based on the averaged profile from global models against all individual profiles from the CLARIS dataset available for the study area.

4 Discussion and Conclusion

Despite the differences between LiDAR and TanDEM-X derived dune and beach morphology and the use of default XBeach parameters, the model results provide valuable information on the occurrence of erosion, the magnitude of maximum retreat (Fig. 3b) and the percentage of subaerial eroded volume to total dune volume (Fig. 3c). However, as the methodology proposed is based on alongshore averaged profiles from global models, there is an intrinsic inability to account for alongshore variability. This variability was significant in the measured profiles but challenging to capture with TanDEM-X topography due to its 12 m resolution and the presence of data inconsistencies at the shoreline. In addition, beach nourishments occurred in the study area, which may have altered the profile, limiting the extent to which a fixed global topo-bathymetry can represent an evolving coastal morphology.

Future work is necessary to confirm the validity of the proposed methodology in multiple sites, which is necessary to evaluate the ability to robustly identify the occurrence and magnitude of erosional events and extend it to locations where high-resolution in-situ data is absent.

References

1. Stutz ML, Pilkey OH (2011) Open-ocean barrier islands: global influence of climatic, oceanographic, and depositional settings. J Coastal Res 27(2):207–222
2. Sherwood CR, Van Dongeren A, Doyle J et al (2022) Modeling the morphodynamics of coastal responses to extreme events: what shape are we in? Ann Rev Mar Sci 14:457–492
3. Spore, N. J., Renaud, A. D., Conery, I. W. & Brodie K. L. Coastal Lidar and Radar Imaging System (CLARIS) Lidar Data Report: 2011–2017 (2019)

4. Kümmerer V, Ferreira Ó, Fanti V, Loureiro C (2024) Storm identification for high-energy wave climates as a tool to improve long-term analysis. Clim Dyn 62(3):2007–2226
5. Fanti V, Ferreira Ó, Kümmerer V, Loureiro C (2023) Improved estimates of extreme wave conditions in coastal areas from calibrated global reanalyses. Communications Earth & Environment 4(1):151

Decoding Storm Response in Geologically Controlled Coastal Barriers Using XBeach

Vincent Kümmerer[1]([✉]), Carlos Loureiro[1,2], and Óscar Ferreira[1]

[1] Centro de Investigação Marinha e Ambiental (CIMA/ARNET), Universidade do Algarve, 8005-139 Faro, Portugal
vkuemmerer@ualg.pt

[2] Geological Sciences, School of Agricultural, Earth and Environmental Sciences, University of KwaZulu-Natal, Durban 4000, South Africa

Abstract. Storm-induced morphological changes in coastal barriers are primarily driven by hydrodynamic forcing. However, in rock-fronted coasts the presence of geological features can modulate storm impacts, but the extent to which these non-dynamic variables influence barrier morphological response to storms is still poorly understood. This study explores the influence of intertidal rock platforms on storm-driven morphological changes. Using the process-based XBeach model, dune erosion during an energetic winter storm was simulated across different schematic cross-shore profiles, incorporating variable intertidal rock platform configurations. The schematic profiles are based on existing barrier profiles in the Outer Hebrides, NW Scotland. Calibration of the model involved adjusting friction coefficients of exposed bedrock surfaces to improve the ability to reproduce observed nearshore wave heights and morphological changes in pre- and post-storm profiles. The results show that higher intertidal rock platforms (above mean sea level) are associated with reduced dune erosion, while lower rock platforms are associated with more extensive dune erosion. This highlights the complex interaction between geological control and storm impacts, with implications for improved understanding of coastal resilience and storm-driven morphosedimentary dynamics in rock-fronted barrier systems.

Keywords: Geological Control · Coastal Barriers · Storm Impacts · XBeach

1 Introduction

Storm-induced morphological changes in coastal barriers are primarily determined by the interaction between the hydrodynamic forcing conditions and the mobile sediment. However, the presence of irregular non-erodible surfaces on rock fronted beaches influences nearshore hydrodynamics and morphodynamic processes [1, 2], resulting in spatially variable morphological changes [3, 4]. Process-based modelling is nowadays commonly used to predict storm-induced morphological changes at unconstrained coastal barriers, but there are still few studies for geologically controlled barriers. Therefore, exploring the influence of intertidal rock platform configurations on morphological responses during storms in such barriers through process-based morphodynamic modelling can help

© The Author(s) 2026
C. Coelho et al. (Eds.): CD 2025, CRL 41, pp. 191–195, 2026.
https://doi.org/10.1007/978-3-032-15473-6_30

to understand spatially varying morphological changes in geologically controlled barrier settings.

Along the coastline of the Outer Hebrides (NW Scotland), a chain of barrier islands developed on a gently sloping undulated bedrock that extends from the coast to ~ 50 km offshore [5]. This coastline is exposed to frequent and extreme coastal storms [6], providing the ideal setting to investigate the understudied role of geological control on storm response in barriers positioned on top of bedrock surfaces. This work investigates storm-driven dune erosion for a range of schematic profiles representative of the Hebridean coastal barriers using the process-based XBeach model.

2 Methods

XBeach was implemented in 1D surfbeat mode to model the morphological response to storms on cross-shore profiles with varying rocky foreshore configurations. Boundary conditions were derived from SWAN, forced by buoy and tide gauge data from a storm event in February 2022 ($H_S \geq 12$ m, $T_p \geq 18$ s). The cross-shore profile was derived from the integration of EMODnet bathymetric data and RTK-DGNSS topographic surveys of pre- and post-storm barrier profiles. The control exerted by the natural bedrock and the schematic rock platforms was numerically implemented by: i) setting the bedrock as an unerodable layer; ii) increasing the wave dissipation through higher bed (c_f) and wave (f_w) friction coefficients in the rock sections; and iii) lowering the breaker index (γ) in the rock sections (Fig. 1a). The sandy part of the profile was set to default values.

The calibration of the bedrock-induced wave dissipation was conducted in two steps. First, the short-wave related friction f_w on bedrock surfaces was calibrated by an optimization routine through the comparison of modeled and observed incident-band nearshore wave heights in the intertidal beach profile. The observed nearshore wave data were obtained from pressure transducer measurements during a spring high-tidal cycle in February 2023. In the second calibration step, 81 combinations of different γ and c_f values on bedrock surfaces were tested for a profile where storm-induced morphological changes were observed during the February 2022 storm event (P_{N1}, Fig. 2). The γ and c_f combination that performed better in reproducing observed morphological changes was validated by comparison of the modelled storm response against observations at two alongshore adjacent profiles. After calibration and validation, four schematic profiles were developed by varying the position of an intertidal rock platform (P_{S1-S4}, Fig. 2). The platform geometry was kept constant, but its elevation was determined according to different tide levels (Fig. 1a), with higher elevations increasing the level of the geological constraint. The schematic cross-shore profiles are representative of existing barrier profiles in the Outer Hebrides (Fig. 1b,c).

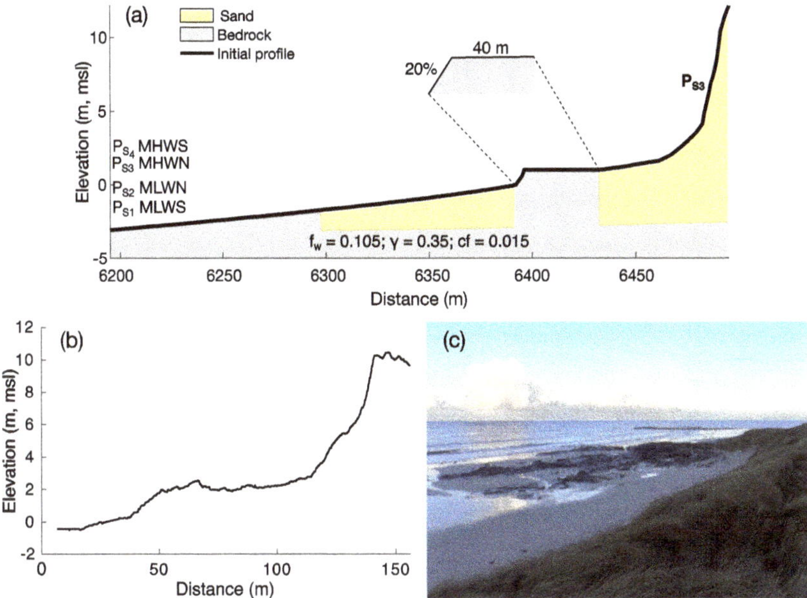

Fig. 1 (a) XBeach profile configuration and geometry of the schematic rock platform, for P_{S3}. (b) Example of a cross-shore profile with an intertidal rock platform in the Outer Hebrides. (c) Photograph of the example intertidal rock platform.

3 Results and Discussion

The predicted incident-band wave heights are in excellent agreement with the observations when f_w is set to 0.105 on the rocky profile sections. In the second calibration step, the combination of $c_f = 0.015$ and $\gamma = 0.35$ for the rock sections resulted in modelled morphological changes that closely matched the observed post-storm profile at P_{N1} and predicted muted morphological response at the adjacent profiles, consistent with field observations. These friction coefficients are within the range of values reported for hydrodynamic modelling with XBeach on reef and rock fronted coastlines [7, 8].

The morphological change predictions obtained using the schematic profiles, show that rock platforms with elevation above the mean sea level (P_{S3-4}) result in dampened dune erosion (Fig. 2), despite the extreme hydrodynamic forcing conditions. No significant reduction in dune erosion is predicted for P_{S1} and P_{S2}, which have rock platforms below this level. The elevation of the intertidal rock platform is found to be a key factor in controlling storm-induced barrier morphological change. This suggests that the relationship between the elevation of the intertidal rock platform and the total water level during the storm is a primary factor affecting storm impacts in rock fronted barrier sectors in the Outer Hebrides.

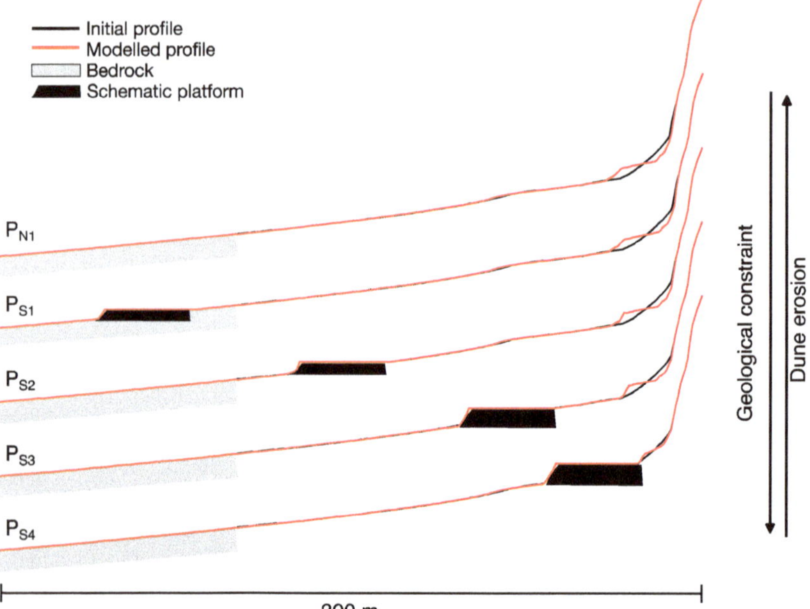

Fig. 2 Modelled storm response of profile P_{N1} (original profile) and the schematic profiles P_{S1}-P_{S4}.

4 Conclusion

This work demonstrates that XBeach can predict spatially variable storm-induced dune erosion during extreme storm events in a highly geologically controlled barrier once wave transformation on bedrock surfaces is calibrated. This enabled testing storm responses of both measured and schematic profiles to disentangle the influence of intertidal rock platforms on barrier morphodynamics during storm events. Model results indicate that increased geological control exerted by intertidal rock platforms at higher tidal elevations (above mean sea level for the tested conditions) contributes to lower storm-induced dune erosion, highlighting the interdependency between dynamic forcing and geomorphological characteristics in modulating storm impacts. Further research with different geological control configurations and forcing conditions will contribute to improved evaluation of the degree of geological control on storm-induced morphological changes of rock-fronted barriers.

References

1. Velegrakis AF et al (2016) Shoreline variability of an urban beach fronted by a beachrock reef from video imagery. Nat Hazards 83:201–222
2. Robinet A, Castelle B, Idier D, Harley MD, Splinter KD (2020) Controls of local geology and cross-shore/longshore processes on embayed beach shoreline variability. Marine Geol (422): 106118

3. Muñoz-Perez JJ, Medina R (2010) Comparison of long-, medium- and short-term variations of beach profiles with and without submerged geological control. Coast Eng 57:241–251
4. Gallop SL, Bosserelle C, Eliot I, Pattiaratchi CB (2012) The influence of limestone reefs on storm erosion and recovery of a perched beach. Cont Shelf Res 47:16–27
5. Cooper JAG, Jackson DWT, Dawson AG, Dawson S, Bates CR, Ritchie W (2012) Barrier islands on bedrock: A new landform type demonstrating the role of antecedent topography on barrier form and evolution. Geology 40:923–926
6. Kümmerer V, Ferreira Ó, Fanti V, Loureiro C (2024) Storm identification for high-energy wave climates as a tool to improve long-term analysis. Clim Dyn 62:2207–2226
7. McCall R, Masselink G, Austin M, Poate T, Jager T (2017) Modelling indicent-band and infragravity wave dynamics on rock shore platforms. In: Coastal Dynamics 2017, vol 144, pp 1659–1669
8. Drost EJF, Cuttler MVW, Lowe RJ, Hansen JE (2019) Predicting the hydrodynamic response of a coastal reef-lagoon system to a tropical cyclone using phase-averaged and surfbeat-resolving wave models. Coastal Eng. (152): 103525

Long-Term Pocket Beach Morphology
at the Romanian Coast - Field Measurements and Numerical Modelling

W. P. Bodde[1(✉)], C. W. T. van Bemmelen[1], P. G. F. Brandenburg[2], and E. Quataert[3]

[1] Witteveen+Bos, Daalsesingel 51C, 3511 SW Utrecht, The Netherlands
`willem.bodde@witteveenbos.com`
[2] Van Oord DMC, Schaardijk 211, 3063 NH Rotterdam, The Netherlands
[3] Deltares, Boussinesqweg 1, Delft 2629 HV, The Netherlands

Abstract. This paper investigates the long-term morphodynamics of a designed beach near Constanta, Romania to assess the design approach. Field measurements, including bathymetric surveys and nearshore wave and current data, were collected from 2015 to 2020. The study utilizes the XBeach numerical model to simulate morphodynamics and sediment loss under various wave conditions. Results indicate that XBeach can predict sediment losses reasonable well and morphological responses over a five-year period, although certain phenomena like swash bar formation and erosion behind structures are not fully captured. Sensitivity analyses highlight the influence of parameters such as horizontal background viscosity, sediment grain size, and wave energy on model predictions of sediment loss. The findings confirm the applicability of XBeach and provide a first step to validate the model for this type of design applications.

Keywords: beach morphodynamics · XBeach · beach design · nourishments

1 Introduction

Along the Romanian coast of the Black Sea a large number of beaches is being rehabilitated. The objective of the project is to increase coastal safety and recreational attractiveness. The design consists of breakwaters and beach nourishments, resulting in a series of semi-protected pocket beaches along the coast. It is aimed at a beach geometry that is in alongshore and cross-shore equilibrium such that maintenance volumes are minimized. To realize the optimal geometry, the relevant processes and timescales for morphological changes in the pocket beaches need to be understood and quantified.

This project requires a novel approach to forecast long term morphological developments and sediment loss in complex geometries. Typically, the empirical Parabolic Bay Shape Equation is applied for equilibrium pocket beach design [1]. Since it provides the equilibrium, variation in shoreline position due to wave conditions and nearshore processes is not taken into account. In previous work it was shown that it is possible to model the morphological response of a pocket beach with process-based numerical

© The Author(s) 2026
C. Coelho et al. (Eds.): CD 2025, CRL 41, pp. 196–202, 2026.
https://doi.org/10.1007/978-3-032-15473-6_31

modelling using Delft3D or XBeach [2, 6]. Since infra-gravity waves play an important role in dry beach and shoreline morphodynamics and retreat in the project, XBeach was applied to assess the response to storm events. It was qualitatively demonstrated that the approach with XBeach led to a stable beach design based on behavior of constructed beaches in an earlier phase of the project [3]. XBeach has been calibrated and validated predominantly for dune erosion and overwash processes during individual storm events [4, 5], but not for morphology of the shoreface in a complex geometry. The phased construction of the rehabilitation project (2012-present) has allowed for gathering data over multiple years which can be used to validate the design approach with XBeach.

The objective of this research is to evaluate the effectiveness of XBeach, quantify uncertainties and relate these to relevant processes.

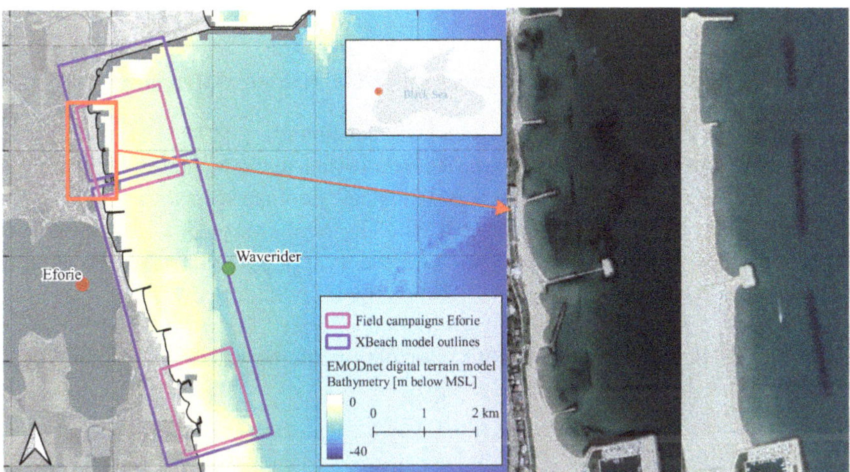

Fig. 1. Left: location field site at Eforie. Right: beach Eforie North before and after construction

2 Materials and Methods

This research considers the beach at Eforie North (Fig. 1), which was completed in 2015. Comprehensive bathymetric surveys of the full area were conducted post-construction in 2015, 2016 and 2020. Field campaigns were conducted to measure nearshore waves and currents. Offshore timeseries of waves for the period 2015–2022 were obtained from a SWAN hindcast and calibrated and validated by nearshore directional waverider buoys. Sediment samples were collected pre-construction at locations along the coastal stretch and the nourishment borrow area for grain size data, resulting in D50 = 0.34mm on average. Over time it is observed that sediment sorting takes place as a large swash bar is formed along the shoreline containing mostly shell fragments.

This paper starts with a description of the wave conditions and observed morphological developments based on the bathymetric surveys since completion in 2015. A XBeach model was then set up for Eforie North and the observations were compared to hindcast results for the two periods between surveys (2015 2016 and 2016–2020), focusing on sediment loss to assess the performance of XBeach for these type of beaches.

Several uncertainties in the approach were identified and quantified. One of the main assumptions of the hindcast model approach is that no losses occur during storm events below a certain wave height threshold. Given restraints in computational time it was decided that $H_{m0} > 1.85$m is an acceptable threshold where the closure depth is close to the structure tips. It is assumed that further reduction would have a limited effect on the net losses. Since 2015 XBeach has been further developed and calibrated resulting in new default settings, referred to as BOI [5]. As part of this research the XBeach code itself was further developed in cooperation with Deltares, resulting in improved model stability and flow patterns. The latest version of XBeach (rev 6122) is currently applied in the project and is assessed in this research, using mostly default parameter settings. The horizontal background viscosity parameter *nuh* was set to 0.8.

3 Results

3.1 System Behavior - Hydrodynamics and Morphological Response

Nearshore wave conditions at the XBeach model boundary (12 to 20m depth) were determined by translating calibrated CFSR offshore data to the project site with a SWAN wave model. The resulting timeseries was validated by comparison with measured wave data at the Waverider buoy. A storm event with $H_{m0} = 3.5$m roughly corresponds with a 1 year return period. The timeseries shows 3 events at or exceeding the RP1 condition and multiple events exceeding $H_{m0} = 1.85$m in the fall/winter period each year. In total about 13 days in period 1 and 44 days in period 2 exceed the wave threshold and are included in the simulated timeseriesAQ. The total incoming wave energy in period 2 is about five times the energy in period 1 (Fig. 2).

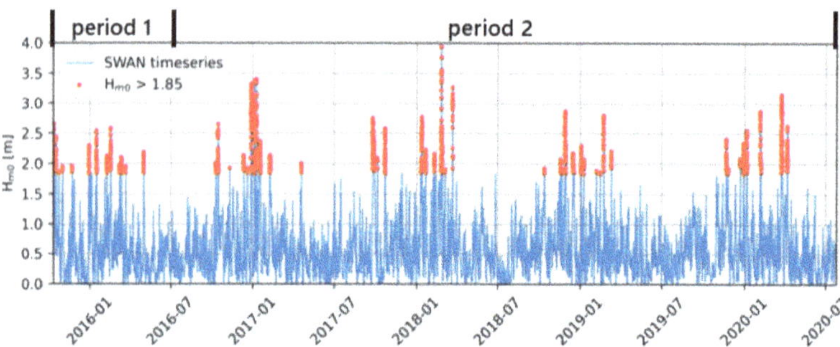

Fig. 2. H_{m0} timeseries at the XBeach model boundary for periods between bathymetric surveys

The bathymetric surveys show a consistent pattern in morphological development across all three pocket beaches. In general, sedimentation and swash bar formation is observed at and above the shoreline, erosion and some sedimentation on the shoreface (2–6 m water depth), erosion focused at and landward of the openings and sedimentation north of the beach cell (Fig. 3). The sedimentation at the shoreline tends to be biased to the north of the pockets, particularly for the two northern beaches, indicating dominant northward transport of sediment. Bed level changes on the shoreface are mainly created by sediment stirring due to breaking waves, interaction of hydrodynamics and structures and wave-driven currents transporting sediment out of the beach cell. Above the shoreline mainly shell fragments are found creating a large swash bar along the entire beach which is a combined effect of aeolian transport and short wave action.

After construction the following is observed: net sediment loss from the beach cells, a stable or even seaward shift of the shoreline and a steep (~1:25) underwater slope, which seems to be the new equilibrium at this location as it persists until 2020 and beyond (see also Fig. 4). This is much steeper than the expected slope if a Dean profile would be assumed. The formation and size of the swash bar was unexpected and is likely to be related to the large shell fragment content in the nourished sand.

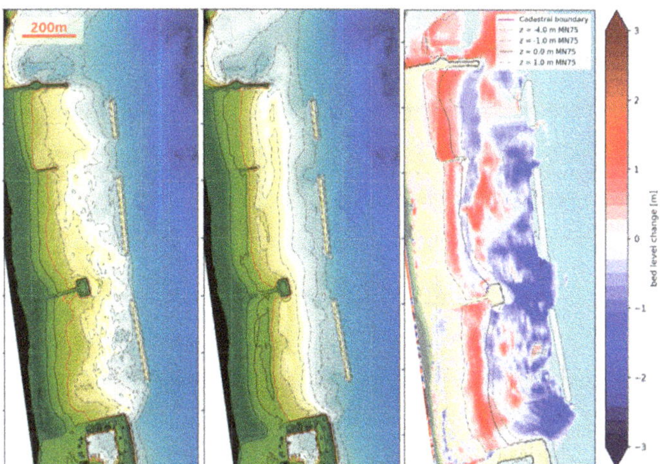

Fig. 3. Bathymetry at Eforie North in 2015 (left) and 2016 (middle) and bed level change (right)

3.2 Model Results - Morphology and Sediment Loss

In the hindcast for the period 2015–2016 at Eforie North, the modelled morphological changes show reasonable agreement with the observed changes (Fig. 4). Particularly the erosion zones inside the openings between the structures are reproduced well. The survey shows a sedimentation zone directly north of the beach cell, which is not accurately reproduced in the model. The model shows sedimentation around the shoreline and profile steepening, though less than observed because sediment sorting and short wave runup are not included in the model. Also the model tends to show less erosion behind the structures whereas in the survey the erosion is more uniform, particularly on the long term. The overall net sediment loss from the beach cell is overestimated by the model by 68% (70,000 vs 41,500 m^3), which is a combined effect of overestimation of erosion under water and underestimation of sedimentation around the shoreline.

Over the second period (2016–2020) the losses are slightly underestimated by the model by 14% (189,000 vs 220,300 m3). The main differences between model and survey are similar to period 1: less accumulation of sediment around the shoreline and less erosion behind the structures in the model versus the observations, see the cross shore profiles in Fig. 4 (profile 4 is behind a structure, 8 is at an opening). This is related sediment stirring due to roller turbulence and currents leading to redistribution behind the structures that is not well represented in XBeach.

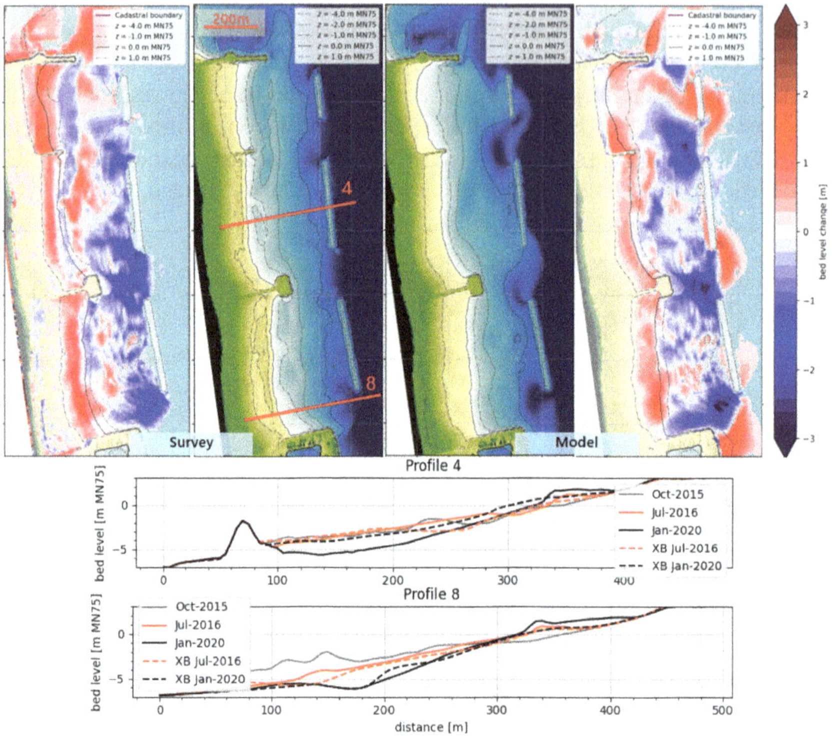

Fig. 4. Top: bathymetry in 2016 and bed level changes compared to 2015 (left panels: survey, right: model). Bottom: cross-shore profiles 2015, 2016 and 2020 comparing survey and model

3.3 Model Results - Uncertainties

The predicted sediment losses are close to the observations, suggesting XBeach is a suitable model for these type of applications. Predicted losses are 14–68% off from the observed losses, depending on the considered period. An explanation for the difference between the periods is that in period 1 the volume related to the formation of the swash bar compared to the erosion at the shoreface was much larger than in period 2, i.e. part of the eroded volume at the shoreface formed the swash bar and did not lead to net loss from the cell. Since the swash bar formation is underestimated in XBeach, in the model a larger part of the eroded volume is lost from the cell, resulting in a larger net loss than observed. In period 2 the in-situ beach profile is closer to the apparent equilibrium and the volume increase at the swash bar is smaller compared to overall sedimentation/erosion, resulting in a better prediction of overall net loss of sediment in period 2.

To further quantify uncertainties, the sensitivity of the calculated losses was investigated for several parameters and model input variations (Table 1). Sensitivities were tested for values in a realistic range and were accounted for in the design process. For example, wave conditions without validation can easily be 10% off which has a large impact on calculated losses. The analysis shows that the calculated losses are most sensitive to the viscosity (*nuh*), grain size and incoming wave energy.

Table 1. Sensitivity of net loss. Values are relative to the base case setting (in brackets)

Period	nuh = 0.1 (0.8)	D_{50} = 0.22 (0.34) mm	H_s > 2.0 (1.85)m	ρ_s = 2650(2500) ρ_w = 1025(1010) kg/m³	H_s,T_p-10%	manning = 0.015 (0.02)
15/16	+ 32%	+ 70%	-15%	-8%	-57%	+ 19%
16/20	+ 21%	-	-	-	-	0

4 Conclusions and Recommendations

The aim of this research is to assess the performance of XBeach for sediment losses from a designed pocket beach and to quantify uncertainties for future applications. The results show that 1) XBeach predicts the sediment losses from the beach cell reasonably well for a period up to 5 years after construction, 2) the assumption that only events above a certain H_{m0} threshold result in sediment loss is acceptable for this case and the main processes that are missing are 3a) sediment sorting and short wave runup leading to swash bar formation and 3b) sediment stirring due to roller turbulence and currents leading to redistribution behind the structures when wave action is limited.

The main uncertainties in the predicted losses result from these missing processes, lack of data to calibrate and validate flow patterns and finetune viscosity, the sediment grain size and the incoming wave energy. As XBeach is typically applied for dune erosion processes, this research provides valuable information regarding the performance of the model for engineering design of underwater slopes in complex geometries to minimize sediment loss. It confirms that XBeach is applicable to similar project sites within the rehabilitation project or elsewhere and it gives an indication which uncertainties to include. We consider this a first step to validate the model for this type of applications. Recommended next steps are a thorough flow calibration, sensitivity analysis and in-depth analysis of sedimentation/erosion patterns for the current case and the application of XBeach for a wider range of cases where survey data is available.

References

1. Hsu JR-C, Yu M-J, Lee F-C, Benedet L (2010) Static bay beach concept for scientists and engineers: a review. Coast Eng 57(2):76–91
2. Daly CJ et al (2011) Morpho-dynamics of embayed beaches: the effect of wave conditions. J. Coastal Res. SI 64(64):1003–1007

3. Langeveld C et al (2017) Constanta Coastline Rehabilitation - Application of advanced models and practical solutions to develop a sustainable low-maintenance coastal defence scheme. Coasts, Marine Structures and Breakwaters 2017. January 2018, 859–868
4. Roelvink D et al (2009) Modelling storm impacts on beaches, dunes and barrier islands. Coast Eng 56(11):1133–1152
5. Coumou L (2022) Validation of dune erosion model XBeach: development of 'BOI Sandy Coasts'
6. Van Bemmelen CWT (2017). Long Term Process-Based Morphological Modelling of Pocket Beaches. Additional Thesis, Delft University of Technology

Identification of Sediment Distribution and Pathways on an Energetic, Sediment-Starved, Embayed Coast

Wassim Seksaf[(✉)], Tim Scott, Gerd Masselink, Nieves G. Valiente, and Daniel Conley

University of Plymouth, Plymouth PL48AA, UK
`wassim.seksaf@plymouth.ac.uk`

Abstract. Sandy beaches and dunes are vital for protecting coastal communities from erosion and flooding, particularly along high-energy, sediment-limited coastlines. This rugged coastline, defined by rocky headlands and sediment-starved systems, requires a comprehensive understanding of sediment dynamics to maintain its natural defenses. This study examines sediment characteristics, dynamics, and connectivity along a 65-km stretch of coastline in southwest England using a comprehensive dataset comprising ground-truth sediment analysis, high-resolution remote sensing (bathymetric and LiDAR), and hydrodynamic modeling outputs. Four sediment clusters, distinguished by grain size and carbonate content, served as fingerprints to track sediment variability. Remote sensing data were used to create a seabed roughness map, differentiating flat sediment areas from rocky reef-dominated zones. By integrating sediment characteristics, geological features, and regional hydrodynamic models, the study identifies seven sediment cells dominated by a strong eastward-northward sediment transport pathway, with sediment settling downstream on the constraining headlands. These results provide a basis for evaluating the impacts of sediment supply and regional transport pathways on the future evolution of sand beaches and dune systems in sediment-starved embayed regions.

Keywords: Sediment connectivity · Headland's geomorphology · Coastal cells

1 Introduction

The southwestern coastline of England (Cornwall) is renowned for its diverse geomorphological features, including rocky cliffs, prominent headlands, small estuaries, and sandy beaches and dunes [1]. The shoreface geology in this region is complex and exhibits considerable variation. Some areas are characterized by large, deep embayments that contain multiple beaches, primarily oriented from west to north. In contrast, other regions feature steep, narrow shorefaces with shallow embayments divided by rocky headlands, most of which face westward. Elsewhere, the coastline is marked by rugged cliffs fronted by sandy beaches, with no distinct embayments. These beaches exhibit a wide range of morphodynamic states, from reflective to dissipative, and are often flanked by dune systems or cliffs, resulting in complex and diverse coastal morphologies [1].

© The Author(s) 2026
C. Coelho et al. (Eds.): CD 2025, CRL 41, pp. 203–209, 2026.
https://doi.org/10.1007/978-3-032-15473-6_32

Understanding sediment supply from offshore and alongshore areas is crucial for predicting the long-term evolution of beaches and dunes [2, 3]. In regions with strong tidal regimes, tidal currents enhance sediment transport both cross-shore and alongshore, with increased sediment mobilization during storm events [4–6].

Recent studies have revealed that sediment transport along the coast of Cornwall predominantly moves northeast, driven primarily by strong tidal currents and wave action [6] through headland bypassing [7]. This process has been investigated using various approaches, including remote sensing, sand tracer experiments, and numerical modelling [e.g., 6,8–14]. However, sediment cell connectivity is still far from being completely understood. This paper examines the characteristics, distribution, connectivity, and dynamics of mobile sediment along a 65-km stretch of high-energy, macro-tidal, sediment-starved, embayed northern coastline of Cornwall (southwest England). We integrate new sedimentological, mineralogical, and geomorphological datasets with established regional hydrodynamic studies [6, 8–10] to enhance understanding of sediment mobility, headland bypassing, and embayment sediment exchange.

2 Data and Methods

In order to address mobile sediment connectivity along the northern coastline of Cornwall, this study provides key insights into sediment characteristics in conjunction with remote sensing data (bathymetry and LiDAR) and hydrodynamic modelling outputs, enabling a detailed understanding of key controls of coastal sediments connectivity.

2.1 Ground-Truth Data

This study utilises ground-truth datasets to analyse sediment characteristics within and between embayments, providing valuable insights into sediment connectivity. 237 sediment grab samples were collected from the dune to the offshore (34 m to -30 m ODN, Fig. 1). Laboratory processing included sieving to isolate the sand fraction for each sample, followed by laser particle size analysis [15] to determine grain size distribution. Carbonate content was estimated using the loss on ignition method [16]. These analysis yielded two key parameters: sediment size and carbonate concentration.

2.2 Surface and Subsurface Data

Two complementary datasets, multibeam bathymetric data and LiDAR, are used to generate a continuous elevation model spanning from offshore waters to the coastal dune systems (Fig. 1). Bathymetric data, collected by the United Kingdom Hydrographic Office (UKHO) between April 2009 and March 2011, is gridded at 2 m spatial resolution, providing high-detail seafloor mapping. LiDAR data collected by the Environment Agency using the OPTECH GALAXY T2000 system in October 2022 supplements inland coverage captured at a 1-m resolution. Both the multibeam bathymetry and the LiDAR datasets offer precise topographic details with positional and elevational accuracies of $\pm$ 0.15 m and $\pm$ 0.1 m, respectively.

2.3 Seabed Roughness Mapping

To map the spatial distribution of sediment surfaces along the coastline, a seabed roughness analysis was performed to differentiate sediment deposits. This method operates on
the assumption that sediment-covered areas exhibit relatively flat surfaces, whereas rocky
areas display greater roughness within the same spatial window [6]. The bathymetry
data and topographic LiDAR were used to calculate roughness, defined as the maximum
detrended standard deviation within a 20-m radius. The resulting roughness metric provides a reliable means of identifying and delineating sedimentary environments along
the coastal seafloor. Sediment areas were delineated based on a 0.2-m threshold informed
by the incorporation of LiDAR data which allowed for detailed mapping of flat sediment
zones.

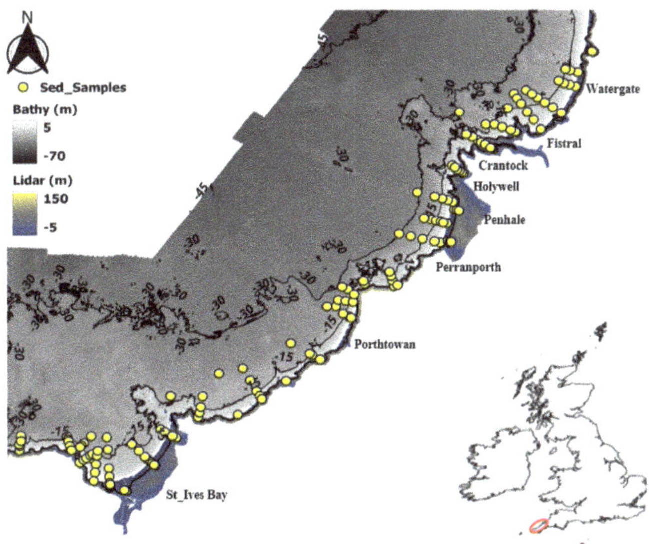

Fig. 1. Location map highlighting the datasets collected along the North Cornish coast. Bathymetric data collected by the UK Hydrographic Office (UKHO) combined with LiDAR data. Sediment
grab samples collected along the north coast from the dunes to the offshore (34m to -34m ODN).

2.4 Embayment Definition and Clustering

The north Cornish coastline is divided into embayments by rocky headlands labeled H1
to H19 (Fig. 2b). Additionally, in order to compare sediment and effectively create a
fingerprint for different sand fractions, a clustering approach was carried out using the
Kmeans algorithm [17].

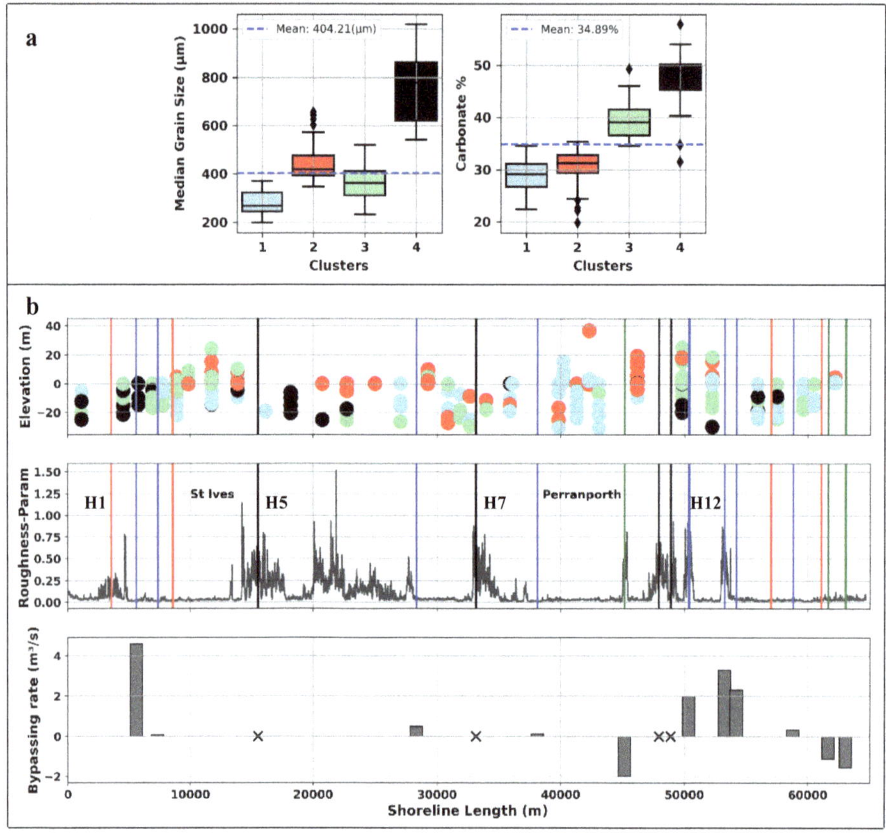

Fig. 2. (a) Kmeans clusters analysis representing the grain size and carbonate content distribution in each of the 4 clusters. (b) Top panel illustrates scatter plot of sediment characteristics; colours represent sediment clusters. Middle panel illustrates modelled headland bypassing rate (m^3/s) for spring low water, uniform sediments and extreme 12-h exceedance wave and tide forcing (281.25° dominant direction) [6]. Bottom panel represents alongshore seabed roughness variation along 600 m offset line from the shoreline. Vertical solid black lines are the modelled constraining headlands with no sediment bypassing, blue solid lines are the modelled open headlands with a northward-eastward sediment transport, green solid lines are modelled open headlands with a southward-westward sediment transport, and the red lines are not considered as headlands by the model during low spring tides [6].

3   Results and Discussion

Four distinct clusters (Fig. 2a) were chosen following the elbow principal. Cluster 1 consists of fine sand with low carbonate content and uniform grains, while Cluster 2 represents coarse sand with low carbonate content and irregular grains. Cluster 3 features fine sand with high carbonate content, marked by carbonate particles like shell fragments, and Cluster 4 comprises coarse sand rich in carbonate materials, including coral and shell fragments. Sediment clusters show the alongshore and the cross-shore variability within each embayment. Whereas seabed roughness along a 600m-offset shoreline reveals reef-dominated areas (high roughness) and flat sediment seabed (low roughness). In addition, sediment bypassing rates indicate northward-eastward transport for positive values, southward-westward for negative, and no bypassing at marked headlands. Headlands are classified as constraining or open based on model outputs [6]. Open headlands (e.g., H2, H3, H8) have low roughness and enable sediment exchange, while constraining headlands (e.g., H5, H7, H11) exhibit high roughness, limiting bypassing and connectivity.

From a modelling perspective, the studied area can be divided into four cells delineated by constraining headlands (crosses in Fig. 2b-bottom panel). However, observations suggest the presence of additional constraining headlands between H9 and H13, likely due to rocky reefs, as indicated by high roughness values (Fig. 2b-middle panel). Furthermore, the embayments in this region exhibit distinct sediment characteristics compared to neighbouring embayments. While the modelling aligns well with observations in larger embayments, field data provide higher-resolution insights into headland characteristics, particularly in sheltered and narrow embayments. By combining modelling and observations, we identified seven main cells along the coast, adding three additional cells: H9 to H10, H11 to H12, and H12 to H13. The distribution of sediment clusters across the defined coastal cells reveals distinct patterns influenced by hydrodynamics and headland morphologies. Larger cells exhibit a consistent trend, where coarser sand with higher carbonate content tends to settle upstream in sediment starved areas represented by high roughness values, while finer sand with lower carbonate content accumulates downstream on the constraining headland, indicative of strong eastward and northward sediment pathways within these cells. In contrast, sheltered cells appear to function as closed systems, likely supplied by offshore sources, as suggested by the cross-shore similarity in sand composition.

The findings from this study align with global studies emphasizing the key role of rocky headlands and hydrodynamic processes in shaping sediment distribution and transport pathways [6, 8, 9]. While constraining headlands are shown to act as sediment traps and limiting connectivity between embayments, open headlands facilitate sediment bypassing and promote alongshore sediment connectivity.

4   Conclusion

The multi-method approach (observations and modelling) provides a robust framework for evaluating sediment transport processes on both large and local scales. The analysis of sediment characteristics and connectivity by integrating ground-truth sediment data, high-resolution remote sensing, and modelled sediment transport revealed the interplay between hydrodynamic forces and headland morphologies. The findings underscore the role of headlands in shaping coastal cells connectivity, with constraining headlands acting as barriers and open headlands facilitating sediment bypassing. The delineation of seven sediment cells along the coastline highlights the complexity of sediment transport pathways, where sediment characteristics vary significantly between larger, connected, and sheltered cells.

References

1. Scott T, Masselink G, Russell P (2011) Morphodynamic characteristics and classification of beaches in England and Wales. Mar Geol 286(1–4):1–20
2. Kinsela MA et al (2022) Mapping the Shoreface of Coastal Sediment Compartments to Improve Shoreline Change Forecasts in New South Wales. Australia. Estuaries and Coasts 45(4):1143–1169
3. Harley MD, Masselink G, Ruiz De Alegría-Arzaburu A, Valiente NG, Scott T (2022) Single extreme storm sequence can offset decades of shoreline retreat projected to result from sea-level rise. Commun Earth Environ 3(1): 112
4. Valiente NG, Masselink G, Scott T, Conley D, McCarroll RJ (2019) Role of waves and tides on depth of closure and potential for headland bypassing. Mar Geol 407:60–75
5. Scott T, Masselink G, Russell P, Castelle B, Dodet G (2016) The extreme 2013/2014 winter storms: Beach recovery along the southwest coast of England. Mar Geol 382:224–241
6. King EV, Masselink G, McCarroll RJ, Scott T, Conley D, Leonardi N (2021) Wave, tide and topographical controls on headland sand bypassing. J Geophys Res Oceans 126(8): e2020JC017053
7. Evans OF (1944) The Relation of the Action of Waves and Currents on Headlands to the Control of Shore Erosion by Groins
8. Valiente NG, Masselink G, McCarroll RJ, Scott T, Conley D, King E (2020) Nearshore sediment pathways and potential sediment budgets in embayed settings over a multi-annual timescale. Mar Geol 427:106270
9. McCarroll RJ, Masselink G, Valiente NG, Scott T, King EV, Conley D (2021) An XBeach derived parametric expression for headland bypassing. Coast Eng 165:103860
10. McCarroll RJ, Masselink G, Valiente NG, Scott T, King EV, Conley D (2018) Wave and Tidal Controls on Embayment Circulation and Headland Bypassing for an Exposed, Macrotidal Site. Journal of Marine Science and Engineering 6(3):94
11. Vieira Da Silva G, Toldo EE, Klein AHDF, Short AD, Woodroffe CD (2016) Headland sand bypassing — Quantification of net sediment transport in embayed beaches, Santa Catarina Island North Shore, Southern Brazil. Marine Geol. 379: 13–27
12. Duarte J, Taborda R, Ribeiro M, Cascalho J, Silva A, Bosnic I (2014) Evidences of sediment bypassing at Nazaré headland revealed by a large scale sand tracer experiment
13. Vieira Da Silva G, Toldo Jr. EE, Klein AHDF, Short AD (2018) The influence of wave-, wind- and tide-forced currents on headland sand bypassing – Study case: Santa Catarina Island north shore, Brazil. Geomorphology 312: 1–11
14. Goodwin ID, Freeman R, Blackmore K (2013) An insight into headland sand bypassing and wave climate variability from shoreface bathymetric change at Byron Bay, New South Wales. Australia. Marine Geology 341:29–45
15. Blott SJ, Pye K (2012) Particle size scales and classification of sediment types based on particle size distributions: Review and recommended procedures. Sedimentology 59(7):2071–2096
16. Heiri O, Lotter AF, Lemcke G (2001) Loss on ignition as a method for estimating organic and carbonate content in sediments: reproducibility and comparability of results
17. Likas A, Vlassis N, Verbeek JJ (2003) The global k-means clustering algorithm. Pattern Recogn 36(2):451–461

Beach Morphodynamic Time Scales Analysis at Hasaki Beach, Japan

Xinyu Chen[1]([✉]), Masayuki Banno[1], and Nobuhito Mori[2]

[1] Port and Airport Research Institute, Yokosuka 239-0826, Japan
xychen.jp@gmail.com
[2] Kyoto University, Uji 611-0011, Japan

Abstract. Understanding beach morphodynamic behavior across different time scales is crucial for developing reliable shoreline models, yet comprehensive analyses are limited by the scarcity of high-frequency, long-term observational data. This study analyzes a 34-year (1986-2019) dataset of daily beach profile measurements from Hasaki Beach, Japan, alongside wave data from the JRA-55 atmospheric reanalysis model. We examined the short-term shoreline response during storm arrival and post-storm conditions, analyzing two-day response patterns under different wave conditions. For decadal-scale analysis, we employed Discrete Wavelet Transform to decompose shoreline variations into different temporal scales. Storm response analysis revealed that erosion continues beyond the initial storm impact despite wave height stabilization, while post-storm recovery shows no significant deviation from stable wave conditions. The wavelet analysis of monthly averaged shoreline positions identified dominant energy concentrations at time scales of 4–8 and 8–16 months, corresponding to seasonal variations. Components exceeding 5-year time scales were found to contribute substantially to overall shoreline variance. After removing this long-term trend, the analysis revealed a marked decrease in shoreline variability after July 2007, with variance reducing from 108.74 m^2 to 34.95 m^2.

Keywords: Beach Morphodynamics · Time Scale · Storm Response · Discrete Wavelet Transform

1 Introduction

Beach morphodynamic behavior can be divided into components that vary from day to year. Understanding the shoreline's behaviors and dominant forcing variants is necessary to develop a more reliable model in shoreline projection. However, studies on these time scales are rare due to the lack of high-frequency and continuous beach observation data. This study utilizes a 34-year (1986 to 2019) dataset from Hasaki Beach to analyze behavior across various time scales.

© The Author(s) 2026
C. Coelho et al. (Eds.): CD 2025, CRL 41, pp. 210–215, 2026.
https://doi.org/10.1007/978-3-032-15473-6_33

2 Dataset

Hasaki Beach is a longshore-uniform open coast located on the east coast of Japan, facing the Pacific Ocean without any shelter (Fig. 1(a, b)). Bathymetric observations have been recorded in the cross-shore direction from 1986 to the present at the Hasaki Oceanographical Research Station (HORS) (Fig. 1(c)). Along the 400-m-long pier, bathymetric data has been measured every 5 m. The measurement frequency was once per day from 1986 to 2011, and increased to once per week from 2012 onwards. The bathymetry is measured by engineers using the plumb-line method, which does not rely on the water level as a proxy for the shoreline as image-derived techniques do, ensures high reliability of the measurements. Following previous research, this study defines the shoreline position as the horizontal position at the high water level (HWL). Wave data is determined from the atmospheric reanalysis model JRA-55 and WAVEWATCH III, which showed good agreement with in-situ wave observations.

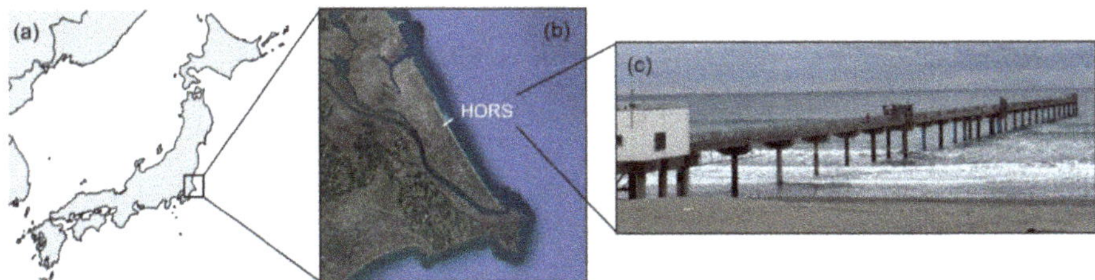

Fig. 1. (a, b) Location of Hasaki beach (Google, n.d.); (c) The pier of Hasaki Oceanographical Research Station (HORS)

3 Analysis and Results

3.1 Daily Variation

This study investigates shoreline response patterns during storm events by analyzing the relationship between significant wave height variations and shoreline position changes. Building upon the ShoreFor model framework [1], which establishes that shoreline evolution is proportional to the deviation from equilibrium wave conditions, we examine shoreline behavior during storm arrival and post-storm recovery phases.

Storm events were classified based on daily changes in significant wave height (ΔHs), with storm arrival defined as $\Delta Hs \geq 1.0$ m and post-storm conditions as $\Delta Hs \leq -1.0$ m (Table 1). Shoreline response was analyzed over two consecutive days for each condition and compared against stable wave conditions (-0.5 m $\leq \Delta Hs \leq 0.5$ m).

Probability density function (PDF) analysis reveals distinct shoreline behavior patterns across different wave conditions (Fig. 2). During storm arrival, the shoreline exhibits pronounced erosion on the first day, evidenced by a significant shift in the PDF toward negative values. This erosional trend persists through the second day, albeit at a reduced rate, despite wave heights stabilizing. This continued erosion under stable wave

Table 1. Wave conditions of storm and the number of cases.

Case	Day 1		Day 2	
	ΔHs (m)	No. of case	ΔH_s (m)	No. of case
Storm arrival	$\Delta Hs \geq 1.0$	253	$-0.5 \leq \Delta Hs \leq 0.5$	87
Post-storm	$\Delta Hs \leq 1.0$	187	$-0.5 \leq \Delta Hs \leq 0.5$	119
Stable wave	$-0.5 \leq \Delta Hs \leq 0.5$	5,139		

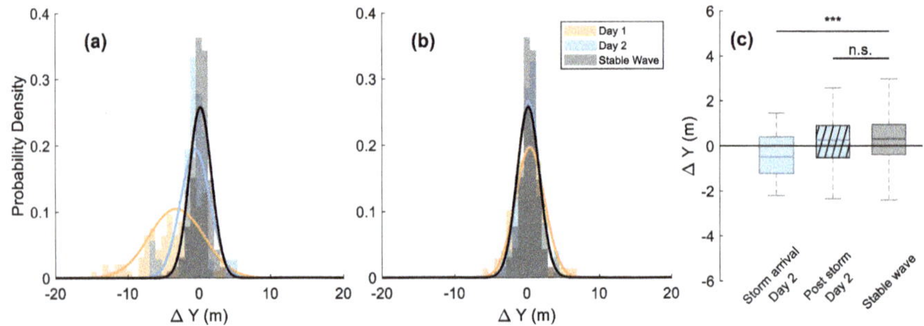

Fig. 2. Histogram and probability density function of shoreline change in (a) storm arrival and (b) post-storm; (c) boxplot comparison of shoreline change.

conditions suggests a lag effect in morphodynamic response, likely due to decreased sediment availability following initial storm impact. Statistical analysis confirms the significance of this persistent erosion (p < 0.001) compared to stable wave conditions.

Post-storm recovery patterns differ markedly from storm arrival response. The PDF analysis shows no significant deviation from stable wave conditions during either the first or second day post-storm (p > 0.05). This finding indicates that while storm-induced erosion exhibits a clear temporal signature extending beyond the initial impact, post-storm recovery processes operate more gradually, with shoreline changes comparable to those observed under stable wave conditions.

3.2 Seasonal to Interannual Variation

To analyze long-term shoreline evolution patterns, we employed the Discrete Wavelet Transform (DWT) method, which offers superior time-frequency localization compared to traditional filtering approaches. The shoreline position data was first averaged monthly to ensure uniform temporal sampling and reduce high-frequency noise. The DWT methodology decomposes the signal through a cascade of filtering operations, as illustrated in Fig. 3, where the original signal (a_0) is progressively filtered into approximation (a_{j+1}) and detail (d_{j+1}) coefficients at each decomposition level j. This multi-resolution analysis enables the separation of shoreline variations across different time scales while preserving signal energy.

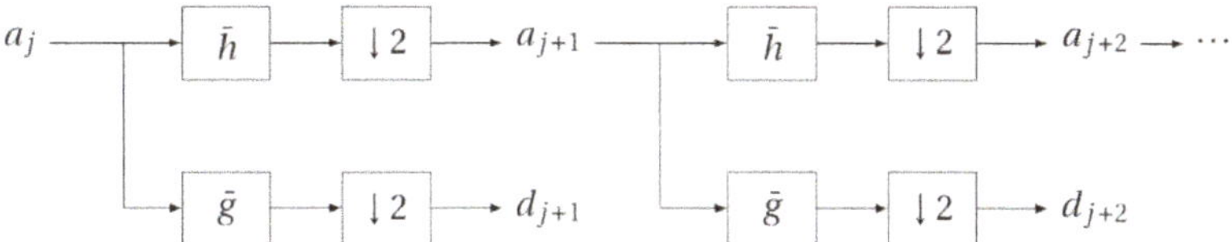

Fig. 3. Conceptualized cascade discrete wavelet transform.

We applied the Biorthogonal-3.9 (bior3.9) wavelet to analyze monthly averaged shoreline positions from 1986 to 2021, with a maximum decomposition level of 8 determined by the dataset length. The variance distribution across decomposition levels, shown in Fig. 4, reveals dominant energy concentrations at levels 2 and 3, corresponding to time scales of 4–8 months and 8–16 months respectively. These peaks align with the previously observed biannual and annual cyclic behavior of Hasaki Beach shoreline positions, which are primarily driven by seasonal wave patterns [2, 3].

The level 5 approximation coefficients, representing variations with time scales exceeding 5.3 years, effectively captured the shoreline's long-term trend (Fig. 5(a)). When compared to a conventional 5-year moving window average, the DWT approach demonstrated superior performance in identifying localized variations, particularly during significant morphological events like the 1991 erosion period. After removing the long-term trend identified by the level 5 DWT residual, the detrended signal revealed a marked reduction in shoreline variability after July 2007, with the variance decreasing from 108.74 m^2 to 34.95 m^2 (Fig5(b)).

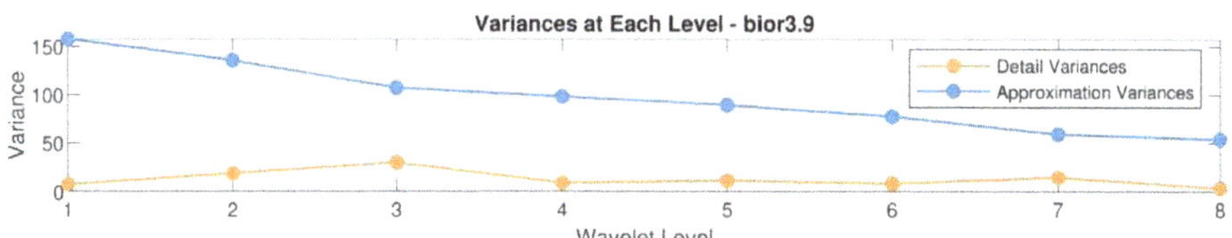

Fig. 4. Variance of decomposed shoreline records at each level by using DWT with bior 3.9.

4   Discussion

4.1   Asymmetry in Erosion and Accretion Processes

Analysis of the two-day wave-shoreline comparison reveals fundamental differences between erosion and accretion processes in both magnitude and temporal scales. During storm events, erosion occurs rapidly but follows a gradual pattern, while significant sediment removal occurs on the first day, the process continues beyond two days with decreasing intensity. This behavior aligns with the ShoreFor model's underlying principle that shoreline change rates are proportional to the magnitude of disequilibrium. However, accretion exhibits markedly different characteristics: the recovery process

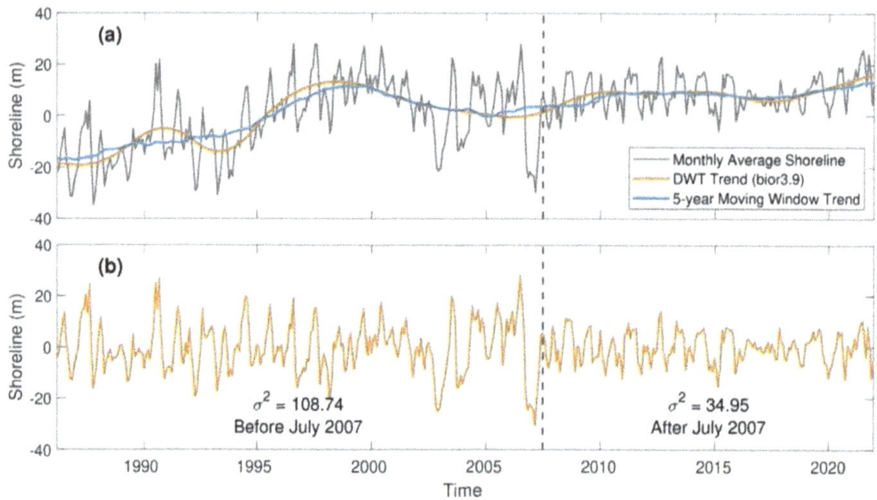

Fig. 5. Comparison of trend extraction methods and shoreline variability analysis: (a) Monthly averaged shoreline position with trends extracted using DWT (bior3.9 wavelet) and 5-year moving window methods; (b) Detrended shoreline fluctuations using DWT method.

proceeds slowly and steadily, showing minimal response to post-storm wave conditions. This asymmetry poses a significant challenge for current equilibrium shoreline models, which typically apply identical time scales to both erosion and accretion processes [1]. The distinct temporal signatures of these processes suggest that future model improvements should incorporate separate parameterizations for erosion and accretion phases to better capture short-term shoreline variations.

4.2 Significance of Long-Term Components

The wavelet analysis reveals that shoreline components with time scales exceeding 5–10 years contribute substantially to overall shoreline variance. This finding raises important questions about the physical processes driving these long-term variations. Conventional equilibrium models like ShoreFor primarily focus on cross-shore sediment transport, yet it remains unclear whether the residual effects of cross-shore processes alone can generate such significant long-term shoreline variations. When these models are calibrated using raw shoreline positions, they may inadvertently attempt to capture long-term components while overlooking important local variations. This suggests the value of frequency-domain preprocessing, such as DWT analysis, to identify dominant processes before model calibration. Such preprocessing could help distinguish between variations driven by cross-shore transport and those potentially caused by other mechanisms, leading to more physically meaningful model parameterizations.

5 Conclusion

Analysis of storm response patterns at Hasaki Beach revealed distinct temporal signatures between erosion and accretion processes, with erosion showing persistent effects beyond initial storm impact while accretion maintains steady rates regardless of wave conditions. These findings challenge the conventional approach of using identical time scales for both processes in equilibrium shoreline models, suggesting the need for separate erosion and accretion parameterizations.

Discrete Wavelet Transform analysis revealed substantial contributions of long-term components to shoreline variance, presenting a critical challenge for model calibration. Direct calibration against unprocessed shoreline positions risks optimizing model parameters to capture these dominant long-term trends while failing to resolve important short-term shoreline variations. Implementation of frequency-domain preprocessing methods therefore emerges as an essential step toward developing more physically representative models of shoreline evolution across multiple temporal scales.

References

1. Davidson MA, Splinter KD, Turner IL (2013) A simple equilibrium model for predicting shoreline change. Coastal Eng 73:191–202
2. Kuriyama Y (2002) Medium-term bar behavior and associated sediment transport at Hasaki, Japan. J Geophys Res Oceans 107(C9):15-1–15-12
3. Eichentopf S, Alsina JM, Christou M, Kuriyama Y, Karunarathna H (2020) Storm sequencing and beach profile variability at Hasaki, Japan. Marine Geol 424

Coastal Management and Risk Assessment

Probabilistic Analysis of Coastal Levee Erosion Using Excess Loading in a Response-Based Framework

Abigail L. Stehno[1,2(✉)], Jeffrey A. Melby[1], Norberto C. Nadal-Caraballo[1], Victor M. Gonzalez[1], and Patrick Lynett[2]

[1] U.S. Army Engineer Research and Development Center, Vicksburg, MS 39180, USA
`Abigail.L.Stehno@usace.army.mil`
[2] University of Southern California, Los Angeles, CA 90089, USA

Abstract. Risk-informed design is typically used for coastal levee projects, which integrate hazards, performance, and consequences. Presently, U.S. practice uses a relationship between overtopping discharge rates and erosion in order to evaluate and design coastal levees. However, excess loading concepts may better characterize physical processes during an erosional event. Herein, high-fidelity modeling and response-based probabilistic methods are combined with an excess loading approach to define levee crest elevations. These response-based methods maintain high statistical and physical fidelities and incorporate uncertainties. An example analysis is used to compare the probabilistic results from overtopping discharge rates and excess loading methods.

Keywords: Levee Erosion · Excess Loading · Stochastic Simulation

1 Introduction

Beginning in the 1980's, coastal levee design incorporated allowable overtopping which was based on overtopping discharge rates (q). These overtopping discharge rates are based on international standards which heuristically relate q to levee erosion damage state, such as start of damage [1, 2], but generally do not account for event duration.

Following Hurricane Katrina, [3] introduced a more physics-based relationship between levee erosion and wave forcing following [4] excess loading concepts. The approach allowed for combined wave and steady overflow. Erosion was related to a loading parameter (velocity, shear stress, and work) in excess of a threshold load from seaside toe to leeside toe. Threshold load values followed [4] and included duration. Limit states were defined for acceptable erosion for plain grass with good, average, and poor cover. These equations were dependent on event duration and intra-storm wave-by-wave timeseries.

Cumulative excess hydraulic load (CEHL) method [5] relates erosional equivalences between velocity and ground cover and removes the dependence on duration but requires wave-by-wave timeseries. Validation of CEHL has been completed using field testing

© The Author(s) 2026
C. Coelho et al. (Eds.): CD 2025, CRL 41, pp. 219–223, 2026.
https://doi.org/10.1007/978-3-032-15473-6_34

with the wave overtopping simulator but does not include combined overtopping and overflow. The individual wave approaches are highly dependent on the details of the incident wave spectrum and the wave-structure interaction which are often not well understood in shallow water and not defined at site-specific locations. Nearshore wave spectra include significant infragravity energy. Spectral transformation in shallow water is not well understood or modeled. Additionally, probabilistic simulation requires thousands of storms and defining individual waves is simply too computationally onerous. A bulk-spectral-parameter cumulative excess work (CEW) approach was proposed [6] that removes the wave-by-wave dependence and compares erosion limit guidelines with those of [4].

The USACE also introduced improved fidelity hydrodynamic modeling and statistical analysis for coastal structure design following Hurricane Katrina. Monte Carlo simulation was used to approximate erosion damage from overtopping discharge [7]. Extremal distribution methods, including the joint probability method, were used to approximate wave and water level hazards for design events in a frequency-based analysis. The USACE advanced estimation of extremal hazards through development of the Coastal Hazards System (CHS) [8]. The CHS framework generates and hosts high-fidelity, high-resolution numerical modeling and probabilistic coastal storm hazard data that spans practical probability (e.g., annual exceedance frequency) and storm parameter space. CHS data has been incorporated into response-based approaches to evaluate overtopping discharge rates on coastal levees [e.g. 9]. Response-based methods increase statistical fidelity over frequency-based approaches [10]. For threshold responses, such as overtopping discharge rates, response-based approaches compute maximum structure response for each individual event in the storm suite then integrate these responses to define the structure response probability. This response-based method maintains multivariate relationships between storm forcing parameters without assuming identical probabilities of the storm responses (e.g., wave height, storm water level) and the structure responses (e.g., overtopping discharge rate). Uncertainties associated with the waves and water levels are applied within the stochastic sampling, and uncertainties associated with the empirical response equations are computed and characterized using confidence levels to the mean value. The software, StormSim, is used by the USACE to integrate high-fidelity CHS data into a computationally efficient probabilistic response-based workflow. The StormSim tool suite computes coastal hazard responses for hundreds of thousands of instances within minutes without reduction in statistical or physical fidelities.

2 Methodology

Excess loading methods were extended by [6] for bulk wave with the relationship:

$$V_{ET}(t) = \sum_{n=1}^{N}(q_n - q_c)\Delta t_n \leq G_F\left(\frac{f_F}{2g\sin\theta}\right) \text{for } q_n > q_c \tag{1}$$

where $V_{ET}(t)$ is cumulative excess water volume (CEV) proportional to cumulative excess work (CEW), N is total number of discrete discharges, q_n is measured load at instance n, q_c is threshold load value, Δt_n is uniform duration associated with each discrete discharge, G_F is erosional limit, f_F is Fanning friction factor, g is gravity, and θ is landward levee slope angle. The erosional limit characterizes grass cover failure where a considerable amount of grass cover is eroded but does not capture levee breaching.

Previously, a response-based probabilistic analysis was computationally demanding for practical application of cumulative excess work methods. In this work, we incorporate CEV from [6] into the computationally efficient StormSim framework to probabilistically evaluate erosional hazard for coastal levee design.

For demonstration purposes, we show a representative coastal levee on the U.S. Gulf Coast. The levee has a crest elevation of 5 m and 1:3 seaside and leeside slopes. Levees in this region are generally subjected to combined overflow and wave overtopping for extreme loading cases. U.S. levees are usually built to a safety level corresponding to the 1% annual exceedance probability (AEP) flood level defined for flood insurance purposes. This is largely because it is expected that flood safety is primarily from evacuation during extreme events. Therefore, design context is flood risk mitigation, and resulting levee crest elevations are near the 1% AEP design water level, where overflow may potentially occur.

For the example, 189 synthetic tropical cyclones from a CHS storm suite [9] with simulated intra-storm time steps of 15 min define the flood hazard. At this representative location, 1% AEP storm water level is 4.2 m, $H_{m0} = 1.2$ m, and associated $T_p = 5.2$ s at a 50% "mean" confidence level.

The CHS storm suite was used in the response-based StormSim framework to demonstrate differences between designs based on q and cumulative excess volume $V_{ET}(t)$ limit states. The start of damage for good quality grass is defined in the Coastal Engineering Manual (CEM) [1] between $q = 9.3 \times 10^{-4}$ m^3/s/m $= 1$ l/s/m and 9.3×10^{-3} m^3/s/m $= 10$ l/s/m. [9] used these limits to define the mean and 90% confidence level q limit states. The q limit states are known to be overly conservative for start of damage on a good quality grass cover layer. The following was used to compute erosional limit states: $G_F = 0.1 \times 10^6$ m^3/s^2 for poor cover, 0.2×10^6 m^3/s^2 for average cover, and 0.5×10^6 m^3/s^2 for good cover, $f_F = 0.015$, and $g = 9.81$ m/s^2 [3, 6]. These values are used in the right-hand side of the equation to define the limit state for each grass cover to the erodibility of the grass cover to the cumulative overflow and overtopping volumes.

3 Results and Discussion

The results herein compare q and CEV for probabilistic levee design and demonstrate how CEV can increase fidelity on levee erosion prediction within a probabilistic context including uncertainty. Figure 1 shows q and CEV hazards at the mean and 90% confidence level and corresponding erodibility limit states as horizontal lines.

In this example, both q limit states are greatly exceeded at a 1% AEP. At an AEP of 1%, the representative design is adequate for average and good grass cover from the CEV method at the mean, but the erosion limit is slightly exceeded for poor grass cover. At the 90% confidence level, the CEV is adequate for good grass cover, is nearly identical to the limit state for average grass cover, and exceeds the limit state for poor grass cover.

We expect CEV to be 4 orders of magnitude greater than q due to summation of overtopping over time but the limit state difference between CEV and q is 6 orders of

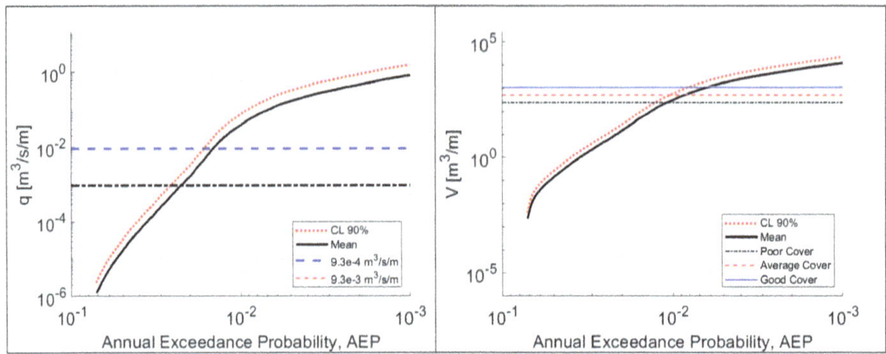

Fig. 1. Hazard curves at the mean and 90% confidence levels (CL) for overtopping discharge rate (left) and CEV (right) for a representative levee.

magnitude. This suggests that the CEV approach design criteria allow approximately two orders of magnitude more total water volume for initiation of damage, depending on grass quality. This is consistent with observations from [5] and [6]. This may lead to differences in designs between approaches.

Overtopping discharge does not give indication of what will happen once the limit state is exceeded, whereas CEV describes erosion for varying grass cover conditions after limit state exceedance. The CEM only heuristically describes start of damage as an entire order of magnitude for overtopping discharge rates; however, these relationships are known to be conservative. The CEV methods have been validated against full-scale testing and field data; however, there is still a poor understanding of some of the complex relationships between the detailed flow characteristics and grass failure, particularly for combined overflow and wave overtopping.

Finally, there has been little research on probabilistic analysis of CEV methods leaving large unresolved gaps in research. The majority of excess loading methods have been implemented in deterministic analysis, leading to large assumptions about the relationship between storm and structure response probabilities and the hydrographs of the events themselves. A probabilistic response-based approach is more likely to accurately characterize risk using CEV methods and is now possible with high-fidelity probabilistic storm databases such as the CHS and the associated probabilistic software such as StormSim.

References

1. USACE (2008) Coastal engineering manual, 2nd edn. EM 1110-2-1100. U.S. Army Corps of Engineers, Washington, DC
2. Van der Meer JW, et al (2018) EurOtop, 2nd edn. www.overtopping-manual.com
3. Dean RG, Rosati JD, Walton TL, Edge BL (2010) Erosional equivalences of levees: steady and intermittent wave overtopping. Ocean Eng 37:104–111
4. Hewlett HWM, Boorman LA, Bramley ME (1987): Design of reinforced grass waterways. CIRIA report 116, Construction and Industry Research and Information Association, London

5. van der Meer JW, Hardeman B, Steendam GJ, Schuttrumpf H, Verheij H (2010) Flow depths and velocities at crest and inner slope of a dike, in theory and with the wave overtopping simulator. In Smith JM, Lynett P (eds) Proceedings of the 32nd international conference on coastal engineering
6. Hughes SA, Thornton C (2015) Tolerable time-varying overflow on grass-covered slopes. J Marine Sci Eng 3:128–145
7. van Ledden M, Lynett P, Resio DT, Powell NJ (2007) Probabilistic design method of levee and floodwall heights for the hurricane protection system in the New Orleans area. In 10th international workshop wave hindcasting forecasting coastal hazard Symposium, U.S. Army Engineer Research & Development Center, Vicksburg MS
8. Nadal-Caraballo NC, Campbell MO, Gonzalez VM, Torres MJ, Melby JA, Taflanidis AA (2020) Coastal hazards system: a probabilistic coastal hazard analysis framework. J Coastal Res 95:1211–1216
9. Melby JA, Massey TC, Stehno AL, Nadal-Caraballo NC, Misra S, Gonzalez VM (2021) Sabine pass to galveston bay, TX pre-construction, engineering and design PED hurricane coastal storm surge and wave hazard assessment: report 1-background and approach. Technical report ERDC/CHL TR-21-15. U.S. Army Engineer Research and Development Center, Vicksburg, MS
10. Stehno AL (2021) Coastal structure overtopping and overflow stochastic simulation method comparison. M.S. thesis. Mississippi College, Clinton, MS

Maldivian Island Typologies for Coastal Adaptation to Future Sea Level Rise

Ahmed Aslam Waheed[1,2(✉)], Gerd Masselink[1], Timothy Poate[1], Lauren Biermann[1], and Mohamed Aslam[1]

[1] University of Plymouth, Plymouth, UK
ahmed.waheed@plymouth.ac.uk
[2] The Maldives National University, Male, Maldives

Abstract. Coral atoll islands, such as those in the Maldives, demonstrate landscapes acutely vulnerable to climate change-driven sea-level rise (SLR) and related coastal hazards. These challenges, including recurrent flooding, shoreline erosion, and salinization of freshwater lenses, directly threaten the sustainability of human habitation and livelihoods. Building on insights from previous studies, atoll islands continue to exhibit inherent dynamism, offering paths for adaptive responses that will complement ecological resilience with engineered solutions. This research employs clustering to classify Maldivian islands based on critical variables, including demographic density, wave energy exposure, and the extent of coastal defences. The analysis reveals six distinct island typologies, each characterized by specific socio-environmental challenges and opportunities for adaptation. Findings stress the critical need for tailored adaptation frameworks specific for islands reinforcing the importance of avoiding generalized approaches, promoting the equitable allocation of resources to enhance resilience across diverse island contexts. The findings of this study maybe useful for shaping adaptive policies in atoll nations facing similar climate challenges worldwide.

Keywords: Coral Atoll Vulnerability · Sea-Level Rise (SLR) · Wave Exposure Adaptation · Maldives Future Habitability

1 Introduction

Coral atoll islands rank among the most vulnerable landscapes to climate change-driven sea-level rise (SLR) and associated coastal hazards. These low-lying islands, composed of biogenic sediments atop reef platforms, face ongoing challenges such as flooding, coastal erosion, and freshwater salinization, all of which threaten their long-term habitability [1, 2]. The Maldives, characterized by islands with elevations barely exceeding 1–2 m above mean sea level, depicts the vulnerabilities of atoll nations. Coastal hazards, including storm surge impacts and wave-driven inundation, are expected to intensify under future SLR scenarios, posing significant risks to both ecosystems and human habitation [2, 3].

Contrary to the perception of atoll islands as static and geologically passive, emerging evidence highlights their dynamic nature. Processes such as vertical accretion and

C. Coelho et al. (Eds.): CD 2025, CRL 41, pp. 224–229, 2026.
https://doi.org/10.1007/978-3-032-15473-6_35

lateral migration, driven by wave overwash and sediment redistribution, reveal the inherent adaptability of these islands to environmental changes [1, 4]. This natural resilience provides a compelling alternative to the prevailing reliance on expensive grey infrastructure, such as seawalls and breakwaters. Nature-based solutions—such as coral reef restoration and sustainable sediment management—are increasingly recognized for their dual role in mitigating risks and supporting socio-ecological systems [5]. Early warning systems can help to prepare for and respond to impending storm-driven flood events. In the long-term, future scenarios of flooding events enable coastal communities and managers to plan and implement adequate risk-reduction strategies [6].

Recent research draw attention to the variability of atoll island vulnerabilities, shaped by factors such as reef morphology, wave exposure, and socio-economic conditions [2, 7]. For example, Amores et al., (2021) demonstrated that variations in wave energy and reef geometry significantly influence flood risk timelines and adaptation priorities across islands. These findings emphasize the necessity of tailored adaptation frameworks that classify islands based on their resilience and adaptive capacity.

This study tries to classify inhabited islands of the Maldives, developing a robust classification system to evaluate their adaptation potentials. By integrating historical shoreline changes, remote sensing data, and geospatial analyses, the research attempts to pave way for adaptation options that can be tailored through a better understanding of oceanographic, environmental, social, and economic contexts of islands.

2 Methods

The study used an extensive database of 72 attributes, which was originally developed and expanded from the framework created by Duvat & Magnan, (2019). The original database focused on assessing coastal geomorphology, vulnerability, and island dynamics in the context of climate change. It incorporated diverse variables, including geomorphological features, vegetation coverage, shoreline positions, and climate hazard exposure, derived from historical records, aerial photographs, satellite imagery, and geomorphic surveys. Building upon this foundation, the database was updated to address the specific socio-environmental and hazard-related factors pertinent to the inhabited islands of the Maldives. The enhanced database integrated recent population data (2014–2024), a decadal analysis (2014–2024) of island changes derived from Landsat-8 imagery, and documentation of land reclamation efforts from Environmental Impact Assessments (EIA) and the Maldivian government budget and expenditure records.

The updated database provides a structure for evaluating geospatial, demographic, geomorphic, and anthropogenic characteristics of all the inhabited 180 Maldivian islands. It covers a wide range of attributes across six domains, including island area, decadal changes in size, shoreline modifications, artificial land expansion, reef-based marine alterations, and benthic/geomorphic features. Geospatial attributes, such as island area in 2024 and 2014, were derived using Landsat 8/9 imagery, while official geospatial boundaries for 2024 were obtained from www.onemap.mv. Demographic data from 2006, 2014, and 2022 provided insights into population dynamics, including density metrics and resource pressures, sourced from the Maldives National Bureau of Statistics.

Artificial land expansion and shoreline changes were analysed using Landsat imagery, with details on reclamation size, sediment volumes, and elevation gathered from

EIA reports and corroborated by the **Maldives Budget Book 2024**. Coastal protection types, such as longitudinal and transversal protections, were documented through combined analyses of EIA reports and Google Earth imagery. Reef-based marine modifications, including dredged harbours, boat channels, and sediment removal activities, were similarly mapped using Google Earth and supplemented by data on sea-level rise-related incidents reported by the Maldives National Disaster Management Authority.

Wave-induced flooding hazards were assessed using the methodology of Amores et al., (2021), incorporating significant wave heights, reef profiles, and island-specific morphological characteristics to estimate flooding risks under varying sea-level rise scenarios. This modeling approach facilitated the identification of vulnerable areas and informed adaptation strategies. Additionally, benthic, and geomorphic attributes, such as rock, rubble, sand coverage, coral/algae cover, seagrass extent, and lagoon features, were derived from the **Allen Coral Atlas**, **Google Earth imagery**, and EIA reports done in the Maldives.

Data processing involved standardizing key attributes, including population density, Wave Exposure, and Shore Protection Length, using the Standard Scaler function from Python's sklearn. [10] preprocessing module to ensure comparability. K-means clustering algorithm [11] was employed to partition the islands into six clusters, guided by the Elbow Method to determine the optimal number of clusters. Principal Component Analysis (PCA) [12] was used to visualize the clustering results for a clear interpretation of the patterns and relationships among the islands.

3 Results Discussion

This study applied hierarchical clustering to group Maldivian islands based on population density, wave exposure, and shoreline protection length. After normalizing variables to eliminate scale biases, six distinct clusters were identified using the elbow method, which determined the optimal number of clusters. Principal Component Analysis (PCA) highlighted key dimensions of variance, with the first two components explaining most of the variation and enabling clear visualization of cluster separations. For example, Cluster 1 featured densely populated islands with moderate shoreline protection, while Cluster 6 included the heavily urbanized Greater Malé Region with extensive coastal defences (Fig. 1 and Table 1).

The cluster analysis revealed distinct socio-environmental characteristics. Islands in Cluster 1, with the highest population density, face significant socio-economic risks, necessitating alternative adaptation options. Cluster 3, the most wave-exposed group, highlights the immediate nature of coastal protection or planned retreat and more creative engineered plus nature-based solutions like coral reef restoration to mitigate risks. Cluster 5, notable for its higher coastal protection, demonstrates substantial investments in engineered infrastructure but features the need for ongoing maintenance and integration with ecological solutions. In contrast, Cluster 6, with its high population density and naturally low wave exposure, requires urban planning to address land use pressures sustainably.

These findings demonstrate the spatial heterogeneity of vulnerabilities and adaptation needs across the Maldives. Tailored strategies will be essential to address specific

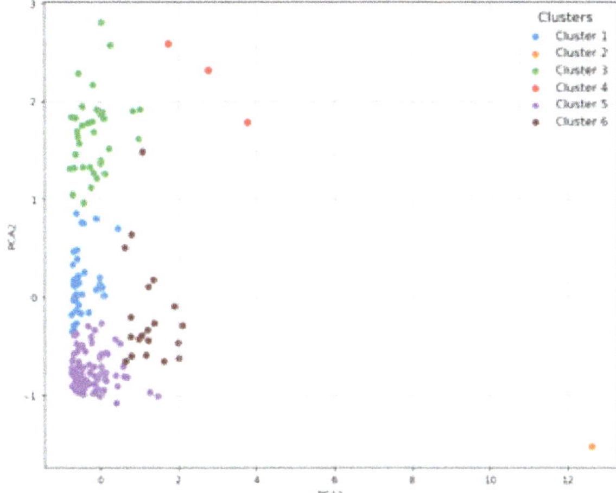

Fig. 1. Principal Component Analysis (PCA) Plot Showing Cluster Separation. Visualization of the clustering results based on PCA, illustrating the separation of islands into six distinct clusters. Each cluster, represented by varying characteristics based on key attributes such as population density, wave exposure, and shore protection. The axes represent the first two principal components (PCA1 and PCA2), which capture most of the variance in the dataset.

Table 1. Socio-Environmental Characteristics of Maldivian Island Clusters Based on Population Density, Wave Exposure, and Coastal Protection

Feature	Clusters					
	1	2	3	4	5	6
Population Density	High	Moderate	Low	Moderate	Moderate	Highest
Wave Exposure	Low	Low to moderate	Highest	Low	Moderate to high	Low
Coastal Protection	Moderate (30–40%)	Least (>20%)	Moderate (30–40%)	Close to fully protected (<80%)	Partially Protected (40–60%)	Close to fully protected (<80%)

risks and opportunities for each cluster. This also means investments to be prioritised based on vulnerabilities of islands. High-density clusters, such as Clusters 1 and 6, require strategies that integrate resilient infrastructure with sustainable urban planning. For wave-exposed islands in Cluster 3, greater investments are needed to alleviate coastal damage. However, in the long term, coastal retreat or relocation should be considered as part of a phased adaptation pathway for clusters 1, 3 and 6. Ecosystem-based solutions, such as coral reef restoration and mangrove rehabilitation, are essential for mitigating hazards and supporting biodiversity and hence the economy. For islands in Cluster

5, ongoing maintenance and upgrades to grey infrastructure may also be necessary, complemented by hybrid approaches that integrate ecological elements.

Moderate clusters, such as Clusters 2 and 4, reflect mid-level vulnerabilities requiring a balanced mix of coastal protection investments and socio-ecological interventions. These clusters demand resource-efficient strategies tailored to their specific risks and opportunities.

4   Conclusion

This study highlights the spatial and socio-environmental diversity of Maldivian islands, stressing the need for tailored adaptation strategies. Through a simple cluster analysis of population density, wave exposure, and coastal protection measures, distinct island groups with varying vulnerabilities and adaptation needs were identified.

The findings emphasize the need to move beyond one-size-fits-all approaches, advocating for cluster-specific measures aligned with the unique socio-ecological dynamics of each group. This approach ensures optimized resource allocation, enhanced climate resilience, and equitable outcomes across the Maldivian archipelago, offering valuable guidance for adaptation planning in other low-lying island nations facing similar climate challenges.

References

1. Beetham E, Kench PS (2018) Predicting wave overtopping thresholds on coral reef-island shorelines with future sea-level rise. Nat Commun 9(1). https://doi.org/10.1038/s41467-018-06550-1
2. Storlazzi CD, et al (2018) Most atolls will be uninhabitable by the mid-21st century because of sea-level rise exacerbating wave-driven flooding. Sci Adv 4(4). https://doi.org/10.1126/sci adv.aap9741
3. Masselink G, Beetham E, Kench P (2020) OCEANOGRAPHY coral reef islands can accrete vertically in response to sea level rise. https://www.science.org
4. Masselink G, McCall R, Beetham E, Kench P, Storlazzi C (2021) Role of future reef growth on morphological response of coral reef islands to sea-level rise. J Geophys Res Earth Surf 126(2). https://doi.org/10.1029/2020JF005749
5. Perry CT, Kench PS, Smithers SG, Riegl B, Yamano H, O'Leary MJ (2011) Implications of reef ecosystem change for the stability and maintenance of coral reef islands. Glob Chang Biol 17(12):3679–3696. https://doi.org/10.1111/j.1365-2486.2011.02523.x
6. Winter G, et al (2020) Steps to develop early warning systems and future scenarios of storm wave-driven flooding along coral reef-lined coasts. Front Mar Sci 7. https://doi.org/10.3389/fmars.2020.00199
7. Hoeke RK, McInnes KL, Kruger JC, McNaught RJ, Hunter JR, Smithers SG (2013) Widespread inundation of Pacific islands triggered by distant-source wind-waves. Glob Planet Change 108:128–138. https://doi.org/10.1016/J.GLOPLACHA.2013.06.006
8. Amores A, et al (2021) Coastal flooding in the Maldives induced by mean sea-level rise and wind-waves: from global to local coastal modelling. Front Mar Sci 8. https://doi.org/10.3389/fmars.2021.665672
9. Duvat VKE, Magnan AK (2019) Rapid human-driven undermining of atoll island capacity to adjust to ocean climate-related pressures. Sci Rep 9(1). https://doi.org/10.1038/s41598-019-51468-3

10. Fabianpedregosa FP, et al (2011) Scikit-learn: Machine Learning in Python Gaël Varoquaux Bertrand Thirion Vincent Dubourg Alexandre Passos Pedregosa, Varoquaux, Gramfort et al Matthieu Perrot. http://scikit-learn.sourceforge.net
11. Macqueen J. Some methods for classification and analysis of multivariate observations
12. Jollife IT, Cadima J (2016) Principal component analysis: a review and recent developments. Royal Society of London. https://doi.org/10.1098/rsta.2015.0202

Simple Physics-Based Rip Current and Shore-Break Wave Hazard Predictors for Beaches in Southwest France

Bruno Castelle[1]([✉]), Jeoffrey Dehez[2], David Carayon[2], Sylvain Liquet[3], and Jean-Philippe Savy[4]

[1] UMR EPOC, Univ. Bordeaux/CNRS/Bordeaux INP, Pessac, France
bruno.castelle@u-bordeaux.fr
[2] ETIS, INRAE, Cestas-Gazinet, France
[3] Météo-France, Toulouse, France
[4] SMGBL, Messanges, France

Abstract. Sandy beaches are attractive yet potentially dangerous environments due to physical hazards in the surf zone. The most severe natural hazards are rip currents and shore-break waves, which form under varying wave, tide, and morphological conditions. This study introduces two simple, physics-based models to forecast channel rip current flow speed V and shore-break wave energy Esb. These models were applied to La Lette Blanche, a high-energy meso-macro-tidal beach in southwest France, where both physical hazards coexist. Lifeguard-perceived hazard data collected hourly from July to August 2022 during patrol hours (11 AM–7 PM) were used to calibrate the models. This data also informed a 5-level hazard scale, from 0 (no hazard) to 4 (maximum hazard). The models accurately predict hazard levels, accounting for tidal and wave influences. Requiring only basic beach morphology metrics, this approach offers a promising tool for forecasting surf-zone hazards on beaches with minimal morphological data and available wave forecasts.

Keywords: Rip currents · Shore-break waves · Surf-zone hazards · Hazard forecasting · Beach morphology

1 Introduction

The prediction of natural hazards like flash floods, wildfires, and hurricanes is critical for protecting lives, property, and ecosystems [1, 2]. Significant advances have been made in forecasting atmospheric and hydrological hazards, but surf-zone hazards, such as rip currents and shore-break waves, have received comparatively less attention. However, for instance in the U.S., rip currents were the third-leading cause of weather-related fatalities between 2012 and 2021, following heat waves and flooding (US Department of Commerce). Unlike many hazards linked to extreme events, surf-zone incidents often occur during fair weather, typically on warm and sunny days (*e.g.*, [3, 4]).

© The Author(s) 2026
C. Coelho et al. (Eds.): CD 2025, CRL 41, pp. 230–235, 2026.
https://doi.org/10.1007/978-3-032-15473-6_36

Rip currents are narrow, fast-flowing seaward currents that form near the shoreline, often in channels between sandbars [5]. These currents can pull swimmers offshore, leading to drownings caused by exhaustion or panic. The activity of channel rips, one of the most common rip types, intensifies under shore-normal waves and low tides. Shore-break waves, in contrast, are plunging waves breaking near steep beach faces, causing injuries like spinal trauma from wave impact or shallow diving [6]. Lifeguards in southwest France report increased shore-break wave hazards during long-period waves and high tides [4], but research on these hazards is limited due to challenges in measuring wave energy and impact. Current models for surf-zone hazards mainly focus on rip currents and fail to account for shore-break waves, underscoring the need for more comprehensive approaches [*e.g.*, 7].

This study presents two physics-based models for predicting rip current and shore-break wave hazards, validated at La Lette Blanche beach in southwest France [4]. Using a 5-level hazard scale, these models integrate lifeguard observations and environmental data, offering a practical framework for forecasting surf-zone hazards across diverse beaches.

2 Study Site and Data

La Lette Blanche beach (Fig. 1a), in southwest France, is representative of the region's open-coast beaches with its intermediate, double-barred morphology. The beach lies in a meso-macrotidal zone. It faces high-energy North Atlantic waves, with a summer mean wave height of 1.1 m and a 9-s peak wave period. Rip currents flowing through the inner-bar rip channels (Fig. 1b), prevalent during shore-normal wave conditions and mean low tide, present significant drowning risks along this coast. Shore-break waves (Fig. 1c), often causing injuries during high tides on steep profiles, are another critical physical hazard (*e.g.*, [4]). In summer (July-August), lifeguards monitor the beach and typically set up supervised bathing zones away from rip currents. These zones are adjusted throughout the day based on tidal changes, with hazard levels communicated via a color-coded flag system: green for safe, yellow-orange for dangerous, and red for prohibited bathing.

In summer 2022, a beach safety experiment collected diverse data, including lifeguard hazard assessments, surf-zone drifter measurements, environmental monitoring and beachgoer surveys [4, 8]. Lifeguards rated rip current and shore-break wave hazards hourly on a 5-level scale. Wave and tide conditions were estimated using numerical hindcasts, validated against local buoy data, and included significant wave height, wave period and direction, and tidal elevation. A preliminary analysis revealed increased rip current hazards during larger, longer-period waves, while shore-break hazards were higher under long-period waves and large tidal range [4].

Fig. 1. (a) Location map of La Lette Blanche beach (green bubble). Photographs in southwest France of the two major surf-zone hazards with (c) rip currents (Ph. Observatoire de la Côte de Nouvelle-Aquitaine) and (d) shore-break waves (Ph. SMGBL).

3   Hazard Forecast Models

Two hazard forecast models were developed. For the sake of conciseness, below we only provide a brief summary of the underlying assumptions and required free parameters of the model followed by a short description of the calibration strategy. For a comprehensive description of the models and calibration approach, the reader is referred to [9].

3.1   Rip Current

Rip current hazard is linked to rip flow speed, which can be estimated using an idealized rip-channeled beach model where breaking waves drive rip currents through the deeper channels. These currents result from alongshore pressure gradients caused by wave breaking variability over the sandbar and channel. This pressure gradient drives feeder currents converging at channels and turning offshore as rips. Wave set-up is simplified as $S \approx 0.16Hs$ where Hs is the significant wave height. Rip flow speed (V) depends on wave set-up differences (pressure gradient) across the bar (Sb) and channel (Sc) linked to wave height decay due to depth-induced breaking. Based on a simplified depth-induced breaking energy dissipation model and the approximated depth-averaged alongshore momentum balance relating flow velocity to pressure gradients across and idealized bar/rip morphology, a simple model computes V as a function of a breaker parameter, sandbar elevation, channel depth, and tide elevation as parameters.

3.2   Shore-Break

A physics-based model was developed to estimate shore-break wave hazard. The model uses the dimensionless Irribarren parameter Irr as a proxy for wave breaking type, with shore-break wave energy (Esb) computed as Irr times a proxy of incident wave energy. An idealized Dean profile models the beach, with a sandbar mimicked by a terrace elevation affecting wave energy dissipation. The model computes *Esb* by accounting for tide elevation, wave breaking offshore across the terrace, and remaining wave energy at the shore. It requires four parameters: Dean profile parameters, breaker parameter and terrace elevation.

3.3 Calibration

A two-step approach was used to model rip current velocity (*V*) and shore-break wave energy (*Esb*) based on lifeguard-perceived data (*RHl, HSl*). First, simulations were run to calibrate parameters by maximizing the correlation between *V* (*Esb*) and *RHl* (*HSl*) during summer 2022. Next, *V* and *Esb* values were sorted and thresholds computed to match a 5-level hazard scale. The accuracy of the models was assessed using confusion matrices and compared with lifeguard data to evaluate prediction of high-hazard days.

4 Results and Discussion

The rip-current model achieved a Pearson correlation *r* of 0.77 for hourly predictions of rip-flow speed (*V*) against lifeguard-perceived rip-current hazard (*RHl*), with an accuracy of 0.55 in classifying hazard levels. After merging low (*RHm* = 0,1) and moderate-to-high hazard levels (*RHm* = 2,3,4), the accuracy improved to 0.83, with an F-Score of 0.81. The model effectively captured tidal modulation and wave energy influence, with the highest rip-current hazard observed at lower tidal elevations and higher wave energy (see the left-hand panels of Fig. 2). The daily mean correlation for rip-current hazard (*RHm* vs. *RHl*) was 0.82, indicating strong performance in predicting high-hazard days.

For shore-break wave hazard, the model achieved a Pearson correlation *r* of 0.71 between modelled shore-break wave energy (*Esb*) and lifeguard-perceived hazard (*SHl*) on an hourly basis, with an accuracy of 0.54. After merging low and moderate-to-high hazard levels, the accuracy rose to 0.83 with an F-Score of 0.72. The model correctly reflected the maximized shore-break wave hazard during higher tidal stages (see the right-hand panels of Fig. 2). The daily mean correlation between modelled shore-break hazard (*SHm*) and lifeguard-perceived hazard (*SHl*) was 0.74, with slight underestimation of the highest hazard days (*e.g.*, July 13 and 26).

Overall, the models demonstrated good performance in capturing the variations in both hourly and daily hazard predictions for both rip-current and shore-break wave conditions. These two models were compared to previous works, including [10, 11], with key differences: *e.g.*, our rip-flow model is simpler, not distinguishing between surf-zone conditions. The models were calibrated based on lifeguard-perceived hazard data, with good performance (accuracy and F-Score > 0.8) for predicting moderate to high hazard hours. For shore-break waves, the hazard model showed similar accuracy. The models provided daily and hourly hazard forecasts, with daily mean rip-current hazard highly correlated (*r* = 0.82) with lifeguard assessments. A simple wave factor, *Wf* [12], was also identified as a useful tool for predicting high rip-current hazard days, but with much less skill for high shore-break wave hazard days. The models can be applied broadly, but further validation is needed across different beach morphologies and wave climates.

Combining rip-current and shore-break models offers comprehensive hazard insights for beach safety management.

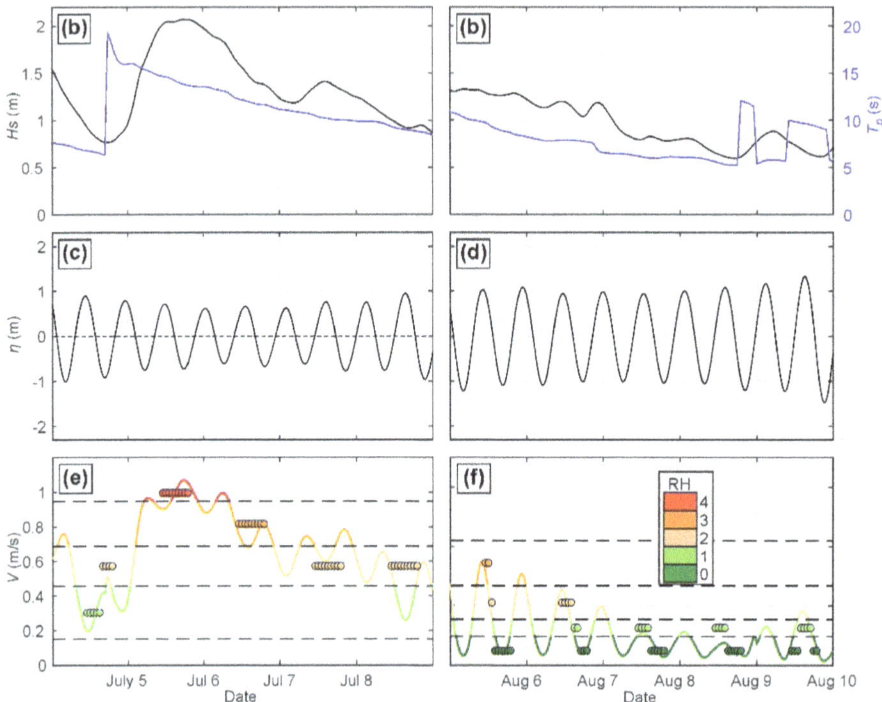

Fig. 2. Time series of (a,b) significant wave height in 10-m depth Hs and peak wave period Tp; (c,d) astronomical tide level η; (e,f) modelled hazard. Zoomed onto a 5-day period. Left-hand panels: rip-current hazard with rip-current flow speed V and its corresponding hazard level (colored); Right-hand panels: shore-break wave hazard with shore-break wave energy Esb and its corresponding hazard (colored). Lifeguard-perceived hazards are indicated but the colored bubbles.

Acknowledgements. This study received financial support from Project SWYM (Surf zone hazards, recreational beach use and Water safetY Management in a changing climate) funded by Région Nouvelle-Aquitaine and the French government in the framework of the University of Bordeaux's IdEx "Investments for the Future" program/RRI Tackling Global Change. We warmly thank the SMGBL (Syndicat Mixte de Gestion des Baignades Landaise), and particularly Stéphanie Barneix and the La Lette Blanche lifeguards who were on duty during the summer of 2022. We are also thankful to the Vielle Saint-Girons council for providing technical support and access to the lifeguard facilities.

References

1. Merz B, et al (2020) Impact forecasting to support emergency management of natural hazards. Rev Geophys 58:e2020RG000704

2. Bates PD, et al (2021) Combined modeling of US fluvial, pluvial, and coastal flood hazard under current and future climates. Water Resourc Res 57:e2020WR028673
3. Stokes C, Masselink G, Revie M, Scott T, Purves D, Walters T (2017) Application of multiple linear regression and Bayesian belief network approaches to model life risk to beach users in the UK. Ocean Coast Manag 139:12–23
4. Castelle B et al (2024) Environmental controls on lifeguard-estimated surf-zone hazards, beach crowds, and resulting life risk at a high-energy sandy beach in southwest France. Nat Hazards J Int Soc Prevent Mitigat Nat Hazards 120:1557–1576
5. Castelle B, Scott T, Brander R, McCarroll R (2016) Rip current types, circulation and hazard. Earth Sci Rev 163:1–21
6. Puleo J, Hutschenreuter K, Cowan P, Carey W, Arford-Granholm M, McKenna K (2016) Delaware surf zone injuries and associated environ-mental conditions. Nat Hazards 81:845–867
7. Scott T, Masselink G, Stokes C, Poate T, Wooler A, Instance S (2022) A 15-year partnership between UK coastal scientists and the international beach lifeguard community. Cont Shelf Res 241:104732
8. Dehez J, Lyser S, Castelle B, Brander RW, Peden AE, Savy JP (2024) Investigating beach-goer's perception of coastal bathing risks in southwest France. Nat Hazards 120:13209–132302
9. Castelle B, Dehez J, Savy J-P, Liquet S, Carayon D (2025) Physics-based forecast modelling of rip-current and shore-break wave hazards. Nat. Hazards Earth Syst. Sci. Discuss. https://doi.org/10.5194/nhess-2024-168 (in review)
10. Casper A, Nuss ES, Baker CM, Moulton M, Dusek G (2024) Assessing NOAA rip-current hazard likelihood predictions: comparison with lifeguard observations and parameterizations of bathymetric and transient rip-current types. Weather Forecast 39:1045–1063
11. Moulton M, Elgar S, Raubenheimer B, Warner JC, Kumar N (2017) Rip currents and alongshore flows in single channels dredged in the surf zone. J Geophys Res Oceans 122:3799–3816
12. Scott T, Masselink 520 G, Austin MJ, Russell P (2014) Controls on 0 rip current circulation and hazard. Geomorphology 214:198–215

RoadRAT - Regional Risk Assessment Tool for Coastal Roads in a Changing Climate

Caroline Hallin[1(✉)], Anna Adell[1], Björn Almström[1], Aart Kroon[2], and Magnus Larson[1]

[1] Lund University, John Ericssonsv 1, 221 00 Lund, Sweden
caroline.hallin@tvrl.lth.se
[2] University of Copenhagen, Øster Voldgade 10, 1350 Copenhagen, Denmark

Abstract. In this paper, a new framework – RoadRAT - for analyzing the probability of inundation, wave runup, and erosion is applied within the context of risk management for coastal roads. The analysis focus on a case study in south Sweden and consists of risk identification, calculation of probability, and discussion of consequences and mitigation methods. The results show that future risks are highly dependent on the future greenhouse gas emissions.

Keywords: Coastal flooding · Wave runup · Storm erosion · Sea level rise

1 Introduction

A new framework for analyzing the probability of inundation, wave runup, storm erosion and long-term coastal change has recently been developed [1]. The framework is called RoadRAT (Risk Assessment Tool for Roads) and is intended to assist road managers in planning and managing climate change adaptation of coastal roads. In this paper, the RoadRAT framework is applied to coastal roads along the south coast of Sweden and the results are discussed in relation to risk management.

There exist numerous definitions of risk. The International Standard [2] defines risk as the "effect of uncertainty on objectives". Risk is typically expressed in terms of risk sources, potential events, their consequences and their likelihood. In case of a coastal road, the source can be the sea, the potential event a storm surge flooding the road, consequences damage to the road or delays, and likelihood can be expressed as the probability of flooding.

Risk management is a methodology that consists of risk assessment and risk treatment to avoid or reduce negative impacts. The risk assessment is the overall process of risk identification, risk analysis, and risk evaluation [2]. The risk assessment process can be guided by the following questions [3]: (i) What can happen? (i.e., What can go wrong?) (ii) How likely is it that it will happen? (iii) If it does happen, what are the consequences? The questions are answered for a case study in the following sections, followed by a discussion about possible risk treatment strategies.

C. Coelho et al. (Eds.): CD 2025, CRL 41, pp. 236–242, 2026.
https://doi.org/10.1007/978-3-032-15473-6_37

1.1   Study Site

The study site is located on the south coast of Sweden (Fig. 1) and encompasses the coastal roads within the municipalities of Trelleborg, Skurup, and Ystad. The management of the roads is the responsibility of either the Swedish Transport Administration, the municipalities, or private road associations.

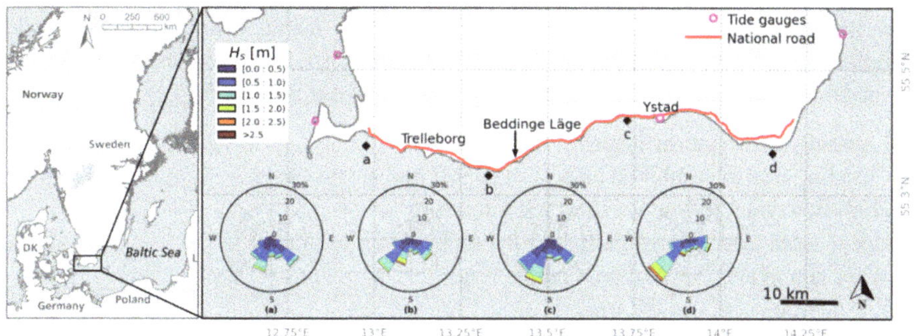

Fig. 1. Overview of study area. The wave roses are derived from hindcasted data at 10 m depth at the locations indicated with the corresponding letters (a–d).

The coast faces the Baltic Sea, which is a marginal sea with brackish water and negligible astronomic tide. The coast consists mostly of sand or gravel beaches with glacial or post-glacial origin. The area is subject to a small post-glacial uplift of 0.08 mm/year, which is outpaced by a sea level rise of about 3.4 mm/year [4]. Most of the coastline is in balance, but parts of the coast are more dynamic, and gradients in the longshore transport cause alternating erosion and accretion, especially in the eastern part of the study area.

Important sources of coastal risks are storm surges and waves. Within the study area, there were several water level gauges (Fig. 1), but no detailed information about the wave climate. The first step in the analysis was, therefore, to set up a wave model for the area [5]. The results revealed that there is a large longshore variability of the nearshore wave climate (Fig. 1).

Most wave energy approaches the coast from the western direction, but in some years, the eastern wave-energy component has dominated [5]. These changes are strongly correlated to the North Atlantic Oscillations (NAO-index). Long-term studies of wind climate directions dating back to the 16th century revealed a long-term wind climate variability that can explain the ongoing re-orientation of the coastline that causes coastal erosion in part of the study area [6].

The future development of the wind directions are uncertain and the projections are inconclusive. However, the inter-annual and inter-decadal variability of the wave climate will likely continue in the future [7]. More certainly, the increasing elevation of the mean sea level is expected to accelerate in the future. For the IPCC scenario SSP5-8.5, the median projection for relative sea level rise is 0.27 m, 0.80 m, and 1.29 m until 2050, 2100, and 2150, respectively [8]. The SSP5-8.5 scenario is the most extreme among the scenarios that the IPCC categorizes as likely—median projections until 2150 range from 0.59 m (SSP1-1.9) to 1.29 (SSP5-8.5). The still water level with 100 years return level is 1.45 m above normal sea level and there is a complex relationship between waves and surge levels [9].

2 What Can Happen?

The risk identification is based on exploring possible negative impacts on the road. Douglass and Webb [10] present a comprehensive list of the impacts of coastal processes on roads. The following are assessed as relevant for the study site:

- Flooding due to storm surge
- Flooding due to combined effects of runoff and storm surge
- High flow velocities associated with flooding
- Local scour due to flow, including the weir-flow mechanism
- Wave runup and wave impact on the road
- Washed up debris on the road
- Overwash deposits of sand
- Erosion due to long-term coastal evolution
- Storm erosion

   This study focuses on the impact of coastal processes. Thus, pluvial and river flooding should be treated in separate assessments. The drivers of the studied impacts are flooding due to storm surges, wave runup reaching the road, and erosion due to long-term and short-term changes in morphology.

   On October 20–21, 2023, the study area was affected by a storm surge in combination with large waves that were caused by strong eastern winds during the storm "Babet". The still water level in the study area reached 1.2–1.35 m above normal sea level and the significant wave heights have been hindcasted to 2–3 m.

   The roads within the study area were impacted by storm erosion and washed-up debris and seaweed (Fig. 2). In one of the locations, a small road owned by a private road association was completely washed away, probably due to a combination of long-term coastline evolution and storm erosion. On the east coast, adjacent to the study area, overwash deposits of sand had to be removed from roads and bike paths following the storm event.

Fig. 2. A private road was washed away in Beddinge läge (left panel). Debris was washed up on a parking lot in Kåseberga (right panel). Photos: Hallin, C.

3   How Likely is It that It Will Happen?

RoadRAT is a recently developed framework for calculating the probability of inundation, wave runup, and erosion impacting coastal roads (Fig. 3). The calculations apply to the present climate based on historical data of waves and water levels. Predictions of the probability of impact in the future should consider the historical trends in coastline evolution as well as future development due to sea level rise.

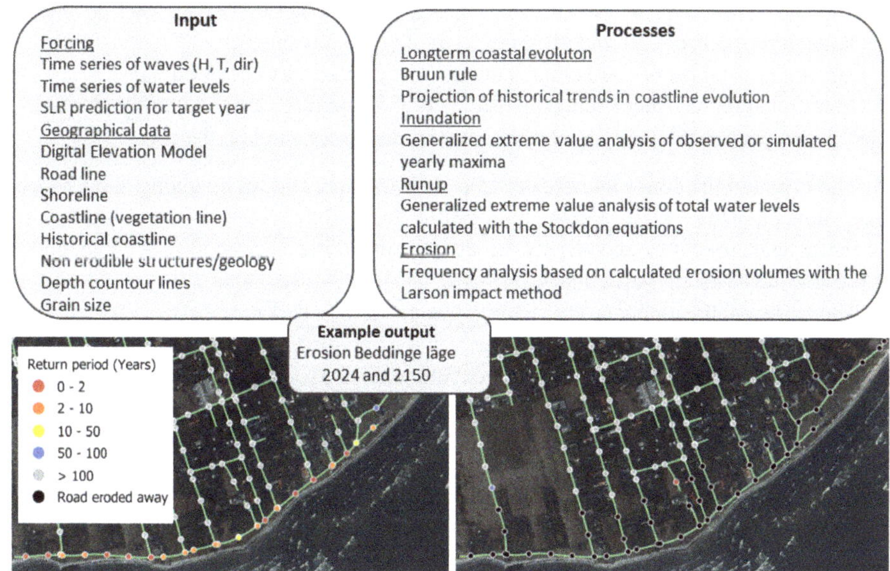

Fig. 3. Overview of input and included processes in RoadRAT and an example of results.

RoadRAT was applied to the national road (Fig. 1) along the coast [1]. The results showed that at present, a few hundred meters of the road had an annual exceedance probability (AEP) of the impact of inundation, wave runup, and erosion impacting the road larger than 1%, i.e. return periods ranging from 1–100 years. In the future, until the year 2150, several kilometres of the road are calculated to be exposed to processes with AEP exceeding 1% (inundation: 11 km, runup: 8 km, erosion: 37 km). In the cities of Trelleborg and Ystad, inundation due to storm surges has the largest impact, while outside the cities, the long-term coastal evolution and storm erosion are the main threats against the road. The results showed large differences depending on the assumed SSP scenarios. Comparing the SSP1-1.9 scenario to the SSP5-8.5 scenario, the length of the road impacted by a 100-year event decreased by 80%, 60%, and 50% for inundation, runup, and erosion, respectively.

On the seaward side of the national road are more exposed roads that the municipalities and private road associations manage. Figure 3 displays an example of a road network in Beddinge läge and the probability of impact due to erosion in present and future conditions.

4   If It Does Happen, What are the Consequences?

The consequences of the impact of coastal processes can include damage to the road, permanent or temporary operational disruptions, and damage to vehicles and road users. This project has focused on the likelihood of impact and not on its consequences. But in general terms, the processes that are analyzed – storm surges, wave runup, and erosion as a result of long-term coastline changes and storm erosion – can give rise to the following consequences [10]:

– Damage to roads due to undermining or damage to pavement
– Damage to road users and vehicles due to wave impact and/or flooding
– Operational disturbances requiring the road to temporary or permanently close
– Reduced accessibility or speed due to water, debris, or sand on the road

The extent of the consequences depends on how often and for how long disturbances occur, how busy the road is, and the possibilities for traffic diversion. Roads where there are no alternative routes and where the disruptions affect accessibility for evacuation and rescue services are particularly sensitive.

The costs of the impacts are mainly due to delays and repair. In addition, there are costs for climate adaptation measures, such as coastal protection and relocation of coastal roads. In the future, the development of traffic and transport patterns influence the consequences. Thus, both the probability of impact and the consequences may change with time.

The results from the RoadRAT-analysis could be linked to models that describe the consequences of the impact of coastal processes on coastal roads [11, 12].

5   How to Adapt Coastal Roads to a Changing Climate?

Overarching adaptation strategies for risk mitigation along coastal roads are: manage and maintain; increase redundancy; protect; accommodate; and relocate [10]. When choosing between these different strategies, it is necessary to consider the public road network as one entity and not divided among different road owners. For some of the roads within the study area, relocation will likely be needed in the future due to the impact of long-term coastal evolution,. However, in the shorter term *no-regret* solutions, such as nourishments could be used as a temporary mitigation method [13].

In response to the damage during the storm Babet, hard protection, such as rock revetments, are being installed at several places within the study area. Considering the negative downdrift effects of such measures on sandy coastlines [14], nourishments could be used as a complement to create hybrid structures [15]. A drawback is that hybrid structures are complicated to design and require detailed information about the sediment dynamics.

In conclusion of the discussed questions, the RoadRAT framework is an efficient and robust tool for analyzing identified risks and evaluate adaptation strategies. In other study areas, the framework can be extended and modified depending on possible impact and appropriate descriptions of the physical processes.

Acknowledgements. This study was funded by the Swedish Transport Administration (TRV 2019/96299).

References

1. Hallin C, Adell A, Almström B, Kroon A, Larson M (2025) RoadRAT a new framework to assess the probability of flooding, waverunup, and erosion impacting coastalroads. Coast Eng 199:104741
2. International Standard: ISO 31000:2018(E) Risk management – guidelines (2018)
3. Kaplan S, Garrick BJ (1981) On the quantitative definition of risk. Risk Anal 1(1):11–27
4. Madsen KS, Høyer JL, Suursaar Ü, She J, Knudsen P (2019) Sea level trends and variability of the Baltic Sea from 2D statistical reconstruction and altimetry. Front Earth Sci 7:1–12
5. Adell A, Almström B, Kroon A, Larson M, Uvo CB, Hallin C (2023) Spatial and temporal wave climate variability along the south coast of Sweden during 1959–2021. Reg Stud Mar Sci 63(103011):1–11
6. Larson M, Fredriksson C, Hanson H (2016) Changing wind properties in south Sweden. VATTEN – J Water Manag Res 72(2):117–128
7. Soomere T (2023) Numerical simulations of wave climate in the Baltic Sea: a review. Oceanologia 65(1):117–140
8. Hieronymus M, Kalén O (2020) Sea-level rise projections for Sweden based on the new IPCC special report: the ocean and cryosphere in a changing climate. Ambio 49(10):1587–1600
9. Hanson H, Larson M (2008) Implications of extreme waves and water levels in the southern Baltic Sea. J Hydraul Res 46(sup2):292–302
10. Douglass SL, Webb BM (2020) Highways in the coastal environment. Hydraulic Engineering Circular No. 25, FHWA-HIF-19- 059. Fairhope, AL (2020)
11. Van der Meer JW, Allsop NWH, Bruce T, De Rouck J, Kortenhaus A, Pullen T, Schüttrump (2018) Manual on wave overtopping of sea defences and related structures. An overtopping manual largely based on European research, but for worldwide application. EurOtop
12. Pregnolato M, Ford A, Wilkinson SM, Dawson RJ (2017) The impact of flooding on road transport: a depth-disruption function. Transp Res D Transp Environ 55:67–81
13. Fredriksson C, Almström B, Hanson H, Larson M, Persson O (2017) Estimation of required beach nourishment volumes along the south coast of Sweden during 2017–2100. VATTEN   J Water Manag Res 73:77–84

14. Nawarat K, Reyns J, Vousdoukas MI, Duong TM, Kras E, Ranasinghe R (2024) Coastal hardening and what it means for the world's sandy beaches. Nat Commun 15(10626)
15. Adell A, Kroon A, Almström B, Larson M, Hallin C (2024) Observed beach nourishment development in a semi-enclosed coastal embayment. Geomorphology 462(109324):1–13

Decision-Support Tools for Beach Management in Context of Climate Change

Céline Trmal[(✉)] and David Criado

Cerema Méditerranée, Aix-en-Provence, France
`celine.trmal@cerema.fr`

Abstract. In order to adapt to climate change, local authorities are encouraged to develop management strategy for their coastline. Thus, the Toulon Provence Méditerranée Metropolitan authority has joined forces with Cerema to initially carry out a hydro-sedimentary diagnostic and an assessment of current beach management. Then a beach classification methodology was developed based on various criteria. For urban beaches, multi-criteria analysis is a decision-making tool aimed at prioritizing action in terms of solutions to be implemented, by rationalizing investment levels for the public body, according to an intervention schedule (short-term/medium-term/long-term). The criteria involve the evolution of the coastline of the beaches and the future permanent submersion of nearby stacks, economic attractiveness and environmental value. Depending on the value of the synthetic index calculated, families of beach management solutions emerge, ranging from nature-based solutions for low-index beaches to reinforcement solutions for high-index beaches in the short term. In the medium and long term, the relocation of stakes exposed to retreating coastlines and rising sea levels is essential to increase beach resilience.

Keywords: Coastal management · Shoreline retreat · Sea level rise · Climate change · Beaches

1 Introduction

1.1 Call for Partners Integrated Management Coastline

Adapting coastal areas to the multiple pressures they face is a real challenge, which will be enhanced by the impacts of climate change and global warming, particularly rising sea levels [1].

To meet this challenge, Cerema and the Association Nationale des Elus du Littoral (National Association of Coastal Elected Officials) launched a call for partners at the end of 2019, with the aim of supporting volunteer coastal territories in a dynamic that will enable them to address their local issues, identify practical long-term solutions, and contribute to bringing the land-sea interface to life in a virtuous way.

The Toulon Provence Méditerranée Metropolitan authority applied for the project in order to improve understanding of hydro-sedimentary process of its coastline, identify and prioritize vulnerable sectors and adapt current coastal management methods to climate change.

C. Coelho et al. (Eds.): CD 2025, CRL 41, pp. 243–248, 2026.
https://doi.org/10.1007/978-3-032-15473-6_38

1.2   The Aims of the Project

The Toulon Provence Méditerranée metropolitan area boasts a highly varied coastline, ranging from pocket beaches between rocky capes to long sandy beaches bordering low-lying lagoon systems (see Fig. 1).

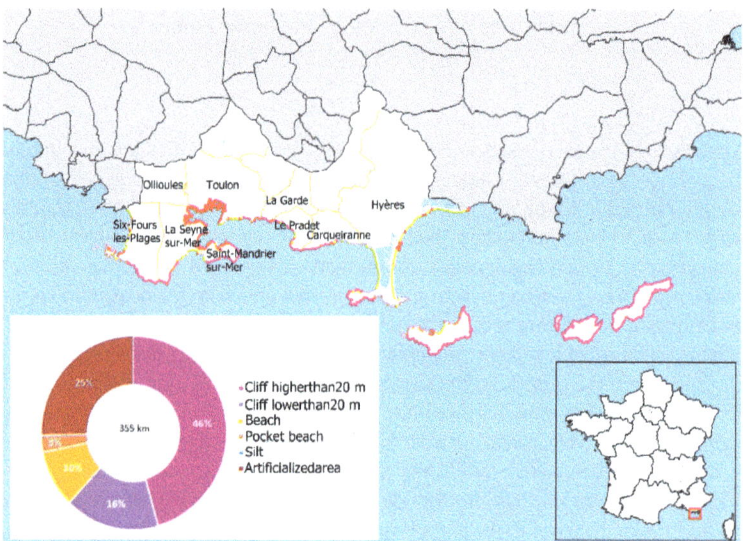

Fig. 1. Localization of the Toulon Provence Méditerranée Metropolitan area.

This coastline has been the subject of many experiments in beach management over the last few decades, including the deployment of geotubes in the early 2000s, and armourstone removal operations to allow beaches to retreat…

At its creation in 2018, the Toulon Provence Méditerranée Metropolitan authority took over the responsibility of several more or less advanced municipal projects, projects often disconnected despite certain similarities.

In order to standardize practices across the area and prioritize public action, in a context of changing marine forcing due to climate change, implementing an integrated beach management strategy is becoming essential.

2   The Coastline Diagnostic

As a first axis, a bibliographic analysis was carried out to describe the physical functioning of the coastline. It has been possible to bring together knowledge that has been scattered across academic and engineering studies for decades.

Interviews with all coastal municipalities were carried out in order to provide a shared vision of these findings, and to obtain information on the current beach management method.

New data was produced in order to have a homogeneous and updated analysis of the past evolution of the coastline. It allowed to identify future sensitive points.

This diagnosis was presented in the form of summary sheets on the scale of the nine homogeneous hydro-sedimentary units. Interactive maps containing the data collected and produced have been developed to facilitate the appropriation of the project stakeholders.

3 Construction of the Decision Support Tool

The coastal diagnostic has identified numerous sensitive sectors, some today and a few in the future. It was decided to concentrate the work on the coastal areas with beaches.

The decision support tool presented here aims to homogenize beach management solution in urban areas, with the common thread being the preservation of a beach in the future.

Global methodologies exist to evaluate coastal solution performance, but few are based on multi-criteria analyses [2, 3]. However, those tools are not sufficiently mature to be implemented in a changing regulatory and physical context. It was therefore decided to develop a tool that would enable an enlightened choice of families of management solution adapted to the beach under study, taking into account its medium- and long-term evolution and the regulatory framework.

3.1 Scoring Method of Beaches

The proposed decision-support tool is based on a scoring method designed to calculate a synthetic index.

The criteria used are relatively simple but objective, each criterion being broken down into sub-criteria not specified here. Then a weighting is given to each criterion to calculate the synthetic index:

- Beach width (mobility space) – Weighting 30%
- Building issues, infrastructures, economic stakes that are difficult to move and exposed to shoreline retreat and flooding – Weighting 35%
- Building issues, light infrastructures, economic stakes that are easier to move, exposed to shoreline retreat and flooding – Weighting 15%
- Economic stakes of the coastal fringe – Weighting 10%
- Environmental value of the beach – Weighting 10%.

The synthetic index makes possible to qualify the beaches according to the following levels:

- High index: score higher than 50
- Medium index: score between 30 and 50
- Low index: score lower than 30.

The index is calculated over the short/medium term and medium/long term (see Fig. 2). In fact, the first three criteria depend on the evolution of the beach shoreline and coastal hazards, evolution largely linked to the sea level rise due to climate change.

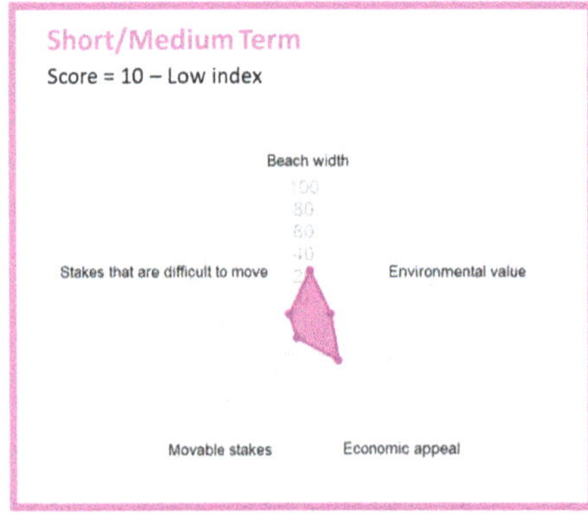

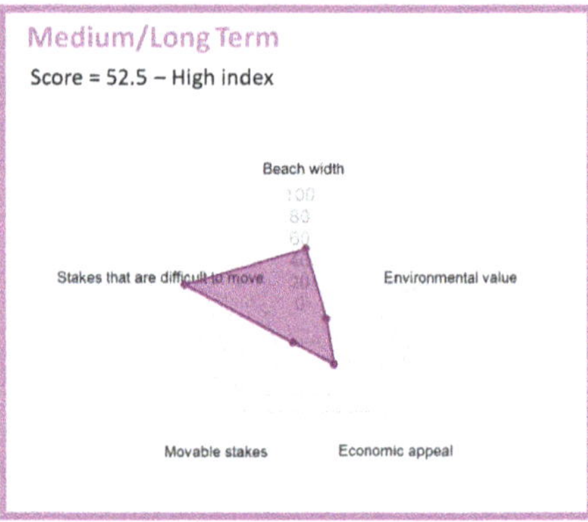

Fig. 2. Example of the scoring of a beach situated closed to an urbane area.

The scoring method was tested on a limited sample of beaches, so as not to adapt the scoring to predefined actions.

To adapt to the regulatory context, natural beaches (i.e. outside urban areas) are rated only on their environmental value, using a grid that aims to identify anthropogenic pressures, leading to families of solutions designed to reduce them, and thus improve their resilience to climate change.

3.2 Implications of the Index

As written, the index is calculated over the short/medium term and medium/long term. Both temporalities are important and necessary to understand how the exposure of the beach and its stakes will evolve over time. Thus, informed decisions can be taken immediately but especially in a more distant temporality, to give time to think and adapt responses as early as possible to make the beach and coastline more resilient.

Depending on the values of the synthetic index, families of beach management solutions emerge, among four large families:

- F1: Hard Engineering Solutions, such as seawalls, armourstone, groynes, breakwaters
- F2: Soft Solutions, introducing a gradation between:

  o F2a: Soft Control Solutions, such as geotubes and beach nourishment (with exogenous sand).
  o F2b: Soft Management Solutions, such as the remobilization of sandy stocks and the management of Posidonia banquettes.

- F3: Support for natural processes – regulation and renaturation, such as the non-management of Posidonia banquettes, de-artificialization, and the diminution of human frequentation.
- F4: Reduction of exposure and vulnerability of assets, such as the release of beach crests.

A flowchart (see Fig. 3) is proposed to aim at choose the solution families adapted to the synthetic index. Note that over time the synthetic index doesn't decrease. In the context of that study, families 1 and 2a will be preconized only at short term, because they can't be sustainable.

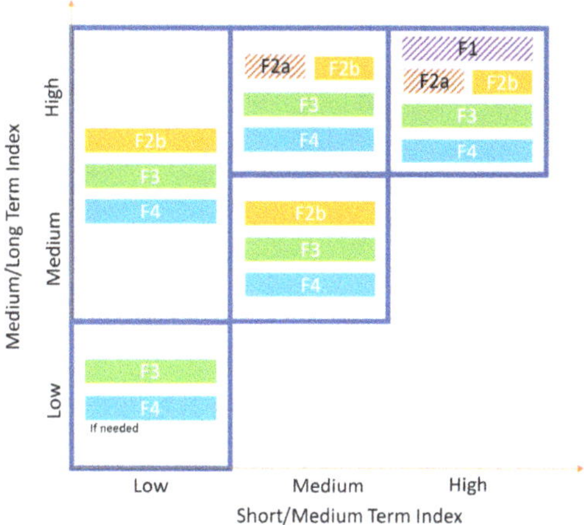

Fig. 3. Flowchart of the decision-support tool

4   Results

This decision-support tool is currently being implemented on around a hundred beaches in the Toulon Provence Méditerranée Metropolitan area, a third of which are located in urban or highly urban areas, and two thirds in natural areas. It is accompanied by a best practice guide detailing solutions families.

Management and action are possible on all urban beaches. Given the current regulations, it is not recommended to turn to hard solutions, only two beaches could use them while integrating a reflection on reducing the vulnerability of the stacks in the medium/long term. The number of beaches increases to six when we include the soft control solutions.

In conclusion, this tool has the advantage of applying the same methodology to all the beaches selected, and prioritizing public action on a metropolitan scale.

The implementation of this tool will require regular reassessments of the criterion. In particular the predominant one, taking into account the beach width evolution, is calculated with regards of past beach management and do not include possible in the future. The evolution of coastal structures and beaches with sea level rise is technically difficult to assess. The establishment of a coastal observatory is essential for future decision-making.

Finally, this decision-making tool does not replace the impact studies to be carried out on a hydro-sedimentary cell scale once a family of solutions has been identified.

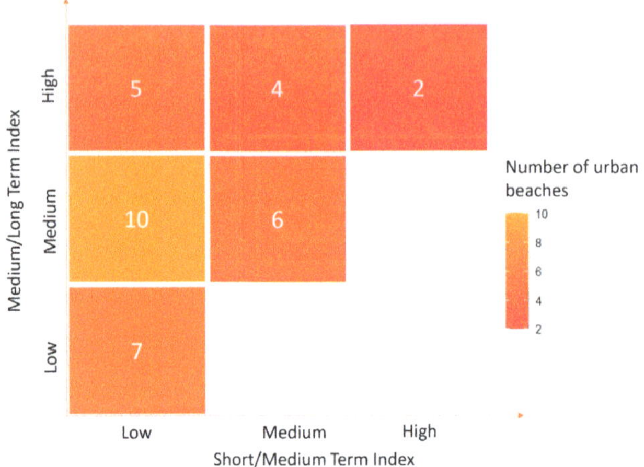

Fig. 4. Number of urban beaches on each category

Acknowledgment. We would like to acknowledge the support of the Toulon Provence Méditerranée metropolitan authority for this partnership, particularly Olivier Le Neannec, Yohan Guérin and Dorian Roggero, as well as technical committee members.

References

1. IPCC (2021) Climate Change 2021: The Physical Science Basis. Contribution of Working Group I to the Sixth Assessment Report of the Intergovernmental Panel on Climate Change
2. Kumar P et al (2021) An overview of monitoring methods for assessing the performance of nature-based solutions against natural hazards. Earth-Sci Rev 217:103603. https://doi.org/10.1016/j.earscirev.2021.103603
3. Philippenko X (2023) L'adaptation au changement climatique dans un territoire côtier: l'archipel de Saint-Pierre-et-Miquelon. https://theses.hal.science/tel-04588557

Beyond Coastal Hazards and Physical Susceptibility: A Sustainable Coastal Vulnerability Index (SCVI) for the County Cork Coastline, Ireland

Emma Chalençon[1,2]([✉]), Fiona Cawkwell[1], Michael O'Shea[2], and Jimmy Murphy[2]

[1] Department of Geography, University College Cork, Cork T12 K8AF, Ireland
emma.chalencon@ucc.ie
[2] MaREI Centre, Environmental Research Institute, University College Cork, Ringaskiddy, Cork P43 C573, Ireland

Abstract. Coastal regions are experiencing increasing pressure from climate change and human activity, with County Cork, Ireland, particularly at risk due to its high percentage of population living near the coast. Recognizing the limitations of traditional Coastal Vulnerability Indices (CVIs), which emphasize physical susceptibility while neglecting broader socioeconomic and environmental factors, this study proposes a novel framework for assessing coastal vulnerability.

Building upon recent reviews of CVIs, this research integrates a diverse range of variables spanning physical, sociocultural, economic, as well as environmental dimensions into a Sustainable Coastal Vulnerability Index (SCVI). Independent sub-indices, incorporating both exposure to coastal hazards and the capacity of communities and ecosystems to adapt, offer a nuanced perspective on vulnerability when combined. This integrated index moves beyond traditional assessments focused solely on hazards and physical susceptibility, delivering a broader evaluation of the interconnected systems shaping coastal vulnerability and resilience, beyond an anthropocentric approach. It emphasizes the need for interdisciplinary research and collaboration, leveraging insights from social and natural sciences. The variables are easy to determine, making it adaptable to the specific needs and priorities of local ecosystems and communities. The SCVI provides a foundation for more informed decision-making, supporting strategies that balance environmental, social, and economic considerations to enhance the long-term sustainability of coastal regions.

Keywords: Coastal · Vulnerability · Sustainability · Transdisciplinary

1 Introduction

Human activities, coupled with the impacts of a changing climate, rising sea levels and increased storm intensity, exert pressure on coastlines worldwide, altering coastal dynamics. Coastal erosion and flooding can ultimately cause habitat and ecosystem loss, damage infrastructure, and disrupt social, environmental and economic systems [1]. At

C. Coelho et al. (Eds.): CD 2025, CRL 41, pp. 249–254, 2026.
https://doi.org/10.1007/978-3-032-15473-6_39

both the international and European levels, the Intergovernmental Panel on Climate Change (IPCC) and the European Union (EU) urge governments to develop vulnerability assessments and implement strategies to reduce exposure to climate variability while enhancing adaptation [2]. Two challenges can be identified when developing vulnerability indices: defining the systems under consideration, which is closely tied to the chosen definition of vulnerability; and selecting appropriate variables [3].

The concept of vulnerability is widely applied in coastal sciences; however, it remains a contested concept due to the often ambiguous and poorly understood relationships between its components [4]. Even within the IPCC, the definition of vulnerability has evolved significantly [5], becoming a function of susceptibility to harm and capacity to cope and adapt [6] with the transition from a vulnerability to a risk framework occurring in 2014 (AR5). This transition recognizes that a significant proportion of impacts are triggered by hazardous events [6]. Recent reviews have highlighted the erroneous designation of many Coastal Vulnerability Indices (CVIs) [7]. The earliest CVIs were primarily focused on coastal hazards and physical susceptibility [8]. Many studies still follow this methodology, however, by failing to integrate socioeconomic factors they provide an incomplete picture, neglecting the broader social and economic contexts that shape coastal vulnerability. CVIs are decision-making tools that rely on simplified representations of complex processes. To maintain their scientific robustness, it is crucial to provide clear and transparent disclosure of their foundations and limitations [9]. The selection of variables and the datasets used to represent them is seldom discussed in detail. While data and methodological constraints are unavoidable, failing to identify and properly flag their influence on the results can lead to misinterpretation and inappropriate policy decisions [4].

Finally, reviews highlight the importance of customizing and tailoring approaches to suit the specific context of each study [4]. With 61% of its population living within 5 km of the coastline and the highest proportion of residents living within 100 m of the coast among all Irish counties [10], Cork County is considered "particularly vulnerable" [11]. Caloca-Casado [2] explains that previous vulnerability analyses in Ireland were primarily focused on socio-economic scenarios, but by adapting Gornitz's methodology [8], Caloca-Casado proposes a high-resolution, index-based approach to assess the relative vulnerability of the coast to sea-level rise and flooding, incorporating six to eight parameters that address coastal hazards and physical susceptibility. Based on Caloca-Casado's recommendations [2] a national CVI is being developed by Geological Survey Ireland, however by focusing on the County Cork coastline, this research demonstrates the advantages of adopting a regional, holistic, and transdisciplinary perspective for conducting integrated vulnerability assessments.

2 Going Beyond Coastal Forcings and Physical Susceptibility

Coastal vulnerability assessments consider a variety of aspects that can be classified into three components: physical, geological, and socioeconomic [12]. Physical and geological processes are usually at the center of assessments, with coastal forcing variables used to evaluate the likelihood of coastal hazards, and geological parameters used to quantify coastal systems' resistance. Socioeconomic variables are typically used to assess the social and economic value of resources and the necessity of protecting them [8]. The parameters that fit within these broad headings, are here reframed to create four sub-indices, with some variables responsible for determining the degree of exposure and others the degree of adaptive capacity within the coastal system (Fig. 1). Following the recommendations of Rocha et al. [7], physical and geological processes form the Coastal Hazard Index and a Physical Susceptibility Index. In this research the Coastal Hazard Index is defined as the combination of coastal forcing parameters, including Relative Sea Level Rise (rSLR) and wave climate, as well as coastal responses, with shoreline change rates and the probabilities of flooding and storm surge extents (Fig. 1). The Physical Susceptibility Index incorporates six parameters with tidal range, slope, elevation, distance to the shoreline, geomorphology and groundwater susceptibility to intrusion. The inclusion of coastal defenses within the Physical Susceptibility Index is recommended, however such a database is not yet available in Ireland.

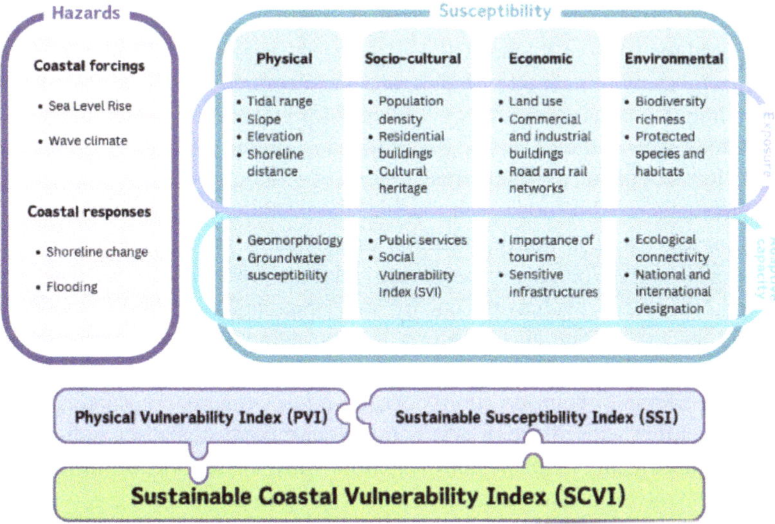

Fig. 1. Proposed Sustainable Coastal Vulnerability Index adapted from Rocha et al. [7].

Many studies have highlighted that the omission of demographic and economic factors in coastal vulnerability indices considerably limits the assessment of vulnerable areas by focusing solely on coastal susceptibility [7, 13]. Bukvic et al. [4] critique the tendency to justify a disciplinary focus by emphasizing the limitations of socio-economic data in terms of availability, resolution, and reliability, rather than addressing the challenges of adopting a more holistic approach that encompasses vulnerability more broadly. When socioeconomic variables are included, they are largely economic in nature. The integration of socio-cultural susceptibility, which is linked to demographic and cultural structures, social inequities, awareness, or preparedness is rare, even though it drives the ability of communities to adapt and be resilient [14].

This research separates socio-cultural and economic susceptibility, diverging from the approach of Rocha et al. [7]. By isolating socio-cultural susceptibility, the study emphasizes the importance of social equity, cultural preservation, and community preparedness in shaping resilience and adaptive capacity, complementing rather than combining these factors with economic considerations. This distinction aligns with the concept of sustainability, which views the social and economic dimensions as distinct yet interdependent poles that must be balanced to achieve sustainable outcomes [1]. The Socio-Cultural Index developed in this study includes population density, residential buildings, cultural heritage sites, public services, and the Irish Social Vulnerability to Environmental Hazards Index (ISVEHI) [15] (Fig. 1). The Economic Susceptibility Index includes land use, business and industrial buildings, rail and road networks, tourism, and sensitive infrastructure or licensed facilities, such as wastewater treatment plants (Fig. 1). This index captures the economic assets at risk, providing a comprehensive understanding of the economic dimension of coastal vulnerability.

3 Making CVIs Sustainable

Going beyond the work of Rocha et al. [7], this study argues that, in addition to socioeconomic factors, environmental variables must be integrated to provide a comprehensive understanding of coastal vulnerability. The need for vulnerability assessments is driven by institutions such as the IPCC and the European Commission, which emphasize sustainability by recognizing the interconnectedness of economic, social, and environmental dimensions that should be addressed together [16]. While sustainability is often referenced as an aspiration in the introduction of studies, it is rarely implemented when designing CVI methodologies or selecting variables. This absence of environmental factors in CVI reviews, such as the one by VanZomeren and Acevedo-Mackey [12], is concerning for studies that aim to inform local sustainable coastal management [17]. Incorporating ecosystems, biodiversity and conservation status into a CVI is complex [18]. The management of conservation sites often requires maintaining natural processes, which may conflict with planning efforts to preserve ecosystems in a non-equilibrium state, but it cannot justify the exclusion of environmental factors. The environment should also not simply be evaluated in terms of contributing to or mitigating geophysical and socioeconomic vulnerability, but as a vulnerable system in itself [4].

To address the need for a more comprehensive and sustainable CVI, this research incorporates an Environmental Susceptibility Index composed of parameters sourced from the Irish National Parks and Wildlife Service. First, biodiversity richness accounts for the variety of species and habitats present, recognizing that higher biodiversity contributes to ecological resilience. Second, the presence of protected species and habitats highlights areas of conservation importance, ensuring that vulnerable or endangered species are factored into susceptibility assessments. Third, the level of ecological connectivity measures the integrity and continuity of ecosystems, which are critical for maintaining natural processes and species migration. Finally, the strength of conservation designations reflects the legal and policy protections in place, acknowledging that stronger designations can enhance resilience. Together, these parameters enable a more robust evaluation of environmental vulnerability, ensuring that ecosystems and biodiversity are given due consideration alongside social and economic factors in the CVI.

4   Addressing Both Exposure and Adaptive Capacity

In order to align with the IPCC's new vulnerability definition, as a function of suscepti-bility to harm and capacity to cope and adapt [6], both exposure and adaptive capacity need to be considered for each element of the susceptibility system (physical, socio-cultural, economic, and environmental). This requires a thorough evaluation of both the exposure or assets that are vulnerable to damage from hazards, and the adaptive capacity or ability to adjust and cope with these hazards (Fig. 1).

The Physical Susceptibility Index, with factors such as tidal range, slope, elevation, and shoreline distance, highlights exposure, while geomorphology and groundwater susceptibility reflect the adaptive capacity of the physical system. Together with the coastal forcings and responses, these form the Physical Vulnerability Index. Sociocul-tural susceptibility, where population density, residential buildings, and cultural heritage are key assets at risk, and public services and social vulnerability provide measures of society's capacity to cope with coastal hazards within the Socio-Cultural Index. In the Economic Index, land use, commercial and industrial buildings, and transport infras-tructure represent assets at risk, while dependencies on tourism and sensitive or licensed infrastructures offer insights into secondary risks and adaptive capacity. Finally, in the Environmental Index, biodiversity, protected species, and habitats that are particularly vulnerable are identified, whereas ecological connectivity and conservation designations signal the capacity of ecosystems to adapt to coastal hazards and change. Together, the components within these three indices form the Sustainable Susceptibility Index.

By identifying relevant variables for each system and combining the five subindices into a Sustainable Coastal Vulnerability Index (SCVI) within a Geographical Information System (GIS), a comprehensive understanding of the coast's vulnerability can be gained, and more effective strategies for mitigation and adaptation devised.

5   Conclusion

The Sustainable Coastal Vulnerability Index (SCVI) proposed for the Cork coastline represents a significant advance over traditional CVIs. By incorporating insights from Rocha et al. [7], separating socio-cultural and economic factors and further integrating environmental variables, the SCVI ensures a comprehensive assessment where suscep-tibility extends beyond mere exposure to also consider adaptive capacity, aligning with the updated IPCC vulnerability definition and offering a more holistic understanding of coastal vulnerability to better inform local sustainable coastal management.

References

1. IPCC (2018) Global Warming of 1.5°C: IPCC Special Report on Impacts of Global Warming of 1.5°C above Pre-industrial Levels in Context of Strengthening Response to Climate Change, Sustainable Development, and Efforts to Eradicate Poverty, 1st edn. Cambridge University Press
2. Caloca-Casado S (2018) Coastal vulnerability assessment of Co. Dublin and Co. Wicklow to impacts of sea-level rise (Ph.D. thesis). National University of Ireland, Maynooth

3. Oloyede M, Williams A, Benson N (2022) A review of coastal vulnerability assessment methods. Int. J. Climate Change: Impacts Responses 14(2):95–103

4. Bukvic A, Rohat G, Apotsos A, de Sherbinin A (2020) A systematic review of coastal vulnerability mapping. Sustainability 12(7):2822

5. Estoque RC, Ishtiaque A, Parajuli J, Athukorala D, Rabby YW, Ooba M (2023) Has the IPCC's revised vulnerability concept been well adopted? Ambio 52(2):376–389

6. GIZ and EURAC (2017) Risk Supplement to the Vulnerability Sourcebook – Guidance on how to apply the Vulnerability Sourcebook's approach to the new IPCC AR5 concept of climate risk

7. Rocha C, Antunes C, Catita C (2023) Coastal indices to assess sea-level rise impacts - a brief review of the last decade. Ocean Coastal Manag 237:106536

8. Gornitz V (1991) Global coastal hazards from future sea level rise. Palaeogeogr Palaeoclimatol Palaeoecol (Glob Planet Change Sect) 89:379–398

9. Harik G, et al (2017) Implications of adopting a biodiversity-based vulnerability index versus a shoreline environmental sensitivity index on management and policy planning along coastal areas. J Environ Manag 187:187–200

10. Central Statistics Office (CSO) (2016). https://www.cso.ie/en/releasesandpublications/ep/p-cp2tc/cp2pdm/pd/. Accessed 08 May 2024

11. Cork County Council: Cork County Council Climate Adaptation Strategy 2019–2024 (2019). https://www.corkcoco.ie/sites/default/files/2021-11/cork-county-council-climate-adaptation-strategy-2019-2024-pdf.pdf. Accessed 08 May 2024

12. VanZomeren C, Acevedo-Mackey D (2019) A review of coastal vulnerability assessments: definitions, components, and variables. Engineer Research and Development Center (U.S.)

13. Mclaughlin S, Cooper JAG (2010) A multi-scale coastal vulnerability index: A tool for coastal managers? Environ Hazards 9(3):233–248

14. Li K, Li GS (2011) Vulnerability assessment of storm surges in the coastal area of Guangdong Province. Nat Hazards Earth Syst Sci 11(7):2003–2010

15. Fitton JM, O'Dwyer B, Maher B (2022) Developing a social vulnerability to environmental hazards index to inform climate action in Ireland. Irish Geogr 54(2):157–180

16. Borrell, J.: Social and environmental sustainability is a global public good. In EEAS (2024). https://www.eeas.europa.eu/eeas/social-and-environmental-sustainability-global-public-good_en. Accessed 18 Dec 2024

17. Myers MR, et al (2019) A multidisciplinary coastal vulnerability assessment for local government focused on ecosystems, Santa Barbara area, California. Ocean Coastal Manag 182:104921

18. McLaughlin S, McKenna J, Cooper JAG (2002) Socio-economic data in coastal vulnerability indices: Constraints and opportunities. J Coastal Res 36:487–497

Overview of New Advances in Modeling Sediment Transport Caused by Tsunami Waves

Ernesto Guerrero Fernandéz[(✉)] and Yong Wei

Cooperative Institute for Climate, Ocean and Ecosystem Studies, University of Washington, John M. Wallace Hall, 3737 Brooklyn Ave NE, Seattle, WA 98105, USA
ernestog@uw.edu

Abstract. In this work we provide an overview of recent developments at the NOAA Center for Tsunami Research (NCTR) for sediment modeling in the framework of tsunamis. We describe a new numerical approach of the well-known shallow-water equations with an additional equation for the transport of the moving bed. In this approach, the eigenvalue structure of the model is leveraged into the path-conservative numerical discretization, resulting in a fast, weakly-coupled and accurate numerical approach. We present some successful comparisons with survey data. Additionally, we explore other potential capabilities of the model, like tide prediction or historical tsunami recreation.

Keywords: Tsunami modelling · shallow-water · sediment transport · GPU implementation

1 Introduction

Natural hazards, such as tsunamis, are among the most destructive phenomena in the geophysical sciences, causing widespread damage to both natural landscapes and human infrastructure. These events often trigger significant sediment transport and deposition, which in turn influence the hydrodynamic behavior of the affected areas. Understanding and predicting sediment transport during extreme events is therefore critical for hazard assessment, risk mitigation, and environmental management.

Sediment transport modeling has traditionally relied on coupling hydrodynamics with morphodynamics through the evolution of the sediment bed, often referred to as the moving bed function (see [1] or related references for sediment modeling [1–3]). This coupling introduces a range of challenges, including the handling of non-conservative product terms and the design of robust numerical schemes capable of maintaining physical accuracy. The shallow-water equations (also known as the Saint-Venant equations, [4]) form the foundation for many sediment transport models, providing an efficient framework for simulating large-scale geophysical flows. By depth-averaging the governing equations, the computational complexity is significantly reduced, making these models particularly suitable for studying large domains, such as coastal regions affected by tsunamis.

© The Author(s) 2026
C. Coelho et al. (Eds.): CD 2025, CRL 41, pp. 255–260, 2026.
https://doi.org/10.1007/978-3-032-15473-6_40

Incorporating sediment dynamics into shallow-water models requires extending the system with transport equations that account for the sediment bed's evolution. A variety of approaches have been developed to model sediment flux, including empirical formulations such as the Meyer-Peter and Müller equation (see [5]) or Van Rijn's formulations [6], among others. However, the choice of sediment flux model introduces additional uncertainties due to the difficulty of accurately characterizing sediment properties, such as grain size, density, porosity and others. Consequently, achieving a balance between physical accuracy and uncertainties remains a core challenge in sediment transport modeling.

Numerical methods for coupled morpho-hydrodynamic models face unique obstacles, particularly in preserving critical properties like the well-balanced property, i.e., the ability to maintain equilibrium solutions without introducing numerical noise (see [7]). While fully coupled solvers provide more stability by considering the complete eigenstructure of the system, they are computationally more expensive. Alternative approaches, such as weakly coupled solvers, aim to reduce computational costs while maintaining sufficient accuracy and stability.

This work comprehends research published in [8] and extends it to new use cases and capabilities. We present here some of these results, including a morphological comparison with survey data available for the 2011 Tohoku tsunami event. We also present early results on model simulation of Discovery Bay, where existing records of past tsunamis can be found. Similarly, we also explore tidal forcing in the model, showing that it can potentially be used to simulate general coastal inundation phenomena related to Sea Level Rise.

2   Model Discussion

In this section, we provide a concise overview of the governing principles and numerical approach associated with the shallow-water model. This discussion is intended to outline key aspects rather than offers a detailed mathematical treatment. Readers interested in a rigorous exploration of the topic can refer to [8].

The model focuses on describing fluid flow dynamics in shallow water using a system of equations defined over two spatial dimensions and time. These equations incorporate variables such as the total water depth, horizontal velocity components, and bathymetry, which represents the underwater topography. The system also accounts for the effects of gravity and friction, along with sediment transport processes.

To simplify the analysis, the two-dimensional system can be reduced to a one-dimensional version that retains the essential features of the original model. This is possible thanks to the invariance under rotations of the original model. The advantage of this approach is that it allows us to design numerical approaches that are simpler than their two-dimensional counterparts.

Sediment transport plays a crucial role in this model, as it influences bed evolution and flow patterns. The transport rate is typically described using empirical formulas derived from experiments with granular, non-cohesive materials. Several well-known formulations, such as those proposed by Grass, Meyer-Peter and Müller, Van Rijn, Nielsen, Kalinske, and Einstein, provide frameworks for calculating sediment movement based on factors like shear stress and sediment properties.

A generalized formula for bedload sediment transport has also been proposed, offering flexibility to represent different transport behaviors by adjusting parameters. These parameters account for porosity, shear stress, and sediment properties, allowing the formula to adapt to various flow and sediment conditions.

Previous research has shown that the use of the Darcy-Weisbach friction formula ensures the unconditional hyperbolicity of the system, making it a preferred choice over alternatives like the Manning formula that can potentially lead to non-hyperbolic behavior.

To numerically approximate the system, the model is discretized using a finite volume method within a path-conservative framework. This approach divides the computational domain into uniform cells, averaging the solution over each cell to track changes over time. The process is designed to handle both the hydrodynamic and morphodynamic components of the system while maintaining stability and accuracy.

The hydrodynamic component focuses on fluid flow variables such as water depth and velocity. It then uses the standard shallow-water eigenstructure to evolve the numerical scheme using a traditional HLL scheme, which is written under the Polynomial Viscosity Method (PVM) in the path-conservative framework.

The morphodynamic component addresses the sediment transport required for bed evolution. It uses an upwind type of numerical approach but leverage the information from the hydrodynamic calculations to ensure consistency between the two components. This coupling minimizes errors and prevents spurious oscillations that can arise from treating the components independently.

One of the key advantages of this numerical approach lies in its reliance on the shallow-water wave speeds, that simplifies computations without compromising the numerical accuracy. The framework also supports second-order accuracy and is well-balanced, so it can preserve the steady state corresponding to a constant free surface, which is crucial for realistic simulations.

3 Validation

We now briefly present some relevant numerical simulations where the model results are compared with available survey data. In particular, the 2011 tsunami event in Tōhoku is especially relevant since there is solid record of its morphological impact on Crescent City (California, U.S.).

Wilson et al. [9] examined morphological changes caused by the 2011 Tōhoku tsunami by comparing two surveys conducted before and after the event, on February 21, 2010, and March 30, 2011. These surveys were carried out by the United States Army Corps of Engineers (USACE), the National Oceanic and Atmospheric Administration (NOAA), the U.S. Geological Survey (USGS), as well as several private companies and local harbor authorities.

For the boundary conditions, we utilize the source data provided by Grilli et al. [10]. That work shows a strong correlation between the numerical results and the observed data for several tide gates and DART buoys. We discretize the domain with 501 elements in the x direction and 361 in the y direction, ensuring that the discretization points align with those of the bathymetry source file. The sediment properties are defined by the following parameters: a relative density of 2.4, a typical grain size of 0.1 mm, a bottom friction coefficient of $f = 2 \times 10^{-2}$, a porosity of $\varphi = 3/5$, and a critical threshold $\theta_a = 1 \times 10^{-3}$.

The results are shown in Figs. 1. The numerical outcomes are compared with those presented by Wilson et al. [9], which highlight the changes in bathymetry following the tsunami event. As shown in Fig. 1, there is a strong correlation between the observed data and the numerical results. The general morphological trends, including significant erosion and deposition near the ends of both jetties, likely resulting from strong currents, are well captured.

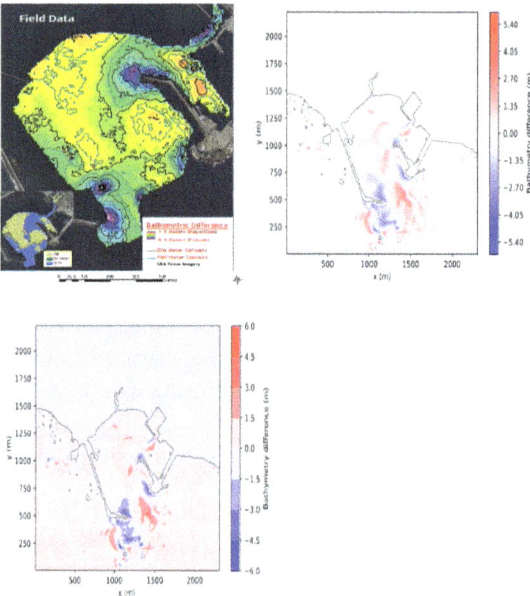

Fig. 1. 2011 Tsunami impact on the morphology in the harbor of Crescent City, California, USA. Left: Wilson et al. [9] results. Middle: Numerical simulation. Right: Numerical simulation using low resolution for sediment changes.

4 Future Applications

This approach also allows for other applications in the field of coastal modeling and inundation. In particular, it allows for high resolution modeling of tides by enforcing tidal boundary conditions along the domain interface. Due to the high-performance implementation, this approach can be computed extremely fast. Figure 2 shows the modeled tides at the Garibaldi (Oregon, USA) for three days (left) and one month (right), respectively, compared to a tide prediction model.

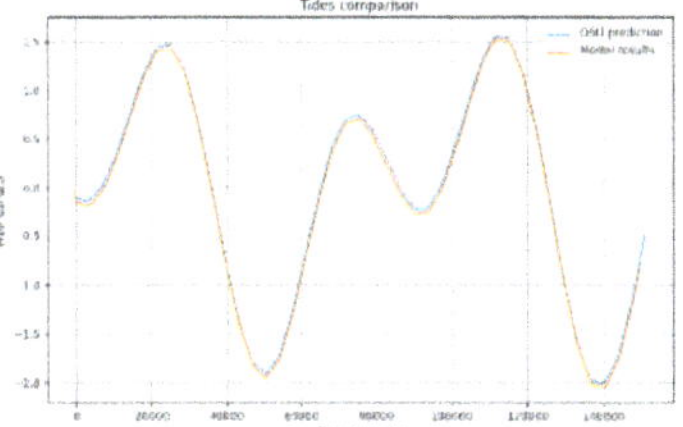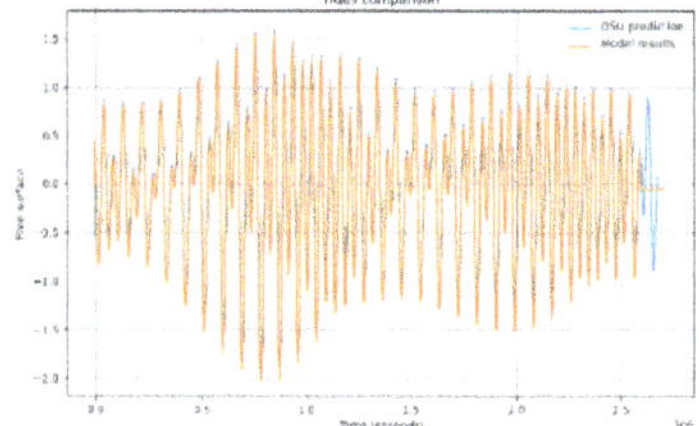

Fig. 2. Tide evolution at Garibaldi (Oregon State, U.S.) for three days (left) and one month (right)

Additionally, preliminary work is currently underway to simulate historic tsunami events in Discovery Bay, (Washington, USA) through the comparison with historical records. Figure 3 shows an example of model-simulated sediment deposition and erosion (right) resulting from the tsunami genereated from a crustal fault in the Puget Sound, Washington, USA, compared with historical records (left).

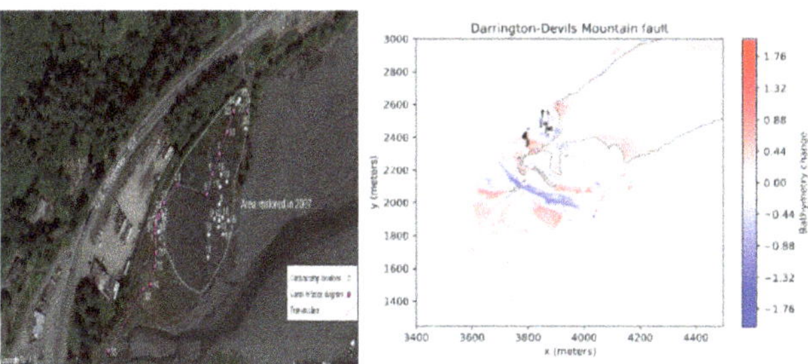

Fig. 3. Discovery Bay deposits (left) and numerical simulation with sediment changes (right).

5 Conclusions

In summary, the present model study provides a framework for analyzing shallow-water dynamics and sediment transport processes. The numerical implementation ensures stability, accuracy, consistency, and speed of the model compuatation, making it suitable for complex simulations. The validation using real-world data, such as the 2011 Tōhoku tsunami, further reinforces the reliability and utility of the model. Furthermore, the robust and efficient implementation of the model allows us to expand its original intent to other geophysical phenomena.

References

1. Castro Díaz M, Fernández–Nieto E, Ferreiro A (2008) Sediment transport models in Shallow Water equations and numerical approach by high order finite volume methods. Comput Fluids 37(3):299–316
2. Teeter AM, et al (2001) Hydrodynamic and sediment transport modeling with emphasis on shallow-water, vegetated areas (lakes, reservoirs, estuaries and lagoons). Hydrobiologia 444(1):1–23

3. Castro Díaz M, Fernández-Nieto E, Ferreiro A, Parés C (2009) Two-dimensional sediment transport models in shallow water equations. A second order finite volume approach on unstructured meshes. Comput Methods Appl Mech Eng 198(33):2520–2538
4. De St. Venant B (1871) Théorie du mouvement non permanent des eaux, avec application aux crues des rivières et à l'introduction des marées dans leur lit. Academic de Sci. Comptes Redus 73(99):148–154
5. Meyer-Peter E, Müller R (1948) Formulas for bed-load transport. In: IAHSR 2nd meeting, Stockholm appendix 2. IAHR
6. van Rijn LC (1984) Sediment transport, part I: Bed load transport. J Hydraulic Eng 110(10):1431–1456
7. Cordier S, Le M, Morales de Luna T (2011) Bedload transport in shallow water models: Why splitting (may) fail, how hyperbolicity (can) help. Adv Water Resourc 34(8):980–989
8. Guerrero Fernández E, Díaz MC, Wei Y, Moore C (2024) Modeling sediment movement in the shallow-water framework: A morpho-hydrodynamic approach with numerical simulations and experimental validation. Ocean Model 192:102445
9. Wilson R, Davenport C, Jaffe B (2012) Sediment scour and deposition within harbors in California (USA), caused by the March 11, 2011 Tohoku-oki tsunami. Sedimentary Geol 282:228–240. The 2011 Tohoku-oki tsunami
10. Grilli ST, et al (2013) Numerical simulation of the 2011 Tohoku Tsunami based on a new transient FEM co-seismic source: Comparison to far- and near-field observations. Pure Appl Geophys 170(6):1333–1359

Coastline Evolution as a Result of Three Decades of Dynamic Coastal Preservation

Evelien Brand[1], Quirijn Lodder[1,2], and Niels van Kuik[1(✉)]

[1] Rijkswaterstaat, Griffioenlaan 2, 3526 LA Utrecht, The Netherlands
Niels.van.Kuik@rws.nl
[2] TUDelft, Stevinweg 1, 2628 CN Delft, The Netherlands

Abstract. The coastline is pro-actively maintained in the Netherlands to prevent negative effects of erosion. Since the 1990's significant nourishment efforts have been carried out for this purpose. Here, the beach topography of a relatively undisturbed stretch of coast, with the exception of nourishment efforts, was investigated to understand the effect of three decades of coastal preservation. The evolution of the beach volume, coastline position, and coastline orientation were compared before and after the implementation of the coastal preservation policy. Structural erosion of the coastline was put to a halt thanks to regular nourishments. On average, the coastline built out, which was proven necessary to maintain the most erosive parts of the coast. Especially adjacent to breakwaters/groynes and nourishments hotspots an expansion of the coastline can be observed. It can thus be concluded that coastline maintenance is achieved with regular nourishments and that sediment accumulates in the beach zone at certain locations along-shore.

Keywords: sand nourishment impact · coastal management · coastline orientation · beach morphology

1 Introduction

Coastal erosion threatens flood safety and other functions of the coast globally. A dynamic coastal preservation policy was implemented in the Netherlands in 1990. This policy aims to prevent chronical land loss while, given the coastal functions, allowing for coastal dynamics where possible [1]. Sand nourishments were adopted as the primary means of maintaining the coast [2]. It has often been concluded that sand nourishments are an effective solution to coastal retreat [3]. The cumulative effect of repeated nourishments, however, is lesser studied and poorly understood. In this study, the effect of three decades of nourishments on the coastline position, coastline orientation, and beach volume is evaluated for a stretch of the Dutch coast. This practice-based evidence helps understanding the relation between coastal dynamics and nature based coastal maintenance strategies.

© The Author(s) 2026
C. Coelho et al. (Eds.): CD 2025, CRL 41, pp. 261–266, 2026.
https://doi.org/10.1007/978-3-032-15473-6_41

2   Approach

Beach volume, coastline position and coastline orientation were calculated from yearly cross-shore topographic transects [4]. The beach volume is calculated as the volume around the mean low water line, from the dune foot to the same elevation difference below the low water line (i.e., approximately from 3 m above to 5 m below mean sea level) [5]. The coastline position is the weighted average of this volume and thus represents the low water line. The coastline orientation is the angle in the coastline position between neighbouring transects. Variability in the coastline position was determined to understand the effect of local protrusions of the coastline on the coastline evolution. This variability was calculated as the standard deviation in coastline orientation over 1.5 km of coastline.

Indicators were compared for 1970, 1990, and 2020 to understand the effect of the nourishment strategy that originates from 1990. Indicators were averaged over three years (e.g. 1989, 1990, and 1991) to exclude short-term variations, such as the effects of individual nourishments.

In this study, we focus on only a part of the Dutch coast: a stretch of the straight coast (Scheveningen to Den Helder) and one of the barrier islands (Texel), which are separated by a tidal inlet. This part of the coast is relatively unaffected by large-scale morphological developments and is therefore suitable for the aim of this study.

3   Evolution of the Coast

Small-scale nourishments were already implemented before the start of regular coastline maintenance in 1990, which resulted in an stable coastline position with a median change of 0.1 m/year (Fig. 1). Without these small-scale nourishments, the coastline would have shown a net retreat [6]. Regardless, 25% of the coast showed significant coastline retreat of 2.0 m/year or more. Since the start of coastline maintenance, the coastline builds out with 1.6 m/year on average (1990–2023). Nowadays, only 0.5% of the coast shows significant coastline retreat of 2.0 m/year or more.

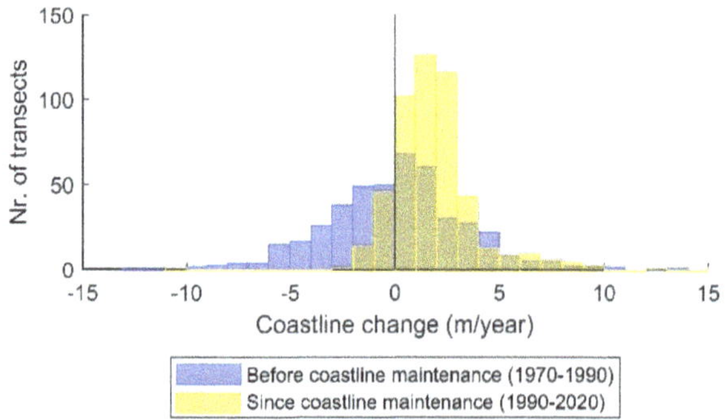

Fig. 1. Histogram of the net coastline change in the period before and after the start of the national nourishment policy.

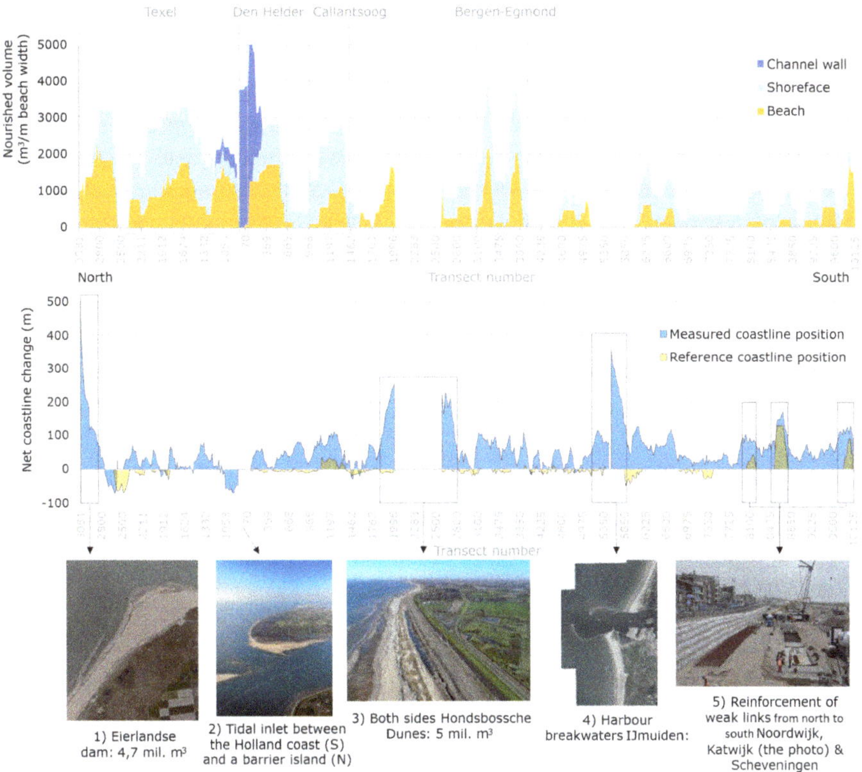

Fig. 2. Total nourishment volume between 1990 and 2020 (top) and changes in coastline position since the start of coastline maintenance (middle) from the north of Texel to Scheveningen. The most frequently nourished areas are indicated in grey. Important features are highlighted (bottom). Photo sources: 1 & 4 beeldmateriaal.nl; 2, 3 & 5 Rijkswaterstaat.

Coastline expansion occurs along the entire coast but the coastline evolution varies alongshore (Fig. 2). In total, 132 mln. m^3 of sand was added in the investigated part of the coast between 1990 and 2020, of which a part was added through reinforcements. After a reinforcement, the coastline is often maintained at a more seaward position to preserve the reinforcement (indicated in yellow, Fig. 2). The coastline expansion that can be observed at reinforced towns Noordwijk, Katwijk, and Scheveningen is thus intentional. A total of 45 mln. m^3 of sand has accumulated between 1990 and 2020 on the stretch of coast shown in Fig. 2. Approximately 10 mln. m^3 of this sand was trapped by the harbour breakwaters of IJmuiden and the large groyne 'Eierlandse dam' (Fig. 2), resulting in a coastline expansion of several hundreds of meters south of these dams.

The remainder of the nourished volume is transported to the dunes, alongshore, offshore, or into the Wadden Sea through the tidal inlet between Texel and Noord-Holland. In earlier studies this transport was estimated to be approximately 3.7 mln. m^3/year [7], resulting in a total of 111 mln. m^3 of sand that would be transported into this inlet in the period 1990–2020. It must be noted that not all of this sand is actually nourished sand: even without the long-term nourishment strategy, sand would have been imported into the Wadden Sea, but without nourishments this would have resulted in coastal retreat of other areas.

Nourishment hotspots (total nourished volume >2000 m^3/m) are marked in grey. At Texel and Den Helder the coastline position was maintained with regular nourishments. At Callantsoog and Bergen-Egmond the coastline was built out into the sea. Along the central Dutch coast there are several coastal towns where regular nourishments directly benefit the safety against floodings (Callantsoog) or socio-economic developments (Bergen and Egmond). Coastline maintenance happens more strictly at these coastal towns. Coastline expansion in these locations can likely be related to this strict maintenance of the coastline in the most critical transects. Expansion also occurs at the beaches surrounding these coastal towns due to the lateral spreading of nourished sediments.

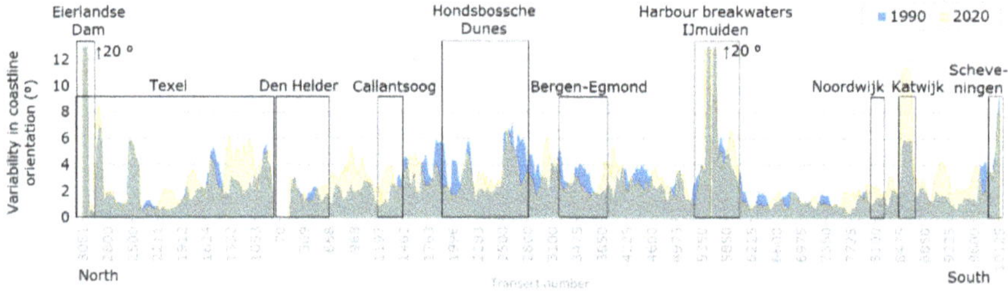

Fig. 3. Variability in coastline orientation pre- and post-coastline maintenance. Several coastal towns and important features are highlighted by the boxes.

It can be assumed that locations where the coastline protrudes further into the sea experience more erosion and thus require greater nourishment efforts. Due to lateral spreading of the nourished sediment, the coastline supposedly gradually becomes less protruded, possibly reducing nourishment efforts. Changes in the coastline orientation were investigated to test this assumption.

The most distinctive protrusions of the coastline are observed near breakwaters/groynes (Fig. 3). Furthermore, a protrusion of the coastline (>5° variation in coastline orientation over 1.5 km) is observed at some of the coastal towns (Katwijk, Scheveningen, Bergen, and Callantsoog). At others a seaward protrusion of the coastline cannot or can hardly be distinguished (Egmond and Noordwijk).

On average, the same amount of variation in coastline orientation is observed nowadays as in 1990 (Fig. 4). The assumption that the coastline tends to become straighter is partly true though. At original protrusions of the coastline, i.e. stretches of the coast with a relative high variability in coastline orientation (right side in Fig. 4) the coastline has become straighter. Here, a negative change in the coastline variability is observed in general (Fig. 4). However, at other locations the coastline is protruding seaward more nowadays than in 1990. At reinforcements that have occurred between 1990 and 2020 the coastline is often locally built out on purpose (e.g. Scheveningen and Katwijk). New protrusions can also be linked to repeated nourishment efforts that result in a seaward

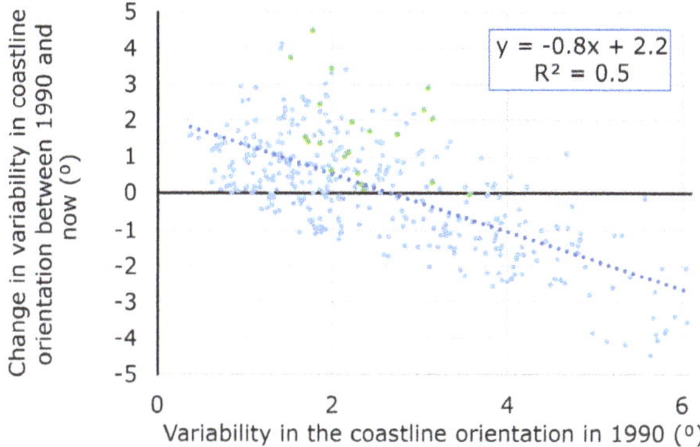

Fig. 4. Change in variability in coastline position compared to the original variability. The green dots represent sites with increased protrusion due to reinforcements or repeated nourishments.

migration of the coastline, such as observed at Bergen and Egmond. These specific sites are marked in green in Fig. 4.

4 Conclusions

With the current Dutch strategy of pro-active nourishments, structural retreat of the coastline was put to a halt in the investigated part of the Dutch coast. The coastline actually builds out on average, which was proven necessary to maintain the coastline in the most erosive parts. Coastline expansion is especially strong near breakwaters/groynes and adjacent to nourishment hotspots. Nourishment hotspots exist for example where the coastline is intentionally built out and maintained at a more seaward position for food safety. It can thus be concluded that coastline maintenance is achieved with regular nourishments and that sediment accumulates in the beach zone at certain locations alongshore.

References

1. Ministerie van Verkeer en Waterstaat: Kustverdediging na 1990: Beleidskeuze voor de Kustlijnzorg. Tweede Kamer 1989–1990, 21 136, nrs. 5–6 (1990)
2. Brand E, Ramaekers G, Lodder Q (2022) Dutch experience with sand nourishments for dynamic coastline conservation – an operational overview. Ocean Coast Manag 217:1–17
3. De Schipper MA, Ludka BC, Raubenheimer B, Luijendijk AP, Schlacher TA (2021) Beach nourishment has complex implications for the future of sandy shores. Coast Eng 2(1):70–84
4. Rijkswaterstaat, JARKUS (2023). https://data.overheid.nl/dataset/27b66c3e-0b1b-485f-8fa2-5bdea40a6250
5. Hillen R, de Ruig JHM, Roelse P, Hallie FP (1991) De basiskustlijn, een technisch/morfologische uitwerking. Nota GWWS 91.006; RWS Dienst Getijdewateren, Den Haag
6. Groenendijk FC (1997) Zand voor Nederland, Een analyse vanaf 1964 en een extrapolatie tot 2010 van het zandvolume in de Jarkus zone. Rapport RIKZ-95.003

7. Taal M, et al (2023) Sedimentbehoefte Nederlands kustsysteem bij toegenomen zeespiegelstijging. Rapportnummer 11207897-002-ZKS-0004

Numerical Model Assessment of Temporary Mobile Shore Protection Modules on an Erosional Transgressive Coast

Gary A. Zarillo[1]([✉]) and Joe Farrell[2]

[1] Florida Institute of Technology, Melbourne, FL 32934, USA
zarillo@fit.edu
[2] Resolve Marine Group, Fort Lauderdale, FL, USA

Abstract. Over the past two decades the southeast coast of the U.S., and especially the east coast of Florida, has been subject to rising coastal ocean sea levels at rates three times higher than the 100-year trend due to the slowing of the Gulf Stream. This, combined with frequent tropical cyclones, limited local sources of terrigenous sands, and rapid development of human infrastructure, has transformed a once stable barrier island setting into an environment of erosion and coastal transgression (Zarillo, 2023). In this situation, erosion hotspots may appear, evolve through a life cycle, and then disappear and reestablish at other locations. Thus, this work aims to evaluate the potential of portable and temporary breakwaters as a treatment for chronic shore erosion and erosion hotspots transgression. This study evaluates the potential of Coastal Recovery Shoal Modules (CRSMs) as a mobile temporary or permanent solution to mitigate erosion. Two configurations of CRSM-based breakwaters were modeled using hydrodynamic and sediment transport simulations calibrated to field data. Results demonstrate that module arrangements can reduce sand volume loss and wave energy, particularly when crest elevations are optimized. This approach shows promise as a sustainable shore protection.

Keywords: Coastal Processes Model · Breakwater Configuration

1 Introduction

The east central Florida coast faces ongoing challenges from shoreline erosion, exacerbated by storm impacts, limited beach fill retention, and highly variable erosion rates. The comprehensive beach preservation plan incorporates measures such as sand bypassing, but erosion persists. This study introduces Resolve Marine Coastal Recovery Shoal Modules (CRSMs) as an innovative mobile shore protection structure. CRSMs are submersible structures engineered to attenuate wave energy and promote sediment deposition. This analysis integrates observational data and numerical modeling to predict their effectiveness along an 8-km shoreline segment. Key objectives include evaluating configurations for erosion control and optimizing module design for minimal environmental impact.

C. Coelho et al. (Eds.): CD 2025, CRL 41, pp. 267–273, 2026.
https://doi.org/10.1007/978-3-032-15473-6_42

2 Study Area and Physical Setting

The U.S. central Florida coast characterized by a barrier island system consisting of mixed terrigenous and carbonate sediments overlying coquina rock formations. The area includes significant reef outcrops influencing wave dynamics and sediment transport. Coastal ocean sea levels have been rapidly rising over the past two decades and are strongly influenced by seasonal and interannual Gulf Stream dynamics (Zarillo, 2023). Tropical storms and hurricanes frequently impact the east Florida coast..

3 Methods

The modeling scheme to assess the potential performance of the shore and reef modules is the Coastal Modeling System (CMS), which is widely used for hydrodynamic, wave and morphodynamic numerical modeling in U.S. coastal settings. The CMS is an integrated 2D numerical modeling system for simulating waves, current, water level, sediment transport and morphology change at coastal inlet and entrances. Model tests of the alternatives were based on measured topography and verified wave hindcast conditions. Model hydrodynamics calibrated with measured data.

3.1 Model Set up

Grid Development. The CMS uses Cartesian grids with a telescoping or Quadtree option for the Flow model to provide more spatial resolution. Grids are developed using the SMS interface. The telescoping grid allowed for refinement to be optimized with small cell sizes to be placed in areas of interest or areas where rapid changes can be expected while leaving larger refinement to offshore regions. Figure 1 shows the grid coverage within Indian River County, FL including grid refinement levels from a base size of 160 m by 160 m to the minimum cell size of 5 m by 5 m along the shoreline. Model bottom topography surfaces were generated using a combination of available beach profiles, multi-beam swath bathymetry data and bottom topography collected from the NOAA Coastal Relief Model.

Model Boundary Conditions and Model Verification. The wave model is driven using a spectra derived from hind cast data developed from Wave Watch III data (WW3; Tolman 2010) and winds derived from data measured onsite. The CMS-Flow model was driven using a combination of tidal constituents and offshore water levels derived from measured data in order capture the influence of the Gulf Stream. Model performance is evaluated by comparing model results with measured data through qualitative (graphical) and quantitative (statistical) methods. The variables used in the hydrodynamic calibration included: wave height, water surface elevation, and current velocity. Verification of the morphological change model was performed independently by a comparison of measured vs. model topographic changes.

Model Test Configurations. Two configurations of module-based breakwaters were applied. One consisted of 100 m long breakwaters constructed from a series of modules spaced along an approximate 8 km of shoreline in water depths of 3.5 to 4 m (Fig. 2A).

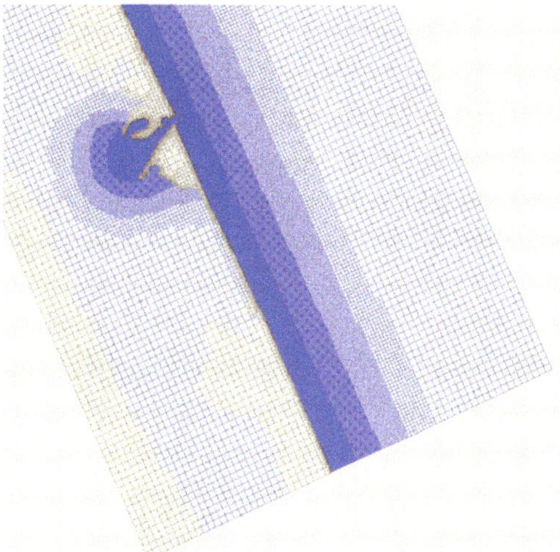

Fig. 1. CMS Flow computational grid illustrating high resolution along the shoreline.

Module segments were placed between reef rock segments to reduce wave energy and prevent excessive offshore sand losses beyond the active shoreface. The crest elevation of modules in this configuration was set at a low profile at a freeboard of about 2 m below the still water level or in high profile near the still water level at a freeboard of about 0.5m. Configurations included 100 m segments of modules to form a continuous breakwaters (Fig. 2A) and segmented groups of modules extending alongshore for 300m (Fig. 2B). Alternative 1modeltest consisted of low-profile segmented modules, Alternative 2 specified high-profile segmented modules, and Alternative 3, a line of 300m high-profile breakwaters, as shown in Fig. 2B. Each alternative was subjected to a model run representing 6 months of real time during 2013.

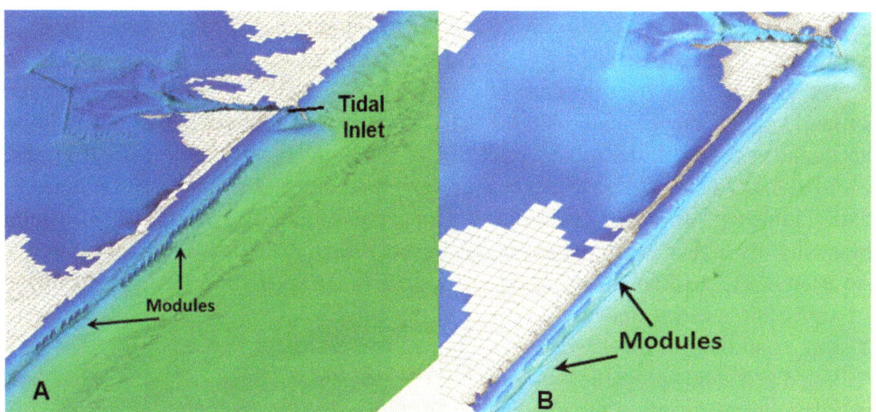

Fig. 2. Continuous lines of modules(A) on the upper shoreface (Alternatives 1 and 2) and 300m long groups of modules (B) forming a segmented breakwater configuration

4 Results

For each of the 6-month model runs of Alternatives 1, 2, and 3, a comparison was made between predicted topographic change with and without the modules represented in the model grid. Figure 3A shows the sand volume calculation cells associated with the Coastal Recovery Shoal Modules seen in Fig. 2A. Figure 3B shows the sand volume calculation cells associated with the 5 breakwaters nested in the model topography, as shown in 2B.

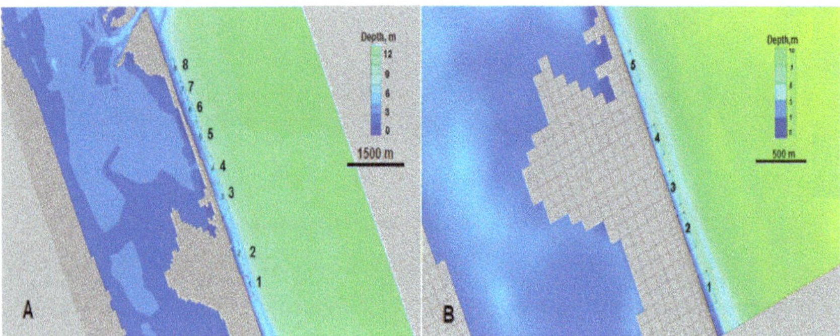

Fig. 3. Location of volume calculation cells 1 – 8 within the CMS model grid on the upper shoreface (A) as seen in Fig. 2A and within the fields of 300 m long module breakwaters(B) as seen in Fig. 2B

4.1 Alternative 1

Alternative 1 includes 30 set of modules set into the shoreface (Fig. 2A In this alternative the modules form low profile breakwaters having a length of about 100 m. The base elevation of the modules is between 3.5 and 4 m below sea level. The crest of the modules extends about 2 m above the base elevation leaving about 1.5 to 2 m of freeboard above meters, the free board over the crest of the crest below Table 2 compares the calculated net sand volume change under Alternative 1 with the predicted volume change without the modules paced in the model grid. Six of the eight sand volume cells (Fig. 2A) recorded lower sand volume losses change compared to the existing conditions and two sand volume cells recorded sand volume gains. Two of the cells recorded sand volume losses slightly higher compared to the existing condition. Alternative 2 increased the top elevation of the module-constructed breakwaters so the crest elevation extended about 0.5 above the mean water level. Table 2 compares alternative model results with the existing condition. In this case, having limited freeboard above the mean water level. Predicted and sand volume losses were reversed to gain in all but two cells, which still recorded improvements relative to the no structure existing condition (Table 1).

Table 3 compares the predicted net sand volume change for Alternative 3, which includes 300 m long CRSM high-profile breakwaters integrating shore protection over longer segments of the coast in this configuration (Fig. 2B). Predicted sand volume

Table 1. Predicted sand volume changes for Alternative 1

Sand Volume Cell	Alternative 1: m^3 (low profile)	Existing: yds^3	Difference: m^3	Percent Improvement
Cell 1	−3,279	−2,960	−329	−11%
Cell 2	5,066	2,265	2,801	124%
Cell 3	41	−6,045	6,086	101%
Cell 4	−10,757	−10,119	−638	−6%
Cell 5	−18,222	−21,562	3,340	15%
Cell 6	−433	−7,487	7,054	94%
Cell 7	−2,954	−9,297	6,343	68%
Cell 8	−9,264	−13,376	4112	31%

Table 2. Predicted sand volume changes for Alternative 2

Sand Volume Cell	Alternative 2: m^3 (High profile)	Existing: m^3	Difference: m^3	Percent Improvement
Cell 1	6,665	−2,960	9,625	325%
Cell 2	11,543	2,265	9,278	410%
Cell 3	7,991	−6,045	14,036	232%
Cell 4	2,669	−10,119	12,788	125%
Cell 5	−18,109	−21,562	3,453	16%
Cell 6	4,158	−7,487	11,645	156%
Cell 7	4,285	−9297	13,582	146%
Cell 8	−3195	−13376	10,181	76%

losses for all but cell 5 were more than 10,000 cubic meters over the 6-month model period. However, placing the modules in a series of 300 m long breakwaters with a high profile and in water depth of about 3.5 m substantially reduced predicted sand losses and reversed predicted losses in cell 1 to a gain. The model prediction also showed a nearly 100% gain in sand volume within Cell 5. Only in Cell 4 were sand volume losses not measurably improved.

Table 3. Predicted sand volume changes for Alternative 3

Sand Volume Cell	Alternative 3: m^3 (High profile)	Existing: m^3	Difference: m^3	Percent Improvement
Cell 1	4556	−22,794	27,350	120%
Cell 2	−2789	−14,000	11,211	80%

(*continued*)

Table 3. (continued)

Sand Volume Cell	Alternative 3: m^3 (High profile)	Existing: m^3	Difference: m^3	Percent Improvement
Cell 3	−7098	−10,917	3,819	35%
Cell 4	−21970	−22,349	379	3%
Cell 5	11,541	5870	5,671	97%

5 Summary and Conclusions

The purpose of this work is to evaluate the potential of the Coastal Recovery Shoal Modules as a temporary to permanent treatment for chronic shore erosion and erosion hotspots. Two configurations of module-based breakwaters were applied, one consisted of a series of modules arranged in continuous breakwaters. The crest elevation of modules in this configuration was set at either about 2 m below the still water level or near the still water level. The second configuration includes five 300 m module-based breakwaters placed along an approximate 3.5 km section of shoreline landward of reef outcrops. The crest of the breakwater modules in this configuration was set in a high-profile model having a freeboard of about 0.5 m.

Model hydrodynamics were calibrated with measured data. Model results showed that the Coastal Recovery Shoal Modules can be arranged in a configuration that will slow or reverse sand volume losses from the upper shoreface. The model tests also show that the performance of the modules for wave energy attenuation and sand volume retention is sensitive to the elevation of the crest of the structure. Model test showed that when high profile modules are arranged in continuous breakwaters or in breakwater segments 300 m in length, a substantial reduction of sand loss from the upper shore may result. This study demonstrates the potential of Coastal Recovery Shoal Modules as a sustainable intervention for shoreline erosion.

Reference

Zarillo GA (2023) Inter-annual sea level change along a barrier island coast. Front. Environ. Sci., 22Sec. Interdisc Climate Stud 11. https://doi.org/10.3389/fenvs.2023.1107458

Coastal Texas Mega Project: Performance and Resiliency of a Dune-Beach System Against Extreme Storms

Himangshu S. Das[(✉)], Kelly A. Burks-Copes, and Robert C. Thomas

U.S. Army Corps of Engineers, Galveston, TX 77550, USA
himangshu.s.das@usace.army.mil

Abstract. The Texas Gulf coast is a complex system with numerous narrow inlets and broad back bays with intricate river and bayou networks. Due to its complex settings, the Texas coast is susceptible to extreme compound flooding due to fluvial, pluvial, and surge during hurricanes and tropical storms. The Texas coast is also one of the United States most dynamic regions in terms of population and economic growth. The objective of the Coastal Texas Study was to improve our capabilities to prepare for, resist, recover and adapt to extreme events.

A "Multiple Lines of Defense" strategy was used in the Coastal Texas study to design cost-effective, environmentally friendly solutions that will reduce risks of storms impacting the coastal communities and restore important wildlife habitat at the same time. This paper specifically addresses the feasibility of nature-based solutions such as a dune-beach system along Galveston Island and Bolivar Peninsula to reduce storm surge impacts and shoreline recession. The numerical model SBEACH (Storm-Induced BEAch Change) has been used to simulate the cross-shore response of representative existing beach profiles and alternative configurations while exposed to RSLC and extreme storm events. Evaluation of alternatives is based on storm-induced performance and sediment budget requirements to identify improvements that optimize resiliency of the coastline to reduce vulnerability to infrastructures.

Keywords: Coastal management · Beach Nourishment · Nature Based Improvements · Adaptation to climate change

1 Introduction

The Gulf of Mexico coastal region is vulnerable to periodic storm surges and rainfall extremes. Since 1900, hurricanes striking in this region have killed 6,000 to 12,000 people and caused tremendous economic damage to infrastructure. The socio-economic risk from these extreme events is projected to grow worse in the coming decades due to urban development and climate change. In 2015, the U.S. Army Corps of Engineers (USACE), in partnership with the Texas General Land Office (GLO), began to explore viable solutions for coastal storm risk management and ecosystem restoration along the Texas coast. The study is known as the comprehensive Coastal Texas Study (USACE, 2021). Detailed maps, graphics, and reports are located at http://coastalstudy.texas.gov.

© The Author(s) 2026

C. Coelho et al. (Eds.): CD 2025, CRL 41, pp. 274–278, 2026.

https://doi.org/10.1007/978-3-032-15473-6_43

The Texas Gulf coast is a complex system with numerous narrow inlets and broad back bays with intricate river and bayou networks. Due to its complex settings, the Texas coast is susceptible to extreme compound flooding due to fluvial, pluvial, and surge during hurricanes and tropical storms. Five of the top six largest tropical cyclone rainfall totals in the continental U.S. have occurred in Texas. The Texas coast is also one of the United States most dynamic regions in terms of population and economic growth. According to the Bureau of Economic Geology, twenty five percent of the population and thirty three percent of the economic resources of Texas are located along the 360-mile-long coast. The objective of the Coastal Texas Study was to improve our capabilities to prepare for, resist, recover and adapt to extreme events.

A "Multiple Lines of Defense" strategy was used in the Coastal Texas study to design cost-effective, environmentally friendly solutions that will reduce risks of storms impacting the coastal communities and restore important wildlife habitat at the same time (Fig. 1). With a focus on redundancy and robustness, the first objective is to keep the storm surge in the Gulf (Gulf Defense). Then, a combination of water from Galveston Bay and Gulf surge that could overtop the Gulf defenses must be addressed through a second line of defense called the Bay Defenses.

Fig. 1. Multiple Line of Defense System along the Texas Coast.

The upper Texas coastline has been receding rapidly on an average 2 feet per year to over 20 feet per year in some regions (HDR, 2014). Severe storm events in the past along with relative sea level changes (RSLC) have contributed to this shoreline recession which causes billions of dollars of damage. This paper specifically addresses the feasibility of nature-based solutions such as a dune-beach system along Galveston Island and Bolivar Peninsula to reduce storm surge impacts and shoreline recession. Evaluation of alternatives is based on storm-induced performance and sediment budget requirements to identify improvements that optimize resiliency of the coastline to reduce vulnerability to infrastructures.

2   Methodology

Various numerical models were used to better understand the potential environmental impacts and enable decision makers to evaluate the largest feasibility decision in USACE history. The integration of a variety of modeling tools, along with qualitative evaluations, was a critical part of the assessment of environmental impacts to the health and ecology of the unique Galveston Bay complex. A coupled surge and wave modeling system was used to quantify the surge and wave hazards, informing economic and design analyses (USACE, 2021). A suite of 660 tropical cyclones were developed to span the full range of tropical storm hazards and a joint probabilistic model framework including historical storms was also employed.

The numerical model SBEACH (Storm-Induced BEAch Change) has been used to simulate the cross-shore response of representative existing beach profiles and alternative configurations while exposed to RSLC and extreme storm events. SBEACH model results were used for comparison of pre and post-storm performance existing and alternative design profiles. Alternative profile configurations modeled include varying quantities, dimensions, and shapes of dune-berm systems along with natural vs. enhanced dune core. The primary performance evaluation criteria were upland damage mitigation along with cost indicators such as estimates of initial & re-nourishment volume requirements. The depth and duration of overtopping has been considered for damage evaluation. We assessed that sediment requirements for initial construction range between 100 and 160 cubic yards per linear foot depending on site conditions for a total of approximately 50-million cubic yards for the entire length of the study site. Borrow source identification and lifecycle analyses are also performed to indicate periodic nourishment and maintenance needs.

3   Results

3.1   Coastal Storm Modeling

Coupled surge and wave modeling system was used to quantify the surge and wave hazards, informing economic and design analyses (USACE, 2021). A suite of 660 tropical cyclones were developed to span the full range of tropical storm hazards and a joint probabilistic model framework including historical storms was also employed. Of those 660 storms, 170 were selected to best capture the Galveston area storm hazard and minimize computational time. New model meshes were developed from the most recent, highest resolution survey data available.

Coastal models were validated against historical storms. Annual exceedance probabilities (AEP) were then computed for the range 1 to 0.0001 for peak storm water level and spectral significant wave height. Structural design and crest elevation were based on the 1% AEP water level combined with wave runup and the appropriate overtopping limit state.

3.2   Beach and Dune System Design

This project includes about 45 miles of beach and dunes system. The proposed beach template (Fig. 2) was designed to reduce coastal storm risk by preventing inundation and breach during all but the largest of storms and to dramatically reduce total inflow during storms that exceed breach conditions. SBEACH model was used to evaluate storm response and optimize dune and beach design. It was determined that 39 million Cubic Yards (MCY) of sand would be needed for initial construction.

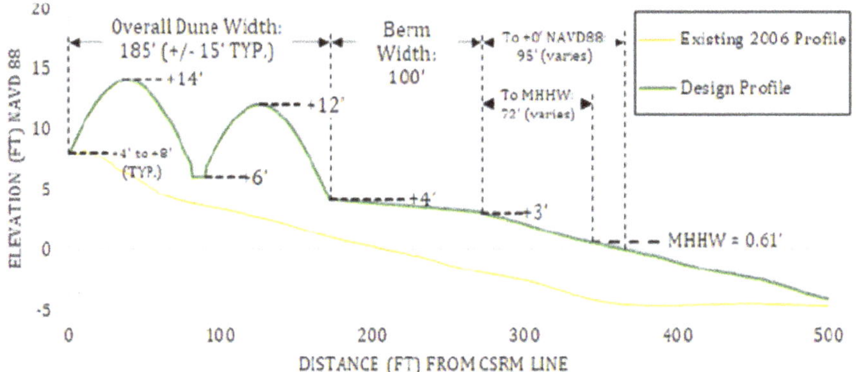

Fig. 2. Typical dimensions of the beach design profile.

Impacts of climate change were included in evaluation of alternatives. The greatest impact on plan selection driven by climate change was a result of Relative Sea Level Change (RSLC). Per guidance it is required to evaluate three RSLC curves; low, based on the observed rate at nearby water level stations, intermediate following modified National Research Council (NRC) Curve I, and high following modified NRC Curve III. Later beach morphology was modeled with a Monte Carlo life cycle simulation to help understand beach maintenance required after large storms (USACE 2021). The model showed that switching from a single dune to a dune field reduced the number of times the system would need to be renourished over the 50-year life. Eliminating bedform smoothing in the model reduced both the single dune and dune field renourishment needs. This resulted in a range of total nourishment requirements from 19.32 MCY to 24.95 MCY, depending on the RSLC scenario.

4 Next Steps

The first construction cost, excluding inflation, for the Coastal Texas Project is over $30B. The $20M study evaluated hazards, economics, and the environment sufficiently to make an investment decision, but much work remains to be completed to safely design this Mega Project. Some key remaining coastal analyses include full integration of gate operations into the coastal numerical model, physical modeling of the gates and pump stations, additional ship simulation, enhanced environmental modeling and data collection, and higher fidelity beach morphology and coupled storm risk analysis.

As we embark on design of this generational project, USACE and our partners at the Gulf Coast Protection District and Texas General Land Office are seeking to engage the best minds in the industry to innovate and deliver. We have already begun to engage universities and other government organizations to plan innovation and delivery. After the project is authorized and funds are appropriated, we will be engaging industry partners to begin the innovation, design, and construction process.

References

HDR (2014) Beach and Shoreline Changes Along the Upper Texas Coast: Recovery from Hurricane Ike, HDR Engineering Inc., Project Number 166742, pp 2–61
USACE (2021) Coastal Texas Protection and Restoration Feasibility Study: Coastal Texas Flood Risk Assessment: Hydrodynamic Response and Beach Morphology, ERDC/CHL TR-21-11

Assessing Compound Flooding Hazards in Estuaries by Integrating a Climate Emulator and a Hybrid Metamodel

Jared Ortiz-Angulo[1(✉)], Paula Camus[1], Beatriz Pérez-Díaz[1], Laura Cagigal[1], Mirian Jiménez[1], Sonia Castanedo[1], Andrea Pozo[2], Luis Cea[3], Juan F. Farfán[3], and Fernando Méndez[1]

[1] Geomatics and Ocean Engineering Group, University of Cantabria, Santander, Spain
jared.ortizangulo@unican.es
[2] University of Canterbury, Christchurch, New Zealand
[3] University of A Coruña, A Coruña, Spain

Abstract. Compound flooding in estuaries results from the interaction of multiple forcings, including tides, storm surges, waves, precipitation and river discharge. These events present significant challenges for dynamic modeling due to their computational requirements. To address this, we present a hybrid statistical-dynamical framework for efficiently assessing compound flooding hazards.

The framework incorporates a new version of the climate-based emulator TESLA, which generates synthetic extreme multivariate events by linking them to weather types that represent atmospheric pressure patterns. A multiscale approach considering Regional Weather Types (RWT) and Local Weather Types (LWT) captures the spatial variability of oceanographic and land drivers. TESLA is coupled with a hydrodynamic metamodel, enabling fast predictions of water levels and inundation under various forcing conditions while significantly reducing computational demands.

Using flood proxies based on water volume, the framework identifies events associated with specific return periods. High-resolution simulations of these events provide detailed assessments of compound flooding hazards.

Applied to the Asón River estuary in northern Spain, this approach demonstrates its ability to analyze historical climate variability, predict flood risks, and define mitigation strategies. It offers an efficient tool for exploring low-probability, high-impact scenarios in estuarine environments under current and future climate conditions.

Keywords: TESLA · Compound · Flooding · Hybrid · Metamodel

1 Introduction

The statistical distribution of compound flooding extents at estuaries is the result of the non-linear interaction and co-occurrence of multiple oceanographic, hydrologic, geologic and meteorological forcings (e.g. astronomical tides, monthly mean sea-level, large-scale storm surge, dynamic wave set-up, river discharge, surface runoff).

© The Author(s) 2026
C. Coelho et al. (Eds.): CD 2025, CRL 41, pp. 279–283, 2026.
https://doi.org/10.1007/978-3-032-15473-6_44

To address this complex problem and to manage with models that are highly computationally expensive, we propose a hybrid statistical-dynamical framework approach that combines: (a) a new version of the climate-based emulator TESLA [4]; (b) a metamodel based on hydrodynamics simulations; (c) the use of a flooding proxy to analyze compound flooding events; (d) a statistical characterization of the proxy based on return periods; and (e) high-resolution modelling to assess compound flooding hazards.

The proposed framework has been tested at the estuary of the river Asón, located in the northern Spain. The Santoña Estuary forms a wide bay at the mouth of the aforementioned river, located in the northern Spain. The estuary is characterized by its large area (about 18.7 km^2), its complex bay-like morphology, the presence of large areas of relatively well-preserved tidal flats and mudflats (although with a high degree of anthropization), and the presence of residual rocky reliefs rising above the bay.

2 Motivation

Many coastal settlements are located in estuaries where flooding can result from oceanographic sources (e.g. tides, storm surges and waves), but also from river discharge (fluvial) and direct surface runoff (pluvial). These drivers are typically causally linked through associated storms, and may therefore occur simultaneously or in close succession, exacerbating the flooding impact. Historical observations of these multivariate drivers causing compound flooding represent only a small fraction of the possible range of conditions capable of causing adverse consequences. Therefore, a climate-based emulator is needed to generate synthetic extreme multivariate events (tides, storm surge, precipitation and river discharge) that allow to explore low probability combinations of oceanographic and fluvial forcing conditions.

On the other hand, a numerical hydrodynamic model is required to simulate the compounding effects of spatially distributed interactions between river discharge and downstream sea level in tidal channels and estuaries. However, these models are typically too computationally expensive to dynamically simulate the new full parameter space of synthetic oceanographic and pluvial/fluvial conditions that can cause compound flooding.

3 Methodological Framework

In this study, a hybrid statistical-dynamical framework is proposed to efficiently predict spatially varying estuarine water levels and inundation areas for assessing compound flooding hazard, see Fig. 1.

A new version of the climate-based emulator TESLA [1] is used to generate new compound flooding events by identifying their relationship with a set of weather types representing different large-scale atmospheric pressure patterns. The incorporation of drivers involved in compound flooding events in the climate-based emulator requires different spatial scales of the predictors to better explain oceanographic and land drivers predictands. To address this problem, a multiscale approach is proposed, that allows incorporating different areas for each predictor ensuring they are spatially defined to adequately characterize the behavior of the predictand they explain: Regional Daily

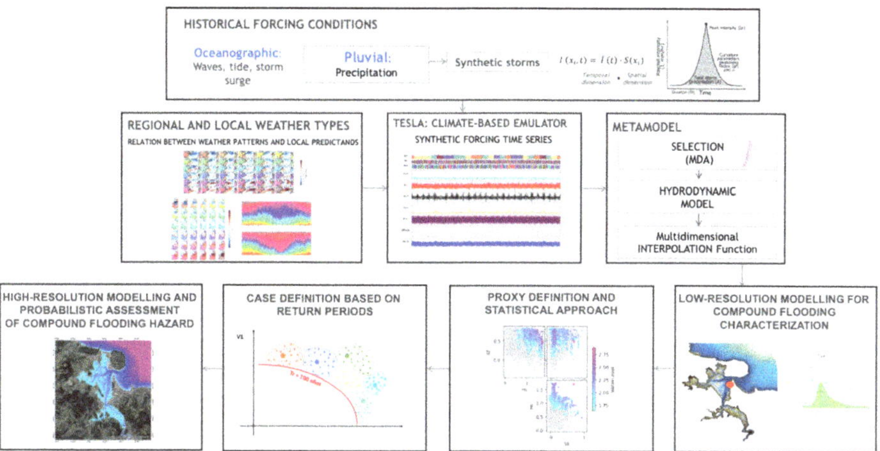

Fig. 1. Methodological framework for assessing compound flooding hazards in estuaries.

Weather Types (RWT) and Local Daily Weather Types (LWT), see Fig. 2a & Fig. 2.b. The use of weather types allows for analyzing the historical climate variability and provides insights into the spatial and temporal variability in the region, see Fig. 2.c.

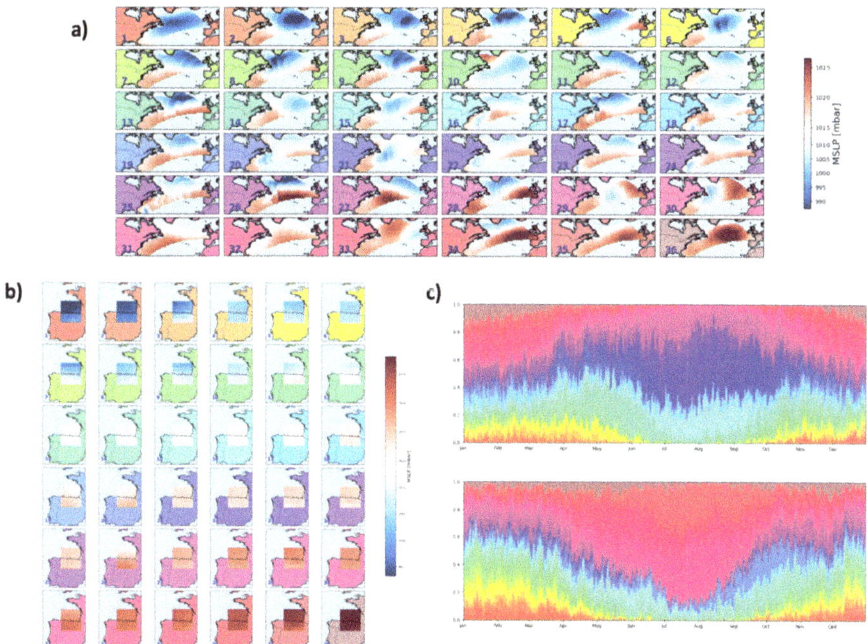

Fig. 2. Multiscale weather types defined in the new version of the climate-based emulator: a) Regional Weather Types (RWT) and b) Local Weather Types (LWT) c) Seasonality of the RWT (upper) and seasonality of the LWT (lower).

Furthermore, the storm parametrization included temporal and spatial dimensions: temporal characteristics are defined by peak intensity, curvature parameters, total storm precipitation, and duration (see Fig. 3.a), while spatial characteristics are captured by the principal components of dimensionless rainfall. The precipitation assigned to the synthetic storms will be transformed into river discharge.

This climate-based emulator is coupled to a metamodel, similarly to [2], based on hydrodynamic simulations, in this case, to emulate water levels and flood inundation under any combination of tidal, storm surge and river discharge conditions. One of the most important aspects to ensure the efficiency of the metamodel is the parameterization of the problem, which involves defining a set of representative variables for the process to be modeled. In this case, discharge and meteorological tide hydrographs are considered to define the typical evolution of these dynamics during extreme events, astronomical tidal cycles of various ranges characteristic of the study site, and the time lag that may exist between discharge and water level, see Fig. 3.b. This fast and efficient method to predict estuarine hydrodynamics is based on selection, and interpolation techniques [3] reducing the computationally intensive cost of classical dynamical modeling. See Fig. 3.c.

Based on the flood maps obtained from the metamodel, a proxy based on water volumes has been defined, which is representative of compound flooding events. Statistical characterization is performed using the proxy values associated with each event, calculating desired return periods from the proxy values obtained from the synthetic series. Applying a clustering technique on the proxy, focus shifts to events that result in flooding associated with a given return period. Using a high-resolution model, the most probable combination of climate drivers producing the target event can be simulated, allowing the assessment of compound flooding hazards.

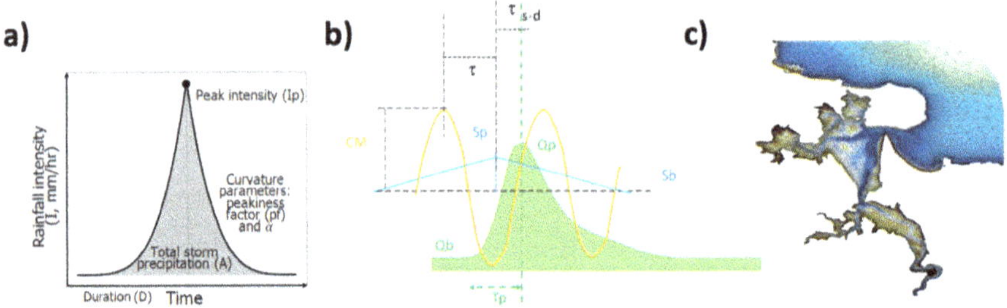

Fig. 3. a) Proposed parametrization of the storms for its application to the climate-based emulator; b) proposed parametrization of the problem for its application to the compound flooding metamodel; and c) mesh considered for the application of the dynamic modelling at the estuary of the river Asón.

Acknowledgements. This project is partially funded by MyFlood (PLEC2022-009362 – MCIN/AEI/https://doi.org/10.13039/501100011033 and EU Next GenerationEU/PRTR), the Office of Naval Research under the ESTCP Project, HyBay (PID2022–141181OB-I00, MCIN/AEI/https://doi.org/10.13039/501100011033/FEDER, EU) and BahiaLab (funded by the Autonomous Community of Cantabria and by the European Union Next Generation EU/PRTR).

References

1. Anderson R, Cagigal A, Mendez R (2019) Time-varying emulator for short and long-term analysis of coastal flood hazard potential. J Geophys Res Oceans 124(12):9209–9234
2. Anderson R, Mendez B, Erikson O'N, Merrifield R, Cagigal, M (2021) Projecting climate dependent coastal flood risk with a hybrid statistical dynamic model. Earth's Future
3. Camus M, Medina R (2011) A hybrid efficient method to downscale wave climate to coastal areas. Coast Eng 58(9):851–862
4. Cagigal L, Rueda, A, Ruggiero P, Merrifield MA, Montano J, Mendez FJ (2020) A multivariate, stochastic, climate-based wave emulator for shoreline change modelling. Ocean Modell 154

Advancing the Use of Spatial Data in Implementing Adaptive Management to Support Coastal Resilience

Karinna Nunez[✉], Pamela Mason, Tamia Rudnicky, Christine Tombleson, Catherine Duning, Jessica Hendricks, Zhonghui Lv, Evan Hill, Jack Graulich, Daniel Schatt, and Karen Duhring

Virginia Institute of Marine Science, William & Mary, Gloucester Point, Williamsburg, VA 23062, USA
karinna@vims.edu

Abstract. This article emphasizes the innovative integration of spatial data, remote sensing, machine learning, and community science to enhance adaptive management strategies aimed at supporting coastal resilience. By advancing the use of spatial analysis, this study addresses the pressing challenges posed by sea-level rise and shoreline erosion. The research focuses on mapping key shoreline features and coastal habitats in Virginia, USA, utilizing photo-interpretation and machine-learning techniques to update shoreline inventories efficiently and accurately. This novel approach not only improves the VIMS Shoreline Management Model (SMM) but also incorporates community science through a mobile application, allowing local residents to contribute valuable data on shoreline conditions and coastal habitats. By fostering community engagement and leveraging real-time information, decision-makers are better equipped to implement adaptive coastal management practices. Ultimately, this study highlights the critical importance of comprehensive spatial data and community involvement in developing effective strategies for enhancing coastal resilience in the face of climate change.

Keywords: Spatial Data · Remote Sensing · Shoreline Management · Coastal Resilience · Community Science

1 Introduction

Sea level rise is significantly impacting coastal communities and natural resources, causing increased shoreline erosion and changes to intertidal habitats such as marshes and beaches. The persistence of natural shoreline habitats depends on their ability to migrate landward into adjacent lands. To enhance coastal resiliency, communities must address multiple needs through carefully planned projects that leverage co-benefits. Effective community resilience requires maintaining up-to-date data for coastal analysis and modeling, as well as expanded monitoring efforts to support adaptive management. Incorporating community science into this research is valuable, as it can provide real-time, localized data on shoreline changes, habitat shifts, and erosion patterns. By engaging

C. Coelho et al. (Eds.): CD 2025, CRL 41, pp. 284–290, 2026.
https://doi.org/10.1007/978-3-032-15473-6_45

local residents in data collection and monitoring activities, researchers can gather more comprehensive and frequent observations, while also fostering community awareness and involvement in coastal resilience efforts. The Virginia Institute of Marine Science (VIMS), Center for Coastal Resources Management (CCRM), has developed shoreline inventories (https://www.vims.edu/ccrm/research/inventory/) that are crucial for supporting decision-making and modeling efforts. Given the rapid changes in coastal conditions due to climate change, there is an urgent need to have updated coastal inventories to aid localities in preparing for climate-induced changes.

This study aims to enhance coastal inventory data through an adaptive management approach. It leverages cutting-edge technology, satellite data, and community science to accomplish four main objectives: 1) applying machine learning techniques to update and expand shoreline condition inventories based on locality needs; 2) upgrading the Shoreline Management Model (SMM) [1] to support the implementation of Natural and Nature-based features (NNBFs) for shoreline erosion control, 3) mapping marsh migration corridors under various sea level rise scenarios to aid in conservation and restoration decisions, and 4) developing and implementing a community science application for citizens and local government to contribute to maintaining current shoreline condition information in Virginia.

2 Methodology

2.1 Study Area

Shorelines along Virginia's coast, USA were selected for shoreline inventory update, SMM upgrade, and mapping of marsh migration corridors. Virginia is a state located in the Mid-Atlantic region of the USA, along the shores of the Chesapeake Bay, which serves as the destination for major tributary rivers. The state's diverse landscape encompasses a wide spectrum of geophysical, biological, and anthropogenic features. These range from low-energy marsh environments with organic sediment-dominated shorelines to high-energy sandy beaches and dune systems facing the Atlantic Ocean and Chesapeake Bay. Virginia's coastline varies from pristine, unmanaged natural shores to heavily developed and commercially managed waterfronts [2].

2.2 Integrated Data Application Process

This study employed an integrated data application approach (Fig. 1) that combines various data sources and analytical techniques to enhance adaptive coastal management. The process integrates spatial data, advanced modeling, and community science, to support coastal management decision-making through the generation and update of shoreline inventories and the development of recommendations for best management practices (BMPs) that contribute to coastal resilience. Baseline data and model outputs were incorporated into a decision-support tool designed for data sharing and visualization. The webtool, Virginia Coastal Resources Tool (VCRT) was developed based on stakeholders' input including decision-makers, coastal planners, state and local agencies, and NGOs.

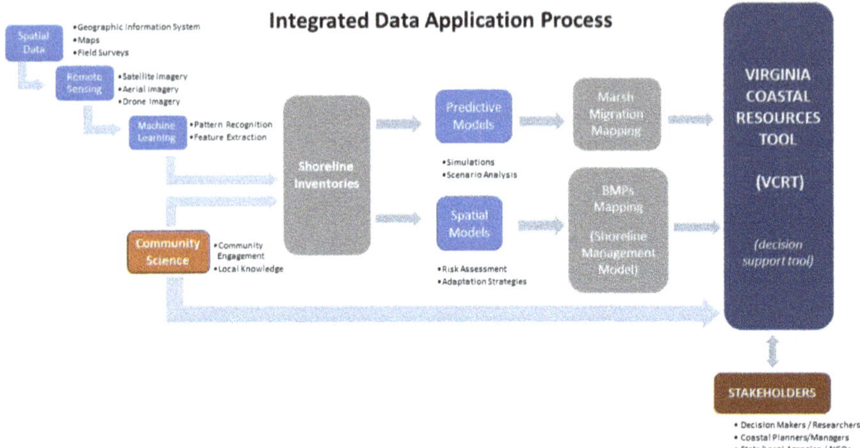

Fig. 1. Integrated data application process for implementing adaptive management to support coastal resilience.

Enhanced Shoreline Inventories

Traditional methods for assessing shoreline conditions are time-consuming and inefficient for capturing rapid changes. This study employed machine learning techniques [3, 4], specifically adapting U-Net architecture, to predict and map shoreline features from high-resolution imagery (National Agriculture Imagery Program [NAIP] and Virginia Base Mapping Program [VBMP]). This approach allows more rapid identification of vulnerable areas and prioritizes resilience projects.

Shoreline Management Model (SMM)

Upgrade of the VIMS Shoreline Management Model (SMM) [1] was conducted to support the implementation of Natural and Nature-based features (NNBFs). The SMM is a geospatial model that recommends strategies for tidal shoreline erosion control and identifies suitable locations for living shorelines. This model processes available information about shoreline conditions through decision flow charts to generate shoreline stabilization recommendations. The SMM's recommendations are based on a comprehensive understanding of shoreline dynamics, including natural shoreline responses to environmental conditions, effects of human interventions on coastal areas, cumulative impacts of traditional shoreline stabilization methods, and best practices derived from extensive field observations. The model incorporates various shoreline factors, such as the presence of natural buffers (e.g., tidal marshes, beaches, riparian forests), bank height, nearshore bathymetry, fetch, existing defense structures, and proximity of upland development. For this study, the SMM upgrades included using wave energy data modeled by SCHISM [5] for erosion risk assessment, and adding variables such as RTE species, tiger beetles, proximity to navigational channels, and oyster structure recommendations for living shorelines, enhancing the model's ability to provide comprehensive coastal management solutions.

Marsh Migration Corridors

A multi-model approach was employed to map potential marsh migration corridors under two sea-level rise scenarios: 0.6 m and 1.2 m, representing mid-century and end-of-century projections, respectively. This method integrates outputs from four established marsh migration models applicable to the study area: Sea Level Affecting Marshes Model 5.0 (SLAMM), Integrated Valuation of Ecosystem Services and Tradeoffs (InVEST), NOAA Sea Level Rise Viewer's Marsh Migration Mapping Method, and Evolution of Tidal Marsh Distribution (VIMS). The consolidated output, Marsh Migration Corridor Envelope (MMCE) [6, 7], identifies current upland areas that could potentially convert into marsh habitats under the specified sea-level rise scenarios. The MMCE is presented as a raster file with a 30-m pixel resolution, illustrating the degree of consensus among the four models' projections. This comprehensive approach aims to inform conservation and restoration efforts.

Community Science

A community-science application to update shoreline conditions was developed using ESRI Survey 123. The mobile App, *MapMyShore* (Fig. 2), was developed based on localities' needs and stakeholders' input (including federal, state, and local government agencies, and NGOs). Features that can be surveyed with *MapMyShore* include: coastal erosion protection structures, cultural features, natural shoreline buffers, Phragmites, recreational/navigational features, riparian land use, shoreline bank height, and slope. Accessible via smartphones, the App allows community members to contribute local knowledge and validate remote sensing data. This App contributes to community cohesion, fosters community engagement, standardizes information sharing, supports local decision-making, and raises awareness about coastal hazards and resilience.

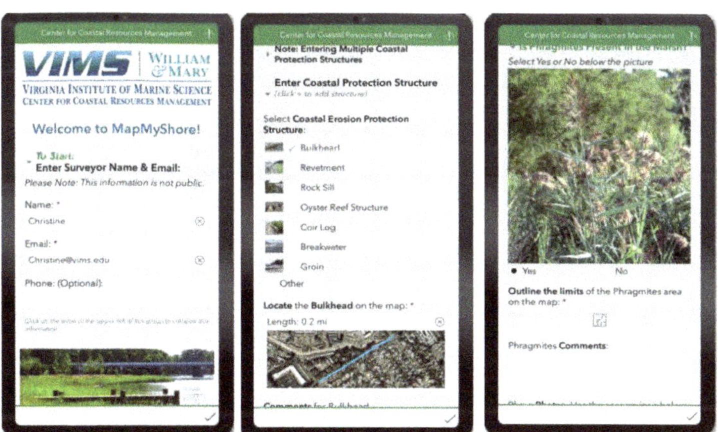

Fig. 2. Selected screenshots of the Community Science App *MapMyShore*

3   Coastal Decision Support Tool for Adaptive Management

A coastal decision support tool, the Virginia Coastal Resources Tool (VCRT) (Fig. 3) was developed using ArcGIS Experience Builder. This web platform was also developed based on feedback received from different stakeholder groups in order to address locality needs. This tool supports data sharing and visualization of all the baseline and analytical data derived from the integrated data application approach: shoreline conditions (e.g., bank height and slope, shoreline structures), natural buffers (tidal marshes, beaches, dunes), marsh migration corridors, and best management practices (BMPs) derived from the upgraded SMM v. 6.0. The VCRT includes two map viewers and five dashboards, depicting spatial and tabular data, as well as several statistics by locality and river system in Virginia. The *MapMyShore* Community Engagement Dashboard displays the data collected through the App and shows the statistics of the different categories surveyed by the community: https://www.vims.edu/ccrm/research/inventory/map-my-shore/. On this site, users can access the App tutorial and associated documentation. GIS files in the VCRT are displayed with color palettes designed to meet the needs of users with visual impairments, including those with color vision deficiencies. All the App documentation displayed in the VCRT, including the tutorial and glossary, has been developed and checked for accessibility.

Fig. 3. Virginia Coastal Resources Tool (VCRT) main page - (https://cmap22.vims.edu/VACoas talResourcesTool/)

The VCRT's integrated approach ensures that coastal management decisions benefit from both scientific analysis and community input, making it an invaluable resource for addressing complex coastal challenges and enhancing overall coastal resilience.

Since this integrated data approach was implemented and shared with stakeholders, the VCRT has received 95,275 hits – 3,648 unique visitors (01/01/24–01/13/25).

4   Conclusion

Accurate and current data are essential for making informed decisions that protect our shorelines against the impacts of climate change and human activity. The application of analytical data, such as SMM outputs, underscores the critical role of updated spatial and modeled data in effective coastal management and enhancing coastal resilience. Furthermore, incorporating community science into this process enriches our understanding of local conditions and fosters greater public engagement in coastal conservation efforts. By leveraging both scientific expertise and community insights, we can develop more robust strategies that not only address immediate challenges but also promote sustainable practices for the future. The methodologies developed in this study are scalable and applicable to other coastal regions, facilitating the integration of climate projections into state-level coastal protection strategies. It enables collaboration among diverse stakeholders to enhance coastal resilience and adaptation efforts. This study exemplifies a proactive and comprehensive approach to coastal management, leveraging innovation and community engagement to effectively manage and protect natural and cultural resources in the face of climate change impacts.

References

1. Nunez K, Rudnicky T, Mason P, Tombleson C, Berman M (2022) A geospatial modeling approach to assess site suitability of living shorelines and emphasize best shoreline management practices. Ecol Eng 179:106617
2. Hobbs CH, Krantz DE, Wikel GL (2015) Coastal processes and offshore geology. In: Bailey C (ed) Geology of Virginia, college of William and Mary, pp 1–44
3. Lv Z, Nunez K, Brewer E, Runfola D (2023) pyShore: a deep learning toolkit for shoreline structure mapping with high-resolution orthographic imagery and convolutional neural networks. Comput Geosci 171
4. Lv Z, Nunez K, Brewer E, Runfola D (2023) Mapping the tidal marshes of coastal Virginia: a hierarchical transfer learning approach. GISci Remote Sens 61(1)
5. Zhang YJ, Ye F, Staney EV, Grashorn S (2016) Seamless cross-scale modeling with SCHISM. Occan Model 102:64–81. https://doi.org/10.1016/j.ocemod.2016.05.002
6. Nunez K, Rudnicky T, Hendricks J, Duning C, Hill E, Graulich J, Lv M, Gregory S, Angstadt K (2025) GIS data for 2025 Gloucester County, Virginia: shoreline inventory, coastal natural buffers, and marsh migration corridors. Virginia Institute of Marine Science, William & Mary. https://doi.org/10.25773/b8xf-be95
7. Mitchell, M., Nunez, K., Tombleson, C., Herman, J. (2023). A marsh multi-model approach to inform future marsh management under accelerating sea level rise. Ecological Solutions and Evidence. https://doi.org/10.1002/2688-8319.12285

BlueMath-Hub: A Cloud-Based, Open-Source, Python Framework with Interactive Notebooks for Statistical Analysis and Simulation of Coastal Climate Hazards in a Changing Climate

Laura Cagigal[1]([✉]), Valvanuz Fernandez-Quiruelas[1], Fernando Méndez[1], Javier Tausia[1], Jared Ortiz-Angulo[1], Alba Ricondo[2], Paula Camus[1], Antonio S. Cofino[3], Dylan Anderson[4], Peter Ruggiero[2], Meredith Leung[5], Mark Merrifield[6], John Marra[7], Borja G. Reguero[8], David Gutierrez-Barcelo[8], Ron Hoeke[9], Emilio Echevarria[9], José A. A. Antolinez[10], Giovanni Coco[11], Brad Murray[12], and Jayantha Obeysekera[13]

[1] University of Cantabria, Santander, Spain
laura.cagigal@unican.es
[2] Oregon State University, Corvallis, OR, USA
[3] Institute of Physics of Cantabria (IFCA, CSIC-UC), Santander, Spain
[4] US Army Corps of Engineers, Washington, D.C., NC, USA
[5] National Center for Atmospheric Research, Boulder, CO, USA
[6] Scripps Institution of Oceanography, San Diego, CA, USA
[7] NOAA NCEI Regional Climate Services, Honolulu, HI, USA
[8] University of California Santa Cruz, Santa Cruz, CA, USA
[9] CSIRO, Aspendale, Australia
[10] TU Delft, Delft, Netherlands
[11] University of Auckland, Auckland, New Zealand
[12] Duke University, Durham, NC, USA
[13] Florida International University, Miami, FL, USA

Abstract. Addressing global challenges such as coastal hazards and climate change requires innovative tools capable of analyzing complex environmental drivers, including waves, storm surges, and cyclones, across varying scales. These tools are vital for predicting floods, assessing risks, and planning adaptive responses. BlueMath-Hub has been developed as a global collaborative initiative to provide accessible, customizable solutions for both researchers and practitioners. It aims to simplify the use of advanced statistical and numerical models, fostering creative and scalable approaches in coastal science and engineering.

BlueMath, the core of this platform, is an open-source repository of Python tools accessible via a cloud-based Jupyter Hub environment. It integrates statistical methods and numerical model wrappers within a modular framework. The system includes: (a) BlueMath-Toolkit, providing tools for data mining, interpolation, and model integration; (b) BlueMath-Statistical Downscaling, focusing on extreme events and generalized models; (c) BlueMath-Hybrid Downscaling, combining statistical and numerical approaches for optimized solutions; and (d) BlueMath-Climate Services, supporting integrated applications such as compound flooding assessments.

© The Author(s) 2026
C. Coelho et al. (Eds.): CD 2025, CRL 41, pp. 291–295, 2026.
https://doi.org/10.1007/978-3-032-15473-6_46

BlueMath is continuously evolving, with its tools already applied in research, publications, and training. By lowering barriers to entry and enabling collaborative workflows, BlueMath-Hub supports the development of innovative solutions to mitigate the impacts of a changing climate.

Keywords: Coastal-hazards · Toolbox · Statistical Models · Repository · Collaboration

1   Significance and Motivation

In the face of increasing global challenges such as coastal hazards and climate change, the use of robust statistical and numerical analysis tools is essential. Tools that facilitate the analysis of multivariate met-ocean climatic drivers (e.g., waves, storm surges, tropical and extratropical tropical cyclones) acting at multiple spatial and temporal scales are key for predicting flooding events, producing risk assessments, or planning for adaptation measures. The development of applications for analyzing coastal hazards in a changing climate demands not only accessibility to such tools but also the flexibility to combine them seamlessly to generate valuable insights and creative solutions.

BlueMath-Hub arises in this context as a collaborative platform of many research groups and universities around the world working together to democratize the access to advanced models and services, empowering both researchers and non-specialists to generate customized, complex solutions. To the best of our knowledge, it is the first tool developed for this purpose in the field of coastal science and engineering. With ongoing contributions and the constant development of BlueMath, BlueMath-Hub promotes collaboration and innovation among scientists while enabling a more resilient future through easily accessible, customizable, and scalable solutions.

2   BlueMath Framework

BlueMath (https://geoocean.unican.es/bluemath.html) is an innovative, open-source repository of Python-coded tools. The repository is also accessible from the muti-user cloud-based Jupyter Hub environment, leveraging shared computational resources and the user-friendly interface of Jupyter Notebooks. BlueMath integrates state-of-the art statistical techniques and high-fidelity numerical model wrappers in a system framework that includes from standalone models with straightforward applications to hybrid model-tool combinations that produce comprehensive climate services (Fig. 1).

The individual modules of BlueMath (Fig. 2) are designed to be highly adaptable to a wide range of needs and applications:

a) **BlueMath-Toolkit**: Provides fundamental tools including data mining (e.g., dimensionality reduction, clustering techniques), interpolation (e.g., radial basis functions), and wrappers for numerical models such as SWAN, SWASH, Delft3D, and SFINCS. It also allows to generate statistical analyses of waves, tides, and tropical cyclones.

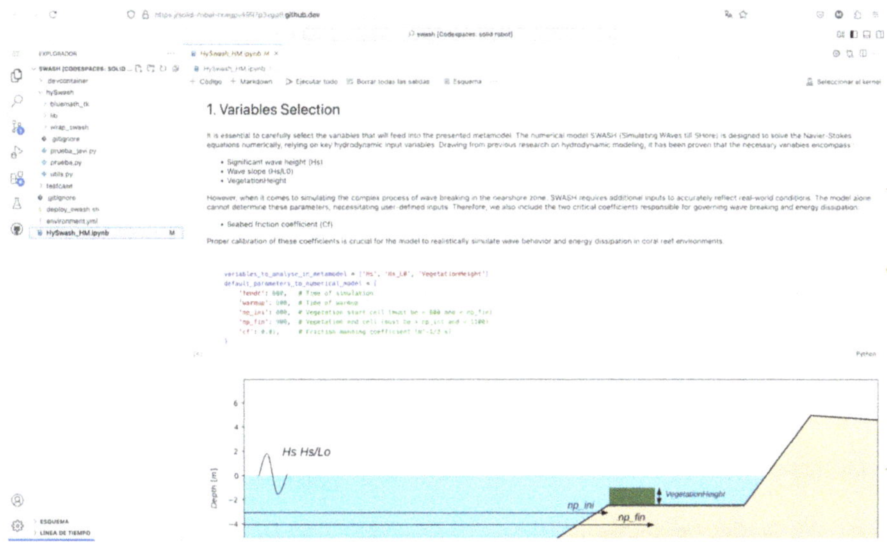

Fig. 1. Hybrid downscaling example implemented in BlueMath

b) **BlueMath-Statistical Downscaling**: Facilitates extreme value analysis, weather-typing, and generalized linear models. This allows to stablish simple downscaling methods as well as the ensemble of complex statistical models to build climate based emulators of metocean parameters in any location around the world (Anderson et al., 2019).

c) **BlueMath-Hybrid Downscaling**: Integrates statistical and numerical models to optimize computational efficiency. It includes the implementation of hybrid models like metamodels (e.g., HySwash (Ricondo et al., 2024)) and additive approaches (e.g., Binwaves (Cagigal et al., 2024), GreenSurge (Pérez-Diaz et al., 2025)).

d) **BlueMath-Climate Services**: Enables the creation of integrated climate applications by combining functionalities from previous modules. Examples include the TESLA climate emulator (Anderson et al., 2019), as well as detailed workflows of the integration of these tools on already implemented compound flooding and risk assessment projects.

The collaborative platform is undergoing continuous development and improvement. Notably, many of its tools have already been successfully applied in publications (e.g., Cagigal et al., 2024; Ricondo et al., 2024; Pérez-Diaz et al., 2025), the development of research projects, and the implementation of hands-on courses. The BlueMath framework aims to enhance collaboration among researchers by reducing the duplication of coding efforts and improving the replicability of published studies and methodologies. This is expected to significantly increase the potential influence of studies, facilitating easier and more transparent collaborations. Ongoing efforts will ensure that BlueMath remains a valuable and up-to-date resource for both researchers and practitioners, enriched by community testing, feedback, and contributions.

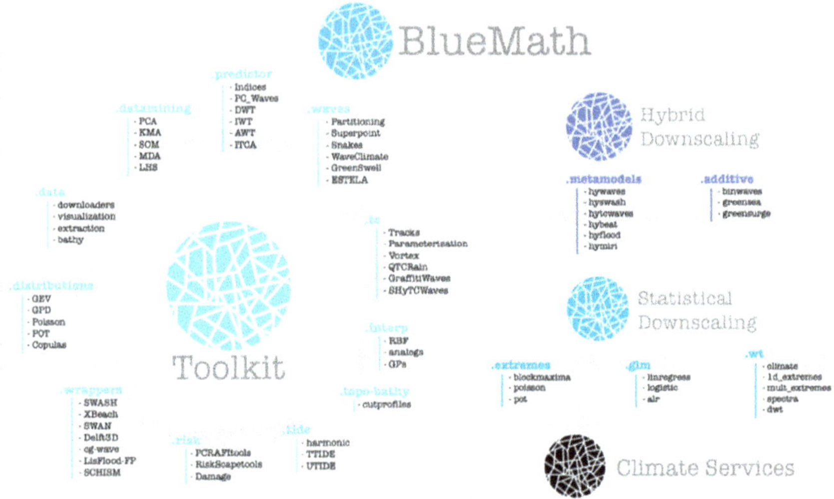

Fig. 2. BlueMath modules with functions to be implemented

Acknowledgements. This project is partially funded by the Office of Naval Research under the MURI and ESTCP projects; MyFlood (PLEC2022-009362 – MCIN/AEI/10.13039/501100011033 and EU Next GenerationEU/PRTR) ; HyBay (PID2022-141181OB-I00 - MCIN/AEI/10.13039/501100011033 y FEDER, UE) ; Perfect-storm (2023/TCN/003 - Gobierno de Cantabria/FEDER, UE) and CE4WIND (CPP2022-010118 MCIN/AEI/10.13039/501100011033 and EU Next GenerationEU/PRTR). LC acknowledges the funding from the Juan de la Cierva – Formación FJC2021-046933-I/ MCIN/ AEI/10.13039/501100011033 and the European Union "NextGenerationEU"/ PRTR.

References

Anderson D, Rueda A, Cagigal L, Antolinez JAA, Mendez FJ, Ruggiero P (2019) Time-varying emulator for short and long-term analysis of coastal flood hazard potential. JGR Oceans 124(12):9209–9234

Cagigal M, Ricondo A, Gutierrez-Barcelo D, Bosserelle C, Hoeke R (2024) BinWaves: an additive hybrid method to downscale directional wave spectra to nearshore areas. Ocean Model 189:102346

Ricondo A, Cagigal L, Pérez-Diaz B, Méndez FJ (2024) HySwash: a hybrid model for nearshore wave processes. Ocean Eng 291:116419

Pérez-Díaz B, Cagigal L, Castanedo S, Fernandez-Quiruelas V, Méndez F (2025) GreenSurge: an efficient additive model for predicting storm surge induced by tropical cyclones. Coast. Eng. 197:104691. ISSN 0378-3839. https://doi.org/10.1016/j.coastaleng.2024.104691

The Role of Machine Learning, Incorporating Wind Data, to Predict Coastal Wave Overtopping

Michael McGlade[1]($\boxtimes$), Nieves G. Valiente[1], Jennifer Brown[2], Christopher Stokes[2], and Timothy Poate[1]

[1] University of Plymouth, Plymouth PL4 8AA, UK
michael.mcglade@plymouth.ac.uk
[2] National Oceanography Centre, Southampton SO14 3ZH, UK

Abstract. Anthropogenic climate change is increasing wave overtopping frequencies. Reliably predicting overtopping is important for minimising casualties and economic losses. Empirical formulas, such as EurOtop, can provide accurate overtopping predictions based on physical model experiments. EurOtop requires nested regional process-based models to provide the input of nearshore coastal conditions. These models are computationally expensive, restricting their application in operational forecasting. EurOtop excludes wind speed and direction; significant variables influencing overtopping. This study explores the role of training random forests, using annual overtopping observations, to predict overtopping at two study sites: Dawlish and Penzance (UK). We compare the performance of random forests with a EurOtop based model, while investigating the performance of including hourly measured wind speed and direction. The random forests estimated overtopping and non-overtopping with approximately 80% and 95% accuracy, respectively. Including wind speed and direction enhanced the random forests' predictive performance (R2 = > 0.70, F1 = > 0.80). Random forest outperformed the predictive precision and accuracy made by EurOtop at Dawlish (F1 = 0.81, F1 = 0.65, respectively) and Penzance (F1 = 0.86, F1 = 0.07, respectively). Random forests generated predictions rapidly (<4 s) and generalised well, demonstrating great potential for implementation in different overtopping hotspots worldwide, provided overtopping observation availability.

Keywords: Machine-Learning · Overtopping · EurOtop · Flooding

1 Introduction

By 2050, approximately 300 million individuals globally may experience high-frequency coastal flooding [1]. Globally aggregated annual overtopping hours have increased by almost 50% over the last two decades [2]. Predicting wave overtopping along seawall defences is challenging given the number of variables influencing flooding. Such variables include wind speed (U_{10}), wind direction (U_{10} Dir), significant wave height (H_s), freeboard, wave direction (D_m), and mean period (T_m). These variables interact and

C. Coelho et al. (Eds.): CD 2025, CRL 41, pp. 296–302, 2026.
https://doi.org/10.1007/978-3-032-15473-6_47

can change significantly spatially and temporally. The EurOtop manual estimates overtopping by considering various conditions, such as hydraulic variables (e.g. H_s) and structural parameters (e.g. crest freeboard), and then calculates overtopping discharge by accounting for sloped and vertical seawall structures [3]. Many process-based models rely on using the EurOtop manual for estimating overtopping. These models often incur errors due to inadequate spatiotemporal resolutions, bathymetry, forcing data, and seawall geometries [4]. Wind effects are also oversimplified due to insufficient data, and although several studies have demonstrated the effects of wind on wave overtopping [5, 6], current widely used empirical parameterisations still neglect these. Machine learning, specifically random forests, may offer new insights to predict overtopping rapidly and reliably. Random forests consist of multiple decision-tree ensembles, analysing different sections of the training dataset, which then formulate together a final prediction [7]. Training random forests over several storms may reliability predict overtopping, derived from learning correlations from historical overtopping events [8]. Including U_{10} and U_{10} Dir in the random forests may also improve such predictive performances. This study examines the accuracy and precision of training random forests, using field observational overtopping data, along with the corresponding wave and wind conditions, to predict overtopping occurrence and frequency (number of overtopping waves per 10 min) in Dawlish and Penzance. We compare the importance of accounting for wind conditions and compare the random forest performance with EurOtop predictions following the process-based approach used in a forecasting system called SWWEP OWWL [9] which is operational across the study region.

2 Methods

2.1 Study Sites and Overtopping Measurements

Wave overtopping was investigated in Dawlish ($50°35'01''$ N, $3°27'52''$ W) and Penzance ($50°07'10''$ N, $5°32'15''$ W), southwest England (Fig. 1). Both locations experience frequent overtopping; for example, in February 2014, the Dawlish seawall collapsed, causing disruption amounting to £50 million [10]. The Dawlish seawall reaches 5.64 m above Ordnance Datum, and the Penzance wall reaches 3.7 m. The seawall structures differ, with Dawlish consisting of a recurved section for deflecting waves, while Penzance is slightly curved ($35°$). The WireWall apparatus, developed by the National Oceanography Centre (NOC), was permanently deployed at Dawlish and Penzance seawalls from March 10th 2021 to March 17th 2022, and November 16th 2021 to March 15th 2022, respectively [11, 12]. The systems were programmed to run for 6 h every tide centered over predicted high tide. At Dawlish, two WireWall systems were adjacently deployed to create a cross-shore transect (50 cm apart); rig 1 at the seawall crest and rig 2 landward close to the railway line. At Penzance, two WireWall systems were deployed with alongshore separation, rig 1 was exposed to SE & SW winds (Eastern seawall location), and rig 2 was more sheltered, exposed to SW winds (western seawall location) (Fig. 1). The corresponding wind conditions (i.e., U_{10} and U_{10} Dir) were recorded using local weather stations, presenting wind data every 10 min. Wave buoys recorded the wave conditions, computing corresponding wave conditions every 30 min. WaveRadar REX, a marine monitoring system, recorded the water level every 10-min and 15-min for Dawlish and Penzance, respectively.

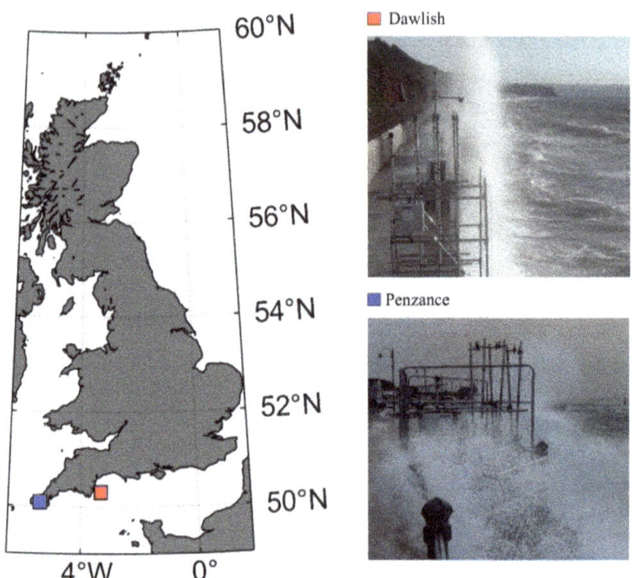

Fig. 1. Study sites for Dawlish (red) and Penzance (blue) illustrating Wirewall setup.

2.2 Machine Learning Development

Wind, wave, water level, and WireWall overtopping observations were processed in 10-min intervals. WireWall overtopping occurrence was classified binary (yes/no) as well as using the mean number of overtopping events computed in 10-min windows. A binary and regression random forest classifier estimated overtopping occurrence and frequency. We selected a random forest given its efficiency in handling limited data, its ability to rank feature variable importance, and its capability of rapidly generating predictions through its ensemble decision-tree structure. The data was partitioned into testing (20%) and training (80%) ratios. 10-fold cross-validation determined this optimised test-to-train ratio. Systematic grid searches analysed the hyperparameter tuning metrics. For the binary model, the threshold was adjusted to optimise the precision and recall. The optimal threshold configuration was 0.43. L2 regularisation addressed overfitting; selected given the few features. The random forest consisted of 100 decision trees, max depth = 10, and sample splits equaling = 10. Variable importance was determined by SHapley Additive exPlanations (SHAP) [13].

EurOtop predictions were generated using the same wave and tidal observations as the machine learning model. These conditions were transformed into the coast using the parametric breaker dissipation model, transforming wave conditions from the buoy to Dawlish and Penzance seawalls, with adjustments for foreshore slope, toe mound effects, and impulsive breaking, following the approach of SWEEP OWWL [14]. EurOtop overtopping volumes exceeding 0.1L/s/m were classified as overtopping (1), and discharge volumes below this threshold as non-overtopping (0), reflecting pedestrian safety thresholds in the EurOtop manual.

3 Results and Discussion

3.1 Variables Influencing Overtopping

Random forests demonstrated strong predictive performances for estimating overtopping occurrences and frequency in Dawlish and Penzance (F1 > 0.80, R^2 > 0.70) (Fig. 2c,d). H_s and freeboard are the most important variables influencing overtopping (Fig. 2a,b). Larger H_s corresponds to more energetic wave conditions, increasing wave runup, and the probability of waves reaching the seawall crest [15]. Lower freeboard conditions reduce the barrier height that waves must overcome, increasing overtopping likelihood. U_{10} and U_{10} Dir are also important (Fig. 2a, b). Excluding U_{10} and U_{10} Dir decreased the random forest performance for estimating overtopping occurrences, particularly at rig 2 in Penzance (F1 = 39% without wind; Fig. 2d). Omitting wind variables reduces the random forest's ability to estimate the number of overtopping events in Dawlish (R^2 = 0.54 and R^2 = 0.19, respectively) (Fig. 2c).

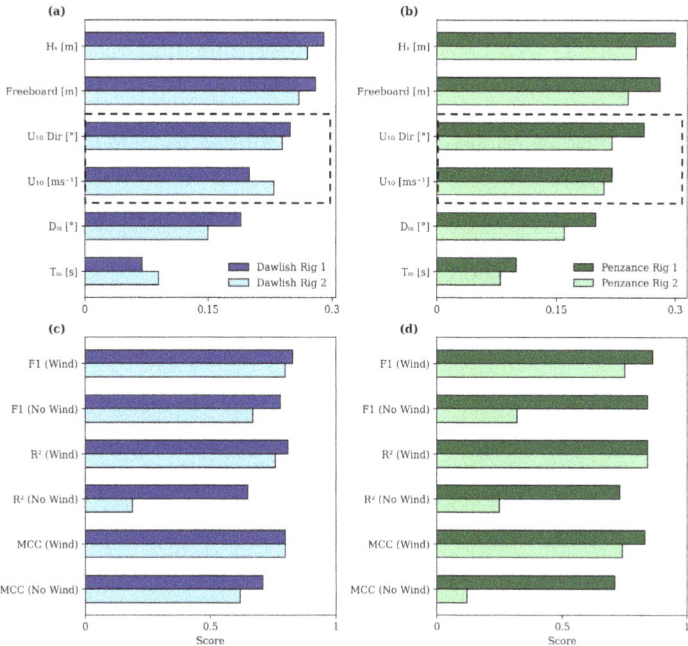

Fig. 2. Random forest variable importance metrics for (a) Dawlish and (b) Penzance comparing with and without wind variables for (c) Dawlish and (d) Penzance.

3.2 Comparing Machine Learning with EurOtop

The random forest yielded high-performance metrics for estimating overtopping in Dawlish and Penzance (F1 > 0.80), demonstrating low instances of false positives and negatives (precision > 80%) (Table 1). Random forests indicated high performances for estimating non-overtopping events (F1 = 0.98). Calculating a single prediction by the random forests was under 4 s. The SWEEP OWWL demonstrated low to moderate performance for estimating overtopping in Dawlish (F1 = 0.65) with very low predictive performances in Penzance (F1 = 0.07). The high accuracy from the random forest could be attributed to its inherent design, aiming to reduce predictive model bias and variance. Random forests sample the training dataset with replacement, meaning each decision tree specialises in different dataset sections. Random forests rank feature variables of importance (Fig. 1), focusing less on redundant unnecessary features, reducing model complexity, bias, and miss-predictions. However, the random forests did indicate several false positives and negatives which require further analysis.

The SWEEP OWWL bases its predictions for overtopping discharge using EurOtop, which is a neural network configuration. Neural networks are powerful machine learners, robust to high noise and variance. The likelihood of SWEEP OWWL's poor performance likely attributed more to the challenges of addressing data imbalances, rather than its configurations of not handling noisy data. Across the WireWall deployment, overtopping was recorded at approximately 2–3%, whereas non-overtopping was >97%, indicating significant class imbalances. Neural networks require extensive training, significantly more than random forests, given the number of neurons, hidden layers, and bias values requiring adjustment for neural networks. This neural network may not be trained enough on the minority overtopping class, likely causing this poor performance. The high performance on the non-overtopping class supports this, indicating that because many cases were non-overtopping, this sufficed the training requirements for the neural network; however, the neural network is likely overfitting the data to non-overtopping correlations, causing the low predictive performance for the overtopping class. Sweep OWWL does not consider wind parameters in its overtopping predictions, with many studies emphasising the importance of influencing wind, which influences H_s, alters Dm, and affects the landward transport of sea spray.

Table 1. SWEEP OWWL performance against the random forest for Dawlish and Penzance.

SWEEP OWWL					Random Forest		
Location	Phase	F1	Recall	Precision	F1	Recall	Precision
Dawlish	Overtopping	0.65	0.73	0.59	0.81	0.77	0.86
Penzance	Overtopping	0.07	0.04	1	0.86	0.82	0.89

4 Conclusion

This study demonstrates the high predictive performance of random forest models to estimate overtopping. Incorporating wind variables significantly improved random forest predictions. This study calls for further research to contribute additional guidance to EurOtop to improve the prediction of wind-related conditions. Our study reveals that random forests can significantly outperform SWEEP OWWL, which bases its predictions for overtopping discharge on EurOtop. Random forest demonstrated great potential for implementation in different hotspots worldwide, given its strong generalisation performance for both study locations, providing overtopping observations are available. Ensuring appropriate generalisation to different locations, this random forest would need training to predict overtopping on various seawall structures.

References

1. Kulp A, Strauss H (2019) New elevation data triple estimates of global vulnerability to sea-level rise and coastal flooding. Nat Commun 1(1):4835–4844
2. Almar R et al (2021) A global analysis of extreme coastal water levels with implications for potential coastal overtopping. Nat Commun 12(3775):3770–3775
3. EurOtop. http://www.overtopping-manual.com. Accessed 11 Dec 2024
4. Buccino MD, Leo A, Tuozzo L, Calabrese D (2023) Wave overtopping of a vertical seawall in a surf zone: a joint analysis of numerical and laboratory data. Ocean Eng 1(288):116114–116120
5. Van Der Van Gent MRA, Bijl RJ, Wolters G, Wüthrich D (2024) The maximum influence of wind on wave overtopping at seawalls with crest elements. Emerald Publishing 1(9999):1007–1021
6. Di Leo A, Dentale F, Buccino M, Tuozzo S, Pugliese C (2022) Numerical analysis of wind effect on wave overtopping on a vertical seawall. Water 14(23):3875–3891
7. Rong G et al (2020) Rainfall induced landslide susceptibility mapping based on Bayesian optimized random forest and gradient boosting decision-tree models - a case study of Shuicheng County. China Water 12(11):3055–3066
8. Naeem S, Ali A, Anam S, Ahmed M (2023) An unsupervised machine learning algorithm: comprehensive review. Int J Comput Digit Syst 1(1):1–12
9. Stokes K, Poate T, Masselink G, King E, Saulter A, Ely N (2021) Forecasting coastal overtopping at engineered and naturally defended coastlines. Coast Eng 1(164):103820–103827
10. Dawson D, Shaw J, Gehrels WR (2016) Sea-level rise impacts on transport infrastructure: the notorious case of the coastal railway line at Dawlish, England. J Transp Geogrpahy 1(51):97–109
11. Brown J et al (2023) Coastal wave overtopping: new nowcast and monitoring technologies. Coast Eng Proc 37(1):1–15
12. Yelland J et al (2024) A system for in-situ, wave-by-wave measurements of the speed and volume of coastal overtopping. Commun Eng 2(9):1–9
13. Deb D, Smith R (2021) Applications of random forest and SHAP tree explainer in exploring spatial (in)justice to aid urban planning. Int J Geo-Inf 10(9):629
14. Janssen T, Battjes J (2007) A note on wave energy dissipation over steep beaches. Coast Eng 54(9):711–7116
15. Salauddin M, Pearson M (2019) Wave overtopping and toe scouring at a plain vertical seawall with shingle foreshore: a physical model study. Ocean Eng 1(171):286–299

Beach Morphodynamic Response to Tropical Storms Using a Spatial-Temporal Surrogate Model. Case Study in Martinique

Nico Valentini[1]([✉]), Yann Balouin[1], Clement Bouvier[2], Jeremy Rohmer[3], and Deborah Idier[3]

[1] BRGM, University of Montpellier, Montpellier, France
n.valentini@brgm.fr
[2] BRGM, Bordeaux, France
c.bouvier@brgm.fr
[3] BRGM, Orleans, France
{j.rohmer,d.idier}@brgm.fr

Abstract. Tropical cyclones (TCs), especially intense hurricanes, pose a significant threat to coastal areas due to the destructive forces of storm surges and wind-generated waves. These forces can cause severe coastal erosion and flooding within a few hours. The present study aims to address the limitations of computational cost of morphodynamic models and data scarcity by developing a surrogate model to better address the morphodynamic beach response to storms event. Few studies have focused on developing surrogate models specifically dedicated to temporal coastal beach erosion forecast. A great subset of TC is selected from a synthetic database, considering time-dependent evolution of the oceanographic variables. Secondly, a 2d XBeach model is used to compute the morphological responses to selected storms. Thirdly, we adapt and test an Artificial Neural Network-based surrogate model to mimic XBeach at low computational cost. Results are presented at the study site of Le Carbet Beach, Martinique, Lesser Antilles.

Keywords: Beach Morphodynamics · Tropical storms · Caribbean · Xbeach · Deep Neural Network

1 Introduction

The Caribbean region is significantly impacted by climate extremes, with Martinique, in the Lesser Antilles archipelago, particularly vulnerable to coastal erosion driven by tropical cyclones (TC). These events generate destructive forces such as storm surges and wind-driven waves, causing severe erosion and flooding over short periods. Shoreline retreat projections rely on accurate evaluations of sediment contributions during extreme events, yet the relationship between weather extremes and erosion in Martinique remains inadequately understood, also due to limited observational data. In such cases, numerical morphodynamic models like XBeach (Roelvink et al., 2015) become crucial tools for predicting coastal morphology evolution during storms. However, their computational intensity often limits their utility for real-time forecasting or early-warning systems.

© The Author(s) 2026
C. Coelho et al. (Eds.): CD 2025, CRL 41, pp. 303–309, 2026.
https://doi.org/10.1007/978-3-032-15473-6_48

To overcome this limitation, data-driven methods like Artificial Neural Networks (ANNs) offer a promising alternative, enabling faster predictions by integrating numerical model outputs with statistical or machine learning techniques (an approach named "surrogate modelling"). ANNs are particularly effective for solving complex, non-linear systems (e.g., López et al., 2017; Van Gent et al., 2007) but often rely on synthetic datasets due to the scarcity of long-term beach erosion observations. Synthetic data generated using process-based models like XBeach have been employed in few studies to predict coastal dynamics, such as changes in dune geometry during storms (Santos et al., 2019) and dune erosion volume (Athanasiou et al., 2022). Gharagozlou et al. (2022) further developed a surrogate to predict beach responses to storm impacts.

This study addresses these challenges by introducing an ANN-based surrogate model to predict morphodynamic beach responses to hurricanes at reduced computational costs. Using a synthetic TC database, the XBeach model simulates morphological responses to selected storms, forming the basis for training the model. This approach integrates the strengths of numerical modeling and machine learning to improve understanding and forecasting of coastal erosion in hurricane-prone regions like Martinique.

2 Methods and Study Area

As part of the INTERREG CARIBCOAST project (https://www.carib-coast.com/), BRGM developed an approach leveraging the STORM database (Bloemendaal et al., 2020), a simulation-based archive of tropical cyclone tracks in the Caribbean basin built on the IBTrACS historical dataset (Knapp et al., 2018). For the present study, 685 tracks were selected from STORM, representing 1000 years of cyclonic activity near Guadeloupe and Martinique. Cyclone parameters along these tracks (e.g., maximum wind speed, pressure, radius to maximum winds) were used to compute wind speed, using a parametric approach (Holland et al., 2010), wave fields (using the spectral wave model WW3; Abdolali et al., 2021), and water levels (using the hydrodynamic model UHAINA; Filippini et al., 2023). The models were validated using historical hurricanes such as 1928 Okeechobee hurricane, Hugo (1989), Dean (2007), Irma and Maria (2017). The XBeach model (2d) was applied to five beaches sites, calibrated using storm-specific parameters like wave dissipation, breaker index, and sediment transport factors. Inspired by Adeli et al. (2023), who utilized deep neural networks (DNNs) for storm surge predictions, this study incorporates a ConvLSTMs (Convolutional Long Short-Term Memory networks) and an Encoder-Decoder architecture to predict beach morphodynamics. ConvLSTMs handle spatial and temporal learning, while the Encoder-Decoder approach supports sequence-to-sequence prediction. The encoder, consisting of three convolutional layers, processes input storm data (e.g., size, intensity, location), compressing it into a concise representation via a ConvLSTM cell. The decoder then upscales this representation using pixel shuffling to predict topo-bathymetric changes. Residual connections further enhance performance, effectively addressing limited data variability, as demonstrated in Adeli et al.'s storm surge models.

The methodology is here presented at Le Carbet Beach located on the island of Martinique (Fig. 1a) which has been triggered over the last ten years by severe tropical storms. Le Carbet is a 2.4-km-long beach located in the north-west side of Martinique

(Fig. 1). The submerged beach is characterized by a slope of around 10%, exposed to a low energy climate, with an average significant wave height of only 0.3 m and peak period of 9.2 s and a microtidal range, maximum amplitude of 0.7 m at a semidiurnal cycle. In its southern part, the beach exhibits a high level of morphodynamic seasonal variability (10 m throughout the year, Bouvier, et al., 2023).

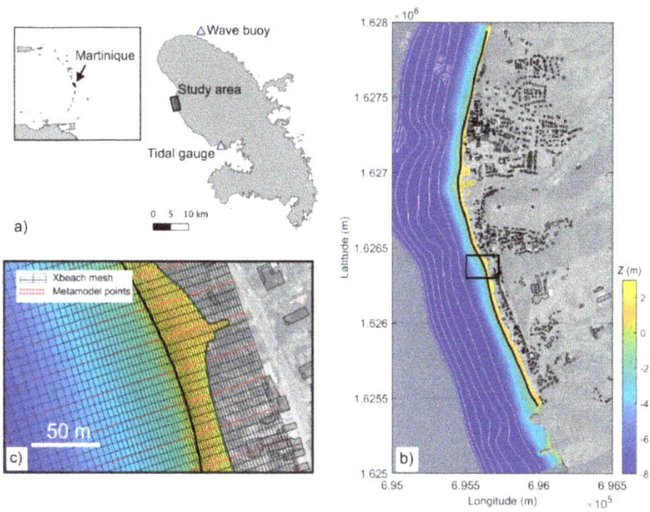

Fig. 1. (a) Map of the study area, showing the location of a directional wave buoy and tidal gauge. (b) Detailed high-resolution bathymetry (Z) of Le Carbet Xbeach model. (c) Close-up view of a selected area, illustrating the curvilinear grid employed by the XBeach morphodynamic model (black) and the computational points utilized by the surrogate (red).

3   Results

A total of 685 XBeach 2D simulations of synthetic tropical cyclones, along with their associated morphological responses, carried out using a cluster server. Each simulation utilized 60 cores on a cluster, with one day (24 h) of simulation requiring approximately 30 min cpu wall-clock time. The total simulation durations ranged from 12 to 82 h. The XBeach simulations provide process-based estimates of beach responses at over 110 cross-shore locations, with higher resolution focused on the subaerial section of the beach (min. 1.5 m). The predicting skill of this method was validated by hindcasting Hurricane Maria, where the modeled storm-induced morphological changes showed satisfactory agreement with historical observations (Fig. 2).

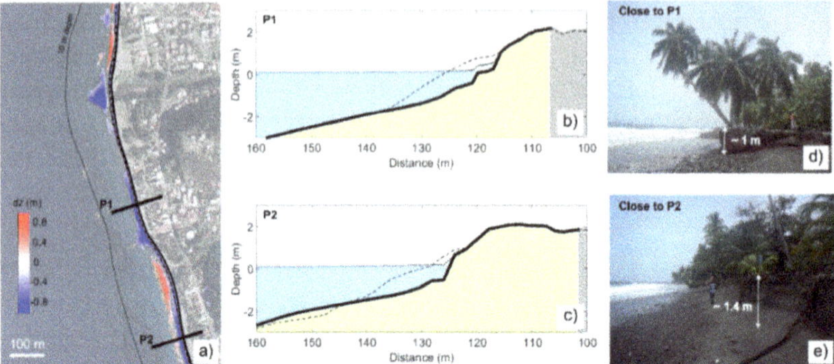

Fig. 2. Beach response to Hurricane Maria in September 2017 at Le Carbet beach. (a) Simulated 2D erosion-deposition patterns generated by the calibrated XBeach model, showing the locations of profiles P1 and P2. (b, c) Cross-shore morphological evolution by Xbeach along P1 and P2, with the dashed line representing the initial beach morphology. (d, e) The observations maximum beach scarps, distribution of those beach scarps are well predicted by Xbeach.

One over 4 Xbeach transects is chosen in order to reduce model complexity, so a total of 96 locations in the cross-shore and 48 in the alongshore are used for training the DNN model, (Fig. 1c). A total of 640 randomly selected storms are used for the calibration of the neural network emulator, and an additional 40 storms are used as a test sample for its validation. The inputs are the classic parametric meteo-marine variables with significant wave heights (Hs), peak period (Tp), peak direction (Dp) and sea level (SeaLev), with peaks Hs ranging from 1 to 17 m, at 50 m offshore depth. The time evolution for all these four parameters is used as input to the neural network. For both the input and the output of the emulator, 82 time-steps (1 h each) are used. This range is chosen to encompass the maximum duration of TCs across the geographic domain of interest. The specific structure of the input sequences for the storm imposes to apply data standardization using the mean and standard deviation. The output label data also undergoes normalization, using a hyperbolic function to center the data around zero. This normalization helps the model to converge more quickly when using the standardized data (Adeli et al., 2023). The model is trained for 20,000 epochs with a learning rate of 10^4, using L2 norm loss function and the Adam optimizer with mini-batch gradient descent. Figure 3(a-b) shows an example of XBeach-simulated versus surrogate-predicted morphological response at Carbet beach for a category 4 tropical cyclone, with a maximum offshore Hs of ~10 m, at 20 m depth. The 2D spatial distribution of the last erosion-deposition pattern correlates well with the final profiles (R2 = 0.95, RMSE ~0.4 m). Additionally, for the 40 validation tropical cyclones the Beach Erosion Volume (BEV) is computed as the volume of the eroded emerged beach (above the 0 m bathymetric contour) over time. Figure 3(d) presents a scatterplot with related statistics showing the comparison between XBeach-simulated and surrogate-model-predicted BEV, plus Hurricane Maria. The colors represent the maximum Hs for each TC, while point size increases with time steps. The statistics appended on the figure show the alongshore median error values averaged across the 40 TCs, with min–max values in brackets. The black stars correspond to the final time step and the black line represents the 1:1 line. As shown, the morphological evolution predictions made by the convolutional recurrent neural network are highly accurate across all tested cases, with a substantial correlation (average Brier Skill Score, BSS > 0.95) between predictions and true responses. Only two out of 40 TC cases have low BSS and Root-Mean-Square-Difference (RMSD) values (minimum BSS of 0.67 and RMSD of 10), still meeting good classification criteria according to Van Rijn et al. (2003). The surrogate model provides around 500x speedup (single core- >30 s) over XBeach (multi-core HPC). Ongoing work aims to optimize the model's architecture, enhance stability, and improve performance, particularly through strategies such as early stopping and regularization techniques (L1 and L2) that introduce penalty terms to the loss function.

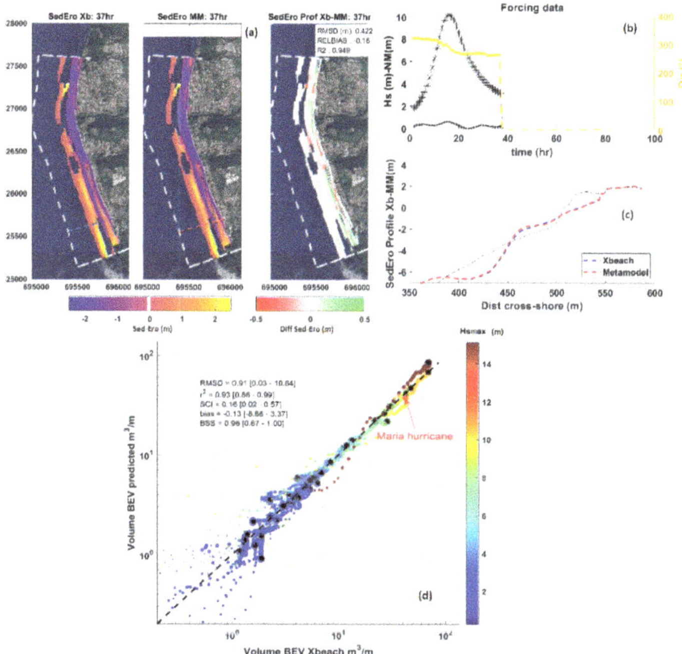

Fig. 3. (a) Xbeach-simulated vs. surrogate morphological response at Le Carbet beach, distribution of elevations and elevations' differences across the domain for category 4 TC, Input TC wave parameters at (b). (c) cross-shore comparison of both responses at last time step. (d) Scatterplot between the XBeach (x-axis) and DNN (y-axis) predicted Beach Erosion Volume (BEV) for the validation set.

4 Conclusions

This research developed a surrogate model using a convolutional recurrent neural network (ConvLSTM) with an encoder-decoder architecture to predict beach morphodynamics over time, utilizing a database of simulated synthetic tropical cyclones. The model captures both spatial and temporal correlations within the data, improving predictions of the erosion-deposition pattern based on storm-driven parameters. Unlike previous studies that focused on dune erosion or beach erosion at isolated profiles, this approach allows for morphodynamic predictions across multiple profiles within a large 2D domain, enhancing accuracy. The accuracy of the model itself depends also on the initial calibration of the morphodynamic model, which affects the beach response to hurricanes. Although validated with synthetic TCs and Hurricane Maria, further testing on other historical cyclones is needed. Additionally, the model uses classical wave parameters like Hs, Tp, and Dp, which may not fully capture the complexity of wave climates with multiple spectral modes. The absence of initial topo-bathymetry in the metamodel further simplifies the approach, ongoing works are used to assess the model's sensitivity to this variable, as pre-storm bathymetry is known to have a significant effect on event-scale beach evolution. Despite these limitations, the key innovation of this research is its ability to generate spatio-temporal predictions across a large geographic area, unlike other models that focus on aggregated predictions. This model advances our ability to predict and understand the impact of severe weather events on coastal landscapes in Martinique and the broader Caribbean region.

References

Adeli E, Sun L, Wang J, Taflanidis AA (2023) An advanced spatio-temporal convolutional recurrent neural network for storm surge predictions. Neural Comput Appl 35:18971–18987. https://doi.org/10.1007/s00521-023-08719-2

Athanasiou P, Van Dongeren A, Giardino A, Vousdoukas M, Antolinez JAA, Ranasinghe R (2022) Estimating dune erosion at the regional scale using a meta-model based on neural networks. Nat Hazards Earth Syst Sci 22:3897–3915. https://doi.org/10.5194/nhess-22-3897-2022

Bloemendaal N, Haigh ID, De Moel H, Muis S, Haarsma RJ, Aerts JCJH (2020) Generation of a global synthetic tropical cyclone hazard dataset using STORM. Sci Data 7:40. https://doi.org/10.1038/s41597-020-0381-2

Bouvier C, Njakam A, Robinet A, Azorakos G (2023) Smartphone based shoreline monitoring: an intersite comparison in lesser Antilles. In: Coastal Sediments 2023: The Proceedings of the Coastal Sediments 2023, pp 255–266

Filippini AG et al (2024) An operational discontinuous Galerkin shallow water model for coastal flood assessment. Ocean Model 192:102447

Holland GJ, Belanger JI, Fritz A (2010) A revised model for radial profiles of hurricane winds. Mon Weather Rev 138(12):4393–4401

Gharagozlou A, Anderson DL, Gorski JF, Dietrich JC (2022) Emulator for eroded beach and dune profiles due to storms. JGR Earth Surface 127:e2022JF006620. https://doi.org/10.1029/2022JF006620

Santos VM, Wahl T, Long JW, Passeri DL, Plant NG (2019) Combining numerical and statistical models to predict storm-induced dune erosion. JGR Earth Surface 124:1817–1834. https://doi.org/10.1029/2019JF005016

Van Rijn LC, Walstra DJR, Grasmeijer B, Sutherland J, Pan S, Sierra JP (2003) The predictability of cross-shore bed evolution of sandy beaches at the time scale of storms and seasons using process-based Profile models. Coast Eng 47:295–327. https://doi.org/10.1016/S0378-3839(02)00120-5

Storm Impact Assessment for Scotland's Wave-Dominated Sedimentary Coasts

Nicola Horsburgh[1(✉)], Carlos Loureiro[2,3], Martin Hurst[4], Alistair Rennie[4,5], and Andrew Tyler[1]

[1] Biological and Environmental Sciences, Faculty of Natural Sciences, University of Stirling, Stirling FK9 4LA, Scotland
nicola.horsburgh@stir.ac.uk
[2] CIMA-ARNET, Faculty of Science and Technology, University of Algarve, 8005-139 Faro, Portugal
[3] School of Agricultural, Earth and Environmental Sciences, University of KwaZulu-Natal, Durban 4000, South Africa
[4] School of Geographical and Earth Sciences, University of Glasgow, Glasgow G12 8QQ, Scotland
[5] NatureScot, Inverness IV3 8NW, Scotland

Abstract. Along wave-dominated sedimentary coasts, risks posed by chronic coastal erosion and rising sea levels during this century have received much attention. However, extreme wave runup during large storms are still poorly characterised at large spatial scales, and can contribute significantly to coastal flooding and erosion. This creates an important challenge for coastal management and adaptation planning. This work presents a national-scale storm impact assessment for Scotland's wave-dominated sand and gravel coasts. It uses readily available national and European datasets to model extreme total water levels that include storm wave runup, tides, storm surge and changes in mean sea level. Future impacts are assessed by integrating likely climate-driven changes to storm forcing and coastal morphology. Outputs estimate the likely storm impact regime at 50 m alongshore transects in Scotland's natural sand and gravel coasts for the present day, 2050 and 2100, considering climate change scenarios RCP4.5 and RCP8.5. Results indicate 65%, 11% and 4% of transects currently experience collision, overwash and inundation as the most severe storm impact. Towards 2100, transects in overwash and inundation increase to 20% (22%) for RCP4.5 (RCP8.5), but local increases vary considerably between coastal cells. This study contributes to a strategic, national scale understanding of coastal risk that supports the spatial prioritisation of areas in need of proactive adaptation solutions.

Keywords: Coastal Adaptation · Flooding · Erosion · Wave Runup

1 Introduction

Coastal zones are dynamic interfaces between the sea and land, and are exposed to hazards that include coastal erosion and flooding due to extreme water levels from tides, storm surge and waves. Despite these risks, coasts are economically and socially

© The Author(s) 2026
C. Coelho et al. (Eds.): CD 2025, CRL 41, pp. 310–316, 2026.
https://doi.org/10.1007/978-3-032-15473-6_49

important and globally an estimated 898 million people currently live in the low-elevation coastal zone, with numbers projected to increase this century under all shared socio-economic pathways [1].

To assess the potential risk to coastal communities, significant effort has been made to characterise coastal change [2] and extreme water levels [e.g. 3] and how it may evolve this century. These studies model extreme water levels at the global scale as the combined effect of tides, storm surge and wave setup, but exclude wave-induced swash. This omission is necessary due to the absence of detailed topographic data, specifically beach slope, which is not available at sufficient resolution in global datasets. In Scotland, wave contributions to extreme water levels are considered for limited sections of the coastline, but whole coast assessments currently only include sea level rise (SLR), tide and storm surge contributions [4]. At the national scale, there is thus a lack of understanding of storm-induced wave runup, which can make a significant contribution to total water levels along exposed coasts and drive substantial morphological changes to the beaches and dunes [5]. This leaves a critical gap in the current understanding of storm-driven coastal risk, which hampers adaptation planning, especially along non-urban coasts, where coastal communities and infrastructure are often located directly behind sand or shingle barriers that act as the first line of defense during storms.

This work presents a national-scale storm impact assessment for Scotland's wave-dominated sand and gravel coasts. By modelling extreme total water levels that include storm wave runup, the objectives are to understand a) where natural sand and shingle barriers are vulnerable (resilient) to present-day storm impacts, b) where coastal assets behind natural barriers are most at risk of storm-driven flooding, and c) how these risks may change towards 2050 and 2100. This improved understanding of coastal storm impacts and the role of natural barriers in coastal protection can then highlight areas where proactive adaptation planning should be prioritised.

2 Study Area and Methods

Scotland's coastline is highly complex and varied, with a heavily indented configuration on the Atlantic coast versus more linear stretches on the North Sea coast. Sediment ranges from fine sand to cobbles, with sandy barrier systems, mixed and composite sand-gravel beaches, and pure gravel beaches. Storm exposure also varies significantly, as the west coast and islands face extreme Atlantic storm systems, while adjacent coastlines on the Irish Sea are more sheltered. The wave climate on the east coast of Scotland tends to be less extreme than in the west, but extratropical cyclones over the North Sea can cause extreme storm surges and coastal flooding when coincident with high tides.

The storm impact assessment was performed for the entire coast of Scotland and included open coastlines with sand or gravel barriers, where wave action (as opposed to tidal inundation) is the predominant hydrodynamic driver of beach morphology. Sedimentary accumulations on rocky and highly engineered coasts were excluded from the assessment, as runup on these beaches fall outside the scope of the models used.

To provide a conceptual understanding of potential storm impacts at the national scale, the storm impact scale framework of Sallenger [6] was used to identify the storm impact regimes of swash, collision, overwash or inundation. The assessment was implemented within a geospatial framework, where each stretch of coastline was populated with shore-normal transects, spaced 50 m apart. Along each transect, the local topography profile was extracted, extreme total water levels were computed, and compared with the toe and crest elevations of the most seaward coastal barrier to estimate the transect's storm regime (Fig. 1). The study comprised of 12,798 transects, equivalent to 640 km (~3.6%) of Scotland's coastline.

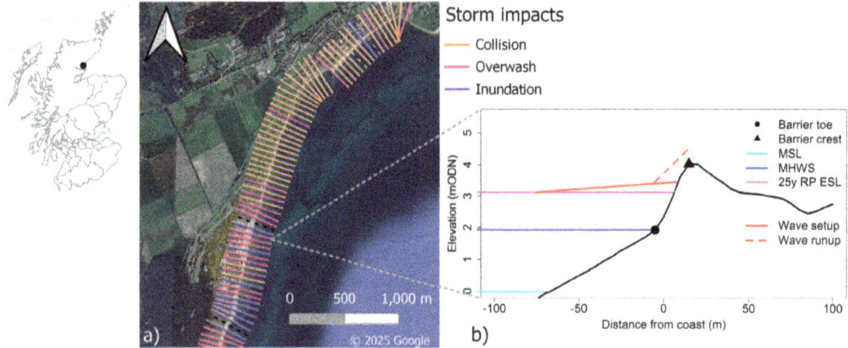

Fig. 1. a) Present day storm regimes for Golspie, Highland b) Transect beach-barrier morphology and water levels, predicting overwash. Map produced in QGIS 3.26 (www.qgis.org). Map data: Google © 2025. Administrative boundaries (level 3) from the GADM database (gadm.org, v3.6).

Extreme total water levels (TWL) were computed as the sum of extreme still water levels (ESL) and extreme wave runup, and this was done for the present day and climate scenarios RCP4.5 and RCP8.5 in 2050 and 2100. ESL comprised of tides and storm surge, uplifted by 95th percentile projected SLR (Table 1). To build a synthetic storm, the 25-year return period ESL was chosen to represent a large but likely storm. For storm wave conditions, the 99th percentile deepwater significant wave height and peak period were used. Using wave and beach slope data, the 2% exceedance of extreme wave runup ($R_{2\%}$) on sandy, composite and mixed sand/gravel beaches was modelled with the empirical parameterisation of Stockdon et al. [7], while $R_{2\%}$ on pure gravel beaches was modelled with a gravel-specific parameterisation by Poate et al. [8].

To assess potential storm impacts for mid- and end of century, future extreme TWL were compared with future barrier morphology, adjusted for SLR with a conceptual model that used historic coastal change as a first-order indicator of sediment budget.

Table 1. Primary data sources for storm impact assessment.

Data	Description	Source
Extreme still water levels	25-year return levels of tide and storm surge, base year 2017	Met Office Hadley Centre
Mean sea level	95th percentile SLR for RCP4.5 and RCP8.5 in 2020, 2050 and 2100	Met Office Hadley Centre
Wave climate	99th percentile H_s and T_p for the present day, mid-century and end-century	Copernicus Climate Change Service
Elevation	National digital terrain model, 5 m resolution	Ordnance Survey
Assets	Roads, railway lines and property	Ordnance Survey

3 Results and Discussion

Along Scotland's wave-dominated sedimentary coasts, 65%, 11% and 4% of coastal transects are modelled to experience collision, overwash and inundation as the most severe present-day impact regime. Towards 2100, transects in overwash and inundation are modelled to increase to 20% (22%) for RCP4.5 (RCP8.5), which suggests storm impacts will intensify towards the end of the century, regardless of emissions scenario.

Given the complexity of the Scottish coastline, storm impacts vary considerably around the coast (Fig. 2). For example, cells 3 and 11 are projected to experience the largest proportions of overwash and inundation for the present day (24% and 40%, respectively), while cell 2 shows the greatest present-day resilience to storm impacts, with 85% of transects in collision and only 8% in overwash or inundation. All cells show an increase in overwash and inundation towards the end of the century, under RCP4.5 and RCP8.5. Under RCP8.5, cells 3, 6 and 7 are predicted to have proportionately the largest increases in more severe impact regimes (Fig. 2).

Areas of overlap between low-lying areas of high asset density and overwash or inundation regimes indicate hotspots of greatest storm-related coastal flood risk (Fig. 3). Locations of regime upshift, where the modelled storm regime for RCP8.5 in 2100 is more severe than for the present day, suggest that for many currently high-risk asset areas, storm impacts are expected to increase towards the end of the century (Fig. 3). At the national scale, these high-risk areas represent 808 transects (~40 km of coastline) and provide a strategic understanding of where to target further local-scale assessments.

While these results provide a first order indication of hotspots in need of coastal adaptation, there are a range of limitations when working at large spatial scales, mainly related to input data and methodology. By using national elevation data, the analysis uses a single snapshot of barrier morphology that varies in acquisition date around the coast. Furthermore, medium horizontal resolution and accuracy of the national DTM adds uncertainty to extracted barrier toe and crest elevations.

314   N. Horsburgh et al.

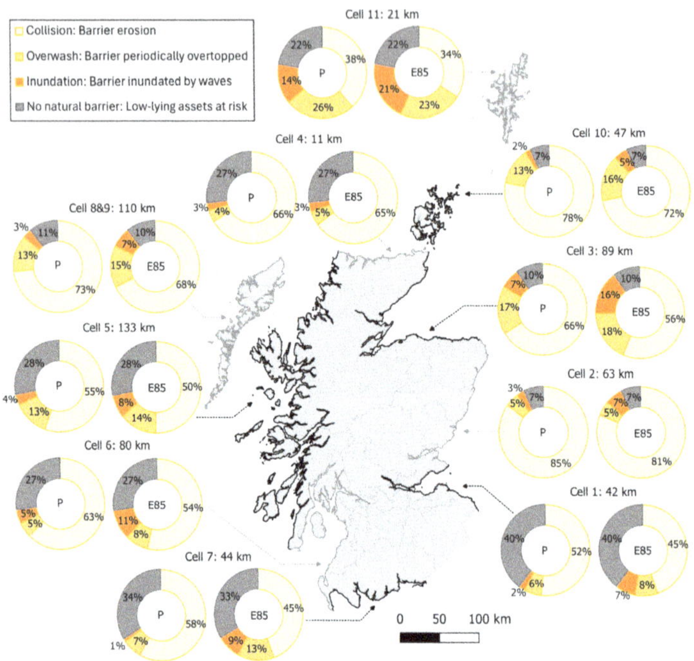

Fig. 2. Percentage of transects in each storm regime per coastal cell (alternating black/grey sections of coastline). Length of coastline studied/cell. P = present day, E85 = end-century RCP8.5. Map produced in QGIS 3.26 (www.qgis.org). Administrative boundaries (level 3) from the GADM database (gadm.org, v3.6).

Extreme still water levels combining tide and storm surge are provided at 2 km spacing around the coast from state-of-the-art national models. By using a 25-year return period ESL, the model uncertainty is small (order of ±10 cm), and thus does not contribute significantly to analysis uncertainty. However, there is large uncertainty regarding the amount of SLR that will happen this century, with the difference between modelled 5th to 95th percentile SLR in 2100 in the order of 0.5 m for Scotland. Therefore, the choice of SLR percentile has a large bearing on total water levels and storm impact regime predictions. Here, we only consider 95th percentile projected SLR in each climate scenario, as a precautionary high-impact estimate of TWL.

Finally, there is still large uncertainty regarding the future changes in the storm wave climate for the North Atlantic. An increase in winter-mean wave height, variability (larger extremes) and periodicity has been observed over the last 70 years, but it is not yet clear if this can be attributed to climatic change [9]. Most existing climate models, including the regional climate model used to produce the wave data used in this study, do not predict a notable increase in wind speed (and in turn wave height) for the North Atlantic in the XXI century [9, 10]. However, if the observed trend of increasing wave height continues in the next decades, wave climate projections used in this analysis likely underestimate wave contributions to future TWL.

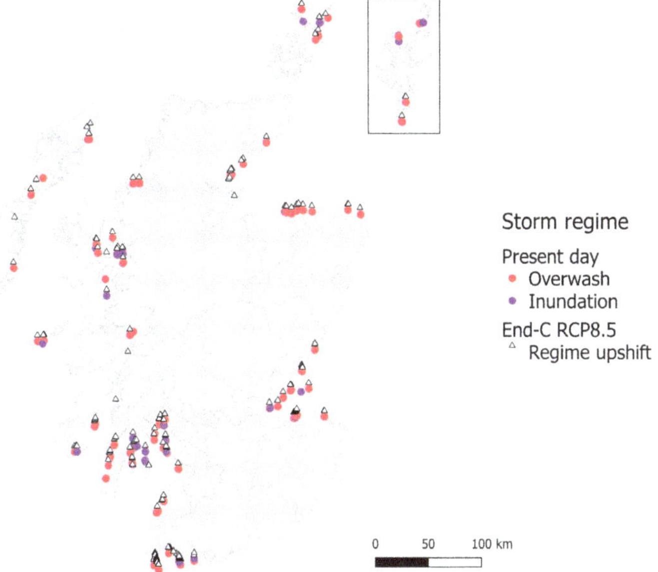

Fig. 3. Areas with low-lying assets and high impact storm regimes. Map produced in QGIS 3.26 (www.qgis.org). Administrative boundaries (level 3) from the GADM database (gadm.org, v3.6).

4 Conclusion

This study presents the first national storm impact assessment for Scotland that includes storm wave runup. It outlines areas of increasing coastal risk for this century, with large increases in high impact storm regimes predicted for sectors of the Scottish coast. The study identifies where natural barriers fronting coastal assets are currently vulnerable to storm-driven flooding, and where impacts are likely to increase in the future. This contributes to a strategic understanding of coastal hazards and supports the spatial prioritization of areas in need of further local assessment and adaptation planning.

References

1. Reimann L, Vafeidis AT, Honsel LE (2023) Population development as a driver of coastal risk: current trends and future pathways. Cambridge Prisms Coast. Futures 1:e14
2. Mentaschi L, Vousdoukas MI, Pekel JF et al (2018) Global long-term observations of coastal erosion and accretion. Sci Rep 8(1):12876
3. Jevrejeva S, Williams J, Vousdoukas MI et al (2023) Future sea level rise dominates changes in worst case extreme sea levels along the global coastline by 2100. Environ Res Lett 18(2):024037
4. SEPA (2023) Coastal Flooding Summary: Methodology and Mapping, pp 1–29
5. Stockdon HF, Long JW, Palmsten ML et al (2023) Operational forecasts of wave-driven water levels and coastal hazards for US Gulf and Atlantic coasts. Commun Earth Environ 4(1):169
6. Sallenger AH (2000) Storm impact scale for barrier islands. J Coastal Res 16(3):890–895

7. Stockdon HF, Holman RA, Howd PA et al (2006) Empirical parameterization of setup, swash, and runup. Coast Eng 53(7):573–588
8. Poate TG, McCall RT, Masselink G (2016) A new parameterisation for runup on gravel beaches. Coast Eng 117:176–190
9. Castelle B, Dodet G, Masselink G et al (2018) Increased winter-mean wave height, variability, and periodicity in the Northeast Atlantic over 1949–2017. Geophys Res Lett 45(8):3586–3596
10. Tobin I, Jerez S, Vautard R et al (2016) Climate change impacts on the power generation potential of a European mid-century wind farms scenario. Environ Res Lett 11(3):034013

Assessing Coastal Hazards Using Multivariate Statistical Analysis: Examples on Two Soft Cliffs in France and Croatia

Olivier Cohen[1]($\boxtimes$), Kristina Pikelj[2], and Emmanuel Blaise[1]

[1] Lab. of Oceanology and Geosciences, Université du Littoral Côte d'Opale, Dunkirk, France
`olivier.cohen@univ-littoral.fr`
[2] Faculty of Science, Department of Geology, University of Zagreb, Zagreb, Croatia

Abstract. In this paper, a multivariate statistical method to assess coastal erosion hazards is tested on two soft cliffs located in France and Croatia. This method requires two 3D digital elevation or surface models at two dates from which cliff edge and toe positions, mean slope and volume are extracted and compared in time. The potential hazard of the cliffs is synthesised and classified using a statistical classification tool (clustering analysis).

Keywords: Coastal cliffs · Hazards · Geographical Information System · Clustering Analysis

1 Introduction

Assessing the risk of erosion hazard on coastal cliffs is a complex task. Most studies use the values or rates of cliff retreat as the main parameter to estimate the intensity of coastal hazards. This is understandable, as the only information available in the long term (e.g. several decades) comes from historical maps, vertical aerial photographs or satellite images, on which the successive cliff edges are digitised and their development is then calculated [1, 2].

Over the last 25 years, 3D topography tools and methods (airborne or terrestrial LIDAR, stereophotogrammetry) have become popular in the scientific community of geoscientists and engineers while their availability has increased [3, 4]. They are very useful to determine the horizontal evolution of the coastline and to study the shape, evolution and morphodynamics of the coast [5] and to assess the natural hazards associated with coastal erosion.

In this methodological paper, we present preliminary results showing how a multivariate cluster analysis using four evolutionary parameters derived from aerial imagery and digital elevation models can help to categorise zones with different levels instability hazard potential. This simple method can be easily applied by stakeholders who want to map coastal risk using 3D data and free software (GIS and statistical analysis package). Two different sites were selected to test this method: The Petite Pointe aux Oies (Wimereux) in France and the Vrgada cliff in Croatia. Both are endangered soft cliffs,

C. Coelho et al. (Eds.): CD 2025, CRL 41, pp. 317–323, 2026.
https://doi.org/10.1007/978-3-032-15473-6_50

318   O. Cohen et al.

but they have different shapes and morphodynamics and they develop in different environmental setting. The sites were studied on two different time scales: medium-term for
the Petite Pointe aux Oies cliff (April 2008 to July 2024) and short-term for the Vrgada
cliff (April to June 2023).

2 Study Sites

The Petite Pointe aux Oies is located on the French coast of the English Channel, south
of the Dover Straits, in the Hauts-de-France region (Fig. 1). It is one of the rare soft
cliffs in the region. The coastal section studied here is ~450 m long, where longshore
geomorphologic variations of the coast are visible. The lithology consists of Pleistocene
clay and sand with a layer of Cretaceous limestone at the base [6], outcropping to the north
and dipping to the south. The northern part of the Petite Pointe aux Oies is characterized
by a limestone cliff that develops with rockfalls. The middle part is characterized by
periodic rotational landslides [7] where the sand and clay layers are thicker (3 to 4 m of
sand, 11 m of clay). The average height of the cliff is 13.1, m but it varies from ~10 m
in the north, 24.3 m in the center to 5.6 m in the south. The slope also changes along the
cliff: it is steeper in the north and the south (~30°) and more gentle in the central part
(~20°). The limestone cliff retreats slowly (e.g. 0 to 1.7 m erosion of the cliff edge from
2008 to 2024), but the sand and clay cliff erodes rapidly (max. – 28.2 m).

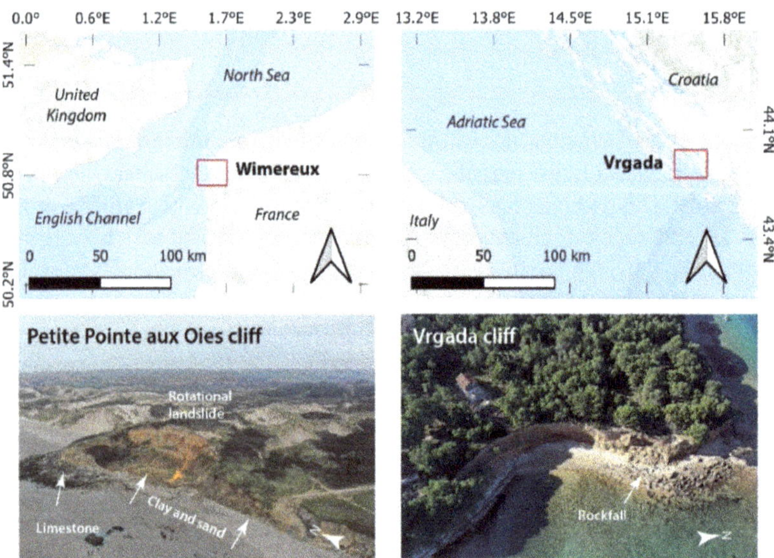

Fig. 1. Locations and aerial photographs of the study sites.

The second site is located on the Vrgada Island in the central part of the Croatian
Adriatic (Fig. 1). The length of the studied part of the cliff is 100 m and of more homogeneous geomorphology compared to the French site. Its lithology comprises Pleistocene

aeolian-alluvial clastites. In the sector shown in the Fig. The cliff is particularly steep (almost 90°). Its average height is 11.5 m and varies between 9.3 and 14.3 m. It is susceptible to erosion and is subject to massive episodic rockfalls [8, 9]. During the study period, the cliff edge remained stable except in the middle part where transects 16 and 17 (see below) retreated by 2.7 and 2 m, respectively, where rockfalls occurred (aerial photograph on Fig. 1).

3 Datasets and Methods

The 3D topographical measurements were carried out twice on each site. The first survey for the Petite Pointe aux Oies was conducted in April 2008 using airborne LIDAR. The Digital Elevation Model (DEM) with a spatial resolution of 0.5 m was provided by the French government. The second survey of the Petite Pointe aux Oies (July 2024) and the two surveys of the Vrgada cliff (April and June 2023) were carried out by the authors using UAVs and Structure-from-Motion (SfM) stereophotogrammetry to calculate high resolution (5 cm) planar orthomosaics and Digital Models (DMs). At both sites, the horizontal and vertical accuracies of the DMs are within 2 and 4 cm with the Root Mean Square Error less than 4 cm.

The data set required for the statistical analysis was created using a free Geographic Information System (QGIS 3.32) and some of its extensions.

First, the QGIS extension Station Lines was used to calculate regularly spaced transects perpendicular to the shoreline. It calculates the transects along a baseline. This baseline must be approximately parallel to the shoreline. In Wimereux, the 17 transects are equidistant by 20 m. In Vrgada, the 25 transects are closer together (5 m). These transects are used for data extraction.

Subsequently, the edges and toes of the cliffs were carefully digitised at each time point on the high-resolution aerial photographs and DMs. The cliff edge is the upper break of the slope, while the cliff toe is the lower break. In certain places the cliff toe was identified as the boundary between debris cones or blocks and beach.

Cliff edges and toes evolution were estimated by measuring the distances between the two data positions along each transect. The greater the erosion of the shorelines, the greater the resulting hazard.

The change in slope is the third parameter to be considered. The faster these changes are, the more unstable the cliff is. This instability can be part of the hazard. The mean slope can change even if the edge and the toe of a soft cliff remain stable: for example, the shape of the slope can change like a rotational landslide by becoming more concave [7]. qProf, another QGIS extension, was then used to calculate the average slope along each transect at each time point. The change in slope comparing the two points in time was then calculated.

qprof also helps with the extraction of cumulative distances and heights along the transects and thus enables the calculation of the volumes of the topographic profiles. As with the slopes, the greater the changes the volume changes, the more unstable the cliff.

A two-way clustering analysis was performed using the latest version of Past (4.17), a free statistical software. It provides a classification based on the four dynamic parameters (edge and toe, slope and volume evolution) and estimates their relative influence. In each phase of the calculations, the Euclidean distances between the individual statistical units (transects) were calculated. In our case, these are 4D distances (one dimension for each parameter). The shorter the distance, the more similar the transects are. The units are gradually grouped into classes. The cluster analysis is presented in a dendrogram showing the steps of progressive combination (Fig. 2). The discretization threshold is set to a distance chosen to obtain the desired number of classes: 4 classes for the Petite Pointe aux Oies and 3 classes for the Vrgada cliff.

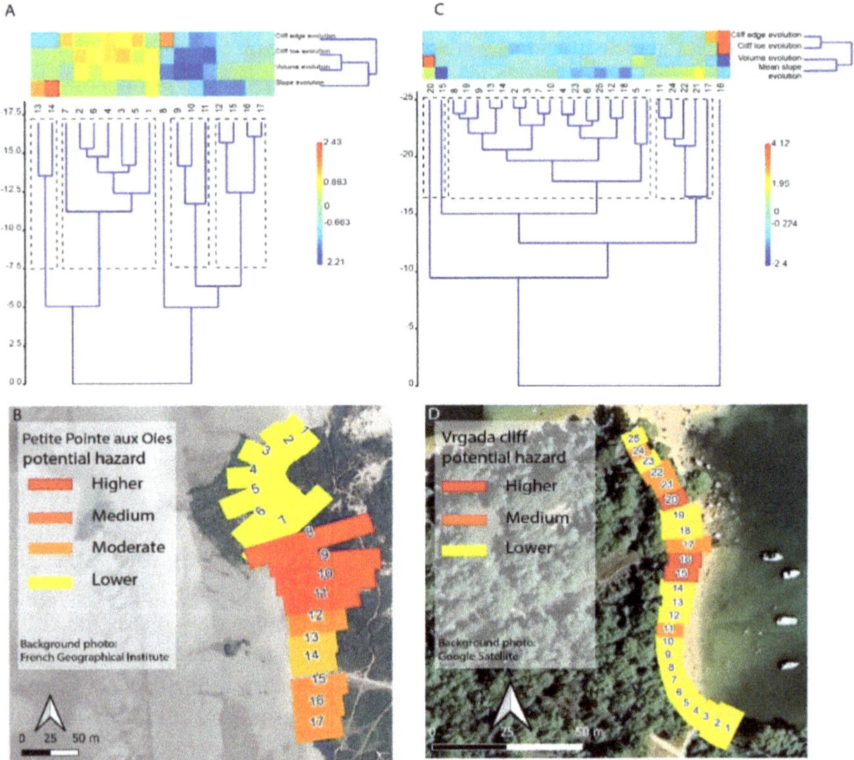

Fig. 2. Clustering Analysis (A and C) with potential hazard classification maps (B and D) for the two study sites.

4 Results

The discretization threshold was set to −7.5 to create 4 groups (rectangles in dashed lines in Fig. 2A). Transects 12, 15, 16 and 17 are the first to be grouped together as they are the most similar. The second group consists of transects 13 and 14, the third group consists of transects 1 to 7 and the fourth group consists of transects 9 to 11 (with the last group being clearly different from the others). Transect 8 is an "outsider" that was merged with the two groups on the right of the dendrogram at the end and it was included in the high hazard group (Fig. 2B). The colour table above the dendrogram helps with a more detailed analysis. Each line refers to a parameter of the clustering analysis, the colours to the influence of the parameter on the transect. Red and dark blue stand for a strong influence, green for a very low influence (0 on the standardised scale on the right side of the dendrogram). The greater the range of colours, the greater the diversity. For example, transects 9 to 11 experienced both a fair amount of coastal erosion and a fair volume reduction, but the change in slope was milder; they were therefore included in the high hazard class. Transects 1 to 7 showed few changes in the four parameters and were therefore classified as low hazard compared to the other transects. Transects 13 and 14 remained fairly stable, with the exception of their slope, which increased (orange and red colours); they were classified as moderate hazard.

The colour table of Vrgada shows more homogeneous colours, as the geomorphological gradients are low and the cliff remained stable overall during the short study period (Fig. 2C). However, transects 16, 17 and 20 are clearly different from the others: on a transects 16 and 17, the cliff collapsed, while the cliff edge and cliff toe retreated, as mentioned in the study area description. A discretization threshold of -16 was set to create 3 groups. In Vrgada, the southern part of the site appears to be at lower hazard even though the cliff is higher. The northern part of the site shows greater variability, mixing transects at low, medium and high hazard transects (Fig. 2D).

5 Discussion and Conclusion

On the Petite Pointe aux Oies, the central and southern parts are at moderately to severely endangered. A large rotational landslide has already been detected in the middle. It occurs where the cliff is highest and where the thicker sand and clay outcrops are located. There is a less instability to the south. The case of transects 13 and 14 should be further investigated: why does the hazard decrease for these two transects even though they are located in an area of medium hazard? Do they represent a transition to a southern zone where the hazard should be lower but it is not, because the cliff in the southern part also consist of an upper and obviously unstable layer of an anthropogenic fill material? The northern part is less vulnerable. This greater stability could be related to the fact that the limestone outcrop becomes thicker from south to north and protects the upper and thinner layers of sand and clay.

In the case of the Vrgada cliff, the result showed that the highest hazard level is along transects 15 to 17, where a rockfall recently occurred. Considering the slopes and heights, which are quite similar along the entire cliff, it seems that other parameters, which were not considered in this analysis, played a more important role for this slope. In the case of the Vrgada cliff, such parameters could be a localised weakness of the rock mass that is the result of the detached part of the cliff. It was suspected that the heavy rainfall in May 2023 was the trigger for the rockfall. After the rockfall, the slope and height of the newly formed cliff face remained largely unchanged, i.e. a steep and vertical slope with an unchanged cliff height. The only exception to slope is the collapse material that is eventually washed away. The evolution of the Vrgada cliff is therefore quite simple and includes a retreat of the cliff over time without any significant change in slope and height.

GIS and statistics are powerful spatial analysis tools to characterise cliff instability and hazards. Further research should be carried out, e.g. over different periods of time, to check whether the hazard zones detected remain the same with the same intensity or change in time and space.

Acknowledgements. The measurements on the French cliffs have been carried out and financed in the framework of the National Observation Service DYNALIT (https://www.dynalit.fr/).

References

1. Dornbusch U (2022) Holocene and historic rates of rock coast erosion – a discussion focussed on southeast England. Mar Geol 449:106817
2. Letortu P et al (2020) The potential of Pléiades images with high angle of incidence for reconstructing the coastal cliff face in Normandy (France). Int J Appl Earth Obs Geoinf 84:101976
3. Eitel JUH et al (2016) Beyond 3-D: the new spectrum of LiDAR applications for earth and ecological sciences. Remote Sens Environ 186:372–392
4. Cohen O, Héquette A (2020) Recent advances in coastal survey techniques: from GNSS to LiDAR and digital photogrammetry - examples on the northern coast of France. In: Tiefenbacher J, Poreh D (eds) Spatial variability in environmental science - patterns, processes, and analyses. IntechOpen, London: 91–112
5. Medjkane M et al (2018) High-resolution monitoring of complex coastal morphology changes: cross-efficiency of SfM and TLS-based survey (Vaches-Noires cliffs, Normandy, France). Landslides 15:1097–1108
6. Buchwald T, Antoine P (2024) Historique des travaux et perspectives de recherche sur deux séquences pléistocènes originales du Boulonnais : Wimereux Pointe aux Oies et Wissant Carrière du Fart (littoral du Pas-de-Calais, France). Géomorphologie Relief Process Environ 30:27–47
7. Blaise E, Cohen O, Trentesaux A, Bourdu-Devulder D (2022) Evolution récente des falaises du Bas Boulonnais (Pas-de-Calais) : l'apport des nouvelles techniques de mesure et de la restitution en 3D. In: Actes des 17èmes Journées Nationales Génie Côtier - Génie Civil, Chatou. Editions Paralia, pp 267–276
8. Pikelj K, Furčić N (2020) Impact of cliff erosion on marine sediment composition - indication of local coastline evolution (Vrgada Island, Croatia). In: Eighth international symposium "monitoring of Mediterranean coastal areas. Problems and measurement techniques", pp 462–468
9. Banak A, Pikelj K, Lužar-Oberiter B, Kordić B (2021) Sedimentary record of Pleistocene aeolian - alluvial deposits on Vrgada Island (eastern Adriatic coast, Croatia). Geol Croat 74:137

Risk Assessment for Rocky Cliffs: Application to the Coastal Area of Lagoa, Southern Portugal

Óscar Ferreira[1]([✉]) and Carlos Loureiro[1,2]

[1] FCT/CIMA/ARNET, University of the Algarve, 8005-139 Faro, Portugal
`oferreir@ualg.pt`
[2] School of Agricultural, Earth and Environmental Sciences, University of KwaZulu-Natal, Durban 4000, South Africa

Abstract. Coastal areas are highly vulnerable to extreme events, particularly in rocky coastal environments where cliff collapse poses a significant risk to human occupation. This study presents a simplified method for assessing coastal risk at the top of rocky cliffs, integrating hazard and exposure into a single risk index. The methodology is based on the Coastal Risk Assessment Framework and incorporates hazard (inland extent of cliff mass movements) and exposure (land use) indices to calculate a Coastal Index (Ci). This approach was applied to the coastline of the municipality of Lagoa (Algarve, southern Portugal), which is characterised by indented rocky cliffs, pocket beaches, and subject to high touristic pressure. Risk maps were developed to determine the spatial variability, revealing a predominantly low to moderate risk along the coastline. However, 22% of the study area was identified as having high to very high coastal risk, particularly in areas with urban or touristic development close to the cliff edge. The method provides a straightforward tool for coastal management, enabling the identification of risk hotspots and the optimization of resource allocation for risk reduction.

Keywords: Mass Movements · Coastal Cliffs · Hazard · Land Use · Coastal Index

1 Introduction

Coastal areas are often densely populated while also exposed to extreme events that cause several hazards, the combination of which can lead to the occurrence of coastal risks. In rocky coasts, a key risk is the potential impact of cliff collapse on the existing human occupation. In contrast to sandy coasts, hazard zones cannot be defined based on the simple projection of shoreline change rates into the future, as even in cases of rocky coasts with low or very low average retreat rates, there is potential for occasional medium to large-scale mass movements to occur (Del Río and Gracia, 2009). Consequently, alternative approaches must be employed for cliffed coasts, reflecting the episodic occurrence of mass movements.

The majority of the existing studies that identify potential impacts of cliff movement on human occupation characterise the hazard, vulnerability and/or risk based on composite indexes (e.g. Del Río and Gracia., 2009; Nunes et al, 2009; Bergillos et al., 2022).

C. Coelho et al. (Eds.): CD 2025, CRL 41, pp. 324–329, 2026.
https://doi.org/10.1007/978-3-032-15473-6_51

These hazard indexes are typically the result of a combination of variables representing indirect factors (e.g. lithology, structure, wave exposure, cliff protection) that determine the alongshore variability of the hazard. However, they do not consider the cross-shore expression of the hazard as a function of the mass movement inland extension associated with different return periods. Other authors have defined the spatial distribution of the hazard based on statistical methods (e.g., Marques et al., 2011, 2013; Marques, 2018), encompassing a set of predisposing factors for cliff failure associated with the cliff's geology and geomorphology. The number of studies that address both the alongshore and cross-shore variability of the hazard remains limited, with many focusing on specific areas with detailed datasets (e.g. Penacho and Marques, 2023). Furthermore, there is a paucity of literature integrating both hazard and exposure to obtain the spatial distribution of the risks for the occupation of cliff tops (e.g., Del Río and Gracia, 2009).

The present study proposes a simplified risk assessment method for areas at the top of rocky cliffs, integrating hazard (as a function of the probability of cliff mass movements' inland extension) and exposure (as a function of the land use). The test area was the coastline of Lagoa council (Algarve, southern Portugal), an indented rocky coastline with pocket beaches, diverse land uses and subject to high touristic pressure (Fig. 1).

2 Methods

The methodology employed in this study is an adaptation of the Coastal Risk Assessment Framework (CRAF) proposed by Viavattene et al. (2018). Indexes related to hazard (H_i) and exposure (E_i) were used to calculate a coastal index (C_i), which serves to quantify the risk along the top of rocky cliffs. The application of these indexes to the coastal area of Lagoa council (Fig. 1) enabled the generation of risk maps, which illustrate the spatial variability of the risk. The C_i is obtained as follows:

$$C_i = (H_i * E_i)^{1/2} \tag{1}$$

The H_i ranges from 1 to 5, depending on the maximum inland extension of the cliff mass movements associated with different return periods (based on Teixeira, 2006), and also takes into account a legally established protection distance of 200 m from the cliff top, which is defined based on the potential for sinkhole collapse (Table 1). Return period estimations by Teixeira (2006) are based on the analysis of two datasets of mass movement width (inland extension) recorded along 46 km of rocky cliffs carved on Miocene calcarenites along the central Algarve coast. These comprise a nine-year-long field inventory (1995–2004) and a 44-year-long catalogue obtained from the analysis of aerial photographs (1947–1991). A total of 155 events with maximum mass movement widths of at least 4 m were considered for the statistical analysis. Power-law distributions were fitted to determine the relationships between return period and mass movement width (see Teixeira, 2006), thereby enabling the delineation of hazard zones according to the designated return periods intervals (Table 1).

The E_i is determined heuristically as a function of various land use types, defined in the 2018 national land use categorisation for Portugal (DGT, 2019), with a minimum mapping unit size of 1 ha, and a minimum spacing between mapping lines equal or greater than 20 m. The exposure, as a function of the land use classes, was also classified from 1 to 5 according to the severity of the possible impacts on people, economic activities and society (examples in Table 1). The resulting C_i index values obtained were divided into 4 classes from low to very high risk (Table 1).

Fig. 1. Location of the Lagoa municipality coastal area. The white diamond represents the location of the study area within the Iberian Peninsula (right panel), and the red rectangle refers to the area shown in Fig. 2, near Carvoeiro town. Image source: Google Earth, CNES/Airbus, June 2020.

Table 1. Hazard and exposure classes and assigned values as a function of the distance from the cliff top and land use. Risk classes as a function of the calculated C_i value.

Return period (years) - Distance from cliff top (m)	H_i	Land Use (examples)	E_i	Risk	C_i
[0–10] - [0, 23]	5	Continuous urban fabric	5	Very High	[4, 5]
]10–25] -]23, 35]	4	Discontinuous urban fabric	4	High	[3, 4[
]25–50] -]35, 48]	3	Road network and car parks	3	Moderate	[2, 3[
]50–100] -]48, 67]	2	Parks and gardens; Agricultural areas	2	Low	[1, 2[
> 100 -]67, 200]	1	Bare rock; Wastelands; Sparse vegetation	1	No risk	0

3   Results and Discussion

The cross-shore distribution of the hazard is the same for the entire study area (Fig. 2a), as the hazard classes are a function of the landward retreat due to the mass movements, as determined by Teixeira (2006). This assumption considers the same likelihood of occurrence of mass movements for the entirety of Miocene rocky cliffs along the central Algarve. Therefore, the variability of the C_i for the coast of Lagoa council depends mainly on the variation in land use. The study area encompasses a wide range of land uses, being dominated by areas with very low exposure (bare rock), but including also moderate exposure (discontinuous urban) along sections of the cliff top area. However, there are areas of touristic agglomeration with consolidated urban uses, often along cliffs near to pocket beaches, which determine very high exposure levels (Fig. 2b). The resulting risk classes along the cliff edge are predominantly low to moderate (78%). Nevertheless, there are several risk hotspots (high to very high risk classes), which together represent 22% of the 13.5 km of coastline analysed. These include areas occupied by small towns (e.g. Carvoeiro; central part of Fig. 2c) or tourist resorts (e.g. Tivoli Carvoeiro Algarve Resort; eastern part of Fig. 2c). In total, 10 hotspots were identified, all near or at the edge of the cliffs, with coastline lengths ranging from about 100 m to 500 m.

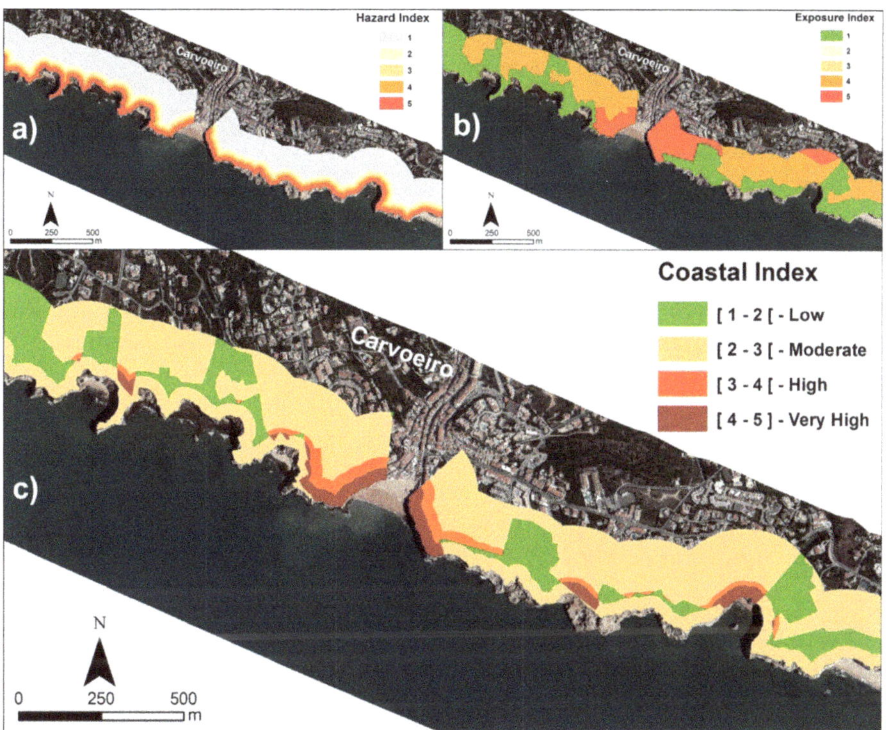

Fig. 2. Coastal risk for the Carvoeiro town, including: a) hazard index, b) exposure index and c) coastal index. Aerial image from 2018 provided by DGT.

Methods based on indexes are the most common approach used to assess hazard and risk along rocky cliff coasts. However, most of them do not provide an analysis of risk based on the probabilistic assessment of hazards, and are limited in their ability to provide a temporally constrained analysis that better supports priority setting. These methods often require the calculation and combination of a large number of variables to derive a composite risk index (for example, 11 physical and 6 socioeconomic variables proposed by Del Rio and Gracia, 2009), which can hinder their use in areas with scarce data. The relative importance of each variable in the final risk assessment is often unknown, and weighting factors have be defined for each variable used in each application, increasing the complexity and reducing the transferability of the methods.

The method presented here provides an integrated risk zonation of coastal areas adjacent to the top of rocky cliffs, allowing the definition of critical (priority) areas. It prioritises the cross-shore differentiation of the hazard as a function of the return period of the hazardous event (mass movement), while the exposure is considered using a simple but integrative parameter (land use). This method provides a straightforward and efficient approach for evaluating and ranking risk levels in rocky coastal regions with similar geomorphological and hydrodynamic characteristics. Therefore, it is a useful and simple tool for coastal management purposes, allowing the optimisation of resources and a rapid decision-making process regarding the areas that should be prioritised for coastal management interventions. The method can be easily applied to other rocky coastal areas, provided that hazard zones are or can be readily defined and land use maps can be meaningfully converted into exposure classes.

4 Conclusions

This work presents a novel yet simple method to characterise and rank coastal risks at the top of rocky cliffs, integrating hazard and exposure in a straightforward index. The application of the method to the coastline of Lagoa council allowed the identification of 10 coastal hotspots (high and very high risk) near the cliff edge, mostly associated with villages or touristic resorts. This approach can be used for management purposes and to optimise the use of resources in rocky coastal zones.

References

Bergillos RJ, Rodriguez-Delgado C, Medina L, Fernandez-Ruiz J, Rodriguez-Ortiz JM, Iglesias G (2022) A combined approach to cliff characterization: cliff stability index. Mar Geol 444:106706

Del Río L, Gracia FJ (2009) Erosion risk assessment of active coastal cliffs in temperate environments. Geomorphology 112:82–95

DGT: Especificações técnicas da Carta de Uso e Ocupação do Solo (COS) de Portugal Continental para 2018. Relatório Técnico. Direção-Geral do Território, Lisbon (2019)

Marques F (2018) Regional scale sea cliff hazard assessment at sintra and cascais counties Western Coast of Portugal. Geosciences 8:80

Marques F, Matildes R, Redweik P (2011) Statistically based sea cliff instability hazard assessment of Burgau-Lagos coastal section (Algarve, Portugal). J Coastal Res SI64:927–931

Marques F, Matildes R, Redweik P (2013) Sea cliff instability susceptibility at regional scale: a statistically based assessment in the southern Algarve, Portugal. Nat Hazard 13:3185–3203

Nunes M et al (2009) Hazard assessment in rock cliffs at Central Algarve (Portugal): A tool for coastal management. Ocean Coast Manag 52:506–515

Penacho N, Marques F (2023) A statistical approach for sea cliff hazard and setback lines mapping in the northeast of Peniche Portugal. Geomorphology 430:108669

Teixeira S (2006) Slope mass movements on rocky sea-cliffs: a power-law distributed natural hazard on the Barlavento Coast, Algarve Portugal. Continental Shelf Res 26:1077–1091

Viavattene C, Jiménez JA, Ferreira Ó, Priest S, Owen D, McCall R (2018) Selecting coastal hotspots to storm impacts at the regional scale: the coastal risk assessment framework. Coast Eng 134:33–47

Optimizing Climate Change Adaptation Pathways: A Probabilistic Approach

Pablo Alonso-Alguacil[1]([✉]), Víctor Collado[2], Roberto Mínguez[2], Mirian Jiménez[1], Beatriz Pérez-Díaz[1], Fernando Méndez[1], and Sonia Castanedo[1]

[1] Universidad de Cantabria, 39005 Santander, Spain
`pablo.alonsoalguacil@unican.es`
[2] Universidad Carlos III, 28903 Madrid, Spain

Abstract. The complex interactions between freshwater and oceanic systems make estuarine environments particularly vulnerable to climate change. Addressing these challenges requires adaptable solutions that can evolve over time, as one-time actions are insufficient for such complex systems. This study presents a novel methodology to identify optimal adaptation pathways by combining a climate emulator with an optimization algorithm, while accounting for uncertainties in current knowledge and future projections. The methodology is exemplified with the analysis of flood risk at a local factory due to fluvial and coastal hazards. The climate emulator provides synthetic time series of the forcing agents that are long enough to accurately considered all plausible combinations of the variables. A medium resolution hybrid model is employed to select annual maxima events based on a key indicator (local total water level). Representative events are then selected with a clustering algorithm to simulate them with a high-resolution model to evaluate the flood risk. By integrating long-term variability, uncertainty, and precision modeling, this approach offers a flexible and robust framework for climate adaptation planning in dynamic and vulnerable systems.

Keywords: Adaptation Pathways · Climate Change · Risk-Assessment

1 Introduction

Estuarine environments are areas that are often exploited for industrial and/or recreational activities due to the diverse ecosystem services they offer. These environments are characterized by complex interactions between freshwater and oceanic systems, making them particularly vulnerable to climate change, which affects precipitation patterns and sea levels. Stakeholders require tools to accurately assess the risks associated with their activities and the efficiency of potential adaptation measures. However, traditional methodologies that are based on a one-time solution are inadequate for addressing the uncertainty inherent in climate change processes. To overcome this gap, recent scientific advances propose the establishment of decision-making pathways that are adaptable to the future evolution of all agents involved [8, 9].

C. Coelho et al. (Eds.): CD 2025, CRL 41, pp. 330–337, 2026.
https://doi.org/10.1007/978-3-032-15473-6_52

The objective of this study is to propose a methodology for establishing the optimal adaptation pathway, considering the variability and uncertainty of the factors that determine risk. This novel approach has been applied to the San Martin de la Arena estuary (Northern coast of Spain) where industrial activities are threatened by flood events due to rising sea levels. Specifically, the methodology is demonstrated through an evaluation of the flood risk at a local factory and the assessment of a levee as an adaptation measure.

2 Methodology

The proposed methodology aims to integrate into a single framework the probabilities of different agents leading to extreme flood events (astronomical tide, storm surge, and river discharge), as well as the uncertainties in sea level rise projections. This approach comprises the following phases:

2.1 Parametrization of the Forcing Agents

The forcing dynamics need to be correctly parametrized to ensure that all events can be correctly simulated with the appropriate hydrodynamic models, in this case, Delft3D [5] for the low-resolution simulations (50~10 m) and SFINCS [4] for the high-resolution simulations (5~1 m). To do so, the river discharge has been characterized as per the SCS Unit Hydrograph [2] (Peak discharge Q_p, Base discharge Q_b and Peak Period T_p), and water levels as the addition of the storm surge SS and the tidal range TR, as well as an offset between the river discharge peak and the astronomical tide peak, *tau*, as represented in Fig. 1.

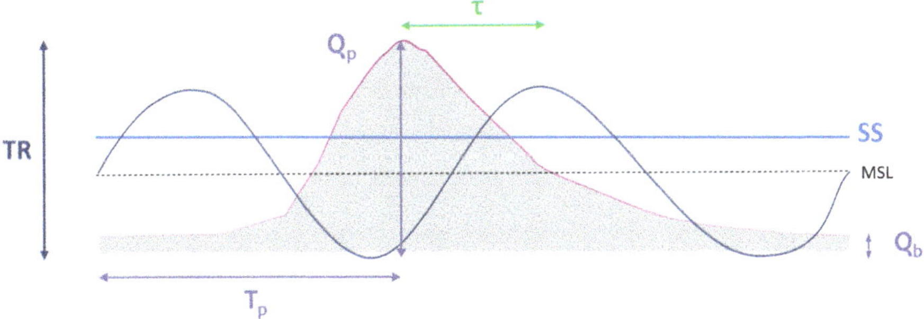

Fig. 1. Parametrization of the forcing variables

2.2 Climate Emulator

Observational records are generally too short for long-term risk assessment, particularly for a multivariate problem, as all relevant combinations of flood drivers (sea levels and discharge data) are not covered. Climate emulators that reproduce synthetic time-series of those drivers are an alternative that can provide plausible extreme events that have not been registered. For such purposes, Anderson et al., (2019) introduced the TESLA methodology that has been employed to create synthetic time-series of river discharges and sea levels [1].

2.3 Hybrid Downscaling and Cluster Analysis

The low-resolution hybrid model is then used to assess the severity of extreme events under any combination of the aforementioned agents. Combining key indicators (proxy), such as the total water level, that measure the severity of an event, the emulator and a clustering algorithm, it is possible to select representative annual maxima events ξ_c^j to be simulated with a high-resolution model (Fig. 2). These events can be defined as the mean distances of all cluster members ξ_i. The size of the cluster $|C_j|$ can be extrapolated to probabilities, w_j, as a fraction of the total number of events $|S|$.

$$\xi_c^j = \frac{1}{|C_j|} \sum_{\xi_i \in C_j} \xi_i; j = 1, \ldots, K \tag{1}$$

$$w_j = \frac{|C_j|}{|S|}; j = 1, \ldots, K \tag{2}$$

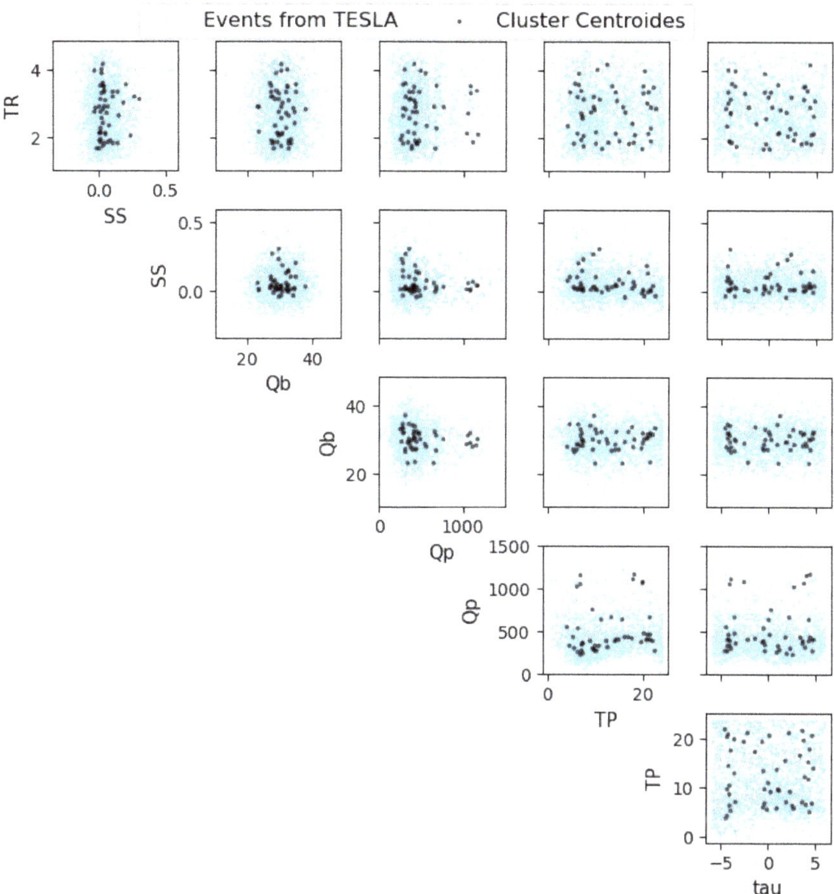

Fig. 2. Cluster analysis of annual maxima

2.4 Probabilistic Optimization of Dynamic Adaptation Pathways

The high-resolution results are then used to calculate damages for each cluster representative by crossing the flood information with exposure and vulnerability data. Exposure data and assets costs are based on the factory's estimates and the damage functions have been obtained from the Global flood depth-damage functions database [3].

Since each cluster representative is associated with a probability, it is possible to characterize the probabilistic distribution of damages, or other metrics such as the Value at Risk (VaR), maximum potential loss at a confidence level, or the Conditional Value at Risk (CVaR), average losses above the VaR (see Fig. 3), which are the focus of the optimization problem.

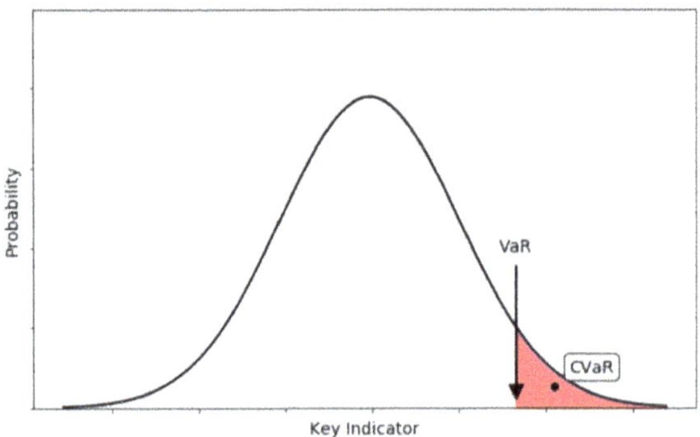

Fig. 3. Definitions of the CVaR and VaR,

These metrics are only valid assuming static climate and without any adaption measures. In order to approach the problem from a probabilistic and dynamic perspective, the representative events need to be modeled under different sea level rise conditions and adaptation measures. As a result, each cluster has associated flood maps and damages for different adaptation scenarios, see Fig. 4.

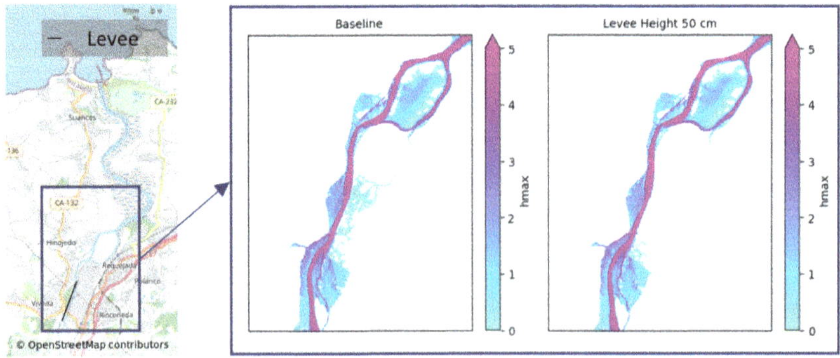

Fig. 4. Example of flood maps associated with a cluster centroid. On the left, a map of the baseline situation and on the right, a map with the application of a 50 cm high levee.

The optimum adaptation pathway can be obtained exploring the interactions between hydrodynamic forcing, climate change impacts (sea level rise) and adaptation measures (levee height, changes in exposure/vulnerability) using an optimization algorithm [6, 7].

This algorithm will return the strategies that minimize the chosen risk metric (CVaR, VaR, et) while taking into account the uncertainties of sea level rise, and climate variability. This approach ensures that the desired level of protection is achieved and implemented at the appropriate time, giving stakeholders the flexibility to adapt as new information becomes available and uncertainty diminishes (Fig. 5).

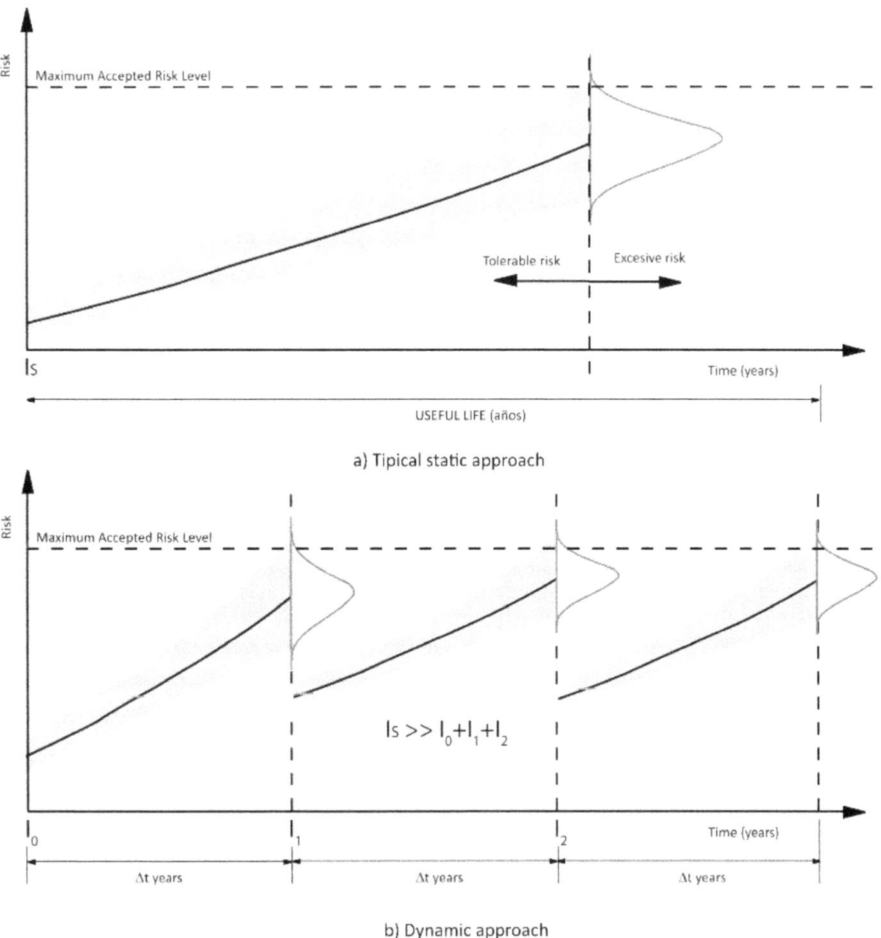

Fig. 5. Comparison between a traditional approach (A), where stakeholders aim to maintain the system at the desired risk level throughout its lifetime, and a dynamic approach (B), where decisions are made at multiple points, allowing stakeholders to incorporate new knowledge and adapt as time progresses.

Acknowledgments. The development of this work is framed within the project Methodologies for risk assessment of the industrial sector and adaptation to climate change in the estuarine environment (MyEst) that has the support of the Biodiversity Foundation of the Ministry for Ecological Transition and Demographic Challenge, through the Call for grants for the implementation of projects that contribute to implement the National Plan for Adaptation to Climate Change (2021–2030); and the HyBay project (PID2022-141181OB-I00 – MCIN/AEI/10.13039/501100011033 and FEDER, EU). P. Alonso-Alguacil acknowledges the Spanish Ministry of Science and Innovation for funding his contract (PREP2022-000316). This work is part of the R&D project "Algorithms for Stochastic Optimization Using Data-driven and Learning Analysis (ASTRAL)" with code PID2023-151013NB-I00, funded by MCIN/AEI/10.13039/501100011033 and the European Union Next Generation EU/PRTR.

References

1. Anderson D, Rueda A, Cagigal L, Antolinez JAA, Mendez FJ, Ruggiero P (2019) Time-varying emulator for short and long-term analysis of coastal flood hazard potential. J Geophys Res Oceans 124(12):9209–9234. https://doi.org/10.1029/2019JC015312
2. Bhunya PK (2011) Synthetic unit hydrograph methods: a critical review. Open Hydrol J 5(1):1–8. https://doi.org/10.2174/1874378101105010001
3. Huizinga J, de Moel H, Szewczyk W (2017) Global flood depth-damage functions. Methodology and the database with guidelines. EUR 28552 EN. https://doi.org/10.2760/16510
4. Leijnse T, Van Ormondt M, Nederhoff K, Van Dongeren A (2021) Modeling compound flooding in coastal systems using a computationally efficient reduced-physics solver: Including fluvial, pluvial, tidal, wind- and wave-driven processes. Coast Eng 163:103796. https://doi.org/10.1016/j.coastaleng.2020.103796
5. Lesser GR, Roelvink JA, Van Kester JATM, Stelling GS (2004) Development and validation of a three-dimensional morphological model. Coast Eng 51(8–9):883–915. https://doi.org/10.1016/j.coastaleng.2004.07.014
6. Rockafellar RT, Uryasev S (2000) Optimization of conditional value-at-risk. J Risk 2(3):21–41. https://doi.org/10.21314/JOR.2000.038
7. Rockafellar RT, Uryasev S (2002). Conditional value-at-risk for general loss distributions
8. Toimil A, Losada IJ, Camus P, Díaz-Simal P (2017) Managing coastal erosion under climate change at the regional scale. Coast Eng 128:106–122. https://doi.org/10.1016/j.coastaleng.2017.08.004
9. Toimil A, Losada IJ, Hinkel J, Nicholls RJ (2021) Using quantitative dynamic adaptive policy pathways to manage climate change-induced coastal erosion. Clim Risk Manag 33:100342. https://doi.org/10.1016/j.crm.2021.100342

Coastal Hazard Assessment on Indonesia's Outermost Small Islands

Rima Harahap[1,2(✉)], Gerd Masselink[1], and Sarah J. Boulton[1]

[1] University of Plymouth, Plymouth 4 8AA, UK
rima.gusrianaharahap@plymouth.ac.uk
[2] Kalimantan Institute of Technology, Balikpapan, Indonesia

Abstract. Coastal hazard assessment is important for the sustainable development and resilience of small islands. The selection of coastal hazard is often dominated by the impact of climate change, rather than integrating natural hazards induced by geophysical activity. In this study, a multi-hazard approach was used and categorized into hazards associated with climate change – specifically sea-level rise and storm wave – and geophysical hazards, namely earthquakes, tsunamis, and volcanoes. This study is the first to address the coastal multi-hazard on the outermost small islands of Indonesia, which refer to the islands located at the edge of the country's territory. The study carries out a multi-hazard assessment to determine hazard level of each island, involving the quantification of hazard index for every island. The result identifies the full range of multi-hazard levels across the outermost small islands represented in the island dataset. Finally, this approach can be applied to address coastal hazard for outermost small islands in other regions, helping to prioritize disaster risk reduction that suitable for each island.

Keywords: Coastal · Multi-hazard · Outermost · Small · Island

1 Introduction

Small islands are recognized for their high level of biodiversity as well as their geographical remoteness and economic dependencies [1]. Critical aspects of small islands, such as health, settlement, culture, economy, infrastructure, food, and water security have been impacted due to the increasing threats of climate change and natural disasters [1–4]. With the number of islands reported reaching 17,508, Indonesia is listed as one of the largest archipelagic countries in the world [5]. Numerous small islands are spread across the country, with 111 identified as the outermost small islands (Fig. 1) [6]. The term "outermost small islands" refers to the islands located at the edge of the country's territory, situated far from development centres and experience limitations in terms of infrastructure, economic, and human resource development [7].

The selection of coastal hazards is often dominated by the impact of climate change [2, 8–11], rather than integrating natural hazards induced by geophysical activity. In this study, a multi-hazard approach was used and categorized into hazards associated with

C. Coelho et al. (Eds.): CD 2025, CRL 41, pp. 338–344, 2026.
https://doi.org/10.1007/978-3-032-15473-6_53

climate change, specifically sea-level rise and storm wave, and geophysical hazards, namely earthquakes, tsunamis, and volcano eruptions. Coastal zones exposed to these hazards can have devastating effects particularly on small island communities. Therefore, the assessment of multi-hazard is required to produce a comprehensive hazard assessment, particularly on outermost small islands.

This study aims to perform a coastal hazard assessment for outermost small islands of Indonesia, involving the quantification of multi-hazard index for each island. This approach provides a clear and straightforward assessment, making it accessible for understanding unique characteristics and distinct triggers associated with coastal hazard. Furthermore, the same approach can be applied to address coastal hazard for small islands in other regions and prioritizing disaster risk reduction that suitable for each island.

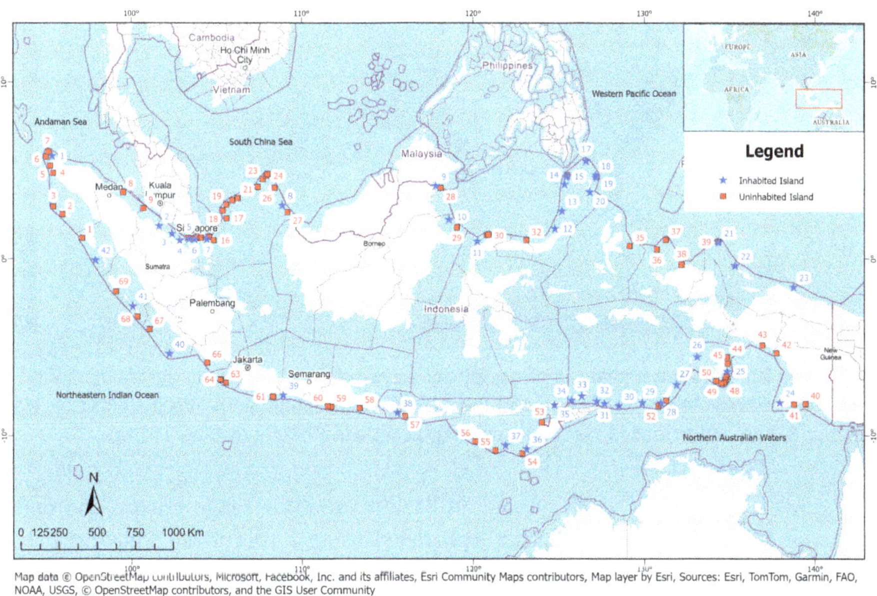

Fig. 1. Outermost small islands of Indonesia serve as markers of the nation's outer boundaries and influence its territorial extent. There are 42 inhabited outermost small islands and 69 uninhabited outermost small islands.

2 Data Collection and Methodology

This study focused on 42 inhabited outermost small islands spread across various maritime regions. Five islands are located close to the Andaman Sea, four in the South China Sea, fifteen in the Western Pacific Ocean, five in the Northeastern Indian Ocean, and fourteen in the Northern Australian Waters region (Fig. 1). These outermost small islands play a key role in national security and international relations. Therefore, the development of these islands is highly prioritized, and the infrastructure on inhabited outermost islands is being increasingly improved.

Hazard data was obtained from national and global datasets produced by various institutions. The trend of sea-level was obtained from NASA's Sea Level Portal [12] and specifically selected for the region around the island. The storm wave hazard was assessed by analysing ERA-5 hourly reanalysis dataset from Copernicus Climate Data Store [13], calculating the 99th percentile of significant wave height (SWH) spanning 33 years (1990–2022) of swell and wind wave data (Fig. 2). Geophysical hazards utilized a map from Indonesian Centre for Volcanology and Geological Hazard Mitigation (PVMBG), which provides information on areas prone to earthquakes and tsunamis [14]. Meanwhile, the database of active volcanoes in Indonesia was obtained from the Global Volcanism Program [15] and the hazard criteria was determined by grouping the islands based on the proximity to the volcanoes.

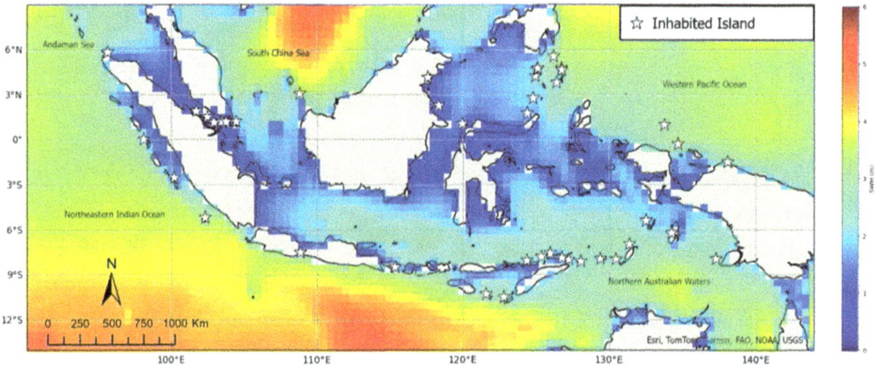

Fig. 2. The ERA-5 dataset analysis of significant wave height in Indonesia during the period of 1990–2022. Islands facing the Indian and Pacific Oceans predominantly reach significant wave height of 3–5 m, while in the inner parts, the significant wave height reaches 1–2 m.

The workflow began with defining all hazard variables, range criteria and ranks. The score for each variable was then calculated using a root mean square equation [16], producing the hazard index for each island. The hazard indices were classified into different categories by setting threshold values for various levels, according to the natural Jenks classification in ArcGIS. This classification is widely used within GIS packages and designed to minimize the variance between each class with uneven data value [17]. The classification reflected the index level of each island, namely Very Low, Low, Medium, High, and Very High. Table 1 shows the breakdown of hazard variables and ranks.

Table 1. Hazard Variables and Ranks

Variables	Sub-variables	Range	Rank	Score
Sea-level trend	Sea-level trend (1992–2017)	<3.5 mm/yr	Very Low	1
		3.52–4.0 mm/yr	Low	2

(continued)

Table 1. (*continued*)

Variables	Sub-variables	Range	Rank	Score
		4.0–4.5 mm/yr	Moderate	3
		4.5–5.0 mm/yr	High	4
		>5.0 mm/yr	Very High	5
Storm waves	99th percentile of SWH (1990–2022)	<1.0 m	Very Low	1
		1.0–1.5 m	Low	2
		1.5–2.0 m	Moderate	3
		2.0–2.5 m	High	4
		>2.5 m	Very High	5
Earthquakes	Earthquake intensity scale	< V MMI	Very Low	1
		V - VI MMI	Low	2
		VI - VII MMI	Moderate	3
		VII - VIII MMI	High	4
		> VIII MMI	Very High	5
Tsunamis	Tsunami wave height modelling at the coastline	<1.0 m	Very Low	1
		1.0–3.0 m	Low	2
		3.0–5.0 m	Moderate	3
		5.0–7.0 m	High	4
		>7.0 m	Very High	5
Volcanic activity	The score is calculated by summing the values assigned to volcanoes based on their distance from the island - Value 0: If no volcano within 100 km of the island - Value 1: If volcano located 50–100 km from the island - Value 2: If volcano located within 50 km of the island	<1	Very Low	1
		1–2	Low	2
		3–4	Moderate	3
		5–6	High	4
		>6	Very High	5

3 Result and Discussion

The local assessment of sea-level trend reveals that the islands in the northern part of Papua facing the Western Pacific Ocean are identified as having the highest sea-level trend, reaching 5.05 mm/yr which is higher than the average sea-level trend in Indonesia of 4.68 mm/yr. Meanwhile, the islands facing the Indian Oceans predominantly have a very high storm wave rank. The highest recorded storm wave was found in the outer island facing the Indian Ocean, reaching 3.35 m for the 99th percentile of SWH during 1990–2022. The potential for earthquakes and tsunamis is widespread across the islands, except for those near the South China Sea and southern Papua. Meanwhile, the volcanic hazard is relatively low, due to most islands being far from the volcanoes.

Figure 3 illustrates the geographical representation of coastal hazards in Indonesia's outermost small islands. Three islands have been identified with a very high level of multi-hazard: Makalehi and Kawaluso – two outermost small islands in the northern part of Indonesia – and Nusa Penida in the southern part. These islands have a very high potential of geophysical hazards including earthquake, tsunami, and volcanic activity, as well as a high sea-level trend and storm waves. The high and moderate levels of multi-hazard are concentrated in the eastern part of Indonesia, while the islands near the South China Sea have the lowest average hazard index, relatively safe from natural disasters.

Multi-hazard assessment is one of the important variables in assessing coastal risk, along with the integration of exposure and vulnerability element. This multi-variable, approach is useful due to the need for policy development, where cross-sectoral collaboration of each risk variable is essential to produce a long-term risk analysis. However, the limited availability of small island's databases often emerges as a key challenge. Finally, the results of this study show that each island has unique characteristics and distinct triggers associated with coastal hazards, providing insights for prioritizing factors specifically tailored to disaster risk reduction.

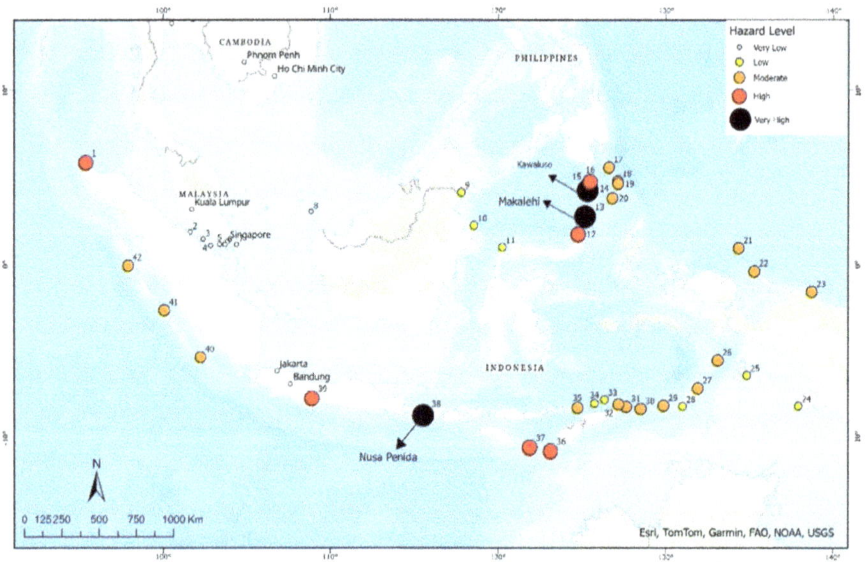

Fig. 3. The geographical representation of the coastal hazard levels. The circles on the map represent the coastal hazard levels, ranging from Very Low to Very High.

4 Conclusion

The multi-hazard assessment was conducted to determine the hazard level of 42 inhabited outermost small islands of Indonesia. The methodology quantified multi-hazard index for each island and was represented in the island dataset. This assessment highlighted the varied multi-hazard levels across the islands, reflecting their unique hazard's driver that necessitate different focus in risk reduction. This study addresses a gap in current research, serving as a foundation for further investigation into coastal risk on outermost small islands, especially considering the distinct challenges they face compared to more densely populated or economically valuable small islands.

Acknowledgement. This work was conducted under the funding of the Indonesia Endowment Fund for Education (LPDP) as a part of PhD Scholarship at University of Plymouth.

References

1. IPCC.: Climate Change 2022: Impacts, Adaptation and Vulnerability. Contribution of Working Group II to the Sixth Assessment Report of the Intergovernmental Panel on Climate Change. Cambridge University Press, Cambridge (2022)
2. Doorga JRS (2022) Climate change and the fate of small islands: the case of Mauritius. Environ Sci Policy 136:282–290
3. Griffen S, Robinson S-A (2023) (Un)just post-disaster mobilities in small island developing states: revisiting the patterns and outcomes of three major environmental disasters in the Caribbean. Int J Disaster Risk Reduc 97
4. Martin PCM et al (2018) Responding to multiple climate-linked stressors in a remote island context: the example of Yadua Island. Fiji. Climate Risk Manag 21:7–15
5. Andréfouët S et al (2022) Indonesia's 13558 islands: a new census from space and a first step towards a one map for small islands policy. Marine Policy 135
6. Presidential Decree of the Republic of Indonesia Number 6 of 2017, https://peraturan.bpk.go.id/Details/57400/keppres-no-6-tahun-2017, Accessed 23 Nov 2023
7. Isnansetyo A et al (2020) Vulnerability index analysis of Bepondi island as a reference for small and outer islands management in Indonesian. In: The 3rd International Symposium on Marine and Fisheries. EDP Sciences
8. Bunting EL, Theuerkauf EJ (2024) What is a coastal hazard? Perceptions of critical coastal hazards amongst decision makers in communities across the Great Lakes. Ocean Coastal Manag 251
9. Giardino A et al (2018) Coastal hazard risk assessment for small islands: assessing the impact of climate change and disaster reduction measures on Ebeye (Marshall Islands). Reg Environ Change 18(8):2237–2248
10. Stalhandske Z et al (2024) Global multi-hazard risk assessment in a changing climate. Sci Rep 14(1):5875
11. Zhang Y et al (2024) A multi-hazard framework for coastal vulnerability assessment and climate-change adaptation planning. Environ Sustain Indic 21
12. NASA sea level portal, https://sealevel.nasa.gov/data-analysis-tool, Accessed 4 Dec 2023
13. Copernicus climate data store, https://cds.climate.copernicus.eu, Accessed 6 Nov 2023
14. Portal Mitigasi Bencana Geologi, https://vsi.esdm.go.id/portalmbg, Accessed 7 Dec 2023
15. Global volcanism program, https://volcano.si.edu, Accessed 27 Dec 2023

16. Gornitz VM et al (1994) The development of a coastal risk assessment database: vulnerability to sea-level rise in the U.S. southeast. J Coastal Res (Special Issue No. 12):327–338
17. Smith MJD, Goodchild MF, Longley PA (2018) Geospatial Analysis: A Comprehensive Guide to Principles Techniques and Software Tools. The Winchelsea Press, Leicester

Regional Application of Coastal Engineering Resilience Index

Rusty Permenter[1(✉)], Catie Dillon[1], and Martin Schultz[2]

[1] USACE ERDC CHL, Vicksburg, MS 39180, USA
rusty.l.permenter@usace.army.mil
[2] USACE ERDC EL, Vicksburg, MS 39180, USA

Abstract. The US invests hundreds of millions of dollars annually into beach nourishment and other coastal storm risk management efforts. Evaluation of the vulnerability of beach communities requires metrics that are broadly applicable in a range of coastlines. Indices that measure these vulnerabilities provide decision makers with the information required to prioritize infrastructure investment in beach communities. The Coastal Engineering Resilience Index(CERI) [2] is a metric that combines multiple parameters that describe a sandy beach's ability to withstand storm impacts. This study applies the CERI index to predict the relative resistance of varied beach cross-sections to storm attack selected from locations of interest (LOI) across the East and Gulf Coast. These profiles are perturbed for greater variability in initial profile conditions and subsequently evaluated through the cross-shore morphology model CSHORE. A comparison of CERI and other indices with the CSHORE modeling results was performed through application of the Kendall rank correlation coefficient.

Keywords: Hazards · Risk · Planning

1 Introduction

Beach nourishments are a globally employed method to mitigate erosion along sandy shorelines that provide a critical buffer for valuable oceanfront properties against storm damage and contribute to local economies that rely on tourism. However, the investment in beach nourishment is a continuous cycle, as the placed sediment is often eroded through natural processes. Since 2018, the United States has allocated over a billion dollars [9] for nourishment initiatives. Coastal development coupled with climate change and sea level rise expose more infrastructure and property to coastal storm hazards each year, with the demand for nourishment rising accordingly.

In the United States, the justification for investing in regional or local beach nourishment plans involves a comprehensive cost-benefit analysis, which require significant time, data, and resources. Evaluating numerous potential project areas would be cost-prohibitive, resulting in the need for simplified metrics to calculate coastal storm hazards for project and regional applications. Traditional evaluation metrics parameterize the beach profile in terms of the dimensions of the dune and berm profile and include

C. Coelho et al. (Eds.): CD 2025, CRL 41, pp. 345–350, 2026.
https://doi.org/10.1007/978-3-032-15473-6_54

protective elevation (PE), protective width (PW), and volume density (VD). PE is the average elevation across the berm, dune slope, and dune crest, PW is the width from the landward dune crest to the seaward edge of the berm, and VD is the total volume in the dune and berm available to be eroded. Protective width has been used to evaluate beach vulnerability in past studies [3] using the label buffer width(BW). The Coastal Engineering Resilience Index is another evaluation metric adapted from the traditional methods to include other characteristics related to vulnerability of the beach profile relative to storm attack. This study investigates how well these indices correlate with the ability of dune and berm profiles to resist storm-induced changes in volume at shorelines on the United States' Eastern seaboard and Gulf Coast. The 1-dimensional time-averaged cross-shore morphology model CSHORE [4] was used to simulate changes in dune and berm volume caused by 2, 10, 25, 50, and 100 year storm events selected from USACE Coastal Hazard System. The correlation between each index and the percent dune and berm volume remaining were evaluated through the application of Kendall's tau-b correlation coefficient [10].

1.1 CERI Index

CERI's initial formulation focused on a simplified trapezoidal profile similar to that used in the USACE Beach-fx model [8], as shown in Fig. 1. The profile simplifies a beach cross-section into a flat upland, a constant landward and seaward dune slope with a flat dune crest, and a flat berm with a constant foreshore slope.

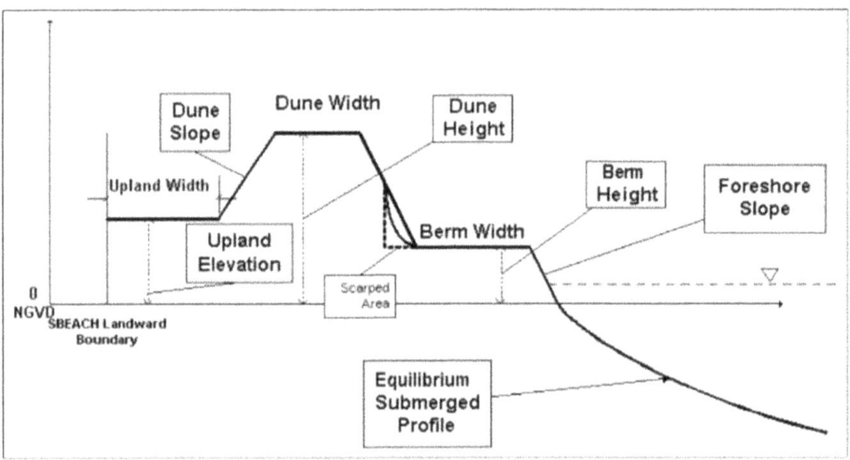

Fig. 1. Idealized Beach Profile from Beach-fx [8]

The CERI index consists of five parameters measuring the protective elevation, protective volume, protective width, crest freeboard, and wave runup as shown in Eqs. 1–6, respectively. PE, PW, s, MR, H_D, H_{MHW}, MS, WR and CF_0 represent protective elevation, protective width, percent fines, maximum recession, dune elevation, mean high water level, maximum storm surge, wave runup, and characteristic freeboard, respectively. The protective width consists of the lateral distance from the landward edge of the idealized dune crest to the seaward end of the berm. The protective elevation consists of the average elevation over the same span. Protective width and protective elevation each occur in two of the five CERI parameters, resulting in dependence among the individual CERI components.

Each factor is normalized using a characteristic value appropriate to the site represented by the subscripted naught values. The a and b terms represent the protective elevation and protective volume contributions. Fines do not contribute to the protective volume factor, and are removed through the (1-s) term. The c value represents the ability of the beach to withstand a 10-yr event without complete erosion of the berm and dune. The d value gives the freeboard of the dune relative to the maximum surge of a 10-yr event, and the e value gives the storm runup relative to a 10-yr event.

A potential obstacle in evaluating a beach profile with CERI is the determination of the characteristic values. When applied at the regional scale, these terms will vary across geography and hazard exposure types, resulting in values for certain areas that will not apply in others. Additionally, as the component values of CERI are not independent of one another, interpreting the resulting indices values is complicated compared to the traditional methods which have physical interpretations.

$$CERI = a + b + c + d + e \tag{1}$$

$$a = \frac{PE}{PE_0} \tag{2}$$

$$b = \frac{PE * PW * (1 - s)}{PE_0 \bullet PW_0} \tag{3}$$

$$c = \frac{PW - MR_{10}}{PW_0} \tag{4}$$

$$d = \frac{H_D - (MS_{10} + H_{MHW})}{CF_0} \tag{5}$$

$$e = \frac{WR_0}{WR_{10}} \tag{6}$$

2 Methods

2.1 Profile Selection

For each LOI, 1000 ideal dune profiles were randomly generated by varying berm width, dune width, and dune elevation from characteristic ranges provided in previous Beach-fx studies. Dune and foreshore slopes were sampled from a range represented by their mean values ± 15%. Offshore profiles used a Dean [1] profile calculated with A values determined based on the local grain size. The 1000 profiles are selected to give a wide range of potential starting CERI values for each LOI.

2.2 Index Calculation

BW, PE, VD and CERI were calculated for each of the 1000 initial profiles at each LOI. CERI indices were calculated for all initial profiles with maximum recession values determined by both the Dean-Kriebel [6] and Komar [5] maximum berm recession methods. For this study the maximal value of PE and PW across all 1000 profiles were used for their respective characteristic values. The MR_{10} was determined using either the Kriebel and Dean or Komar methodologies on a 10-yr event. USACE Coastal Hazards System(CHS) [7] provided location-specific values for the 10-yr surge that was applied as the MS_{10} value. The characteristic freeboard CF_0 is the maximum of all dune elevations for each site. All values were applied using mean high water as the datum.

2.3 Storm Selection and CSHORE Parameters

CHS data was obtained for all areas of interest, and storms were selected to intersect with the collision regime to allow for the most interaction between the storm and the profile. The storms were selected based on the 2-, 10-, 25-, 50-, and 100-yr return periods for the linear addition of storm surge and wave amplitude. This procedure allowed the team to focus on storms primarily occurring in the collision regime of the various profiles while avoiding storm events insufficient to cause profile damage or too severe for appreciable differences between profiles to be evident.

Typical CSHORE parameters were applied across all LOIs, with D_{50} values assigned by LOI. The suspension efficiency due to wave breaking was 0.002 with a value of 0.004 for the suspension efficiency due to bottom friction. A suspended load parameter of 0.5 was used. The IPROFL factor in CSHORE was set to 1.1 to avoid the initial smoothing typically implemented in the model. The IPROFL value determines if the model applies a movable bed and rounds the initial profile prior to evaluation of morphology change. The application of smoothing results in implied erosion even without any storm impact to the dune or berm. The grid cells were spaced at 1m across the whole profile.

2.4 Kendall Correlation of Indices

The correlation between each index and the percent dune and berm volume remaining after the storm was evaluated for each storm event and LOI using Kendall's tau-b statistic, which reflects the concordance between two sets of values. A perfectly concordant set of variables results in a Kendall tau-b correlation value of 1, and a perfectly discordant set of variables results in a value of -1. A value of 0 indicates there is no association between the variables.

3 Results

The rank correlation statistics for every storm event at every LOI are shown in Fig. 2. In the majority of cases VD provided the best indication of the percent volume remaining on the beach. CERI_K99 and PE typically demonstrated the lowest levels of concordance, and in some cases exhibited discordance. The low levels of concordance for CERI_K99 are attributed to the use of the Komar methodology for calculating the maximum recession variable, MR_{10}. The maximum recession calculated using Komar in some profiles partially contributed to a negative CERI value, implying severe vulnerability to erosion and canceling out the majority of benefits from the other beach characteristics represented within the CERI equation. Solely applying PE to predict relative beach strength omits any information about the width of the beach, reducing the value of the index. Furthermore, VD performs better than both BW and PE since it contains both BW and PE within its calculation. BW also appears to be a better indicator of the overall profiles resistance to erosion as compared to PE, indicating that width may be a more important factor in beach profile vulnerability than elevation. The Dean Kriebel methodology for calculating max recession within the CERI index provided reasonable results as evidenced by Fig. 2, with concordance values generally higher than both BW and PE.

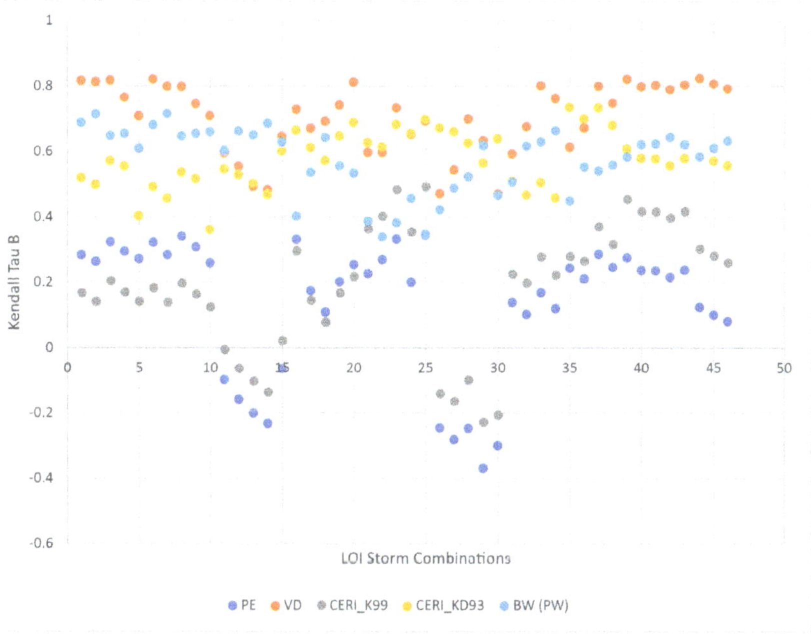

Fig. 2. Rank Correlation for CERI and other metrics

4 Conclusion

The CERI_KD93 and VD indices provide the greatest concordance across the LOIs. CERI_KD93 provides greater utility in certain locations, but typically VD provides a stronger correlation with the percent volume remaining within CSHORE modeling results. The Kendall tau-b score for PE ranged from 0.47 to 0.83, and is outperformed by the BW scores. VD outperforms both PE and BW due to the inclusion of both parameters within the VD formulation. For East Coast and Gulf Coast beaches, the Komar maximum recession parameter gives large values, outweighing the rest of the values in the CERI formula. The low concordance and occasional discordant values for CERI_K99 demonstrate this reduction in CERI's predictive value when the Komar maximum recession formulation is applied.

References

1. Dean R (1991) Equilibrium beach profiles: characteristics and applications. J Coastal Res 53–84
2. Dong Z, Elko N, Robertson Q, Rosati J (2018) Quantifying beach and dune resilience using the coastal resilience index. Coastal Eng Proc 1(30): https://doi.org/10.9753/icce.v36.papers
3. Chambers DM (2019) Coastal resilience metrics from Beach-fx. ERDC/CHL CHETN-VI-49. Vicksburg, MS: U.S. Army Eng Res Dev Center. 12 https://doi.org/10.21079/11681/31462
4. Johnson B, Kobayashai N, Gravens M (2012) Cross-shore numerical model CSHORE for waves, currents, sediment transport and beach profile evolution. Technical Report ERDC/CHL TR-12-22. Vicksburg, MS: U.S. Army Engineer Research and Development Center, Coastal and Hydraulics Laboratory. September. 159 (2012)
5. Komar P, McDougal W, Marra J, Ruggiero P (1999) The rational analysis of setback distances: applications to the Oregon coast. Shore and Beach 67(1):41–49
6. Kriebel D, Dean R (1993) Convolution method for time-dependent beach-profile response. J Waterw Port Coast Ocean Eng 119(2):123–229
7. Nadal-Caraballo N, Campbell M, Gonzalez V, Torres M, Melby J, Taflanidis A (2020) Coastal hazards system: a probabilistic coastal hazard analysis framework. In: Malvárez G, Navas F (eds) Global Coastal Issues of 2020. Journal of Coastal Research, Special Issue No 95, pp 1211–1216. Coconut Creek (Florida), ISSN 0749-0208
8. Rogers C, Jacobson K, Gravens M (2009) Beach-fx user's manual: Version 1.0. Technical Report, No. ERDC SR-09-6. Vicksburg, MS: USACE
9. WCU (Western Carolina University), Program for the study of developed shorelines. http://beachnourishment.wcu.edu/, Accessed 14 Jan 2025
10. Schultz, M., Permenter, R., Dillon, S.: An Investigation Into the Correlation Between Selected Coastal Protection Indices and Percent Residual Dune and Berm Volumes Following Coastal Storms (2025). ERDC/CHL TR-25-35. Vicksburg, MS: U.S. Army Eng Res Dev Center. https://dx.doi.org/10.21079/11681/49994

Future Projections of Inundation Risk
in the Three Major Bays in Japan Considering
Society Change

Shion Yamamoto[1]([✉]), Takuya Miyashita[2], Tomohiro Yasuda[3], and Nobuhito Mori[2]

[1] Graduate School of Science and Engineering, Kansai University, Suita, Japan
k723812@kansai-u.ac.jp
[2] Disaster Prevention Research Institute, Kyoto University, Uji, Japan
[3] Department of Civil, Environmental and Applied Systems Engineering, Kansai University,
Suita, Japan

Abstract. Sea level rise and intensified storm surges due to climate change are assumed to increase inundation risks in coastal areas. While climate change adaptation policies are being developed in Japan, the long-term changes in inundation risk have not yet been quantified. This study estimated potential inundation risk areas by considering sea level rise and future population changes for three major bays in Japan under different SSP scenarios. Additionally, storm surge simulations for Typhoon Jebi in Osaka Bay were conducted under high tide conditions, incorporating sea level rise, to estimate future changes in the inundated area and the storm surge-affected population. The results indicated that the affected population was greater than that of changes in hazard intensity. A significant increase in impacts after 2020 was observed only under the high emission scenario with SSP5–8.5 in Tokyo Bay. This analysis highlighted that differences between scenarios for the population in high-risk areas concerning storm surges become more significant after 2060.

Keywords: Sea Level Rise · Population Change · Inundation Risk. SSP scenarios · Storm Surge

1 Introduction

Climate change is expected to increase the risk of inundation in coastal areas through sea level rise and intensified storm surges. In Japan, the impacts of climate change are becoming increasingly apparent, as demonstrated in Typhoon Jebi, which caused significant storm surge damage along the Osaka Bay coast. In future climates, the number of typhoons is projected to decrease, while their intensity is expected to increase, leading to higher storm surge risks. Japan's population is heavily concentrated in low-lying coastal areas, making these regions particularly vulnerable to extensive damage from storm surge inundation. However, the combined effects of future population changes and increased disaster hazard intensity on long-term inundation risks remain insufficiently understood. This study aims to quantify changes in storm surge risks under different

C. Coelho et al. (Eds.): CD 2025, CRL 41, pp. 351–357, 2026.
https://doi.org/10.1007/978-3-032-15473-6_55

SSP (Shared Socio-economic Pathways) scenarios and estimate temporal changes in the population of potential inundation areas. The analysis focuses on the three major bays of Japan: Osaka Bay, Ise Bay, and Tokyo Bay. Storm surge simulations using Typhoon Jebi are also conducted for Osaka Bay to assess future inundation risks. By comparing simplified estimates with detailed storm surge inundation calculations performed using numerical models, the study highlights the importance of reducing the affected population through levee functions and other mitigation measures.

2 Sea Level Rise, Future Population Change and Tides

As projections of sea level rise vary by SSP scenario and model, it is essential to consider sea level rise specific to each SSP scenario when assessing coastal areas. The sea level rise projections used in this study are based on the sea-level change datasets from IPCC AR6 [1]. Figure 1 shows the projected sea level rise along the Pacific coast of Japan, averaged across 468 coastal points, for the period from 2020 to 2100. The projections are presented for each SSP scenario.

The SSP is a scenario used in the IPCC AR6 to assess the impact of socio-economic development on climate change. However, the SSP datasets are only available for five global regions, limiting their applicability to detailed regional analyses. To address this, Japan version of the SSP was developed to reflect the country's unique socio-economic outlook. Population projections for each SSP scenario, developed by the Japanese SSP Development Team at the National Institute for Environmental Studies (NIES), were used in this study [2]. The population trends for these areas under different scenarios are illustrated in Fig. 2.

Table 1 provides tide levels for each bay. The mean sea level is based on the 5-year average from 2017 to 2021 [3], while the high water level is derived from the 5-year average from 2018 to 2022 [4]. The factors contributing to the maximum tide levels in each bay are Typhoon Jebi (2018) for Osaka Bay, Typhoon Vera (1959) for Ise Bay, and Typhoon Tip (1979) for Tokyo Bay [5].

3 Methods

3.1 Estimation of Changes in Inundation Risk for the Three Major Bays

In this study, potential storm surge inundation hazard areas due to sea level rise were estimated for Osaka Bay, Ise Bay, and Tokyo Bay. Sea level rise resulting from climate change was added to specific reference tide levels, and areas at lower elevations were identified as potential inundation areas. Four water levels were used as the basis for estimating the impacted area and population: 1) 0 m elevation (T.P. $\pm$ 0 m), 2) Mean Sea Level (MSL), 3) High Water Level (HWL), and 4) Highest High Water Level (HHWL), respectively. The affected population was calculated based on the population within the potential inundation areas. These calculations were performed at five-year intervals from 2020 to 2100 for each scenario to analyze the characteristics of population trends in the three major bays.

Table 1. Tide levels for three major bays.

	MSL	HWL	HHWL
Osaka Bay	0.192	0.872	3.29
Ise Bay	0.092	1.185	3.89
Tokyo Bay	0.046	0.902	2.03

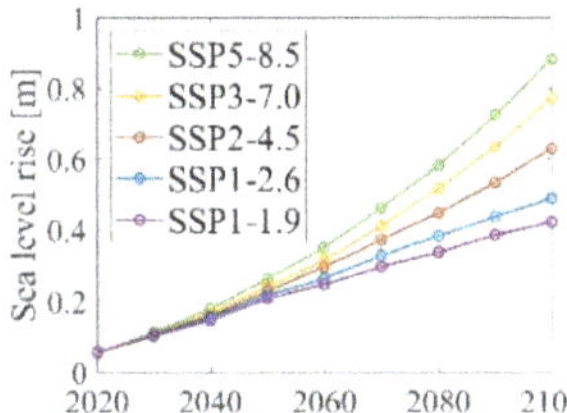

Fig. 1. Projected sea level rise for each SSP scenario for the Pacific coast of Japan.

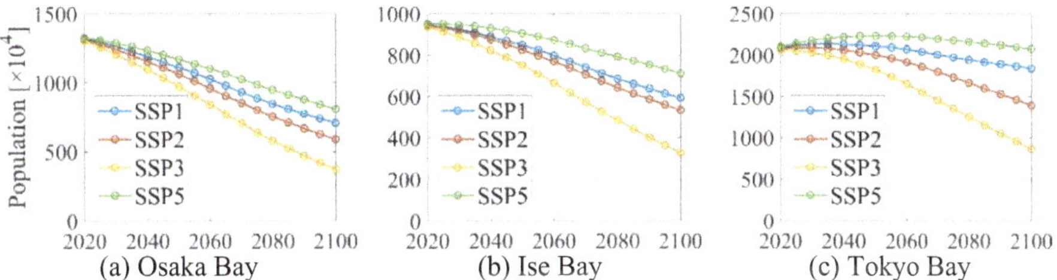

Fig. 2. Population trends in the coastal areas of the three major bays based on the second edition of the Japanese SSP population scenarios.

3.2 Storm Surge Inundation Calculations for Osaka Bay

Storm surge inundation calculations were performed for the deep bay area of Osaka Bay to estimate the inundation area and affected population for the years 2020, 2060, and 2100, and to assess changes over time. The inundation area was determined based on inundated meshes for which population density data were available. The storm surge calculations use JAGURS, a numerical model based on the nonlinear long wave equation. Typhoon Jebi, which caused the highest recorded storm surge in Osaka Bay, was used as the base case, with the initial water level set at HWL. Bathymetry and elevation data were consistent with those in Sect. 3.1, and the calculations employed nested grids, refining the horizontal spatial resolution from 2430 m to 90 m meshes. Meteorological data, including hourly atmospheric pressure and wind speed during Typhoon Jebi, were downscaled using the WRF meso-weather model [6]. To evaluate storm surge hazards under future climate conditions, Typhoon Jebi was intensified using the 100-year probability values of central pressure and wind speed projected in future climates.

4   Storm Surge Hazard Change Analysis Considering Sea Level Rise for Each Climate Change Scenario

Figure 3 shows the annual change in the affected area due to sea level rise for the three major bays under different climate change scenarios, based on simplified estimations. Among the three major bays, the affected area of Ise Bay is significantly larger. This indicates that Ise Bay is already at high risk of inundation in 2020 due to its extensive low-lying areas. Additionally, the difference between the affected areas at MSL and HWL is larger for Ise Bay than for the other two bays, further highlighting its vulnerability to sea level rise.

The annual trends in affected areas show that for all three bays, differences between scenarios remain small until approximately 2060. After 2060, these differences become more pronounced due to accelerated sea level rise in higher emission scenarios. While the increase in affected areas caused by tidal level differences varies across bays, there are no significant changes in the annual trend patterns for any of the bays.

Figure 4 illustrates the annual changes in the affected population due to sea level rise and population trends under different climate change scenarios for the three major bays. While Ise Bay has the largest affected area, Tokyo Bay has the highest affected population. This is attributed to Tokyo Bay's large overall population and high population density in low-lying areas. Although Ise Bay has a substantial affected area, its affected population is smaller than that of Tokyo Bay and is comparable to Osaka Bay.

The annual changes in the affected population across scenarios are compared. Trends vary among bays, but SSP5–8.5 consistently shows the largest affected population, followed by SSP1–2.6. While SSP1–2.6 represents a smaller sea level rise scenario, it still has the next largest affected population after SSP5–8.5. This indicates that while changes in hazards due to sea level rise influence the affected population, the impact of population trends is more significant.

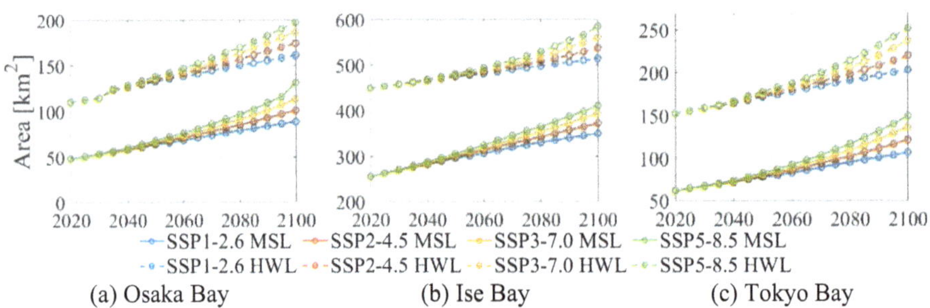

Fig. 3. Changes in affected area of the three major bays

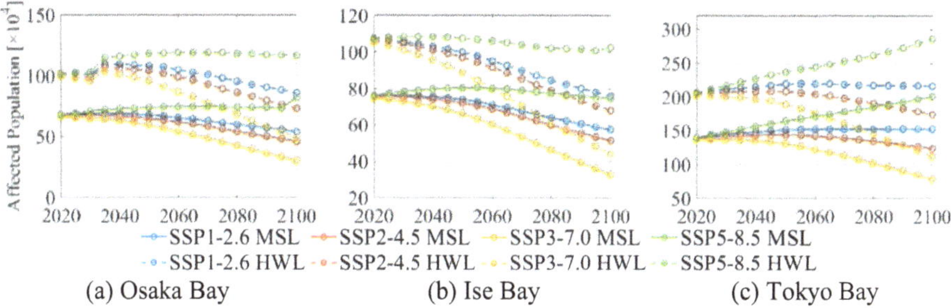

Fig. 4. Changes in affected population of the three major bays

5 Storm Surge Hazard Change Analysis Considering Sea Level Rise for Each Climate Change Scenario

Figure 5 compares the affected areas from simple estimation and storm surge inundation calculations, while Fig. 6 compares the affected populations. (a) and (b) show the results of the simple estimation under different tidal conditions, (c) to (e) present storm surge inundation calculations. In the storm surge calculations, (c) and (d) represent scenarios without levees, while (e) is the condition with levees. The original intensity of Typhoon Jebi is used for (c), whereas intensified Jebi under future climate conditions is used for (d) and (e).

Although the external force conditions for HHWL in the simple estimation (a) and for no levees in the storm surge calculation (c) are the same, the values differ significantly. This discrepancy arises because the simple estimation includes areas that are not actually inundated. Simple estimation considers only elevation data associated with population distribution and does not account for seawater dynamics, leading to overestimation of the affected area.

Comparing (c) and (d), which include the impact of typhoon intensity under future climate scenarios, both the affected area and population increase under 2 °C and 4 °C scenarios. However, these increases are relatively small compared to the difference between the scenarios themselves. This suggests that sea level rise and population change have a greater impact on coastal areas than typhoon intensity.

Comparing (d) without levees and (e) with levees, the affected population is significantly reduced when levees are functional. Furthermore, the scenarios with the largest affected population vary depending on whether levees are in place, highlighting their critical role in mitigating storm surge impacts.

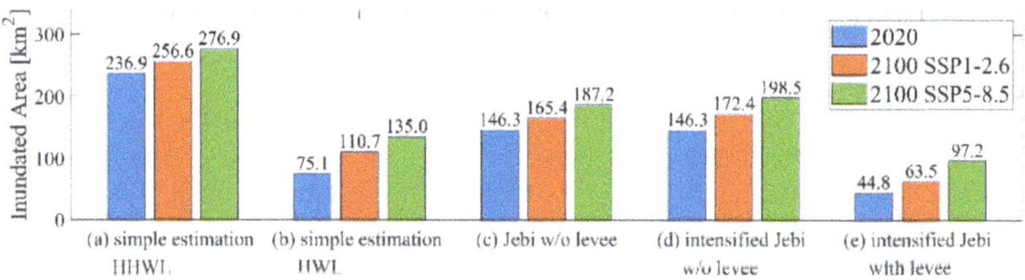

Fig. 5. Comparison of affected areas from simplified estimation and storm surge calculations

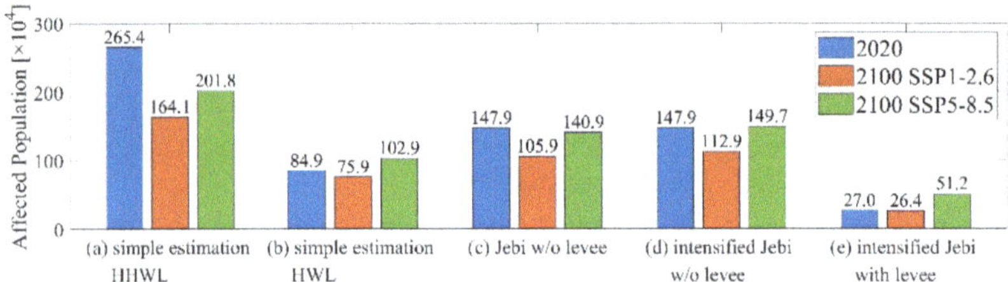

Fig. 6. Comparison of affected population from simplified estimation and storm surge calculation

6 Conclusions

For all three major bays, the differences between scenarios tend to increase after 2060. This is because the differences in sea level rise and population change among the scenarios increase after this period. However, each bay has unique topographic characteristics and exposure changes (e.g., population trends), making it essential to conduct separate inundation assessments to evaluate future trends in impact changes accurately.

The simple estimation method can generally capture the trend of affected population changes in cases without levees in storm surge inundation calculation. However, when compared to calculations that incorporate levee functions, significant differences in the affected area and population are observed. This highlights the necessity of performing detailed storm surge inundation calculations to obtain accurate estimates of the affected area and population.

This study conducted storm surge inundation calculations using meteorological data from Typhoon Jebi. However, due to the high level of uncertainty, risk assessment based on the single typhoon simulation is insufficient. To comprehensively assess future inundation risks, storm surge calculations using data from multiple typhoons are required.

References

1. Garner GG et al (2021) IPCC AR6 Sea Level Projections. Version 20210809. https://doi.org/10.5281/zenodo.5914709. Accessed 2 Mar 2024
2. National Institute for Environmental Studies, Climate Change Adaptation Information Platform (A-PLAT). Japanese shared socio-economic pathways population scenarios by mesh, 2nd edition. https://adaptation-platform.nies.go.jp/socioeconomic/population.html. Accessed 28 Jan 2024
3. Japan Meteorological Agency, List of Tide Table Listing Locations (2023). https://www.data.jma.go.jp/kaiyou/db/tide/suisan/station.php. Accessed 18 Dec 2023
4. Japan Meteorological Agency, Tides for each year. https://www.data.jma.go.jp/gmd/kaiyou/db/tide/gaikyo/nenindex.php. Accessed 18 Dec 2023
5. Japan Meteorological Agency, List of Highest Tide Levels. https://www.data.jma.go.jp/gmd/kaiyou/db/tide/list2.html. Accessed 18 Dec 2023
6. Fujiwara K, Takemi T, Mori N (2023) Response of Intensity and Structure of Typhoon Jebi (2018) before Landfall to 2-K and 4-K warmed future climates in dynamical downscaling experiments. SOLA 19:142–149. https://doi.org/10.2151/sola.2023-019

Assessment of Coastal Erosion Hazard in View of Climate Change: Case Study Camposoto Beach, Spain

Thet Oo Mon[1,2]($\boxtimes$), Theocharis Plomaritis[2], Juan Montes[1], Javier Benavente[1], Laura Del Rio[1], Cristina Montes[2], and Michalis Vousdoukas[3]

[1] Department of Earth Sciences, Faculty of Marine and Environmental Sciences, University of Cádiz, Cádiz, Spain
`thet.oo@uca.es`
[2] Department of Applied Physics, Faculty of Marine and Environmental Sciences, University of Cádiz, Cádiz, Spain
[3] Department of Marine Science, University of the Aegean, Mytilene, Greece

Abstract. Coastal risks, potentially amplified by climate change, are highly uncertain and can impact all coastal environments. Effective coastal risk management demands simple, adaptable methods that capture nearly accurate coastal dynamics, allowing for continuous improvement and integration of new data and approaches. This research validated a regional method for assessing climate change related coastal erosion hazards, by applying high-resolution data and a combination of simple process-based transfer functions at Camposoto Beach (Spain). Higher erosion hazard levels agree with dune erosion areas and historical erosion trends assessed in previous studies in the area. Test cases considering sea-level rise indicate higher coastal erosion hazards in areas with narrow berms and low dune profiles, aligning with current erosion rates depicted by studies of different methodology. The proposed method provides a more detailed perspective on potential hotspots compared to similar regional studies and offers better comparability among different sections and climate change scenarios.

Keywords: Coastal Risks · Climate Change · Erosion · Storms

1 Introduction

Coastal dynamics are an ensemble of complex processes, driven by a variety of interacting inputs and environmental factors that act over diverse time periods and across varying spatial scales. Growing concern and impacts from climate change-induced extreme events call for effective risk assessment and management measures regarding natural disasters in coastal areas. Understanding and reliable assessment of coastal hazards, associated risks and potential impacts of climate change constitute critical information to be derived from state-of-the-art research methods in coastal dynamics. This research aims to develop a computationally optimized and robust methodology for preliminary regional assessment of coastal erosion hazards, considering climate change scenarios and

© The Author(s) 2026
C. Coelho et al. (Eds.): CD 2025, CRL 41, pp. 358–363, 2026.
https://doi.org/10.1007/978-3-032-15473-6_56

using high-resolution topobathymetric information. The proposed method was validated in Camposoto Beach (SW Spain).

2 Study Area

Camposoto Beach is located in San Fernando (Cádiz province) on the southwest Atlantic coast of the Iberian Peninsula. The study area extends along 11 km in a NNW–SSE direction to the end of Sancti Petri sand spit and belongs to the Bay of Cadiz Natural Park. Dominant wind direction in the study area is from the West-SW and East-SE directions and longshore drift is directed Southeastwards. Average significant wave height is 1 m (up to 5 m in most energetic conditions), with associated periods of 5–6 s and dominant waves from the West. Tides are of meso-tidal and semidiurnal with mean neap and spring tidal range of 1.20 to 2.96 m [1].

Camposoto is a sandy beach with average slope of 0.02–0.06 [2], backed by a low dune ridge with sparse vegetation and intermittent washover fans. The dune elevation varies between 3 and 10 m above MSL, reaching its maximum towards the south. The inland area transitions from marshland in the north to a road in the middle and then to higher dunes with salt marshes and a tidal channel in the south. The beach widens towards the road, plateauing in the middle part with decreasing dune height. The southernmost part has a wider berm and higher dune heights than the others and suffers less anthropic pressure. Previous research in the central part of the study area showed varying erosion trends, with net dune foot retreat between 11.26 m and 20.44 m over 11 years in about 400m middle section [3] and up to 45 m over 60 years [4].

3 Methodology

In order to assess coastal hazard along cross-shore profiles at Camposoto beach, a methodology based on phase 1 of the regional-scale coastal risk assessment framework (CRAF) [5, 6] was applied and validated (Fig. 1).

Wave and surge projections across 6 coupled model of European Joint Research Centre (JRC) for RCP4.5 and RCP8.5 scenarios [7] were used as input. Regional sea-level rise data (median values), derived from [8] based on the IPCC report [9], were incorporated. Topobathymetric information of the profiles was obtained from a detailed 2-m resolution Digital Elevation Model (DEM) provided by REDIAM [10]. Results were analyzed across two decadal timescales: 2010–2050 and 2051–2100, to assess temporal variations. Storms were identified using a Peak Over Threshold (POT) analysis with defined threshold parameters [11]. Events exceeding the 95th percentile of Hs for over 12 h with at least 72 h of intermittence were considered individual storms [12]. Peak wave parameters for each storm were transformed into nearshore wave inputs accounting for refraction and shoaling. Dune foot retreat on transects separated 20m apart was calculated applying time dependent Convolution method [13] using nearshore wave and surge data of the storms and profile properties from DTM.

Hazard levels for 10-, 50-, and 100-year return periods were derived from retreat values using the Generalized Extreme Value (GEV) distribution [11, 14]. The following cases were analyzed: wave only storms erosion (Baseline) and two distinct cases for sea-level rise (SLR) induced retreat.

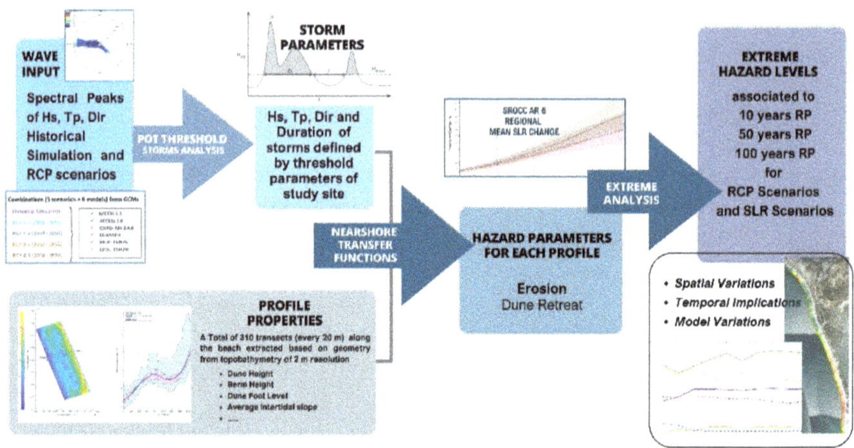

Fig. 1. Process diagram showing detailed methodology applied for hazard assessment.

- By adding retreat values calculated using convolution method, assuming a SLR increment and its duration until the end of century as a single event (Case 1)
- Total net change from the first case as a linear rate added to wave-only hazard levels until the target period (Case 2).

4   Results and Discussion

Dune retreat hazard levels of only 100-year return period for the test cases are shown here: Spatial variations of wave only retreat levels (assuming that the morphological parameters can maintain their values and the beach can keep up with sea-level rise, Fig. 2) and results of SLR induced retreat test cases (Figs. 3 and 4).

High erosion hazard (medium to high values closer to the end of road in Fig. 2) aligns with dune erosion areas obtained by field measurements in [3] and historical erosion trends in [15]. Areas of low or medium (1 to 3 Hazard levels) dune retreat coincide with high erosion trends in [15] and high shoreline erosion rates in [2]. Values are lower at the end of littoral cell transport to the south near the channel, except a few profiles. Estimated dune retreat by the convolution method initiates with wave parameters (Hs) are comparable to the dune erosion threshold obtained by [11].

For both wave and SLR induced retreat cases, some transects northward and the middle section close to the road exhibit higher hazard levels due to relatively lower dunes and narrower berms, whereas lower retreat close to the southern tip correlates with higher dunes and wider berms.

Baseline case suggests that end century RCP 4.5 scenario shows higher variation with higher maximums, which influence higher average hazard (Fig. 2), while RCP 8.5 demonstrates rather low range of deviation from historical simulation. This gives a range of dune retreat approximately agreeing with current erosion rates in the area of around

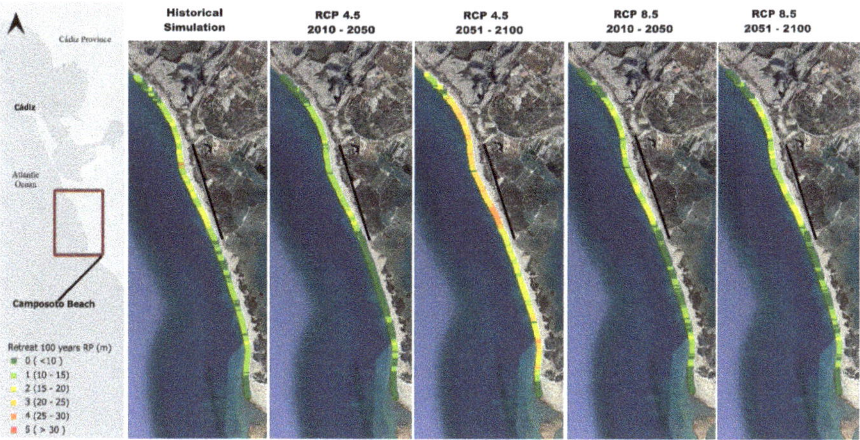

Fig. 2. Spatial variation of 100 year RP Hazard levels (dune foot retreat in m) for RCP scenarios along the transects in study area: Wave Only Test case(Black line represents location of the road).

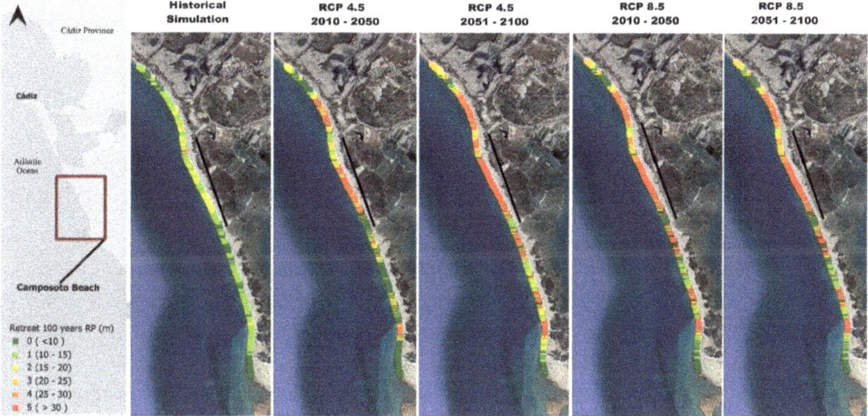

Fig. 3. Spatial variation of 100 year RP Hazard levels (dune foot retreat in m) for RCP scenarios along the transects in study area: SLR Test Case 1.

0.75 to 1.5 m/year [15]. SLR induced retreat test cases (Figs. 3 and 4) highlight potentially higher hazard levels in areas where one or more profile parameters are susceptible (narrow berms and lower dunes in the middle section) in contrast to fully developed dunes in the southern tip.

These changes give a perspective on spatial variation regarding possible climate change induced hazards. Current or historical hazard seem to be comparable to wave only scenario. However, there is high uncertainty to conclude which processes are major drivers for those: sediment deficit (due to lower discharge of sediment upstream), SLR or both. There is an opportunity to explore the joint influence of many processes such

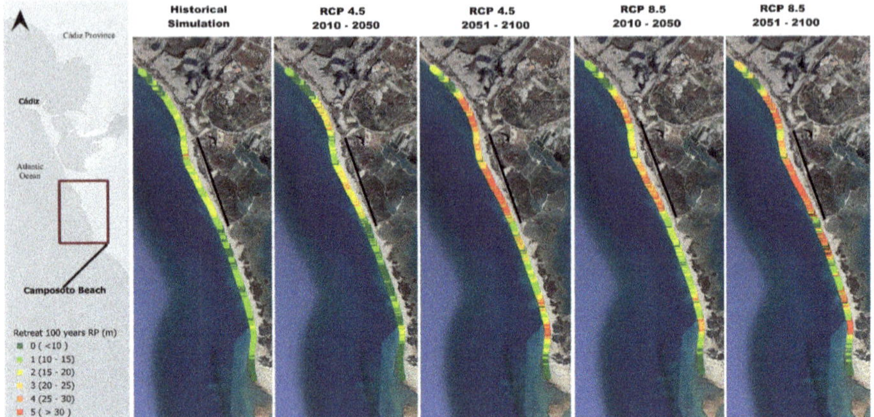

Fig. 4. Spatial variation of 100 year RP Hazard levels (dune foot retreat in m) for RCP scenarios along the transects in study area: SLR Test Case 2 (A linear rate of change of end century Retreat from case 1 is added).

as longshore drift, sediment budget, equilibrium state of the profile and which is more significant using methods as ShoreTrans [16] and ShorelineEvol [17].

5 Conclusion

Overall, spatial variation across profiles is comparable to historical trends. The highest hazard levels of 100 years RP considering only wave scenario suggest high erosion patterns similar to the historical ones revealed in existing studies. Significant changes in many profiles were found when SLR scenarios were introduced, especially for profiles with limited berm width and dune height. Monitoring and data collection over time provided valuable information for the validation of erosion values due to historical storms. Validation of future scenarios is a challenge, but comparison of different cases and approaches that combine storm forcing and morphological parameters can provide valuable information about the future trends of erosion risk.

Acknowledgements. This research was funded by the CRUNCH project (FEDER-UCA18-107062) by Regional Andalucian government and European Union and the CRISIS project (PID2019-109143RB-I00) by Ministry of Science and Innovation. It is also a contribution to PAI research group RNM-328.

References

1. IGN (Instituto Geográfico Nacional) La Red de Mareógrafos, https://www.ign.es/web/gds-la-red-mareografos, Accessed 17 May 2021
2. Del Río L, Gracia FJ, Benavente J (2013) Shoreline change patterns in sandy coasts. A case study in SW Spain. Geomorphology 196:252–266

3. Montes C, Benavente J, Puig M, Montes J, Talavera L, Plomaritis T (2024) High-resolution interannual evolution of the dune toe at a mesotidal barrier (Camposoto Beach, SW Spain). J Mar Sci Eng 12(5):718
4. Puig M (2016) Análisis de las variables que controlan la evolución de la costa a corto-medio plazo: Aplicación a la Bahía de Cádiz. (Universidad de Cádiz)
5. Viavattene C, Jiménez JA, Ferreira O, Priest S, Owen D, McCall R (2018) Selecting coastal hotspots to storm impacts at the regional scale: a coastal risk assessment framework. Coast Eng 134:33–47
6. Plomaritis TA, Ferreira Ó, Costas S (2018) Regional assessment of storm related overwash and breaching hazards on coastal barriers. Coast Eng 134:124–133
7. Vousdoukas MI, Mentaschi L, Voukouvalas E, Verlaan M, Feyen L (2017) Extreme sea levels on the rise along Europe's coasts. Earth's Futur 5(3):304–323
8. IPCC Sea Level Projection Tool, https://sealevel.nasa.gov/ipcc-ar6-sea-level-projection-tool?psmsl_id=209&data_layer=scenario, Accessed 01 Dec 2024
9. Lee J-Y et al (2021) Future global climate: scenario-based projections and near- term information. In: Climate Change 2021: The Physical Science Basis. Contribution of Working Group I to the Sixth Assessment Report of the Intergovernmental Panel on Climate Change. Cambridge University Press, Cambridge, United Kingdom and New York, NY, USA, pp 553–672
10. REDIAM (Red de Información Ambiental de Andalucía, Junta de Andalucia), https://www.juntadeandalucia.es/medioambiente/portal/acceso-rediam, Accessed 21 Nov 2022
11. Del Río L, Plomaritis TA, Benavente J, Valladares M, Ribera P (2012) Establishing storm thresholds for the Spanish Gulf of Cádiz coast. Geomorphology 143–144:13–23
12. Harley M (2017) Coastal Storm Definition. Coast. Storms, pp 1–21
13. Kriebel DL, Dean RG (1993) Convolution method for time-dependent beach-profile response. J Waterway, Port, Coastal, Ocean Eng 119:204–226
14. Kotz S, Nadarajah S (2000) Extreme Value Distributions Theory and Applications. Imperial College Press, London
15. Puig M, Del Rio L, Plomaritis T, Benavente J (2014) Influence of storms on coastal retreat in SW Spain. J Coast Res 70:193–198
16. McCarroll RJ et al (2021) A rules-based shoreface translation and sediment budgeting tool for estimating coastal change: ShoreTrans. Mar Geol 435:106466
17. De Santiago I et al (2021) Impact of climate change on beach erosion in the Basque Coast (NE Spain). Coast Eng 167:1–14

Numerical Modeling of Tsunami Caused by Submarine Landslide Using FDM-MPM Coupled Analysis

Yota Enomoto[1(✉)], Naoto Kihara[2], Chiaki Tsurudome[2], and Taro Arikawa[1]

[1] Chuo University, Bunkyo-Ku, 1-13-27, Tokyo, Japan
`a17.p7cj@g.chuo-u.ac.jp`
[2] Central Research Institute of Electronic Power Industry, AbikoAbiko 1646, Chiba, Japan

Abstract. This study proposes a novel numerical model of fluid-soil interaction to investigate tsunami generation caused by submarine landslides. The approach integrates CADMAS-SURF3D/2F (CS2F) for free-surface flow modeling with Anura3D's saturated soil solver. Material Point Method (MPM) was used to capture the complex interaction between soil and fluid. Experiments involving a sandy slope with an impermeable layer were conducted where slope failure was triggered by excess pore water pressure. By supplying water from the bottom of the slope, the effective stress in the lower sand layer was reduced, which resulted in a rapid collapse and subsequent wave generation. The numerical model reproduced the formation of an arc-shaped shear zone and the swift propagation of excess pore water pressure leading to landslide initiation. Experiments demonstrated slip-type failures triggered by elevated pore pressure beneath the vinyl sheeting, which acted as an impermeable layer, whereas simulations revealed slump-type failures governed primarily by dead weight. This difference underscores the importance of modeling fluid flow effects after initial slope failure. Overall, the developed model successfully simulated the key physical processes of submarine landslide tsunamis. The presented model is expected to offer insights into their complex mechanisms and provide a basis for improved coastal hazard assessments.

Keywords: Submarine landslide tsunami · Fluid-soil interaction · Material Point Method

1 Introduction

Submarine landslides and the resulting tsunamis have the potential to cause significant damage to coastal areas. Since submarine landslide tsunamis can reach coastal regions more quickly than fault-generated tsunamis, accurate prediction by numerical analysis is crucial for mitigating damage. Previous studies report that submarine landslides can occur even on seabed whose slopes are gentler than the angle of repose. The key trigger for such landslides is liquefaction, which results in water films between sediment layers of differing permeability [1] and uniforms excess pore water pressure beneath impermeable layers [2].

© The Author(s) 2026
C. Coelho et al. (Eds.): CD 2025, CRL 41, pp. 364–370, 2026.
https://doi.org/10.1007/978-3-032-15473-6_57

There are mainly three numerical approaches for analyzing tsunamis generated by submarine landslides: (1) directly inputting topographic changes, (2) employing a two-layer flow model that solves the bidirectional motion of both the landslide body (lower layer) and seawater (upper layer) where the landslide chape can change, and (3) using the Grilli and Watts (2005) [3] model, which derives an prediction law for the initial wave amplitude due to a landslide based on hydraulic experiments and numerical analysis. However, few studies rigorously address the internal ground behavior in submarine landslides, and there is no unified numerical model for submarine landslides and tsunami generation which captures their mechanism. In this study, a numerical model is developed to reproduce the physical processes from landslide initiation to tsunami generation. The experiments of submarine landslide tsunamis are conducted and the model is validated through comparison with experiments.

2 Method

2.1 Experimental Set up

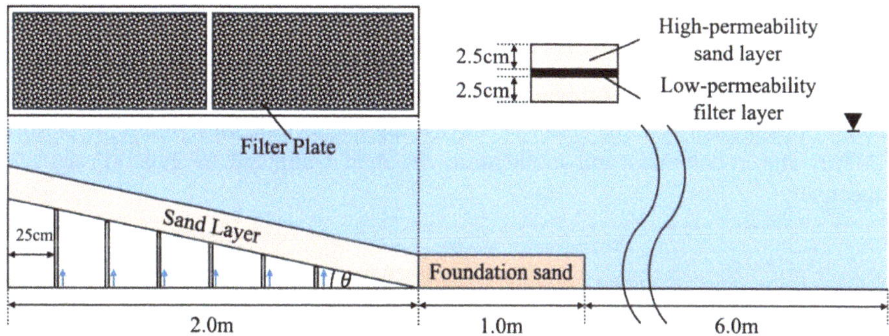

Fig. 1. Experiment condition

In the experiment, a sandy slope with an impermeable layer was prepared in a channel to model a landslide by applying excess pore pressure from the bottom [2]. The experimental channel measures 37 m in length, 0.5 m in width, and 1.0 m in height. A perforated slope with a slope angle of 15°, a horizontal length of 2.0 m, and a width of 0.5 m, was installed inside the channel. Eight hoses were connected to the perforated section of the slope to supply water and generate excess pore pressure. A filter layer was placed on the slope to ensure uniform application of excess pore pressure. Crushed stones, polypropylene perforated plates, and felt fabric were laid from bottom to top inside the filter layer (see Fig. 1). A sand matrix was formed on the filter layer using sand with a mean grain size of 0.1 mm according to the following steps:

1. Sand is sprinkled on the filter layer to create a 2.5cm thick lower layer by water pluviation method.

2. To represent an impermeable layer, a vinyl sheet (2.0 m long, 0.5 m wide, 1 mm thick) is submerged and placed parallel to the lower-layer slope.
3. Sand is again sprinkled on the vinyl sheet to create a 2.5cm thick upper layer by water pluviation method.

Water level variations during landslide were measured by detecting the water-surface edge in high-speed camera images.

2.2 Development of FDM-MPM Coupled Method for Soil-Fluid Interactions

In this study, a coupled model is developed using CADMAS-SURF3D/2F (CS2F) proposed by Arikawa et al. (2019) [4] as the fluid analysis model, and Anura3D's two-phase saturated soil solver developed by the Anura3D MPM Research Community [5] as the soil analysis model. CS2F employs the Volume of Fluid (VOF) method to capture free-surface flow, while Anura3D's governing equations are described in detail in the Anura3D scientific manual.

In the coupled method, CS2F transmits fluid pressure along the ground surface, while Anura3D updates the positions of soil particles. Specifically, CS2F passes the time-step size and surface fluid pressure to Anura3D, and Anura3D returns porosity information to CS2F. To maintain stability, Anura3D operates with a smaller time-step size than CS2F when they exchange information at regular intervals.

In this section, the calculation procedure of Anura3D is mainly described.

Mass and Momentum Interpolation. Each particle's mass and velocity are interpolated onto the background mesh (nodes) using shape functions (linear or uniform GIMPs). The nodal mass and momentum are then computed by Eqs. (1) and (2), respectively:

$$m_g^n = \sum_{i=1}^{n_p} S_{i,g} m_p \tag{1}$$

$$M_x^{n-1/2} = \sum_{i=1}^{n_p} S_{i,g} v_p^{n-1/2} \tag{2}$$

where m is the particle mass, n_p is the number of particles, S is the shape function, M is the momentum, v is the velocity. The subscripts g and p represent the quantities at the node and at the particle, respectively, and n represents the computation step.

Porosity Calculation. For each cell, the volume occupied by the particle cluster (defined by particle centers) is evaluated. This porosity information is then sent to CS2F each time step.

Equation of Motion. The background mesh is deformed by solving the equations of motion. The forces acting on each node are given by Eq. (3), where b_x is determined by the pressure from CS2F, based on the normal vector and the area of influence:

$$f_{x,I}^n = \sum_{i=1}^{n_p} \left[S_{i,I} b_x + S_{i,I} m_i g_x + \left(\tau_{xx,i} \cdot \frac{\partial S_{i,I}}{\partial x_g} + \tau_{xy,i} \cdot \frac{\partial S_{i,I}}{\partial y_g} + \tau_{xz,i} \cdot \frac{\partial S_{i,I}}{\partial z_g} \right) V_i^n \right] \tag{3}$$

where f is the force acting on the grid point, g is the acceleration of gravity, τ is the stress, V is the volume.

Water Pressure Application. Water pressure on each mesh is applied as a surface load based on Eqs. (4)–(9) including calculations for Jacobian $|J|$ and the surface normal vector $\boldsymbol{n}$.

$$\int_{S_2^e} N_i t \, ds = \int_{S_2^e} N_i p \boldsymbol{n} \, ds = \int_{-1}^{1} \int_{-1}^{1} N_i p \boldsymbol{n} |J| \, d\xi \, d\eta \tag{4}$$

$$r = \begin{Bmatrix} x(\xi, \eta) \\ y(\xi, \eta) \\ z(\xi, \eta) \end{Bmatrix} = \begin{Bmatrix} N_1(\xi, \eta) X_1 + N_2(\xi, \eta) X_2 + \cdots \\ N_1(\xi, \eta) Y_1 + N_2(\xi, \eta) Y_2 + \cdots \\ N_1(\xi, \eta) Z_1 + N_2(\xi, \eta) Z_2 + \cdots \end{Bmatrix} \tag{5}$$

$$r_1 = \frac{\partial r}{\partial \xi} = \frac{\partial}{\partial \xi} \begin{Bmatrix} N_1(\xi, \eta) X_1 + N_2(\xi, \eta) X_2 + \cdots \\ N_1(\xi, \eta) Y_1 + N_2(\xi, \eta) Y_2 + \cdots \\ N_1(\xi, \eta) Z_1 + N_2(\xi, \eta) Z_2 + \cdots \end{Bmatrix} \tag{6}$$

$$r_2 = \frac{\partial r}{\partial \eta} = \frac{\partial}{\partial \eta} \begin{Bmatrix} N_1(\xi, \eta) X_1 + N_2(\xi, \eta) X_2 + \cdots \\ N_1(\xi, \eta) Y_1 + N_2(\xi, \eta) Y_2 + \cdots \\ N_1(\xi, \eta) Z_1 + N_2(\xi, \eta) Z_2 + \cdots \end{Bmatrix} \tag{7}$$

$$n = \frac{r_1 \times r_2}{r_1 \times r_2} \tag{8}$$

$$|J| = r_1 \times r_2 = \sqrt{r_1^2 r_2^2 - (r_1 \cdot r_2)^2} \tag{9}$$

where S_2^e is the boundary surface of element, N_i is the shape function, p is surface pressure, ξ amd η is parametric coordinates and $r(\xi, \eta)$ is position vector.

Particle Update. After solving the equations of motion, particle positions and velocities are updated using Eqs. (10) and (11), where x is the particle position, a is the acceleration.

$$x_p^{n+1} = x_p^n + \Delta t \cdot \sum_{I=1}^{n_g} S_{i,g} v_g^{n+1/2} \tag{10}$$

$$u_p^{n+1} = u_p^n + \Delta t \cdot \sum_{I=1}^{n_g} S_{i,g} a_g^{n+1/2} \tag{11}$$

By iterating these steps, the coupled analysis progresses over time.

2.3 Numerical Analysis Condition

A water pressure of 20 kPa was imposed as a boundary condition on the slope for the first 1.0 s of the simulation. An impermeable boundary was implemented between the upper and lower layers to replicate the vinyl sheet used in the experiment (see Fig. 2). Both the grid size for fluid analysis and the background mesh size for soil analysis were set to 1 cm. Table 1 lists the parameters and soil properties.

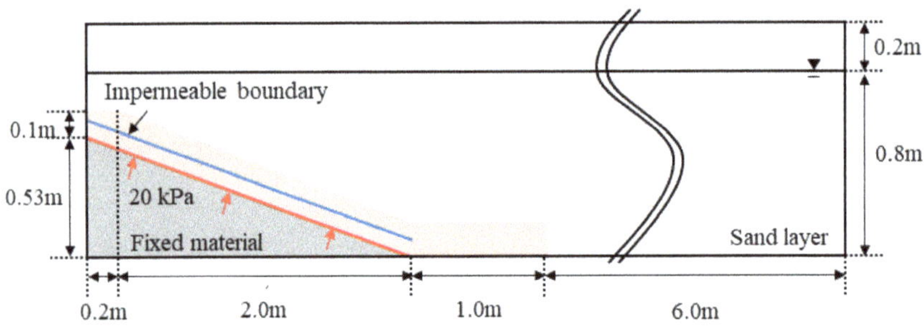

Fig. 2. Computational domain

Table 1. Table captions should be placed above the tables.

Heading level	Example
Initial porosity [-]	0.45
Density of solid [kg/m^3]	2650
Density of liquid [kg/m^3]	1000
Permeability [m/s]	4.0×10^{-7}
Bulk module liquid [kPa]	45000
Dynamic viscosity of liquid [kPa · s]	1.0×10^{-6}
Material model solid	Mohr-Coulomb
Effective Poisson ratio [-]	0.2
Effective Young modules [kPa]	5000
Effective friction angle [°]	32
Effective cohesion [kPa]	0

3   Results and Discussion

Figure 3 illustrates the time evolution of (a) plastic deviatoric strain, (b) effective vertical stress, and (c) the change in effective vertical stress from the first calculation step, respectively. The dashed line denotes the final ground surface position after the landslide. As time progressed, an arc-shaped shear zone formed, and excess pore water pressure rapidly propagated by seepage to the top of the slope, which eventually triggered a collapse. Consequently, the effective stress decreased in the sand layer at the bottom of the impermeable boundary, which caused the slope to lose its bearing capacity. The upper sand layer then slides and initiates a landslide. Figure 4 shows the spatial distribution of water-level variation induced by the landslide. This figure corresponds to the moment of maximum negative water-level variation at the top of the landslide mass. In the experiment, the landslide configuration conformed to the slide-type described by Watts et al. (2005) [3], whereas the numerical simulation yielded a slump-type failure. During the experiment, water flow induced by excess pore pressure remained even after the landslide commenced and propagated along the slope. The water flow might impart more energy than simple collapse under the soil's weight. In contrast, in the numerical model, only water pressure was imposed from the slope. Once the soil mass began sliding, no additional water flow was applied. Thus, the slope collapse proceeded mainly under self-weight, resulting in a slump-type failure.

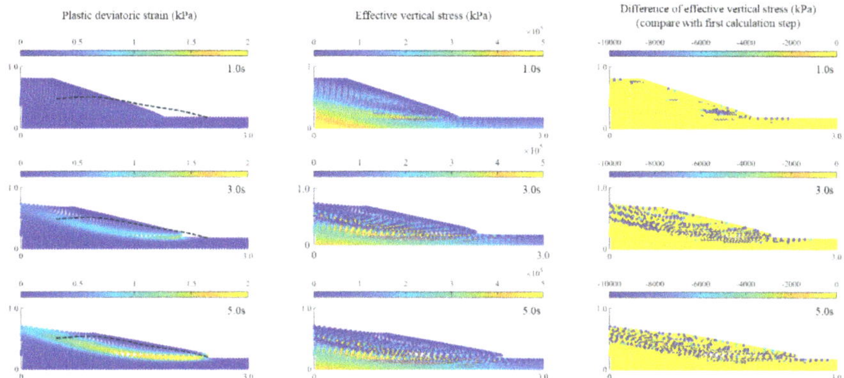

Fig. 3. Time evolution of (a) plastic deviatoric strain, (b) effective vertical stress, and (c) difference effective vertical stress compares with first calculation step in the numerical analysis.

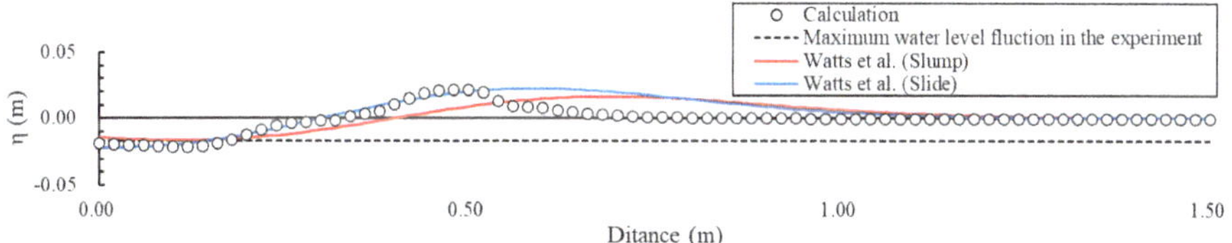

Fig. 4. Spatial distribution of water level fluctuation due to landslide occurrence.

References

1. Kokusho T, Kojima T (2002) Mechanism for postliquefaction water film generation in layerd sand. J Geotech Environ Eng 128(2):129–137
2. Yasui S, Iwai H, Kimura M, Cho H (2020) An experimental study on landslide kinematic on s submarine slope with an impermeable layer. J Japan Soc Civil Eng Ser. A2, 76(2):I_313-I_323. (In Japanese)
3. Grilli ST, Watts P (2005) Tsunami generation by submarine mass failure: 1 modeling, experimental validation, and sensitivity analysis. J Waterw Port Coast Ocean Eng 131(6):298–310
4. Arikawa T, Chida Y, Seki K, Takagawa T, Shimosako K (2019) Development and applicability of multiscale multiphysics integrated simulator for Tsunami. J Disaster Res 14(2):225–234LNCS
5. Anura3D MPM Research Community (2023) Anura3D Version 2023 Source Code, www.anura3d.com

Topanga Lagoon Restoration – Shoreline Morphology

Zhanxian Wang[1]([✉]), Qing Wang[2], and Weixia Jin[2]

[1] Moffatt & Nichol, Raleigh, NC 27609, USA
zwang@moffattnichol.com
[2] Moffatt & Nichol, Costa Mesa, CA 92626, USA

Abstract. The Topanga Lagoon Restoration project aims to expand the current Topanga Lagoon habitat area and return the lagoon toward its historic footprint, improve estuarine hydrologic function and water quality, enhance sea level rise and coastal erosion resilience, and integrate public and emergency access while not negatively affecting the quality of the Topanga Point surf break.

Unintended negative consequences to the surrounding shoreline caused by the restoration efforts are of concern. To avoid these consequences, a shoreline morphology study is requested to further understand the effects of restoration alternatives.

Two-dimensional hydrodynamics, waves, sediment transports, and morphological change numerical modeling were performed to investigate potential impacts on adjacent beaches and surf quality due to the restoration of Topanga Lagoon.

Keywords: Beach Morphology · Numerical Modeling · Coastal Restoration

1 Introduction

Topanga Lagoon is the lagoon of lower Topanga Creek where it empties into the Pacific Ocean in Los Angeles County, California. Topanga Creek drains an 18-square-mile watershed in the Santa Monica Mountains to the Pacific Ocean. It conveys flood flows to the lagoon during rain events and low flows during dry weather.

Once more than 30 acres, the historic lagoon area has been encroached over the years with buildings and hard-scape infrastructure and shrunk to approximately 1 acre in size currently. The beach adjacent to the lagoon is an important regional coastal access and recreation location, hosting millions of visitors per year. Topanga Lagoon hosts important natural resources such as the federally endangered tidewater goby and federally endangered Southern California steelhead trout [1, 5]. Remnants of the Historic Topanga Ranch Motel are also located in the lagoon area.

The Topanga Lagoon Restoration project (as shown in Fig. 1) aims to expand the current Topanga Lagoon habitat area and return the lagoon toward its historic footprint, improve estuarine hydrologic function and water quality, enhance sea level rise

C. Coelho et al. (Eds.): CD 2025, CRL 41, pp. 371–377, 2026.
https://doi.org/10.1007/978-3-032-15473-6_58

and coastal erosion resilience, and integrate public and emergency access while not negatively affecting the quality of the Topanga Point surf break.

There are three action design alternatives that are currently being considered for the restoration project in which the Pacific Coast Highway bridge is lengthened, material is excavated from the lagoon, and the existing Lifeguard building and Helipad are relocated farther inland to allow the site to be more resilient to sea level rise. The excavated materials are being considered placing nearshore as beneficial use.

Unintended negative consequences to the surrounding shoreline caused by the restoration efforts are of concern. To avoid these consequences, a shoreline morphology study is requested to further understand the effects of several action items of the project.

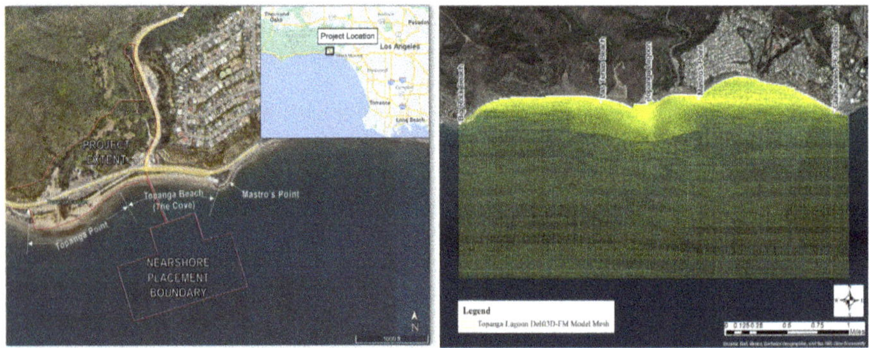

Fig. 1. (left) Topanga Lagoon project location; (right) flexible model mesh

2 Morphology Modeling

2D hydrodynamics, waves, sediment transports, and morphological change numerical modeling were developed using the open-source Delft3D Flexible Mesh Model suite (Delft3D-FM) developed by Deltares [3] due to its flexible meshes along with enhanced features than its predecessor Delft3D. The morphological model flexible mesh is presented in Fig. 1. Model calibrations were conducted based on publicly available data for water levels, currents, waves, and bathymetry changes. Due to space limitations, only the morphology calibration results are presented herein.

2.1 Topography and Bathymetry

Topographic and bathymetric data from several sources were compiled and processed to cover the entire computational domains. All datasets were adjusted to the model vertical datum - NAVD88. The 2009 USACE Topobathy Lidar DEM was used to develop the initial bathymetry for the morphology model calibration. In the offshore region where the Lidar DEMs are absent, the Santa Monica 1/3 arc-second DEM developed by NOAA was used.

2.2 Sediments

Four different sediment classes were included in the current modeling efforts: sand (0.4 mm), fine pebble (3 mm), medium pebble (12.5 mm), and cobble (160 mm). These sediment sizes were selected based on the composition of nearshore surface grabs along two cross-sections at the Topanga Beach area and borings in the Topanga Lagoon restoration areas as shown in Fig. 2. These sediments were sampled in 2022.

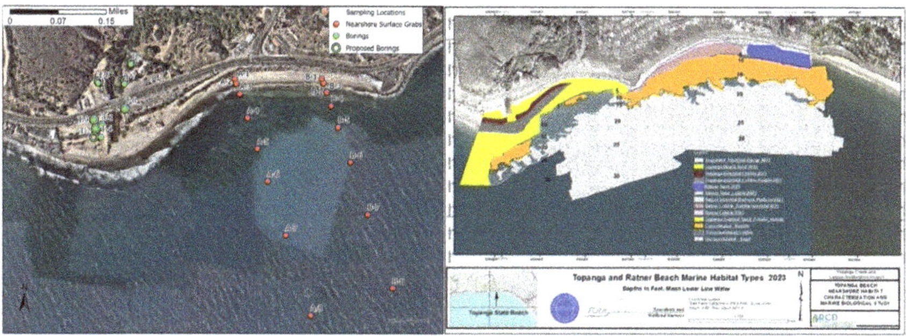

Fig. 2. (left) Sediment sampling locations; (right) Marine habitat types [2].

Detailed information about initial sediment size distributions and their thickness on the seabed in the model domain are lacking. They were developed for model calibration based upon:

- Sediment distributions from nearshore surface grab samples (Fig. 2).
- Topanga State Park seafloor habitat characterization studies from Coastal Resources Management, Inc. [2] as presented in Fig. 2.
- Outside of the CRM study area, only sediment grain size of 0.4 mm was assumed available in limited nearshore area based on aerial photos in Google Earth.
- Trial-and-error iterations based on engineering judgement together with discussions with CRM staff who performed the habitat survey.

The sediment transport formula of van Rijn 2007 for the non-cohesive sediments was applied for this study.

2.3 Wave Model Boundary Conditions

The USACE WIS hindcast data offshore of the project location from 1980 to 2021 were used to develop the wave boundary conditions. Most of the offshore waves come from directions between Southwest and West.

Modeling long term sediment transport and the resulting coastal morphology with Delft3D-FM model using a real-time series of waves as input would lead to unsustainably long run times. To avoid this problem, the wave data at the model boundary were numerically analyzed to derive a limited but representative set of wave conditions to be used as input into the morphological model. The goal is to reduce the wave classes as much as possible without losing much accuracy in the morphological development of these waves compared to the full wave time series. In this study, the wave class selection for the wave climate schematization with multiple classes was based on the OPTI-method [4]. Nine representative waves as listed in Table 1 were selected based on the OPTI analysis.

Delft3D-FM also adopted a technique called "morphological time scale factor" (morfac) whereby the speed of the changes in the morphology is scaled up to a rate that it begins to have a significant impact on the hydrodynamic flows. Variable morphological time scale factors (as presented in Table 1) were used for the representative waves in this study.

Table 1. Representative waves selected from OPTI-method and morfac.

Significant wave height (ft)	Peak wave period (s)	Mean wave direction (°N)	morfac
2.1	14.1	201.1	10.4
2.0	14.6	210.4	12.8
2.0	14.6	220.1	12.7
4.0	13.6	230.7	5.4
2.2	13.4	250.0	12.4
4.2	11.1	250.6	4.9
11.0	13.6	244.8	1.2
4.4	10.8	259.9	5.3
7.5	10.3	259.3	8.3

2.4 Flow Model Boundary Conditions

Along the west and east lateral open boundaries of the flow model, Newmann boundary conditions were applied. For the morphology model calibration, it was assumed no inflow from Topanga Creek since the Creek is mostly dry and the creek sediment contribution is very limited except during large fluvial storm events. At the south offshore open boundary, tidal water level boundary conditions were used. A one-day peak average spring tidal cycle was selected and repeated. This is a conservative approach for morphology modeling due to the largest tidal range captured in this peak spring tidal cycle.

2.5 Morphology Calibration Results

It is an advantage to use good quality lidar data for morphological model calibration because of the continuous coverage. USACE NCMP topobathy lidar data collected in 2009 and 2014 are available in the project area. However, after carefully examined the lidar data, it seems that the 2014 lidar bathymetric data has some issues. Eventually the 2014 lidar elevations were lowered by 1.3 ft for the model calibration purpose.

Figure 3 presents the total volume changes per unit area between the model results after 5 years and lidar. Volume changes from both original and lowered 2014 lidar data relative to 2009 lidar are included. The volume changes based on the original 2014 lidar indicate accretion along the whole coastline both east and west side of the Topanga Lagoon (+30 CY/ft on average). On the contrary, the volume changes from the lowered 2014 lidar show erosion on both sides of the lagoon. The model results are in good agreement with the lowered 2014 lidar data.

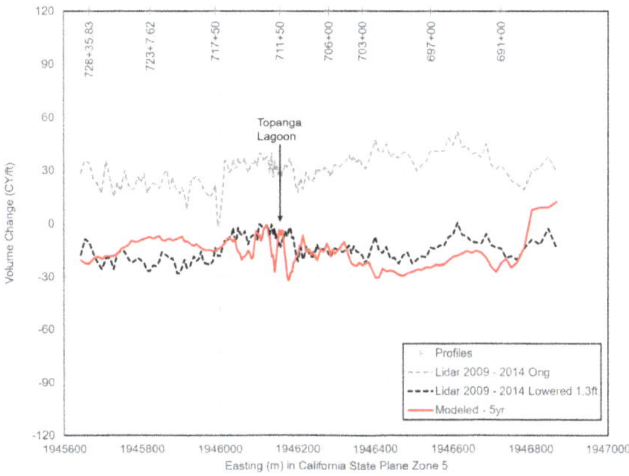

Fig. 3. Volume changes near Topanga Lagoon above -24 ft-NAVD (modeled vs lidar).

3   Restoration Alternative Analysis

Four Topanga Lagoon restoration alternatives were developed and studied for the project. These four restoration alternatives are: Alternative 1 – No project/no build, managed decline; Alternative 2 – Maximum lagoon habitat, removal of motel; Alternative 3 – Limited lagoon habitat expansion, retention of motel; and Alternative 4 – Maximum managed retreat, partial motel retention.

For the restoration alternative analyses, the lagoon was assumed initially breached with a straight 40-ft wide channel connecting to the Pacific Ocean. Three Topanga Creek flow conditions were considered: (1) dry-weather condition (no flow and sediment input); (2) 10-yr return period fluvial flow condition; (3) 100-yr return period fluvial flow condition. The fluvial flow and sediment inputs were assumed to occur at the beginning of the model simulation.

Due to space limitations, only the model results for Alternative 1 and Alternative 2 are presented herein. The bathymetry differences (as shown in Fig. 4) due to initial alternative lagoon grading are excluded in the difference maps to isolate and demonstrate the morphological changes over time caused by environmental forces.

Under dry weather conditions without lagoon sediment input, the morphological changes in Alternative 2 are similar to Alternative 1. This result is to be expected since all gradings are above + 14 ft-NAVD88, except a pilot channel under the proposed PCH bridge which will not have any impact under the dry weather conditions, and no grading south of the current beach berm.

After the 10-yr RP fluvial storm, the shoreline differences occurred between 0 to -5 ft-NAVD88 contours in the Topanga Point area and the magnitudes are less than 0.4 ft. The modeling results indicate that there is erosion in the pilot channel after the storm event, which could lead to west breach of the lagoon mouth. A west breach is considered favorable for surfing.

After the 100-yr RP fluvial storm, the existing cobbles offshore of the lagoon were transported in slightly different directions due to the flow field differences during the 100-yr RP fluvial storm between alternatives. Thus, slightly different channel orientation was generated. Alternative 2 cuts through the cobble berm slightly to the west compared to Alternative 1, which could lead to west shift of the thalweg and breach. The magnitudes of morphological differences for the nearshore area are less than 1ft.

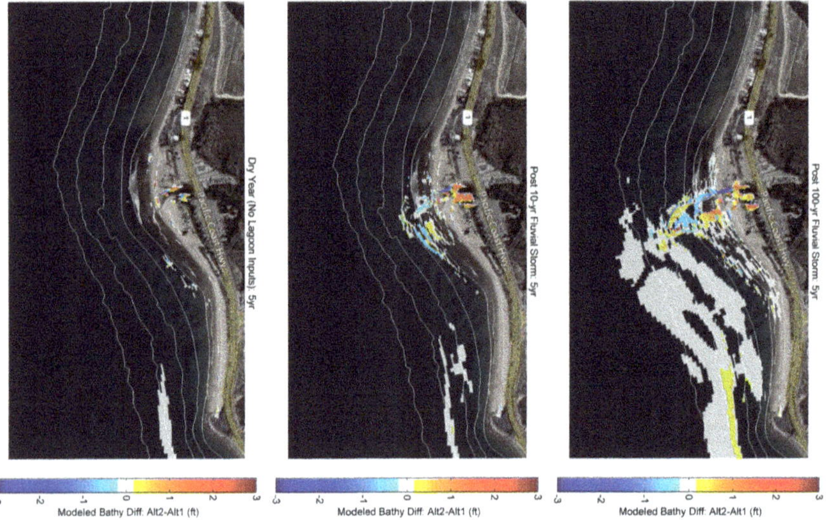

Fig. 4. Predicted morphology differences between Alternatives 1 and 2 after 5 years.

Acknowledgements. We appreciate the opportunity given by the Resource Conservation District of Santa Monica Mountains to take on the study of the Topanga Lagoon Restoration project. Special thanks also must be given to Integral Consulting LLC, our project partner and other funding agencies and shareholders for their support and insightful information.

References

1. CalTrout: Santa Monica Mountains steelhead habitat assessment, final project report, 18 January 2006
2. Coastal Resources Management, Inc.: Preliminary draft Topanga State Park seafloor habitat characterization and marine biological - second year studies June-July 2023, sediment reuse project for Topanga Lagoon. Prepared for Resource Conservation District of the Santa Monica Mountains. 10 September 2023

3. Deltares: D-Flow flexible mesh computational cores and user interface, user manual (2023)
4. Mol A.C.S.: R&D coastal hydraulic engineering reduction of wave conditions OPTI manual. Research report H4959.10. WL|Delft Hydraulics, the Netherlands (2007)
5. National Park Service: Santa Monica Mountains national recreation area general management plan, July (2002)

Coastal Wetlands for Coastal Resilience and Climate Adaptation

Use of Sub-Pixel Imagery Classification to Assess Salt Marsh Plants' Adaptation

A. Rita Carrasco[1]([⊠]), Alexandra Astori[2], and Katerina Kombiadou[1]

[1] Centre for Marine and Environmental Research (CIMA) I Aquatic Research Network (ARNET), University of Algarve, Campus of Gambelas, 8005-139 Faro, Portugal
azarcos@ualg.pt
[2] University of Milan-Bicocca, Milan, Italy

Abstract. The inherent heterogeneity of coastal wetlands and the small size of halophytic plants present challenges in accurately sensing plant species, even with very high-resolution satellite imagery. This study used sub-pixel imagery classification methods on high spectral and spatial resolution imagery from Worldview-3 to predict plant species distribution in a mesotidal coastal wetland system. The predicted sub-pixel fractional abundance of plant species is discussed for three targeted wetland categories in the Ria Formosa lagoon: naturally evolving patches, patches modified by human activities, and patches affected by coastal squeeze.

The Random Forest Regression algorithm was proven to be highly effective in unmixing the spectral signal of halophytic vegetation, enabling the retrieval of plant species distribution (7 plant species). To train the algorithm, field observations were used to classify satellite images. Differences in band feature importance for key species and bare soil were observed across the various sites. The comparison of species distribution between sites suggests that, in addition to biotic factors, other environmental influences likely affect ecological succession; therefore, large-scale mapping approaches based on remote sensing should be undertaken with caution. The results are important for understanding the diverse ecological behavior of marsh plants within the same system and highlight the variability in plant reflectance and the need for ground truthing when sensing plant cover from satellite data.

Keywords: salt marsh · satellite imagery · species fractional cover · reflectance · environmental pressures

1 Introduction

The temporal and spatial evolution of coastal wetlands is complex and challenging to forecast and made even more complicated by changing climatic conditions [1]. While remote sensing offers many clear benefits for marsh monitoring [2], greater care should be taken when generalizing patterns of ecosystem service provision to larger scales (e.g., blue carbon assessments). Satellite imagery has shown a strong potential for acquiring comprehensive and long-term observations in extensive areas, enabling the characterization of habitat change in various spatial and temporal scales [3].

© The Author(s) 2026
C. Coelho et al. (Eds.): CD 2025, CRL 41, pp. 381–387, 2026.
https://doi.org/10.1007/978-3-032-15473-6_59

Image classification methods applied to salt marshes have been developed for and applied to multi- and hyperspectral remote sensing data in a diverse set of biomes worldwide, to determine vegetation zonation and species abundance. Tracking such ecological variables is important to conclude about the health of wetlands and their adjustments to climate change, particularly to sea-level rise effects and limited sediment supply [4]. This study uses very high spectral and spatial resolution imagery (Worldview-3, WV3), combined with spectral unmixing methods, to assess their ability to identify the subpixel fractional cover of marsh species, focusing on zones of different physical, ecological, and anthropogenic conditions.

2   Methods and Data

2.1   Satellite Imagery and Field Data

The research was conducted in Ria Formosa, one of the main coastal lagoons of Portugal. Subpixel classification methodologies, such as the Random Forest Regression (RFR; [4]) were employed to assess their accuracy in predicting marsh plant fractional abundance in challenging marsh environments from very high-resolution EO products (WorldView-3, WV3, pansharpened imagery with pixel of 0.5x0.5 m). WV3 satellite data from May 2023 was used for the subpixel plant's fractional abundance prediction within three salt marsh sites, i.e., Site 1 - patches evolving naturally, Site 2 - patches highly affected by human activities, and Site 3 - patches facing coastal squeeze (Fig. 1). At Site 1, a 'natural' development of ecological succession is observed, with vegetation distributed in bands/zones according to topography and the wide accommodation space; sediment is characterized by a heterogeneous composition, predominantly sandy mud, with smaller contributions of muddy sand and sand. At Site 2, a modified ecological succession is observed, where vegetation zonation is interrupted over areas of high active sedimentation (burial); sediment is mainly composed of sandy mud and muddy sand, with contributions from slightly gravelly sandy mud and mud. At Site 3, due to the limited accommodation space, species are distributed in bands but with some degree of mixing of plants, especially between the medium and low marsh; sediment is homogeneous and predominantly composed of sandy mud.

The RFR algorithm was written in Python, using the open Scikit-learn package [5]. Training/testing data were collected at the three sites during 2023–2024, using RTK-DGPS and visual assessment of plant species cover. The marsh plant species present in each of the three sites determined the classes used (increased by one for the bare soil). The importance of each WV3 spectral band (i.e., feature importance used in the classification; [4]) in key plant species was extracted based on the WV3 bands.

2.2   Performance Analysis

To estimate model performance, the root-mean-square error (RMSE) and the coefficient of determination of the prediction (R^2) between predicted and observed fractional abundance were calculated for each class for both training and testing populations.

Fig. 1. Views of the test-sites: (A) Site 1, (B) Site 2, and (C) Site 3.

3   Results and Discussion

The proposed method was successfully tested against field observations. RFR proved to be highly effective in unmixing halophytic vegetation (average train population R^2 > 0.83 and test population overall accuracy of RMSE < 5.12 and R^2 > 0.83; Table 1), allowing for reliable predictions of species coverage (see Fig. 2); some of the plant species had to be grouped in pairs to improve the prediction results. The predicted and observed percentages for the classes (plant species) closely align with a 1:1 prediction line at all three sites, especially for Sites 1 and 2 (Fig. 2A and Fig. 2B). Still, Site 3, which corresponds to a wetland patch facing coastal squeeze, shows lower accuracy (higher values of RMSE) than the other two sites, mostly for the classes *Spartina maritima* and the pair *Salicornia fruticosa* and *Sarcocornia perennis* (Fig. 2C and Fig. 2D). This is likely due to the shortened development of the ecological succession and mixture of species observed at this location, which leads to a 'noisy' signal and should not be deemed as a diminished predictive ability of the algorithm.

Table 1. Accuracy assessment (averaged values) for the RFR in training and testing (RMSE; R^2).

Sites	$RMSE_{test}$	R^2_{test}	R^2_{train}
Site 1	4.58	0.90	0.89
Site 2	4.42	0.83	0.83
Site 3	5.12	0.87	0.88

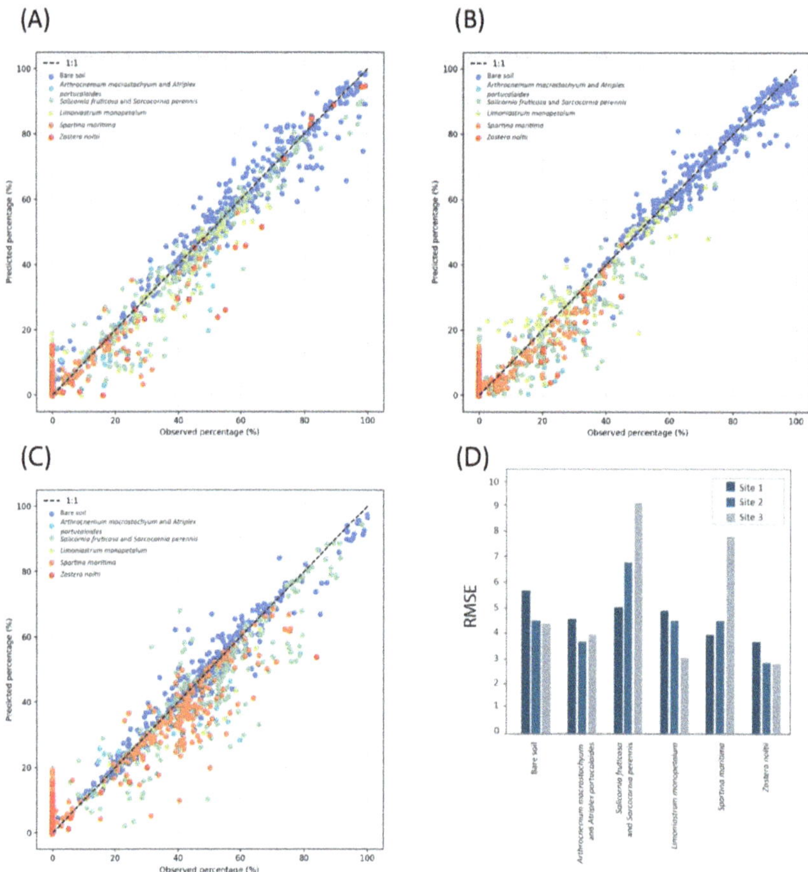

Fig. 2. Model skill (observations *vs.* predicted probabilities in each class) for Site 1 (A); Site 2 (B); Site 3 (C); and a comparison of the root-mean-square error (RMSE) for each class (bare soil, *Arthocnemum macrostachyum* and *Atriplex Portucaloides*, *Salicornia fruticosa* and *Sarcocornia perennis*, *Limoniastrum monopetalum*, *Spartina maritima*, and *Zostera noltii*) between sites (D).

The results indicate variability in band importance (feature importance shown in Fig. 3) for key species, including bare soil. Notably, the findings suggest that NIR1 (near-infrared 1, 765–901 nm) is critical for bare soil (corroborating the results of [4]). Greater variability in the NIR1 band feature importance for bare soil among the sites suggests differences in sediment composition and texture that are likely to impact ecological zonation (as observed in recent experiments of [6]). In the low marsh, for the pair *Salicornia fruticosa* and *Sarcocornia perennis*, and the *Spartina maritima*, the critical band is the NIR2 (near-infrared 2, 856–1043 nm; Fig. 3). Variability in the NIR2 band importance reveals differences in water presence, likely linked to variations in the elevation of these plant species. In the tidal flat, for *Zostera noltii* is the Red (0.64–0.67 nm; Fig. 3). Variability in this band importance implies deviations in chlorophyll absorption regions [7] associated with differences in coverage and biomass between the monitored sites.

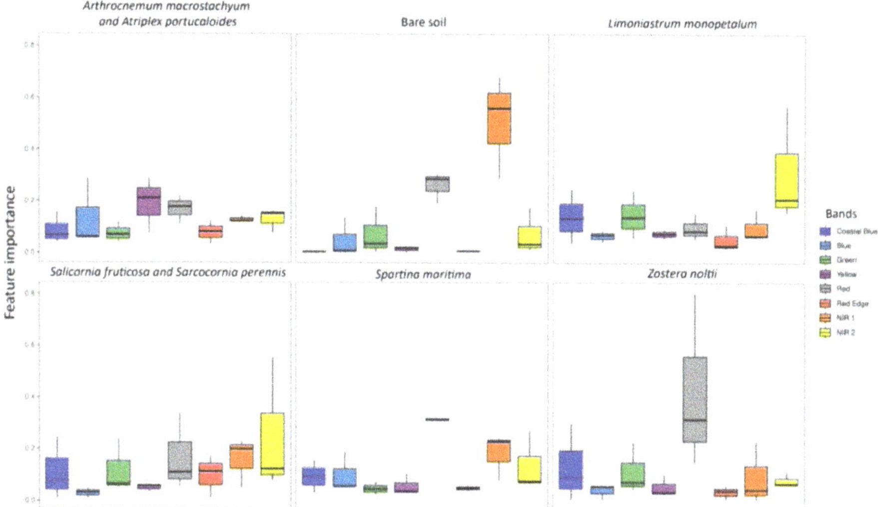

Fig. 3. Feature importance of each band for *Arthrocnemum macrostachyum* and *Atriplex Portucaloides*; bare soil; *Limoniastrum monopetalum; Salicornia fruticosa* and *Sarcocornia perennis*; *Spartina maritima*; and *Zostera noltii*, in the three sites.

The referred variability suggests that ecological succession develops differently within the same coastal system due to various natural and human pressures. The study raises concerns about the representativeness of training data and the upscaling of mapping results in wetland systems that may experience different adaptations to environmental drivers. This has already been suggested by other authors, such as [8], in the classification of UAV images in mesotidal wetlands.

4 Concluding Remarks

Improving the accuracy of remote sensing for mapping wetlands is in high demand because it offers an effective way to accurately inventory plant species and provides essential baseline data for climate change mitigation strategies. This study applied sub-pixel imagery classification methods to examine the distribution of salt marsh plants to environmental characteristics. The application of the RFR approach proved skillful in unmixing halophytic plants in mesotidal marshes, enabling accurate predictions of percentage cover for each species. When comparing sites, differences in band importance for key species were observed, indicating variability in vegetation coverage and density. This interspecific variability may be attributed to differences in life cycles or the plants' adaptive responses to the unique environmental factors at each site.

The study makes two significant contributions: first, it addresses the challenges of large-scale mapping using satellite imagery, and second, it provides insights into how environmental conditions within the same basin system may influence the development of salt marsh plants.

Acknowledgments. The work contributes to projects TraceLands (ESA-ID PP0090200) and DEVISE (https://doi.org/https://doi.org/10.54499/2022.06615.PTDC), funded by the European Space Agency (ESA) and the Fundação para a Ciência e Tecnologia (FCT).

A. R. Carrasco and K. Kombiadou were supported by the individual contracts https://doi.org/https://doi.org/10.54499/CEECINST/00052/2021/CP2792/CT0007 and https://doi.org/https://doi.org/10.54499/CEECINST/00146/2018/CP1493/CT0011, respectively, and had the support of the funding attributed to CIMA (https://doi.org/https://doi.org/10.54499/LA/P/0069/2020) and the Associate Laboratory ARNET (https://doi.org/https://doi.org/10.54499/LA/P/0069/2020), all funded by the FCT. The authors would like to thank Zhicheng Yang for kindly making available the code on the Random Forest Regressor.

References

1. Campbell A, Wang Y (2019) High spatial resolution remote sensing for salt marsh mapping and change analysis at fire Island national seashore. Remote Sens 11
2. Blount TR, Carrasco AR, Cristina S, Silvestri S (2022) Exploring open-source multispectral satellite remote sensing as a tool to map long-term evolution of salt marsh shorelines. Estuarine Coastal Shelf Sci 266
3. Lopes CL, Mendes R, Caçador I, Dias JM (2019) Evaluation of long-term estuarine vegetation changes through Landsat imagery. Sci Total Environ 653:512–522
4. Yang Z, D'Alpaos A, Marani M, Silvestri S (2020) Assessing the fractional abundance of highly mixed salt-marsh vegetation using random forest soft classification. Remote Sens 12:1–25
5. Scikit-learn package. RandomForestRegressor. https://scikit-learn.org/stable/modules/generated/sklearn.ensemble.RandomForestRegressor.html. Accessed 20 May 2024
6. Stoorvogel MM, Temmerman S, Oosterlee L, Schoutens K, Maris T, van de Koppel J et al (2024) Nature-based shoreline protection in newly formed tidal marshes is controlled by tidal inundation and sedimentation rate. Limnology and Oceanography
7. Lopes CL, Mendes R, Caçador I, Dias JM (2020) Assessing salt marsh extent and condition changes with 35 years of Landsat imagery: Tagus estuary case study. Remote Sens Environ 247
8. Routhier M, Moore G, Rock B (2023) Assessing spectral band, elevation, and collection date combinations for classifying salt marsh vegetation with unoccupied aerial vehicle (UAV)-acquired imagery. Remote Sens 15

Salt Marsh Hydrodynamics to Inform Beneficial Use of Dredged Material for Coastal Resilience

Conrad Davis[1]([✉]), Anna Wargula[1], Liliana Velásquez-Montoya[1], Tori Tomiczek[1], Monica Chasten[2], and Lenore Tedesco[3]

[1] United States Naval Academy, Annapolis, MD 21402, USA
conrad.p.davis25@gmail.com
[2] U.S. Army Corps of Engineers, Philadelphia, PA 19107, USA
[3] The Wetlands Institute, Stone Harbor, NJ 08247, USA

Abstract. As salt marshes struggle to keep up with sea level rise, natural resource managers and agencies are increasingly utilizing nature-based solutions and exploring Beneficial Use of Dredged Material (BUDM) for marsh restoration. A United States Army Corps of Engineers (USACE) BUDM project in cooperation with The Wetlands Institute and the State of New Jersey at the Scotch Bonnet Island wetlands, in Stone Harbor, NJ, USA provided a unique opportunity to measure hydrodynamic and morphologic conditions in two marsh channels (North and South) before, during, and immediately after the material placement to better understand impact to these parameters. This paper focuses on pre-placement observations of tides, flow velocities, and pre-placement channel bathymetry and presents a methodology to estimate volumetric flow at channel transects, with a future goal of evaluating change in tidal prism as a potential consequence of BUDM. Using SONAR survey, satellite imagery, and walking survey measurements, area models at channel transects were created. From Nov.-Dec. 2023, with the ebb and flood stages of tide, the North and South Channels both saw phase-averaged peak discharge of -2.9 m^3/s but inflow of 3.8 m^3/s and 2.0 m^3/s respectively.

Keywords: Engineering With Nature · Marsh restoration · Volumetric flow

1 Introduction

Salt marshes play an integral role along coastlines as buffers against storm surge and wave energy and critical habitats for shorebirds, fish, and other fauna. As sea level rise or sediment deficits surpass the ability of salt marshes to morphologically adapt, increased inundation is resulting in vegetation die-off on marsh platforms [1]. Consequently, coastal salt marshes are prone to interior erosion, drowning, and land loss resulting from conversion of marsh to mudflats [2]. The United States Army Corps of Engineers (USACE) is promoting Beneficial Use of Dredged Material (BUDM) from local dredging projects as a method to combat this land loss while enhancing wetland habitat [3]. In the Fall of 2024, Scotch Bonnet Island wetlands in Stone Harbor, New

© The Author(s) 2026
C. Coelho et al. (Eds.): CD 2025, CRL 41, pp. 388–394, 2026.
https://doi.org/10.1007/978-3-032-15473-6_60

Jersey, USA (Fig. 1), was restored using BUDM, where sediment was hydraulically pumped onto a partially contained marsh area to increase its elevation. By occupying a portion of the tidal prism of Scotch Bonnet Island wetlands, this BUDM initiative has the potential for hydrodynamic impact in existing marsh channels. Salt marsh channels serve as pathways for nutrient and sediment exchange [4]. In order to better inform future BUDM placement in salt marsh systems and provide quantitative data linking salt marsh hydrodynamics to morphological evolution, this research aims to correlate changes in channel hydrodynamics to the BUDM placement. Observed time-series of water level and flow velocity were used to estimate volumetric flows. This paper describes the methodological approach employed to estimate volumetric flows based on observed hydrodynamic and morphological data prior to the BUDM project.

2 Study Site and BUDM Specifications

Scotch Bonnet Island wetlands measures approximately 22.2 hectares and contains two tidal channels, designated herein as North (N) and South (S). The marsh platform is primarily colonized by high and low vigor smooth cordgrass (*Sporobolus alterniflorus*). From 1941 to 2019, Scotch Bonnet Wetlands lost 34% of its marsh area (12.1 hectares, 30 acres), and the island is projected to lose an additional 33% (8.1 hectares, 20 acres) of marsh area by 2050 as a result of vegetation die-off [5]. The BUDM placement that took place in Scotch Bonnet Island wetlands in the Fall of 2024 encompassed an approximate area of 4.8 hectares (12 acres), outlined in red in Fig. 1. Biodegradable coir logs reduced unwanted spread of material. The targeted increase in elevation was 0.15–0.37 m (0.5–1.2 ft) to bring marsh elevation up to 0.85–1.07 m (2.8–3.5 ft) relative to the North American Vertical Datum of 1988 (NAVD88), with the goal of stabilizing the marsh platform, infilling expanding pools, and reversing marsh conversion to mudflat. The BUDM placement occupied the lowest-lying areas of the marsh, including upper portions the North Channel (Fig. 1).

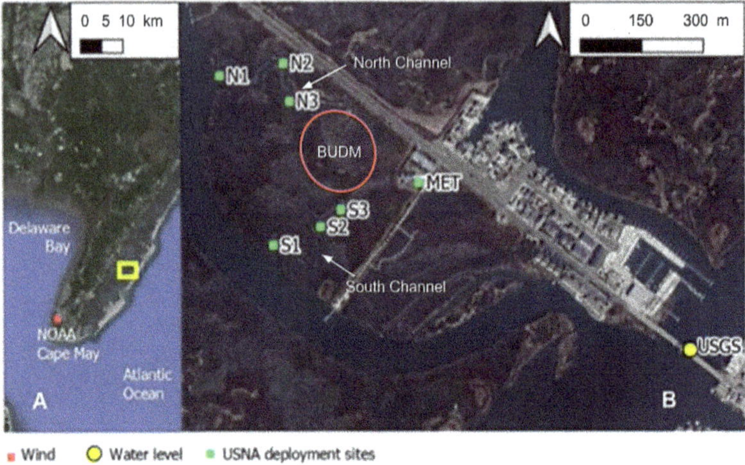

Fig. 1. A. BUDM study site location in New Jersey, USA; B. Detail of Scotch Bonnet Wetlands. N1–3 and S1–3 - locations of water level and flow velocity observations in North and South channels, respectively; MET - location of atmospheric pressure measurements. USGS – location of near site tidal gage operated by the United States Geological Survey.

3 Methods

3.1 Water Level and Channel Velocity

Pre-BUDM water levels and velocities were collected at 3 sensor locations per channel, N1–3 and S1–3 (Fig. 1) between Nov. and Dec. 2023. Sensors were placed approximately mid-cross section of each channel. Water level was measured using Onset HOBO U20 absolute pressure gauges. Atmospheric pressure, recorded by a pressure gauge in air (MET, Fig. 1), was subtracted from absolute pressure readings by sensors in water, from which depth could be calculated using an assumed density from fixed temperature and salinity conditions. Water level was corrected to a common datum (NAVD88) using an instantaneous offset from USGS Gauge 01411360, located outside Scotch Bonnet in the Great Channel (USGS, Fig. 1). Flow velocity was measured using Lowell Tilt Current Meters (TCM-1 and TCM-4). Velocity readings were truncated according to submersion requirements using depth from the co-located pressure gauge. Pressure loggers recorded at 10-min intervals, while velocities recorded in 20-s bursts at 10-min intervals. Bursts were averaged to obtain final readings. Velocity components were rotated into along- and cross-channel velocities using principal axes. The mean cross-channel velocities were small (typically < 19% of the mean along-channel velocity) and are neglected in this study.

3.2 Volumetric Flow

Volumetric flow was used as a metric to estimate the spatiotemporal variability of hydro-dynamics across the six measurement locations in the North and South Channels prior to the BUDM project. Volumetric flow Q is the product of flow cross-sectional area A and velocity V:

$$Q = AV \tag{1}$$

In the channels, A is not constant as the water level rises and falls with the tide. Instantaneous A can be derived from a constructed channel transect at sensor locations, "sliced" to obtain cross-sectional area corresponding to instantaneous water level relative to a datum (NAVD88). V is the measured along-channel velocity.

Three sources of information were used to define transects at N1–3 and S1–3 to compute the values of A: (1) SONAR survey of bathymetry, (2) satellite imagery for channel slope widths, and (3) walking survey measurements for channel widths and elevations. Figure 2 shows an overhead view of where different measurements were used. Measurement sources are also annotated in Fig. 3. (1) was carried out at a mid-high tide. Cross-channel depth was collected using a Remotely Operated Vehicle (ROV) fitted with a PPK GPS and SONAR (Fig. 2). Depth was sampled using SONAR at 30 Hz and corrected for slant range associated with the ROV's list angle. PPK GPS position was sampled at 1 Hz. Depths collected were assigned linearly interpolated positions by time. Depth was referenced to NAVD88 by subtracting depth and the vertical offset between the PPK GPS unit from measured PPK GPS elevation. (2) was sourced from Google Earth. (3) was a collection of on-land laser rangefinder measurements, bank inclination measurements, and positions and elevation from a PPK GPS. Total channel width was measured between the transitions from marsh platform to channel bank slope, defined by low and high vigor grass on either side of the transect. Slope was measured using an inclinometer on a stiff 2-m rod, laid over the slope. For all transects, a PPK GPS point on the interior side of the channels and approximate perpendicular azimuth created the defining transect line.

Fig. 2. A: Transect measurements and B: ROV driving transect.

4 Results

4.1 Transects' Cross-Sectional Area A for N1 and S1

Channel transects were completed at N1–3, S1, and S3. For brevity, N1 and S1 (channel mouths) are discussed here. Gaps in elevation data along transect widths are reflected in low, mid, and high area estimates of channel cross sectional area A. The low estimate was obtained by assuming the elevation profile rose vertically from the last known survey depth to the calculated bank height. The high estimate was obtained by dropping vertically from bank height to last known survey depth. The mid estimate assumed the profile was linear between bank height and last known survey depth. Figure 3 shows the completed transect profiles for locations N1 and S1, which were used to produce total area models to support volumetric flow estimates.

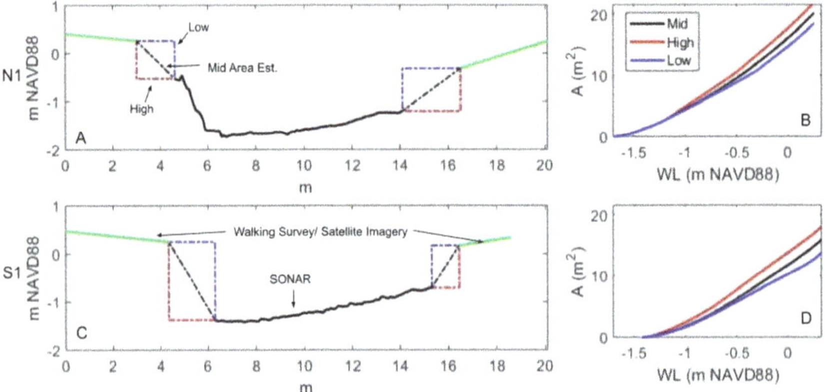

Fig. 3. A, C: N1 and S1 composite transects; B, D: Low, mid, and high area estimates for N1 and S1.

To calculate A with respect to water level, elevations were linearly interpolated every 5 mm along the transect width. Cross-sectional area corresponding to water level was calculated using a trapezoidal area estimates that ignored points above incrementing step elevations.

At overtopping elevation (the elevation where water would "spill" out of the channel) N1 low, mid, and high area estimates are 18.41 m², 20.17 m², and 21.36 m², respectively, whereas S1 low, mid, and high area estimates are 13.67 m², 15.71 m², and 17.81 m² respectively. For Q calculation, the mid area estimate was assumed for N1 and S1. When water level was above overtopping elevation (0.24 m and 0.34 m NAVD88 for N1 and S1 respectively), Q was calculated using the maximum mid area estimate.

4.2   Water Level, Velocity, and Volumetric Flow Q

Sample time series of water levels and along-channel velocities at N1 and S1 are shown in Fig. 4 (A, B, D, E), where a positive and negative velocities correspond with flood and ebb tides, respectively. Water levels were dominated by the M2 tide, with a tidal amplitude of 0.6 m. Time varying currents showed tidal asymmetry, with the transition from maximum flood to maximum ebb taking approximately 3 h and the transition from maximum ebb to maximum flood taking approximately 9 h.

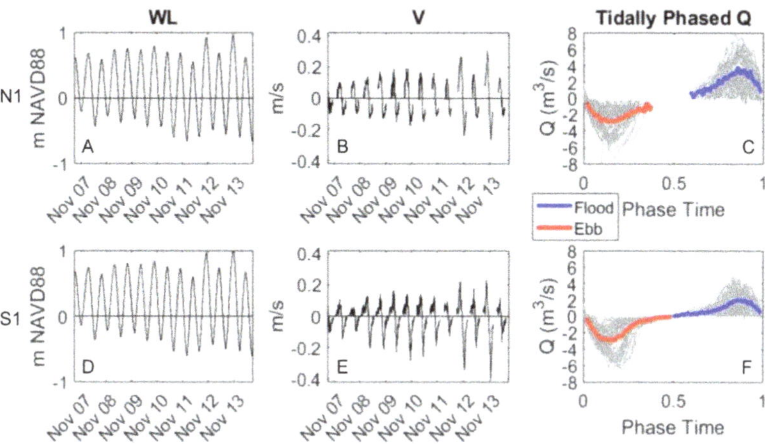

Fig. 4. Water level (WL), velocity (V) and, volumetric flow (Q) at N1 (A, B, and C) and S1 (D, E, and F)

Tidally phased averaged volumetric flows for Nov. 06 to Dec. 05, 2023 are shown in Fig. 4 (C, F). The tidal phase (high tide to high tide using water level data) for each tidal cycle was divided into 1% increments (for a 12.42-h M2 tidal phase, approximately 7.5 min), with individual tidal cycles shown in the gray curves. The red/blue curves are the result of phase averaging across tidal cycles. Maximum discharge and inflow occur when the channel is overtopped for both N1 and S1. The noticeable discontinuity in Q near mid phase (low tide) at N1 is a result of insufficient tilt meter submersion to resolve flow velocity.

5   Discussion and Conclusion

Vegetation, marsh platform instability, and periodic submersion of tidal channels presented unique challenges to obtaining the physical measurements necessary to compute cross-sectional area A of N1–3 and S1–3 transects. These challenges resulted in data gaps when combined with SONAR data. At overtopping elevation, high and low area estimates for N1 are within ±9% of mid area estimate, and for S1 within ±13%. N1 mid area estimate is 28% larger than S1 mid area estimate. Using area estimates, N1 and S1 exhibit volumetric tidal cycle asymmetry. Flood peak inflow exceeds ebb peak discharge in N1, while ebb exceeds flood at S1.

Although volumetric flow estimates rely on uniform flow assumptions, they can be used to understand baseline channel flow conditions pre-BUDM project. Water levels and velocities during and post BUDM were collected Sept. 2024 through Jan. 2025 at N1 and S1 (Fig. 1). Future work will compare pre-, during-, and post-BUDM placement periods of water level, velocity, and volumetric discharge to determine changes in channel hydrodynamics concurrent with BUDM. Additionally, exterior tidal and wind forcing will be time-correlated to water level and velocity in the channels. Such correlations can inform the dominant factors in high flow magnitude events, especially those with erosive potential for existing channels and applied dredge material.

Acknowledgements. The authors thank The Wetlands Institute staff and USNA Hydromechanics Laboratory staff for their assistance with instrument deployments. The project was supported by funding from the US Army Corps of Engineers through project number W25PHS32555958. Any opinions, findings, conclusions, or recommendations expressed in this material are those of the authors and do not necessarily reflect the views of USNA or the DoD.

References

1. Schuerch M, Spencer T, Temmerman S et al (2018) Future response of global coastal wetlands to sea level rise. Nature 561:231–234
2. FitzGerald DM, Hughes Z (2019) Marsh processes and their response to climate change and sea-level rise. Ann Rev Earth Planetary Sci 47(1):481–517
3. Piercy C, Welp T, Mohan R (2023) Guidelines for How to Approach Thin-Layer Placement Projects. ERDC/EL SR-23-4. Vicksburg, MS: U.S. Army Engineer Research and Development Center. p 179
4. Geng L, D'Alpaos A, Sgarabotto A, Gong Z, Lanzoni S (2021) Intertwined eco-morphodynamic evolution of salt marshes and emerging tidal channel networks. Water Resourc Res 57(11):e2021WR030840
5. USACE Philadelphia District. Scotch Bonnet Marsh Enhancement Project. https://www.nap.usace.army.mil/Missions/Civil-Works/Coastal-Dredging-Beneficial-Use/Scotch-Bonnet-Project/. Accessed 12 Jan 2025

Oyster Reefs as a Nature-Based Solution for Coastal Protection: An Experimental Study

Hani Ghasemi[1], Victoria Jakaus[2], Brian Rice[2], John O'Sullivan[1], Ali Dastgheib[3,4], Paul Brooks[2], and Md Salauddin[1(✉)]

[1] UCD Dooge Centre for Water Resources Research, UCD School of Civil Engineering, and UCD Earth Institute, University College Dublin, Belfield, Dublin 4, Ireland
`md.salauddin@ucd.ie`
[2] UCD School of Biology and Environmental Science and UCD Earth Institute, University College Dublin, Belfield, Dublin 4, Ireland
[3] Department of Coastal and Urban Risk and Resilience, IHE Delft Institute for Water Education, Delft, Netherlands
[4] IMDC, Antwerpen, Belgium

Abstract. Coastal regions, which are home to over 40% of the global population, are increasingly threatened by climate change effects, such as rising sea levels, storm surges and coastal erosion. Traditional hard engineering solutions, such as seawalls and breakwaters, help mitigate these risks but often lead to ecological degradation and require high maintenance. Nature-based solutions (NbS), such as oyster reefs, have emerged as sustainable alternatives that provide both ecological restoration and coastal protection. This study examines how the integration of oyster-covered artificial reef balls can attenuate wave energy under controlled laboratory conditions. We report on a comprehensive set of wave flume experiments, where we tested four different configurations: no reef balls, plain reef balls, and reef balls with low and high oyster densities. The results demonstrated a direct correlation between oyster density and wave attenuation. High-density oyster configurations reduced wave energy by up to 42%, significantly outperforming other setups, especially under high-energy wave conditions. The findings of this study highlight the potential of oyster reef balls for sustainable coastal protection and provide valuable insights for designing resilient coastal defense systems that can adapt to changing climate conditions.

Keywords: Artificial Reef Balls · Climate Change · Coastal Ecosystems · Ecological Engineering · Wave Attenuation

1 Introduction

Coastal regions play a crucial role in economic, ecological, and social systems, with over 40% of the global population living within 100 km of the shore [1]. However, infrastructure, habitats and communities in coastal areas are increasingly threatened by the impacts of global climate change, including rising sea levels, stronger storm surges, and coastal erosion. The Intergovernmental Panel on Climate Change (IPCC) emphasizes

C. Coelho et al. (Eds.): CD 2025, CRL 41, pp. 395–402, 2026.
https://doi.org/10.1007/978-3-032-15473-6_61

the urgent need for effective adaptation and mitigation strategies to safeguard coastal assets and populations [2]. Traditionally, hard-engineering solutions such as seawalls and breakwaters have been used to address these challenges. While they effectively reduce wave energy, they often lead to ecological degradation and require high maintenance [3, 4]. To this end, nature-based solutions (NbS), such as oyster reefs, seagrass beds, salt marshes, and mangroves, have emerged in recent years as promising alternatives for coastal protection. These solutions integrate ecological restoration with protective measures, offering benefits such as natural wave attenuation, enhanced biodiversity, and improved ecosystem resilience [5, 6].

Oyster reefs, in particular, have garnered attention for their multifaceted benefits. They not only mitigate wave energy and stabilize sediments but also improve water quality through filtration, supporting marine biodiversity [6, 7]. Artificial reef structures, such as reef balls, have been widely deployed in habitat restoration and coastal protection projects. These systems, when integrated with oysters, form hybrid solutions that deliver ecological and hydrodynamic benefits [8]. However, the influence of oyster density on wave attenuation remains underexplored, posing challenges to optimizing design for maximum efficiency.

Here we examine the effectiveness of artificial reef balls paired with oysters at different densities in a controlled laboratory setting. By simulating various wave climates and water depths, the research aims to measure wave attenuation performance across configurations with no oysters, low-density oysters and high-density oysters. The findings will contribute to the development of sustainable, nature-based coastal defense strategies and enhance discussions on ecological engineering and coastal management.

2 Experimental Setup

The experiments were conducted in a two-dimensional wave flume (20 m in length, 1 m in width, and 0.6 m in depth) located in the hydraulics laboratory in the School of Civil Engineering at University College Dublin, Ireland. The channel is equipped with a piston-type wave generator that can produce both regular waves, characterized by their periodic and uniform behavior, and irregular waves that simulate the stochastic and unpredictable characteristics of natural sea states. To ensure precise and comprehensive measurements of wave dynamics, two arrays of wave gauges, each consisting of calibrated sensors, were strategically positioned along the length of the flume. A schematic representation of the experimental setup is provided in Fig. 1.

The artificial reef balls used in this study were modeled after those deployed along the Irish coastline in Malahide, north Dublin. These reef balls are hemispherical structures made from marine-grade concrete, designed to mimic natural reef systems. Each reef ball has a diameter of 0.6 m and a height of 0.28 m, resulting in a total surface area of approximately 1.1 m2. The design features four circular openings, each 0.1 m in diameter, which are strategically positioned to enhance water flow and facilitate oyster colonization.

To replicate practical deployment scenarios, two reef balls were arranged longitudinally in the flume (Fig. 2). This setup simulates a multi-unit system typically used in coastal restoration projects, allowing the study to assess the cumulative impacts on wave attenuation.

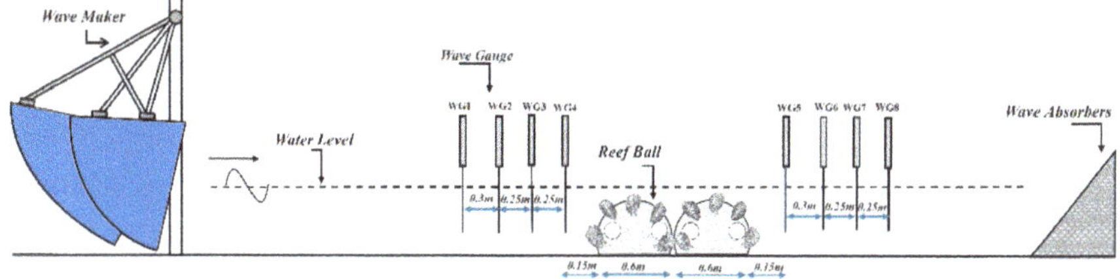

Fig. 1. Schematic of the test set-up

Four configurations were tested to evaluate the effect of oyster colonization density on wave attenuation:

- Baseline (No Reef Ball): A control condition with no structures.
- Reef Ball Only (R): Two reef balls without oyster coverage.
- Low-Density Oysters (R + L): Two reef balls sparsely populated with oysters.
- High-Density Oysters (R + H): Two reef balls densely covered with oysters.

Fig. 2. Configurations of reef balls with different oyster densities tested within this study

The oysters used in this study were whole specimens originally collected from Tralee Bay in County Kerry, Ireland. A total of 50 samples were selected, with a median size of 0.091 m in length, 0.06 m in width, and approximately 0.04 m in thickness. The density configurations of the oysters were based on the European Native Oyster Restoration Guidelines [9], representing varying stages of habitat development for the UK and Ireland. The tested configurations included:

- Low Density (R + L): 5–10 individuals/m^2;
- High Density (R + H): > 20 individuals/m^2.

The densities described reflect different stages of oyster habitat development, ranging from sparse clusters to dense, mature reef structures. When placing oysters, we took into account the surface area and openings of the reef ball to mimic the natural attachment patterns typically observed in both natural and restored habitats.

Incident significant wave heights (H_i) ranged from 0.04 m to 0.10 m, as detailed in Table 1. Wave periods were tested between 0.72 and 2.07 s to represent varying levels of wave steepness, while the water depth was consistently maintained at 0.30 m. All tests were performed using a JONSWAP wave spectrum with a peak enhancement factor of 3.3 for around 1000 waves. Experimental conditions were specifically designed to evaluate the effectiveness of reef ball configurations with varying levels of oyster attachment ranging from no oysters to low and high-density coverage under different wave heights, steepness.

Table 1. Incident wave conditions tested within this study

Experimental Configurations	Test Series	Water Depth (m)	Incident Wave Heights (m)	Wave Steepness	Periods (s)
Baseline (No Reef Ball)	1A	0.30	0.04, 0.06, 0.08, 0.10	0.015	1.31,1.6, 1.85,2.07
				0.05	0.72,0.88, 1.01,1.13
Plain Reef Ball (No Oyster Attachment)	2A	0.30	0.04, 0.06, 0.08, 0.10	0.015	1.31,1.6, 1.85,2.07
				0.05	0.72,0.88, 1.01,1.13
Reef Ball with Low-Density Oyster Cover	3A	0.30	0.04, 0.06, 0.08, 0.10	0.015	1.31,1.6, 1.85,2.07
				0.05	0.72,0.88, 1.01,1.13
Reef Ball with High-Density Oyster Cover	4A	0.30	0.04, 0.06, 0.08, 0.10	0.015	1.31,1.6, 1.85,2.07
				0.05	0.72,0.88, 1.01,1.13

3 Results and Discussions

The study analyzed wave attenuation performance across five experimental configurations: Baseline (No Reef Ball), Reef Ball Only (R), Low-Density Oysters (R + L) and High-Density Oysters (R + H). Wave attenuation was evaluated using the wave transmission coefficient (K_t), which is the ratio of transmitted wave height (H_t) to incident wave height (H_i), as shown in Eq. 1:

$$K_t = \frac{H_t}{H_i} \tag{1}$$

The results show the variation of the wave transmission coefficient (Kt) with wave period for reef ball configurations with different oyster densities under tested wave conditions (see Fig. 3). Under low wave steepness conditions (swell waves), the baseline configuration exhibited minimal wave attenuation approximately of 2%. In comparison, the plain reef ball (R), tested in Series 2A, showed a significant improvement with 14% attenuation. The addition of oysters further enhanced performance: low-density (R + L) configurations reached 17% and high-density (R + H) achieved 26%. For high wave steepness conditions (storm-like waves), the baseline configuration provided a modest 4% attenuation. The plain reef ball (R) attenuated around 28%, and the addition of oysters again demonstrated incremental improvements: Low-density (R + L) reached 30%, and high-density (R + H) achieved the highest attenuation of 42%. These results emphasize the capacity of reef balls integrated with densely populated oysters to dissipate wave energy, particularly under storm wave conditions.

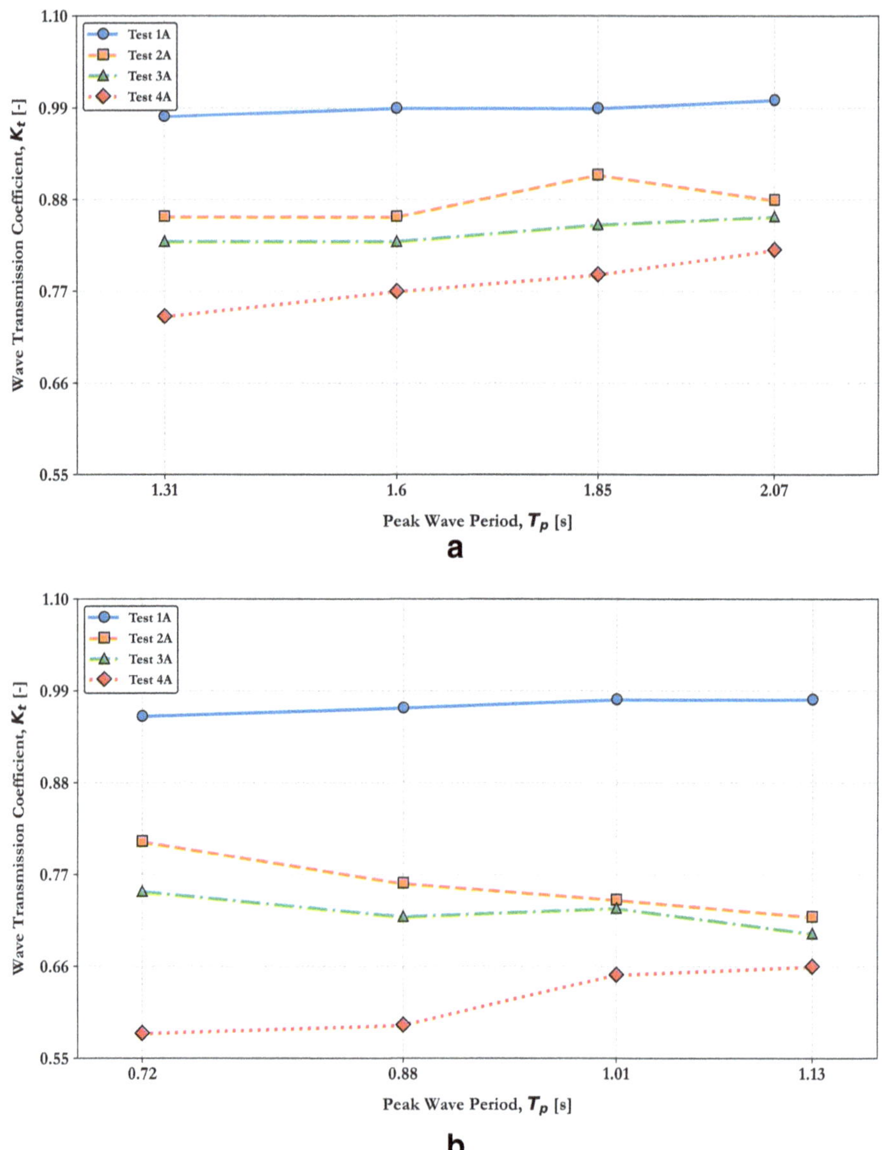

Fig. 3. Wave transmission coefficient (Kt) for reef ball configurations with different oyster densities: (A) swell waves and (B) storm waves

4 Conclusions

The results of this experimental study show that higher oyster densities significantly improve the ability of reef balls to reduce wave energy, especially under high-energy conditions. By integrating natural and structural elements, this approach not only supports marine ecosystems but also strengthens coastal protection. This simple and effective solution offers a sustainable way to manage and protect coastlines in the face of climate change.

Acknowledgements. This study was funded under the Irish Environmental Protection Agency (EPA) research program 2021–2030, through the research project "Enhancing Blue Carbon and Ecological Services through Nature-Based Solutions: Integrated Restoration in Irish Coastal Waters (2023-NE-1224)". The EPA research program is a Government of Ireland initiative funded by the Department of Communications, Climate Action and Environment. It is administered by the EPA, which has the statutory function of coordinating and promoting environmental research.

References

1. OECD (2021) Adapting to a changing climate in the management of coastal zones. OECD, Pari. https://doi.org/10.1787/b21083c5-en
2. IPCC. Intergovernmental Panel on Climate Change. https://www.ipcc.ch/. Accessed 13 Jan 2025
3. Salauddin M, O'Sullivan JJ, Abolfathi S, Pearson JM (2021) Eco-engineering of seawalls—an opportunity for enhanced climate resilience from increased topographic complexity. Front Mar Sci 8:674630. https://doi.org/10.3389/fmars.2021.674630
4. André C, Boulet D, Rey-Valette H, Rulleau B (2016) Protection by hard defence structures or relocation of assets exposed to coastal risks: contributions and drawbacks of cost-benefit analysis for long-term adaptation choices to climate change. Ocean Coast Manag 134:173–182. https://doi.org/10.1016/j.ocecoaman.2016.10.003
5. Temmerman S, Meire P, Bouma TJ, Herman PMJ, Ysebaert T, de Vriend HJ (2013) Ecosystem-based coastal defence in the face of global change. Nature 504(7478):79–83. https://doi.org/10.1038/nature12859
6. Beck MW et al (2011) Oyster reefs at risk and recommendations for conservation, restoration, and management. Bioscience 61(2):107–116. https://doi.org/10.1525/bio.2011.61.2.5
7. Grabowski JH et al (2012) Economic valuation of ecosystem services provided by oyster reefs. Bioscience 62(10):900–909. https://doi.org/10.1525/bio.2012.62.10.10
8. Perricone V, Mutalipassi M, Mele A, Buono M, Vicinanza D, Contestabile P (2023) Nature-based and bioinspired solutions for coastal protection: an overview among key ecosystems and a promising pathway for new functional and sustainable designs. ICES J Mar Sci 80(5):1218–1239. https://doi.org/10.1093/icesjms/fsad080
9. zu Ermgassen P et al (2021) European native oyster habitat restoration monitoring handbook. https://researchportal.port.ac.uk/files/49260381/European_Native_Oyster_Habitat_Monitoring_Handbook_WEB_Final.pdf. Accessed 19 Jan 2025

First Observations of Saltmarsh Cliff Erosion Mechanics During Extreme Storm Conditions in a Full-Scale Wave Flume

J. R. M. Muller[1]($\boxtimes$), B. W. Borsje[1], J. J. van der Werf[1,2], D. Dermentzoglou[3], A. Antonini[3], P. W. J. M. Willemsen[2,4], V. G. Mason[5,6], and S. J. M. H. Hulscher[1]

[1] University of Twente, Enschede, The Netherlands
j.r.m.muller@utwente.nl
[2] Deltares, Delft, The Netherlands
[3] Delft University of Technology, Delft, The Netherlands
[4] Wageningen University and Research, Wageningen, The Netherlands
[5] Royal Netherlands Institute for Sea Research (NIOZ), Yerseke, The Netherlands
[6] Utrecht University, Utrecht, The Netherlands

Abstract. Saltmarshes provide many vital ecosystem services, like wave dampening and carbon sequestration. These services are under pressure since saltmarshes are eroding worldwide. The erosion mechanics of saltmarshes are not well understood, especially during extreme storm conditions. Full-scale flume experiments were conducted on a real-life transplanted Dutch Wadden sea saltmarsh with a common brushwood dam protecting the marsh edge, to test the erosion resistance during extreme wave conditions. Little erosion occurred on the marsh platform during 40 h of exposure to various wave conditions, with most of the significant erosion occurring at the seaward edge of the marsh. Erosion was initiated by damaging and uprooting of the vegetation, exposing the substrate and leading to gradual erosion. Uprooting events frequency and erosion rates were higher during high wave conditions and water depths ($H_{m0} = 2$ m and $d_m = 4$ m). The brushwood dam did not protect the saltmarsh from eroding at the cliff. Our results demonstrate the erosion mechanics under extreme storms conditions and that uprooting may lead to instability of the saltmarsh on long-term.

Keywords: Saltmarsh · erosion · storm conditions · flume experiment

1 Introduction

Intertidal saltmarshes offer vital ecosystem functions, such as reduction of wave loads during storms, habitat for coastal species and carbon sequestration. However, many of these coastal saltmarshes are eroding globally [1]. Experimental and field-based observations on saltmarsh erosion have been made for short-term storm events time scale to decadal time scale (e.g. [2, 3]). An off-set in sediment dynamics and wave energy between the mudflat and saltmarsh can lead to cliff initiation at the marsh edge and subsequent lateral erosion of the saltmarsh. These cliffs can grow up to 0.5–1 m in

© The Author(s) 2026
C. Coelho et al. (Eds.): CD 2025, CRL 41, pp. 403–408, 2026.
https://doi.org/10.1007/978-3-032-15473-6_62

height, reaching the high marsh in meso- to macrotidal environments depending on the local tidal levels [4, 5]. The lateral retreat of the saltmarsh cliff is shown to be most severe when water levels and cliff elevations are similar, occurring during medium to strong storm events [6]. However, storm surge levels and wave heights in meso-tidal environments (e.g. the Wadden Sea, The Netherlands) are higher than the marsh bottom level. Detailed observations on the near-bed velocities, the interaction with vegetation and their root-systems and saltmarsh erosion mechanics under extreme storm conditions typical for the design of Dutch coastal protection (e.g. 10000 years return period) are missing.

In this research, we aim to identify the erosion mechanics of a saltmarsh cliff under extreme storm conditions and quantify the effect of soil characteristics and vegetation on the erosion rate. We do this through full-scale wave flume experiments on a real-life saltmarsh, simulating extreme wave conditions with a maximum significant wave height of $H_{m0} = 2$ m and a maximum inundation depth on top of the marsh of $d_m = 4$ m.

2 Methods

A series of wave flume experiments were performed in the 300 m long, 8 m wide and 9.5 m high Delta flume at Deltares. The physical model represented a typical mudflat-saltmarsh-sea dike profile (Fig. 1). A foreshore was made of non-erodible sand cement with a gentle slope ranging from 1:9 to 1:45. A real-life saltmarsh was created by harvesting blocks (2 m long, 2 m wide and 0.7 m deep) at a field site at the Dutch Wadden sea coast during Autumn. Blocks were harvested from a high marsh at both a landward and at a seaward location, with the seaward location having a higher sand fraction. The dominant vegetation species at the site was *Elymus athericus*. The blocks were placed next to each other in the flume to form two saltmarsh transects with a total length of 70 m. The first block was placed on top of the foreshore, creating a ca. 0.6 m high saltmarsh cliff. Initially, the cliff was protected by a brushwood dam during 39 experiments. The brushwood dam was removed during the final 3 experiments to test cliff erosion. At the end of the model, a non-erodible dike was built with a slope of 1:3.6, typical for a Dutch coastal sea dike.

Wave scenarios ranged from intermediate to extreme storm conditions. Wave-states with offshore significant wave heights ranged between $H_{m0} = 0.75$ to 2.0 m with a peak wave steepness ranging between $s_0 = 0.02$ and 0.04. The offshore water depth ranged between $d_0 = 4.4$ to 6.9 m, corresponding to a water depth on top of the marsh of $d_m = 1.5$ to 4.0 m, respectively. In total the test setup was exposed to various wave forcings for 40 h. During the experiments, the vegetation was gradually damaged. To measure the effect of a reduced standing biomass on the erosion rate of the saltmarsh, the experimental program was repeated for different vegetation states, categorized as 'healthy', 'damaged', and finally 'removed' by mowing the remaining vegetation.

Saltmarsh surface elevation measurements were taken with FARO 3D laser scanner between wave scenarios when the flume was drained. Erosion volumes were calculated by aligning and subtracting elevation scans. In this analysis, surface scans measured after removing the vegetation are used to avoid measurement errors due to standing vegetation covering the actual saltmarsh bottom (ca. last 20 h of exposure). Wave height and period were measured by wave gauges at both offshore and halfway at the marsh (x = 40 m) locations.

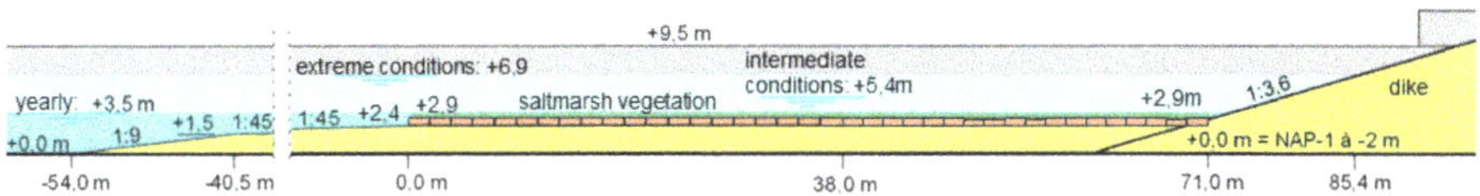

Fig. 1. Schematic cross view of the saltmarsh model in the Delta flume (courtesy of Deltares).

3   Results

During the experiments, when the standing vegetation and root systems were still intact, no significant erosion of the marsh platform was observed (Fig. 2). However, as the experiment progressed, the cumulative wave forcing lead to gradual damaging and subsequent removal of the above-ground vegetation. Slow removal of cohesive soil in between the roots led to less attachment of the vegetation to the soil. Once enough material was removed in combination with strong wave scenarios ($H_{m0} = 2$ m), large areas of vegetation were uprooted (ca. 1×1 m^2) exposing the soil beneath (Fig. 2, Fig. 3). Uprooting of the vegetation occurred sporadically along the saltmarsh both on the seaward and landward harvested transects. After removal of the roots, surface erosion rates were observed higher compared to sections with roots still intact.

Most significant erosion was measured in the first 20 m from the cliff edge (Fig. 2). Large scour holes developed at the front of both transects, contributing the most to the overall erosion volumes. The presence of the brushwood dam did not prevent the erosion of the saltmarsh cliff. The cliff continued to erode more rapidly and the saltmarsh started to erode vertically. After removal of the brushwood dam, the initial straight cliff was further scoured into a smooth sloped surface (Fig. 3).

Erosion rates after vegetation removal were further analyzed for the first 20 m from the cliff (Fig. 4). The total erosion between the experiments after mowing of the vegetation and the end of the experiment was limited, ranging between 1 to 1.5 m^3 during a total of 20 h of wave forcing. Slightly higher erosion raters were measured at the seaward harvested transect (i.e. more sandy). Higher erosion rates were measured after experiments with higher wave conditions. An increase in wave steepness did not show a significant change in erosion rates. The increase in erosion rates due to the removal of the brushwood dam is lower compared to the effect of a higher wave height (Fig. 4, hours 8 to 20).

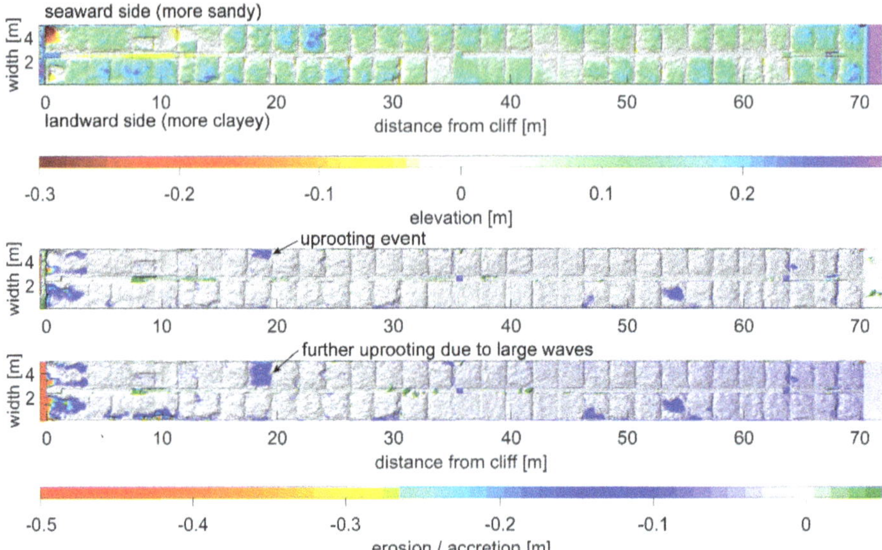

Fig. 2. Initial profile after mowing of the vegetation (top) with the upper transect ($y = 4$ m) harvested landward (i.e. higher clay fraction) and the lower transect ($y = 2$ m) harvested seaward (i.e. higher sand fraction) and the change in bathymetry after 4 h of additional testing (middle) and at the end of the experiment after 20 h of testing (bottom).

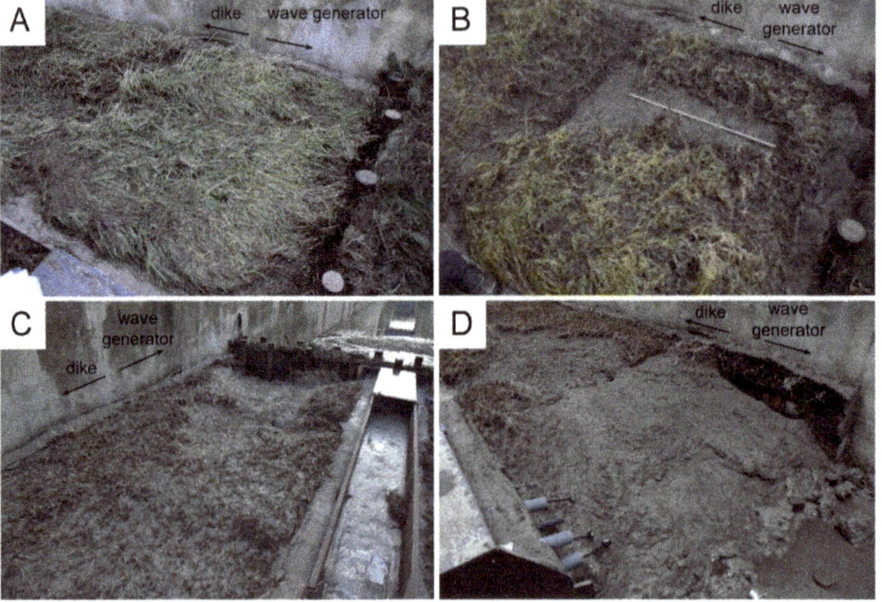

Fig. 3. Erosion at the saltmarsh cliff at the start of the experiment (A), after 17 h of testing (B), after 24 h of testing and mowing of vegetation (C) after removal brushwood dam and the end of the test program (D) (courtesy of Deltares)

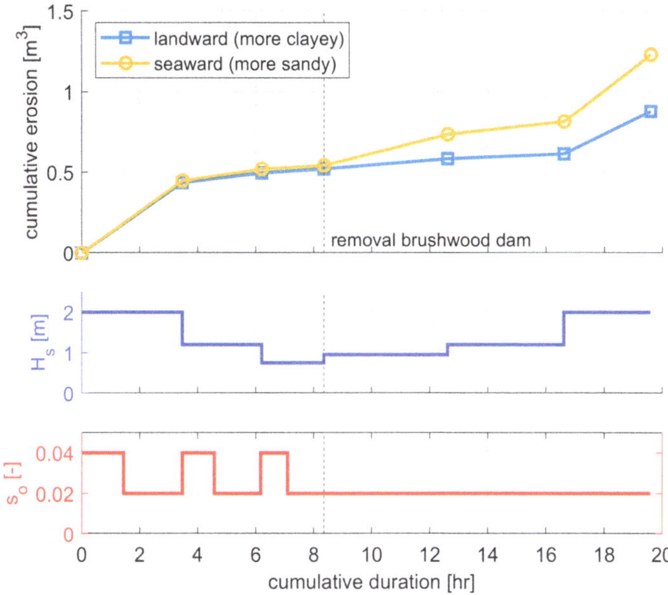

Fig. 4. Erosion volume of first 20 m saltmarsh after removal vegetation, the offshore significant wave height (H_{m0}) and wave steepness (s_0) over the cumulative duration of exposure in hours.

4 Discussion and Conclusion

Limited erosion was measured for both landward and seaward harvested saltmarsh transects during over 40 h of exposure to extreme wave conditions. Previous research showed similar limited erosion during wave conditions up to $H_{m0} = 1$ m [7]. Our results extend these observations to wave conditions up to $H_{m0} = 2$ m and inundation depth up to 4 m. Most of the significant erosion occurred at the first 20 m of the marsh transect. Large scour holes emerged at the first block behind the brushwood dam, contributing the most to the overall measured erosion volumes. The removal of the brushwood dam showed a small increase in erosion rates (Fig. 4, between hour 8 to 12), where higher wave heights were needed to further erode the more consolidated exposed substrate at this location (Fig. 4, hour 16 to 20).

Under high waves, strong near-bed velocities led to continuous damaging of the vegetation and eroding soil particles from the root mat. Only after sufficient detachment of the root mat and the rerunning of high wave conditions (i.e. $H_{m0} = 2$ m), large areas of vegetation roots were subsequently ripped out. These observations suggest the existence of a threshold near-bed velocity for uprooting to take place. Similar observations of uprooting have been made after hurricanes making landfall at coastal wetlands [8]. The removal of the root mat and the loss of the complete vegetation (above and below-ground biomass) means that vegetation will need several seasons to grow back to its former state. With the exposed substrate eroding at a faster rate, uprooting during extreme events can leave locations on the saltmarsh vulnerable for erosion during subsequent lower storm conditions.

Our results demonstrate the irreversible effect of extreme storm conditions on saltmarshes. Although short term erosion is limited, severe uprooting causes a weaker substrate, which takes time to be vegetated and regain its former strength. Further linking these observations to hydrodynamics and soil properties will allow for a better quantification of saltmarsh erosion under extreme wave conditions.

Acknowledgements. This work is part of the 'Living Dikes – Realising resilient and climate-proof coastal protection' project NWA.1292.19.257, funded by the Netherlands Organisation for Scientific Research (NWO). The authors want to thank M. Klein Breteler, V. van Bergeijk, G. Mourik and all lab technicians from Deltares for the management, design, construction and measurements during these experiments and support in post-processing of the data.

References

1. Deegan L et al (2012) Coastal eutrophication as a driver of salt marsh loss. Nature 490:388–392
2. Francalanci S, Bendoni M, Rinaldi M, Solari L (2013) Ecomorphodynamic evolution of salt marshes: experimental observations of bank retreat processes. Geomorphology 195:53–65
3. Feagin RA, Lozada-Bernard SM, Ravens TM, Möller I, Yeager KM, Baird AH (2009) Does vegetation prevent wave erosion of salt marsh edges? Proc Natl Acad Sci 106(25):10109–10113
4. Allen JRL (1989) Evolution of salt-marsh cliffs in muddy and sandy systems: a qualitative comparison of British West-Coast estuaries. Earth Surf Process Landforms 14:85–92
5. Van der Wal D, Wielemaker-van den Dool A, Herman PMJ (2008) Spatial patterns, rates and mechanisms of saltmarsh cycles (Westerschelde, The Netherlands). Estuar Coast Shelf Sci 76:357–368
6. Leonardi N, Ganju NK, Fagherazzi S (2016) A linear relationship between wave power and erosion determines salt-marsh resilience to violent storms and hurricanes. Proc Natl Acad Sci 113(1):64–68
7. Spencer T et al (2016) Salt marsh surface survives true-to-scale simulated storm surges. Earth Surf Process Landforms 41:543–552
8. Priestas AM, Mariotti G, Leonardi N, Fagherazzi S (2015) Coupled wave energy and erosion dynamics along a salt marsh boundary, hog Island Bay, Virginia, USA. J Marine Sci Eng 3(3):1041–1065

Enhancing Resilience and Biodiversity: Adaptation Measures in the Elbe Estuary

Johannes Pein[(✉)] and Joanna Staneva

Institute of Coastal Systems Modelling and Analysis, Helmholtz-Zentrum Hereon,
Max-Planck-Street, 1, 21502 Geeshacht, Germany
`johannes.pein@hereon.de`

Abstract. Estuaries are among the most biologically productive and intensively used marine areas in the world. Their diversity and the number of interrelationships and conflicts of use make their sustainable management difficult. In the face of climate change, however, "business as usual" scenarios are becoming less and less viable. This is particularly true for the Elbe estuary of the North Sea, which is the lifeline of the Hamburg metropolitan region and the Port of Hamburg. Although scientists and stakeholders recognise the need for action in the face of high environmental and economic costs and increasing flood risks, there is no widely accepted strategy for a sustainable, resilient future for the estuary. Based on stakeholder workshops and publications, as well as the scientific literature, we have analysed several adaptation scenarios. We adopt the strategy of adapting the estuary geometry and use a numerical model for cross-sectoral analysis of system responses.

Keywords: Estuaries · Human intervention · Climate change · Sustainable adaptation · Resilience. (Maximum 5 words)

1 Introduction

Estuaries have been attractive to settling and economic use for time immemorial thanks to their high biological productivity and convenient accessibility. Acting as a vital connection between inland regions and the ocean, they serve as ideal transport routes. These advantages have led to estuaries being modified by humans for their own purposes since at least the Middle Ages to optimise navigability and other economic uses, therefore causing profound changes in the physical and ecological characteristics. Today, these transformed landscapes fulfil both economic and ecological roles, yet balancing these interests remains a formidable challenge, often prioritising economic gains at the expense of ecological integrity.

Recent decades have witnessed unprecedented climate and environmental changes, prompting widespread acknowledgment among scientists, policymakers, and society that current trajectories are unsustainable. Of particular concern is ongoing and projected sea level rise (SLR), necessitating a critical reassessment of past interventions to devise sustainable adaptation options for the future. While many estuaries of the North Sea are

C. Coelho et al. (Eds.): CD 2025, CRL 41, pp. 409–415, 2026.
https://doi.org/10.1007/978-3-032-15473-6_63

under the combined pressure of climate change and human activities, the Elbe estuary stands out due to its centuries-long history of human intervention and notably high eutrophication load. Historical disputes over river engineering privileges date back to the 16th century, and since the 1970s, regular channel deepening has accelerated physical transformations and sparked extensive scientific and societal debates. Consequently, what was once Europe's largest inland delta with weak tides has evolved into an almost macrotidal waterway strongly influenced by marine forces. The upper end of the Elbe estuary, where a tidal weir functions as a total reflector to the tidal wave, receives high levels of algae from the Middle Elbe, which is considered the federal waterway with the highest algae concentrations in Germany.

The concept of estuaries as socio-ecological systems (SES) is complex and still in its early stages of definition [1]. In contrast to fisheries SES, it is challenging to define resource units for estuaries. Although fishing continues in the Elbe estuary, the resource that is actually the water surface or water volume (space) of free-flowing brackish water. There is a consensus among scientists and stakeholders that the system is in an unsustainable state. The stakeholder grouping has not yet reached consensus on sustainable adaptation measures. The so-called 'Forum Tide-Elbe'[1], which discussed a dozen adaptation options such as the creation of polders or the opening of the historic 'Alte Süderelbe' arm of the river, has achieved some public impact. However, ultimately, the proposed measures were not pursued due to cost-effectiveness and efficiency concerns in the eyes of the participants. This can be attributed to the selection and size of the measures, which were opportunistic and did not exceed a maximum size of 150 ha. In contrast, [2] calculated that approximately 4000 ha of shallow water area would need to be connected to the harbour area in order to mitigate the summer oxygen minimum.

The SES 'Elbe Estuary' is at an impasse. Human interventions in the system have taken place over centuries. The deepening campaigns in recent decades have not been compensated for by designated river engineering measures, although compensation measures are a long-accepted principle in other infrastructural sectors. The governance system is not capable of producing sustainable regulation. However, stakeholder experts are aware of suitable adaptation strategies at a fundamental level [3]. Drawing upon these principles and knowledge we have developed four potentially efficient sustainable adaptation options that largely transcend the concrete adaptation options that have been put forward so far. To this end, we developed two adaptation options based on the principle of reducing tidal currents and two based on the principle of increasing the tidal prism. We implemented these in the numerical model of the Elbe estuary and comparative simulations were carried out under present and future projected conditions. In the following section, we will describe the modelling approach and the model areas, including the adaptation options. Subsequent sections will discuss the simulation results and a final evaluation of the results.

---

[1] https://www.forum-tideelbe.de/

2 Methods

2.1 Model

Researchers have successfully employed the semi-implicit cross-scale hydroscience integrated system model [4] to model both idealised and realistic estuarine domains, addressing research questions in hydrodynamics, sediment dynamics and ecology [5, 6, 7]. The model reliably predicts water level, horizontal currents, vertical exchange and tracer transport [8]. In the coupled configuration used in this study, the SCHISM predicts the transport and diffusion of ecosystem state variables and provides salinity, water temperature and bottom stresses to the ecosystem module [7].

2.2 Forcing

The forcing data comprise historical hydrodynamic, atmospheric and hydrological reanalyses and observations. To force the Elbe estuary model we use an intermediate model domain of the southern North Sea that is driven by CMEMS AMM7 hydrodynamics [9] (CMEMS product id: NWSHELF_MULTIYEAR_PHY_004_009). Elbe river discharge and nutrient load data were obtained from public sources and in-house measurements of Helmholtz-Centre Hereon (see details in [8]). Atmospheric forcing is represented by Cosmo-EU model reanalysis provided by the German Weather Service [10]. Model calibration and validation under historic forcing conditions of 2012–2013 is described in [8]. In order to benchmark the sustainable adaption options against the major effects marine climate change, we have expanded the current conditions to a climate experiment with increased sea levels and higher global temperatures, simulating conditions expected under an extreme climate scenario like RCP8.5. This is achieved by adding by increasing sea levels by 1 m at the open boundaries of the forcing model. Temperatures are increased by 4 °C at the boundaries. In the remainder we label the climate experiment as year "2100".

2.3 Model Domain and Adaptation Options

The reference model domain convers the outer estuary which has part of the German Bight and Elbe estuary proper until the tidal weir at Geesthacht. The Elbe estuary model has been calibrated and validated for forcing conditions of 2012–2013, for details see [8].

The sustainable adaptation options assessed in the paper are as follows:

- A) Adding Middle Elbe river (AME): This option includes removal of the present tidal weir which is 'reconstructed' as an active tidal weir that partially closes upon flood in the mid-estuarine section (km 695, see Fig. 1). The reference domain is enlarged by adding part of the Elbe river topography (Fig. 1a).
- B) Adding Süder-Elbe (ASE): This option keeps the present tidal weir but includes the reconnection of an historic channel branch (Fig. 1b).
- C) Polder (POL): Construction of a large (4800 ha) flood polder downstream of the port of Hamburg connecting to the estuarine main channel via the lower part of the Alte Süderelbe channel.

- D) *Leitdamm* (LDAM): A training wall is constructed to shield the mouth of the Elbe estuary from the flood arriving from the western direction in the German Bight.

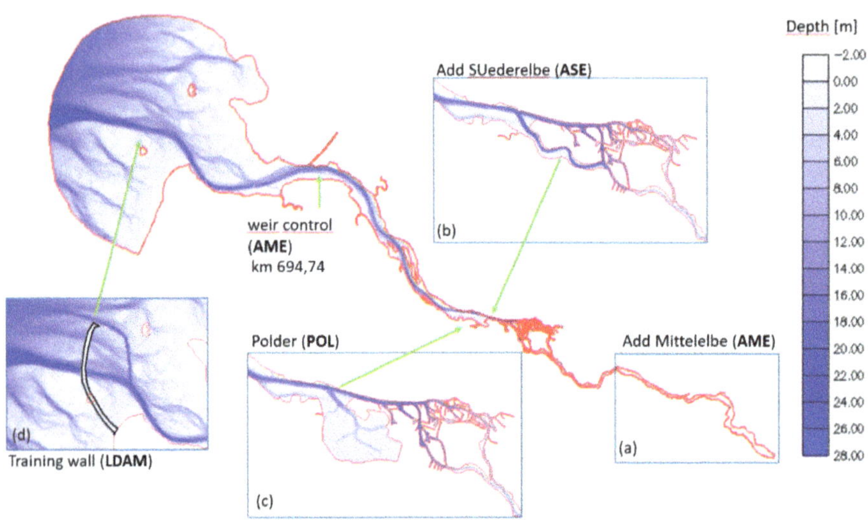

Fig. 1. Elbe estuary model domain bathymetry (center) with zoom-in views showing sustainable adaptation options. This image has been adopted from Fig. 1a in Pein et al. (2025), https://doi.org/https://doi.org/10.1016/j.ocemod.2024.102467.

3   Results and Discussion

The general functioning of the coupled physical-biogeochemical dynamics is given in [8]. Applying 1m of SLR and a 4 °C warming the open boundaries of the forcing model results in warming and increase of salinity intrusion in the Elbe estuary model domain (Fig. 2a, b). The increased water depth and subsequently increased celerity of the tidal wave and tidal pumping [11] lead to increased suspension of particulate organic nitrogen/phytodetritus in the port region and the shallow upper estuary (Fig. 2c). This region is consequently particularly impacted by decreasing bottom oxygen levels against the background of globally decreasing oxygen levels owing to the increase of biogeochemical process rates under the temperature increase (Fig. 2d).

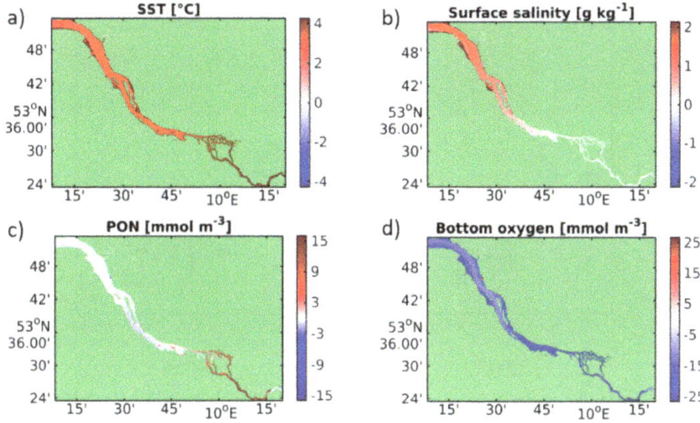

Fig. 2. Impact of sea level rise and warming in the Elbe estuary.

The proposed sustainable adaption options reveal different resilience potential under the climate change experiment. Here we focus on the statistical distribution of a major driver of oxygen depletion, the phytodetritus/particulate organic nitrogen (PON) and dissolved oxygen in the deepened freshwater section of the Elbe estuary including the port of Hamburg (Fig. 3). In the reference year of 2012, the reference geometry ("business-as-usual"), the AME and LDAM adaption scenarios reveal the highest median concentrations of PON (Fig. 3a). Under the future scenario, the PON concentration increase in the reference domain while they decrease in the AME and LDAM scenarios. The ASE and POL scenarios show no significant changes. These results indicate that the latter scenarios largely buffer the change of kinetic forcing due to SLR.

Median oxygen levels equal ~ 220 mmol/m3 in the reference run in the historic period with ASE and LDAM adaptation scenarios being close to that value (Fig. 3c). AME scenarios shows much lower oxygen levels with median less than 180 mmol/m3. The POL scenario reveals median oxygen levels of about 230 mmol/m3. This means that even under historic forcing conditions the ensemble of adaptation scenarios diverges clearly from the default geometry, with the active flood control scenario revealing undesirable response from the ecological viewpoint.

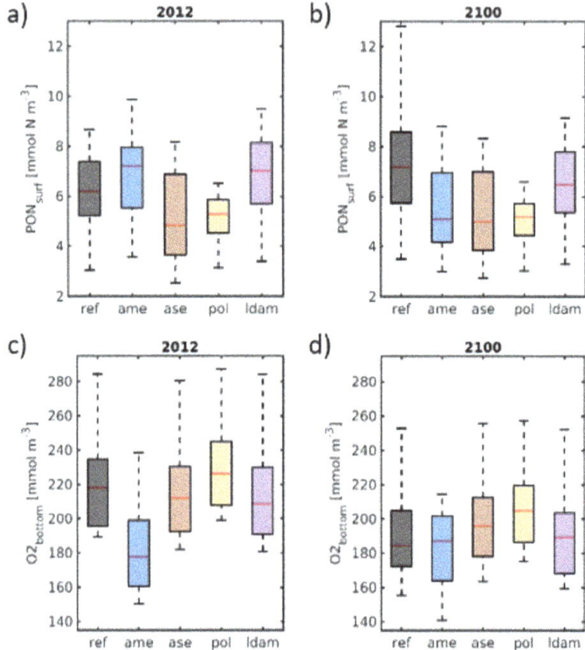

Fig. 3. Response to climate change experiment of reference case and sustainable adaptation scenarios (2).

The climate experiment leads to a general deterioration of oxygen levels in the reference case and LDAM scenario while it remains at low levels in the AME scenario (Fig. 3d). ASE and POL scenario reveal the highest dissolved oxygen levels demonstrating resilience to the major impacts of climate change, i.e. global warming and SLR.

4 Conclusions

We used a coupled physical-biogeochemical numerical model of the Elbe estuary to study the prospect of sustainable adaptation of the estuarine geometry under historic forcing conditions and under a climate change experiment. The business-as-usual scenario and scenarios aiming to reduce flood currents demonstrated the least ability to preserve oxygen levels under climate change. Adaptation options following the principle of increasing tidal prism showed better ability to cope with impacts of climate change revealing the highest oxygen levels of the model ensemble. This leads to the conclusion that reduction of the tidal forcing may have adverse consequences in a deepened and eutrophied estuary because the water column is deprived of the tidal currents mixing oxygen from the surface towards the bottom. Increasing the intertidal area or channel diameter in the upper estuary, i.e. enhancing tidal prism, improves local mixing locally in the area where low oxygen levels typically occur. It seems thus that this kind of adaptation would be sustainable in the face of climate change. In a parallel study we have demonstrated that the "passive" or "nature-based" adaptation options decrease also tidal range and siltation in the port area [12].

References

1. Author F (2016) Article title. Journal 2(5):99–110
2. Author F, Author S (2016) Title of a proceedings paper. In: Editor F, Editor S (eds) CONFERENCE 2016, LNCS, vol 9999. Springer, Heidelberg, pp 1–13
3. Author F, Author S, Author T (1999) Book title, 2nd edn. Publisher, Location
4. Author F (2010) Contribution title. In: 9th International Proceedings on Proceedings, pp 1–2. Publisher, Location
5. LNCS Homepage. http://www.springer.com/lncs. Accessed 21 Nov 2016

Blue Transitions in the Black Sea: Multi-actor Forums to Advance a Sustainable Blue Economy

Lydia Papadaki[(✉)], Ebun Akinsete, and Phoebe Koundouri

Sustainable Development Unit, Athena RC, School of Economics and ReSEES Research Laboratory, Athens University of Economics and Business, UN SDSN Global Climate Hub, Athens, Greece
lydia.papadaki@athenarc.gr

Abstract. The Blue Economy, encompassing coastal, oceanic, and sea-related economic activities, is crucial for sustainable global development. The Black Sea, located between Asia and Europe, has significant potential for expansion. DOORS Black Sea, an EU-funded initiative, aims to revitalize the Black Sea by fostering "blue economy" opportunities through collaboration among industry, academia, and local communities, addressing climate change and human activities' effects on the marine ecosystem. Multi-Actor Forums (MAFs) facilitate the collaboration of diverse national stakeholders from Georgia, Romania, Bulgaria, and Turkey in order to assist scientists in the prioritisation of Black Sea issues, with an emphasis on innovations to address gaps and blue economy policies. This method also contributes to the co-design of the region's System of Systems, which provides the necessary datasets for researchers to address environmental challenges and advance the blue economy. The results from the first round of MAFs show the sectors which should be prioritized in the Black Sea and the most significant challenges per country that need to be put at the forefront of the public dialogue.

Keywords: Multi-Actor Forums · Living Labs · Co-creation · Blue Economy · Black Sea

1 Introduction

Approximately half of the European populace resides within 50 km of the sea, with a majority of the population residing in urbanizations along the coast (Eurostat, 2011). The Sustainable Blue Economy, which establishes the conditions for sustainable governance and implements programs and instruments, significantly influences the prosperity of the Black Sea by encompassing the sustainable utilization of marine and aquatic resources. (European Commission, 2022). The Black Sea Blue Economy, encompassing aquaculture, tourism, shipping, and fisheries, has the potential to generate significant development, employment, and innovation. However, it relies on sustainable marine resource management. The region faces political, socio-economic, and environmental challenges, threatening its economic potential and biodiversity. Regional initiatives align with European Green Deal objectives to transition to a sustainable Blue Economy.

© The Author(s) 2026
C. Coelho et al. (Eds.): CD 2025, CRL 41, pp. 416–423, 2026.
https://doi.org/10.1007/978-3-032-15473-6_64

The Common Maritime Agenda (CMA) for the Black Sea and the Black Sea Strategic Research and Innovation Agenda (SRIA) were two flagship strategies that were introduced in 2019 with the objective of enhancing sustainability and enhancing governance and cooperation in the Black Sea (Connect Black Sea, 2019; European Commission, 2019). SRIA has four central aspirations for the Black Sea, including addressing the fundamental research challenges, developing products, solutions, and clusters, constructing critical support systems and research infrastructures, and fostering education and capacity development. CMA is in pursuit of healthy marine and coastal ecosystems, a competitive, innovative, and sustainable blue economy for the Black Sea, and the promotion of investment in the Blue Economy of the Black Sea (Connect Black Sea, 2023). The Black Sea SRIA's Implementation Plan was officially unveiled in 2023. Conversely, the CMA has been actively implemented since its adoption, with annual stakeholder conferences and meetings to advance its objectives.

Living Labs, as defined by ENoLL are *"open innovation ecosystems in real-life environments based on a systematic user co-creation approach that integrates research and innovation activities in communities and/or multi-stakeholder environments, placing citizens and/or end-users at the centre of the innovation process"* are particularly well suited to anchor this process in real word systems (ENoLL, 2024). Living labs are centered around stakeholders, who are broadly defined as all parties that are actively involved in or influenced by the innovation processes facilitated by the Living Lab environment. They are involved in co-creation processes that promote innovation that is specifically designed to address real-world requirements and challenges. (Schuurman et al., 2015). Using the Quadruple Helix Approach, the stakeholders who participate in a Living Lab are selected from the following categories: Industry and Business; Government and Policy makers; Civil Society, NGOs & Associations; and Research and Academia. They work together to ensure that solutions are both socially pertinent and technologically feasible (Hossain et al., 2019; Schäpke et al., 2018). This multi-stakeholder engagement improves the efficacy of Living Labs by harmonizing disparate perspectives, leveraging local knowledge, and addressing specific community priorities in a participatory manner (Schäpke et al., 2018).

The living labs can be used for the co-development of innovation pathways towards sustainability, taking into account local priorities and needs. This present paper is looking at how a certain type of Living Labs, the Multi-Actor Forums (MAFs), are used to support the transition towards a sustainable Blue Economy in the Black Sea through the coordinated implementation of 4 distinct workshops in four Black Sea countries, namely, Bulgaria, Georgia, Romania and Turkey, with a common goal of prioritizing Blue Economy sectors in each country and identifying the challenges that are associated with these sectors.

2 Methodology

Stakeholder mapping is a crucial process that identifies and analyzes individuals, groups, and organizations with a vested interest or influence in the DOORS project. It aids in developing communication strategies, prioritizing engagement efforts, and fulfilling their needs. Stakeholder mapping was conducted for each MAF country individually from July to September 2022, identifying stakeholders from the Quadruple Helix operating within 17 established and emerging industries, as defined by the (European Commission, 2021). Stakeholders were mapped and analyzed using an 'influence/interest' matrix (Fig. 1), where 'influence' refers to the stakeholder's power and capacity to effect change and 'interest' refers to their likelihood to engage in activities relevant to the case study focus. The map was validated by an external expert. The core group selected for workshops was those with significant influence and interest, while the lower right and upper left quadrants included those with low interest and influence. This methodology helps identify stakeholders who are most relevant to the study and are more likely to be involved in research endeavors. Identifying stakeholders willing to dedicate time and effort to the research process is also possible through the examination of 'Interest' (Brugha & Varvasovszky, 2000; Mendelow, 1981).

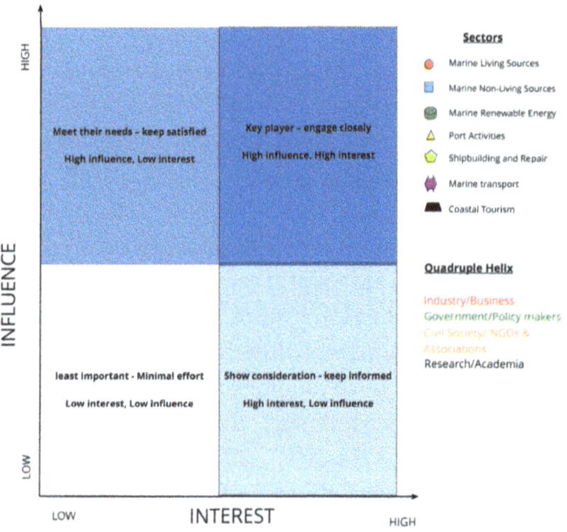

Fig. 1. Influence-Interest Matrix

The principal objective of the first round of MAFs was to establish Blue Economy industries as national priorities and to map the needs and challenges for each priority sector using the PESTLE framework. During the workshops, an overview of the DOORS project was delivered to the participants, encompassing both its own objectives and those of the MAF. The MAFs were comprised of two core exercises. In the first, the "Established" and "Emerging" BE sectors, defined by (Caribbean Development Bank, 2018), were used to set national priorities in each of the four BS countries. The second exercise employed the PESTLE (Political, Environmental, Social, Technological, Legal, Economic) approach to identify and analyze the challenges that are associated with the top five prioritized sectors. This activity incorporated their perspectives and information to enhance our understanding of the gaps and needs.

3   Results and Discussion

The Blue Economy priority sectors for the Black Sea, including fishing and marine research and development, are based on the results of the 1st MAF for Bulgaria, Georgia, Romania, and Turkey (Table 1). The region accounts for 37.4% of the complete FAO-designated fishery territory, with Bulgaria and Romania having fishery interests in the Black Sea. In 2019, revenues and GVA were expected to reach €10.5 million and €7.4 million, respectively (European Commission, 2022). The Black Sea region is experiencing an increase in tourism and sustainability concerns, including inadequate visibility, limited accessibility during high demand periods, and environmental strains. The Tourism 4.0 for the Black Sea initiative, co-financed by the European Maritime and Fisheries Fund, aims to equip local stakeholders with the necessary information to understand current trends, visitor impact, and movement patterns. The marine and coastal tourism sectors in Bulgaria, Georgia, and Romania generate substantial value, while Georgia, Turkey, and Romania are experiencing an increase in marine aquaculture cultivation. Aquaculture productivity in the Black Sea region has steadily increased over the past few years, with the region producing over 700,000 tonnes of farmed seafood in 2019 (FAO, 2023).

Romania and Turkey have recognized Ocean Renewable Energy as a critical economic sector, with progress being made in the commercial viability of floating offshore wind in deep waters and hazardous environments. This technology has the potential to unlock previously untapped markets in the Mediterranean Sea, Black Sea, and Atlantic Ocean. Meanwhile, Georgia and Bulgaria have designated maritime and port sectors as areas of paramount significance, with the transport sector contributing an average of 145 million euros to their GDP over the past two decades (The World Bank, 2020; Trading Economics, 2023). BlueInvest, a EU-financed initiative, aims to enhance investment and innovation in sustainable technologies for the blue economy by providing financial readiness and access to early-stage enterprises, small and medium-sized enterprises (SMEs), and scale-ups. These results align with the Common Maritime Agenda for the Black Sea and the Black Sea Strategic Research and Innovation Agenda, which define the roles of these two sectors in the Blue Economy. The SRIA supports food systems research in fields such as biotechnology, fisheries, recruitment, stock assessment and sustainability, and marine protected areas (MPAs). However, shipbuilding and offshore oil/gas, which are prioritized in the MAFs but underemphasized in the CMA and SRIA, are considered critical industries for regional economies and should be included in these approaches.

The initial round of MAFs has also highlighted the challenges faced by the region, including environmental degradation and pollution (Table 2). The four Black Sea states face significant obstacles at the national level, such as lack of investments in coastal infrastructure, inadequate renewable energy, and insufficient funding for scientific research. Georgia faces issues with marine litter and waste management, lack of a beach quality award system, and insufficient funding for research. Turkey and Romania prioritize marine litter and waste surveillance and control, with Romania emphasizing digitalization. The need for interdisciplinary, international, and institutional collaboration is crucial for promoting the adoption of the Blue Economy in the Black Sea region.

Table 1. Black Sea Blue Economy sectors prioritization

Sector	Definition	BE Sector	Black Sea countries
Capture Fisheries	The practise of obtaining naturally occurring living resources in both freshwater and marine environments in a sustainable manner	Established	Bulgaria, Georgia, Turkey, Romania
Marine R&D	The activities centred around the advancement of technology, knowledge, and capabilities pertaining to marine environments, encompassing oceans, seas, and other aqueous bodies	Established	Bulgaria, Georgia, Turkey, Romania
Marine & Coastal Tourism	The provision of tourism-related services in and around littoral or marine environments, which support the local community's sustainable development	Established	Bulgaria, Georgia, Romania
Marine aquaculture	The practise of aquaculture and farming with the intention of minimising any adverse effects on the purity of air, water, and soil	Emerging	Georgia, Turkey, Romania
Shipping/ Ports	The operations related to maintaining a sustainable maritime transport ecosystem, encompassing terminal services and the conveyance of passengers and cargo via water	Established	Bulgaria, Georgia
Ocean Renewable Energy	The production of pure and renewable energy from natural sources, such as wind, wave, tidal, and solar, at sea, offshore, on land, and in close proximity	Emerging	Turkey, Romania
Shipbuilding	The goods and services necessary for the construction, upkeep, restoration, and repair of vessels used for ecologically conscious maritime transportation	Established	Turkey
Marine Business Services	The commercial activities that rely on water and are associated with marinas and other vessel service operations	Established	Bulgaria

(continued)

Table 1. (*continued*)

Sector	Definition	BE Sector	Black Sea countries
Offshore oil & gas	The extraction of gas and hydrocarbons from submerged sources	Emerging	Turkey

Table 2. Challenges in the Black Sea region as presented in the first round of MAFs

POLITICAL	ENVIRONMENTAL	SOCIAL
• Geopolitical instability • Lack of collaboration between all state institutions • Need for regional cooperation and intersectoral synergies • Lack of international cooperation • Insufficient political will • Lack of vision and long-term planning • Lack of a Black Sea brand at local and regional level • Lack of compliance with political obligations towards the EC	• Pollution and environmental degradation • Seawater quality • Imbalance of aquaculture sustainability and overfishing • Climate change's impact on biodiversity	• Need for training and capacity building in all Blue Economy sectors (limited human resources available) • Lack of job opportunities • Public awareness on aquaculture sector • Need for connectivity through cultural and natural heritage, between neighbouring countries
TECHNOLOGICAL	LEGAL	ECONOMIC
• Need for marine research • Need for initiatives, such as a beach quality award system (e.g., Blue Flag) • Lack of adequate infrastructure • Lack of advanced technologies (e.g., for monitoring the fishing vessel in the region) • Need for digitization and creation of a unitary database • Need for use of non-polluting technologies	• Harmonization of national and EU legislation • Specific and integrated legislations drafted clearly and concisely, with the elimination of legislative loopholes • Lack of implementation of existing regulations • Lack of carrying out adequate monitoring and control • High bureaucracy • Corruption	• Need for state aid for fisheries and aquaculture sectors • Need for removing the tax barriers for the maritime sector • Lack of quotas in the fishing industry • Financial crisis • High international competition in marine products • Lack of Investments

4 Conclusions

This study examined the utilization of a specific type of Living Labs, the Multi-Actor Forums (MAFs), to facilitate the transition to a sustainable Blue Economy in the Black Sea. This is achieved through the coordinated implementation of four distinct workshops in four Black Sea countries, namely, Bulgaria, Georgia, Romania, and Turkey. The shared objective is to identify the challenges associated with Blue Economy sectors in each country and to prioritize them. The results show that there is significant regional consensus on sectors such as capture fisheries, marine research and development, aquaculture, and tourism. These sectors are crucial for economic development and ecological sustainability, but they are often underrepresented in regional policies like the Common Maritime Agenda (CMA) and the Strategic Research and Innovation Agenda (SRIA). Sustainability is a prevalent issue across all sectors, including the reduction of tourism's environmental impact, the shift to renewable energy, and the promotion of circular economy ideas.

The policy landscape has both strengths and deficiencies, with some countries connecting their policies with EU frameworks to promote innovation, while others encounter bureaucratic bottlenecks, legal loopholes, and uneven execution. Investment in strategic vision, governance changes, and optimized bureaucratic procedures is essential. Enhanced public awareness and developing technical knowledge are also crucial for fostering private sector and community engagement in the Blue Economy. Economic limitations and technological inadequacies continue to pose significant obstacles to the Blue Economy. Restricted access to venture capital, inadequate initial investment, and deficient infrastructure hinder entrepreneurial development. Technological deficiencies in digitalization and data management impede efficiency and decision-making. Establishing resilient finance structures and promoting an innovative culture via government-sponsored grants, subsidies, and tax incentives can mitigate these problems and propel sustained advancement.

Cross-sectoral insights highlight the importance of inclusion, regional cooperation, and strategic long-term planning. Public engagement, youth participation, and community-led projects are essential for harmonising local priorities with regional objectives. Enhancing cross-border collaborations, disseminating exemplary practices, and using shared expertise can augment regional influence. Innovations focused on sustainability, especially in renewable energy and circular economy practices, should remain fundamental to regional strategy. Integrating these initiatives with global frameworks like the European Green Deal and the UN Sustainable Development Goals will enhance the resilience of the Black Sea Blue Economy.

Acknowledgments. This paper is an output of the science project DOORS. DOORS has received funding from the from the European Union's Horizon 2020 Framework Programme for Research and Innovation under grant agreement No 101000518.

References

Brugha, R., Varvasovszky, Z.: Stakeholder analysis: a review. Health policy plan. **15**(3), 239–246 (2000). https://doi.org/10.1093/heapol/15.3.239

Caribbean Development Bank (2018). Financing the Blue Economy: A Caribbean Development Opportunity. In. https://www.caribank.org/publications-and-resources/resource-library/thematic-papers/financing-blue-economy-caribbean-development-opportunity

Connect Black Sea (2019) The SRIA|Black Sea Connect. http://connect2blacksea.org/the-sria/

Connect Black Sea (2023) Black Sea SRIA Implementation Plan. http://connect2blacksea.org/wp-content/uploads/2023/08/23062023-Black-Sea-SRIA-Implementation-Plan.pdf

ENoLL (2024) Living Labs - ENoLL. https://enoll.org/living-labs/#living-labs

European Commission (2019) Common Maritime Agenda (CMA) for the Black Sea. https://black-sea-maritime-agenda.ec.europa.eu/

European Commission (2021) COM (EU) 2021/240 COMMUNICATION FROM THE COMMISSION TO THE EUROPEAN PARLIAMENT, THE COUNCIL, THE EUROPEAN ECONOMIC AND SOCIAL COMMITTEE AND THE COMMITTEE OF THE REGIONS on a new approach for a sustainable blue economy in the EU Transforming the EU's Blue Economy for a Sustainable Future. https://eur-lex.europa.eu/legal-content/EN/TXT/?uri=COM:2021:240:FIN

European Commission (2022) The EU blue economy report 2022. Publications Office of the EU. https://op.europa.eu/en/publication-detail/-/publication/156eecbd-d7eb-11ec-a95f-01aa75ed71a1/language-en

Eurostat. Population in coastal regions. Statistics Explained. European Commission (2011). https://ec.europa.eu/eurostat/statistics-explained/index.php?title=Population_in_coastal_regions

FAO (2023) Recent boom in aquaculture under threat in the Black Sea region | General Fisheries Commission for the Mediterranean (GFCM). Food and Agriculture Organization of the United Nations. https://www.fao.org/gfcm/news/detail/en/c/1506130/

Hossain M, Leminen S, Westerlund M (2019) A systematic review of living lab literature. J Clean Prod 213:976–988. https://doi.org/10.1016/J.JCLEPRO.2018.12.257

Mendelow, A.L.: Environmental Scanning--The Impact of the Stakeholder Concept (1981). ICIS 1981 Proceedings. 20. https://aisel.aisnet.org/icis1981/20

Schäpke N et al (2018) Jointly experimenting for transformation? Shaping real-world laboratories by comparing them. GAIA - Ecol Perspect Sci Soc 27:85–96. https://doi.org/10.14512/GAIA.27.S1.16

Schuurman D (2015) Bridging the gap between Open and User Innovation? : exploring the value of Living Labs as a means to structure user contribution and manage distributed innovation. http://hdl.handle.net/1854/LU-5931264

Schuurman D, De Marez L, Ballon P (2015) Living Labs: a systematic literature review. http://scholar.google.be/

The World Bank. Blueing the Black Sea Program (BBSEA) overview. The World Bank (2020). https://www.worldbank.org/en/region/eca/brief/blueing-the-black-sea-program-bbsea

Trading Economics. Georgia GDP from Transport. TradingEconomics.com (2023). https://tradingeconomics.com/georgia/gdp-from-transport

A Paleoenvironmental Reconstruction Study in a River Mouth, Southeastern Brazilian Coastal Region

Marlon Carlos França[1,2(⊠)], Fernando Augusto Borges Silva[2],
Nisya Robelly Cardoso Pantoja[2], Angela Esmeralda Cely Torres[2],
Luke Ortiz-Whittingham[3], Giseli Modolo Vieira Machado[4], Pablo Azevedo Rocha[4],
Renata de Castro Ribeiro França[1], Emuobosa Orijemie[5],
and Ana Carolina Lima Coutinho Gomes de Vasconcelos[1]

[1] Federal Institute of Espírito Santo, Piúma, ES 29285000, Brazil
`marlon.franca@ifes.edu.br`
[2] Federal University of Pará, Belém, PA 66075110, Brazil
[3] Amherst College, Amherst, MA 01002, USA
[4] Federal University of Espírito Santo, Vitória, ES 29075910, Brazil
[5] University of Ibadan, Ibadan 200001, OY, Nigeria

Abstract. The purpose of this work was to identify and correlate vegetational and sedimentary changes with sea level fluctuation events and climatic changes that have occurred over the last two thousand years at the mouth of the Jucu River (ES), southeastern Brazil. Thus, a sedimentary core was collected with a depth of 190 cm and was used to perform pollen, sedimentary, and C-14 dating analyses. The results showed the formation of four zones. The first zone started around 2212 cal years BP, with sandy sediments, which is indicative of high energy flow. The analysis of the palynological profile indicated the presence of herbaceous vegetation. In the second and third zones, which correspond to the period of ±2210 to ±460 cal years BP, there was a predominance of silt-sandy sediment, with the installation of the mangrove. The fourth zone considers the period of ±460 cal years BP until the present, marked by the presence of silt-clay sediment and predominance of herbaceous vegetation. Furthermore, the presence of *Laguncularia* and *Rhizophora* remained almost continuous during the fourth zone.

Keywords: Climate Change · Holocene · Mangrove · Sea Level · Sedimentology

1 Introduction

Climate change and Atlantic sea-level oscillations have produced impacts on coastal ecosystems along the Western South Atlantic Margin during the late Holocene [1]. The coastal zone is largely controlled by complex interactions involving gradients of tidal oscillation, river discharge, littoral currents, sediment, and nutrient supply [2]; these interactions influence mangroves.

© The Author(s) 2026
C. Coelho et al. (Eds.): CD 2025, CRL 41, pp. 424–430, 2026.
https://doi.org/10.1007/978-3-032-15473-6_65

Mangroves occur broadly on the Brazilian coast [2]. Land-ocean interaction controls these forests' evolutionary development, and their expansion is determined by topography, sediment geochemistry, and current energy conditions. This ecosystem is highly adaptive, with plants tolerant of extreme environmental conditions such as high salinity, anoxia, and constant water inundation [3]. This adaptability has allowed mangroves to withstand environmental change throughout the Holocene and become a marker of great importance for scientific analysis of coastal change [4].

Along the Brazilian coast, mangroves are found from the extreme northern coast in the Oiapoque River (04°20′N) to Laguna (28°30′S) on the southern coast [2].

Previous studies of pollen, biogeochemistry, and sedimentary records along the Brazilian coast have demonstrated that multi-proxy analysis can provide important information about coastal vegetation history [1].

Therefore, this work aims to explore the environmental history in the southern region of Espírito Santo State, focusing on the processes that led to the establishment of mangroves during the late Holocene. Our research concentrates on the development of vegetation near the boundaries between mangrove and coastal herbaceous areas, where sensitive vegetation changes related to relative sea level (RSL) and tidal water salinity are expected. This approach is based on the integration of multi-proxy analysis.

2 Materials and Methods

For this study thirty-eight samples were analyzed for pollen content, forty samples for granulometry, and three samples for C-14 dating from a sediment core collected from the area occupied by a mangrove ecosystem and herbaceous vegetation using a peat sampler. The core was X-rayed to identify sedimentary structures.

2.1 Background of Study Area

The study area is located at the mouth of the Jucu River that flows over Quaternary fluviomarine and paludal deposit into the coastal plain of the southeastern Brazilian littoral (Fig. 1), State of Espírito Santo. The coastal plain of the Jucu River has a maximum width of about 7 km and a length of about 5 km. This region is characterized by stretches with rocky embayments, forming a very indented coastline, where small Quaternary coastal plains and small rivers are present [5]. This coastal region is influenced by the Atlantic Ocean with semidiurnal micro-tides (tidal range <2 m), and water salinity between 9 and 34 ‰. The Jucu River has a maximum and minimum outflow of 42 and 15 m3 s-1 [6].

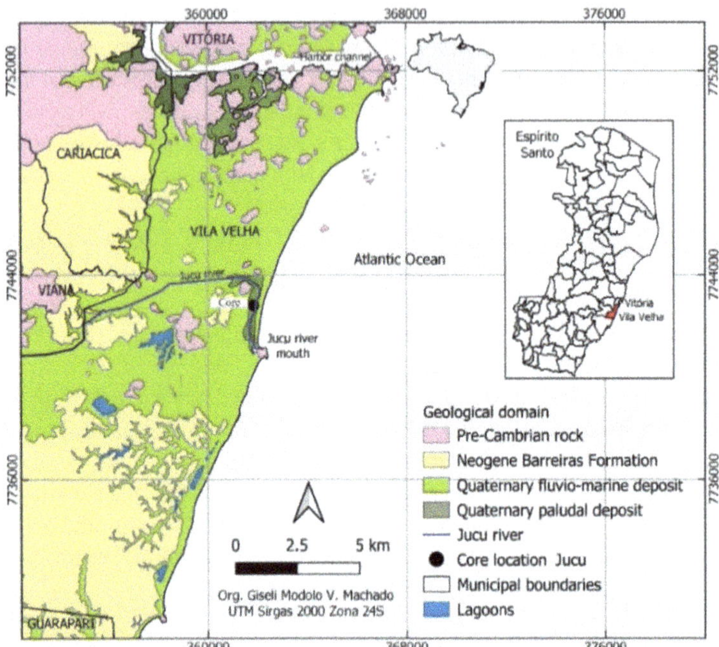

Fig. 1. Study area. Location of sediment core (Jucu) and geological domain of Espírito Santo State, southeastern Brazil.

2.2 Radiocarbon Dating

Samples were checked and physically cleaned under the stereomicroscope. The residual material for each sample was then extracted with 2% HCl at 60 °C for 4 h, washed with distilled water until neutral pH was reached, and dried at 50 °C. The organic matter from the sediment was analyzed by Accelerator Mass Spectrometry (AMS) at the Center for Applied Isotope Studies (CAIS), University of Georgia (UGA), USA. Radiocarbon ages are reported in years before AD 1950 (years BP) normalized to $\delta^{13}C$ of -25‰ VPDB and in cal years BP, 2σ [7].

2.3 Granulometry Analysis

The sediment grain size distributions were determined following the methods of [8], and the graphics were constructed using the SYSGRAN software [9] for sand (2–0.0625 mm), silt (62.5–3.9 μm), and clay fractions (3.9–0.12 μm). Facies analysis included descriptions of color, lithology, texture, and structure. The sedimentary facies were codified according to Miall [10].

2.4 Palynological Analysis

All samples were prepared using standard analytical techniques for pollen including acetolysis [11]. Sample residues were placed in Eppendorf microtubes and kept in a glycerol gelatin medium. Morphological descriptions [12] were consulted for the identification of pollen grains and spores. Software packages TILIA and TILIAGRAPH were used to calculate and plot pollen diagrams.

3 Results

The results for this work consist of radiocarbon data (Table 1), pollen data, and sedimentary features, since at least 2210 cal years BP from a current estuary area of the Jucu River mouth, southeastern Brazil. The sediment core presents light brown muddy, and sandy silt. This deposit presents massive, cross-laminated, parallel-laminated, and heterolithic bedded.

3.1 Radiocarbon Dates and Sedimentation Rates

Radiocarbon dates for the core are shown in Table 1. The result of radiocarbon dating of the sedimentary core samples ranges from 2210 cal years BP to 460 cal years BP. The results revealed the following sedimentation rates: 0.42 mm/year (165–163 cm), 44 mm/year (92–90), and 0.23 mm/year (12–10). This core presents an alternation for the sedimentation rates. Although the rates are non-linear between the dated points, they are within the vertical accretion range of mangrove forests from southeastern Brazil [1].

Table 1. Radiocarbon dating results for Jucu core according to Calib 7.1.

Laboratory number	Depth	C-14 (years BP, 1σ)	Calibrated range (cal years BP, 2σ)	Median of calibrated range (cal years BP)	Sedimentation rates (mm/years)
UGAMS#39907	12–10	375 ± 24	427–502	460	0.23
UGAMS#39908	92–90	414 ± 25	450–516	480	44.0
UGAMS#39909	165–163	2248 + 30	2156–2268	2210	0.42

3.2 Lithofacies and Palynomorphs Association

The core presents light brown muddy and sandy silt with an upward increase in grain size to 62–60 cm depth. After 62–60 cm depth occurred a decrease of sandy silt until the surface. These deposits are massive, cross-laminated, parallel-laminated, and heterolithic bedded. The texture, description of sedimentary structures, and pollen results allowed the identification of four zones.

The first zone consists of sandy sediments with massive sand (Sm), coarse sand, and cross-laminated fine-grained sand (facies Sc) until at least around 2210 cal years BP. This deposit corresponds to the bottom section of the core. The pollen assemblage of this zone is characterized by the predominance of herbaceous pollen (34–72%). No pollen of the mangrove was counted.

The second zone consists basically of sand and silt sediments (±88%), coarse sand (facies Sm), and a low percentage of clay (±12%), between around 2210 and 480 cal years BP. The pollen assemblage is mainly characteristic of mangroves beginning with the presence of *Laguncularia* (9%), *Rhizophora* (2%), and *Avicennia* (2%).

The third zone occurs between around 480 and 470 cal years BP, between 115 and 45 cm in depth. It mainly consists of silt-sandy sediments, with around 56% silt and approximately 32% sand. Clay sediments were much lower around 12%. The pollen assemblage of this zone is characterized by the predominance of herbaceous pollen (46–55%). A mangrove occurs in this zone (3–7%), mainly represented by *Laguncularia* (6%), and *Rhizophora* (1%).

The deposit of the fourth zone consists of silt-clay sediments since at least around 460 cal years BP. The bottom of this zone was marked by silt (±71%), followed by clay sediments (±18%), while sand sediments were around ± 10%. The pollen assemblage is characterized by mangrove pollen presence as well around 2–7%, mainly represented by *Laguncularia* (2–8%), and *Rhizophora* (±2%). The herbaceous pollen (42–53%) was mainly represented by Poaceae (20–27%), Cannabaceae (2–18%), Asteraceae (±10%), Moraceae (5–9%), Amaranthaceae (±6%), and Cyperaceae (±4%). Pollen of shrubs (14–34%), and trees (20–36%) are also registered, represented by Fabaceae (3–10%), Anacardiaceae (±9%), Euphorbiaceae (3–8%), Rubiaceae (±2%), Apocynaceae (5–15%), Ericaceae (9–12%), Combretaceae (±11%), Myrtaceae (5–7%), and Ilex (±2%); furthermore, Arecaceae is found around 15%. Samples from this zone contain marine microfossils such as foraminifera (±2%).

4 Interpretation and Discussion

The data suggest two phases and four pollen zones of wetland development and vegetation close to the mouth of the Jucu River on the central coast of Espírito Santo State, southeastern Brazil. During the first phase, the area was colonized by herbs, palms, and shrubs. Sandy sediments, massive sand, coarse sand, and cross-laminated fine-grained sand were present, and flow energy oscillated. The area typically registered high energy and current influence until around 2210 cal years BP, probably showing a high marine influence between the mid-to late-Holocene around the mouth of the Jucu River. The second phase was marked by mangrove beginning and establishment, with hydrodynamic energy decreasing, resulting in a mud deposition, probably due to the mangrove presence; this phase was also marked by a sea-level drop during the late Holocene. For instance, a change in lithofacies characteristics from coarse sand to fine sand indicates a change from high energy to low energy (a drop in sea level).

5 Conclusion

Palynological, sedimentary, and C-14 dating data obtained from a sedimentary core sample collected at the mouth of the Jucu River, State of Espírito Santo, southeastern Brazil (ES), were used to identify vegetation dynamics and sedimentary dynamics, correlated with RSL fluctuations and climate changes that occurred during the late Holocene. Until around 2210 cal years BP the RSL was above the current position and the climate was more humid, influencing hydrodynamic factors and more intense rainfall, which were not favorable for the establishment of mangroves during this period, but by herbaceous pollen. After this time there was a predominance of silt-sandy sediment and the installation of mangroves, initially colonized by *Laguncularia*. Later there was the installation of *Rhizophora* and *Avicennia*, occurring in a non-expressive way. From ±460 cal years BP to the present herbaceous vegetation was predominant and the number of spores decreased possibly evidencing a drier climate. During this period, the appearance of *Laguncularia* and *Rhizophora* pollen grains occurred almost continuously, which can be explained by the predominance of silt-clay sediment. In this context, the installation and development of the mangrove forest at the mouth of the Jucu River was possible mainly due to the variation in the RSL and climate fluctuations. However, more studies are needed to understand modern mangrove dynamics, according to new scenarios for climatic changes and sea-level rise.

References

1. França M et al (2016) Millennial to secular time-scale impacts of climate and sea-level changes on mangroves from the Doce River delta. Southeastern Brazil Holocene 26(11):1733–1749
2. Schaeffer-Novelli Y, Cintrón-Molero G, Soares M, De-Rosa T (2000) Brazilian mangroves. Aquat Ecosyst Health Manage 3(4):561–570
3. Vannucci M (2001) What is so special about mangroves? Braz J Biol 61(4):599–603
4. Blasco F, Saenger P, Janodet E (1996) Mangroves as indicators of coastal change. CATENA 27(3–4):167–178
5. Mahiques M, de Souza L (1999) Shallow seismic reflectors and upper Quaternary Sea Level Changes in the Ubatuba region, São Paulo State. Southeastern Brazil. Revista Brasileira de Oceanografia 47(1):01–10
6. Deina M, Bastos A, Quaresma V (2011) Variação Morfológica do Cordão Litorâneo Associado À Foz do Rio Jucu (ES). Revista Geografares 9:203–230
7. Reimer P et al (2013) IntCal13 and marine13 radiocarbon age calibration curves 0–50,000 years cal BP. Radiocarbon 55(4):1869–1887
8. Wentworth CK (1922) A scale of grade and class terms for clastic sediments. J Geol 30(5):377–392
9. Camargo M (2006) Sysgran Um Sistema de Código Aberto para Análises Granulométricas do Sedimento. Revista Brasileira de Geociências 36:371–378
10. Miall A (1978) Lithofacies Types and Vertical Profile Models in Braided River Deposits: A Summary. In: Miall AD (ed) Fluvial Sedimentology. Canadian Society of Petroleum Geologists, Calgary, pp 597–604
11. Faegri K, Iversen J (1975) Textbook of pollen analysis, 3rd edn. Hafner Press, New York
12. Colinvaux P, Oliveira P, Patiño J (1999) Amazon Pollen Manual and Atlas. Harwood Academic Publishers, Amsterdam

Shoreline Stabilization by Coastal Mangroves During Extreme Wave Events

Noelle Poovey[(⊠)] and Tori Tomiczek

United States Naval Academy, Annapolis, MD 21412, USA
pooveynoelle@gmail.com, vjohnson@usna.edu

Abstract. Sea level rise and the escalation of intense storms has increased the occurrence of coastal flooding and erosion in recent decades. Engineers, community planners, and military base commanding officers around the world are seeking creative solutions to mitigate the damage done by these processes to nearshore structures and naval bases. One such solution is mangrove forests, which are known to reduce wave heights and, subsequently, their effects, including wave loading and wave-driven erosion. However, their effect on sediment transport and potential for erosion control, particularly during short-term, extreme wave conditions, has not been well quantified. This research analyzed sediment erosion and accretion in a *Rhizophora* mangrove forest model during simulated storm conditions, using a 1:8-scale physical model. Mangrove roots changed the sediment transport patterns compared to the baseline case. This study provides data necessary for communities to develop and implement nature-based solutions such as mangrove forests to mitigate coastal flood hazards.

Keywords: Sediment Transport · Erosion · Nature-Based Solutions · Physical Model

1 Background

Resilient and sustainable solutions are needed to prevent shoreline retreat and protect coastal structures vulnerable to erosion and scour. Mangroves have been identified as a nature-based alternative for shore protection. However, previous field and laboratory experiments have focused on wave height reduction by mangroves, with less focus on erosion. Few experiments have linked hydrodynamic parameters such as flow velocity to sediment transport or erosion reduction [1].

Accretion rates in mangroves have mainly been studied on time scales of one year or more. For example, Krauss et al. [2] compared vertical accretion among various mangrove root types in the field, and Lovelock et al. [3] analyzed sea level and sediment field data over a four-year period to evaluate whether sediment accumulation among mangroves shore their root growth could match sea level rise, which was shown true in areas containing fine sediment. However, engineers lack quantitative data for wave effects on mangroves during shorter-term acute events period, such as a hurricane.

© The Author(s) 2026
C. Coelho et al. (Eds.): CD 2025, CRL 41, pp. 431–437, 2026.
https://doi.org/10.1007/978-3-032-15473-6_66

While previous studies have provided insight about sediment accretion in mangrove root systems, relationships between incident wave and storm surge conditions, erosion or accretion, and shoreline stabilization by mangroves are still not well understood. Therefore, this study addressed the research question of where erosion and deposition occur among mangrove roots in storm wave conditions through a physical model experiment in the United States Naval Academy's (USNA) Hydromechanics Laboratory.

2 Experimental Methods

This experiment measured sediment accretion and erosion in the presence or absence of a realistic representation of a mangrove trunk-root system subjected to strong, regular wave conditions. Evaluations of sediment profile change were taken throughout the wave trials and final equilibrium beach profiles were compared for configurations with or without mangroves.

2.1 Physical Model Setup

Table 1 presents experimental-scale and full-scale parameters of the experiment. Constructed mangrove models at large (1:8) geometric scale were installed in the wave flume, each with a 1.5 cm, brass trunk diameter and 25 cm trunk height. The model root diameter was 0.4 cm and was made from semi-flexible galvanized steel wire. The models were constructed in previous experiments [4] to replicate *Rhizophora* stilt roots based on a mature mangrove forest. This experiment was performed in the USNA Hydromechanics Laboratory, utilizing the Sediment Tank in the Coastal Laboratory (Fig. 1). The tank is 10.7 m long and 1.0 m wide, with sediment that can be shaped to an initial beach profile. The sediment had a D50 of 0.31 mm, which corresponds to a full-scale D50 of 2.5 mm and an ASTM Classification of coarse sand. The forest was installed on an area of 1 m x 1.8 m on the sediment slope, and the mangrove models were placed in an alternating grid pattern to create a stem density of 12 trees/m^2 in the flume.

Table 1. Scaled and actual parameters for the experiment

Parameter	Lab	Actual
Wave Height	0.07 m	0.56 m
Wave Period	1.25 s	3.54 s
Forest Density	12 trees/m^2	0.1875 trees/m^2
Water Depth	0.35 m	2.80 m
Storm Length	40 min	113 min
Forest Area	1.0 m × 1.8 m	8.0 m × 14.6 m
Mangrove Trunk Diameter	1.5 cm	12 cm
Mangrove Root Diameter	0.4 cm	3.2 cm

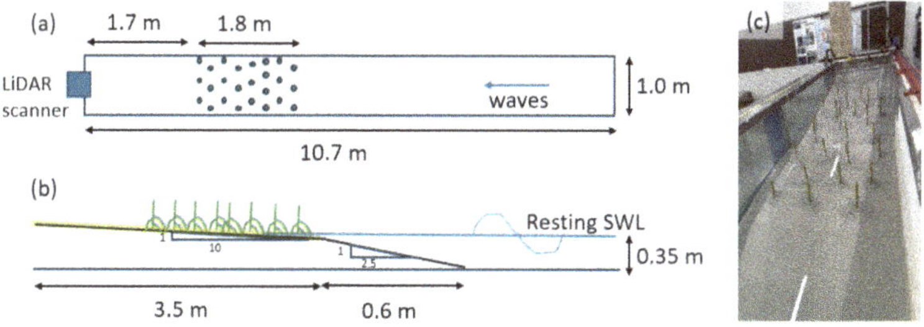

Fig. 1. (a) Plan and (b) profile schematics of experiments (not to scale) showing location of mangroves and sediment in testing section; (c) model mangroves installed in wave flume.

2.2 Hydrodynamic Conditions

Experiments were conducted using repeated sets of regular wave conditions. The water depth in the flume was 0.35 m relative to its fixed base. The model-scale wave amplitude and period were 0.07 m and 1.25 s, respectively (Table 1), corresponding to a 1.12 m wave height and 3.54 s wave period at full scale. These waves are representative of storm conditions in sheltered estuarine areas and were the largest waves that could be safely run in the flume. These waves created breaking conditions in the first two rows of mangroves, as shown for the mangrove and bare (no mangrove) conditions in Fig. 2.

Waves were run for 40 min in 5-min increments to reduce reflection, with the final increment being 10 min to reach an equilibrium condition, visually identified when minimal changes were observed in subsequent profiles for the bare condition. Waves were stopped for at least 15 min between trials to allow the wave energy to dissipate in the tank.

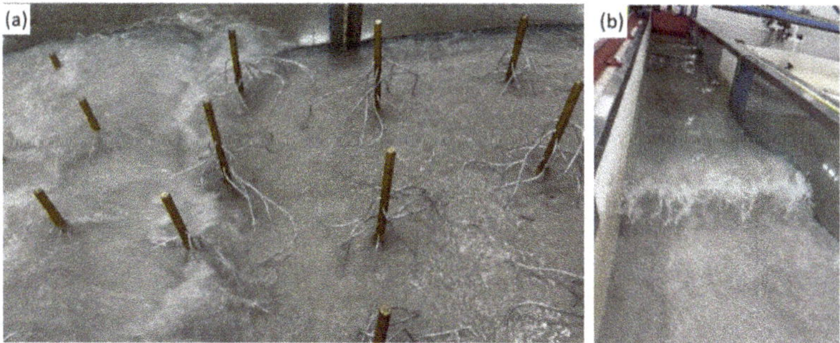

Fig. 2. (a) Profile view of waves breaking through the modeled forest; (b) Front view of waves breaking in the bare condition.

2.3 Morphology Measurements

Data collection and analysis was done using a LiDAR beach profile scanning system. Excel and other analysis tools were used to explore the relationships between wave conditions and sediment erosion/accretion. The LiDAR scanner combined a ToughSonic50-21s-232 Ultrasonic sensor for horizontal distance with a mobile, 50-inch Acuity AR700 laser displacement sensor for vertical distance. A trolley fixed in the lateral and vertical dimensions was rolled over top of the flume to obtain a profile of the sediment heights along the beach. The sampling rate was 10 Hz; data was taken for each scan in 70 s along the first ~ 5.2 m of the flume, corresponding to one data point every 0.86 cm. Before each scan, the laser was re-zeroed to account for differences in the laser's transmittance in air and water, using a fixed starting point for the laser and at the water line. Voltage-based calibration constants were used. The average calibration constant marking the starting point for the laser at the highest, most landward point of the beach was 0.418 V. However, the calibration constant marking the waterline on the beach shifted as the beach morphology changed, ranging from 2.638 V to 2.738 V.

2.4   Experimental Procedure and Data Processing

For each condition (mangrove or bare), the sediment tank was raked and smoothed to achieve the initial beach profile, with a 1:10 slope on the landward side and a 1:2.5 slope on the seaward shelf. Initial laser scans were taken along five transects of the sediment tank's width, approximately 20 cm apart. Wave increments were run until the equilibrium condition was reached, confirmed by comparing centerline transect laser scans after each increment (Fig. 3). Upon reaching equilibrium, five transects were measured to determine the final beach profile for each condition. For the mangrove condition, profile transects were scanned both with the mangroves in place and after removing the mangroves for comparison to the bare condition.

3   Results

Figure 3 shows the before and after centerline transect plots for both the bare (green and red, respectively) and mangrove (purple and blue, respectively) conditions, along with the intermediate trials (grey) for the bare condition. Both conditions reached similar equilibrium beach profiles, with an offshore sandbar, trough at the site of initial wave breaking, and erosion in the forest area. However, the sandbar shifted slightly seaward for the mangrove condition.

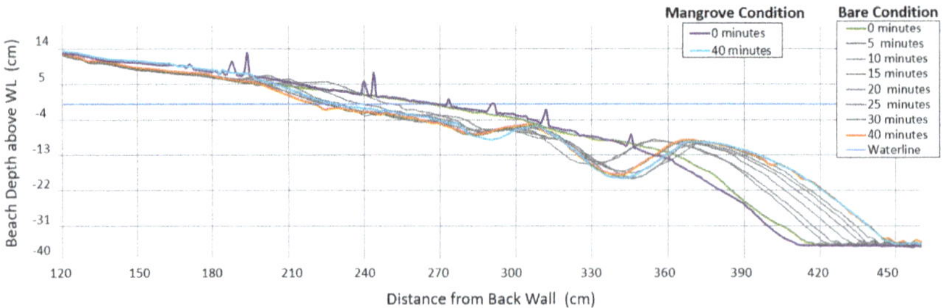

Fig. 3. Initial and final beach profile comparisons for mangrove and bare configurations, with intermediate scans for the bare configuration trials shown in gray.

Difference plots (Fig. 4) were created by subtracting the initial beach scans (0 min) from the final beach scans (40 min) for each condition. Final beach scans were zeroed by subtracting the average change in the shoreward-most 1.5 m of each scan from each condition, which did not experience wave action. Figure 4 shows that the mangrove condition lost less sand in the shoreward portion of the forest but experienced more erosion in the seaward portion before the trough. Bumps in the mangrove difference plot (Fig. 4) and the initial mangrove condition (Fig. 3) are due to the laser hitting the mangrove trunk and roots in the initial profile. Qualitative differences in sediment movement were observed during the experiments; for example, a more enhanced scour edge at the waterline was observed in the mangrove condition compared to the bare condition.

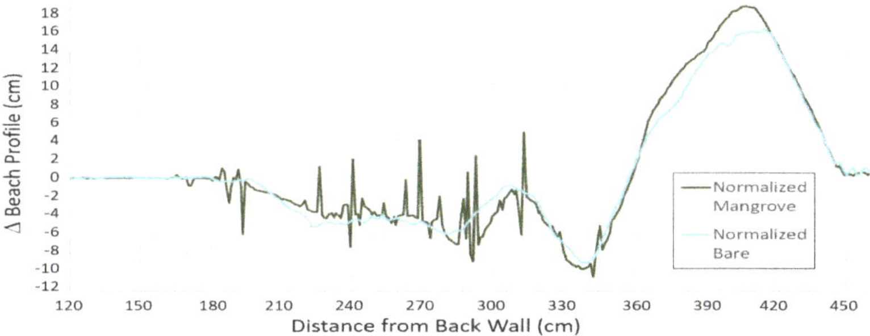

Fig. 4. Normalized difference plots for the bare and mangrove conditions.

4   Discussion

Mangroves were observed to retain sediment in the shoreward eroded area compared to the bare condition, but in some areas of the mangrove forest, more scour was experienced than for of the bare condition. This could be due to enhanced turbulence and varied wave velocities as the waves transit around the mangrove trunks and roots.

While this is one of the first studies to provide insight on sediment transport under storm waves and time scales, results must be interpreted with caution. Mangrove models did not include a canopy, only considered primary roots, and used metal wire to represent prop roots. Mangrove branches and leaves may be important in wave height and erosion mitigation during conditions when the trunks and roots are submerged. A real *Rhizophora* root system may have greater friction and blockage area if smaller roots are considered. In addition, although the scaled sediment is realistic for many beaches, the scaled particles were coarser-grained and non-cohesive, which might behave differently than fine grained material expected in real mangrove forests. These idealizations could have led to enhanced scour around roots and minimized the contribution of mangrove roots in trapping sediment during wave action.

Due to time constraints, the equilibrium profile was determined after 40 min of waves, and only one wave condition was tested. This condition caused wave breaking in the two seaward-most rows of mangroves, covering most of their trunks and all of their root system (Fig. 1). Crosswise reflection in the tank led to some diagonal wave breaking and may have contributed to the trough formation in the forest area. This led to asymmetrical edge effects and increased the amount of sediment movement on the flume's right side. A longer time period for equilibrium or varying wave conditions could lead to different patterns of erosion or accretion in the forest.

5 Conclusion

This experiment investigated sediment accretion and erosion around mangrove root structures in comparison with a bare condition. While extreme differences in the equilibrium mangrove and bare profiles were not observed for the wave conditions tested here, it was clear that sediment moved differently and demonstrated more scour in some regions and more accumulation in others, especially at the shoreward portion of the mangrove forest. Future work will compare volumetric sediment change along the scanned profiles, utilizing irregular storm waves or other regular waves, varying the forest species and density, and adding organic material to the sediment. Ultimately, results could allow the development of an empirical relationship between wave strength, wave attenuation through mangroves, and shoreline erosion, determining sediment transport rates which could be scaled up or down for application in the field.

By establishing and quantifying the benefits of mangrove structures, this experiment provides results on the usefulness of mangrove forests in communities and naval bases in subtropical regions. Further, it can also be used as a framework for testing other types of vegetation or nature-based solutions that could be used on both naval bases and civilian infrastructure in colder climates.

Acknowledgements. The authors thank Mr. John Balano, Mr. Bill Beaver, Dr. James Gose, Mr. Adam King, and Mr. Michael Stanbro of the USNA Hydromechanics Laboratory for their invaluable assistance and expertise in experimental setup. This work was supported in part by funding from the National Science Foundation through Grant 2110262. Any opinions, findings, conclusions, or recommendations expressed in this material are those of the authors and do not necessarily reflect the views of the NSF or USNA.

References

1. Horstman EM, Bryan KR, Mullarney JC, Pilditch CA (2016) Model versus nature: Hydrodynamics in mangrove pneumatophores. In: Proceedings of 35th Conference on Engineering https://icce-ojs-tamu.tdl.org/icce/index.php/icce/article/view/8254/pdf
2. Krauss K, Allen J, Cahoon D (2003) Differential rates of vertical accretion and elevation change among aerial root types in Micronesian mangrove forests. Estuar Coast Shelf Sci 56(2):251–259. https://doi.org/10.1016/S0272-7714(02)00184-1
3. Lovelock C et al (2015) Sea level and turbidity controls on mangrove soil surface elevation change. Estuar Coast Shelf Sci 153:1–9. https://doi.org/10.1016/j.ecss.2014.11.026
4. McMann D, Keck J, Tomiczek T, Lomonaco P, Cox D, Libby M (2024) Small scale experiments' ability to augment large scale laboratory testing for designing nature-based and hybrid solutions for coastal flood hazard mitigation. In: Proceedings of the 9th International Conference on Physical Modeling in Coastal Engineering https://doi.org/10.59490/coastlab.2024.697

The Sensitivity of the Ebb Dominance in Tidal Creeks of a Microtidal Mangrove Forest

Rik Gijsman[1]([✉]), Erik Horstman[1], Sabine Engel[2,3], Daphne van der Wal[1,4], Rob van Zee[1,5], Jessica Johnson[3,6], Matthijs van der Geest[7], and Kathelijne Wijnberg[1]

[1] University of Twente, Enschede, The Netherlands
r.gijsman@utwente.nl
[2] STINAPA National Parks Foundation, Kralendijk, The Netherlands
[3] Mangrove Maniacs, Kralendijk, The Netherlands
[4] Royal Netherlands Institute for Sea Research, Yerseke, The Netherlands
[5] Nelen and sSchuurmans, Utrecht, The Netherlands
[6] Coastal Dynamics, Kralendijk, The Netherlands
[7] Wageningen Marine Research, Den Helder, The Netherlands

Abstract. Tidal creeks in mangrove forests dissect the mangrove-vegetated forest platforms. Due to the difference in resistance to tidal flows between the mangrove-vegetated platform and the tidal creeks, tidal flows in tidal creeks are typically ebb-dominant. As a result, tidal creeks maintain their depth by the net export of sediments. However, with small tidal ranges, the ebb-dominance of tidal creeks may be smaller and their self-scouring may fall short, possibly resulting in net accretion in the creeks when excess sediment is available. In this study, we performed field measurements to quantify tidal flow velocities through tidal creeks of a microtidal mangrove forest in Bonaire, a small island in the Caribbean Netherlands. We found that the ebb-dominance of the tidal creeks was sensitive to spatial variations in the mangrove forest resistance as well as to temporal variations in tidal range on spring-neap tidal timescales. These findings imply that tidal creeks in microtidal mangroves are not necessarily ebb-dominant, and this may enhance their susceptibility to sediment infilling, the clogging of creeks and the resulting degradation of mangrove forests.

Keywords: Mangroves · Tidal Creeks · Tidal Flows · Microtidal

1 Introduction

Intertidal mangrove ecosystems have resilient capacities that allow them to adapt to changing environmental conditions. For instance, the attenuation of hydrodynamic forces in mangrove forests allow for increased sediment deposition and support forest dynamics and survival (Temmerman et al., 2023; Gijsman et al. 2024a). Mangroves could thus continue to provide their essential ecosystem services to coastal communities despite ongoing climatic changes such as sea level rise. Mangrove forests typically consist of vegetated upper intertidal platforms that are dissected by tidal creeks. These tidal

© The Author(s) 2026
C. Coelho et al. (Eds.): CD 2025, CRL 41, pp. 438–443, 2026.
https://doi.org/10.1007/978-3-032-15473-6_67

creeks form low-resistance conduits for the transport of water and sediments, processes that are essential for the hydrological connectivity in mangroves and the morphological development of mangroves (Horstman et al., 2018).

Tidal flows through tidal creeks in mangroves are typically ebb-dominant (Furukawa and Wolanski, 1996, Horstman et al., 2021). Ebb-dominant tidal flows mean that the flow speed during the ebbing stage of the tide exceeds the flow speed during the flooding stage of the tide. The ebb-dominance in tidal creeks is caused by the difference in resistance between the mangrove-vegetated areas and the tidal creeks. Particularly with the onset of the ebbing phase of the tide (i.e. at high slack tide), the higher resistance of the mangrove-vegetated areas causes that the tidal flows are rerouted through the tidal creeks (Furukawa and Wolanski, 1996). Due to the ebb-dominance in tidal creeks of mangroves, the tidal creeks remain at depth by self-scouring (Horstman et al., 2015).

The ebb-dominance in tidal creeks increases with the volume of water flowing through mangrove-vegetated areas and thus its tidal range. Hence, in microtidal mangroves, the ebb-dominance of tidal creeks may be limited, possibly making them susceptible to infilling when an excess of sediment is available. Presently, little quantitative knowledge is available on tidal flows and sediment transport through tidal creeks in microtidal mangrove settings, making assessments of their vulnerability challenging. This study presents field measurements of tidal flow velocities through tidal creeks of a microtidal mangrove forest in the Caribbean Netherlands in the southern Caribbean Sea. We evaluate the ebb-dominance and its sensitivity on a spring-neap tidal timescale in order to obtain a better understanding of the observed siltation in these tidal creeks.

2 Methodology

This study quantifies water levels, tidal flow velocities and sediment transport in tidal creeks of a mangrove forest located in Lac Bay, a coastal lagoon in Bonaire (Caribbean Netherlands), see also Fig. 1a (Gijsman et al., 2024b,c). Here, we focus on the hydrodynamic processes. Between January and May 2022, water levels were measured throughout the lagoon with Onset HOBO Water Level Loggers (Fig. 1b). Flow velocities were measured simultaneously in the tidal creeks with Lowell Instruments LLC Tilt Current Meters (TCM) (Fig. 1b,c,d).

The measurements were used to investigate the water level gradients across the mangrove forest and the associated flow velocities through the tidal creeks. The ebb-dominance in the tidal creeks was determined by dividing the maximum tidal flow speed during the outflowing ebb phase of each tide by the maximum tidal flow speed during the inflowing flood phase of that same tide, see Eq. 1:

$$A_{ebb} = \frac{|\mathbf{u}_{max,ebb}|}{|\mathbf{u}_{max,flood}|} \tag{1}$$

Fig. 1. Mangrove forest in Lac Bay in Bonaire (Caribbean Netherlands). (A) Caribbean island of Bonaire. (B) Lac Bay mangrove forest and field measurement stations. (C) Picture of a tidal creek in the mangroves of Lac Bay (D) Picture of a deployed TCM in the mangroves of Lac Bay.

3   Results

In Lac Bay, the tide is diurnal with an average tidal range of 0.21 m and a maximum of 0.44 m. The tidal wave is substantially attenuated across the mangrove forest (Fig. 2a). In the north of the mangrove forest, at station N, the tidal range decreased to on average 0.03 m (maximum of 0.11 m). In addition to the tidal attenuation on a daily timescale, a clear spring-neap tidal variation was present. On the landward side of the mangrove forest (stations W, N and E), the water level gradually increased towards spring tides, and afterwards gradually decreased towards neap tides (Fig. 2a). Moreover, the water levels in the west of the mangrove forest dropped further towards and after neap tides than in the north and east.

The flow speeds in the tidal creeks were driven by the concurrent water level gradients across the mangrove forest. Flow speeds in the east creek increased to maxima of about 0.1 m/s and were ebb dominant (Fig. 2b,c). Remarkably, in the west creek, flow speeds sometimes exceeded 0.3 m/s and were flood dominant (Fig. 2b). On a spring-neap tidal timescale, the ebb/flood dominance varied as well. At greater tidal ranges, the flood peak flow speed typically increased. In the east creek, the ebb peak flow speed did not increase as much, resulting in a more flood dominant asymmetry during spring tides. In the west creek, the ebb peak flow speed did increase as well, resulting in a slightly more ebb dominant asymmetry during spring tides (Fig. 3).

Fig. 2. Timeseries of hydrodynamic measurements in Lac Bay. (A) Water levels measurements of several spring-neap cycles (B) Flow speeds in the east and west creeks with tidal maxima and minima. (C) Ebb dominance in the east and west tidal creek. Vertical axis on logarithmic scale.

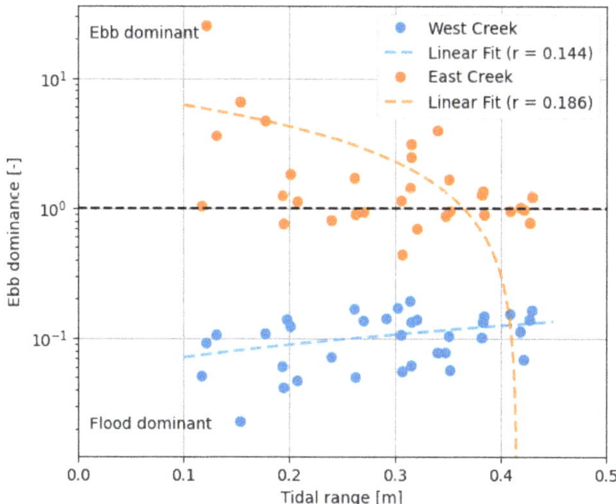

Fig. 3. Relation between the tidal range and the ebb dominance of the tidal creeks in the east and west. Vertical axis on logarithmic scale.

4 Discussion and Conclusion

The presence and dynamics of the tidal creeks that typically dissect mangrove-vegetated areas are driven by asymmetries in tidal flows. While tidal creeks are typically ebb dominant in macrotidal mangroves, providing them their self-scouring maintenance capacity, little is known about tidal flow asymmetries in microtidal mangroves. In this study, we investigated the sensitivity of the ebb-dominance of tidal flows in creeks of microtidal mangroves in Lac Bay, a coastal lagoon in the southern Caribbean. The field measurements revealed a substantial attenuation of the tidal amplitude across the mangrove forest. The tidal flows through the tidal creeks of the mangrove forest, both their speed and direction, were driven by the resulting water level gradients across the mangrove forest.

While the east creek showed mostly ebb-dominant tidal flows, the west creek was flood-dominant. Hence, the ebb-dominance that would typically be induced by local variations in tidal flow resistance in mangrove forests was overruled by a larger scale difference in tidal flow resistance on the mangrove forest scale. As a result, a tidal flow circulation is induced throughout the mangrove forest in Lac Bay, where relatively more water enters the mangrove forest through the west tidal creek than through the east tidal creek. In addition to these spatial differences, the ebb/flood dominance of the tidal creeks in the mangrove forest in Lac Bay also showed sensitivity to spring-neap tidal variations. At increasing tidal ranges, the flood peak flow speed increased and the tidal creeks were more flood dominant. During neap tides, when the tidal ranges were small, the tidal creeks were more ebb dominant. This sensitivity is caused by the presence of a permanently-inundated basin at the back of the mangrove forest in Lac Bay.

Overall, our results show that the ebb dominance of tidal creeks in microtidal mangrove forests is sensitive to both spatial and temporal variations. The observed ebb-dominance in microtidal settings is much smaller than in previous studies in macrotidal mangroves (e.g. Furukawa and Wolanski, 1996; Horstman et al., 2021). In microtidal mangroves, this ebb-dominance of the tidal creeks is found sensitive to spatial variations in the flow resistance induced by the mangrove forest, and thus also to changes in the tidal creek configuration over time. In addition, the ebb-dominance is sensitive to spring-neap tidal variations. These results imply that the typical ebb-dominance in tidal creeks of macrotidal mangrove forest is not necessarily applicable to microtidal mangroves, and this may play an important role in the susceptibility of the mangrove forest in Lac Bay to sediment infilling, tidal creek clogging, and the observed deterioration of the most landward mangrove forest areas.

Acknowledgements. This study was part of the Mangrove-RESCUE project, funded by the Dutch Research Council (NWO-Grant no. 15899). The field measurements were partly funded by the Ecology Fund of the Royal Netherlands Academy of Arts & Sciences (KNAW - Grant no. 705/2022). MvdG was funded by the Netherlands Ministry of Agriculture, Fisheries, Food Security and Nature within the framework of the policy supporting research theme of the Caribbean Netherlands through project no. BO-43-117-007. The field measurements were conducted with approval of STINAPA, the National Parks Foundation of Bonaire. The authors are grateful for the field support of the volunteers of the Mangrove Maniacs and the rangers of STINAPA. We thank Body Ouwersloot and Femke van der Drift (Wageningen Marine Research) for servicing the Onset HOBO Water Level Loggers. The authors thank Daniel Morillo (Kadaster Bonaire) and

Emre Ozturk (University of Twente) for their support in setting up the RTK-GPS device. Peter van Breugel (NIOZ Royal Netherlands Institute for Sea Research) is acknowledged for the sediment sample analysis. Furthermore, we would like to show our gratitude to Tjeerd Bouma (Utrecht University and NIOZ Royal Netherlands Institute for Sea Research) for providing the Teledyne ADCP and TCM sensors.

References

Furukawa K, Wolanski E (1996) Sedimentation in mangrove. Forests 1:3. https://doi.org/10.1023/A:1025973426404

Gijsman R et al (2024) Mangrove forest drag and bed stabilisation effects on intertidal flat morphology. Earth Surf Proc Land 49(3):1117–1134. https://doi.org/10.1002/esp.5758

Gijsman R et al (2024b) The Importance of Tidal Creeks for Mangrove Survival on Small Oceanic Islands [Unpublished Manuscript]

Gijsman R et al (2024c) Field measurement data of hydrodynamic and morphological processes in the mangrove forest of Lac Bay, Bonaire, Caribbean Netherlands. Zenodo. https://doi.org/10.5281/zenodo.13904523

Horstman EM, Dohmen-Janssen CM, Bouma TJ, Hulscher SJMH (2015) Tidal-scale flow routing and sedimentation in mangrove forests: combining field data and numerical modelling. Geomorphology 228:244–262. https://doi.org/10.1016/j.geomorph.2014.08.011

Horstman EM et al (2018) The dynamics of expanding mangroves in New Zealand. In: Threats to Mangrove Forests. Coastal Research Library, Vol. 25, eds. C. Makowski and C. Finkl (Cham: Springer). https://doi.org/10.1007/978-3-319-73016-5_2

Horstman EM, Bryan KR, Mullarney JC (2021). Drag variations, tidal asymmetry and tidal range changes in a mangrove creek system. Earth Surf Process Landf 1–19. https://doi.org/10.1002/esp.5124

Temmerman S, Horstman EM, Krauss KW, Mullarney JC, Pelckmans I, Schoutens K (2023) Marshes and mangroves as nature-based coastal storm buffers. Ann Rev Mar Sci 15(1):95–118

Integrated Salt Marsh Cliff Erosion Modelling: Evaluating Different Approaches

S. Dzimballa[1(✉)], V. Kitsikoudis[1], P. W. J. M. Willemsen[2,3], E. O. Folmer[4], B. W. Borsje[1], I. Y. Georgiou[5], M. C. Bregman[5], and D. C. M. Augustijn[1]

[1] Water Engineering and Management, University of Twente, Enschede, The Netherlands
s.dzimballa@untwente.nl
[2] Hydrology and Environmental Hydraulics Group, Department of Environmental Sciences, Wageningen University and Research, Wageningen, The Netherlands
[3] Department of Ecosystems and Sediment Dynamics, Deltares, The Netherlands
[4] Aeria Solutions Ltd., Squamish, BC, Canada
[5] The Water Institute, New Orleans, LA, USA

Abstract. Lateral erosion of marsh cliffs is a key driver of salt marsh loss but remains poorly understood and underrepresented in numerical models. This study investigates cliff erosion processes using two modeling approaches—subgrid1D and 2D depth-averaged—applied to the Wierum salt marsh in the Dutch Wadden Sea. Both approaches simulated cliff erosion over 9 months and were validated with drone-derived elevation data. Results indicate that monthly-averaged wave conditions yield more accurate lateral retreat predictions than yearly averages, as they better capture short-term variability in wave energy. Both models identify areas without erosion in regions without cliffs, demonstrating good alignment with field data. The subgrid1D model provided closer agreement with observed retreat patterns, while the 2D model, which assumes constant erosion across the cliff face, tended to overestimate retreat. Discrepancies between model results and measurements in areas of high erosion suggest the need to account for additional spatial processes beyond the wave impacts considered here.

Keywords: Salt marsh · Lateral retreat · Cliff erosion · Numerical modelling · Drone data

1 Introduction

Salt marshes are coastal ecosystems that provide wave dampening, support biodiversity, and contribute to carbon sequestration. Because of their importance, it is a problem that they are increasingly threatened by rising sea levels and more frequent storms, resulting in marsh loss. An important driving mechanism of marsh loss is lateral erosion, caused by the erosion of salt marsh edges through cliff erosion. This process, although recognized as a significant driver of marsh loss [1, 2] remains poorly understood and is rarely included in numerical models of salt marsh development [3–5]. Previous empirical studies on lateral marsh erosion have identified a linear relationship between wave power and cliff

C. Coelho et al. (Eds.): CD 2025, CRL 41, pp. 444–450, 2026.
https://doi.org/10.1007/978-3-032-15473-6_68

erosion rates e.g., [2, 6–9]. Efforts to model lateral erosion have incorporated a range of spatial and temporal scales, from basin-wide assessments over millennia [5] to a 1D transect spanning days [3]. Those models used the linear relationship between wave power and erosion rate to simulate lateral retreat. Despite these advancements, critical gaps persist. Integrated modelling approaches at the marsh scale are still lacking, as is validation of existing models using field data. This study addresses these gaps by comparing existing cliff erosion modelling methods within a case study of a salt marsh in the Dutch Wadden Sea.

2   Methods

2.1   Study Area

The Wadden Sea is a large intertidal area protected by barrier islands. The Wierum salt marsh, located in front of an 8–10 m high dike (Fig. 1), spans approximately 830 m laterally and is up to 150 m wide. The tidal range at Wierum is, on average, 2.3 m. The salt marsh is predominantly colonized by the species *Puccinellia*, and has 20–60 cm high cliffs at its seaward edge.

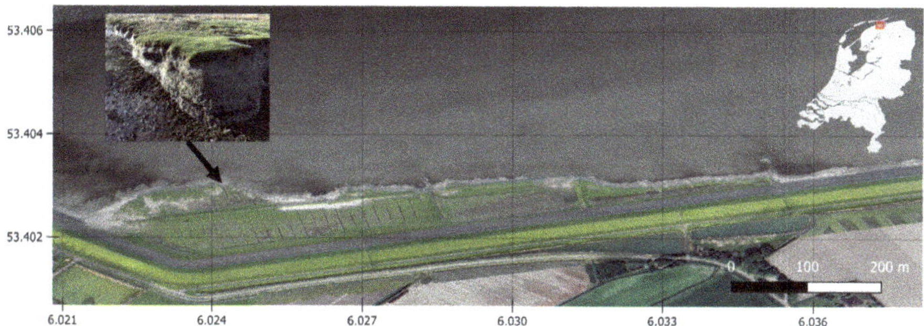

Fig. 1. Drone image of the Wierum salt marsh from 24-11-2022, including a photo of the cliff.

2.2   Hydrodynamic Model

The model domain is based on the Wierum salt marsh, using a drone-derived DEM from 24-11-2022 and existing laser altimetry data [10] for the bathymetry. Drone images were processed into DEMs using structure-from-motion photogrammetry, and this data is used for model validation. The grid resolution is 10×5 m (cross-shore $\times$ longshore) offshore, 5×5 m in a transition zone on the mudflat, and 1×5 m from the salt marsh edge landwards. At the seaward boundary, existing hourly water level data [10] and monthly averages of wave conditions (height, period, and direction) [10] are applied for the simulation period, which runs from 24-11-2022 to 05-09-2023. Hydrodynamics are computed using Delft3D Flexible Mesh (DFM). The DFM module D-Flow calculates tidal flow with 2D unsteady depth-averaged shallow water equations [13], while D-Waves, based on the SWAN spectral wave model, simulates wave propagation hourly. Vegetation is assumed above 1.6 m NAP (Normaal Amsterdams Peil) with a uniform density of 6500 stems/m2, stem diameter of 0.005 m, and height of 0.1 m, representing *Puccinellia* [13]. In DFM, vegetation effects are incorporated as increased roughness based on plant dimensions [14].

2.3 Cliff Erosion Models

A cliff erosion model is integrated with DFM to simulate the lateral retreat of the salt marsh cliff under wave impact. The erosion framework is based on field studies and modeling approaches, where the cliff erosion rate (E) is linearly proportional to the incoming wave energy flux (P) with an erodibility coefficient derived from data (α_w = 1 m3/(yr·W)) [2, 3, 5, 7, 8] (Eq. 1).

$$E = \alpha w \cdot P \tag{1}$$

The cliff erosion models are coded in Python and coupled with DFM every hour (3600 s). During coupling, wave conditions and water levels are used to compute lateral retreat. Two deterministic approaches are employed: the 2D depth-averaged and sub-grid1D models. The 2D depth-averaged model, based on Mariotti & Canestrelli [5] and The Water Institute [15], calculates erosion rates using wave energy flux at each coupling timestep. A bookkeeping method tracks retreat over time, enabling erosion smaller than the grid cell length. The sub-grid1D model, adapted from Bendoni et al. [3], uses a sub-grid cell size of 1 cm and considers the water level at the cliff. Erosion rates decrease for emergent parts of the cliff when water is at the cliff and cease entirely when water levels exceed the cliff height. Both models apply erosion to vegetated cells fronted by bare cells with a minimum 10 cm bed level difference. Erosion percentages are computed every 30 days, and if more than 80% of a cell's erodible volume is lost, erosion updates are communicated to DFM. In this case, 50% of the eroded sediment is redistributed to non-vegetated neighbouring cells (including the eroded cell) in equal volumes, while the remaining 50% is removed from the system [5]. Vegetation in the eroded cell is set to zero. To evaluate the impact of averaging boundary conditions on cliff erosion modeling, two scenarios are simulated using the subgrid 1D model: one with monthly averaged wave conditions and another with wave conditions averaged over the entire simulation period (referred to as yearly averaged). For the 2D depth-averaged model, only monthly averaged wave conditions are applied.

3 Results and Discussion

The lateral retreat of the marsh edge is more pronounced when using monthly-averaged wave conditions across the entire marsh (Fig. 2a). The model predicts that 80% of the cliff erosion occurs in January, coinciding with elevated wave heights, longer wave periods, and a prolonged period of water levels reaching the cliff (Fig. 3). This extended water exposure, combined with larger and longer waves, increases wave thrust [3], significantly contributing to cliff retreat. This finding aligns with previous studies in Delaware Bay, USA [8], which emphasize the importance of inundation duration in marsh edge erosion. The notably lower erosion observed when considering yearly-averaged waves underscores the need to account for short-term variations of boundary conditions in numerical salt marsh models [16]. Incorporating even more detailed wave conditions (e.g., daily or weekly) could further enhance prediction accuracy.

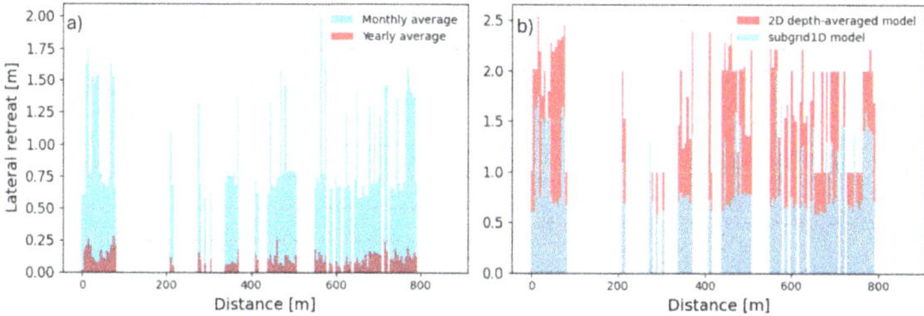

Fig. 2. Lateral retreat of the salt marsh cliff between 24-11-2022 and 05-09-2023 for a) Waves averaged over a year, and waves averaged over a month using the subgrid1D model; b) 2D depth-averaged and subgrid1D model with monthly averaged waves.

In this study, which simulated moderate hydrodynamic conditions without significant storm events, erosion rates were relatively high compared to the long-term average of 0.9 m/year [11]. This agrees with field data from previous research that highlights the importance of lower wave power in salt marsh lateral erosion [9]. It was found that high-magnitude events have a reduced influence on lateral erosion when water levels exceed the cliff face [7], suggesting that the 2D depth-averaged modelling approach may overestimate lateral erosion during storm events.

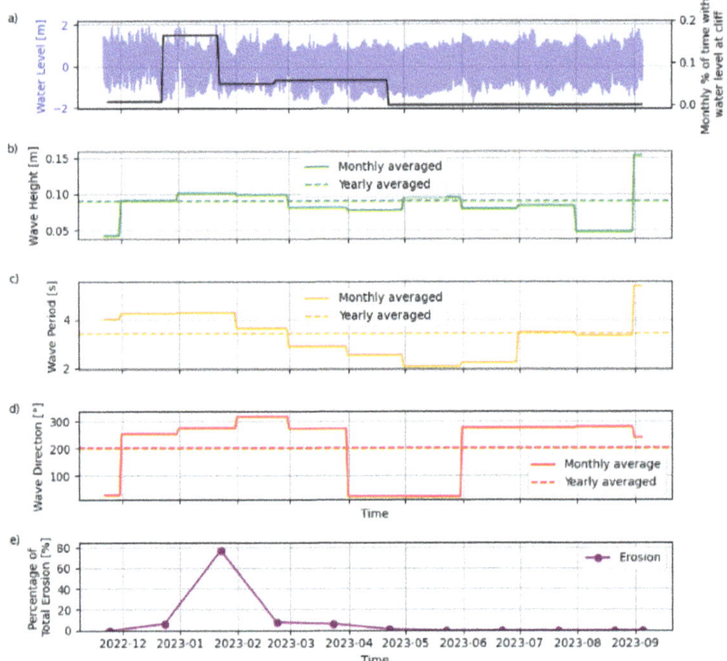

Fig. 3. a) Water levels and monthly period of water level at the cliff; b) Yearly and monthly boundary wave heights; c) Wave periods; d) Wave directions; e) Monthly lateral erosion percentage.

Notable differences in lateral retreat are observed when comparing the 2D depth-averaged and subgrid1D models (Fig. 2b). The 2D model generally predicts greater lateral retreat, as it assumes constant erosion across the entire cliff face and includes erosion for water levels above the cliff. In contrast, the subgrid1D model uses a water-level-dependent erosion rate [3] and excludes erosion above the cliff, which explains the observed differences.

When comparing the simulated lateral erosion for both models with the observed erosion derived from the drone-based DEMs, it is apparent that the subgrid model generally underestimates lateral retreat, while the 2D depth-averaged model tends to overestimate (Fig. 4). The subgrid1D model provides a closer match to the observed retreated area with monthly-averaged wave conditions, with about 121 m2 less erosion. In contrast, the 2D model shows about 221 m2 more erosion than observed.

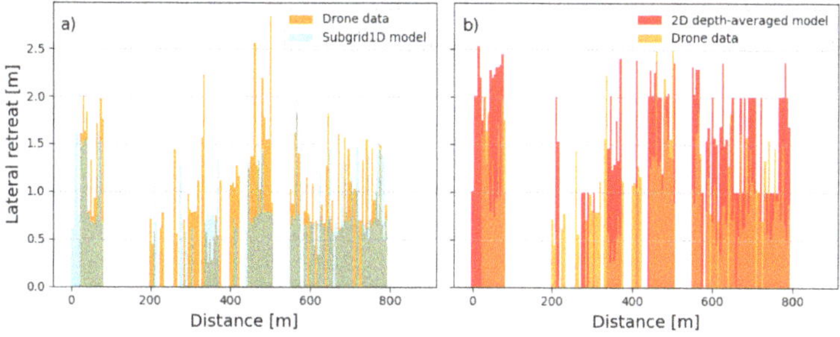

Fig. 4. Lateral retreat observed between 24-11-2022 and 05-09-2023 and modelled with the a) subgrid1D model and b) the 2D depth-averaged model over lateral marsh length (distance).

Both modelling approaches show a relatively good match with observed lateral retreat, particularly in identifying areas without cliffs, visible in areas not showing erosion. However, discrepancies in representing areas of high erosion highlight spatial variations not captured, indicating that further calibration of alpha could be needed and that there are unaccounted-for processes and/or possible deviations from the linear relationship between erosion rate and wave power. Local variability in erosion, both vertically at the cliff and spatially along the marsh, noted by Van der Wal et al. [9], emphasizes the complexity of these systems, while WinklerPrins et al. [17] highlight the challenges in addressing short-term spatial erosion variability with simple models. Therefore, incorporating non-linear erosion-wave power relationships [17, 18] and robust calibrations for marsh-specific conditions could enhance model performance [18].

4   Conclusions

This study demonstrates the value of numerical modelling in understanding the lateral erosion of salt marsh cliffs. The subgrid1D model more accurately captured observed erosion under moderate conditions, while the 2D depth-averaged model overestimated lateral retreat. Monthly wave conditions were more effective than yearly averages in reproducing erosion dynamics, highlighting the importance of shorter-term variability. However, spatial variability in erosion not represented in the models indicates the need to consider additional spatial factors, such as sediment properties. Future improvements should integrate non-linear erosion relationships and site-specific calibrations to better predict salt marsh retreat under diverse conditions.

References

1. Leonardi N, Fagherazzi S (2014) How waves shape salt marshes. Geology 42(10)
2. Marani M, D'Alpaos A, Lanzoni S, Santalucia M (2011) Understanding and predicting wave erosion of marsh edges. Geophys Res Lett 38(21)
3. Bendoni M, Georgiou IY, Roelvink D, Oumeraci H (2019) Numerical modelling of the erosion of marsh boundaries due to wave impact. Coastal Eng 152
4. Valentine K, Mariotti G (2019) Wind-driven water level fluctuations drive marsh edge erosion variability in microtidal coastal bays. Continental Shelf Res 176(2)

5. Mariotti G, Canestrelli A (2017) Long-term morphodynamics of muddy backbarrier basins: fill in or empty out? Water Resour Res 53(8)
6. Bendoni M, Mel R, Solari L, Lanzoni S, Francalanci S, Oumeraci H (2016) Insights into lateral marsh retreat mechanism through localized field measurements. Water Resour Res 52(2)
7. Leonardi N, Ganju NK, Fagherazzi S (2016) A linear relationship between wave power and erosion determines salt-marsh resilience to violent storms and hurricanes. Proc Natl Acad Sci USA 113(1):64–68
8. Schwimmer RA (2001) Rates and processes of marsh shoreline erosion in rehoboth bay, Delaware, U.S.A. J Coastal Res 17(3)
9. van der Wal D, van Dalen J, Willemsen PWJM, Borsje BW, Bouma TJ (2023) Gradual versus episodic lateral saltmarsh cliff erosion: Evidence from Terrestrial Laser Scans (TLS) and Surface Elevation Dynamics (SED) sensors. Geomorphology 426
10. RWS. (n.d.). Data Rijkswaterstaat. https://rijkswaterstaatdata.nl/
11. Siemes RWA, Borsje BW, Daggenvoorde RJ, Hulscher SJMH (2020) Artificial structures steer morphological development of salt marshes: a model study. J Marine Sci Eng 8(5)
12. Deltares: User Manual D-Flow Flexible Mesh (2020)
13. Marin-Diaz B, et al (2023) Using salt marshes for coastal protection: effective but hard to get where needed most. J Appl Ecol 60(7)
14. Baptist MJ (2005) Modelling Floodplain Biogeomorphology. PhD dissertation TUDelft
15. The Water Institute (2022) Partnership for Our Working Coast: A Community-Informed Transdisciplinary Approach to Maximizing Benefits of Dredged Sediment for Wetland Restoration Planning at Port Fourchon, Louisiana. Baton Rouge, LA
16. Dzimballa S, Willemsen PWJM, Kitsikoudis V, Borsje BW, Augustijn DCM (2025) Numerical modelling of biogeomorphological processes in salt marsh development: Do short-term vegetation dynamics influence long-term development? Geomorphology 471
17. WinklerPrins L, Lacy JR, Stacey MT, Logan JB, Stevens AW (2024) Seasonality of retreat rate of a wave-exposed marsh edge. J Geophys Res Earth Surf 129(7)
18. Houttuijn Bloemendaal LJ, FitzGerald DM, Hughes ZJ, Novak AB, Georgiou IY (2023) Reevaluating the wave power-salt marsh retreat relationship. Sci Rep 13

Adjusting the Critical Shields Parameter Within Salt Marshes Based on Vegetation-Induced Resuspension

T. J. van Veelen[1,2]($\boxtimes$), H. M. Nepf[2], S. J. M. H. Hulscher[1], S. Dzimballa[1], and B. W. Borsje[1]

[1] University of Twente, PO Box 217, 7500 AE Enschede, The Netherlands
`t.j.vanveelen@utwente.nl`

[2] Massachusetts Institute of Technology, 77 Massachusetts Avenue, Cambridge, MA 02139, USA

Abstract. Salt marshes are vegetated coastal habitats with current- and wave-driven transport of fine sediment particles. Vegetation-induced turbulence reduces the critical velocity for sediment resuspension within salt marsh canopies, which is not captured by transport models developed for bare beds. Therefore, a comprehensive flume experiment was conducted to quantify the reduction of the critical velocity for resuspension in wave-current flows and to provide recommendations for sediment transport models. The experiments with artificial *Spartina Anglica* vegetation and two non-cohesive grain sizes under combined wave-current flows showed that vegetation reduced the critical velocity by 35–64% compared to a bare bed depending on a ratio of the current and wave velocity. These results were not sensitive to variations in stem density and grain size. We showed that the reduction in critical velocity was equivalent to a reduction of 57–88% of the critical Shields parameter depending on the same ratio of the current and wave velocity. We recommended that the so-called vegetation-adjusted critical Shields parameter could be implemented in sediment transport models to include vegetation-induced resuspension.

Keywords: Salt marshes · Sediment resuspension · Shields parameter · Vegetation · Suspended sediment transport

1 Introduction

Salt marshes are intertidal coastal habitats that consist of sediments and vegetation. They are commonly found in estuaries and along sheltered coasts, where they are exposed to currents and waves. Salt marshes are important for coastal protection thanks to their capacity to attenuate incoming waves [1]. They are also dynamic habitats that grow and erode over time [2], driven by the suspended transport of fine particles and affecting the coastal protection that salt marshes provide. Therefore, it is critical that we understand and can model sediment transport dynamics within salt marshes.

© The Author(s) 2026
C. Coelho et al. (Eds.): CD 2025, CRL 41, pp. 451–458, 2026.
https://doi.org/10.1007/978-3-032-15473-6_69

Sediment transport models developed for bare beds (e.g. [3]) only partially apply to salt marshes. Vegetation increases the turbulent kinetic energy, which can elevate shear and lift forces on sediment particles [4]. Flume experiments with pure current [5–7] and pure wave flows [8, 9] showed that this led to a reduction in the critical velocity for resuspension. Using these results, Tinoco & Coco [6, 9] proposed a vegetation-adjusted critical Shields parameter which could model the reduction in the critical velocity for resuspension in pure currents and wave flows, but they did not study the application to simultaneous wave-current flows that typically occur on salt marshes.

Therefore, we derived a formulation for the vegetation-adjusted Shields parameter based on flume experiments with artificial *Spartina Anglica* canopies in combined wave-currents flows. We studied three vegetation densities and two non-cohesive grain sizes. Using the experimental data, a vegetation-adjusted Shields parameter is derived as function of a ratio between the current and wave velocity.

2 Methodology

2.1 Flume Experiments

Experiments were designed to mimic emergent salt marsh canopies with *Spartina Anglica* vegetation. Rigid wooden dowels with diameter $b_v = 6.4$ mm and height $h_v = 300$ mm were fitted in 1 m-long baseboards forming a staggered grid with stem densities of $n_v = 0$ (bare), 263 (sparse), 526 (medium), or 1053 (dense) stems/m^2. One baseboard was constructed with a recess of 10 mm which was filled with artificials sediments after the dowels were in place. The sediments were glass spheres with a mean diameter of $d_{50} = 70$ or 35 µm, representing two grain sizes commonly found on salt marshes [10]. Two normal baseboards and one sediment baseboard in the middle formed the vegetation canopy, which was placed in the wave-current flume of the Massachusetts Institute for Technology (Fig. 1). The flume measured 24 m in length, 0.38 m in width, and 1 m in height. The water depth was set at 0.22 m for all tests. The flume was fitted with a piston-type wave maker which generated regular waves and a variable-speed pump which controlled flow circulation in the same direction as wave propagation. Furthermore, one acoustic Doppler velocimeter (ADV) and two optical backscatter sensors (OBS) were installed to measure near-bed velocities and suspended sediment concentrations respectively. At the start of each experiment, the sediment bed was flattened and a current velocity below the critical velocity for resuspension was set by the pump. Then, the wave height was incrementally increased until resuspension was measured by at least one OBS. The critical wave velocity amplitude $U_{w,cr}$ was derived from the corresponding ADV signal. The critical current velocity $U_{c,cr}$ was defined as the depth-averaged velocity and was obtained from the pump. This procedure was repeated for wave-current conditions, pure wave conditions (no flow recirculation), and pure current conditions, for which the current velocity rather than wave height was increased. An overview of the experiments was listed in Table 1.

Table 1. List of experimental conditions.

Grain size	Vegetation densities tested	Number of conditions tested per density	Total number of conditions tested
$d_{50} = 70$ μm	Bare, sparse, medium, dense	6	24
$d_{50} = 35$ μm	Bare, medium	5	10

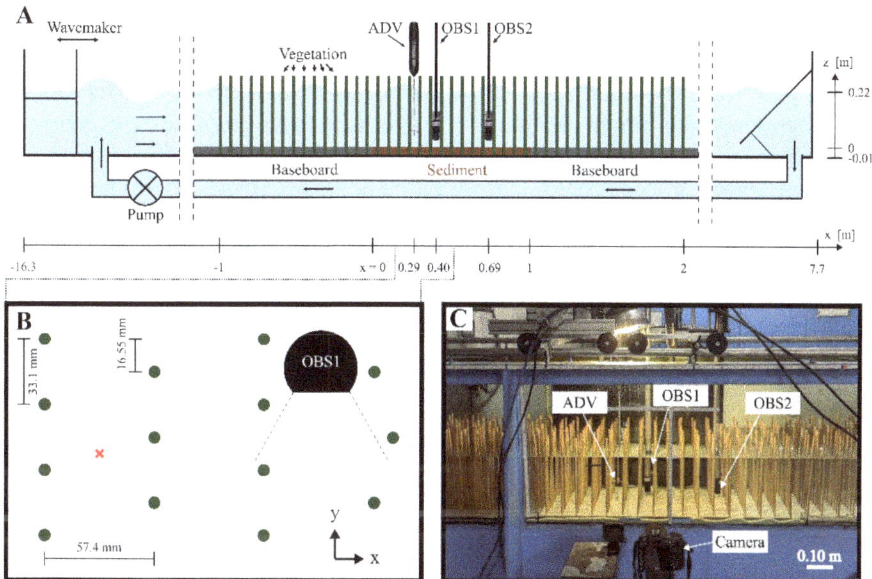

Fig. 1. (a) Side view of the experimental setup, figure not to scale. (b) Top view of the vegetation and instrument locations between $x = 0.26$ m and 0.43 m in a medium density canopy. The green circles are vegetation stems. The red cross marks the location of the ADV measurement. (c) Photo of the side view of the experimental setup. Source: van Veelen et al. [11] with minor edits.

2.2 Vegetation-Adjusted Shields Parameter

The measured critical current velocity $U_{c,cr}$ and critical wave velocity amplitude $U_{w,cr}$ were used to calculate a combined critical wave-current velocity

$$U_{wc,cr} = \sqrt{U_{c,cr}^2 + U_{w,cr}^2} \tag{1}$$

and a wave-current velocity ratio

$$\xi = \frac{U_{c,cr}^2}{U_{c,cr}^2 + U_{w,cr}^2}, \tag{2}$$

which represented the relative contributions of currents and waves to $U_{wc,cr}$. $\xi = 0$ denotes pure wave conditions, $\xi = 0.5$ marks equal current and wave velocities, and $\xi = 1$ represents pure current conditions. $U_{wc,cr}$ is fitted as function ξ to the experimental data for bare beds, resulting in $U_{wc,cr,bare}$, and for vegetated beds, resulting in $U_{wc,cr,veg}$. Following Tinoco & Coco [6, 9], the vegetation-adjusted critical Shields parameter is defined as

$$\theta_{cr,veg} = \theta_{cr,bare} \frac{U_{wc,cr,veg}^2}{U_{wc,cr,bare}^2}, \tag{3}$$

in which $\theta_{cr,bare}$ is the critical Shields parameter on a bare bed for the same sediment.

3 Results

3.1 Critical Velocity

Our results showed that vegetation reduced the critical velocity for sediment resuspension in wave-current flows for both grain sizes (Fig. 2). For the coarser grain diameter of 70 μm (Fig. 2a), the presence of vegetation had a much greater effect on the critical velocity than stem density of the canopy. Therefore, a single threshold was fitted for vegetated beds using the data from all three stem densities. Vegetation reduced the critical velocity the most in pure current flows with a 64% reduction. The smallest reduction of 35% was observed in pure wave flows. The reduction for combined wave-current was within this 35–64% range.

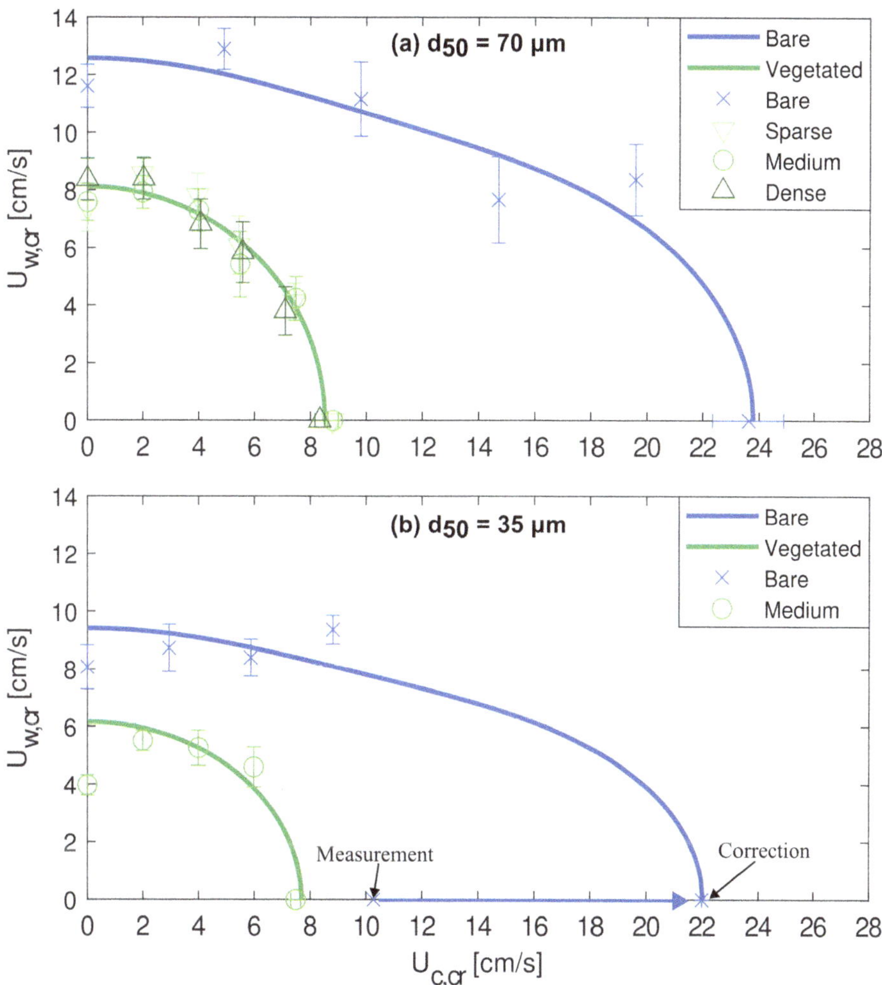

Fig. 2. The critical velocity for sediment resuspension on bare (blue) and vegetated (green) beds for grain sizes (a) $d_{50} = 70$ µm and (b) $d_{50} = 35$ µm. The markers and error bars denote the measured critical velocities and their standard deviations respectively. The solid lines are fitted thresholds. For the finest grain size (b), the bare bed pure current measurement was replaced with a value obtained from Soulsby [3], as discussed in text and marked by the arrows in (b).

For the finest grain diameter of 35 µm (Fig. 2b), the measured pure current critical velocity was $U_{c,cr} = 10.3$ cm/s on a bare bed, which was significantly smaller than $U_{c,cr} = 19\text{–}27$ cm/s measured in previous experimental studies for a similar grain size [3, 5, 12]. Furthermore, $U_{c,cr}$ should only respond weakly to a change in grain size on a bare bed [3], whereas it is reduced by more than 50% in our experiments. Therefore, we replaced the measured bare bed pure current critical velocity with $U_{c,cr} = 22$ cm/s based on Soulsby [3]. Using the replaced value, vegetation reduced the critical velocity for the 35 µm grains by a similar ratio as for the 70 µm grains with a 64% reduction under pure current flows and a 35% reduction under pure wave flows.

3.2 Adjusted Shields Parameter

The reduction in critical velocity by vegetation compared to a bare bed was expressed by an equivalent reduction in the critical Shields parameter defined by Eq. 3 (Fig. 3). The critical Shields parameter for a bare bed $\theta_{cr,bare}$ was reduced by 57–88% by vegetation-induced resuspension within a *Spartina Anglica* canopy. This reduction was nearly identical for both grain sizes, showing that our results can be applied across a range of grain sizes typically found on salt marshes. Like the critical velocities (Sect. 3.1), the greatest reduction in the critical Shields parameter was observed for pure current flows, the smallest reduction was found for pure wave flows. The observed reductions for pure currents and pure waves were comparable with those found by Tinoco & Coco [6, 9]. The vegetation-adjusted Shields parameter could be applied in sediment transport models for *Spartina Anglica* marshes to include vegetation-induced resuspension. The adjusted critical Shields parameter only addressed resuspension and did not include other effects such as root binding and cohesiveness.

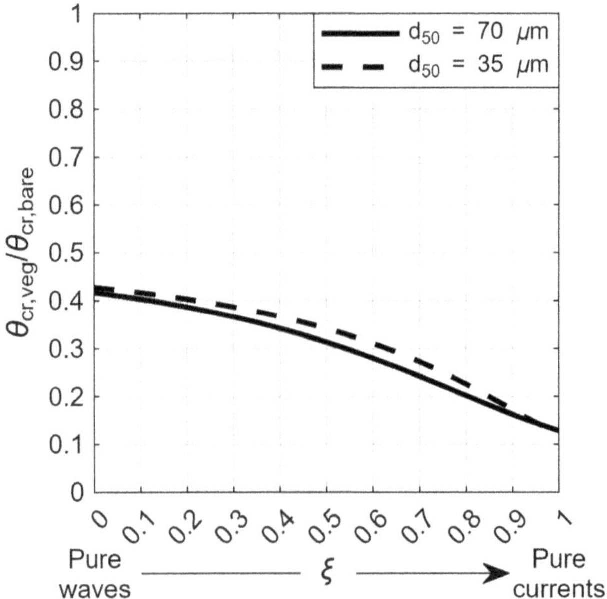

Fig. 3. Reduction in the critical Shields parameter due to vegetation-induced resuspension.

4 Conclusions

Our experimental results showed that bare bed models for sediment resuspension over-estimated the critical velocity for resuspension in salt marsh canopies under combined wave-current conditions. Artificial vegetation that was inspired by *Spartina Anglica* reduced the critical velocity by 35–64% depending on a ratio of current and wave velocities. This reduction was the same for both non-cohesive grain sizes with diameters of 70 and 35 μm. Furthermore, the critical velocity was not sensitive to stem density across the range of densities tested, suggesting that the presence of vegetation was more important than stem density for sediment transport models. We recommended that the vegetation-induced resuspension could be implemented in sediment transport models through a vegetation-adjusted critical Shields parameter. The adjusted Shields parameter was 57% to 88% lower than a bare bed critical Shield parameter depending on the ratio of currents and waves. The implementation of the vegetation-adjusted Shields parameter in sediment transport models is expected to improve predictions of suspended sediment transport within saltmarsh canopies.

References

1. Möller I, Kudella M, Rupprecht F, Spencer T, Schimmels S (2014) Wave attenuation over coastal salt marshes under storm surge conditions. Nat Geo 7(10):727–731
2. Fagherazzi S, Mariotti G, Leonardi N, Canestrelli A, Nardin W, Kearney WS (2020) Salt Marsh dynamics in a period of accelerated sea level rise. JGR: Earth Surf 125(8): e2019JF005200
3. Soulsby RL (1997) Dynamics of marine sands. Thomas Telford Publishing
4. Stoesser T, Kim SJ, Diplas P (2010) Turbulent flow through idealized emergent vegetation. J Hydr Eng 136(12):1003–1017
5. Liu C, Shan Y, Nepf H (2021) Impact of stem size on turbulence and sediment resuspension under unidirectional flow. WRR 57(3):e2020WR028620
6. Tinoco RO, Coco G (2016) A laboratory study on sediment resuspension within arrays of rigid cylinders. Adv Water Res 92:1–9
7. Yang JQ, Chung H, Nepf HM (2016) The onset of sediment transport in vegetated channels predicted by turbulent kinetic energy. Geoph Res Let 43(21):11261–11268
8. Tang C, Lei J, Nepf HM (2019) Impact of vegetation-generated turbulence on the critical, near-bed wave-velocity for sediment resuspension. WRR 55(7):5904–5917
9. Tinoco RO, Coco G (2018) Turbulence as the main driver of resuspension in oscillatory flow through vegetation. JGR: Earth Surf 123(5):891–904
10. Hu Z, Willemsen PWJM, Borsje BW, Wang C, Wang H, Bouma TJ (2021) Synchronized high-resolution bed-level change and biophysical data from 10 marsh–mudflat sites in northwestern Europe. Earth Sys Sci Data 13(2):405–416
11. Van Veelen TJ, Nepf H, Hulscher SJMH, Borsje BW, The thresholds of sediment resuspension within emergent vegetation under combined wave-current conditions – a flume experiment. Submitted manuscript
12. Widdows J, Pope ND, Brinsley MD (2008) Effect of Spartina anglica stems on near-bed hydrodynamics, sediment erodability and morphological changes on an intertidal mudflat. Mar Eco Prog Series 362:45–57

Green, Grey, or Both? Mangroves, Revetments, and Hybrid Systems to Mitigate Wave Overtopping

Tori Tomiczek[1]($\boxtimes$), Margaret Libby[2], Daniel T. Cox[3], and Pedro Lomónaco[3]

[1] United States Naval Academy, Annapolis, MD 21402, USA
vjohnson@usna.edu
[2] HDR Engineering, Inc., Corpus Christi, TX 78401, USA
[3] Oregon State University, Corvallis, OR 97331, USA

Abstract. We performed a 1:2 scale experiment investigating the performance of green (mangrove), grey (revetment), and hybrid green-grey (mangrove + revetment) systems in mitigating wave overtopping. Six configurations were tested under irregular, regular, and transient (tsunami-like) wave conditions. This paper focuses on the performance of the systems under transient wave conditions, with amplitudes between 0.113 m and 0.228 m and rise times between 4.18 s and 15.40 s. Results indicated that while the conventional (revetment only) configuration reduced total overtopping volumes compared to the baseline (wall-only configuration) by 6% to 30%, the mangrove-only configurations reduced overtopping significantly (by 25% to 54%). Hybrid mangrove + revetment configurations produced the greatest reduction in overtopping, with overtopping reductions of 44% to 83% compared to the baseline configuration. These results show the potential for nature-based and hybrid solutions to provide similar engineering benefits as conventional systems to mitigate future coastal flood hazards.

Keywords: *Rhizophora mangle* · Nature-based Solutions · Engineering With Nature · Physical model · Tsunami-like waves

1 Introduction

As effects of sea level rise and climate change manifest at an alarming rate, coastal communities across the globe are seeking coastal flood hazard mitigation strategies that are effective, robust, and resilient in the face of nonstationary future conditions. Nature-based solutions (NbS), particularly emergent vegetation such as mangroves, have rapidly gained attention in recent years owing to their ability to store carbon, provide crucial habitat, and attenuate waves, leading to other engineering benefits like damage reduction and overtopping mitigation.

© The Author(s) 2026
C. Coelho et al. (Eds.): CD 2025, CRL 41, pp. 459–466, 2026.
https://doi.org/10.1007/978-3-032-15473-6_70

1.1 Coastal Flood Hazards and Performance Metrics for Green and Grey Shore Protection Infrastructure

Mangrove effects on wave height attenuation are well-documented and predicted (*e.g.,* [1]). However, the resulting effects of wave height attenuation on other performance metrics (*e.g.,* wave overtopping, wave runup, and wave forces on near-coast infrastructure) are not well understood, even though these metrics may be of greater interest for engineers and planners to ensure the robustness and resilience of coastal communities. Only a few studies have considered wave height attenuation through mangrove forests and the subsequent effects on wave overtopping [2], wave runup [3], or wave force reduction [4]. Similarly, few direct comparisons of green solutions relative to their grey counterparts such as rock revetments exist, and the combined effects of hybrid green-grey systems (e.g., mangrove + revetment) have not been quantified.

1.2 Wave Height Attenuation and Other Performance Metrics

This study focuses on wave overtopping (Fig. 1), which can create damaging and dangerous conditions on coastal roadways or pedestrian walkways and is an important performance metric for nearshore communities. We evaluated the effectiveness of a mangrove forest, a rock revetment alone, and a combination of the two (hybrid system) to reduce wave overtopping of a vertical wall. We hypothesized that the hybrid system can be modeled as a linear system (Fig. 2). Analogous to light passing through two filters, f_1 (green, mangroves) and f_2 (grey, revetment), the quantity passing through the green filter can be given by $A_1 = A_0 (1\text{-}f_1)$ and through the grey filter by $A_2 = A_1 (1\text{-}f_2)$. Combining the two filters, the quantity passing through a green + grey filter, f_3, is $A_3 = A_0(1\text{-}f_3)$, where $f_3 = (f_1 + f_2 - f_1 f_2)$.

We conducted physical model experiments to investigate whether this analogy holds for the complex process of wave attenuation and overtopping. If the relation holds, this could simplify the use of vegetation in coastal engineering design of hybrid systems because f_1 and f_2 can be calculated from well-known equations.

Fig. 1. Wave overtopping a seawall during coastal flood event in Annapolis, MD, USA.

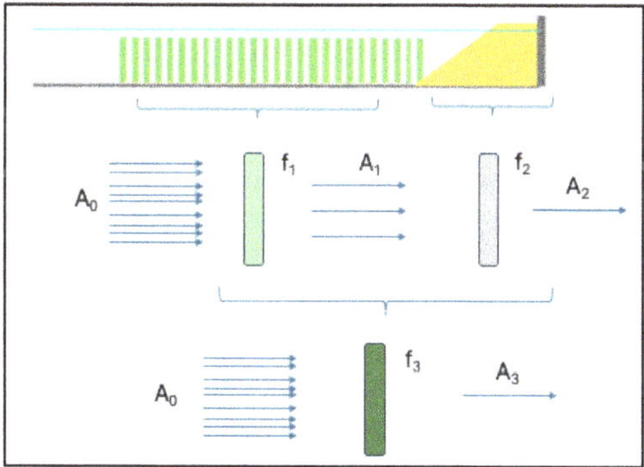

Fig. 2. Filter analogy for linear performance of green-grey systems.

2 Physical Model Experiments

The physical model was constructed in the Large Wave Flume at Oregon State University. Mangrove specimens were the same as used by [5] to model wave height attenuation; individual specimens were constructed of polyvinyl chloride (PVC) trunks with diameter 0.06 m; each trunk had fourteen PEX tubing roots (seven root pairs) with diameter 0.016 m. The specimens were based on a 1:2 scale model of reported *Rhizophora mangle* characteristics in a mature forest in Rookery Bay, FL [6]. The Large Wave Flume measures 3.7 m wide × 107.0 m long; a profile view of the experimental setup is shown in Fig. 3. Six configurations were tested to consider conventional (grey) infrastructure and mangrove forests with realistic densities and cross-shore width. Three configurations were tested with a wall (A: wall with no mangroves, B: wall with a forest with density N, 0.20 tree/m^2, and C: wall with a forest of density 2N), and three configurations were tested with a wall in combination with a revetment (D: wall + revetment with no mangroves, E: wall + revetment with mangrove forest of density 2N, and F: wall + revetment with forest of density 4N). For all configurations with the forest, the forest cross-shore width was 19.60 m.

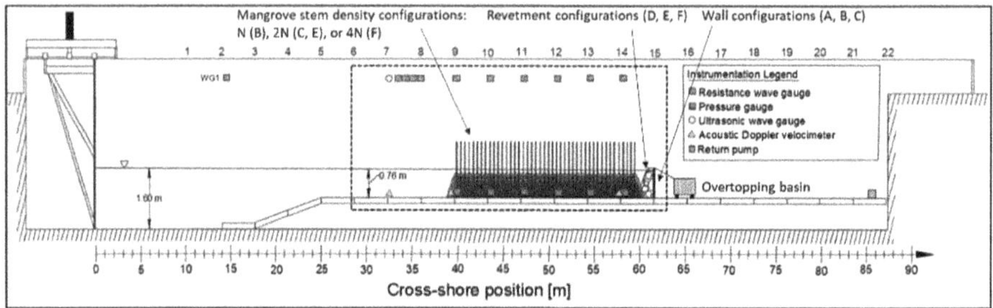

Fig. 3. Profile view of experiments.

Regular, single- and double-peaked irregular wave spectra, and transient wave conditions were tested, with significant wave heights between 0.13 m and 0.30 m and peak periods between 3 s and 8 s. Water surface elevations were measured along the test section using wire-resistance and ultrasonic wave gauges and pressure gauges installed throughout the mangrove forest, and acoustic Doppler velocimeters measured water particle velocities. The water depth was held constant at 1.60 m, giving a water depth at the wall of 0.76 m and a freeboard of 0.09 m. A 0.997-m wide tray directed overtopping into a catchment basin; overtopping was recorded using four load cells and a wire-resistance wave gauge. Overtopped water was returned to the flume during and after tests to maintain a constant still water level along the test section. Single-peaked irregular wave data were analyzed by [2], who found that (1) the lowest-density mangrove forest provided similar mitigation of overtopping discharge compared to the revetment alone, and (2) assumption of linear (independent) performance of the green and grey systems for overtopping reduction was conservative and accurate to within 10%.

This study extends the work of [2] to consider the performance of each configuration under transient (tsunami-like) wave conditions. The transient wave is similar in shape to a solitary wave, but uses the full stroke of the wave machine following an error function time series for wavemaker displacement, resulting in a longer, more realistic tsunami-like wave. Table 1 lists the three transient wave conditions, averaged across all six configurations, measured at wg2, located in the array seaward of the mangrove forest. Wave amplitudes A ranged 0.113 m to 0.228 m and rise times T_{rise}, defined here as the time between when the water surface elevation exceeded 5% of the wave amplitude and the maximum water surface elevation during the wave, ranged 4.18 s to 15.40 s.

Table 1. Transient wave condition, amplitude A, and rise time, T_{rise}, measured at wg2. Values in parentheses indicate standard deviation.

Transient Wave Condition	A [m]	T_{rise} [s]
Tr-WC-1	0.113 (0.004)	15.40 (0.49)
Tr-WC-2	0.153 (0.003)	8.56 (0.13)
Tr-WC-3	0.228 (0.005)	4.18 (0.07)

3 Results and Discussion

Figure 4 shows the total overtopping volume, normalized for the width of the overtopping tray, for each configuration and transient wave condition tested. Blue symbols indicate configurations with the wall, while orange symbols denote configurations with the wall + revetment. As seen in the figure, all protection configurations provided overtopping reduction compared to the baseline (wall only) configuration. However, for the long waves considered here, the green configurations (mangroves only, configurations B and C) were more effective at reducing wave overtopping than the grey configuration (revetment only, configuration D) as indicated in Table 2. These results are consistent with [2], who found that mangroves provided better protection for longer-period random waves compared to the revetment. Doubling the forest density did not result in a linear reduction in overtopping volumes; in general, higher forest density was observed to provide greater benefit for higher-amplitude wave conditions. Combining green and grey infrastructure resulted in the greatest reduction in overtopping, with the lowest overtopping volumes observed for configuration F (wall + revetment + 4N-density forest), with reductions between 64% to 83% (Table 2).

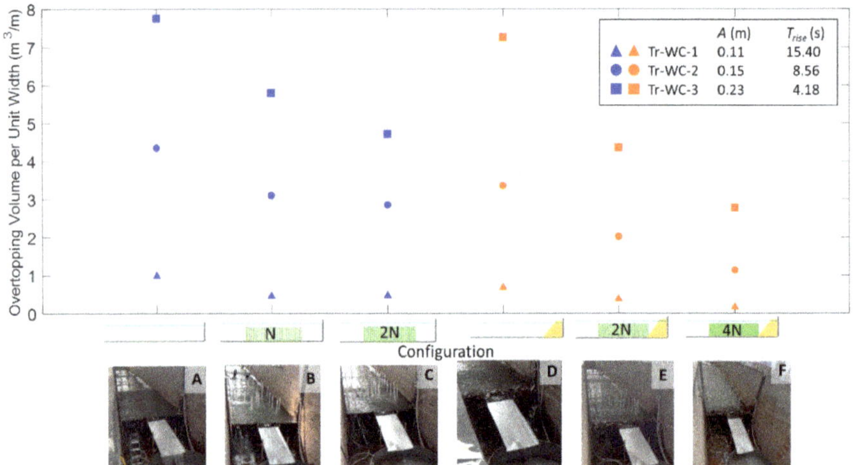

Fig. 4. Overtopping volumes per unit width (m³/m) for transient wave trials across tested configurations: A- wall only; B- wall + N-density forest; C- wall + 2N-density forest; D- wall + revetment only; E- wall + revetment + 2N-density forest; F- wall + revetment + 4N density forest

Table 2. Overtopping volumes per unit width (m³/m) for each transient wave condition. Values in parentheses indicate percent reductions from baseline configuration (A).

Condition	Volume per unit Width [m³/m] (Percent Reduction [%])					
	A (wall)	B (wall + N)	C (wall + 2N)	D (wall + rev)	E: (wall + rev + 2N)	F: (wall + rev + 4N)
Tr-WC-1	0.99 (0)	0.46 (54)	0.47 (52)	0.69 (30)	0.38 (61)	0.17 (83)
Tr-WC-2	4.35 (0)	3.11 (29)	2.86 (34)	3.36 (23)	2.03 (53)	1.14 (74)
Tr-WC-3	7.75 (0)	5.80 (25)	4.72 (39)	7.27 (6)	4.36 (44)	2.79 (64)

4 Conclusions

This paper presents results of a 1:2-geometric scale experiment investigating the overtopping mitigation performance of nature-based (mangrove), conventional (revetment), and hybrid (mangrove + revetment) systems subjected to transient wave conditions. Measured overtopping volumes for each configuration showed that, for the transient (tsunami-like) wave conditions tested in these experiments, the revetment provided some overtopping mitigation, while mangrove configurations significantly reduced the overtopping volumes. The hybrid (green + grey) configurations further reduced overtopping, showing the improved performance of multi-tiered solutions.

These results suggest that nature-based solutions (emergent vegetation) or combinations of vegetation with structural measures can perform at a similar or higher level as conventional structural measures to mitigate overtopping. However, additional work is needed to understand how green and grey systems combine and their respective contribution to other performance metrics (runup, wave height attenuation, erosion mitigation, *etc.*). Relationships between wave height, wave period, and system performance must also be further investigated. As the performance of green, grey, and hybrid systems continues to be quantified, coastal engineers will become more equipped with tools needed to design resilient systems that ensure the vitality of coastal landscapes and communities in the face of climate change and future coastal flood hazards.

Acknowledgements and Data Availability. The authors thank Tim Maddux and Rebekah Miller of the O.H Hinsdale Wave Research Laboratory staff, Duncan Bryant of the U.S. Army Corps of the Engineers for the use of the model mangroves, USNA midshipmen McKenna Brophy, Jordan Keck, Dan McMann, and Nico Valdivieso, and NHERI REU undergraduate researchers Alanna DaRin, Brenna Derby, and Grace Pond for their assistance with the experiments. The project was supported by funding from the National Science Foundation through Grants 2037914, 2110262, and 2110439 and the US Army Corps of Engineers through project number W912HZ2120045. Any opinions, findings, conclusions, or recommendations expressed in this material are those of the authors and do not necessarily reflect the views of the NSF, USACE, or USNA. The data are publicly available on DesignSafe.org as PRJ-4041, see [2] and [7].

References

1. Kelty K, Tomiczek T, Cox DT, Lomónaco P, Mitchell W (2022) Prototype-scale physical model of wave attenuation through a mangrove forest of moderate crossshore thickness: LiDAR-based characterization and Reynolds scaling for engineering with nature. Front Mar Sci 8(780946), 2(5):1–18. https://doi.org/10.3389/fmars.2021.780946
2. Libby M, Tomiczek T, Cox D, Lomónaco P (2024) The sum of the parts: Green, gray, and green-gray infrastructure to mitigate wave overtopping. Coast Eng 194(104615):1–15. https://doi.org/10.1016/j.coastaleng.2024.104615
3. Roldán MMM, Lara JL, Losada IJ (2024) Hybrid solutions for coastal protection combining physical and numerical CFD modeling. In: Proceedings of the 9th International Conference on Physical Modelling in Coastal Engineering (Coastlab24), pp 1–2. TU Delft, Netherlands. https://doi.org/10.59490/coastlab.2024.813
4. Tomiczek T, Mitchell WT, Lomónaco P, Cox DT, Kelty K (2024) Prototype-scale laboratory observations of wave force reduction by an idealized mangrove forest of moderate cross-shore width. Ocean Eng 310(2):118740, 1–17. https://doi.org/10.1016/j.oceaneng.2024.118740
5. Bryant M, Bryant D, Provost L, Hurst N, McHugh M, Wargula A, Tomiczek T (2022) Wave attenuation of coastal mangroves at a near-prototype scale. Engineer Research and Development Center, USACE. https://doi.org/10.21079/11681/45565
6. Novitzky P (2010) Analysis Of Mangrove Structure And Latitudinal Relationships On the Gulf Coast Of Peninsular Florida [MA Thesis]. University of South Florida
7. Libby M, T. Tomiczek D, Cox P (2024) Lomonaco: large-scale physical model study of wave overtopping mitigation by hybrid infrastructure. In: Understanding Hybrid Green-Gray Coastal Infrastructure Processes and Performance Uncertainties for Flood Hazard Mitigation. DesignSafe-CI. https://doi.org/10.17603/ds2-9pqq-qa78

Coasts and Climate Change Effects and Responses

Parameterised Barrier Morphological Evolution for Future Storm Impact Assessment

Carlos Loureiro[1,2(✉)], Nicola Horsburgh[3], and Óscar Ferreira[1]

[1] CIMA/ARNET, Faculdade de Ciências e Tecnologia, Universidade do Algarve, 8005-139 Faro, Portugal
cloureiro@ualg.pt
[2] School of Agricultural, Earth and Environmental Sciences, University of KwaZulu-Natal, Durban 4000, South Africa
[3] Biological and Environmental Sciences, Faculty of Natural Sciences, University of Stirling, Stirling FK9 4LA, Scotland

Abstract. Coastal storm impacts are mainly determined by the interaction of extreme water levels with barrier morphology, particularly the height of the foredune. While future projections of storm hydrodynamic forcing, integrating sea level rise, tides, waves and storm surges, are readily available, predicting future barrier morphology at a decadal scale remains a challenge due to uncertainties in modeling morphodynamic processes. This study proposes a simplified model to predict future foredune morphology at decadal scale and determine storm impacts on coastal barriers, focusing on the critical role of foredune morphology. The model identifies three morphological behavior modes based on the nature of the sediment budget and associated shoreline dynamics, distinguishing between: "keep-up mode" for accreting barriers, "catch-up mode" for stable to mildly accreting barriers, and "give-up mode" for eroding barriers. Application to the Portuguese coast, specifically the eroding barrier of Torrão do Lameiro, demonstrates that even with rising sea level and higher total water levels, storm impacts remain within the same regime by 2050. This simplified model offers a practical approach for considering an evolving foredune morphology, which can support first-order assessments of future storm impacts in changing climate.

Keywords: Barrier Morphology · Foredune · Sediment Budget · Storm Impacts · Coastal Behaviour Modelling

1   Introduction

The impacts of storms on coastal barriers are largely determined by the interaction of extreme water levels with the morphological characteristics of the beach-dune system [1]. The storm-induced extreme water levels determine the severity of the forcing, while the morphology of the beach and, more importantly the dune, regulates the ability of barriers to withstand erosion and prevent flooding. This interdependency of hydrodynamic forcing and morphological setting is thus fundamental for categorising and predicting storm impacts [2].

© The Author(s) 2026
C. Coelho et al. (Eds.): CD 2025, CRL 41, pp. 469–474, 2026.
https://doi.org/10.1007/978-3-032-15473-6_71

When considering future storm impacts in a changing climate, predictions of hydrodynamic forcing integrating sea level rise (SLR), tides, storm surge and waves are nowadays readily available from individual or ensemble climate projections e.g. [3]. These are increasingly used to determine changes in extreme water levels and estimate flooding extents at local to global scales [4, 5]. However, predictions of future barrier morphology remain elusive, as morphodynamic linkages in beach-dune systems across scales are still poorly resolved, particularly when scaling from process observations to decadal morphological changes [6]. For storm impact assessments, this results in an incomplete, but often applied approach, where future extreme sea levels are only compared to present day barrier morphology [4, 7].

Predicting future dune morphology, particularly the height of foredunes, is fundamental for assessing future storm vulnerability in coastal barriers [8]. However, despite recent advances in coastal dune morphological models, both process-based or data-driven, there are still many uncertainties in the numerical modelling of long-term (decadal) coupled beach and dune morphodynamic processes [9]. Therefore, simplified models that describe key evolutionary modes, integrating empirical observations or parameterising observed barrier change behaviours, remain a useful solution for predicting coastal morphological change. The reduced complexity in terms of processes and data used, alongside the low computational requirements make such models particularly useful for long-term and large-scale application, and potentially useful for integration in climate change impact assessments. Aiming to evaluate future storm impacts in coastal barriers, this work proposes a parameterised morphological behaviour model of barrier evolution, exploring its application to the Portuguese coast.

2   Model Framework

For first-order assessment of storm impacts on coastal barriers, such as the well-established Storm Impact Scale outlined by Sallenger [1], the elevation of the foredune toe (D_{low}) and crest (D_{high}) are critical morphological parameters. While these two variables simplify barrier morphology, they represent the key vertical dimensions and are commonly incorporated in simplified or idealized coastal behaviour models. Considering this, to parameterise barrier evolution the proposed modelling framework considers exclusively changes to the height of the toe and the crest of the foredune.

Foredune evolution depends primarily on wind, vegetation and sediment supply [10]. At decadal timescales, and assuming quasi-stationary wave and wind regimes, while excluding human impacts and changes in vegetation cover, the elevation of foredunes is mainly controlled by sediment budget, shoreline change and beach morphodynamic type [11]. Although morphodynamic beach type is variable around a modal state, it can be assumed stable at decadal scales if there are no changes to sediment type. This leaves sediment budget and associated shoreline change as key controls of decadal foredune morphological evolution. Different foredune evolutionary behaviour modes can then be considered for accreting, stable and eroding barriers [11]. Taking a sediment budget conceptual approach to foredune evolution [10] and focussing exclusively on the decadal change of the foredune toe and crest elevation in a context of rising sea levels, three morphological behaviour modes are proposed (Fig. 1). In moderately to highly accreting barriers, indicative of a strongly positive sediment budget, foredunes can keep pace with rising sea levels, increasing their toe and crest elevations at the rate of SLR as the shoreline progrades (*keep-up mode*). In stable to mildly accreting barriers, where a slightly positive sediment budget is necessary to maintain the shoreline position with rising sea level, the foredune toe elevation increases at the rate of SLR, but the foredune crest remains at the same elevation (*catch-up mode*). In this mode, most of the sediment is used to aggrade the beach and also prograde the stoss slope of the foredune, limiting the sediment availability for vertical foredune growth. Finally, in eroding barriers the sediment budget is negative, and the foredune retreats landward, with the crest elevation determined by the topography of the landward dune ridges (*give-up mode*). In this mode, the toe elevation continues to increase at the rate of SLR, as sediment is supplied to the beach from the eroding dunes.

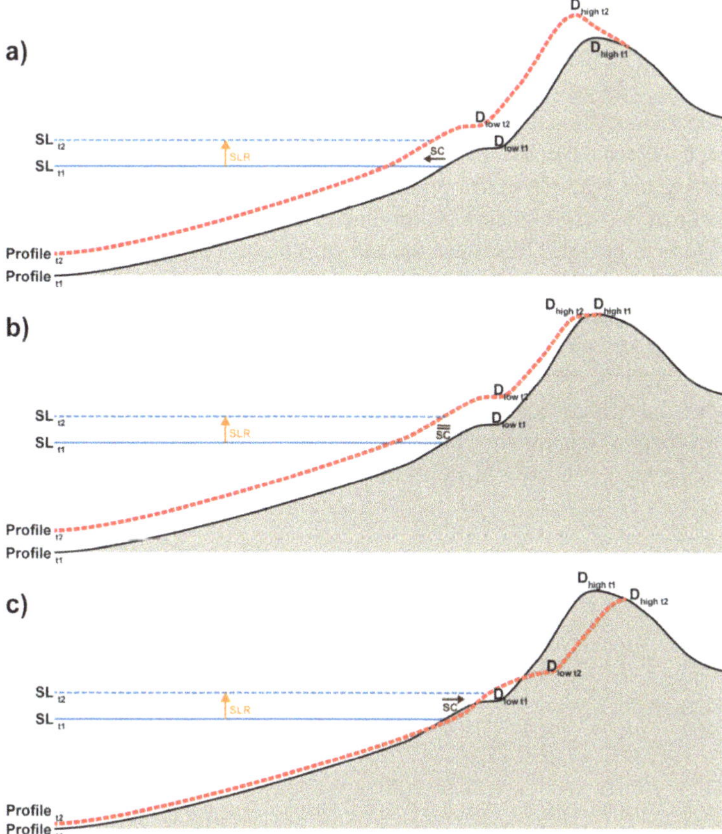

Fig. 1. Schematic overview of change in foredune morphological parameters (D_{low} and D_{high}) in relation to shoreline change (SC) and sea level rise (SLR). a) keep-up mode, b) catch-up mode and c) give-up mode.

Across the three behaviour modes the elevation of the foredune toe increases at the rate of SLR, as the beach profile adjusts vertically to maintain morphodynamic equilibrium with mean sea level [12]. This reflects the fact that, at decadal scales, water levels will be permanently raised by the amount of SLR, but only episodically raised by the storm-induced surge and runup, and foredune toe reflects only the permanent increase in water levels. If the wave regime is not quasi-stationary, changes in the modal wave climate should be incorporated in the trajectory of the foredune toe elevation.

3   Application to the Portuguese Coast

The morphological behaviour model was applied to sandy barriers along the Portuguese coast, using high-resolution LiDAR data of beach and dune systems collected in 2011, combined with multidecadal shoreline change rates from 1958 to 2010 determined by Lira et al. [13]. Storm impact regimes were determined following Sallenger [1] for the present and in 2050 considering the RCP8.5 climate change scenario. Extreme water levels, specifically total water level (R_{high}) and maximum still water levels (R_{low}), were computed for the 100-year return period storm using wave and storm-surge projections from the Copernicus Climate Change Service for the European coastline, alongside the regional IPCC AR6 median sea level projection and considering a mean high tide level.

Considering the example of Torrão do Lameiro barrier, located in the exposed NW coast of Portugal and experiencing a shoreline change rate of −0.5 m/yr, applying the give-up mode to project the 2050 dune toe and crest elevations results in a D_{low} of 7.6 m and D_{high} of 12.6 m, respectively (Fig. 2). The exact crest location is adjusted from the stoss slope to the adjacent peak to better align with dune crest morphology. In this case, even with a projected increase in R_{high} of 1 m and R_{low} of 0.6 m, the total water level remains under the dune crest elevation estimated for 2050, as the retreating shoreline will reach a higher dune ridge located landward of the current foredune. As such, a collision storm impact regime is maintained between the present and 2050, despite the rise in sea level and the increased severity of storm forcing.

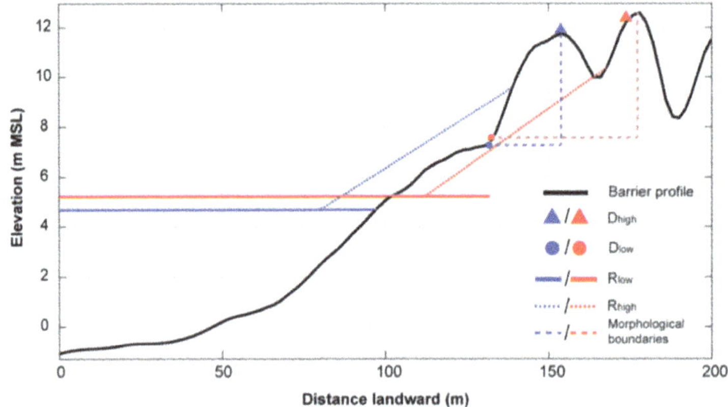

Fig. 2. Overview of present (blue) and RCP8.5 2050 projection (red) of foredune morphological parameters (D_{high} and D_{low}) and and extreme water levels (R_{high} and R_{low}) for Torrão do Lameiro.

4 Discussion

Sandy barriers are highly variable in form and setting, which results in a range of morphological responses to SLR that depend on a variety of processes and variables that cannot be meaningfully captured by a single model [14]. However, at decadal to centennial scales, the nature of the overall sediment budget and associated shoreline change can be used to infer generic barrier and foredune morphological change patterns under positive, neutral and negative sediment budgets [12]. The parameterised morphological behaviour model of barrier evolution proposed in this work incorporates simplified relationships between shoreline and foredune morphological change under rising sea levels, which although not universally valid, attempt to distinguish between common behaviour modes. The reliance on shoreline change and barrier topography metrics makes this approach widely applicable, particularly given the increasing availability of medium resolution global datasets that can be used to obtain, even if with limitations e.g. [15], all the hydrodynamic and morphological parameters considered in the model.

Although the use of on invariant coastal morphology is commonly recognised as a key limitation for assessments of future storm impacts [4, 7], very few studies have incorporated changes to future barrier morphology when modelling the potential impacts of storm events in a changing climate e.g. [16]. As first-order storm impacts in coastal barriers are largely controlled by foredune height [1, 8], predicting the entire barrier profile is not essential. Foredune toe and particularly crest, i.e. D_{high}, are the fundamental variables that must be considered for more confidently determining if future storms are likely to lead to increased collision, overwash and inundation of coastal barriers. Applying the parameterised model to Torrão do Lameiro, a coastal barrier under significant erosion and exposed to more extreme total water levels, indicates that storm impacts will remain within the same regime towards the mid-century. This is due to the presence of a second, inland-located, dune ridge, which highlights the critical importance of landward morphology when evaluating storm impacts in retreating coastal barriers under the give-up mode.

5 Conclusion

This work proposes a new simplified morphological behaviour model to estimate future foredune morphological parameters required for first-order assessments of storm impacts in coastal barriers. This model can be applied with limited information, but addresses a critical limitation in storm impact assessments in a changing climate, which typically evaluate future extreme water levels but do not consider the expected multi-decadal variation in barrier morphology. The application to Torrão do Lameiro in the NW coast of Portugal demonstrates that, despite the increase in total water levels, future storm impacts remain in the collision regime, even for extreme 100-year return period storm conditions.

References

1. Sallenger A (2016) Storm impact scale for barrier islands. J Coast Res 16(3):890–895

2. Leaman C, Harley M, Splinter K, Thran M, Kinsela M, Turner I (2021) A storm hazard matrix combining coastal flooding and beach erosion. Coast Eng 170:104001
3. Jevrejeva S, Williams J, Vousdoukas M, Jackson L (2023) Future sea level rise dominates changes in worst case extreme sea levels along the global coastline by 2100. Environ Res Lett 18(2):024037
4. Thomas J et al (2024) The projected exposure and response of a natural barrier island system to climate-driven coastal hazards. Sci Rep 14:25814
5. Kirezci E et al (2020) Projections of global-scale extreme sea levels and resulting episodic coastal flooding over the 21st Century. Sci Rep 10(1):11629
6. Walker IJ et al (2017) Scale-dependent perspectives on the geomorphology and evolution of beach-dune systems. Earth-Sci Rev 171:220–253
7. Passeri D, Bilskie M, Plant N, Long J, Hagen S (2018) Dynamic modelling of barrier island response to hurricane storm surge under sea level rise. Clim Change 149:413–425
8. Plant N, Thieler R, Passeri D (2016) Coupling centennial-scale shoreline change to seal-level rise and coastal morphology in the gulf of Mexico using a Bayesian network. Earth's Futur 4:143–158
9. Kombiadou K, Costas S, Roelvink D (2023) Exploring controls on coastal dune growth through a simplified model. J Geophys Res Earth Surf 128:e2023JF007080
10. Davidson-Arnott R et al (2018) Sediment budget controls on foredune height: comparing simulation model results with field data. Earth Surf Proc Landf 43:1798–1810
11. Doyle TB, Hesp P, Woodroffe CD (2024) Foredune morphology: regional patter and surfzone-beach-dune interactions along the New South Wales coast. Earth Surf Proc Landf 49:3115–3138
12. Davidson-Arnott R, Bauer B (2021) Controls on the geomorphic response of beach-dune systems to water level rise. J Great Lakes Res 47:1594–1612
13. Lira C, Silva A, Taborda R, Andrade C (2016) Coastline evolution of Portuguese low-lying coast in the last 50 years: an integrated approach. Earth Syst Sci Data 8:165–278
14. Cooper J et al (2020) Sandy beaches can survive sea level rise. Nat Clim Chang 10:993–995
15. Fanti V, Ferreira Ó, Kümmerer V, Loureiro C (2023) Improved estimates of extreme wave conditions in coastal areas from calibrated global reanalyses. Commun Earth Environ 4(1):151
16. Passeri D, et al (2020) The roles of storminess and sea level rise in decadal barrier island evolution. Geophys Res Lett 47:e2020GL089370

Methodological Approach to High-resolution Future Wave Climate Projections Using a Relocatable Model "BinWaves"

Gabriel Bellido Prieto[✉], Laura Cagigal Gil, and Fernando Méndez Incera

Geomatics and Ocean Engineering Group, University of Cantabria, Santander, Spain
bellidog@unican.com

Abstract. Providing computationally efficient and robust high-resolution projections of coastal wave directional spectra is a key challenge in coastal prediction under future scenarios. On one hand, the available wave spectra outputs from CMIP6 global climate models (GCMs) are limited and scarce; on the other hand, dynamic downscaling models are often computationally expensive, making it impractical to explore future scenarios in a probabilistic and detailed manner. To address these issues and overcome existing limitations, the present study proposes a novel and innovative framework to obtain the full directional wave spectra at a specific location using representative bulk parameters. For this purpose, clustering techniques and the analysis of their conditioned probabilities are established and validated over a common historical period. Once the spectra are generated and bias-corrected using a robust directional technique, the hybrid additive model BinWaves [1] is employed to efficiently downscale the waves to the coast, providing highly detailed and reliable information for analyzing probabilistic future impacts under different scenarios.

Keywords: Hybrid model · Directional wave spectra · Spectral correction · Linear wave · Wave projections

1 Introduction

Understanding future changes in wave patterns becomes crucial for effective coastal management. This study focuses on analysing the historical wave climate patterns of the coast of Cantabria and their future changes by applying "BinWaves", an efficient hybrid methodology [1].

"BinWaves" is the first additive hybrid model that propagates waves from deep to intermediate waters taking into account the full complexity of the directional wave spectrum. Recognized for its versatility, this methodology is capable not only of reconstructing high-resolution historical wave data series but also of projecting future wave climates based on climate change projections in an efficient manner. Therefore, this study uses as forcing historical wave data [2] and the results from 140-year wind-wave climate simulations (1961–2100), forced with different CMIP6 Global Climate Models (GCMs) under two different climate scenarios: SSP1–2.6 and SSP5–8.5 [3].

© The Author(s) 2026
C. Coelho et al. (Eds.): CD 2025, CRL 41, pp. 475–479, 2026.
https://doi.org/10.1007/978-3-032-15473-6_72

2 Methodology

2.1 Data Integration and Reconstruction

Firstly, the bulk parameters wave projections used, were corrected by a directional bias adjustment methodology to improve their representativeness in the offshore domain. Afterwards, a relationship between the full directional spectra and the bulk parameters over the historic period was determined using clustering techniques. This approach, based on the K-means algorithm, allows to identify characteristic patterns of the wave spectra and to assign probabilities of occurrence to each of the identified clusters. With this methodology, it was possible to generate representative full wave spectra for the corrected wave projections (see Fig. 1).

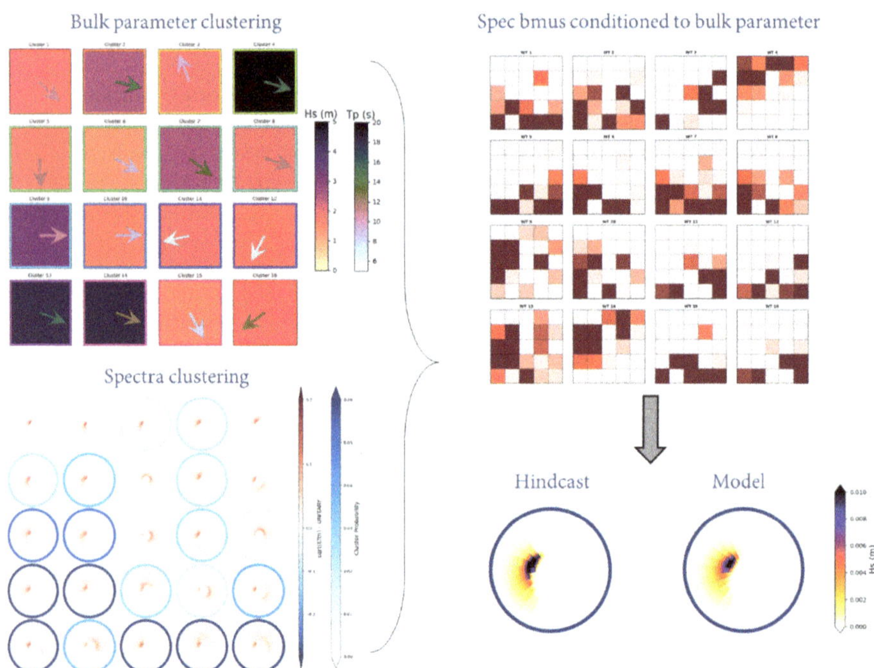

Fig. 1. Flowchart of the clustering techniques to determine wave spectra based on bulk parameters.

2.2 BinWaves Application

The corrected wave spectra have been used as inputs for reconstructing new directional wave time series for each horizon, scenario, and GCM, based on the BinWaves hybrid methodology. Hybrid models combine the computational efficiency of statistical models with the precision and detail of dynamical downscaling models, resulting in faster and more cost-effective wave predictions.

BinWaves involves disaggregating the full directional wave spectrum into its component energy "bins" of discretized frequency and direction. Under the assumption that, as waves propagate in deep waters, non-linear interactions are negligible, it is possible to build a library of pre-run propagation coefficients "BinWaves Library" in the area of interest using the numerical model SWAN.

This library serves as the baseline for the reconstruction of directional wave spectra that represents the wave energy at the boundary, so that the propagated spectra at any point of the domain were obtained by rescaling and aggregating the different energy bins (see Fig. 2).

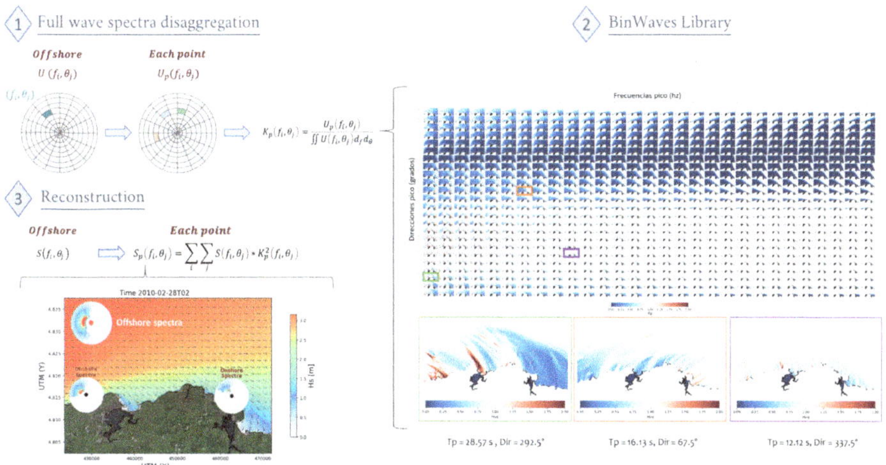

Fig. 2. Flowchart for the BinWaves Methodology.

3   Results

The proposed framework, based on the development of a method to obtain the directional wave spectra from its bulk parameters, combined with the additive model BinWaves allows to reconstruct, in a matter of seconds, high-resolution full directional spectra spatial fields for both the historical period and future wave projections. This method provides an ideal framework to analyse the expected local changes in wave patterns on the coast under different climate scenarios.

By validating the hindcast with regional buoy data in the coast of Cantabria, and applying corrections to future projections, the methodology ensures robust and reliable wave projections, that can be easily replicated in other regions worldwide (see Fig. 3).

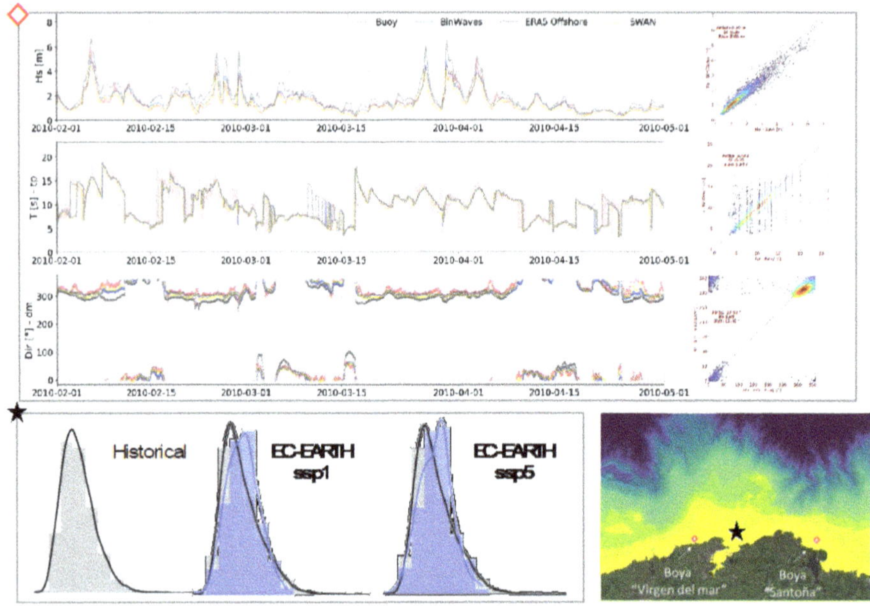

Fig. 3. Hindcast validation and Hs changes in a specific point.

Acknowledgements. This work has been partially funded by the projects MyFlood (PLEC2022–009362 - MCIN/AEI/https://doi.org/10.13039/501100011033 and European Union Next GenerationEU/PRTR), HyBay (PID2022-141181OB-I00, MCIN/AEI/https://doi.org/10.13039/501100011033/FEDER, EU), and PerfectStorm (2023/TCN/003 - Government of Cantabria/FEDER, EU). LC acknowledges the funding from the Juan de la Cierva – Formación FJC2021–046933-I/ MCIN/ AEI/ https://doi.org/10.13039/501100011033 and the European Union "NextGenerationEU"/ PRTR.

References

1. Cagigal L, Mendez F, Ricondo A, Gutierrez-Barcelo D, Bosserelle C, Hoeke R (2024) BinWaves: an additive hybrid method to downscale directional wave spectra to nearshore areas. Ocean Model 189:102346
2. ECMWF (2023) IFS Documentation CY48R1 - Part VII: ECMWF Wave Model, vol. 7
3. Meucci A, Young IR, Trenham C et al (2024) An 8-model ensemble of CMIP6-derived ocean surface wave climate. Sci Data 11 (2024)

A Methodology for Seasonal and Long-Range Forecasting of Shoreline Erosion Hazards Across Australia

Katie Wilson$^{(\boxtimes)}$, Mitchell D. Harley, and Kristen D. Splinter

Water Research Laboratory, UNSW, Sydney 2093, Australia
`katie.wilson@unsw.edu.au`

Abstract. This study presents a novel methodology to forecast shoreline position at the seasonal to five-year time scale using the statistical time series models SARIMAX and SVARX in combination with satellite-derived shorelines (Coast-Sat). 25 years of CoastSat shorelines at Curl Curl Beach in SE Australia were used to test and train the models, with forecast inputs including offshore and nearshore significant wave height and several climate indices relevant to the Australian coastline (ENSO and IOD). Over a five-year forecast horizon, it was found that the SARMIAX and SVARX models showed higher skill than a baseline linear regression model, with RMSE of approximately 10 m across all nine beach transects. The performance of these statistical time series models indicates the potential for the methodology to be extended to large spatial scales, providing valuable long range (i.e. seasonal to multi-year) forecasts of shoreline change to assist coastal managers.

Keywords: Shoreline forecasting · coastal erosion · remote sensing · statistical modelling · risk preparedness

1 Introduction

With coastal population densities growing and climate change increasing the intensity of storms and hazardous sea conditions, new disaster risk reduction tools are needed to prepare for coastal erosion and flooding hazards [1]. Whilst short range forecast driven early warning systems are critical for informing the public of hazardous events, extending the lead time on early warning systems to the monthly or seasonal range allows for more proactive actions by coastal managers, engineers, and other decision makers.

Long range forecasts require inputs which vary at the seasonal scale and beyond due to a lack of wave data for wave-forced models [2]. In the Pacific Basin, studies have identified teleconnections between shoreline and climate variability, namely the El-Nino Southern Oscillation (ENSO) [3, 4], suggesting the use of seasonally varying climate processes to inform seasonal shoreline models will improve skill. Understanding the effects of other climate processes, such as the Indian Ocean Dipole (IOD) on coastal hazards are also important to potentially enhance forecast skill at longer ranges.

© The Author(s) 2026
C. Coelho et al. (Eds.): CD 2025, CRL 41, pp. 480–486, 2026.
https://doi.org/10.1007/978-3-032-15473-6_73

This study utilizes a multi-decadal (25 year) dataset of shoreline variability (using the CoastSat toolbox [5]) across nine cross-shore transects at an embayed beach in SE Australia to develop a novel approach for seasonal to long range forecasting of shoreline position. Three different model types are used for prediction: (1) a baseline linear regression forecast model; (2) a statistically-based time series model (Seasonal AutoRegressive Integrated Moving Average with eXogenous variables – SARIMAX model); and (3) a multivariate statistically-based time series model (Seasonal Vector-Autoregressive model with eXogenous variables – SVARX model). Model skill is assessed for a five-year forecast horizon to evaluate which model and combination of model inputs performed most effectively over these time scales.

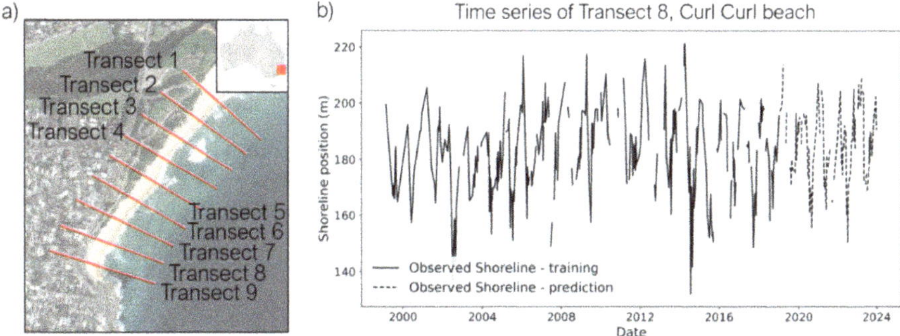

Fig. 1. a) Study area, Curl Curl beach, SE Australia, and b) time series of shoreline position

2 Method

2.1 Data Used

The study used a 25-year dataset of shoreline position across nine transects at Curl Curl Beach, SE Australia (Fig. 1) from the open-source CoastSat toolbox [5], with a frequency of around two observations per month. On the northern end of the beach, the shoreline is exposed to predominantly southerly swells characteristic of SE Australia. The southern end of the beach meanwhile is slightly sheltered from these swells but influenced by northeasterly wind waves caused by sea breezes in the summer months.

ERA5 wave reanalysis data was used to derive offshore significant wave height, with nearshore wave heights calculated at the 10 m depth contour at each transect using the BinWaves downscaling method [6]. A 150-day rolling mean of significant wave height for both nearshore and offshore conditions (hereafter $H_{S,150}$) was used in the models as it had the best correlation (not shown here) to shoreline positions at Curl Curl. Climate indices describing the behavior of ENSO (NINO3.4) and IOD (DMI) were used to characterize interannual controls on the shoreline that could improve forecast skill at multi-year time horizons.

2.2 Selection of Models

A total of eight different models were selected in this study from the three model types (i.e. SARIMAX, SVARX and linear regression) to compare the influence of different combinations of input variables on model forecast skill.

The linear regression model is described in Eq. 1 and simulates a simple linear relationship between shoreline position and wave/climate variables:

$$y_t = \beta_0 + \beta_1 ENSO + \beta_2 IOD + \beta_3 H_{S,150} + \epsilon \tag{1}$$

Two linear regression models were tested here, one including offshore $H_{S,150}$ and one including nearshore $H_{S,150}$.

The SARIMAX model is described in Eq. 2, and includes terms for the relationship between the shoreline and itself (autocorrelation) and the seasonality, along with a linear relationship between shoreline position and wave climate/variables:

$$y_t = SARIMAX(p, d, q)(P, D, Q, s) \tag{2}$$

where p refers to the order of autoregression, d the order of differencing required, q the order of moving average terms, P the order of seasonal autoregression, D the order of seasonal differencing required, Q the order of seasonal moving average terms and s the period of the seasonality. X refers to the exogenous variables ENSO, IOD and nearshore or offshore $H_{S,150}$.

Four SARIMAX models were tested, the first with only shoreline position of the transect, one with shoreline and climate indices, one with shoreline, climate indices and offshore $H_{S,150}$ and the final including shoreline, climate indices and nearshore $H_{S,150}$.

Finally, the SVARX model in Eq. 3 includes terms for the relationship between the shoreline position of interest and previous shorelines at other transects, along with waves and climate.

$$Y_t = B_0 + B_1 Y_{t-1} + B_2 ENSO + B_3 IOD + B_4 H_{S,150} + B_5 month + \epsilon_t \tag{3}$$

where Y and B are vectors/matrices of the coefficients for each variable at each transect, while *month* refers to the seasonality.

Two SVARX models were selected, each including previous shoreline position at every transect, climate indices and either offshore $H_{S,150}$ at each transect or nearshore $H_{S,150}$ at each transect, along with a seasonality component.

The models were trained with a rolling window of testing and validation between 1999 and 2019, predicting monthly average shoreline position at each transect. The final predictions were made with a forecast window from 2019 to 2024 (see Fig. 1b).

2.3 Model Evaluation

The scores used to measure skill at each transect included Root Mean Squared Error (RMSE), RMSE normalized to the range of shoreline change of the observed shoreline during the prediction time period (NRMSE), the coefficient of determination; R^2, and the seasonal mean absolute scaled error (seasonal MASE) to show how the models perform against a seasonal-naïve forecast. Values of Seasonal MASE close to 0 indicate a well performing model, where values of 1 or higher indicate the model performs the same or worse than a seasonal naïve model, respectively.

3 Results

Results of the model skill over the five-year forecast horizon for the eight models (averaged over the nine transects) are summarized in Table 1. This table indicates that the best performing model over this five-year horizon was a SVARX model including ENSO, IOD and nearshore $H_{s,150}$, with an average RMSE of 9.2 m, NRMSE of 0.18, and R^2 of 0.43 and Seasonal MASE of 0.52 across all transects. A SARIMAX model with ENSO, IOD and nearshore $H_{s,150}$ meanwhile performed second best (RMSE = 9.5 m). Conversely the poorest performing models was the linear regression model. Seasonal MASE scores indicate that all models showed improved performance when compared to a seasonal naïve forecast (i.e. Seasonal MASE <1). In each model type, the inclusion of nearshore downscaled $H_{s,150}$ was able to improve model skill (e.g. reducing RMSE by between 0.5 m and 1 m).

Table 1. Comparison of forecast skill scores between models (averaged between transects).

Model Type	RMSE (m)	Normalized RMSE	R^2	Seasonal MASE
Linear regression with ENSO, IOD, offshore $H_{s,150}$	11.5	0.22	0.19	0.65
Linear regression with ENSO, IOD, nearshore $H_{s,150}$	10.5	0.20	0.26	0.57
SARIMA with shoreline position	11.0	0.21	0.13	0.62
SARIMAX with shoreline position, ENSO, IOD	10.9	0.21	0.16	0.61
SARIMAX with shoreline position, ENSO, IOD, offshore $H_{s,150}$	10.0	0.19	0.30	0.55
SARIMAX with shoreline position, ENSO, IOD, nearshore $H_{s,150}$	9.5	0.18	0.36	0.50
SVARX with ENSO, IOD, offshore $H_{s,150}$	10.1	0.19	0.36	0.57
SVARX with ENSO, IOD, nearshore $H_{s,150}$	9.2	0.18	0.43	0.52

At the individual transect level, for the northernmost transect (Transect 2), the SARIMAX model with nearshore $H_{S,150}$ had the highest skill (RMSE = 9.3 m) and all models including nearshore $H_{S,150}$ performed well, suggesting strong teleconnections to average wave climate and some seasonality (Fig. 1).

At the southernmost transect (Transect 8), the SARIMAX model with nearshore $H_{S,150}$ also performed best (RMSE = 9.5 m). However, all the models which included a term for seasonality at this transect performed well, while the two linear regression models performed more poorly, suggesting this transect was dominated by seasonality and possibly seasonally varying wave climate (Fig. 1).

Conversely, at the center of the beach (Transect 5), the SVARX model with nearshore $H_{S,150}$ had higher accuracy, and the models did not capture all the extreme values as closely as the other transects (Fig. 1). All of these transects in the center of the beach showed worse skill metrics for each model, and it is unclear if there is another driver for shoreline change at the interannual scale not captured by the models (Fig. 2).

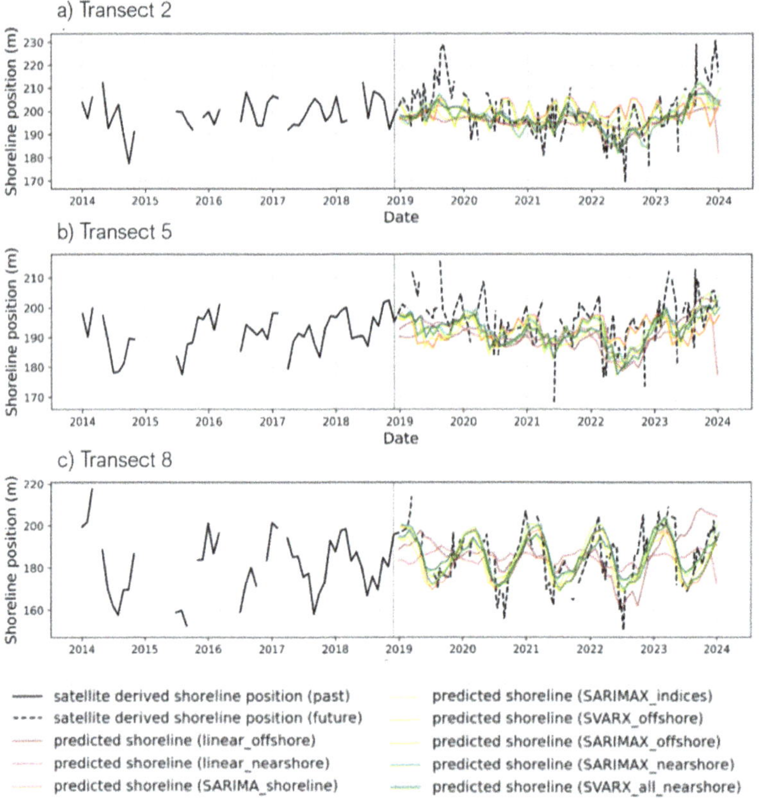

Fig. 2. Shoreline predictions at key transects; a) transect 2 at the north end of the beach showed high teleconnections with $H_{S,150}$, b) transect 5, in the center showed poorer skill scores than those at the northern and southern end and c) transect 8 which showed high levels of seasonality.

4 Discussions and Concluding Remarks

This study utilized a number of statistical time series modelling techniques to forecast shoreline position up to five years into the future, indicating that statistical models (both SARIMAX and SVARX), when trained on satellite-derived shorelines can generate useful projections (RMSE ~10 m) of shoreline change at time horizons of up to five years.

Whilst models that incorporated a term for nearshore $H_{s,150}$ performed better than those including offshore $H_{s,150}$, the reasonable skill exhibited in models with offshore $H_{s,150}$ allow for the potential to hindcast shoreline position in wave dominated shorelines where the bathymetry is not known. Using data-driven models to hindcast or forecast shoreline erosion hazards allows for a regional to national approach and the application of such models to areas with physically or socially vulnerable populations where there is a lack of ground truthing of shoreline information.

Comparing data pre-processing methods such as filtering and smoothing is valuable to balancing prediction of long term trends with extreme events. For example, the models including $H_{S,150}$ captured interannual variability in shoreline position, but had challenges with short term changes, suggesting the smoothed wave data limited the capture of short-term variability from extreme wave events such as East Coast Lows that are common in SE Australia in the austral autumn/winter. This is consistent with [2], which highlighted challenges in trying to capture multiple timescales in shoreline change models and [7] which suggested different forcing in models for different time scales.

Satellite derived shorelines based on lower-resolution LandSat/Sentinel-2 imagery are inherently noisy and their use requires a degree of care when assessing prediction skill. For example, some of the 'observed' extreme events may represent noise e.g. an extreme erosion event observed in transect 5 in May 2021, but not at other transects. As such, models that underpredicted such events may not necessarily be inaccurate.

Applying these techniques to the continental/Australia-wide scale could potentially give decision makers greater insight into likely shoreline changes months in advance – and execute risk mitigation measures such as early warning systems accordingly.

References

1. Turner, I. L., Leaman, C. K., Harley, M. D., Thran, M. C., David, D. R., Splinter, K. D., Matheen, N., Hansen, J. E., Cuttler, M. V. W., Greenslade, D. J. M., Zieger, S., Lowe, R. J.: A framework for national-scale coastal storm hazards early warning. Coastal Engineering, 192. https://doi.org/10.1016/j.coastaleng.2024.104571. (2024)
2. Hunt, E., Davidson, M., Steele, ECC., Amies, JD., Scott, T., Russell, P. Shoreline modelling on timescales of days to decades. Cambridge Prisms: Coastal Futures. 1:e16. https://doi.org/10.1017/cft.2023.5 (2023)
3. Vos K, Harley M, Turner I, Splinter K (2023) Pacific shoreline erosion and accretion patterns controlled by El Niño/Southern Oscillation. Nat Geosci 16(2):140–146
4. Harley M, Turner O, Short A, Ranasinghe R (2010) Interannual variability and controls of the Sydney wave climate. Int J Climatol 30(9):1322–1335
5. Vos, K., Splinter, K. D., Harley, M. D., Simmons, J. A., & Turner, I. L.: CoastSat: A Google Earth Engine-enabled Python toolkit to extract shorelines from publicly available satellite imagery. Environmental Modelling and Software, 122. (2019)
6. Cagigal, L., Méndez, F, J., Ricondo, A., Gutiérrez-Barceló, D., Bosserelle, C., Hoeke, R. BinWaves: An additive hybrid method to downscale directional wave spectra to nearshore areas, Ocean Modelling, Volume 189 (2024)
7. Schepper, R., Almar, R., Bergsma, E., de Vries, S., Reniers, A., Davidson, M., Splinter, K. Modelling Cross-Shore Shoreline Change on Multiple Timescales and Their Interactions. Journal of Marine Science and Engineering. 9(6). (2021)

Climate-Driven Waterline Variability Along the North American West Coast

Marcan Graffin[1,2(✉)], Rafael Almar[1], Erwin W. J. Bergsma[2], Julien Boucharel[1], Sean Vitousek[3,4], Mohsen Taherkhani[5], and Peter Ruggiero[5]

[1] Université de Toulouse, LEGOS (CNES/CNRS, IRD/UT3), Toulouse, France
marcan.graffin@ird.fr
[2] Earth Observation Lab of the French Space Agency (CNES), Toulouse, France
[3] Pacific Coastal and Marine Science Center, U.S. Geological Survey, Santa Cruz, CA, USA
[4] Department of Civil, Materials, and Environmental Engineering, University of Illinois Chicago, Chicago, IL, USA
[5] College of Earth, Ocean, and Atmospheric Sciences, Oregon State University, Corvallis, OR, USA

Abstract. Sandy coasts are highly dynamic environments shaped by a myriad of hydro-sedimentary processes operating across various spatio-temporal scales. From seasonal to centennial timescales, sandy beach dynamics are strongly influenced by climate variability expressed in various forms, including seasonal cycles, climate modes (e.g., El Niño Southern Oscillation (ENSO)), and global warming. Satellite-based methods enable the observation of coastlines with unprecedented spatial and temporal coverage. However, over the past decade, as these methods have matured, coastal change studies have predominantly focused on long-term trends rather than on climate-driven cycles of shoreline change. In this study, we utilize a large dataset of monthly-sampled time series of waterline positions along the Pacific Coast of the United States and Baja California, Mexico. From these time series, we derived information on long-term trends and seasonal cycles of waterline change. Our analysis reveals clear contrasts in seasonal patterns of waterline variability across the study area, with greater amplitudes observed at higher latitudes. A shift in the dominant dynamics is evident in Southern Baja California, where shoreline variability is more influenced by intertropical dynamics and sea-level anomalies than by wave energy modulation, which dominates further north. ENSO is found to induce contrasting responses of the waterline along the study area. During El Niño events, shoreline retreat is observed in subtropical regions, while La Niña winters drive waterline retreat in higher latitudes, with southern regions either remaining stable or advancing.

Keywords: Waterline · Climate Variability · Seasonal cycle · ENSO

1 Introduction

Coastal morphodynamics have long been studied for their critical implications on ecosystems and human activities. Early research relied on in-situ methods to monitor indicators such as waterline and shoreline changes, dune erosion, and beach slopes [1]. While these

© The Author(s) 2026
C. Coelho et al. (Eds.): CD 2025, CRL 41, pp. 487–492, 2026.
https://doi.org/10.1007/978-3-032-15473-6_74

approaches provided valuable local-scale insights, they often failed to capture broader patterns of coastal change, limiting their applicability at regional or global scales [2].

Recent advancements in Earth observation satellites and cloud-based platforms [3] have revolutionized coastal monitoring. These tools enable the automated extraction of shoreline and waterline indicators across large temporal and spatial scales, offering new opportunities to study coastal dynamics at regional, continental, and global levels [4]. However, challenges persist, particularly in ensuring data quality and addressing variability beyond long-term trends. Seasonal and interannual variability, driven by environmental factors such as waves, sea level, and climate modes, often overshadow decadal shoreline trends but remain poorly quantified at larger scales.

ENSO, the dominant source of interannual variability in the global climate system, exerts profound and diverse impacts on coastal systems by reorganizing atmospheric circulation and storm tracks [5]. While large-scale studies have highlighted ENSO's influence on regional-to-global coastal variability, finer-resolution analyses are essential to understanding its regional effects.

In this study, we investigate the coastal response of the North American West Coast (NAWC) to environmental drivers over multiple timescales using a 25-year dataset of satellite-derived waterline positions. Spanning from Washington State to Baja California, the NAWC (see Fig. 1) features diverse oceanic conditions, with sandy beaches dominating much of the coastline. Northern regions, influenced by storm tracks, exhibit strong seasonal wave height variability (up to 4 m), while southern regions experience less pronounced seasonal patterns. Sea-level variability also differs across the NAWC, driven by thermal expansion, storm surges, and ENSO dynamics. These contrasts underscore the complexity of coastal processes in this region, leading to varied and complex responses.

2 Data and Methods

This study leverages advanced remote sensing techniques and satellite imagery (Landsat 5–9 and Sentinel-2) to investigate shoreline variability along the North American West Coast (NAWC). A high-resolution satellite-derived waterline dataset was generated using state-of-the-art band combination methods (SCoWI [6]) and Otsu image segmentation [7], enabling the extraction of waterline positions at a 250-m spatial resolution. The waterline extraction was performed on transects identified as sandy by [8].

To account for the influence of tides on the waterline position, the SDW dataset was corrected using outputs from the FES2022 tide model and local estimations of intertidal beach slopes, following the beach slope estimation approach of [9].

After resampling the data into a monthly-sampled waterline position dataset spanning 1997–2022, we isolated the trend, seasonal cycles, and anomalies of waterline variability. Trends were derived from the time series using the slope of the time series, calculated as the median slope between all pairs of points in the dataset. From the detrended time series, seasonal cycles were calculated as the mean waterline position for each calendar month, capturing recurring monthly patterns of variability. Anomalies were obtained as the residuals after removing the long-term trend and the seasonal cycle from the original time series.

Fig. 1. Map of the North American West Coast (NAWC). The monitored sandy coast is highlighted in red.

Wave data (monthly significant wave height and peak period) and sea-level data were sourced from the ERA5 [10] and DUACS [11] products. ERA5 provides atmospheric and oceanic variables at a 0.5° resolution, including wind and pressure fields, while DUACS offers sea-level anomaly data.

El Niño, neutral, and La Niña winters were identified using the Multivariate ENSO Index (MEI) [12]. Winters between 1997 and 2022 were classified as El Niño if the winter-averaged MEI value exceeded the standard deviation of the time series. La Niña winters were defined as those with a winter-averaged MEI value below the negative standard deviation of the time series. Winters with MEI values within $\pm$ one standard deviation were classified as neutral.

3   Results

Figure 2 shows the three distinct components identified in the time series of waterline position: the long-term trends, the seasonal cycle (both calculated between 1997–2022), and the anomalies of waterline variability.

Long-term trends of waterline change are evenly distributed along the study area, with 55% of the transects showing no trend of erosion or accretion (absolute trend below 0.5 m/yr), while 28% show trends of accretion and 17% show trends of erosion.

Seasonal cycles of shoreline change are not uniform along the study area. In terms of amplitude, we observe a clear latitude-dependent pattern, with beaches in the northern

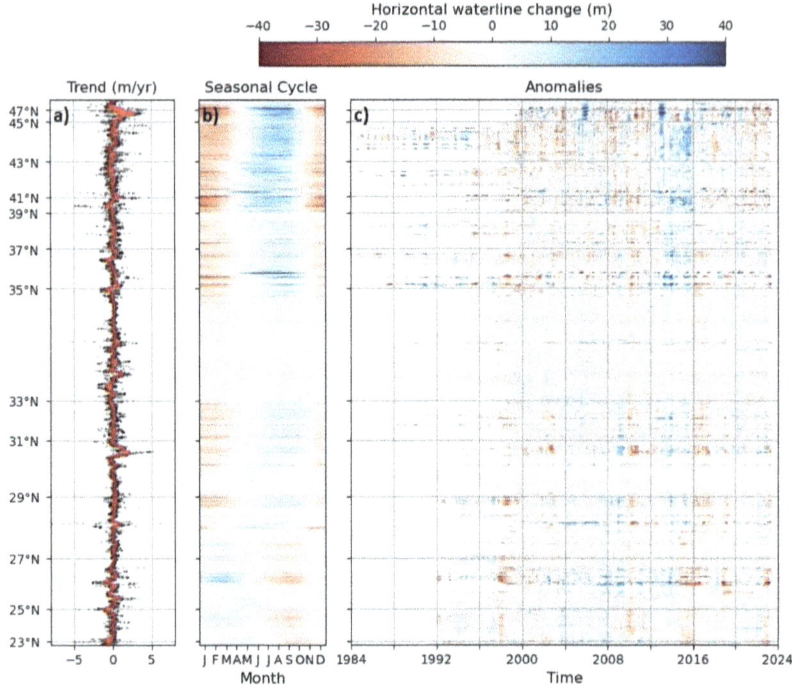

Fig. 2. Components of waterline change along the North American West Coast (NAWC). (**a**) Long-term trends of waterline change from 1997–2022. Black dots show local trends and the red line shows a 30 km running mean over local trends. (**b**) Seasonal cycles of waterline positions, derived from the 1997–2022 time series. (**c**) Waterline position anomalies, obtained after removing trends and seasonal signals from the original time series.

part of the study area (Washington, Oregon) showing more variability over the year than those further south. On average, in the Pacific Northwest (PNW), the excursion of the beach over a year is around 25 m, compared to less than 10 m in Southern California and Baja California. Waterline change along the NAWC is primarily driven by seasonal variations in wave energy. Specifically, waterline positions from Northern Washington state to Northern Baja California evolve jointly with seasonal modulation of wave energy, resulting in accreted beaches in summer, when waves are weakest, and eroded beaches in winter. Further south, in Southern Baja California, sea-level anomalies, influenced by both atmospheric pressure variations and oceanic conditions, play a dominant role in driving seasonal shoreline change, while wave energy remains almost constant throughout the seasons. This results in a waterline position evolving jointly with the regional sea-level anomalies, accreting in winter and spring, and eroding in summer and fall.

On the interannual scale, waterline positions are significantly influenced by climate modes, particularly ENSO. ENSO events modulate oceanic and atmospheric conditions across the Pacific basin, causing deviations from seasonal norms. During El Niño years, the shift in storm tracks typically brings more intense and frequent storms to the North

American West Coast, leading to higher wave energy, sea-level anomalies, and increased shoreline erosion. In contrast, La Niña conditions often shift storm tracks further north, bringing cooler waters and reduced wave energy, which leads to more stable or accreting shorelines, particularly in the southern parts of the study area (see Fig. 3).

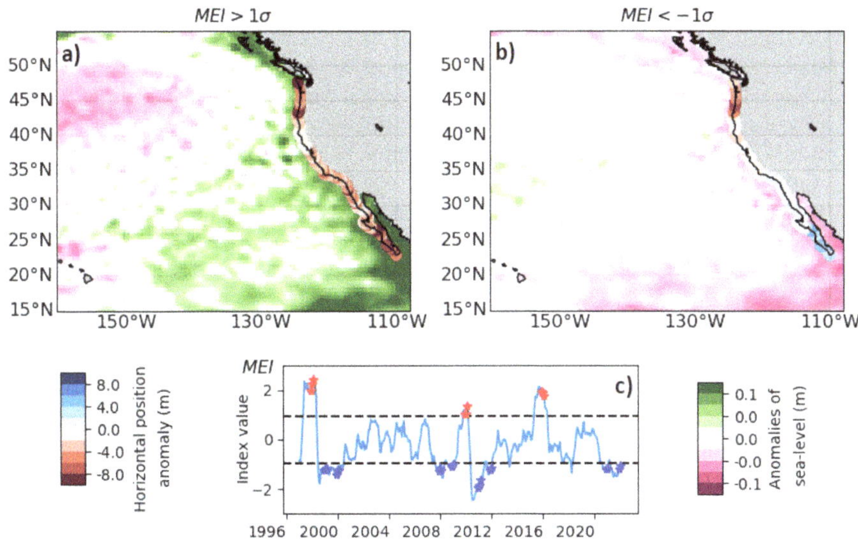

Fig. 3. Sea-level and waterline response to ENSO. **(a,b)** Composites of sea-level and waterline position anomalies for El Niño (a) and La Niña (b) winters between 1997–2022. **(c)** Time series of MEI between 1997–2022, red (blue) star markers indicate El Niño (La Niña) winter months.

4 Conclusion

There are clear latitude-dependent patterns in the variability of waterline change along the NAWC. The seasonal cycles of waterline change exhibit greater amplitude in the PNW than further south. In the PNW and California, shoreline change is primarily driven by seasonal modulations in wave energy, with beaches accreting in summer and eroding in winter. Further south, in Southern Baja California, sea-level anomalies dominate the seasonal shoreline dynamics. On the interannual scale, shoreline change along the NAWC is strongly influenced by ENSO. During El Niño events, the shift in storm tracks leads to increased wave energy and higher sea-level anomalies, resulting in widespread erosion across the NAWC. In contrast, La Niña years bring cooler waters and reduced wave energy to the southern part of the region, leading to recovery and more stable shorelines in Baja California and Southern California, while the northern regions remain stable or experience continued erosion.

References

1. Boak EH, Turner IL (2005) Shoreline definition and detection: a review. J Coastal Res 21(4):688–703

2. Vitousek S, Buscombe D, Vos K, Barnard PL., Ritchie A, Warrick J (2022) The future of coastal monitoring through satellite remote sensing. Cambridge Prisms: Coastal Futures, pp 1–43
3. Gorelick N, Hancher M, Dixon M, Ilyushchenko S, Thau D, Moore R (2017) Google earth engine: planetary-scale geospatial analysis for everyone. Remote Sens Environ 202:18–27
4. Luijendijk A, Hagenaars G, Ranasinghe R, Baart F, Donchyts G, Aarninkhof S (2018) The state of the world's beaches. Sci Rep 8:6641
5. McPhaden MJ, Lee T, Fournier S, Balmaseda MA (2020) ENSO Observations. El Niño Southern Oscillation in a Changing Climate, pp 39–63, American Geophysical Union
6. Bergsma E et al (2024) Shoreliner: a sub-pixel coastal waterline extraction pipeline for multi-spectral satellite optical imagery. Remote Sens 16:2795
7. Otsu N (1979) A threshold selection method from gray-level histograms. IEEE Trans Syst Man Cybern 607 9:62–66
8. Hulskamp R, et al (2023) Global distribution and dynamics of muddy coasts. Nat. Commun. 14:8259
9. Vos K, Harley MD, Splinter KD, Walker A, Turner IL (2020) Beach slopes from satellite-derived shorelines. Geophys Res Lett 47(14)
10. ECMWF: ERA5 hourly data on single levels from 1940 to present. https://cds.climate.copern icus.eu/datasets/reanalysis-era5-single-levels. Accessed 1 Dec 2024
11. CNES: Altimetry product SSALTO/DUACS. https://www.aviso.altimetry.fr/en/data/pro duct-information/information-about-mono-and-multi-mission-processing/ssaltoduacs-mul timission-altimeter-products.html. Accessed 1 Dec 2024
12. Zhang T et al (2019) Towards probabilistic multivariate ENSO monitoring. Geophys Res Lett 46:10532–10540

Coastal Risk Mitigation in a Nature-Based Solution Framework: New Tuscan Integrated Approach in Horizon R4C

Maria Fruzzetti[1]([✉]), Eleonora Duchini[1], Clemente Fuggini[1], Fabrizio Tavaroli[1], Paolo Lovisolo[1], Cecilia Evangelista[1], Ludovica Sartini[1], Federica Vergari[1], Andrea Montanaro[1], Luigi E. Cipriani[2], Andrea Esposito[3,4], and Lorenzo Cappietti[3,4]

[1] Rina Consulting SPA, 16129 Genoa, Italy
`maria.fruzzetti@rina.org`
[2] Region of Tuscany, 50126 Florence, Italy
[3] University of Florence, 50139 Florence, Italy
[4] A-MARE, Joint Laboratory, 50139 Florence, Italy

Abstract. The manuscript describes the design of an artificial restored dune belt on a coastal protected area implemented by means of hybrid modeling approach given by numerical modelling, field monitoring and experimental modelling.

Nearshore waves were estimated by means of a suitable, calibrated numerical wave model on unstructured mesh, while an experimental modelling was used to simulate wave phenomena in shallow water to consider breaking, surf and swash zone phenomena, wave-dune interaction, and over washing processes. A calculation model based on the limit equilibrium approach was employed for artificial dune design. Finally, a real-time weather station was installed on-site to assure the overall stability conditions of the dune.

Keywords: Nature Based Solutions · Numerical Modelling · Waves · Physical Modelling · Extreme Weather Conditions

1 Introduction

The growing frequency of extreme weather conditions attributed to climate change has culminated in a critical juncture that poses a grave threat to our livelihoods, well-being, and the environment that demands immediate and robust responses.

This is where Regions4Climate (R4C), a consortium of visionary partners, emerges as a beacon of hope and innovation. They have embraced a dual mission: to confront the prevailing and projected challenges arising from climate change and to cultivate greater resilience within European communities. The approach is uniquely innovative, socially engaged, and citizen-driven, with a profound emphasis on risk mitigation and natural hazard management. The paramount focus is to develop more resilient regional ecosystems by nurturing cross-sectoral innovations, an endeavor that benefits and safeguards all stakeholders.

C. Coelho et al. (Eds.): CD 2025, CRL 41, pp. 493–499, 2026.
https://doi.org/10.1007/978-3-032-15473-6_75

Central to the agenda of R4C is the concept of risk and one of the critical elements within this comprehensive effort is addressing the risk posed by coastal erosion and inundation.

The present manuscript presents the methodological approach and related activities currently being undertaken to develop a specific Nature-Based Solution (NBS) for a case study in the Tuscany region, Italy, to leverage the value of coastal dunes and natural materials by developing nature-based solutions for managing these coastal areas.

In particular, the design of an artificial restored dune belt implemented by means of hybrid modeling approach will be described.

2   Case Study

The study area located in the Gulf of Follonica is a coastal protected area known as "Sterpaia" managed by the Municipality of Piombino, characterized by high naturalistic and landscape values (relict dune vegetation, plant and animal species of conservation interest) but with high anthropogenic pressures (i.e. coastal tourism, coastal dune erosion and breaching due to tourist load, exotic flora species, wrong beach cleaning techniques) [1].

The coastal strip is characterized by a narrow dune belt, with a maximum thickness of 50 m, with shrubby herbaceous dune vegetation and relict pine forests. The dune belt constitutes an important protection for some humid areas behind the dunes and for important agricultural areas. The study area has experienced several interventions of hydrogeological defense and restoration of the ecological functionality of the coastal dune system for approximately 7,5 km of restored dune belt. A last intervention to restore the dune belt for a total length of approximately 800 m is planned for year 2025 in the framework of the HORIZON Regions4Climate Project [2, 3].

3   Methodology

To assess the design and effectiveness of the artificial dune in controlling dune over washing and salinization of the low-lying coastal area, a hybrid modeling approach was employed combining numerical modelling, field monitoring and experimental modelling, exploiting findings given by the previous experiences of OPERANDUM project [4].

4   Numerical Modelling

To estimate nearshore waves occurring within the Gulf of Follonica, the offshore wave climate was propagated from offshore to nearshore using numerical simulations. More precisely, the numerical wave model SWAN [5, 6] was forced from the offshore area over a rather large domain using the Mediterranean-scale wave hindcast CMEMS MEDSEA_ANALYSISFORECAST_WAV_006_017 database (Copernicus) as boundary conditions.

Furthermore, a nearshore wave buoy was installed at 15 m water depth to collect wave data for 10 months, allowing numerical model calibration and validation.

The computational mesh was designed to cover a sufficiently large area to evaluate the dynamics of wave motion at the subscale in port and coastal areas with an unstructured calculation mesh (Fig. 1), whose spatial resolution varies from a few hundred meters offshore to a few meters below the coast in interest. The bathymetry used is given by the combination of the extraction from the GEBCO database (General Bathymetric Chart of the Oceans) and the bathymetry measured covering the study area.

The boundary wave series were calibrated by comparison with the satellite data available in the area extracted from the Ifremer Cersat website.

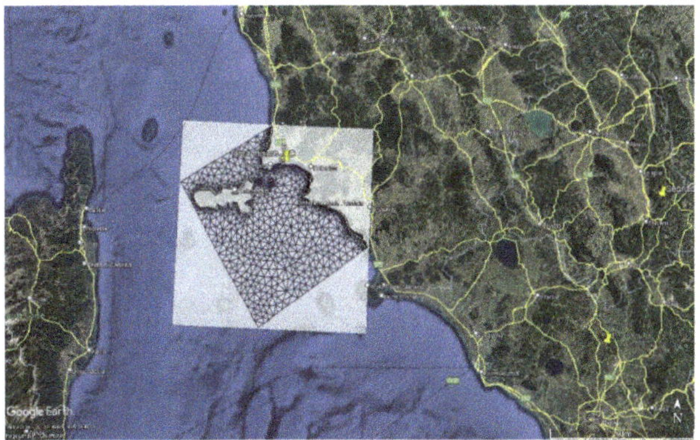

Fig. 1. Computational domain of wave model.

Wave time series comparison for some selected storms and quantiles plots shows a good correlation between wave database and the measurements, except for the peak of the storm only which is not present in the forecast series (Fig. 2).

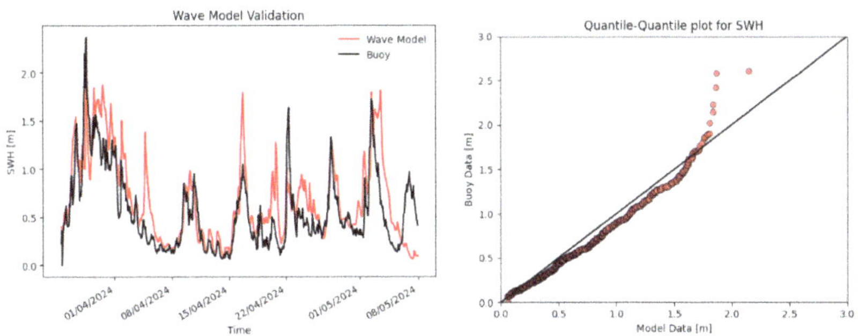

Fig. 2. Left: comparison between Sterpaia wave buoy (black) and model timeseries (red). Right: quantiles plot between model and measured wave data

The wave model was applied to estimate the nearshore wave condition for selected storms associated with defined return periods.

To consider the effects of climate change the hindcasted sea level variation was considered in numerical simulations, as well as extreme wind and waves trends used as forcing.

The results were extracted for each simulation along the several cross-shore transects and were used as boundary conditions for the next physical modelling.

5   Physical Modelling

Once wave conditions were reconstructed by means of numerical modelling, the experimental modelling was used to simulate wave phenomena from 15m water depth to the dune rear, encompassing the breaking zone, surf zone, swash zone, wave-dune interaction, and over washing processes.

The model was tested in a wave-current flumes at the Laboratory of Maritime Engineering (LABIMA) at the University of Florence (Fig. 3) under the action of 27 different random wave conditions with JONSWAP spectrum which is characteristic of the Mediterranean Sea, with a peak factor of 2.4 (Table 1).

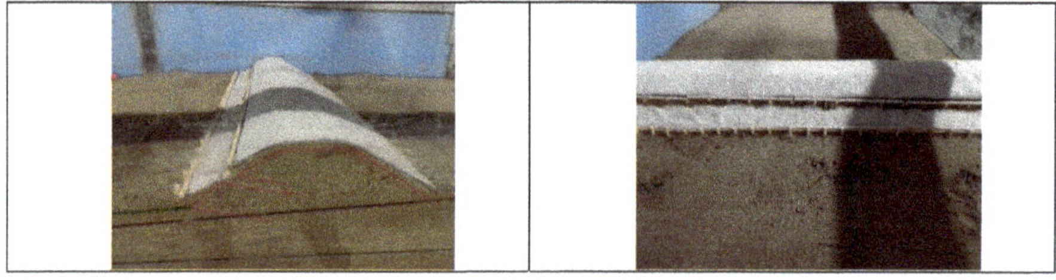

Fig. 3. Dune model. Left: side view. Right: Frontal View.

Table 1. Test code of each tested wave conditions. Significant wave heights (Hs) and peak periods (Tp) are related to 10m water depth, prototype scale.

		Hs [m]						
		1.5	2	2.5	3	3.5	4	4.5
T_p [s]	6	1	2	3	4	5	6	X
	8	7	8	9	10	11	12	13
	10	14	15	16	17	18	19	20
	12	21	22	23	24	25	26	27

All 27 random waves were tested under 5 different sea levels with respect to the local mean sea level. Each wave condition lasted one hour in the model, corresponding to five hours in the prototype, corresponding to 1500 wave periods for wave condition with T = 12 s and 3000 wave periods for wave condition with T = 6s.

During each tested sea state, overwash measurements were carried out providing the number of events in percentage and their magnitude. From statistical analysis of the wave climate in the past 50 years the occurrence of each sea state in percentage and the number of hours per year of overwashing was obtained as key performance parameter for the evaluation of the effectiveness of this NBS in controlling overwashing.

6 Artificial Dune

The design methodology of the artificial dune is based on an engineering approach. The location and the general shape for the dune has been chosen as a function of topographical conditions and regulatory limitations. The inputs for calculations, as topographical, metocean, bathymetric, environmental, geotechnical and seismic conditions, were collected by dedicated surveys and monitoring plans.

Then, a specific calculation model based on limit equilibrium approach (LEM, Fig. 4) was implemented; the model is aimed to the verification of overall stability conditions of the dune, according to the current Italian regulatory system. The verification phase analyzes the possible failure mechanisms for the dune, with reference to the slope stability verifications. Both the slopes of the embankment have been considered for verification, considering different loading conditions, static conditions, in presence of loadings due to the presence of crowd on the embankment or in presence of wave impact, and seismic conditions.

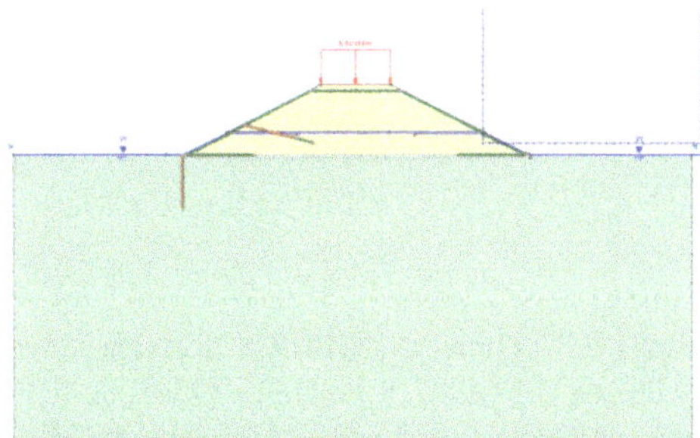

Fig. 4. LEM Model for stability evaluation of the dune

7 Field Monitoring

To ensure a comprehensive analysis of environmental parameters at Sterpaia, a stand-alone real-time weather station (WS) was installed on-site (Fig. 5) to collect wind speed and direction, air temperature and humidity, precipitation levels, solar radiation, and atmospheric pressure. Additionally, a data acquisition and storage unit has been integrated into the battery and data box which collects and stores environmental data at regular intervals every 10 min. These measurements are transmitted to an FTP server hourly, ensuring continuous and up-to-date monitoring, and uploaded to the R4C project dashboard to allow final users (i.e. local authorities, private beach club owners, citizens) to take advantage of them.

Fig. 5. Left: stand-alone weather station installed on Piombino site. Right: battery and data hub.

In addition to the weather station, the wave buoy data collected at 15 m depth in front of the study area as described in Sect. 4 are uploaded to the R4C project dashboard available for final users.

8   Concluding Remarks

With sea level rising and extreme storms becoming more common, coastal communities are highly vulnerable. This methodology provides a framework for future initiatives to enhance coastal resilience and protect communities from the impacts of climate change.

References

1. Bacci M, Corsi S, Lombardi L, GiuntiM (2019) Gli interventi di ripristino morfologico ed ecologico del sistema dunale del Golfo di Follonica (Toscana, Italia): tecniche utilizzate e risultati del monitoraggio. Atti Soc Tosc Sci Nat Mem Serie A 126:41–48. https://doi.org/10.2424/ASTSN.M.2017.29
2. European climate, infrastructure and environment executive agency (CINEA, 2022, Region4Climate, Project: 101093873-R4C-HORIZON-MISS-2021-CLIMA-02
3. Cantegiani C et al (2024) Enhancing coastal resilience: cross-border collaboration and innovation in the Regions4Climate project. Revista de Investigación Marina, UHINAK 2024 30(1):20
4. European Union's Horizon 2020 Research: OPERANDUM (OPEn-air laboRAtories for Nature baseD solUtions to Manage hydro-meteo risks) project (Grant No. 776848)
5. Booij N, Ris RC, Holthuijsen LH (1999) A third-generation wave model for coastal regions, part I, model description and validation. J Geophys Res 104(C4):7649–7666
6. Booij N, Holthuijsen LH, Ris RC (1996) The "SWAN" wave model for shallow water. In: Coastal Engineering Proceedings, vol 1, pp 668–676

Nature-Based Solutions to Mitigate Coastal Erosion: A Modelling Approach Along the Mediterranean Coast

Massimiliano Marino$^{(\boxtimes)}$, Ahmad I. K. Alkharoubi, Luca Cavallaro, Enrico Foti, and Rosaria Ester Musumeci

Department of Civil Engineering and Architecture, University of Catania, Via Santa Sofia 64, 95123 Catania, Italy
massimiliano.marino@unict.it

Abstract. Nature-based Solutions are gaining recognition as viable alternatives to traditional coastal defenses, offering ecological benefits while addressing the challenges of a changing climate. However, current modeling practices often rely on deterministic approaches based on historical storm data, neglecting the dynamic impacts of climate change. The present study introduces a hydro-morphodynamic modeling framework designed to evaluate the efficacy of NbS in reducing coastal erosion under evolving climatic scenarios. Applied to a Mediterranean case study along the Sicily coast (Italy), the framework integrates SWAN and XBeach models to assess three NbS interventions: dune revegetation, seagrass meadow reconstruction and a beach nourishment. The performance of these measures is analyzed under present conditions and projected 4.5 and 8.5 W/m^2 radiative forcing scenarios, demonstrating their potential to mitigate storm-induced erosion while contributing to broader climate resilience strategies.

Keywords: Coastal restoration · Building with Nature · Climate Change · SWAN · XBeach

1 Introduction

Coastal zones have long been essential for human settlement and the utilization of marine resources [1, 2]. However, rapid urbanization and the escalating impacts of climate change, i.e. rising sea levels, intensified storms, and increasing erosion, pose significant threats to these areas. In this regard, coastal ecosystems play a critical role in mitigating coastal hazards by attenuating wave energy and buffering storm surges [3, 4]. These habitats, however, face widespread degradation driven by urban development and resource exploitation, reducing their capacity to provide essential ecosystem services [5, 6]. Nature-based Solutions (NbS) have emerged as scientifically supported alternatives to traditional coastal defenses. NbS leverage natural processes to enhance biodiversity, mitigate climate impacts, and offer a cost-effective and adaptive approach to coastal management [7–10].

© The Author(s) 2026
C. Coelho et al. (Eds.): CD 2025, CRL 41, pp. 500–505, 2026.
https://doi.org/10.1007/978-3-032-15473-6_76

Despite increasing interest, the performance of NbS under future climate scenarios remains insufficiently examined. Existing studies predominantly rely on historical data, limiting their relevance for adaptive planning [11–13].

To address this gap, this study proposes a hydro-morphodynamic modeling framework to evaluate the efficacy of NbS, including dune revegetation and seagrass meadow reconstruction, in reducing storm-induced erosion under present and future climate scenarios. Applied to a Mediterranean case study in Sicily (Italy), the framework integrates climate projections to provide evidence-based insights, supporting the advancement of resilient and adaptive coastal management strategies.

2   Methodology

The case study is the Granelli coast in Sicily, Italy (Fig. 1a), a sandy beach environment in close connection with a nearby coastal lagoon system. The area is subject to an escalating anthropogenic pressure, with the establishment of a small urban area likely being the primary driver of a decadal shoreline erosion trend. Moreover, the area is subject to a rising climatic threat, with storms increasing in frequency and severity, and sea level rise projections predicting a ~6 km^2 permanent flooding by 2100 [14–17].

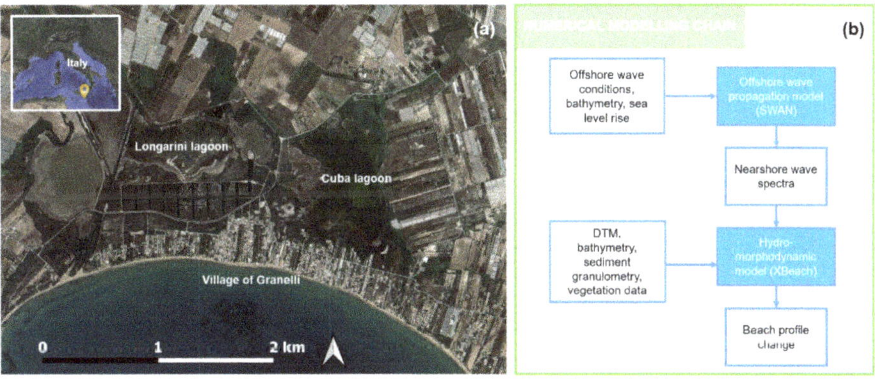

Fig. 1. Aerial view of the Granelli case study (a), scheme of the numerical modelling chain (b).

Local wave conditions (based on present and RCP4.5 and 8.5 projections), and sea level rise data (based on SSP2-4.5 and SSP5-8.5 projections) were used to define what-if scenarios, combining present and future hydrodynamic conditions with NbS based on site-specific vegetation types and distributions. Two time horizons were analyzed, namely 2070 and 2100. Three interventions are investigated, a dune revegetation, which consists of revegetating the dune strip in front of the Granelli village, a seagrass revegetation (SR), which consists of replanting *P. oceanica* seagrass to expand the existing eastern patch, and a sand beach nourishment (BN) are compared with a no-NbS case (NN).

3 Results

Findings indicate that all three interventions mitigate storm-induced erosion across all scenarios. As an example, Fig. 2b, c and d show the beach profiles before (dashed line) and after the storm (continuous line) for the BN intervention and for the S4.5-2070 scenario ($H_s = 8.70$ m, $T_p = 11.77$ s, SLR $= 0.37$ m) along three beach transects, highlighted in Fig. 2a. Instead, Fig. 2e, f and g display the beach profiles before and after the storm for the more severe S8.5-2100 scenario ($H_s = 9.18$ m, $T_p = 12.00$ s, and SLR $= 0.78$ m) along the same transects.

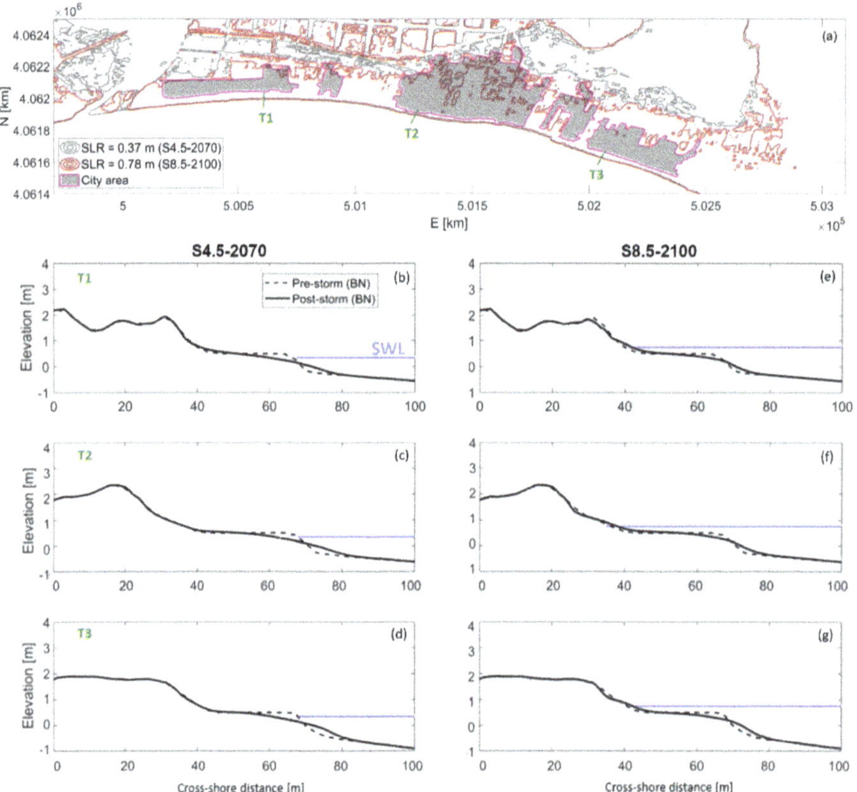

Fig. 2. Pre- and post-storm (modelled) beach profiles. Beach profiles: pre-storm (black dashed line), no-NbS (NN, black continuous line), beach nourishment (BN, orange dotted line) at transects T1 (b, e), T2 (c, f) and T3 (d, g), for S4.5-2070 (b, c and d) and S8.5-2100 (e, f and g) (Color figure online).

From the results for the S4.5-2070 scenario, it is observed that the sand nourishment intervention (BN) in the western part of the area (transect T1, Fig. 2b) determines a post-storm eroded volume of 4.42 m^3/m, reducing the eroded volume for the NN case which is 9.41 m^3/m, corresponding to a reduction of 47%. Moving towards the central part of the beach (transect T2, Fig. 2c), the protective effect of nourishment induces a reduction in eroded volumes as well, with an eroded volume of 4.16 m^3/m compared with 7.83 in the NN case, corresponding to a eroded volumes reduction of 40%. In the eastern part of the domain (transect T3, Fig. 2d), the most affected by the storm, the intervention results in an eroded volume of 5.13 m^3/m, compared to eroded volume of 9.10 m^3/m of the NN case (44% reduction).

Moving to the more critical scenario, S8.5-2100, characterized by a higher sea level rise but a lower wave height compared to the previous scenario, the erosion volumes are lower. In the western part of the domain (transect T1, Fig. 2e), the nourishment intervention determines an eroded volume of 2.11 m^3/m compared with the NN case with an eroded volume of 12.76 m^3/m, corresponding to a reduction of 84%. Along transects T2 and T3, eroded volumes for the NN scenario are 22.32 m^3/m and 18.70 m^3/m, compared with 0.89 m^3/m and 1.11 m^3/m of the BN intervention.

4 Conclusions

The present study underscores the importance of integrating Nature-based Solutions (NbS) into coastal management strategies to address the growing challenges posed by climate change. By employing a hydro-morphodynamic modeling framework, the research demonstrates the potential of NbS, to mitigate storm-induced erosion effectively under both current and projected future conditions.

The findings highlight the critical role of incorporating climate projections into modeling efforts to ensure adaptive and resilient coastal planning. As coastal zones face increasing risks from rising sea levels and intensified storm activity, the evidence presented reinforces the need for sustainable interventions that balance ecological restoration with hazard mitigation.

While the framework was applied to a specific Mediterranean case study, its adaptability makes it a valuable tool for broader application. Future research should aim to refine these methods and expand their application to diverse coastal environments, fostering more comprehensive and globally relevant strategies for climate resilience.

Acknowledgements. This work was funded by the projects: REST-COAST - Large scale RESToration of COASTal ecosystems through rivers to sea connectivity (call: H2020-LC-GD-2020; Proposal no. 101037097),"; National Recovery and Resilience Plan (NRRP), Mission 4 Component 2 Investment 1.3 - Call for tender No. 341 of 15/03/2022 of Italian Ministry of University and Research funded by the European Union – NextGenerationEU. Award number: PE00000005, Concession Decree No. 1522 of 11/10/2022 adopted by the Italian Ministry of University and Research, D43C22003030002, "Multi-Risk sciEnce for resilienT commUnities undeR a changiNg climate" (RETURN) - Cascade funding - Spoke VS1 "Acqua", Concession Decree No. 2812 of 09/01/2024 adopted by the General Director of Politecnico di Milano, "Mitigation and Adaptation in Resilient Coastal and estUarine integrated unitS" (MARCUS), and "VARIO - VAlutazione del Rischio Idraulico in sistemi cOmplessi" of the University of Catania.

References

1. Small C, Nicholls RJ (2003. A global analysis of human settlement in coastal zones. J Coastal Res 584–599

2. Nicholls RJ, Hinkel J, Lincke D, van der Pol T (2019) Global investment costs for coastal defense through the 21st century. World Bank Policy Research Working Paper (No. 8745)

3. Barbier EB, Hacker SD, Kennedy C, Koch EW, Stier AC, Silliman BR (2011) The value of estuarine and coastal ecosystem services. Ecol Monogr 81(2):169–193

4. Vuik V, Jonkman SN, Borsje BW, Suzuki T (2016) Nature-based flood protection: the efficiency of vegetated foreshores for reducing wave loads on coastal dikes. Coast Eng 116:42–56

5. Geijzendorffer I et al (2018) Mediterranean Wetlands Outlook 2: Solutions for Sustainable Mediterranean Wetlands. Tour du Valat

6. Waltham NJ et al (2020) UN decade on ecosystem restoration 2021–2030—what chance for success in restoring coastal ecosystems? Front Mar Sci 7:71

7. Van Zelst VTM et al (2021) Cutting the costs of coastal protection by integrating vegetation in flood defences. Nat Commun 12(1):6533

8. Sánchez-Arcilla A et al (2022) Barriers and enablers for upscaling coastal restoration. Nat Based Solutions 2:100032

9. O'Leary BC et al (2023) Embracing nature-based solutions to promote resilient marine and coastal ecosystems. Nat Based Solutions 3:100044

10. Costa GP, Marino M, Cáceres I, Musumeci RE (2023) Effectiveness of dune reconstruction and beach nourishment to mitigate coastal erosion of the Ebro delta (Spain). J Mar Sci Eng 11(10):1908

11. Guannel G, Arkema K, Ruggiero P, Verutes G (2016) The power of three: coral reefs, seagrasses and mangroves protect coastal regions and increase their resilience. PLoS ONE 11(7):e0158094

12. James RK et al (2021) Tropical biogeomorphic seagrass landscapes for coastal protection: Persistence and wave attenuation during major storm events. Ecosystems 24:301–318

13. Unguendoli S, Biolchi LG, Aguzzi M, Pillai UPA, Alessandri J, Valentini A (2023) A modeling application of integrated nature-based solutions (NBS) for coastal erosion and flooding mitigation in the Emilia-Romagna coastline (Northeast Italy). Sci Total Environ 867:161357

14. Antonioli F et al (2020) Relative sea-level rise and potential submersion risk for 2100 on 16 coastal plains of the Mediterranean sea. Water 12(8):2173

15. Musumeci RE, Marino M, Cavallaro L, Foti E (2023) Does coastal wetland restoration work as a climate change adaptation strategy? The case of the south-east of Sicily coast. Coastal Eng Proc 37:66. https://doi.org/10.9753/icce.v37.events.66

16. Marino M, Nasca S, Alkharoubi A, Cavallaro L, Foti E, Musumeci RE (2025). Efficacy of Nature-based Solutions for coastal protection under a changing climate: a modelling approach. Coastal Eng 197

17. Marino M, et al (2025). Nature-based Solutions as Building Blocks for coastal flood risk reduction: a model-based ecosystem service assessment. Sci Rep 15(1):12070, 104700

Assessing the Role of Ports on Coastal Evolution Through a Coupled Erosion-Flood Modelling Methodology

Moisés Álvarez-Cuesta[✉], Alexandra Toimil, and Iñigo J. Losada

IHCantabria-Instituto de Hidráulica Ambiental de la Universidad de Cantabria, Isabel Torres 15, 39011 Santander, Spain
alvcuestam@unican.es

Abstract. Ports exist at the intersection of human activities and the highly-vulnerable natural environment. However, the interactions with the adjacent coastal areas are usually disregarded. This work presents a modelling chain to assess the role of seaports on coastal flood and erosion hazards. It relies on a nearshore evolution model capable of efficiently reproducing the beachfront evolution at areas affected by coastal infrastructures in the long-term. Morphological changes are then used to update the digital terrain model used in flood simulations. The methodology is demonstrated at a 40 km coastal stretch in the Mediterranean Sea affected by successive enlargements of the port of Valencia (Spain). The impact assessment methodology developed in this work can serve as the basis to support flooding and sediment-related risk assessments and adaptation planning within port-coast systems.

Keywords: Port planning · erosion · flooding coastal management

1 Introduction

Port development and operation are pivotal for economic growth worldwide but significantly influence coastal ecosystems with both positive and negative contributions to coastal erosion, flooding and ecological balance. Changes in coastal erosion or flooding at the vicinity of ports may compromise their societal acceptance, their operations, their safety and increase economic and financial risks. Neglecting the interconnectedness of ports and their surrounding coastal zones can result in incomplete risk assessments and adaptation planning.

There are various approaches to modeling coastal evolution in the presence of coastal structures. Process-based models are utilized to simulate detailed physical processes, including wave dynamics, sediment transport, and coastal morphology over short-term scales [1]. In contrast, reduced-complexity models are dedicated to broader, long-term impacts of coastal infrastructure, such as shoreline evolution, erosion patterns, and sediment distribution, focusing on larger spatial and temporal scales [2, 3].

Despite advances in modelling, process-based models cannot achieve the necessary temporal scale to address long-term planning issues, while reduced-complexity models

© The Author(s) 2026
C. Coelho et al. (Eds.): CD 2025, CRL 41, pp. 506–512, 2026.
https://doi.org/10.1007/978-3-032-15473-6_77

lack the detailed output of their process-based counterparts and may miss some relevant processes like diffraction due to coastal structures. To address this gap, [4] developed IH-LANSloc, a reduced-complexity nearshore evolution model capable of resolving the complete beachfront evolution at complex sandy coastal settings over more than a few years and kilometers. This allows to consider the morphological feedback into flood studies.

Currently, no existing modeling framework rigorously integrates the morphological feedback in presence of coastal structures into the assessment of long-term erosion and flood hazards while accounting for the primary physical processes. In this sense, this paper introduces a suite of models that combines the reduced-complexity nearshore evolution model IH-LANSloc, enhanced to include port-induced diffraction, with a flood modeling suite that captures essential surf-zone hydrodynamic processes and efficiently simulates inland flood propagation using SFINCS [5]. This approach is applied to a large-scale study (~100 km) in the Mediterranean Sea over the period 1990–2020, during which the port of Valencia underwent significant expansion. The combined erosion and flood analysis quantitatively demonstrates the short- and long-term coastal impacts of modifications to the port's outer boundary, highlighting the critical importance of integrated analysis for sustainable port and coastal planning.

2   A Reduced-Complexity Coastal-Port Impact Modelling Chain

The proposed modelling chain consists of two building blocks, the morphological and the flood calculation. First, changes in terrain heights influenced by coastal infrastructure are calculated using IH-LANSloc. In the flood calculation block, morphological changes are applied to the topobathymetry grid, which is used by the flood model SFINCS to assess coastal flooding.

The modelling chain starts by collecting all the necessary inputs including waves and water levels, coastal structures and their temporal evolution, ecosystems if any, shoreline observations and sediment characteristics. Then, coastal drivers are preprocessed for the morphological evolution and the coastal flooding modules. Long-term morphological evolution may be prohibitive from a computational perspective. Hence, reduced wave time series are obtained via input reduction following [6]. In the case of coastal flooding, extreme wave and water level dynamics need to be collected to define the extreme event. Event definition for coastal risk assessments involves characterizing the physical conditions that lead to major damages of exposed assets.

The following step consists of simulating the long-term morphological evolution using a reduced-complexity physics-based morphological model (IH-LANSloc) to solve for the evolution of the nearshore area. The coastal flooding calculation takes as input the topobathymetry evolution, which incorporates the effects of morphological changes on coastal infrastructure and comprises two steps: the surf-zone hydrodynamic modelling and the inland flood propagation. Surf-zone hydrodynamic modelling implies solving the total water level (TWL) at the coast. Subsequently, overland flooding is assessed using SFINCS forced with empirically-derived TWLs. Wave diffraction due to coastal structures is considered in both morphological and flood calculations by applying empirical corrections to nearshore waves according to [7].

3 Demonstration Case

3.1 Study Area and Input Data

The methodology is demonstrated at a 40 km coastal stretch in the Mediterranean Sea affected by the port of Valencia (Spain) as displayed in Fig. 1. Since 1992, the port has experienced some major enlargements that affected its offshore boundary [8]. First, the construction of the southern dike started in 1992 and finished a year later with the purpose of sheltering a basin finished in 1996. The next important change occurred after 2009, when the northern basin was enlarged and external dikes were built in 2010. Since then, the external boundary of the port has not changed.

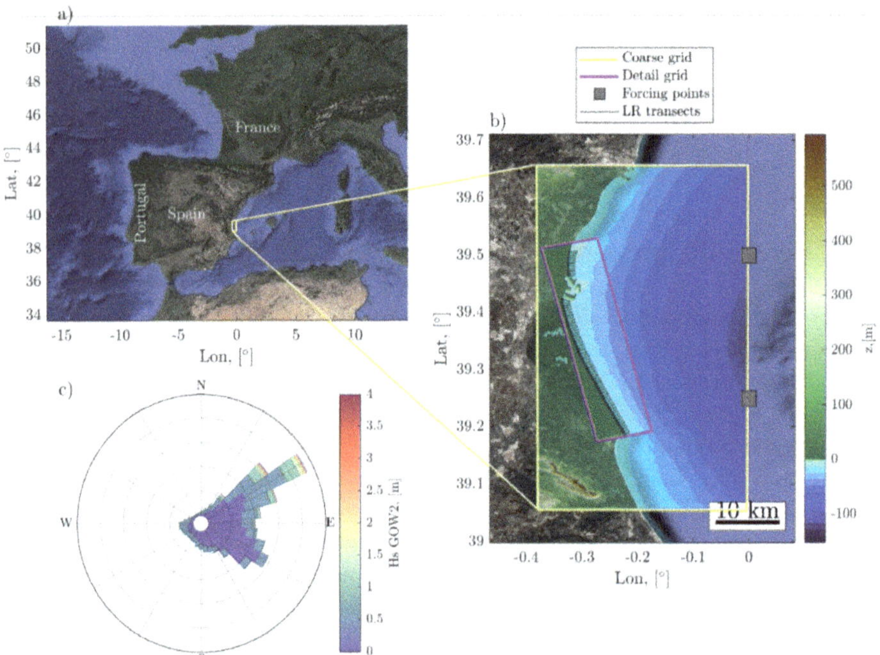

Fig. 1. Description of the study area.

3.2 The Role of the Port Enlargement in Coastal Evolution

In order to assess the role of the port of Valencia in coastal evolution, the flowchart displayed in Fig. 1 is applied. Six topobathymetries are extracted for the years 1990, 1992, 1993, 2009, 2010 and 2020, aligned with changes in external port morphology. To ensure a consistent assessment, the same extreme event, named storm Gloria, is selected as forcing for all modelled topobathymetries. Storm Gloria struck the Mediterranean coast of Spain in January 2019 and lasted 96 h. During this extreme event, significant wave heights exceeded 6.4 m offshore with periods of 12 s coming from the east together with storm surges of more than 0.25 m. This 30-year return period event, which generated major damages all along the Spanish Mediterranean coast, is taken as reference to simulate coastal flooding over the six topobathymetries influenced by coastal infrastructure.

Coastal Erosion

Monthly-equivalent dynamics are obtained from the hourly offshore time series following the input reduction technique. Then, IH-LANSloc solves the evolution of the nearshore by propagating waves to breaking using SWAN [9] and integrating the effect of diffraction empirically. Shoreline evolution is calculated and observations are assimilated to correct for the predicted shoreline and adjust the free model parameters. Shoreline changes, calculated at the 60-m spaced IH-LANSloc profiles, are then transcribed to the detail grid which is used for the subsequent SWAN run.

IH-LANSloc translates shoreline changes to the topobathymetry using the profile translation tool ShoreTrans [10]. Figure 2 shows the initial 1990 topobathymetry and the change in elevation of the nearshore area when significant port infrastructure modifications. It becomes noticeable the construction of the southern marina in 1992 and 1993 and the enlargement of the northern dike after 2009. The construction of the southern dike did not play a significant role in the nearshore evolution. However, after 2010, the erosion trend began to spread southward at an accelerated rate, reaching beaches located 15 km downstream from the Port of Valencia.

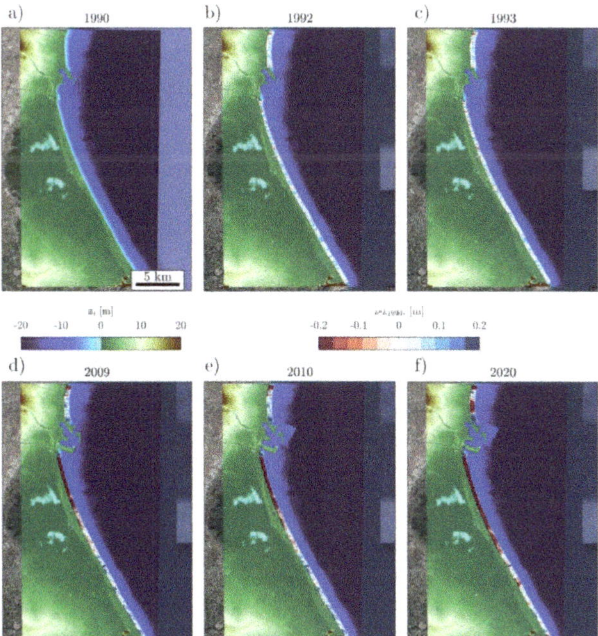

Fig. 2. Topobathymetry changes over the active beach profile after major changes in the port morphology with respect to the initial 1990 topobathymetry.

Coastal Flooding

First, offshore wave and water levels associated to storm Gloria are downscaled to the nearshore by running SWAN dynamically considering the influence of the port evolution and the empirical diffraction corrections. Then, the total water level time series are calculated following the empirical formulation of [11] integrating the wave contribution to set-up and infragravity swash.

SFINCS is then used to derive the flood maps associated to the event Gloria and results are summarized in Table 1. Coastal zonation in Table 1 is carried out considering the morphological behavior of the study site with the purpose of aggregating coastal flooding results. Thus, due to the variability of the nearshore, a consistent seaside boundary is defined starting from the first line infrastructure. Then, by analyzing the results, the relationship between the coastal morphology and flooding can be disentangled. For instance, Zone 1, which corresponds to the northern area of the port, suffers erosion in 1992 and 1993 with respect to 1990, as evidenced in Fig. 3. This situation implies a slight increase of 600 m^2 in the flooded area in 1993 with respect to 1990. Then, Zone 1 receives sediment from the incoming littoral drift and flooding reduces 1200 m^2 in 2010. After 2010, the littoral cell experiences a reorientation due to longshore drift characterized by accretion next to the port and erosion in the north. This is translated into an increase of the coastal flooding generated by storm Gloria with respect to the 2010 situation.

Table 1. Flooded areas associated to storm Gloria at different dates corresponding to major changes in the port morphology.

Flooded area, (ha)	1990	1992	1993	2009	2010	2020
Zone 1	24.1	24.15	24.16	23.92	23.98	24.05
Zone 2	81.55	77.52	77.04	78.78	79.01	81.91
Zone 3	18.23	18.2	18.15	16.99	17.01	16.36

Zones 2 and 3 make up the littoral cell that stretches from the southern section of the port to the Gulf of Cullera. Zone 2 covers the area sheltered by the port, extending to the southern edge of the Albufera coastal lagoon. Zone 3 continues from the Albufera to the Gulf of Cullera. Flooding in this littoral cell is influenced by two contradictory processes: wave sheltering at the lee of the port and coastal erosion/accretion. By analyzing results in Table 1 and comparing them to the morphological evolution it can be noticed how flooding in Zone 3 is progressively reduced since 1990. This zone receives sediment from northern beaches due to predominant sediment drift and this may explain the slight reduction in flooding as more emerged beach area implies more friction resistance to overland flooding. Zone 2 experiences in 1992 and 1993 a slight reduction of the flooding extent of around 4.03 and 4.51 hectares respectively with respect to the 1990 situation. This is explained by the sheltering effect exerted by the construction of the southern marina in the port of Valencia. In this instance, the morphological feedback following the port's modification is not significant because of the short timeframe involved. Then the situation changes and flooding in Zone 2 begins to progressively increase in 2009, 2010 and 2020. Even if TWL is reduced due to the construction of the northern dike in 2009 and 2010, morphological feedback becomes relevant after 1994 (Fig. 3) and erosion of Zone 2 is aggravated after 2010. This fact translated into an increase in flooding extents of 4.87 hectares in 2020 with respect to the 1993 situation and a slight increase of 0.36 hectares with respect to 1990 due to the compensated effects of erosion and wave sheltering.

4 Conclusions

This work presents a modelling chain to assess erosion and flood-related impacts in coastal areas influenced by port. Applied to the Port of Valencia, results show initial wave sheltering but long-term erosion effects. The methodology aids in evaluating port expansion impacts and climate change hazards on coastal areas.

References

1. Karambas TV (2015) Modelling of climate change impacts on coastal flooding/erosion, ports and coastal defence structures. Desalin Water Treat 54(8):2130–2137
2. Robinet A, Idier D, Castelle B, Marieu V (2018) A reduced-complexity shoreline change model combining longshore and cross-shore processes: the LX-Shore model. Environ Model Softw 109:1–16
3. Roelvink D, Reniers A, Van Dongeren AP, De Vries JVT, McCall R, Lescinski J (2009) Modelling storm impacts on beaches, dunes and barrier islands. Coast Eng 56(11–12):1133–1152
4. Álvarez-Cuesta M, Losada IJ, Toimil A (2023) A nearshore evolution model for sandy coasts: IH-LANSloc. Environ Model Softw 169:105827
5. Leijnse T, van Ormondt M, Nederhoff K, van Dongeren A (2021) Modeling compound flooding in coastal systems using a computationally efficient reduced-physics solver: including fluvial, pluvial, tidal, wind-and wave-driven processes. Coast Eng 163:103796
6. Scipione F, De Girolamo P, Castellino M, Pasquali D, Celli D, Di Risio M (2024) Reduced wave time series for long-term morphodynamic applications. Coast Eng 189:104453
7. Dabees MA (2000) Efficient modeling of beach evolution. Queen's University
8. Pardo-Pascual JE, Palomar-Vázquez J, Cabezas-Rabadán C (2022) Estudio de los cambios de posición de la línea de costa en las playas del segmento València-Cullera (1984–2020) a partir de imágenes de satélite de resolución media de libre acceso. Cuadernos de Geografía de la Universitat de València 108–9:79–104
9. Booij NRRC, Ris RC, Holthuijsen LH (1999). A third-generation wave model for coastal regions: 1. Model description and validation. J Geophys Res Oceans 104(C4):7649–7666
10. McCarroll RJ et al (2021) A rules-based shoreface translation and sediment budgeting tool for estimating coastal change: ShoreTrans. Mar Geol 435:106466
11. Stockdon HF, Holman RA, Howd PA, Sallenger AH Jr (2006) Empirical parameterization of setup, swash, and runup. Coast Eng 53(7):573–588

Waves of Sand: Erosion and Accretion Across Multiple Coastal Cells

Patricio A. Catalán[1,5(✉)], Daniel Rojas[2], Maruan Charifeh[3],
and Rodrigo Cienfuegos[4,5]

[1] Departamento de Obras Civiles, Universidad Técnica Federico Santa María, Valparaíso, Chile
patricio.catalan@usm.cl
[2] Geobrugg Chile, Santiago, Chile
[3] 4C Ingenieros Consultores, Viña del Mar, Chile
[4] Departamento de Ingenieria Hidraulica y Ambiental, Pontificia Universidad Católica de Chile,
Santiago, Chile
racienf@ing.puc.cl
[5] CIGIDEN, Santiago, Chile

Abstract. Remote sensing analysis along a section of the coast of central Chile was used to quantify coastal evolution. The results at a single beach were contrasted with simple linear superposition models to identify parameters that could correlate to the observed change, including tectonic-induced vertical change, climatic indices, wave parameters, and riverine discharges. It was found that riverine discharges were significant contributors to this linear model. Next, the coastal evolution analysis was conducted over a large section of the coast, encompassing several watersheds. The shoreline change showed a mesoscale pattern of propagating erosion, that was interspersed by river discharges, although the trends were similar across cells. It is speculated that these observations point to mesoscale and watershed processes to be relevant and ought to be considered in assessing and forecasting coastal evolution.

Keywords: Shoreline evolution · Long-term coastal change · Climate patterns

1 Introduction

It is well known that sandy beaches evolve seasonally and interannually owing to a wide range of forcings, chiefly the wave conditions. If the cycles that undergo these forcings are in a state of quasi-equilibrium beaches would remain stable when averaging over a long-term period. However, changes in forcings due to climate have resulted in model predictions presenting a dire prospect for beaches around the world, where erosion might be the dominant trend. The main driver for these assessments is relative sea level rise but changes in wave climate are relevant (Vousdukas et al., 2020).

Recent studies point to a more complex situation where long-term oscillations such as El Niño cycles or the Pacific Decadal Oscillation might also explain rates of change and trends (Vos et al., 2023). Although Vousdoukas et al. (2020) and others have highlighted

© The Author(s) 2026
C. Coelho et al. (Eds.): CD 2025, CRL 41, pp. 513–518, 2026.
https://doi.org/10.1007/978-3-032-15473-6_78

the exchange across neighboring coastal cells, most analyses analyze coastal change at either the local scale or global scales, with an underlying assumption of independence among coastal cells.

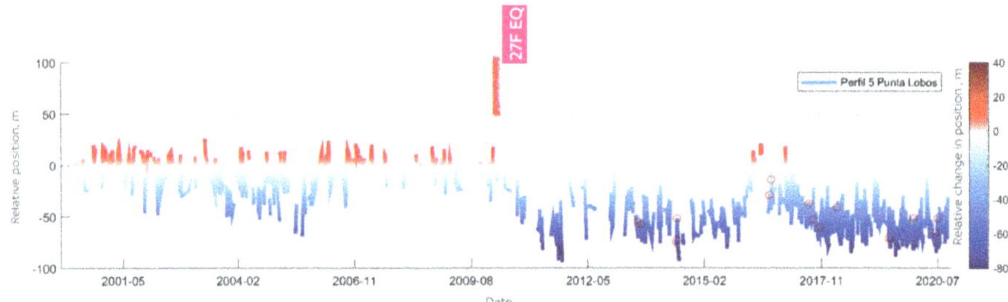

Fig. 1. Coast Sat extracted shoreline position near Punta de Lobos, central Chile. Values are relative to the first data point, thereby showing net change. The downward line marks the occurrence of the Maule Earthquake and Tsunami (Fritz et. Al, 2011)

This has been observed on the coast of central Chile, a 500 km section of coast that is well exposed to the Pacific Ocean wave climate but is also subject to strong El Nino signals, as well as active seismicity and tectonics. This signal can overcome sea level rise due to climate change. Net erosion rates have been reported (Winckler et al., 2023). The focus of this work is to diagnose the long-term evolution along multiple coastal cells and Andean watersheds and to understand its main drivers.

2   Material and Methods

A two-step process was followed. First, using a combination of in situ data, numerical modeling of wave conditions, readily available oceanic indices such as ONI and others, and shoreline tracking using CoastSat (Vos et al., 2023b), the coastal evolution of Punta de Lobos Beach in Central Chile (34.42°S) is characterized during the period 1985–2020. This beach was reported to have changed its erosive state following the 2010 Maule Earthquake and Tsunami. Figure 1 shows the relative shoreline change of a single cross-shore profile near the pivoting section of the beach, that is, the section that showed smaller seasonal variation when exposed to changes of wave direction between summer and winter seasons. Qualitatively, three different patterns are found: First, the period between 2000–2010 shows a balance of accretion and erosion, with a mean shoreline position of about −5m. This period is followed by a rapid retreat, spanning 1.5 years, with net erosion of nearly 50 m. After 2011, the new average profile is located near −60 m, except for the year 2016, when a rapid net accretion was found only to decrease rapidly to the previous state. In terms of variance, both time windows (pre 2010 and after 2011), have similar range, with changes of about 30 m about the mean.

A linear correlation model between coastline position CL and a set of plausible forcings X was tested (Fig. 2).

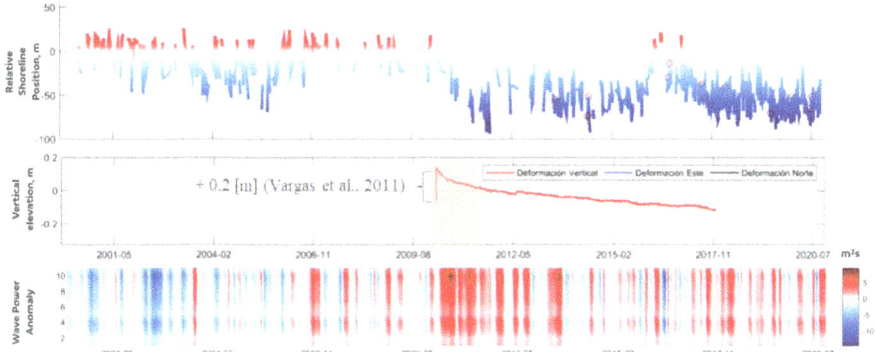

Fig. 2. (Top)Coast Sat extracted shoreline position near Punta de Lobos, central Chile. (Middle) Vertical elevation signal

$$CL = \sum \beta_i X_i \tag{1}$$

where β_i are weights to be determined by least squares regression. The actual shoreline position is used, instead of its rate of change as usually is done in shoreline change models (e.g. Schepper et al., 2021). This supposes that the shoreline reacts instantaneously to the other parameters which could affect the model's physical interpretation, but it is considered a suitable step to identify the dominant parameters. The problem is compounded by differences in acquisition time among images, that leads to a non-uniform temporal distribution to estimate rates of change. Hence, the parameters are sampled to match shoreline position data. The model considered wave-derived parameters, such as wave height H, wave power, wave steepness $H/L,$ and wave direction. Due to lack of long-term coastal monitoring, data were propagation of hindcast data using a unstructured grid domain to shallow waters offshore of the breaker line using SWAN (Booij et al, 1999, using a constant bathymetry. The model was calibrated against a short-term data collection campaign near Punta de Lobos, with a good agreement in reproducing a 3 m. storm.

Central Chile is also subject to significant tectonic activity, that manifests by a constantly evolving land elevation relative to sea level, due the interseismic cycle, which is interrupted by earthquakes that can induce large magnitude, episodic changes during coseimic and post-seismic cycles (Wesson et al., 2015). The Punta Lobos and Pichilemu regions were within the rupture zone of the 2010 Maule Earthquake and in front of a large slip patch. This led to a relatively small vertical uplift of 20–50 cm (Kelson et al., 2012, Vargas et al., 2011). Two additional earthquakes occurred near Pichilemu in 2011, (Ruiz et al., 2011), without significant vertical change. Since then the subsidence rate has been about 20 mm/yr (Orellana et al., 2022). These changes are comparable with the 100-year total relative sea level change in the region (Albrecht and Schaffer, 2016). To account for this in the model, daily vertical data spanning from April 2010 to Nov 2017 from the GNSS station near Pichilemu (−34.391S −72.003W) maintained by the Nevada Geodetic Laboratory was used (http://geodesy.unr.edu) (Fig. 3).

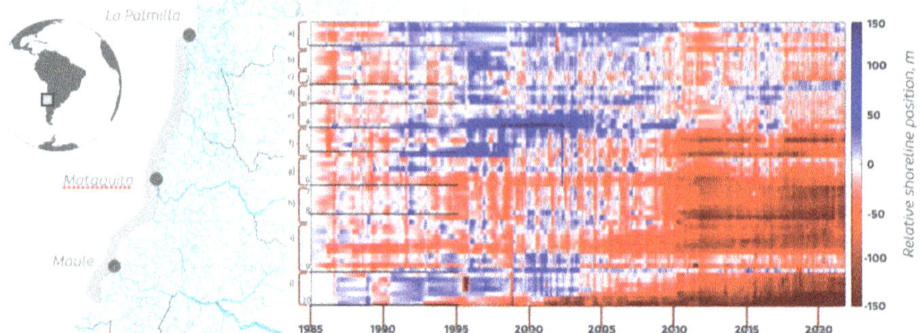

Fig. 3. Locations of interest and time-space map the evolution of the relative shoreline position along the coast. Map shows the hydrographic network. Red denotes erosion while blue is accretion, using as reference the first data point.

Oceanic indices associated with sea level temperature such as Oceanic El Nino Index (ONI) and Pacific Decadal Oscillation index (PDO) were considered. Finally, flow rates measured at a series of rivers and creeks in the vicinity of Pichilemu were also introduced. The rationale was to account for the effect of the mega drought that affected central Chile in the period 2010–2022 (e.g. Garreaud et al., 2019).

All data were normalized in the range 0–1.0, to reduce the effect of different scales. A total of 43 different combinations of parameters were tested, and those with the largest correlation coefficient and minimal root-mean-square were chosen.

3   Results and Discussion

The combination of parameters that yield the best correlation coefficient ($R^2 = 0.63$) and minimal RMSE (19 m) included wave power and wave steepness as the main wave parameters, and land vertical elevation signal, PDO, ONI and the flow rate of four major rivers, even those located more than 50 km from the site. Despite correlation not meaning causation, especially considering some of the assumptions used herein, the fact that incorporating rivers outside this particular coastal cell improved the correlation led us to analyze all beaches along a 100 km stretch of coast with 20 profiles spanning from Maule river (35°31S), to Estero La Palmilla, nearby Punta de Lobos. This section spans multiple coastal watersheds and two Andean rivers.

This analysis does not consider studying the correlation. Instead, it is aimed at assessing qualitatively coastal evolution in a synoptic way. Figure 4 Shows the results on a time-space map, where black lines denote the location of discharge of each river. Each row corresponds to a single beach (one profile per beach), therefore local evolutions are not included. A good qualitative correlation between the river location and trends in the shoreline evolution is found, especially in the southern section of the domain. It appears that erosion patterns begin in the south and propagate northward. They do so slowly at first (1995–2005) only to increase their rate of change in the period 2005–2010. After 2010, the erosive trend is sustained, just as observed in Pichilemu. Traces of this northward propagation of erosion can be observed in northern cells. These interpretations led to some intriguing aspects: First, the rate of change appears to be decoupled from the drought itself, but to another process that is relevant at mesoscale that has not been identified here. Second, these changes appear to have some long-term oscillation since the period 1985–1990 was also characterized by erosion to be followed by an accretional window in 1990–2010.

The corollary is twofold: First, long-term climatic variability affecting sediment supply ought to be included in the analysis of coastal evolution. In the case of Chile, these effects can be augmented by the tectonic setting. Most of the analysis to date has focused on the changes arising from changing wave conditions and appears to have overlooked these effects. Second, trends appear not to be local but rather affect a series of surrounding coastal cells. This implies that the definition of what a coastal cell is and how sediment can bypass morphological features might need revisiting. Also, analysis of shoreline evolution might require to be analyzed at larger scales.

4 Conclusions and Future Work

The analysis of shoreline evolution of a single beach in central Chile was conducted using remote sensing tools. The observations showed a rapid change spanning two years, which coincided with a large earthquake but also with the onset of a megadrought. An analysis that included the flow rate of neighboring rivers as a proxy for sediment supply showed an improved correlation in reproducing the shoreline evolution with a linear model.

The analysis was partially extended to a section of coast encompassing several watesheds. The relation to rivers discharges was observed qualitatively, but also that erosion trends could be identified at mesoscale, with an apparent south to north propagation pattern.

While more work is required to understand the physical drivers of these process, these results suggest that the study of long-term shoreline evolution needs to incorporate effects in the watersheds, as well as having a larger spatial footprint.

References

Albrecht F, Shaffer G (2016) Regional sea-level change along the Chilean coast in the 21st century. J Coastal Res 32:1322. https://doi.org/10.2112/JCOASTRES-D-15-00192.1

Booij N, Ris RC, Holthuijsen LH (1999) A third-generation wave model for coastal regions 1. Model description and validation. J Geophys Res 104:7649–7666. https://doi.org/10.1029/98JC02622

Fritz H et al (2011) Field survey of the 27 february 2010 Chile tsunami. Pure Appl Geophys 168:1989–2010. https://doi.org/10.1007/s00024-011-0283-5

Garreaud RD, Boisier JP, Rondanelli R, Montecinos A, Sepúlveda HH, Veloso-Aguila D (2019) The central Chile mega drought (2010–2018): a climate dynamics perspective. Int J Climatol 40:421–439. https://doi.org/10.1002/joc.6219

Kelson K et al (2012) Coseismic tectonic surface deformation during the 2010 Maule, Chile, mw 8.8 earthquake. Earthq Spectra 28:S39–S54. https://doi.org/10.1193/1.4000042

Ruiz JA, Hayes GP, Carrizo D, Kanamori H, Socquet A, Comte D (2014) Seismological analyses of the 2010 march 11, Pichilemu, Chile Mw 7.0 and Mw 6.9 coastal intraplate earthquakes. Geophys J Int 197:414–434. https://doi.org/10.1093/gji/ggt513

Schepper R et al (2021) Modelling cross-shore shoreline change on multiple timescales and their interactions. J Marine Sci Eng 9:582. https://doi.org/10.3390/jmse9060582

Orellana F, Hormazábal J, Montalva G, Moreno M (2022) Measuring coastal subsidence after recent earthquakes in Chile central using SAR interferometry and GNSS data. Remote Sens 14:1611. https://doi.org/10.3390/rs14071611

Vargas G, Farias M, Carretier S, Tassara A, Baize S, Melnick D (2011) Coastal uplift and tsunami effects associated to the 2010 Mw8.8 Maule earthquake in central Chile. Andean Geology 38. https://doi.org/10.5027/andgeoV38n1-a12

Vos K, Hartley M, Turner IL, Splinter KD (2023) Pacific shoreline erosion and accretion patterns controlled by El Niño/Southern oscillation. Nat Geosci 16(2):140–146. https://doi.org/10.1038/s41561-022-01117-8

Vos K, et al (2023b). Benchmarking satellite-derived shoreline mapping algorithms. Commun Earth Environ 4(1). https://doi.org/10.1038/s43247-023-01001-2

Vousdoukas MI et al (2020) Sandy coastlines under threat of erosion. Nat Clim Chang 10(3):260–263. https://doi.org/10.1038/s41558-020-0697-0

Wesson, RL, Melnick D, Cisternas M, Moreno M, Ely LL, (2015).Vertical Deformation through a complete seismic cycle at Isla Santa Maria, Chile. Nat Geosci. https://doi.org/10.1038/ngeo2468

Winckler P, Martín RA, Esparza C, Melo O, Sactic MI, Martínez C (2023) Projections of beach erosion and associated costs in Chile. Sustainability 15(7):5883. https://doi.org/10.3390/su15075883

The Influence of Climate Change on Compound Coastal Flooding in San Francisco Bay: An Application of a Hybrid Statistical-Dynamical Framework

Peter Ruggiero[1]([✉]), Zhenqiang Wang[1], Meredith Leung[2], Sudarshana Mukhopadhyay[3], Sai Veena Sunkara[4], Scott Steinschneider[3], Jonathan Herman[4], Marriah Abellera[5], and John Kucharski[6]

[1] Oregon State University, Corvallis, OR, USA
peter.ruggiero@oregonstate.edu
[2] NCAR, Boulder, CO, USA
[3] Cornell University, Ithaca, NY, USA
[4] University of California, Davis, CA, USA
[5] US Army Corps of Engineers, Washington, USA
[6] Deltares, Delft, The Netherlands

Abstract. Compound coastal flooding poses significant risks to urban and ecological systems, particularly under changing climate conditions. This paper synthesizes findings from a recent study that leveraged a hybrid statistical-dynamical modeling framework to investigate compound flooding in the San Francisco Bay Area. By generating over 4.3 million hourly total water levels (TWLs) across 100 simulations spanning 500 years, this framework fully quantifies and propagates uncertainties in forcing drivers, such as sea-level rise and river discharge, to predict extreme flooding events. Results reveal that despite the variability in individual forcing parameters, the range of return level events, such as 100-year TWLs, remains low across simulations, underscoring the robustness of the approach. This characteristic highlights the unique nature of compound flooding, where diverse forcing combinations can produce similar flood magnitudes. The framework's computational efficiency allows for the exploration of a broader range of scenarios than traditional dynamical or statistical methods. Key findings emphasize the increasing frequency and magnitude of extreme TWLs under more severe climate scenarios, with implications for adaptation and resilience planning. This synthesis demonstrates the power of hybrid modeling approaches in capturing complex interactions driving compound flooding, providing critical insights for mitigating flood risks in vulnerable coastal regions.

Keywords: Compound Coastal Flooding · Climate Change · Hybrid Modeling

© The Author(s) 2026
C. Coelho et al. (Eds.): CD 2025, CRL 41, pp. 519–523, 2026.
https://doi.org/10.1007/978-3-032-15473-6_79

1 Introduction

Compound coastal flooding due to multiple drivers poses severe threats to coastal communities today, and these threats are expected to increase due to climate change. This study explores the impact of climate change, from the perspective of both sea-level rise and increased river flows, on coastal flooding in San Francisco Bay, CA, USA. This region is characterized by complex interactions between nearshore, estuarine, and river environments [1] with multiple water level drivers such waves, tides, and river flows contributing to extreme total water levels (TWLs). San Francisco Bay has historically experienced coastal flood events resulting in infrastructure damage and economic loss and is becoming more vulnerable to compound flooding [2].

2 Approach

Here we use the hybrid statistical-dynamical approach (Fig. 1) developed by Wang et al. [3, 4] to analyze compound coastal flooding in San Francisco Bay under climate change. By combining a stochastic generator of compound flooding drivers and surrogate models of a high-fidelity hydrodynamic model, the hybrid framework can explore the full range of possible forcing combinations of various drivers [5]. Representative climate change scenarios are defined by linking three sea-level rise scenarios (low, medium, and high impact) and three increased temperature scenarios defining six river discharge scenarios associated with different precipitation rates. The impact of climate change on flooding is investigated in terms of flood magnitude, frequency, and the relative contributions of different drivers.

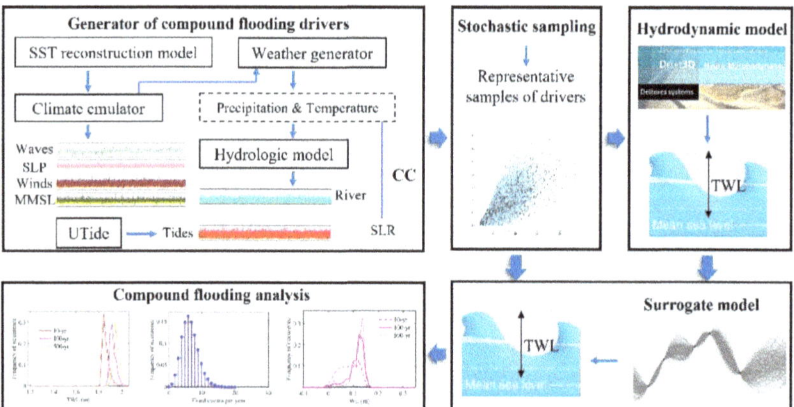

Fig. 1. Hybrid statistical-dynamical framework for compound coastal flood analysis. The flood driver generator combines an SST reconstruction model, a climate emulator, a weather generator, and a hydrologic model. Surrogate models are developed from hydrodynamic simulations to predict TWLS.

3 Results

Results of the hybrid statistical-dynamical framework under the no climate change scenario (C0, Fig. 2), reveals that each of non-tidal forcing drivers (e.g., monthly mean sea level anomalies, waves, sea level pressure anomalies, wind, and river flow) are of similar importance to TWLs and none of them can be ignored when analyzing extremes in San Francisco Bay. Climate change scenarios that account for increases in temperature and associated modifications to precipitation in the watersheds feeding the Bay (C1-C6, Fig. 2), result in significant increases in the relative contribution of river discharge to extreme TWLs (and therefore a decrease in the relative contributions from other drivers). In some scenarios river discharge may account for over 80% of the non-tidal water level signal during extremes, even approximately 20 km seaward of the San Francisco Bay Delta. Exploring the full range of potential climate change impacts on compound flood drivers will be useful in developing adaptation pathways for the region.

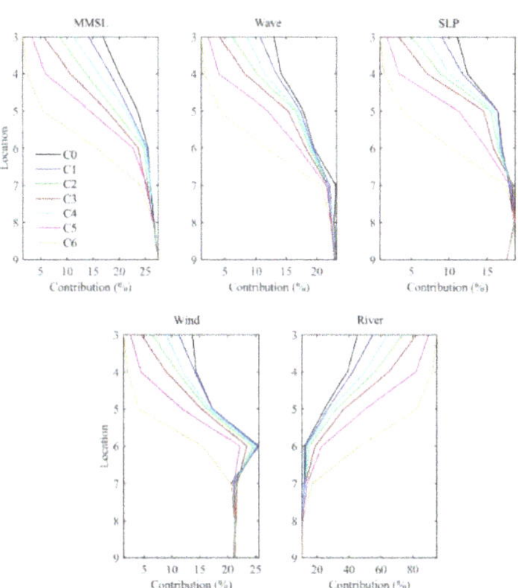

Fig. 2. Spatial distribution of the relative contributions of each driver to nontidal extreme TWLs under different climate scenarios along the centerline of SF North Bay (Stations are separated by approximately 8km). Scenario C0 represents no climate change while scenarios C1-C6 represent six future SLR and temperature change combinations.

Using the hybrid statistical-dynamical approach, this study generated an extensive dataset of hourly total water levels (TWLs), amounting to 4,382,928 TWLs for each of the 100 simulations over a 500-year period. This robust approach ensures that uncertainties in all forcing drivers are fully quantified and propagated, enabling the simulation of compound flood events with high confidence. A specific return level event, such as the 100-year event, is identified by selecting a particular percentile (e.g., the 5th largest TWL). The analysis revealed that the variation in return level events across simulations remains low, even under extreme climate scenarios, owing to the comprehensive quantification of uncertainties (Fig. 3).

Interestingly, the results demonstrate that a wide range of forcing parameter combinations can produce similar magnitudes of compound flooding events. This highlights a unique characteristic of compound flooding, where diverse forcing drivers interact to yield comparable extreme outcomes. Moreover, the high computational efficiency of the hybrid statistical-dynamical approach makes it a powerful tool for capturing extreme TWL events, outperforming purely dynamical or statistical methods. This finding underscores the potential for this framework to advance our understanding of flood risks in complex coastal environments and to support more informed climate adaptation strategies.

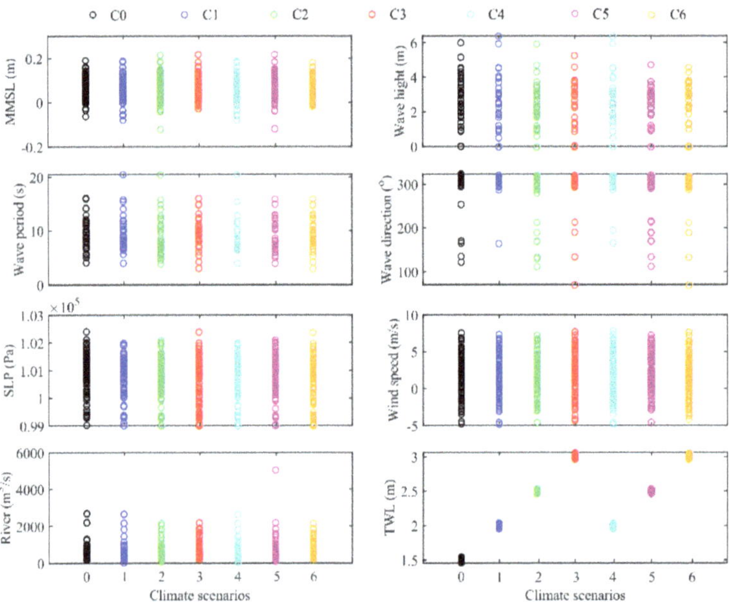

Fig. 3. Distribution of simulated 100-year extreme TWLs and associated forcing drivers at the San Francisco station (near the Golden Gate) across different climate scenarios.

4 Discussion

This study employs a hybrid statistical-dynamical framework to investigate compound coastal flooding under climate change, providing a detailed assessment of flood magnitude, frequency, and driver contributions. By simulating over 4.3 million hourly total water levels (TWLs) across 100 simulations spanning 500 years, the framework ensures that uncertainties in all forcing drivers—such as sea-level rise (SLR), river discharge (RD), and meteorological factors—are fully quantified and propagated.

The results reveal that extreme TWL events, including the 100-year return level, demonstrate low variability across simulations despite significant variability in individual forcing parameters. This finding underscores a fundamental characteristic of compound flooding: diverse combinations of forcing drivers can yield similar flood magnitudes. Such insights emphasize the complexity of interactions among drivers and the challenges of addressing compound flood risks. Additionally, the study demonstrates a significant increase in flood frequency and magnitude under more severe climate scenarios, with SLR and RD becoming the dominant contributors to extreme TWLs, particularly near the Delta.

The hybrid framework's ability to handle extensive datasets with high computational efficiency allows for a comprehensive exploration of forcing combinations, which would be challenging for purely dynamical or statistical methods. This makes the framework a powerful tool for informing climate adaptation strategies and developing pathways for resilience in vulnerable coastal regions like San Francisco Bay. The consistency in results across climate scenarios highlights the robustness of the framework in capturing the complex dynamics of compound flooding.

Acknowledgements. This work was primarily supported by the U.S. Army Corps of Engineers (USACE) via the Anticipating Threats in Natural Systems (ACTIONS) Congressional Interest project. The authors also acknowledge support from the U.S. Department of Defense's Multi-University Research Initiative grant 2033958-N00014-23-1-2862 and Environmental Science and Technology Certification Program grant 2035806-W912HQ24C001 as well as by the Cascadia Coastlines and Peoples Hazards Research Hub, an NSF Coastlines and People Large-Scale Hub (NSF award number 8092103713).

References

1. Nederhoff K, et al (2021) Drivers of extreme water levels in a large, urban, high-energy coastal estuary–A case study of the San Francisco Bay. Coastal Eng 170
2. Cayan DR, Bromirski PD, Hayhoe K, Tyree M, Dettinger M, Flick R (2008) Climate change projections of sea level extremes along the California coast. Clim Change 87:57–73
3. Wang Z et al (2025) A hybrid statistical-dynamical framework for compound coastal flooding analysis. Environ Res Lett 20(1):014005. https://doi.org/10.1088/1748-9326/20/1/014005
4. Wang Z, et al (2025) Compound coastal flooding in San Francisco Bay under climate change. NPJ Nat Hazards 2(3). https://doi.org/10.1038/s44304-024-00057-0
5. Anderson D et al (2021) A hybrid statistical dynamical model incorporating climate variability into future coastal total water levels for flood risk analysis. Earth's Future. https://doi.org/10.1029/2021EF002285

Monsoon-Intermonsoon Influence on Spatio-Temporal Variability of Seabed Sediments Along the Urbanised Coast of Singapore

Stephen Chua[1(✉)], Isaac Lai[1], Benjamin S. Grandey[3], Yu Ting Yan[1], Abang Nugraha[1], Koi Siek[1], Yun Fann Toh[2], and Adam D. Switzer[1,2]

[1] Earth Observatory of Singapore, Nanyang Technological University, Singapore, Singapore
`stephen.chua@ntu.edu.sg`
[2] Asia School of the Environment, Nanyang Technological University, Singapore, Singapore
[3] School of Physical and Mathematical Sciences, Nanyang Technological University, Singa-Pore, Singapore

Abstract. Singapore's urbanized coastal environment is highly influenced by both natural and anthropogenic drivers. However we lack understanding of nearshore sediment deposition patterns and the relationship to these climatic drivers. This research aims to characterize the coastal conditions off Singapore to understand the monsoon-intermonsoon influences on near-offshore seabed sediment in the Singapore Strait. We collected monthly seabed boxcorer samples, CTD water parameters and surface/bottom current at two sampling points per site (i.e., proximal and distal to shore) from September 2023 to August 2024. These sites, namely Tuas (TS2, 5), Marina South (MS1, 4), and East Coast (EC1, 4), represent three unique coastal typologies in the form of highly engineered coast (100% seawall), coastal dam-floodgate and sandy beach (with seaward drainage channels), respectively. We performed particle-size analysis and loss-on-ignition to obtain the grain-size distribution and organic and inorganic matter percentages, respectively. Our results show that climate and current parameters largely correlate with minor temporal lags, but deposition trends show significant spatially and temporally variable site-specific differences, despite the close proximity providing similar oceanographic conditions. Our study suggests decoupling of sediment behaviour from monsoon-driven hydrodynamics at local scales due to coastal modifications. Which underpins the need for field-based sampling of shallow marine sediments due to complex heterogeneity of developed nearshore zones which is difficult to comprehensively model.

Keywords: Sediment · Climate · Monsoon · Coastal development

1 Introduction

Global coastal zones are being rapidly urbanized and are highly vulnerable to erosion by present and future sea-level rise [1] exacerbated by sediment starvation due to geoengineering mitigation measures [2]. We lack adequate understanding of nearshore

C. Coelho et al. (Eds.): CD 2025, CRL 41, pp. 524–530, 2026.
https://doi.org/10.1007/978-3-032-15473-6_80

depositional trends on short climate cycles (e.g., monsoon, ENSO), with assumption of coupled sedimentation response to physicals drivers strengthening and waning during monsoons and intermonsoons, respectively [3]. Currently, there remains a paucity of *in situ* data of how such structures affect sediment deposition of nearshore zones, with implications for coastal resilience and seabed sedimentology/morphology.

Studies have shown highly coupled geochemical shifts associated with monsoon cycles [4] that bring typically strong currents and high precipitation to tropical Southeast Asia (SEA) where up to 40% of the global coastal population reside [5]. SEA countries have developed their coastal regions and reinforced coastlines with manmade structures which altered subtidal sedimentology. Singapore is a highly urbanised city-state located in the approximate core of tropical Southeast Asia, making the Singapore Strait an ideal site for investigating monsoon-intermonsoon influence on nearshore sedimentation.

Here we present an annual record of seabed sediments obtained at monthly resolution to compare sedimentological with oceanographic and meteorological data. We also aim to investigate sediment deposition behaviour along different common coastal typoglogies along urbanized engineered coastlines. This research aims to characterize the coastal conditions off Singapore to investigate the monsoon/intermonsoon influences on nearshore seabed sediment in the Singapore Strait.

2 Study Area

Singapore's southern coastline is highly urbanised and influenced by both natural and anthropogenic drivers. The geography of the Singapore Strait contributes to complex ocean dynamics due to the exchange of water between the Pacific and India Oceans, contributing to complex relationships and interactions at intra-seasonal timescales driven by interannual phenomenon such as monsoons [4]. These relationships have been investigated using hydrodynamic and sediment transport models [6], but to date there exists no in situ observational data for ground-truthing purposes.

Singapore experiences rainfall year-round. Intraseasonal variation exists with Northeast (NE) Monsoon from December to early and the relatively drier Southwest (SW) monsoon from mid-May to mid-September. The intermonsoon periods are generally drier, with Intermonsoon 2 (September to November) generally wetter than the Intermonsoon 1 (March to May) (Fig. 2).

Sampling sites are shown in Fig. 1 representing three unique coastal typologies situated along the ~ 40 km south-southwestern coast of Singapore. These sites, namely Tuas (TS), Marina South (MS), and East Coast (EC), represent three unique coastal typologies in the form of highly engineered coast (100% seawall), coastal dam-floodgate and sandy beach (with seaward drainage channels), respectively. It is noteworthy that these different typologies are located within 20 km on the same coastline providing relatively similar oceanographic conditions.

3 Methodology

We performed sampling and measurements from September 2023 to August 2024 (Fig. 2). Specifically, Tuas sites are TS 2 (188 m from shore) and TS 5 (769m from shore). Marina sites are MS 1 (230 m from shore) and MS4 (4317 m from shore).

526   S. Chua et al.

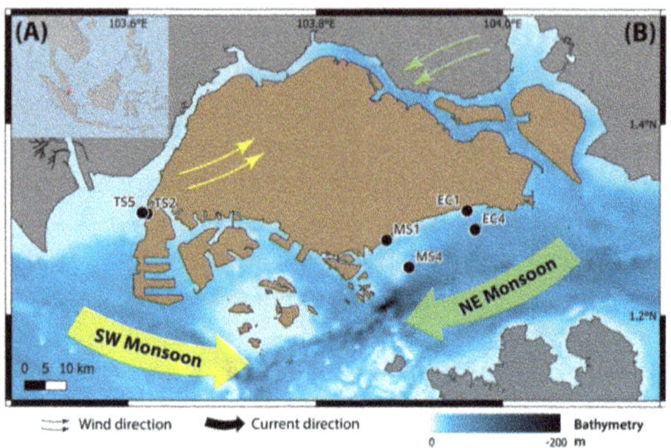

Fig. 1. (A) Location of Singapore within Southeast Asia represented by red dot. (B) Map of Singapore with sampling sites TS, MS and EC marked by black dots. Thick yellow and green arrows represent dominant current direction, and thin yellow and green arrows represent dominant wind direction, during respective monsoon seasons.

East Coast sites are EC1 (495 m from shore) and EC4 (2818 m from shore) (Fig. 1). We collected monthly seabed samples aboard RV Galaxae using a GOMEX25 box-corer. Water column parameters were measured using a Sontek Castaway CTD. Surface and bottom current velocity and direction were measured using a Valeport 106 current meter. Rainfall, atmospheric temperature and wind speeds are obtained from nearest meteorological stations. We performed sediment analysis of seabed sediments through particle-size analysis (PSA) and loss-on-ignition (LOI).

4 Results

Generally, atmospheric and water temperature at Tuas, Marina South and East Coast (Pearson Correlation Coefficient (PCC) = 0.49, 0.66 and 0.70, respectively) show fluctuations associated with seasonal cyclicities (Fig. 2A), with negative correlation between rainfall and atmospheric temperature (PCC = −0.27, −0.84, and −0.70, respectively). Correlation is stronger between wind and surface current speed for TS2 and EC1 with associated seasonal fluctuations (PCC = 0.45 and 0.70, respectively) but weakly correlated for MS1 (PCC = 0.20) (Table 1, Fig. 2).

Surface currents for offshore sites are generally stronger than nearshore sites. Bottom currents for all three sites are generally stronger than surface currents (difference = 0.33 ± 0.15 m/s) particularly during the NE monsoon. Current direction for nearshore and offshore sites are generally similar coincident with seasonal cycles, with shifts toward coast-parallel orientations at offshore sites (Fig. 2).

All three locations generally show finer sediment with lower temporal variability nearshore compared to offshore (Fig. 2B, C), with grain-size peaks observed during different seasons. At Tuas, maximum mean grain size at TS2 (121.16 ± 8.91 um) was recorded in first half of the NE monsoon. In contrast maximum mean grain size at TS5 (227.21 ± 31.3 um) was recorded during the SW monsoon. At Marina South, mean grain

Table 1. Climate and current data for nearshore and offshore sites.

Site (nearshore)	TS2	MS1	EC1
Wind speed (Ave)	5.19 ± 0.8 m/s	7.35 ± 2.80 m/s	11.14 ± 3.24 m/s
Surface current speed (Ave)	0.27 ± 0.09 m/s	0.24 ± 0.07 m/s	0.26 ± 0.1 m/s
Surface current speed (range)	0.13–0.37 m/s	0.11–0.34 m/s	0.13–0.39 m/s
Bottom current speed (range)	0.07–0.63 m/s	0.11–0.47 m/s	0.14–0.55 m/s
Site (offshore)	TS5	MS4	EC4
Surface current speed (range)	0.08–0.48 m/s	0.15–0.54 m/s	0.19–0.33 m/s
Bottom current speed (range)	0.16–0.71 m/s	0.08–0.5 m/s	0.1–0.38 m/s

size at MS1 remained consistently low at 15.33 ± 6.96 um. In contrast MS4 records highly variable mean grain size with peaks recorded in both early NE monsoon at 95.88 ± 0.79 um, and during Intermonsoon 1 to SW monsoon ranging between 66.40 ± 1.79 and 138.15 ± 16.44 um. At East Coast, we observe a reversal with coarser sediment at EC1 recorded during Intermonsoon 1 to early-SW monsoon period ranging between 47.71 ± 0.65 um and 65.42 ± 0.47 um. The offshore EC4 site recorded finer sediments at an average of 11.0 ± 4.76 um. Organic and inorganic percentages show little relationship with seasonal fluctuations, varying from site to site (Fig. 2).

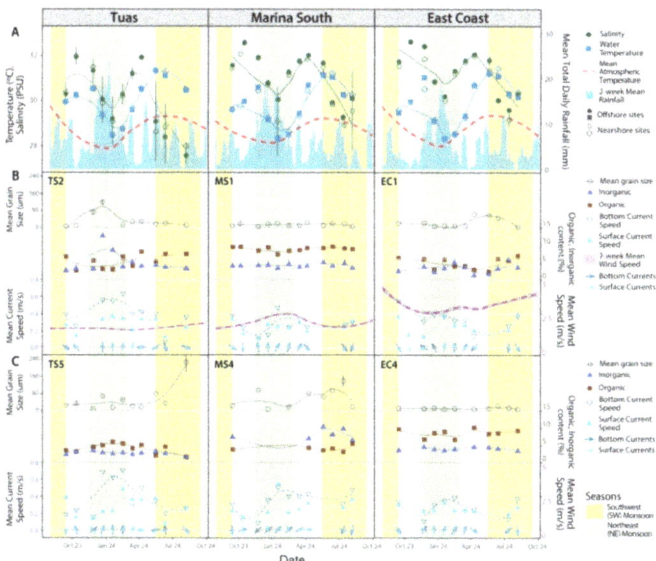

Fig. 2. (A) Climatological, water temperature and salinity for Tuas, Marina South and East Coast (B) Sedimentology, geochemistry and wind and current speed/direction for nearshore sites (C) Sedimentology, geochemistry and wind and current speed/direction for offshore sites.

5   Discussion

Coastal structures disrupt sediment transport balance and introduce new erosion and deposition regimes, introducing local-scale environmental and climatic conditions which can be difficult to model or predict [7]. Our results show high spatio-temporal sedimentation variability that are (1) typology-dependent and (2) decoupled from monsoonal peaks and intermonsoon lulls.

Spatially, we observe variability in sedimentation behaviour. TS 2 and 5 sampling sites, located off concrete seawalls, experience pronounced coarse sediment deposition at both TS2 and TS5, with a more correlative pattern for TS 2 during the NE monsoon coincident with higher rainfall. Although distal from the Johor Strait to the north, interference from tidal currents could cause lower bottom flow velocities [8] resulting in preferential nearshore deposition of highly inorganic coarser sediment due to increased bed friction, and more organic sediments offshore (TS 5) from northern mangrove sources transported by Johor Strait outflow. MS 1 and 4 are located off the Marina Barrage which periodically discharge freshwater during floods, with adjacent coasts protected by seawalls. However, MS 1 display dissimilar coeval sedimentation compared with Tuas with consistently fine sediment of <30 um even during monsoon months, likely due to restricted fluvial output caused by damming. EC 1 and 4, located off a narrow reclaimed sandy littoral beach, have relatively fine sediments in both near-and offshore sites with more organic sediment deposited distally. This could be a combination of improved sediment trapping capabilities in the nearshore zone and preferential west-east sorting of coarse to fine material driven by monsoon-independent residual currents weakening further offshore between Marina South and East Coast (Fig. 3).

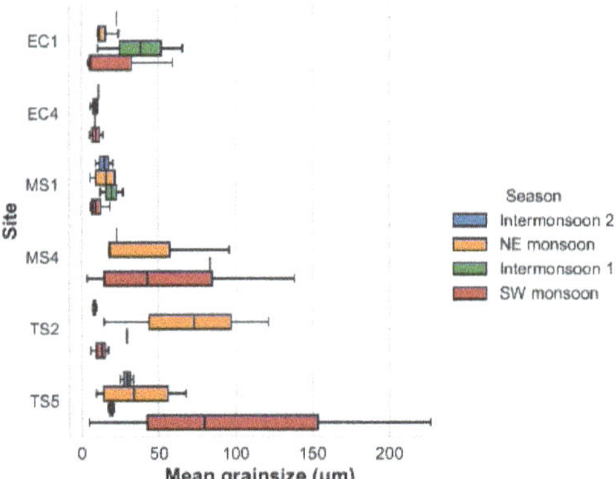

Fig. 3. Boxplots of mean grainsize in one complete monsoons-intermonsoon cycle, showing the median, interquartile range, and range across the samples.

Temporally, it is surprising that except for TS 2 and EC 1, coarser sediments for all sites are deposited during the 'weaker' SW monsoon compared with the NE monsoon (Fig. 2A). This suggest that local factors such as sediment supply and not current strength is the dominant factors controlling deposition. Despite high rainfall during the

NE monsoons resulting in higher suspended loads in fluvial runoff, drainage control via a coastal dam (MS) and concrete channels (EC) potentially resulted in finer and more organic-rich sedimentary outflow due to preferential organic adsorption by clay sediments. This cross-shore sediment transport is reduced distally as more long-shore coast-parallel transport begin to dominate further from the coast transporting coarser material (e.g., shell hash) from more distal sources.

6   Conclusion

Coastal typologies strongly influence sediment dynamics along urban coastlines even at low km-scales. Understanding how water and sediments respond to seasonal changes driven by monsoons is critical for coastal and marine sustainability, as well as provide a case study for other highly urbanised coastal cities within the monsoon belt amidst background climate change in Southeast Asia. Our study suggests decoupling of sediment behaviour from monsoon-driven hydrodynamics at local scales due to coastal modifications. This study underpins the need for field-based sampling of shallow marine sediments due to complex heterogeneity of developed nearshore zones which is difficult to comprehensively model.

Acknowledgements. This research is supported by the National Research Foundation, Singapore, and PUB, Singapore's National Water Agency, under its Competitive Funding for Water Research Initiative. Any opinions, findings, and conclusions or recommendations expressed in this material are those of the author(s) and do not reflect the views of the National Research Foundation, Singapore, and PUB, Singapore's National Water Agency.

References

1. Nicholls RJ, Cazenave A (2010) Sea-level rise and its impact on coastal zones. Science 328(5985):1517–1520
2. Shawler JL et al (2021) The effect of coastal landform development on decadal-to millennial-scale longshore sediment fluxes: evidence from the Holocene evolution of the central mid-Atlantic coast, USA. Quatern Sci Rev 267:107096
3. Behera MR et al (2013) Temporal variability and climatology of hydrodynamic, water property and water quality parameters in the West Johor strait of Singapore. Mar Pollut Bull 77(1):380–395
4. Martin P, et al. (2022) Monsoon-driven biogeochemical dynamics in an equatorial shelf sea: time-series observations in the Singapore strait. Estuarine, Coastal Shelf Sci 270:107855
5. Nicholls RJ et al (2021) A global analysis of subsidence, relative sea-level change and coastal flood exposure. Nat Clim Chang 11(4):338–342
6. van Maren DS, Gerritsen H (2012) Residual flow and tidal asymmetry in the Singapore Strait, with implications for resuspension and residual transport of sediment. J Geophys Res Oceans 117(C4)
7. Allen R, Lacy JR, Stevens AW (2021) Cohesive sediment modeling in a shallow estuary: model and environmental implications of sediment parameter variation. J Geophys Res Oceans 126(9)
8. Sun Y, Eltahir E, Malanotte-Rizzoli P (2017) The bottom water exchange between the Singapore strait and the West Johor strait. Cont Shelf Res 145:32–42

Coral and Nature-Based Reefs: Hydrodynamics and Morphodynamics

A Comparative Hydrodynamic Characterization of the Flow Through Regular and Stochastically Generated Synthetic Coral Reefs Over Flat Topography

Akshay Patil$^{(\boxtimes)}$ and Clara García-Sánchez

3D Geoinformation Research Group, Delft University of Technology, Delft, The Netherlands
a.l.patil@tudelft.nl

Abstract. Coral reefs are vital to marine ecosystems, supporting biodiversity and driving nutrient cycling. Despite significant research on the interaction between surface waves and natural or artificial reefs, the turbulent flow dynamics within coral canopies remain poorly understood due to their intricate geometries. This study addresses this knowledge gap using a turbulence-resolving computational framework based on the volume-penalizing immersed boundary method (vIBM). Comparing the serial, staggered, and stochastic arrangements of various coral roughness types we observe that massive corals and cylinders lead to a similar hydrodynamics response and the effect of dispersive stresses can introduce a large difference when stochastic coral reefs are considered. These observations highlight the importance of better understanding the hydrodynamics of complex coral reef geometries, emphasizing the need for further studies on this aspect of coral reef hydrodynamics.

Keywords: Coral Reefs · Canopy Turbulence · Direct Numerical Simulations

1 Introduction

The hydrodynamics of coral reefs play a central role in their ecological functions and resilience. By dissipating wave energy, promoting nutrient cycling, and supporting marine biodiversity, coral reefs are indispensable to both natural ecosystems and human societies [1]. However, their utility is under severe threat from climate change and anthropogenic stressors [2]. Protecting coral reefs requires a comprehensive understanding of their hydrodynamic processes and concerted global efforts to mitigate threats. A wide range of studies have been conducted to better understand the hydrodynamic response of coral reefs under varying forcing conditions [3–10], to list a few. These studies have provided a deeper insight into the hydrodynamics over coral reefs evidenced by the recent developments [9, 11–13]. In this study, we further the understanding of hydrodynamics over coral reefs by studying the effect of regularly arranged and stochastically arranged coral reefs using a scale-resolving computational framework. The focus here

C. Coelho et al. (Eds.): CD 2025, CRL 41, pp. 533–539, 2026.
https://doi.org/10.1007/978-3-032-15473-6_81

is to better understand and quantify the effect of relatively complex coral reef morphologies (referring to their complex spatial arrangement) compared to typically used serial and staggered arrangements comprising of a single coral species placed in a repeating fashion.

2 Methodology

2.1 Governing Equations and Computational Method

To model the flow through the synthetic coral canopies, we solve the non-dimensional Navier-Stokes momentum equations subject to the incompressibility constraint given by

$$\partial_t u_i + \Gamma \partial_j u_j u_i = \Gamma \left(-\partial_i p + \frac{1}{Re_k^b} \partial_j \partial_j u_i \right) + \cos(t)\delta_{i1} + F_{IBM}, \tag{1}$$

and

$$\partial_i u_i = 0, \tag{2}$$

where x_i are the coordinate directions corresponding to the streamwise, spanwise, and vertical directions respectively, t is time, u_i is the velocity vector, p is the pressure, δ_{ij} is the Kronecker delta, $\Gamma \equiv U_b k_c / \omega$ is the relative-roughness, $Re_b^k \equiv U_b k_c / \nu$ is the roughness-height based wave Reynolds number, and F_{IBM} is the immersed boundary force. The non-dimensional form of the equations is obtained by using the maximum wave orbital velocity (U_b), maximum coral canopy height (k_c), wave frequency (ω), and inertial pressure scaling $(p \equiv p^* / \rho_o U_b^2)$, where p^* is the dimensional pressure and ρ_o is the density of the fluid, and ν is the kinematic viscosity of the fluid. The governing equations are solved using a second-order accurate finite difference method [14] where all terms in the governing equations are discretized explicitly, and the time integration is carried out through a three-step Runge-Kutta scheme using the fractional-step algorithm [15]. The complex roughness is introduced into the computational geometry using a highly scalable signed-distance-field (SDF) generator [16] using a volume-penalizing Immersed Boundary Method (vIBM) [17]. Further details on the computational method can be found in [14].

2.2 Simulation Parameters

The coral geometries are obtained from the Smithsonian Institute coral repository [18] and the stochastic coral reef was generated using two coral species, viz. *Madrepora Formosa* and *Pseudodiploria Strigosa* correspond to the branching and massive coral types that are typically used to represent coral geometries [9]. In this paper, we compare the effect of morphological complexity of coral reef arrangements i.e., serial arrangement, staggered arrangement, and stochastic arrangement. Here morphology refers to the spatial arrangement of the individual corals that constitute the coral reef over a flat topography. Consequently, a total of 8 simulations are carried out, as shown in Fig. 1, which details the arrangement of the cylinder and the coral geometries. Simulations Morph-1 and Morph-2 correspond to the stochastically generated coral reef using the two species of coral geometries. For cases with serial and staggered arrangements, the geometric centre-centre distance in both streamwise and spanwise directions is $S_x = S_y = k_c/1.75$ and the maximum coral canopy height $k_c = 25.5\delta_s$ where $\delta_s \equiv \sqrt{2\nu/\omega}$ is the Stokes' wave boundary layer thickness and $\nu = 10^{-6} m^2/s$. The coral height-based wave Reynolds number is set to $Re_b^k = 2533$ and the wave Reynolds number $Re_w \equiv U_b^2/(\omega\nu) = 5000$. The domain size for the simulations is $20k_c \times 20k_c \times 2.5k_c$ in the streamwise, spanwise, and vertical directions, respectively. All the simulations were run on the Snellius supercomputer on the Genoa partition having 192 CPUs per node. Each simulation was run using 2 full nodes and required 65,520 CPU hours to simulate a total of 110 wave periods of which only the last 50 wave periods are used to calculate the statistics.

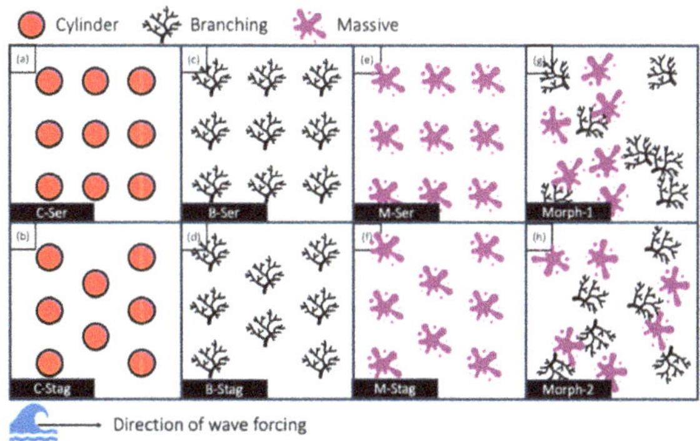

Fig. 1. Morphological setup for the various simulations carried out in this paper along with the names of each of the scenarios. The case names are denoted at the bottom left of each panel, where, C stands for cylinder, B stands for Branching, M stands for Massive, Ser stands for serial, Stag stands for staggered, and Morph stands for stochastically generated morphology.

3   Results and Discussions

Using a triple decomposition of the velocity given by

$$U_i(x_i, t) \equiv \tilde{U}_i(x_i, \omega t) + u_i'(x_i, \omega t) = \langle \tilde{U}_i \rangle(x_3, \omega t) + \breve{U}_i(x_i, \omega t) + u_i'(x_i, \omega t), \tag{3}$$

where the terms on the right-hand-side in order are the plane- and phase-averaged velocity, dispersive velocity, and the turbulent velocity components, respectively. As shown in Fig. 2, for all the cases discussed in this paper, there are no substantial differences observed between the serial and staggered arrangements for a given coral geometry. However, when comparing the branching and massive coral types, there is a substantial difference in the near-wall velocity, where the branching type coral for both arrangements much closely follows the flat-wall Stokes' solution [19]. In contrast, the massive corals that act as large roughness elements are observed to undergo appreciable reduction in the wave velocity for all wave phases considered in this paper. This can be mainly attributed to the difference between the frontal area density (A_f) of the two coral types where the massive coral have a relatively larger frontal area density when compared to the branching type. Figure 2 further elucidates the dependence of the coral type when considering the stochastically generated coral reef which consists of both the massive and branching coral type and is observed to have mean response that is half-way between either of the coral types for all wave phases.

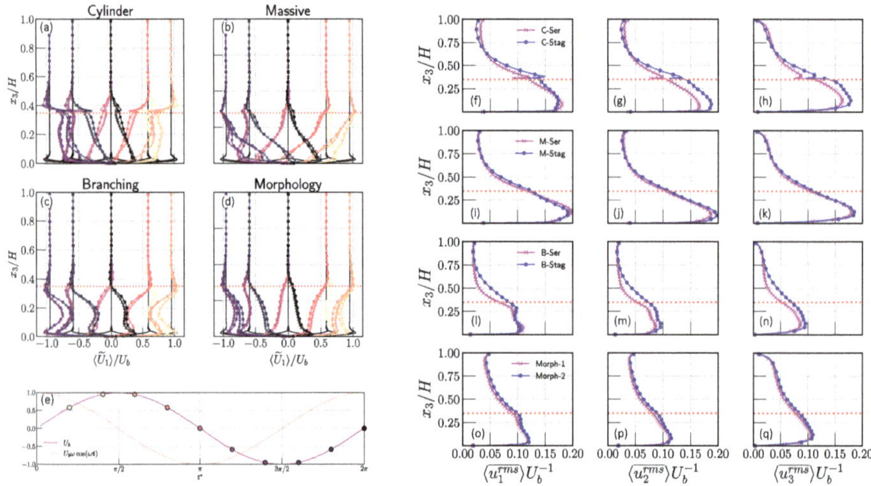

Fig. 2. Plane- and Phase-averaged velocity streamwise velocity profile comparison for the four distinct geometries considered in this paper (panels a - d). Panels (f-q) compare the plane- and time-averaged root-mean-squared (rms) velocity components for all the cases simulated. Panel (e) shows the phase variations of the driving pressure gradient and the velocity with the filled circle markers denote the locations where the data in Panels (a-d) is presented. In Panels (a-d) the line markers correspond to those presented in panels (f-q). The horizontal, red-dashed line marks the location of the top of the coral reef roughness elements for all the cases.

Furthermore, when the plane- and time-averaged response of the turbulent rms velocity components is compared, there is a clear difference between the large A_f coral reef (cylinder and massive) when compared with the relatively small A_f coral reefs (branching and morphology). For all the cases, it is observed that the staggered arrangement led to a suppression of the streamwise velocity component at the expense of the other two when compared to the serial arrangement. As expected, for both the stochastic morphologies considered, there is identical flow response with minor differences that can be attributed to the stochastic nature of the coral reef. This further supports the observations made by [9] where the primary differences are observed for the cylindrical roughness cases that exhibit a sharp shear layer at the top. For all the cases, far away from the bottom wall, the rms velocity profiles agree with each other irrespective of the geometric arrangement except for the cylindrical roughness that shows minor differences.

Figure 3a presents the plane- and time-averaged turbulent kinetic energy (TKE) dissipation rate profiles for the various cases detailed in this paper. The similarities observed in the streamwise velocity are further exemplified when the TKE dissipation rate is compared. Specifically, cases C-Ser, C-Stag, M-Ser, and M-Stag are observed to exhibit a similar overall trend. Since the massive coral has a similar $A_f(x_3)$ as the cylinder, the hydrodynamic response is observed to be similar both at the large and the small scales. Interestingly, the effect of stochastic sampling of the coral geometry does not seem to have a large impact on the overall dynamics and can be modelled using branching type coral reefs, as they exhibit similar hydrodynamic response. As seen in Fig. 3h-i, the TKE dissipation rate is highly localized in the vicinity of large clusters of corals, while there are large patches of the flow domain where the TKE dissipation rate is small in magnitude. Overall, there is a strong influence of the A_f on the small-scale dynamics of the flow suggesting that the dispersive stresses arising out of the triple-decomposition described in Eq. 3, can introduce strong heterogeneity in the flow leading to highly localized dissipation of the TKE. Consequently, further work is needed to accurately quantify and characterize the dispersive stress contribution in such environmental flows.

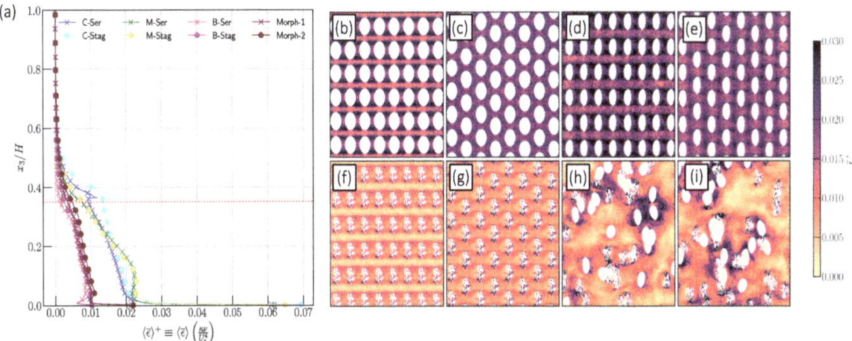

Fig. 3. (a) Plane- and time-averaged turbulent kinetic energy dissipation rate profiles for all the cases. Panels (b-i) show time-averaged turbulent kinetic energy dissipation rate at $x_3/H = 0.2$ for all the cases discussed in this paper.

4 Conclusions

In this paper, we studied the similarities and differences in the hydrodynamic response of synthetically generated coral reefs using organized and stochastic arrangements of individual corals. Our work suggests that cylindrical roughness elements can act as adequate surrogates for massive coral type. Additionally, we observed large differences in the hydrodynamics response between the branching and massive corals primarily due to the differences in the frontal area density between the two. We also observed that using a stochastically generated coral reef with an equal mix of branching and massive coral types can provide similar results in the mean hydrodynamic response when compared to the staggered arrangement of identical branching type corals. A principal finding in this work was the observation of localized zones of TKE dissipation in the stochastic coral reefs motivating further investigations into better characterizing the dispersive stress contribution on the overall dynamics of the flow.

Acknowledgments. This work made use of the Dutch national e-infrastructure with the support of the SURF Cooperative supplemented by the Dutch Research Council (NWO). This publication is part of the project "Multi-fidelity computational modeling of environmental fluid systems" with grant number 2024/ENW/01763969.

References

1. Lowe RJ, Falter JL (2015) Oceanic forcing of coral reefs. Ann Rev Mar Sci 7:43–66. https://doi.org/10.1146/ANNUREV-MARINE-010814-015834
2. Caldeira K, Wickett ME (2003) Anthropogenic carbon and ocean pH. Nature 425(6956):365. https://doi.org/10.1038/425365a
3. Reidenbach MA, Monismith SG, Koseff JR, Yahel G, Genin A (2006) Boundary layer turbulence and flow structure over a fringing coral reef. Limnol Oceanogr 51(5):1956–1968. https://doi.org/10.4319/lo.2006.51.5.1956
4. Stocking JB, Rippe JP, Reidenbach MA (2016) Structure and dynamics of turbulent boundary layer flow over healthy and algae-covered corals. Coral Reefs 35(3):1047–1059. https://doi.org/10.1007/s00338-016-1446-8
5. Reidenbach MA, Koseff JR, Monismith SG, Steinbuck JV, Genin A (2006) The effects of waves and morphology on mass transfer within branched reef corals. Limnol Oceanogr 51(2):1134–1141. https://doi.org/10.4319/LO.2006.51.2.1134
6. Monismith SG (2006) Hydrodynamics of Coral Reefs vol 39, pp 37–55. https://doi.org/10.1146/ANNUREV.FLUID.38.050304.092125
7. Davis KA, Pawlak G, Monismith SG (2021) Turbulence and coral reefs. Ann Rev Mar Sci 13:343–373. https://doi.org/10.1146/ANNUREV-MARINE-042120-071823
8. Lowe RJ, Koseff JR, Monismith SG (2005) Oscillatory flow through submerged canopies: 1. Velocity structure. J Geophys Res Oceans 110(10):1–17. https://doi.org/10.1029/2004JC002788
9. Hamilton JF, Kelley BZ, Monismith SG, Koseff JR (2025) On the use of simplified geometries to represent turbulent flow over coral reefs. J Fluid Mech 1004:A11
10. Lowe RJ et al (2005) Spectral wave dissipation over a barrier reef. J Geophys Res Oceans 110(4):1–16. https://doi.org/10.1029/2004JC002711
11. Ascencio JA, Jacobsen NG, McFall BC, Groeneweg J, Vuik V, Reniers AJHM (2022) Evaluation of implicit and explicit wave dissipation models for submerged and emergent aquatic vegetation. J Coast Res 38(4):807–815. https://doi.org/10.2112/JCOASTRES-D-21-00110.1
12. Jacobsen NG (2024) Spatially averaged, vegetated, oscillatory boundary and shear layers. J Fluid Mech 999:A33. https://doi.org/10.1017/JFM.2024.697
13. Jacobsen NG, McFall BC (2022) Wave-averaged properties for non-breaking waves in a canopy: Viscous boundary layer and vertical shear stress distribution. Coast Eng 174:104117. https://doi.org/10.1016/J.COASTALENG.2022.104117
14. Patil A, Fringer O (2022) Drag enhancement by the addition of weak waves to a wave-current boundary layer over bumpy walls. J Fluid Mech 947:A3. https://doi.org/10.1017/jfm.2022.628
15. Ferziger JH, Perić M, Street RL (2020) Computational Methods for Fluid Dynamics. Springer, Cham. https://doi.org/10.1007/978-3-319-99693-6
16. Patil A, Paranjothi UCK, García-Sánchez C (2025) GenSDF: an MPI-fortran based signed-distance-field generator for computational fluid dynamics applications. SoftwareX 30:102117. https://doi.org/10.2139/ssrn.5042856

17. Scotti A (2006) Direct numerical simulation of turbulent channel flows with boundary roughened with virtual sandpaper. Phys Fluids 18(3):031701. https://doi.org/10.1063/1.2183806
18. Smithsonian Institution. Smithsonian Coral Archive: Digital repository of coral reef data. https://3d.si.edu/corals
19. Stokes GG (1851) On the effect of the internal friction of fluid on the motion of pendulums. Trans Cambridge Phil Soc 3:1880–1905

The Role of Coral Rubble on Modern Coral Reefs

Ana Vila-Concejo[1,2(✉)], Claudia Le Quesne[1,2], Thomas E. Fellowes[1,2,3],
Ana Paula da Silva[1,2], Tristan Salles[1,2], Lachlan Perris[1,2], Lara Talavera[4],
and Maria Byrne[2,5]

[1] Geocoastal Research Group, The University of Sydney, Sydney 2006 NSW, Australia
`ana.vilaconcejo@sydney.edu.au`
[2] Marine Studies Institute, Faculty of Science, The University of Sydney, Sydney 2006 NSW,
Australia
[3] Water Research Laboratory, University of New South Wales, Sydney, Australia
[4] Department of Biology, Geology, Physics and Inorganic Chemistry, University of Rey Juan
Carlos, Madrid, Spain
[5] School Life and Environmental Sciences, The University Sydney, Sydney, Australia

Abstract. Sedimentary deposits in coral reefs represent more than 50% of their composition. The term coral rubble refers to unconsolidated sediment particles originating from the breakdown of the skeletons of coral and other calcifying organisms. Rubble results from destruction and deposition during high-energy events and stabilization can occur naturally facilitating coral regrowth. Under climate change, modern coral reefs are undergoing catastrophic bleaching events leading to production of large amounts of rubble. Here we discuss the nature of the new rubble generated and whether is good or bad for coral reef ecosystems. We argue that coral rubble can be good when it drives constructive processes on coral reefs and islands. Otherwise, bad rubble occurs when deposition impedes corals growth, and/or when where it cannot be transported to constructive rubble environments. For One Tree Island, southern Great Barrier Reef Tropical Cyclone Gabrielle (February 2023) drove rubble transport from the SE reef flat towards the island while causing an overall island erosion of 3%. Additionally, ongoing bleaching in 2024 is currently generating large amounts of rubble, further highlighting the need to understand its implications for reef stability and recovery.

Keywords: Tropical storms · Sediment transport · Coral islands

1 Introduction

Coral reefs are biodiverse ecosystems that fascinate humans; but corals themselves only represent a thin veneer that grows over the reef framework (Fig. 1A). It is the sedimentary deposits, including coral rubble and coralline sand, that fundamentally compose more than 50% of coral reefs [1]. Coral rubble, present in most reefs globally, results from the destructive effects of tropical storms [2, 3]. These events also play a key role in shaping reef geomorphology, building coral islands [4, 5], rubble-dominated reef flats [6], and

C. Coelho et al. (Eds.): CD 2025, CRL 41, pp. 540–544, 2026.
https://doi.org/10.1007/978-3-032-15473-6_82

rubble spits, tracks or ramparts that often form in the same locations [7]. Anthropogenic climate change has intensified coral bleaching events that when combined with tropical storms, can drive an increased in rubble production that can transform biodiverse reefs into rubble-dominated areas (Fig. 1B) hindering coral recruitment and growth [8, 9], and undermining reef resilience.

Recent studies, driven by Australia's Reef Restoration and Adaptation Plan (RRAP), highlight the challenges coral reefs face in recovering over rubble substrates [10] and explore opportunities for stabilising rubble as a Nature Based Solution (NBS) to facilitate coral restoration efforts [11]. Other NBS approaches are being investigated globally, such as dynamic revetments, where large clasts (including rubble) are placed on the supratidal zone of sandy beaches to mitigate coastal erosion [12]. However, significant knowledge gaps remain in our understanding of rubble formation, transport and stabilisation thresholds.

Here we present data from One Tree Island, a cay located on a rubble-dominated reef in the southern Great Barrier Reef. We first quantify the area and volume of the island, and the rubble delivery rates during the Holocene. We then compare them to the changes caused by Tropical Cyclone Gabrielle (February 2023) and with rates of modern rubble generation caused by coral bleaching events.

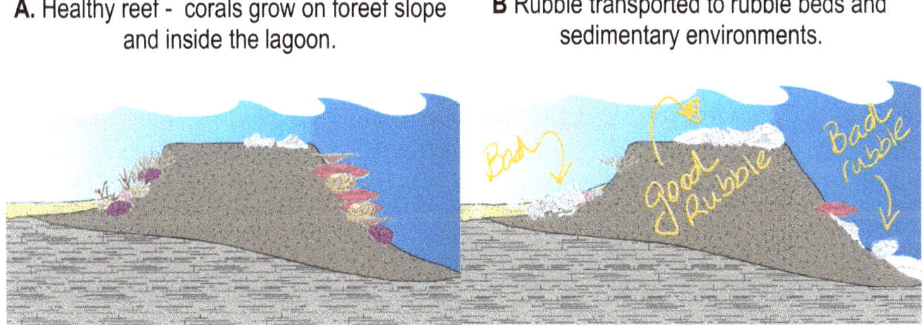

Fig. 1. Healthy reefs under climate change stressors bleach and generate rubble that can be transported by high energy events to rubble beds that might be too mobile for coral to grow (Bad Rubble) or sedimentary environments like rubble ramparts and islands (Good Rubble).

2 Methods

We use different sources of remotely sensed imagery including Unpiloted Aerial Vehicle (UAV) surveys using a DJI Phantom 4 Real Time Kinematic (RTK) Drone with a 20 m/px RGB camera. Additionally, we used a high-resolution LiDAR Digital Terrain Model of One Tree Reef, where the vertical error of the point cloud was ±0.1 m. The UAV data was processed using Agisoft Metashape to produce high-resolution photo orthomosaics and digital elevation models (DEMs). The mosaics and DEMs were imported into ArcGIS Pro and QGIS for further analysis, including areas, volumes and respective net changes. We used wave data from SOFAR spotter buoys deployed on the eastern margin of One Tree Reef in 2022.

3 Preliminary Results and Discussion

The area of One Tree Island is 60,000 m^2, which translates to an approximate volume above MSL (Mean Sea Level) of ~ 210,000 m^3. Previous cores drilled through the rubble flat showed coral rubble reaching a depth of 1.4 m below MSL at One Tree Reef and yielded an approximate age for the rubble flat/island of 4,000 years [13]. Adding this volume of rubble below MSL to the island volume, the total of rubble contained at One Tree Island is ~300,000 m^3. These values indicate an average rubble delivery rate of 75 m^3 per year during the 4,000 years of island formation during the Holocene. One Tree Island grew in area by 7% between 1978 and 2019, emphasizing the importance of large storms (including tropical cyclones) in producing coral rubble that is often delivered to the island via rubble spits or tracks [5]. However, the actual content of rubble and sand at One Tree Reef remains uncertain with recent observations showing more sand than expected on the rubble dominated SE reef flat.

The peak of tropical cyclone Gabrielle in February 2023 coincided with high tide, with average significant wave height (H$_s$) of 3.6 m from the ENE and associated wave power of 103.5 kW/m. A detailed analysis of the rubble tracks present on the SE reef flat offshore One Tree Island, reveals rubble migration towards the island driven by cyclone impacts. However, despite this addition of sediment, the tropical cyclone caused island erosion of the order of 600 m^3, representing a 3% decrease in volume.

The Great Barrier Reef has a significant value of $56 billion; it supports 64,000 jobs and contributes $6.4 billion to the Australian economy [14]. As the steward of this globally significant ecosystem, Australia invests millions of dollars each year in reef management, conservation strategies, and preparations to address future challenges, particularly those related to climate change. The exceptional nature of the sea surface temperature in the waters surrounding the Great Barrier Reef in early 2024 [15] triggered the worst coral bleaching in One Tree Reef's history causing extensive coral mortality [16] and subsequent rubble formation. Understanding the dynamics of these newly formed rubble-dominated habitats is critical for effective reef management. Key questions include whether the rubble stays in place, is transported to form sedimentary deposits, or stabilizes in ways that may support coral growth and ecosystem recovery. Quantifying the impacts of tropical cyclones and recurrent bleaching for coral rubble generation is essential for modelling the future of the Great Barrier Reef. Such information provides critical insights into how these processes influence reef resilience and the ecosystem services that reef provides, enabling informed management and conservation efforts.

Acknowledgements. This research was funded by ARC Discovery 'The GBR in 2100' (DP220101125), Geoscience Australia, UKRI-funded ARISE project (EP/X029506/1). LP funded by RTP USYD scholarship. LT funded by NextGeneration EU funds.

References

1. Blanchon P, Jones B, Kalbfleisch W (1997) Anatomy of a fringing reef around Grand Cayman; storm rubble, not coral framework. J Sediment Res 67:1–16. https://doi.org/10.1306/D42684D7-2B26-11D7-8648000102C1865D

2. Vila-Concejo A, Kench PS (2017) Storms in coral reefs. In: Ciavola P, Coco G (eds) Coastal Storms: Processes and Impacts. John Wiley and Sons, pp 127–144

3. Rasser M, Riegl B (2002) Holocene coral reef rubble and its binding agents. Coral Reefs 21:57–72. https://doi.org/10.1007/s00338-001-0206-5

4. Chivas A, Chappell J, Polach H, Pillans B, Flood PG (1986) Radiocarbon evidence for the timing and rate of Island development, beach-rock formation and phosphatization at Lady Elliot Island, Queensland, Australia. Mar Geol 69:273–287. https://doi.org/10.1016/0025-3227(86)90043-5

5. Talavera L, et al (2021) Morphodynamic controls for growth and evolution of a rubble coral island. Remote Sens 13. https://doi.org/10.3390/rs13081582

6. Thornborough KJ, Davies PJ (2011) Reef flats. In: Hopley D (ed) Encyclopedia of Modern Coral Reefs. Springer, Netherlands, pp 869–876

7. Shannon AM, Power HE, Webster JM, Vila-Concejo A (2013) Evolution of coral rubble deposits on a reef platform as detected by remote sensing. Remote Sens 5:1–18. https://doi.org/10.3390/rs5010001

8. Kenyon TM et al (2023) Coral rubble dynamics in the Anthropocene and implications for reef recovery. Limnol Oceanogr 68:110–147. https://doi.org/10.1002/lno.12254

9. Leung SK, Mumby PJ (2024) Mapping the susceptibility of reefs to rubble accumulation across the Great Barrier Reef. Environ Monit Assess 196:211. https://doi.org/10.1007/s10661-024-12344-4

10. Kenyon TM et al (2020) The effects of rubble mobilisation on coral fragment survival, partial mortality and growth. J Exp Mar Biol Ecol 533:151467. https://doi.org/10.1016/j.jembe.2020.151467

11. Ceccarelli DM et al (2020) Substrate stabilisation and small structures in coral restoration: State of knowledge, and considerations for management and implementation. PLoS ONE 15:e0240846. https://doi.org/10.1371/journal.pone.0240846

12. Bayle PM, Blenkinsopp CE, Conley D, Masselink G, Beuzen T, Almar R (2020) Performance of a dynamic cobble berm revetment for coastal protection, under increasing water level. Coast Eng 159:103712. https://doi.org/10.1016/j.coastaleng.2020.103712

13. Dechnik B, Webster JM, Davies PJ, Braga JC, Reimer PJ (2015) Holocene "turn-on" and evolution of the Southern great barrier reef: revisiting reef cores from the capricorn bunker group. Mar Geol 363:174–190. https://doi.org/10.1016/j.margeo.2015.02.014

14. O'Mahony J, et al (2017) At What Price? The Economic , Social and Icon Value of the Great Barrier Reef. Deloitte Access Economics, Brisbane

15. Henley BJ et al (2024) Highest ocean heat in four centuries places Great Barrier Reef in danger. Nature 632:320–326. https://doi.org/10.1038/s41586-024-07672-x

16. Byrne M, et al (2025) Catastrophic bleaching in protected reefs of the Southern Great Barrier Reef. Limnol Oceanogr Lett

Observations of Wave Setup on a Coral Atoll

Ashley Holsclaw[(⊠)], Mathilde Lindhart, Gerd Masselink, and Mark Davidson

Coastal Processes Research Group, University of Plymouth, Plymouth 4 8AA, UK
ashley.holsclaw@plymouth.ac.uk

Abstract. Atoll island nations (such as the Maldives, Marshall Islands, and Tuvalu) are amongst the most critically vulnerable to climate change due to rising sea levels and subsequent flooding and overwash events. To investigate the wave and water level dynamics on a coral atoll, a six-month dataset consisting of three pressure gauges was collected at three distinct sites on the island of Dhigelaabadhoo in the Republic of the Maldives along with a tide gauge located in a nearby harbor. The observed setup varied across the sites and was found to correlate with the offshore significant wave height and period, as well as tidal water level. The sites with wider reef platforms and steeper forereef slopes were observed to have higher wave setup at the shore and a slightly stronger linear correlation between setup and offshore conditions. Furthermore, the results indicate the potential influence of incident wave direction on alongshore setup gradients.

Keywords: Coral reef · Hydrodynamics · Wave Setup · Infragravity Waves

1 Introduction

Coral atolls are comprised of low-lying islands and shallow reefs situated atop ring-shaped platforms. Surrounding reefs provide vital protection by dissipating incident wave energy through wave breaking and bottom friction [2, 6], but are susceptible to increased flooding and overwash due to sea-level rise [8]. Recent studies have shown that overwash, the process of wave run-up exceeding the beach crest elevation and transporting sediment, plays an important role in atoll adaptation [5]. Wave run-up, the maximum vertical extent of waves onshore, is composed of wave setup and swash. Wave setup refers to the mean water level increase due to wave breaking and is strongly dependent on incident wave height, wave period, reef width, and tidal stage (i.e., water levels). This study presents observations of setup on an atoll island and investigates the morphological and hydrodynamic controls of setup and longshore setup gradients.

2 Methods

2.1 Study Site

Data were collected over a 5-month field campaign from February to August 2023 on the uninhabited island of Dhigelaabadhoo located at the south-western rim of the southern Huvadhu Atoll in the Republic of Maldives. The work was focused on three primary sites located on the Western (W), Southern (S), and Eastern (E) sections of the island to encompass a range of topographic and hydrodynamic variabilities (Fig. 1A).

© The Author(s) 2026
C. Coelho et al. (Eds.): CD 2025, CRL 41, pp. 545–551, 2026.
https://doi.org/10.1007/978-3-032-15473-6_83

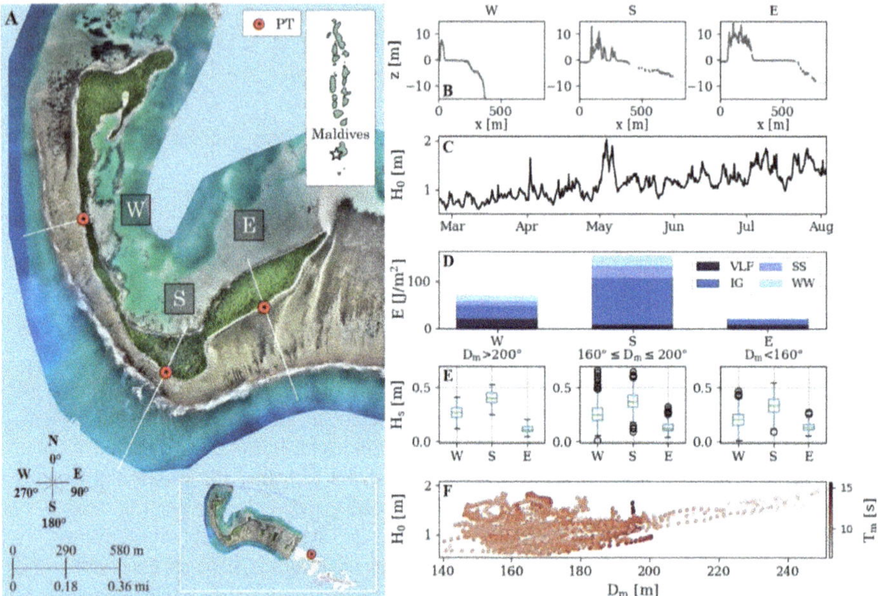

Fig. 1 (A) Map of the site (red circles denote PTs). Imagery from UAV flight. Base maps from OpenStreetMap and OpenStreetMap Foundation (CC-BY-SA). © https://www.openstreetmap.org and contributors. (B) Topographic profiles of each transect. (C) Offshore significant wave height, H_0. (D) Distribution of mean frequency contributions to local wave energy, E. (E) Local significant wave height H_S, at each site partitioned by three equal ranges of mean offshore wave directions representing SW ($D_M > 200°$), S ($160° \leq D_M \leq 200°$) and SE ($D_M < 160°$), and (F) H_0 as a function of D_M with symbol color scaled by offshore wave period, T_M.

2.2 Field Measurements

From February 2023 to August 2024, three RBR Duo pressure transducers (PT) were deployed at the shoreline of each transect (Fig. 1A) sampling at 1 Hz for 34 min (2048 samples) every 2 h. In addition, a Valeport tide gauge was deployed in the nearby Fares-Maathodaa harbor (approx. 3.5 km from Dhigulaabadhoo) sampling every 20 min. In March 2024, topographic and bathymetric data were collected with RTK GNSS and UAV photogrammetry surveying (Fig. 1A, B).

2.3 Offshore Estimates

Estimates for offshore significant wave height (H_0), mean period (T_M), peak period (T_P), and mean direction (D_M, clockwise from 0° N) were obtained from the Copernicus Marine Environment Monitoring Service (CMEMS) Global Ocean Waves Reanalysis model. The 3-hourly time series was resampled to the 2-hourly bursts of the field data.

2.4 Data Analysis

Each pressure record was corrected for atmospheric pressure and adjusted to be relative to the mean water level (MWL) datum. The data were partitioned into two-hour bursts for spectral analysis and calculating burst wave statistics. Significant wave height and mean wave period were estimated from the energy spectra. The spectra were further divided into frequency bands defined as wind wave (WW, $f > 0.2$ Hz), sea swell (SS, $0.2 < f < 0.04$ Hz), infragravity (IG, $0.04 < f < 0.004$ Hz), and very low frequency (VLF, $f < 0.004$ Hz), as well as into low ($h < -0.2$ m), mid ($-0.2 \leq h \leq 0.2$ m), and high ($h > 0.2$ m) tide. Wave setup (η) was estimated for each site per 2-hourly burst as

$$\eta = \langle \zeta \rangle - \langle h \rangle \tag{1}$$

where $\langle \rangle$ indicates the time average over each burst, ζ is the free surface elevation, and h is the tidal water level. Only bursts where the incident wave height was significant enough to produce detectable setup on the reef ($H_0 \geq 1$ m) were analyzed for setup.

Reef platform width and slope was calculated for the region between the toe of the beach and reef crest, and forereef slope was calculated between the reef crest and the 10 m depth contour (i.e., the maximum surveyed depth for sites S and E). Both reef and forereef slopes were estimated using a linear regression fit of the respective profiles.

3   Results

3.1   Offshore Wave Climate

The offshore wave climate was characterized by a dominant S swell (180°) with significant wave heights ranging between 0.6 and 2.0 m and mean periods between 6.4 and 15.6 s. Offshore wave height and period varied according to incident wave direction (Fig. 1F) with larger waves and shorter periods from the SW. Tide range was 1.1 m.

3.2   Reef Characteristics

Despite the proximity, the sites varied in terms of wave exposure (SE-SW, 145°–250°) and topography with cross-shore platform widths (x_{reef}) of 180, 80, and 360 m for W, S, and E, respectively. Furthermore, the reefs varied in terms of forereef and platform steepness, whereas site W had the steepest forereef slope ($\tan\beta_{fr}$), followed by E and S (Table 1). Reef W was the only site to have been surveyed below 10 m depth, where the slope appeared to drastically steepen ($\tan\beta_{fr} = 0.47 \approx 1/2$).

On average, site S was the most energetic, followed by W and E (Table 1). The W and E shores were considerably sheltered from the dominant southerly swell, unlike S.

Table 1. Summary of site characteristics. An overbar (-) indicates the mean.

	x_{reef} [m]	$\tan\beta_{reef}$ [-]	$\tan\beta_{fr}$ [-]	$\overline{H_S}$ [m]	$\overline{T_M}$ [s]
W	180	0.002, ≈1/500	0.06, ≈1/17	0.24	18.3
S	80	0.004, ≈1/250	0.01, ≈1/100	0.35	14.1
E	360	0.0002, ≈1/5000	0.04, ≈1/25	0.13	71.9

3.3 Wave Frequency Composition

On average, all sites were IG-dominated with VLF waves being a large secondary contributor to total wave energy at the sheltered sites, W and E. Combined IG and VLF wave energy constituted 70.7%, 68.3%, and 93.1% of the total wave energy for W, S, and E, respectively (Fig. 2). Consistent with previous observations [1, 7], wave energy was attenuated at low tide. At low tide, VLF waves were the most prevalent at site E and often became more predominant than IG waves.

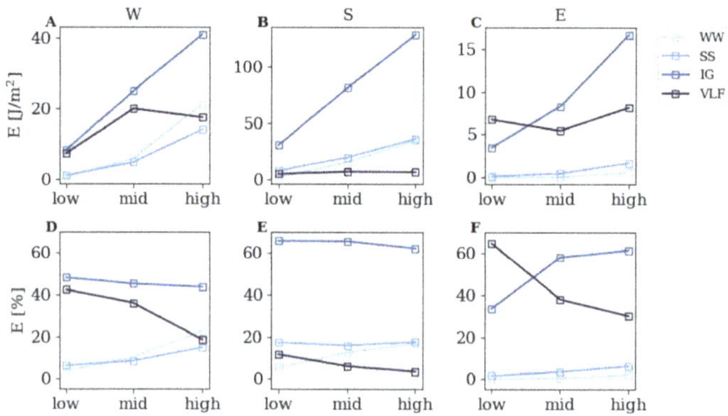

Fig. 2. (A-C) Mean energy and (D-F) contribution to total wave energy for each wave frequency partition at low ($h < -0.2$ m), mid ($-0.2 \leq h \leq 0.2$ m), and high ($h > 0.2$ m) tide.

3.4 Wave Setup Variability

Generally, larger offshore wave heights and longer periods resulted in higher setup on the reef flat (Fig. 3). This trend was strongest for site E, which had the widest, flattest reef flat and steepest forereef, followed closely by site W. Wave setup at the shoreline was found to increase with decreasing water levels (Fig. 3).

Setup was found to be relatively larger at a site when the offshore wave direction was more closely aligned with the site shoreline. That is, waves predominantly coming from the SW (250°) corresponded with elevated setup at W, while waves from the SE (150°) corresponded with elevated setup at E (Fig. 4). The reef flat wave heights similarly varied according to offshore wave direction. While wave heights at W and S were consistently greater than at E, westerly waves resulted in a more substantial difference.

4 Discussion

4.1 Morphological Controls

The physical position and angle of the atoll island shoreline largely determines the site's exposure to incident wave energy which directly relates to onshore wave setup (Fig. 3). The results indicate a stronger positive correlation between incident wave height/period and setup for the reefs with wider platforms ($\geq$ 180 m) and steeper forereefs ($\tan\beta \geq$

$0.04 \cong 1/25$). This morphological combination also functions to support long wave periods (IG/VLF waves), which have been demonstrated to relate to setup [4]. As setup was greater at sites W and E despite being sheltered, it suggests that reef width and forereef slope may be important factors in shoreline setup, coinciding with previous findings [8].

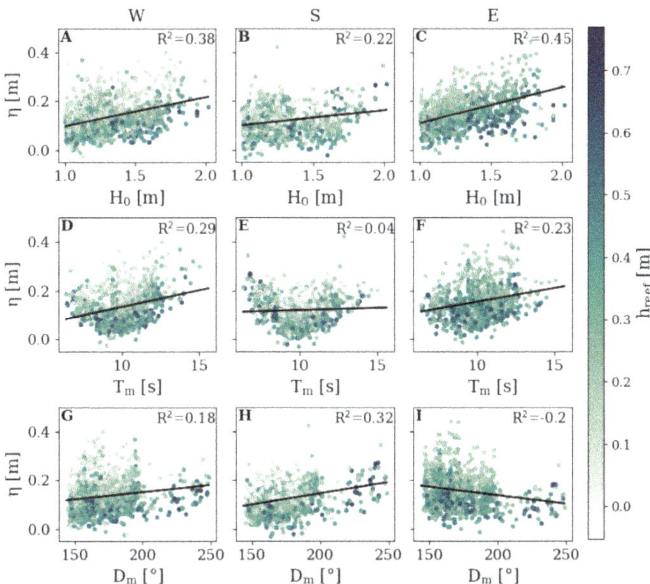

Fig. 3. Setup as a function of offshore (A-C) wave height, (D-F) mean period, and (G-I) mean direction for each site. Color-scale indicates the reef flat water depth.

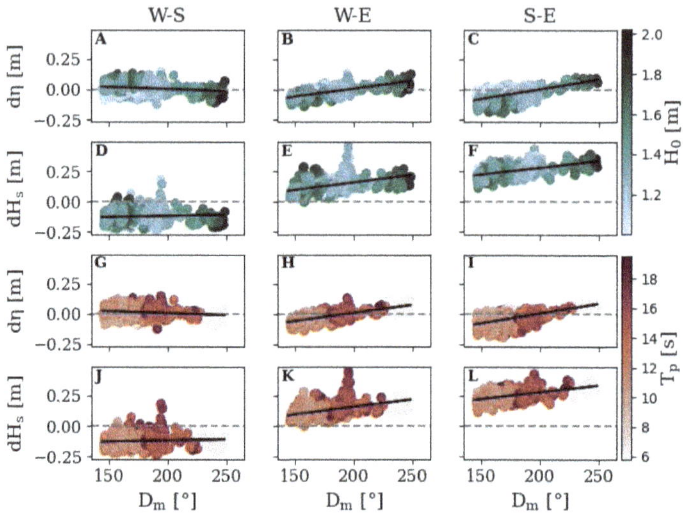

Fig. 4. The difference in (A-C,G-I) wave setup and (D-F,J-L) wave height between sites W-S, W-E, and S-E. Color-scales are offshore (A-F) wave height and (G-L) peak period.

4.2 Hydrodynamic Controls

In agreement with previous works [3, 7, 9, 10], wave setup was observed to increase with lower reef flat water levels. Setup is thought to be linearly related to incident wave height [10], and this relation is apparent for sites W, S, and E ($R^2 = 0.38, 0.22, 0.45$, respectively, with p-values < 0.001 indicating significance) (Fig. 3A,C). However, this correlation is possibly dampened due to the use of model offshore parameters and tidal modulation. Similarly, longer incident wave periods correlated with higher setup at the shoreline (Fig. 3D,F), which may be due, in part, to both long periods and higher wave setup being prevalent at low tide. Results also suggest that wave direction can influence the alongshore variability of shoreline setup (Fig. 4).

5 Conclusion

Setup was observed to depend on a combination of morphological and hydrodynamic factors, particularly reef flat width, forereef slope, incident wave height, and period with setup gradients influenced by wave direction. While this study is limited by data scarcity, it highlights the 3D complexity of wave setup on coral atolls and the need for further study investigation of the effects of alongshore variability on setup gradients.

References

1. Beetham E, Kench PS, O'Callaghan J, Popinet S (2016) Wave transformation and shoreline water level on funafuti atoll, tuvalu. J Geophys Res Oceans 121(1):311–326
2. Ferrario F, Beck MW, Storlazzi CD, Micheli F, Shepard CC, Airoldi L (2014) The effectiveness of coral reefs for coastal hazard risk reduction and adaptation. Nature News

3. Gourlay MR (1996) Wave set-up on coral reefs. 2. Set-up on reefs with various profiles. Coastal Eng 28(1):17–55

4. Hench JL, Leichter JJ, Monismith SG (2008) Episodic circulation and exchange in a wave-driven coral reef and lagoon system. Limnol Oceanogr 53(6):2681–2694

5. Masselink G, Beetham E, Kench P (2020) Coral reef islands can accrete vertically in response to sea level rise. Sci Adv 6(24):eaay3656

6. Munk WH, Sargent MC (1948) Adjustment of bikini atoll to ocean waves. EOS Trans Am Geophys Union 29(6):855–860

7. Pequignet C, Becker J, Merrifield M, Boc S (2011) The dissipation of wind wave energy across a fringing reef at ipan, guam. Coral Reefs 30:71–82

8. Quataert E, Storlazzi C, van Rooijen A, Cheriton O, van Dongeren A (2015) The influence of coral reefs and climate change on wave-driven flooding of tropical coastlines. Geophys Res Lett 42:6407–6415

9. Symonds G, Black K, Young I (1995) Wave-driven flow over shallow reefs. J Geophys Res Oceans 100(C2):2639–2648

10. Tait RJ (1972) Wave set-up on coral reefs. J Geophys Res (1896–1977) 77(12):2207–2211

A Modelling-Based Analysis of Extreme Wave Setup in a Fringing Coral Reef

Axelle Gaffet[1,2(✉)], Xavier Bertin[1], Damien Sous[3], and Emmanuel Cordier[4]

[1] UMR 7266 LIENSs, CNRS/La Rochelle Université, 17000 La Rochelle, France
`axelle.gaffet@univ-lr.fr`
[2] Créocéan, 17000 La Rochelle, France
[3] Université de Pau et des Pays de L'Adour, 2S-UPPA, SIAME, 64600 Anglet, France
[4] OSU-Réunion, UAR 3365, Université de La Réunion, CNRS, IRD, Météo France, 97490 Saint-Denis, La Réunion, France

Abstract. Coastal hazards in tropical islands are likely to increase under the combined effects of sea-level rise, coral reef degradation and the potential intensification of extreme cyclones. In volcanic islands bordered by deep waters, storm surges are primarily driven by atmospheric pressure gradients and the development of a wave setup in the nearshore, which results from wave dissipation. Recent studies on the hydrodynamics of reef lagoon systems have demonstrated that the mean wave-driven circulation can either increase or decrease the wave setup along the shoreline, depending on the system configuration. This study combines the analysis of a unique dataset capturing a paroxysmal event (significant height exceeding 7 m at the breaking point) with phase-averaged numerical modelling to provide further insight on wave setup development in fringing reef environments under extreme conditions. When considering a 2DH modelling approach, the largest wave setup is underestimated by over 40%. Wave setup predictions are partly improved when using fully-coupled 3D runs, due to the representation of the vertical mixing and the horizontal and vertical advections associated with the intense hydrodynamic circulation that takes place in the surfzone.

Keywords: Surf zone hydrodynamics · Wave setup · SCHISM · Coral reef · Extreme wave event

1 Introduction

Coastal hazards in tropical islands are expected to increase under the combined effects of sea-level rise, the degradation of coral reefs and a possible intensification of the most intense cyclones. In volcanic islands bordered by deep waters, storm surges are generally dominated by atmospheric pressure gradients and the development of a local wave setup, the mean increase in water level due to wave breaking, resulting from wave dissipation. Over the last decade, the understanding of the processes controlling the hydrodynamic circulation driven by waves at reef/lagoon environments made important progresses. Thus, several studies showed that the mean wave-driven hydrodynamic circulation can

© The Author(s) 2026
C. Coelho et al. (Eds.): CD 2025, CRL 41, pp. 552–557, 2026.
https://doi.org/10.1007/978-3-032-15473-6_84

either increase (Buckley et al. 2016) or decrease (Sous et al., 2020; Rijnsdorp et al., 2021) the wave setup along the shoreline depending on the configuration of the system reef/lagoon. However, in-situ measurements under paroxysmal conditions are very scarce in the literature, questioning the validity of the findings of these studies under extreme waves. In this context, this study takes advantage of permanent bottom pressure measurements collected to the SW of La Réunion Island (Indian Ocean) to investigate the development of extreme wave setup under the distant-source swells of June, 2022, which drove waves of significant height exceeding 7 m at breaking.

2   Study Site and Field Datasets

2.1   Study Site

The Hermitage fringing reef is located on the western coast of La Reunion Island in the South Western Indian Ocean. It is directly exposed to distant-source swells (DSS) coming from the Southern Ocean and generated by strong atmospheric depressions during the austral winter from April to October. Tides around La Reunion are mixed, mainly semi-diurnal with a Mean Tidal Range (MTR) of 0.37 m (Cordier et al., 2013). The fringing reef is around 500 m wide, with a steep-slope of 1:10.

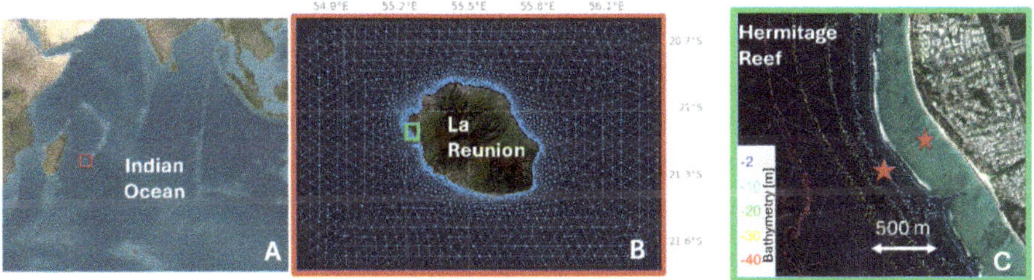

Fig. 1. Location of La Reunion Island in the Indian Ocean (A), satellite view of La Réunion Island with the extension of the model grid (B) and zoom in on l'Hermitage fringing reef (B) with a bathymetric map and the location of and the pressure sensors on the reef (C).

2.2   Field Datasets

Two pressure transducers are located along the central section of the 8 km-long Hermitage fringing reef, near the Hermitage Pass, which divides the reef into two equal units. This data are collected continuously in the scope of the French National Observation Service (SNO) ReefTEMPS (Cordier et al. 2024). The first site is situated on the reef flat at a depth of 0.9 m and the second site is located on the forereef at 12 m depth (see Fig. 1). For this study, bottom pressure data from both sites were analyzed, with a focus on the period from April to August 2022, during which an extreme swell event occurred. Wave parameters were computed from spectral analysis of the bottom pressure measurement using the nonlinear moderately dispersive reconstruction described by Martins et al. (2021).

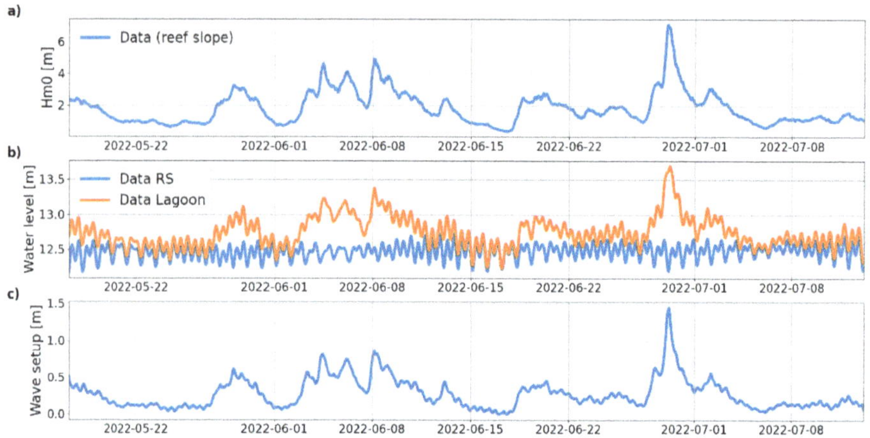

Fig. 2. Field observation data derived from the pressure sensor during the extreme distant-source swell event of June 2022 at the Hermitage fringing reef (La Reunion) for a) Hm0 b) water levels at the reefslope (RS) and in the lagoon and c) wave setup computed as the difference of both.

3 Modelling System

The numerical simulations are based on the 3D unstructured-grid modelling system SCHISM (Semi-implicit Cross-scale Hydroscience Integrated System Model, Zhang et al. (2016)). The model uses the combination of a semi-implicit scheme and an Eulerian-Lagrangian Method to treat the momentum advection, which allows to relax the associated numerical stability constraints. The whole domain is forced with ERA5 wind, corrected using the quantile-quantile approach explained in Gaffet et al. (2024). In the circulation model, tides boundary conditions were extracted from the global tide model FES 2014 (Lyard et al. 2021). The bottom drag coefficient is calculated according to the Manning's law with $n = 0.02$, a value presently too low, which will be discussed later. The hydrostatic solver of SCHISM is fully coupled with the spectral wave model WWM-II (Roland et al. 2012) using the vortex force formalism (VF) proposed by Bennis et al. (2011) and described by Martins et al. (2022). The 3D VF configuration is compared in this study to a 2DH radiation stress (RS) configuration based on Longuet-Higgins & Stewart (1964). The wave breaking, in both configurations, is computed according to the model of Battjes & Janssen (1978) where the breaking index γ is set to 0.75. WWM-II is also forced with ERA5 corrected winds and with time series of directional spectra at the oceanic boundary, extracted from the global application of WAVEWATCHIII described by Gaffet et al. (2024).The spatial grid covers the deep sea bordering La Reunion Island, with a resolution of 10 km refined to 20 m along the Hermitage reef (see Fig. 1-B). The bathymetry is interpolated from DEM produced by the French Hydrographic Service at 100 m (HOMONIM) and 5 m (Litto3D) corresponding to a LiDAR survey.

4 Results

Sea State. The model validation is performed over the month of June 2022. The forereef pressure sensor reveals firstly that the distant-source swell event of June 2022 drove waves with Hs over 7 m (Fig. 2) and peak periods of 20 s. The comparison between forereef and reef-flat sensors shows the development of a wave setup strongly correlated with the incident Hs (R2 = 0.98), peaking at 1.5 m, one of the largest values found in the international literature. Both models are able to capture the variability of the sea state over that period, with Normalized-Root-Mean-Square Error (hereafter NRMSE) values of 19.79% and 20.41%, respectively. They correctly represent the extreme distant-source swell of June 29th, 2022 where H_{m0} reached 7 m, although the wave height peak is underestimated by 1 m.

Wave Setup. The two models correctly represent the wave setup variability with NRMSE of 44% and 34.5% for the 2DH-RS and 3D-VOR configurations, respectively (see Fig. 3). The 2DH-RS configuration shows an underestimation of 42% of the wave setup peak, which is slightly better represented with the 3D-VOR approach. The underestimation of the wave setup peak is partly related to the underestimation of the wave height peak in the model and possible other reasons will be discussed below.

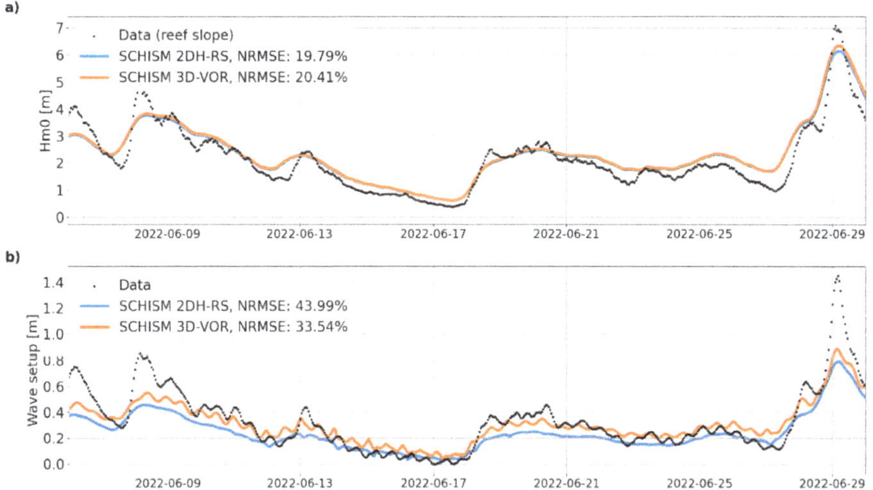

Fig. 3. Observed (black) against modeled in 2DH (blue) and 3D (yellow) a) H$_{m0}$ and b) wave-setup at the Hermitage (La Reunion) in June 2022.

5 Discussion and Conclusions

The numerical model shows correct predictions of wave height and water levels, although with a substantial underestimation of wave setup at the peak event. Wave setup underestimations using 2DH models was already reported by Bruch et al. (2022) next to our studied site, although the underestimation was limited to about 20%.

The underestimation of H_{m0} at the event peak might be related to the wave input forcing as wave modelling in this region of the Indian Ocean is particularly challenging due to the presence of the Agulhas Current, on the way of DSS reaching La Réunion (Marechal et Ardhuin 2021).

In order to estimate the impact of underestimated wave heights at the peak of the event, the spectral forcing was tuned along the open boundary but this experiment only partly improved wave setup predictions. In the model configuration used so far, the bottom stress term in the momentum equations is treated explicitly, which poses a severe time step restriction and forced us to use a weak friction in the circulation model. In reality, the presence of spur and grooves patterns on the reef crest and coral reefs should induce a large bottom friction, which could impede the flushing of the lagoon and hence increase wave setup. This hypothesis is being explored through the implicit method to treat bottom stress in SCHISM proposed by Zhang et al. (2020).

References

Battjes JA, Janssen JPFM (1978) Energy loss and set-up due to breaking random waves. In: Proceedings of 16th Conference on Coastal Engineering, Hamburg, Germany

Bennis A-C, Ardhuin F, Dumas F (2011) On the coupling of wave and three-dimensional circulation models: choice of theoretical framework, practical implementation and adiabatic tests. Ocean Model 40(3):260–272. https://doi.org/10.1016/j.ocemod.2011.09.003

Bruch W, Emmanuel C, France F, Stuart GP (2022) Water level modulation of wave transformation, setup and runup over la saline fringing reef. J Geophys Res: Oceans 127(7). https://doi.org/10.1029/2022JC018570

Buckley ML, Lowe RJ, Hansen JE, Van Dongeren AR (2016) Wave setup over a fringing reef with large bottom roughness. J Phys Oceanogr 46(8):2317–2333. https://doi.org/10.1175/JPO-D-15-0148.1

Cordier E et al (2024) 'ReefTEMPS-OI - The Indian Ocean Island Coastal Ocean Observation Network'

Cordier E, Lézé J, Join J-L (2013) Natural tidal processes modified by the existence of fringing reef on La Reunion Island (Western Indian Ocean): impact on the relative sea level variations. Cont Shelf Res 55:119–128. https://doi.org/10.1016/j.csr.2013.01.016

Gaffet A, Xavier B, Damien S, Héloïse M, Aron R, Emmanuel C (2024) A new global high resolution wave model for the tropical ocean. EGUsphere, 1–26. https://doi.org/10.5194/egusphere-2024-2610

Longuet-Higgins MS, Stewart, RW (1964) Radiation stresses in ~vater waves; a physical discussion, with applications

Lyard FH, Allain DJ, Cancet M, Carrère L, Picot N (2021) FES2014 global ocean tide atlas: design and performance. Ocean Sci 17(3):615–649. https://doi.org/10.5194/os-17-615-2021

Marechal G, Fabrice A (2021) Surface currents and significant wave height gradients: matching numerical models and high-resolution altimeter wave heights in the Agulhas current region. J Geophys Res: Oceans 126(2):e2020JC016564. https://doi.org/10.1029/2020JC016564

Martins K et al (2022) Wave-induced mean currents and setup over barred and steep sandy beaches. Ocean Model 179:102110. https://doi.org/10.1016/j.ocemod.2022.102110

Martins K, Bonneton P, Lannes D, Michallet H (2021) Relation between orbital velocities, pressure, and surface elevation in nonlinear nearshore water waves. J Phys Oceanogr 51(11):3539–3556. https://doi.org/10.1175/JPO-D-21-0061.1

Rijnsdorp DP et al (2021) A numerical study of wave-driven mean flows and setup dynamics at a coral reef-lagoon system. J Geophys Res: Oceans 126(4). https://doi.org/10.1029/2020JC 016811

Roland A, et al (2012) 'A fully coupled 3D wave-current interaction model on unstructured grids. J Geophys Res: Oceans 117(C11). https://doi.org/10.1029/2012JC007952

Sous D, Dodet G, Bouchette F, Tissier M (2020) Momentum balance across a barrier reef. J Geophys Res: Oceans 125(2). https://doi.org/10.1029/2019JC015503

Zhang YJ, Gerdts N, Ateljevich E, Nam K (2020) Simulating vegetation effects on flows in 3D using an unstructured grid model: model development and validation. Ocean Dyn 70(2):213–230. https://doi.org/10.1007/s10236-019-01333-8

Zhang YJ, Ye F, Stanev EV, Grashorn S (2016) Seamless cross-scale modeling with SCHISM. Ocean Model 102:64–81. https://doi.org/10.1016/j.ocemod.2016.05.002

Wave Scattering and Energy Reduction by a Patch of Idealized Coral-Growing Units

Ayrton Alfonso Medina Rodríguez[1], Zhenhua Huang[1(✉)], and R3D Consortium[2]

[1] Department of Ocean and Resources Engineering, School of Ocean and Earth Science and Technology, University of Hawaii at Manoa, Honolulu, HI 96822, USA
zhenhua@hawaii.edu

[2] R3D Consortium University of Hawaii, 2800 Woodlawn Drive Suite 263, Honolulu, HI 96822, USA

Abstract. We report here a numerical study of wave-interaction with an array of idealized coral-growing units, with each unit being a thin-walled, perforated, hollow, cylindrical structure. This study focuses on the effects of four layouts on wave transmission and energy dissipation. Numerical simulations are performed using a Boundary Element Method (BEM), with the energy dissipation by the perforated plates being parameterized in the potential flow solver. Due to wave scattering by 3D units, wave transmission and energy dissipation cannot be evaluated by transmission and reflection coefficients calculated by local wave amplitudes alone. To evaluate the wave transmission and energy dissipation, we use the energy flux across vertical 2D planes in the far field instead. Among the four layouts examined in this study, one layout has units closely arrange in rows, similar to submerged breakwaters, and other layouts all have wider space between units. Results showed that the layout has minor influences in wave transmission and energy dissipation, unless the units are closely arranged in rows. Factors other than layouts should be considered to optimize the energy dissipation by a patch of coral-growing units.

Keywords: Boundary element method · perforated thin-wall structures · wave scattering · wave energy dissipation

1 Introduction

Natural coral reefs can dissipate wave energy and thus can serve as a natural flood defense along coastlines [1]. Many natural-based solutions have been proposed to mitigate coastal flooding and storm damages. In this study, a patch of idealized coral-growing units, which are dual-function units designed to dissipate wave energy and grow corals, is examined to understand their performance in reducing wave energy reaching the shoreline (see Fig. 1 and Fig. 2). The key parameters to be examined will include the dimensions and layouts of the idealized coral-growing unit.

© The Author(s) 2026
C. Coelho et al. (Eds.): CD 2025, CRL 41, pp. 558–563, 2026.
https://doi.org/10.1007/978-3-032-15473-6_85

2 Methodology

Assuming that the wave amplitude is much smaller than the wavelength and water depth, a linear water wave theory is used to study linear hydrodynamic interactions.

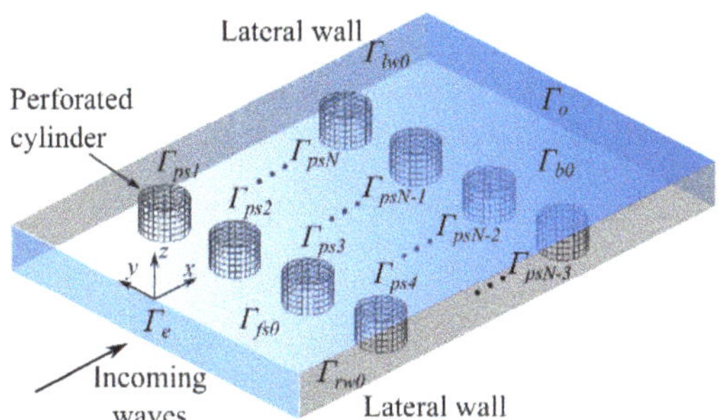

Fig. 1. N perforated cylinders distributed in the wave flume.

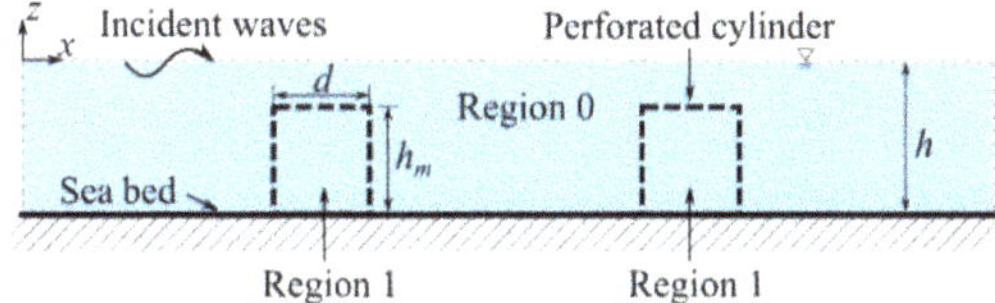

Fig. 2. Lateral view of perforated cylinders with defined regions.

Referring to Figs. 1 and 2, the continuity of mass flux and the pressure drop are the conditions connecting the regions separated by the perforated plates. The pressure drop across a perforated structure can be parameterized by

$$P_1 - P_0 = \frac{\alpha}{2}\rho V_0 |V_0|, \, on(x, y, z) \in \Gamma_{ps} \tag{1}$$

where P refers to the total pressure in the corresponding region, V is the velocity at the perforated structure, α is the loss coefficient and ρ is the water density [2]. The pressure drop can be linearized using Lorentz's principle of equivalent work [3, 4]. Upon linearization and subsequent simplification, the continuity of mass flux and Eq. (1) can be written in terms of velocity potential as [5]:

$$n_0 \nabla \phi_0 = -n_1 \nabla \phi_1 = ikG(\phi_1 - \phi_0), \, on(x, y, z) \in \Gamma_{ps} \tag{2}$$

where the subscripts 0 and 1 refer to the two sides of the plate and G is given by

$$G = \frac{\omega}{\beta k} \tag{3}$$

with β representing the linearized dissipation coefficient defined by

$$\beta = \frac{\alpha}{2}\left(\frac{8}{3\pi}\right)\frac{\int_{\Gamma_{ps}}\left|\frac{\partial\phi_0}{\partial n}\right|^3 ds}{\int_{\Gamma_{ps}}\left|\frac{\partial\phi_0}{\partial n}\right|^2 ds}, \tag{4}$$

In this equation, $|\partial\phi_0/\partial n|$ is integrated along the surface of the perforated structure, denoted by "Γ_{ps}". Similar matching conditions have been used by other authors [5–7] for porous or slotted walls. The loss coefficient α in Eq. (4) can be estimated by

$$\alpha = \left(\frac{1}{\varepsilon c_c} - 1\right)^2, \tag{5}$$

where ε is the plate's porosity and c_c is the contraction coefficient [4, 8, 9]. In Eq. (3), G is the porous effect parameter to be determined iteratively since β is part of the solution.

A multi-domain Boundary Element Method (BEM) was used to solve the boundary value problem (BVP) described in the previous section. The whole domain in Fig. 2 is separated into two regions, one is the water region outside the perforated structures (region 0) and the other is the water region inside all the structures, which is collectively called region 1.

The ratios of the reflected and transmitted energy fluxes through vertical planes to the incident energy flux, K_R and K_T quantify, respectively, the proportion of wave energy reflected and transmitted. According to conservation of energy, the amount of wave energy lost due to wave-interaction with the perforated units can be quantified by an energy loss coefficient K_L defined as [10]

$$K_L = 1 - K_R - K_T, \tag{6}$$

3   Results

Four distinct layout configurations were evaluated, each containing an identical number of structures $N = 24$, see Fig. 3. These structures were distributed in a region within a wave flume measuring 42.5 m in length and 20 m in width. All units shared the same dimensions, with a diameter-to-depth ratio of $d/h = 1.0$ and a submerged height-to-depth ratio of $h_m/h = 0.67$, and the porosity $= 10\%$.

The wave conditions included a water depth of $h = 2.5$ m, a wave height-to-wavelength ratio $= 0.01$, and incident wave period (T) ranging from 5.6 s to 53 s. Figure 4 demonstrates that the row-aligned layout is the most effective configuration for wave energy reduction, followed by the staggered layout. The non-staggered and random layouts perform similarly and are less effective by comparison. The superior performance of the row-aligned layout can be attributed to the minimal gaps between structures within each of the primary rows, which force the water to flow through the array rather than around it. In contrast, the less efficient layouts allow more pathways for water to bypass the structures, reducing their overall effectiveness in dissipating wave energy.

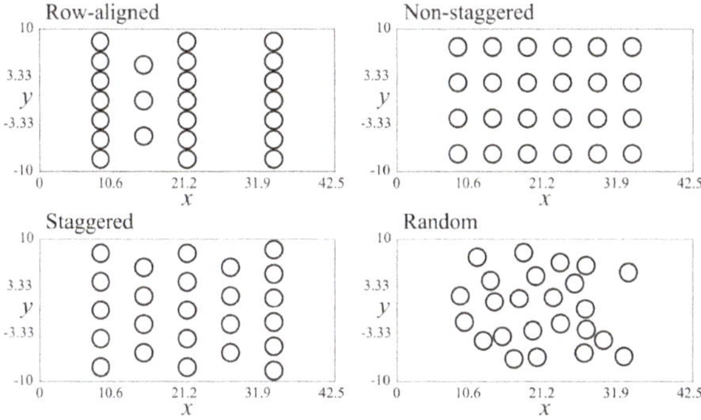

Fig. 3. Four layout configurations examined in the study. The unit for x and y is meter.

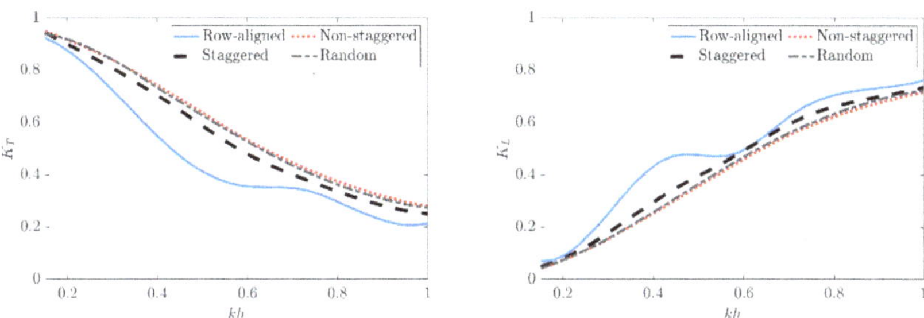

Fig. 4. Comparison of K_T and K_L against kh for the four different layouts.

Figure 5 shows the contour lines of normalized surface displacement within the wave flume for the row-aligned and random layouts. In the row-aligned layout, contour lines in the shoreline-facing region maximum wave amplitude is around 0.45 to 0.47, closely matching the energy transmission coefficient $K_T = 0.214$ shown in Fig. 4. Conversely, the random layout exhibits larger wave amplitude reaching the shoreline ($K_T - 0.274$ from Fig. 4), with concentrations along the wave flume's center line due to the asymmetrical placement of structures.

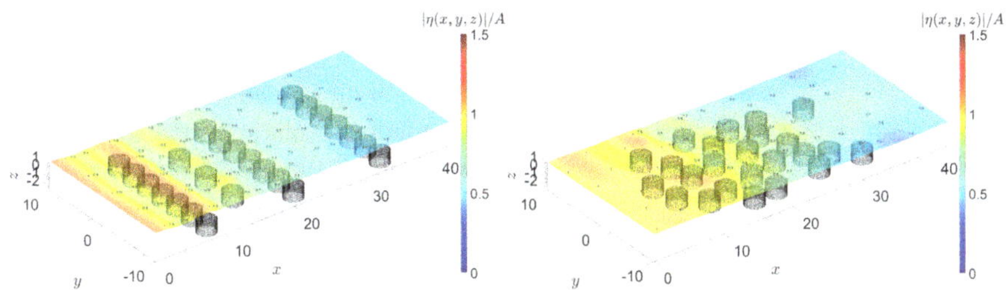

Fig. 5. Contour lines of the non-dimensional free surface displacement for row-aligned and random layouts. In this example, $kh = 1.0$.

Institution	Authors (lab leads are in bold)
Applied Research Laboratory at the University of Hawai'i	**Benjamin A. Jones (Project PI)**, Joshua Levy, Sean Mahaffey, Aricia Argyris, Mark Aruda and Ian Robertson
Department of Ocean and Resources Engineering, University of Hawai'i at Manoa	**Zhenhua Huang (Lead of Technical Area 1)**, Ayrton Medina-Rodriguez, Mert Gokdepe, Brady Halvorson, Jon Chase, Charlotte White, Cami Dillon, Kristian McDonald and Anna Mikkelsen
Hawai'i Institute of Marine Biology, University of Hawai'i at Manoa	**Josh Madin (Lead of Technical Area 2)**, Mollie Asbury, Jessica Reichert, Hendrikje Jorissen, Nina Schiettekatte and Marion Chapeau
	Rob Toonen ((Lead of Technical Area 3), Christopher R. Suchocki, Van Wishingrad, Chris Jury, Daniel Schar, Madeleine Hardt, Claire Lewis, Claire Bardin and Joshua Kualani
	Crawford Drury
	Kira Hughes, Josh Hancock and Carlo Caruso
Ohio State University	**Andrea Grottoli**, Shannon Dixon and Ann Marie Hulver
Florida Atlantic University	**Joshua D. Voss** and Allison Klein
	Siddhartha Verma and Alejandro Alvaro
Makai Ocean Engineering	**Richard Argall**, Kevin Chun, William Hicks, Alex LeBon and John Yeh
Scripps Institution of Oceanography, University of California San Diego	**Aaron Thode** and Oceane Boulais
	Daniel Wangpraseurt, Samapti Kundu, Natalie Levy, Lindsey Badder and Stefan Kolle

4   Conclusion

The research findings show the role of layout configuration in coral-growing units for wave dissipation and energy transmission. The simulation results show that a layout with row-aligned units demonstrates a higher capacity for dissipating wave energy compared to more dispersed layouts. To mimic a natural coral reef, the coral-growing units may not take a row-aligned layout; therefore, optimizing energy reduction of a patch of coral-growing units should focus more on (i) the energy dissipation by individual units and (ii) number of units in a given area. It is important to note that these potential flow-based models have inherent limitations, as it cannot simulate energy loss due to wave breaking. The energy dissipation presented in this study may tend to be on the conservative side.

Acknowledgements. This material is based upon work supported by the Defense Advanced Research Projects Agency under the Reefense Program, BAA HR001121S0012. The views, opinions and/or findings expressed are those of the author and should not be interpreted as representing the official views or policies of the Department of Defense or the U.S. Government. The current members of R3D Consortium are listed in the table below. R3D Consortium Members are part of the DARPA Rapid Resilient Reefs for coastal Defense (R3D) project centered at the University of Hawaii, 2800 Woodlawn Drive, Suite 263, Honolulu, HI, 96822, r3d-consortium@arlcollab. onmicrosoft.com.

References

1. Ferrario F, Beck M, Storlazzi C, Micheli F, Shepard C, Airoldi L (2014) The effectiveness of coral reefs for coastal hazard risk reduction and adaptation. Nat Commun 5:3794
2. Fugazza M, Natale L (1992) Hydraulic design of perforated breakwaters. J Waterw Port Coast Ocean Eng 118:1–14
3. Sollitt CK, Cross RH (1972) Wave transmission through permeable breakwaters. Coastal Eng Proc 1:99
4. Mei CC (1992) The Applied Dynamics of Ocean Surface Waves. World Scientific
5. Huang Z (2006) Method to study interactions between narrow-banded random waves and multi-chamber perforated structures. Acta Mechanica Sinica/Lixue Xuebao 22:285–292
6. Yu X (1995) Diffraction of water waves by porous breakwaters. J Waterw Port Coast Ocean Eng 121:275–282
7. Isaacson M, Premasiri S, Yang G (1998) Wave interactions with vertical slotted barrier. J Waterw Port Coast Ocean Eng 124:118–126
8. Zhu S, Chwang AT (2001) Investigation on the reflection behaviour of a slotted wall. Coast Eng 43:93–104
9. Suh KD, Choi JC, Kim BH, Park WS, Lee KS (2001) Reflection of irregular waves from perforated-wall caisson breakwaters. Coast Eng 44:141–151
10. Yu X, Chwang A (1994) Water waves above submerged porous plate. J Eng Mech

Runup Modeling in Low-Data Coral Reef Environments: Implications for Nesting Sea Turtles

Daniel Dédina[1], Jakob C. Christiaanse[1]([✉]), Floortje Roelvink[2],
Ahmed I. A. Elshinnawy[3], Robert T. McCall[2], Ad Reniers[1], Carlos Duarte[4],
and José A. A. Antolinez[1]

[1] Delft University of Technology, Delft, The Netherlands
j.c.christiaanse@tudelft.nl
[2] Deltares, Delft, The Netherlands
[3] Tanta University, Tanta, Egypt
[4] King Abdullah University of Science and Technology (KAUST), Thuwal, Saudi Arabia

Abstract. Sea turtles are key species in many coastal ecosystems worldwide, particularly coral reef and seagrass habitats. Yet, six of seven species are endangered. Their nests, which incubate in beach sand and rely on specific climatic conditions for egg viability, face significant threats from inundation, for example through wave runup. This paper examines a method to rapidly predict wave runup in low-data coral reef environments, and the implications thereof on the inundation of sea turtle nests. The study uses two metamodels, BEWARE-2 and HyCReWW, to predict wave runup at Ras Baridi, Saudi Arabia, a key nesting site of the Red Sea green turtle population. The models were used to analyze runup events and inundation durations and provide a first estimate of a safe nesting elevation. Despite data limitations, the study provides valuable insights for coastal managers to protect sea turtle nests, suggesting that a 5-year return period runup elevation could serve as a threshold for nest relocation. However, the findings also highlight the importance of more accurate hydrodynamic predictions and the need for in-situ data to validate models and improve conservation strategies.

Keywords: Wave runup · coral reefs · metamodels · sea turtles · nest flooding

1 Introduction

Rising sea levels and increased storm activity due to climate change pose significant threats to coastal ecosystems. This is particularly problematic for endangered sea turtles, who rely on sandy beaches for nesting. Sea turtles lay numerous eggs due to the low survival rate of hatchlings, and specific climatic conditions are crucial for egg viability [1]. The salinity, moisture, and temperature of nests significantly influence hatchling development. Inundation of their nests due to heightened wave runup can lead to egg mortality and altered sex ratios of the hatchlings, potentially destabilizing turtle populations [2, 3]. The eggs cannot survive salt-water inundation for prolonged periods (>12 h), and even short durations (<1 h) can significantly decrease egg viability, particularly at either end of the incubation period (freshly laid or near hatching) [2].

© The Author(s) 2026
C. Coelho et al. (Eds.): CD 2025, CRL 41, pp. 564–569, 2026.
https://doi.org/10.1007/978-3-032-15473-6_86

Mitigation strategies may include relocating nests to higher elevations or enhancing wave energy dissipation through reef preservation, restoration, or artificial reef creation. For example, defining safe nesting elevations ensures nests are moved out of high-frequency flood zones early in the incubation period without compromising the hatchlings' ability to reach the water. These strategies require a thorough understanding of local hydrodynamics, which is challenging in many of the data-scarce environments where sea turtles nest. Coastal managers need quantitative assessments of flooding likelihood along beach elevations to reduce flood risk, which benefits both coastal development and nest management [4].

This study explores the viability of wave runup modeling in data-limited coral reef environments by combining global datasets and metamodels. We aim to equip coastal managers with tools to assess and mitigate flood risks to sea turtle nests in low-data environments like our study area in Ras Baridi, Saudi Arabia.

2 Methods

Ras Baridi is a major green turtle nesting site on the central Saudi Arabian Red Sea coast (Fig. 1a) [5], where local coastal managers have reported flooding of nests. The beach features a fringing coral reef, ranging from 50–200 m in width and is bordered by a steep drop into deep water (Fig. 1b). It is a microtidal environment with a 0.2–0.3 m tidal range and consistent year-round waves of about 1 m.

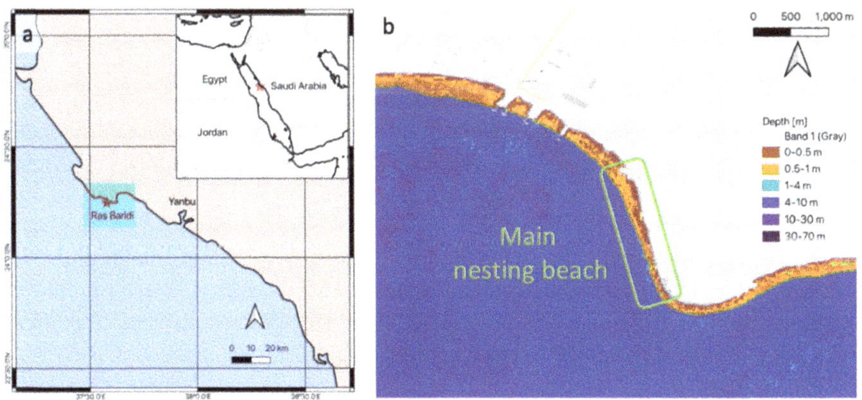

Fig. 1. **A** (a) Geographical overview of the study area, Ras Baridi, in Saudi Arabia and the Red Sea region; (b) bathymetry of the study area derived from the Allen Coral Atlas [10]

We used two different 1D metamodels, BEWARE-2 [6] and HyCReWW [7], to predict wave runup at low computational cost. HyCReWW parameterizes the input bathymetry profile, while BEWARE-2 uses a database of real bathymetry profiles (Fig. 2) to compute the 2% exceedance runup value ($R_{2\%}$). Because these two approaches differ significantly, their results are important to compare.

BEWARE-2 was developed using a training dataset of hydrodynamics and wave runup computed by the XBeach Non-Hydrostatic process-based hydrodynamic model for 440 combinations of water level, wave height, and wave period with 195 representative reef profiles that encompass the natural diversity in real-world fringing coral reef systems. The metamodel functions by matching input profiles to a database of 550 representative (measured) profiles with validated XBeach results and returns a weighted ensemble of the runup (Fig. 2). In the validation stage, the BEWARE-2 modeling system showed a 13% relative root mean square error and 5% relative bias compared to XBeach Non-Hydrostatic [6].

HyCReWW uses a schematic representation of fringing reef geometry through 7 parameters (Fig. 2): the water level, the offshore wave height, the offshore wave steepness, the fore-reef slope, the reef width, the beach slope, and the reef friction factor. The metamodel then interpolates the user input to a database of XBeach runup results through a radial basis function. HyCReWW has been validated with field and laboratory studies and predicts runup with an average root mean square error of 28 cm [7].

As model input, we used 1088 representative sea states, derived from 40-year large scale hindcasts of waves [8] and water levels from 1978–2018 [9] (Fig. 2). Allen Coral Atlas 10-m horizontal resolution bathymetry [10] was used for nearshore bathymetry (Fig. 1b), and ETOPO 2022 Global Relief Model with 15 arc-second resolution was used for offshore bathymetry [11].

The lack of accurate local data led to significant uncertainty in the input parameters for HyCReWW. A sensitivity analysis was therefore conducted to understand the influence of the input parameters on the model output. The sensitivity analysis was also used to screen the parameters to decrease the number of simulations that were run. Based on the sensitivity analysis results, the reef geometry parameters (beach slope, reef width, fore-reef slope and reef friction factor) were found to have a small influence on the results relative to the water level, wave steepness, and significant wave height. The reef geometry parameters were estimated using local data or from literature. We modeled wave runup for 4 different profiles, representing varying reef widths (80–170 m) along the beach, and computed runup values with return periods from 1 to 40 years to assess inundation risk.

The metamodel results provided timeseries data on wave runup, enabling the analysis of nest inundation duration. By examining the distributions of wave runup events exceeding a threshold value, we aimed to understand the flood risk to nests. Specifically, we investigated the duration of runup events and runup heights using HyCReWW and BEWARE-2 models.

The goal was to determine if inundation duration correlated with runup height, allowing nests to be relocated to safer elevations with little chance of being inundated for periods of time that can diminish egg viability. Runup event durations were defined by consecutive runup events with heights of at least 0.48 m and 0.61 m for HyCReWW and BEWARE-2, respectively. These were the minimum runup values modeled by each metamodel due to their input restrictions. Empirical probability density functions (PDFs) were created for inundation durations, and log-normal PDFs were fitted to extract statistical insights.

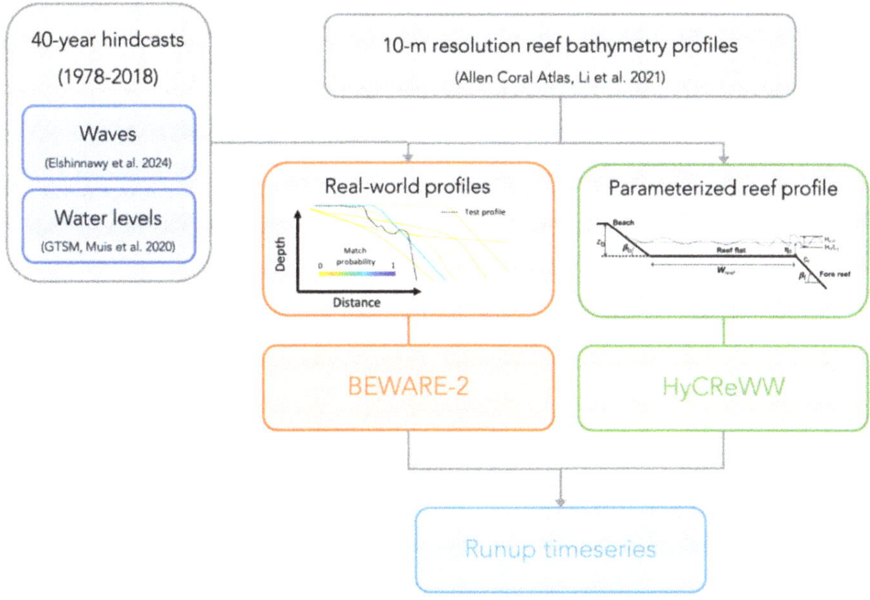

Fig. 2. Flowchart showing the methodology explored in this study. The input data (waves, water levels, and 10-m bathymetry) are used to force the BEWARE-2 and HyCReWW meta-models, which return timeseries of wave runup.

3   Results

The values of runup for different return periods and reef widths are shown in Table 1. They highlight the range of runup values across the beach for extreme events. BEWARE-2 showed a greater sensitivity to reef width than HyCReWW. This appeared to be due to the profile matching process for the reef with of 80 m, which was matched to a profile with a reef width of 50 m, resulting in an overestimation in runup there.

Table 1. Runup ($R_{2\%}$) return period values in meters from HyCReWW and BEWARE 2.0 for two different reef widths (W_r).

Return Period (yrs)	$W_r = 80$ m		$W_r = 170$ m	
	HyCReWW	BEWARE-2	HyCReWW	BEWARE-2
1	0.95	1.35	0.9	0.85
5	1.3	1.75	1.2	1.25
10	1.4	2	1.3	1.3
20	1.9	2.65	1.75	1.8
40	1.95	2.7	1.9	1.8

The HyCReWW results indicated that average $R_{2\%}$ was not well correlated with inundation duration. The median duration of a runup event was 7 h, and the lower and upper quartiles were 4 and 11 h, respectively (Fig. 3). Hence, every nest that gets inundated has a significant risk of reducing the viability its eggs.

A correlation between inundation duration and runup height was also not found in the BEWARE-2 results. Figure 3 shows that the lower quartile was 2 h, and the upper quartile was 10 h, which is apparent in the skew of the PDF towards the lower durations. The results show that the runup events have the ability to reduce the viability of eggs, to a lesser extent than the HyCReWW results.

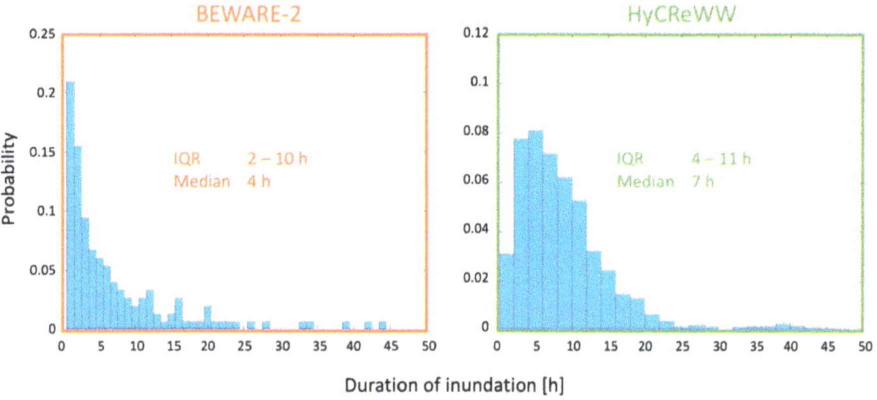

Fig. 3. Weighted histograms of the duration of the runup events modeled with BEWARE-2 (left, orange) and HyCReWW (right, green).

4 Discussion

The duration of the modeled runup events is important to deepening the understanding of flood risk to sea turtles as inundation for 6 h can decreases the viability of eggs by 30% [2]. The median durations of inundation were 4 and 7 h for BEWARE-2 and HyCReWW, respectively. The incubation period of green turtles is about 50–70 days [1], so nests should be placed at an elevation where it is unlikely that the nest will be inundated at all during the 2-month period.

As a conceptual exercise, the 5-year return period runup elevation at Ras Baridi was suggested as a 'safe nesting elevation' for sea turtles at Ras Baridi, with nests below this elevation deemed at risk. Such a threshold can be used in situ to determine which nests to relocate to higher elevations. This was at 1.25–1.75 m (BEWARE-2) and 1.2–1.3 m (HyCReWW) depending on the reef width. BEWARE-2 consistently predicted higher runup than HyCReWW, indicating it could be more suitable for conservative conservation strategies.

This study explored a methodology to estimate wave runup in data-limited coral reef environments using global, low-resolution data and low computational-cost meta-models. In regions where there is no alternative, such approaches may provide first estimates that can, for instance, help inform conservation strategies aimed at protecting vulnerable sea turtle nests from flooding.

References

1. Ackerman RA (1997) The nest environment and the embryonic development of sea turtles. In: Lutz PL, Musick JA (eds) The biology of sea turtles. CRC Press, vol 1, pp 83–106
2. Limpus CJ, Miller JD, Pfaller JB (2021) Flooding-induced mortality of loggerhead sea turtle eggs. Wildl Res 48(2):142. https://doi.org/10.1071/WR20080
3. Ware M, Ceriani S, Long J, Fuentes MMPB (2021) Exposure of loggerhead sea turtle nests to waves in the florida panhandle. Remote Sens 13(14):2654. https://doi.org/10.3390/rs1314 2654
4. Ahles N, Milton SL (2016) Mid-incubation relocation and embryonic survival in loggerhead sea turtle eggs. J Wildl Manag 80(3):430–437. https://doi.org/10.1002/jwmg.1023
5. Shimada T et al (2021) Distribution and temporal trends in the abundance of nesting sea turtles in the Red Sea. Biol Cons 261:109235. https://doi.org/10.1016/j.biocon.2021.109235
6. McCall R, Storlazzi C, Roelvink F, Pearson SG, de Goede R, Antolínez JAÁ (2024) Rapid simulation of wave runup on morphologically diverse, reef-lined coasts with the BEWARE-2 (Broad-range Estimator of Wave Attack in Reef Environments) meta-process model. Nat Hazard 24(10):3597–3625. https://doi.org/10.5194/nhess-24-3597-2024
7. Rueda A et al (2019) HyCReWW: a hybrid coral reef wave and water level metamodel. Comput Geosci 127:85–90. https://doi.org/10.1016/j.cageo.2019.03.004
8. Elshinnawy AI, Lobeto H, Menéndez M (2024) Changing wind-generated waves in the Red Sea during 64 years. Ocean Eng 297:116994. https://doi.org/10.1016/j.oceaneng.2024. 116994
9. Muis S et al (2020) A high-resolution global dataset of extreme sea levels, tides, and storm surges, including future projections. Front Mar Sci 7:512955. https://doi.org/10.3389/fmars. 2020.00263
10. Li J, Knapp DE, Lyons M, Roelfsema C, Phinn S, Schill SR, Asner GP (2021) Automated global shallow water bathymetry mapping using google earth engine. Remote Sens 13(8):1469. https://doi.org/10.3390/rs13081469
11. NOAA National Centers for Environmental Information (2022) ETOPO 2022 15 Arc-Second Global Relief Model

Phase-Resolved Modeling of Surf Zone Wave Transformation Over Idealized Rough Bottoms

Emile Guélard-Ancilotti[1]([⊠]), Damien Sous[1], Denis Morichon[1], Patrick Marsaleix[2], Héloise Michaud[3], and Matthias Delpey[4]

[1] Université de Pau et des Pays de l'Adour, E2S UPPA, SIAME, Anglet, France
`emile.guelard-ancilotti@univ-pau.fr`
[2] CNRS, LEGOS, Toulouse, France
[3] Shom, Toulouse, France
[4] SUEZ, Rivages Pro Tech, Bidart, France

Abstract. Rough rocky seabeds dominate world's coastlines, making accurate modeling of wave transformation in such environments essential for understanding coastal processes and hazards. Wave breaking and friction are critical drivers of wave dissipation over rough seabeds, especially in the surf zone. Phase-resolved wave models, often relying on classical bed shear stress or canopy drag approaches, can fail to capture these processes due to limitations in their physical assumptions, especially for large roughness elements typical of rocky seabeds. The present study adapts the Bulk Canopy Drag (BCD) parameterization, inherited from vegetation and porous media studies, into the 3D non-hydrostatic phase-resolved SYMPHONIE model to simulate wave dissipation over rough seabeds. The model was tested against laboratory experiments (LEGOLAS), using controlled irregular wave forcing propagating over a rough ramp. Results reveal that turbulent drag dominates the outer surf zone, while inertial drag plays a key role in the inner surf zone. A combined optimization of these influences achieved strong agreement with experimental measurements under one irregular wave forcing and a specific bottom configuration. These findings underscore the importance of accurately incorporating both turbulent and inertial contributions into wave dissipation models for the rocky surf zone. The proposed BCD approach provides a promising framework for improving phase-resolved modeling of wave dynamics, with potential applications for more complex field conditions. Future work will aim to extend the approach to diverse roughness configurations and refine empirical coefficients for broader scalability and in situ applicability.

Keywords: Waves · dissipation · roughness · drag · phase resolved modeling

1 Introduction

Rough bottoms account for almost 70% of the world's coastlines [1]. Understanding wave transformation and predicting wave-driven processes and hazards in rocky environments requires accurately representing wave transformation in the surf zone for rough bottoms. Wave breaking and friction are key processes to quantify wave dissipation in models

C. Coelho et al. (Eds.): CD 2025, CRL 41, pp. 570–576, 2026.
https://doi.org/10.1007/978-3-032-15473-6_87

for these types of environments, especially in the surf zone. The focus is placed here on 3D free surface phase-resolved models, which offer promising prospects to describe interaction between natural and artificial structures at the wave scale. Up to date, the description of roughness-induced dissipation in phase-resolved wave models remains mostly based on the classical bed shear stress, often requiring the use of empirical or site-specific bottom drag coefficients. The classical bed shear stress approach, which relies on the assumption of a thin boundary layer, is questionable in rocky environments where large roughness elements can produce significant drag form, especially in the surf zone [2]. An alternative is to describe bottom drag using a canopy-type volume drag, widely applied in vegetation or porous media studies (e.g., [3]). Comparisons between bed stress and canopy drag approaches on very rough bottoms remain very sparse [4]. We will focus here on the canopy approach which has shown promising results on coral reef applications, offering a comprehensive closure of wave and momentum dynamics [5].

The present study aims to develop a parameterization of roughness-induced wave dissipation into 3D non-hydrostatic phase-resolved wave models. In particular, we aim to identify the respective roles played by turbulent and inertial drag components throughout the surfzone. The selected strategy was to adapt the bulk canopy drag (BCD) approach inherited from vegetation and porous media studies (e.g., [3]) to rough seabed and discretized water column. The phase-resolved 3D model SYMPHONIE [6] was tested against a series of recent laboratory experiments on surf zone waves dissipation over a ramp and in the presence of idealized roughness, including very large roughness elements [4].

2 Materials and Methods

2.1 BCD Approach

Initially designed to predict large-scale ocean circulation, the 3D non-hydrostatic version of SYMPHONIE [6] is adapted to perform phase-resolved simulations of surface waves. This model provides a detailed and dynamic representation of the temporal and spatial evolution of the free surface. By contrast to depth-averaged models, the vertical discretization of the water column also allows accounting for the vertical structure of the flow, particularly useful to describe interaction with complex seabeds.

Wave dissipation due to seabed roughness is modeled using the Bulk Canopy Drag (BCD) approach, which introduces an additional drag force into the governing momentum equations. This force captures the effects of solid obstacles in the water column and is inherited from studies on the effects of porous media for non-stationary flow [7]. Its complete expression, expressed as the extended Darcy-Forchheimer equation, is:

$$F_{BCD} = \rho \left(C_D |u| u + C_L u + C_M \frac{\partial u}{\partial t} \right)$$

(1)

where C_D (in m^{-1}), C_L (in s^{-1}) and C_M (adimensional) are coefficients representing turbulent, laminar, and inertial influences, respectively. Empirical determination of these coefficients can be challenging, especially for optimization against observations. The

572   E. Guélard-Ancilotti et al.

Reynolds (Re) and Keulegan-Carpenter (KC) numbers are commonly used to identify the hydrodynamic regimes within the canopy and the dominant components controlling the drag in Eq. 1. These dimensionless numbers are defined as:

$$Re = \frac{u_c l_c}{\nu} \quad KC = \frac{u_c T}{l_c} \tag{2}$$

where u_c is a characteristic velocity, l_c a characteristic length, ν the kinematic viscosity, and T the wave period. In the present study, T is defined as the peak period T_p, l_c as the roughness influence height, equal to four times the standard deviation of the bottom elevation [8], and u_c as the "significant" orbital wave current at the half of this height inside the canopy, equal to the square root of the squared current module multiplied by two. Although blocs width or height can be directly used as the characteristic length in the present study, the influence height was chosen following [9] because it is based on a statistical metric of the topography variability that remains accessible for real rough seabed, in contrast to the geometrical dimensions of an idealized seabed configuration. The influence height is also used to determine the portion of the water column over which the BCD force is applied.

In the presence of large roughness (such as large gravels or rocks) and for the range of Re and KC numbers calculated from the simulation (Fig. 1), [7] recommends focusing on turbulent and inertial influences. Thus, only those two influences were considered.

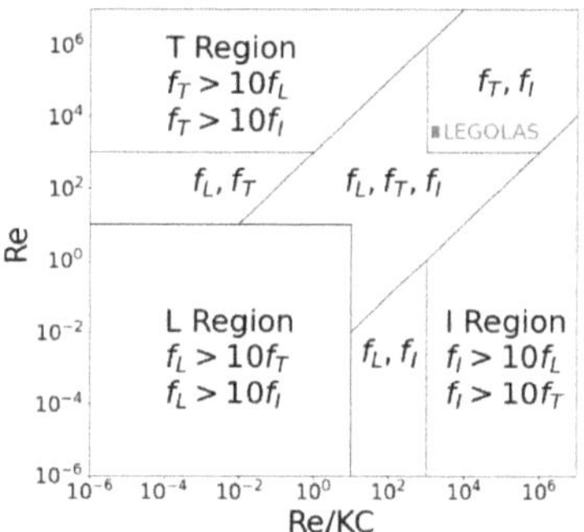

Fig. 1. LEGOLAS study case (green rectangle) in the framework of the Gu & Wang diagram (expressed as [7]) showing the relative influence of each expression of the drag (f_L for the laminar one, f_T for the turbulent and f_I for the inertial) depending on Re and KC

2.2 Experimental Setup

The model developments were compared to the LEGOLAS laboratory experiments [9] conducted in the CASH flume of SeaTech engineering school in Toulon (France). During these experiments, [9] measured variations in water surface elevation using 18 resistive probes, regularly spaced over a linear sloping bed (Fig. 2). In the selected benchmark case, a JONSWAP irregular wave forcing with a T_p of 1.2 s and a H_s of 0.061 m was generated using a piston wave maker. Two seabed configurations were tested here: the smooth one, to validate the representation of wave transformation due to wave breaking only, and the rough one, based on a quincux configuration of 0.03 m cubes, with a roughness influence height of 0.044 m.

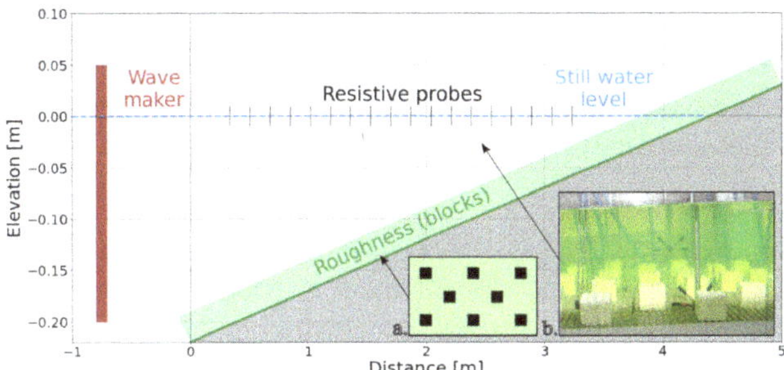

Fig. 2. Cross-shore scheme of the LEGOLAS experimental wave flume with (a.) the chosen blocs planar pattern on the ramp and (b.) a picture of the water column

2.3 Numerical Setup and Analysis

The numerical domain was defined as an 8 m long wave flume (extending by 2 m offshore the experimental channel length to stabilize wave generation), 0.15 m wide and 0.22 m deep. The wave flume presented a linear slope of 1/20 starting 3 m from the wave generator. The horizontal resolution was 0.03 m and the time step was 0.01 s. Ten sigma layers, varying with the bathymetry and the free surface elevation variations, divided the water column. The geometric configuration was fully alongshore uniform, so that the analysis was carried out in a 2DV cross-shore framework. Lateral boundaries were defined as periodic and a flather condition was imposed at the wave maker open boundary.

Numerical sea surface elevation time series were extracted at each probe location and significant wave height (Hs) profiles were computed by integrating spectra at each location over the frequency band of gravity waves ($0.5f_p < f < 3f_p$, where f_p is the peak frequency).

The first step was to calibrate the breaking parameters of the model without roughness-induced drag force against the smooth bottom configuration, to ensure the quality of the wave breaking representation. Once optimized, breaking parameters were kept constant for the rough case study, allowing us to isolate the roughness contribution to frictional dissipation.

Three model configurations for the rough seabed were compared:

- the sole turbulent component, with a constant optimized C_D of 10 m^{-1}. Laminar and inertial drag components were neglected with coefficients values of 10^{-5},
- the sole inertial component, with a constant optimized C_M of 0.5. Laminar and turbulent drag components were neglected with coefficients values of 10^{-5},
- the combination of turbulent and inertial influences, with constant optimized C_D and C_M of 10 m^{-1} and 0.5. The laminar turbulent drag component was neglected imposing coefficient value of 10^{-5} s^{-1}.

Root mean square error (RMSE) and Willmott Index (WI) were computed to compare the results with the corresponding experimental measurements.

3 Results

The model results are compared with experimental cross-shore Hs profiles, as shown in Fig. 3. Four profiles illustrate the impact of different drag influences along the cross-shore profile:

- The blue line represents the cross-shore profile of modelled Hs compared to observations (blue crosses) for the smooth case. A good agreement is observed, with RMSE of 0.0024 m and WI of 0.988, demonstrating the performance of the model in reproducing the sole breaking-induced dissipation.
- The green line shows the cross-shore profile of modelled Hs for the case with turbulent influence solely. When compared to experimental measurements on the rough ramp (orange crosses), the model achieves an RMSE of 0.0041 m and a WI of 0.966.
- The red line represents the cross-shore profile of modelled Hs for the case with inertial influence solely, resulting in an RMSE of 0.0022 m and a WI of 0.992 compared to experimental measurements on the rough ramp.
- Finally, the yellow curve corresponds to the Hs profile obtained by combining turbulent and inertial influences. This configuration provides the best agreement with experimental measurements on the rough ramp, achieving an RMSE of 0.0017 m and a WI of 0.995.

These comparisons demonstrate the specific impacts of the drag components, which vary across the beach profile. Prior to breaking or around the breaking point, waves are large, inducing strong orbital velocities, which should explain the dominant effect of turbulent drag when compared to the inertial component. Toward the end of the ramp, in very shallow waters, the wave height decreases but the wave asymmetry is expected to be strong, which may explain why inertial forces dominate. Further analysis is required to better understand the driving processes. Overall, these results suggest that the combined optimization of turbulent and inertial influences is critical for accurately modeling wave transformation over rough bottom in the surf zone. Moreover, laminar drag did not need to be considered in that scenario, as the model showed good results compared to experimental measurements using only turbulent and inertial drag components. However, it would be important to take it into account for scenarios showing lower Re numbers.

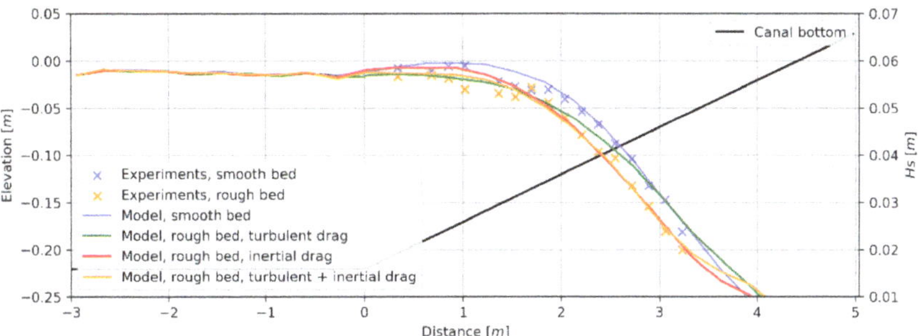

Fig. 3. Experimental and model *Hs* profiles comparison (in colors) for the turbulent and/or the inertial contribution(s) to the drag force along the canal bottom (in black)

4 Conclusion and Perspectives

This study highlights the potential of a 3D non-hydrostatic phase-resolved model, SYM-PHONIE, to simulate wave transformation over rough seabeds in the surf zone. By incorporating the Bulk Canopy Drag (BCD) in the model, we parameterized roughness-induced wave dissipation, accounting for both turbulent and inertial drag influences. The comparison of numerical results with experimental data from the LEGOLAS laboratory experiments demonstrates the robustness of the proposed approach.

The results indicate that turbulent drag plays a dominant role in the outer surf zone, whereas inertial drag becomes more significant in the inner surf zone. This dual influence highlights the necessity of combining both contributions to accurately represent wave dissipation across varying depths in the surf zone. An optimized combination of these factors resulted in a good agreement with experimental measurements for a specific irregular wave forcing and bottom configuration.

These findings provide new insights for improving the representation of wave dissipation in rocky coastal environments, which are still poorly understood in current models. Future work will aim to extend the approach to more complex scenarios, including different roughness configurations and hydrodynamic conditions. Additionally, further investigation is needed to refine the empirical coefficients and explore the scalability of the approach for in situ applications. Although optimized drag coefficients were sufficient to represent turbulent and inertial influences for the chosen forcing and bottom configuration, further improvements could be achieved by expressing these coefficients according a local hydrodynamic expression depending on KC or Re ([4]) and a geometric influence depending on representative topographic metrics ([9]).

References

1. Bird EC (2008) Coastal geomorphology: an introduction. John Wiley & Sons, Hoboken
2. Boles E et al (2024) Bottom stress and drag on a shallow coral reef. J Geo-Phys Res Oceans 129:11
3. Chen H, Liu X, Zou QP (2019) Wave-driven flow induced by suspended and submerged canopies. Adv Water Resour 123:160–172

4. Dealbera S, Sous D (2024) Frictional wave dissipation in macro-roughness environments: A comparison of bottom stress and bulk canopy drag models. Ocean Eng 314:119738
5. Sous D, Dodet G, Bouchette F, Tissier M (2020) Momentum balance across a barrier reef. J Geophys Res Oceans 125(2):e2019JC015503
6. Marsaleix P, Michaud H, Estournel C (2019) 3D phase-resolved wave modelling with a non-hydrostatic ocean circulation model. Ocean Model 136:28–50
7. Van Gent M (1993) Stationary and oscillatory flow through coarse porous media. Commun Hydraulic Geotech Eng 59
8. Chung D, Hutchins N, Schultz M, Flack K (2021) Predicting the drag of rough surfaces. Ann Rev Fluid Mech 53:439–471
9. Dealbera S, Sous D, Morichon D, Michaud H (2024) The role of roughness geometry in wave frictional dissipation. Coast Eng 189:104478

Quantifying Coral Reef Flat Vertical Accretion Rate Using Photogrammetry

Emma Ryan[1]([⊠]), Brendan Hall[1], Kira Brereton[1], and Alexandre Lhériau-Nice[2]

[1] School of Environment, The University of Auckland, Auckland, New Zealand
e.ryan@auckland.ac.nz

[2] Institute of Marine Science, The University of Auckland, Auckland, New Zealand

Abstract. Monitoring reef growth is essential for understanding sea-level rise impacts on reef-lined coasts. However, measuring coral reef vertical accretion in the field is challenging and rarely achieved. Here we present a method for measuring reef flat vertical accretion at sub-mm scale using close-range structure-from-motion photogrammetry. Comparison of repeat digital surface models constructed from field-based and lab-based photos reveal that the accuracy and precision of the method is on average 0.42 mm and 0.181 mm, respectively. We provide an example of how this method can be utilized to measure reef vertical accretion on a reef flat in the Kingdom of Tonga, South Pacific region. Results reveal the reef accreted by 8.42 ± 0.601 mm over the one-year study period. We highlight the need for future research to scale up the methodology spatially and extend the timeframe of analysis in order to track reef response to ongoing sea-level rise.

Keywords: reef flat · structure-from-motion · reef growth · Kingdom of Tonga · reef crest

1 Introduction

Coral reefs are highly vulnerable to climate change impacts, such as mortality from elevated sea surface temperatures, and reduced vertical growth under rising sea levels [1]. Reefs provide vital wave protection for tropical coastlines, but if sea-level rise outpaces reef accretion shorelines may lose this natural barrier, increasing exposure to wave energy. While sea-level rise could theoretically allow more vertical accommodation space for reef growth, the compounding effects of climate change make the future of reefs uncertain [2].

Quantifying vertical accretion rates of coral reefs is a crucial aspect of understanding and predicting reef response to climate change impacts. To date, reef vertical accretion rates have typically been inferred from geological and ecological studies [3–5], rather than directly measured [2]. Data from reef cores show that reefs typically accreted at $5 - 15$ mm/yr over the past few millennia [6]. The lack of *in-situ* reef accretion data is largely due to limitations associated with data collection, particularly at the outer reef edge, where wave processes dominate (often at all tidal stages), making data collection

C. Coelho et al. (Eds.): CD 2025, CRL 41, pp. 577–583, 2026.
https://doi.org/10.1007/978-3-032-15473-6_88

problematic. The deployment of new wave-resistant tools such as the Coral Reef Accretion Frame have been used by one study to quantify reef flat accretion rates with high precision [2]. Close-range structure-from-motion photogrammetry (hereafter referred to as photogrammetry) has potential to provide an alternative, simple and rapid technique for measuring *in-situ* reef accretion [7]. Modelling techniques in photogrammetry convert overlapping features in overlapping photographs to points, generating a 3D point cloud. Interpolation of these points generates a series of triangles to create a 3D Digital Surface Model (DSM) of topography, which is referenced to ground control points and scaled in three dimensions. The use of photogrammetry in reef environments to determine topography and benthic cover has proliferated over the past decade [8–10] to quantify coral growth rates at the individual coral colony-scale [11] and determine benthic ecology and substrate rugosity [12–14]. Despite the potential, no study has attempted to use photogrammetry to quantify coral reef accretion to our knowledge. This paper presents the first study to use photogrammetry to explore quantifying reef flat vertical accretion with sub-mm precision and accuracy.

2   Methods

2.1   Data Collection

Lab Test Data Collection. To determine accuracy and precision of our method, an artificial exposed coral reef plot was established in the laboratory (500 × 520 mm) at the University of Auckland consisting of foam blocks as substrate, with branching and massive coral specimens attached. Between 30 and 40 photographs were taken of the plot using a Canon PowerShot G7 X Mark iii (20.1MP resolution) digital camera. To determine accuracy 3D-printed objects of known dimensions were then added to the artificial reef plot, while everything else remained the same. A subsequent set of photographs was taken to compare with the initial set (see Sect. 2.2) to quantify precision. Automated ground control points (GCPs) were used in this test for scaling and control.

Field Data Collection. Two quadrats (approx. 1000 × 1000 mm) were established at the outer reef flat in July 2023 (labelled RE3 and RE4). GCPs were generated by installing and securing three cross-haired bolts into the reef framework at each quadrat. Automated GCPs were not available for this field campaign. Two 600 mm rulers and one wooden block (180 × 90 × 90 mm) were used in each plot for scaling and orientation control. The wooden block was levelled and oriented north.

Following published methods [6, 11] modified for the intertidal reef flat setting, the two quadrats were photographed in July 2023 at spring low tide when the reef was sub-aerially exposed using the same camera as for the lab test. Between 100 and 200 overlapping photographs were taken of each plot using back-and-forth technique from a range of orientations, ensuring GCPs and scaling and orientation devices were captured. Two sets of photographs were taken at RE4 in 2023, one directly after the other, for quantifying precision. Both quadrats were re-surveyed in July 2024 to establish a one-year dataset of reef surface change. The results from RE3 are presented herein.

2.2 Data Analysis

All photographs were post-processed in Agisoft Metashape Professional. GCPs were placed at the crosshair mid-point for each bolt before building a dense cloud, a DSM and an orthomosaic. To quantify change between two DSMs with the same GCPs (either from the field or lab data), VXelements11 software was used. Heat maps were used to determine field and lab test precision. A grid was added across the heatmap consisting of 25×25 mm cells and elevational change was quantified at the central point within each cell. This process was repeated three times for three areas of interest (AOI), covering different areas of the heatmap and taken from different orientations. The change data were manually extracted for statistical analysis (a total of 1591 data points for the lab data and 1748 for the field). Incorrect values/errors outside of the 1st and 99th percentiles were removed before statistical analysis. Basic statistical analysis (percentiles, averages, room mean square error [RMSE], see Table 1) was then conducted using R 4.3.3. To determine accuracy, the 3D printed objects were used to quantify the difference between the output heatmap 'growth' rates (lab test) with the known dimensions of the objects. We did this manually by pinpointing 50 points across four individual objects (Seven of these pinpoints are shown in Fig. 1A) and generating an average.

3   Results and Discussion

3.1   Accuracy and Uncertainty

The average precision of the photogrammetry approach in measuring upward growth ranged between 0.001 to 0.181 mm respectively (Table 1). This was very similar to the field test, where the average precision ranged from 0.033 and 0.150 mm (Table 1). While outliers of up to 4.33 mm were measured in the tests, 95% of the precision data for both the field and lab tests lie within or very close to ± 1 mm (Fig. 2). For example, the data for AOI 1 (field) lie within −1.095 mm and 0.65 mm (Table 1). All outliers were measured in areas with high structural complexity or in gaps in the reef structure (i.e. areas where the model was not well-constructed due to lack of point cloud data), indicating this approach is less suitable for determining change in cryptic reef spaces.

Accuracy of the method was on average −0.42 mm, indicating that this method tends to underpredict 'accretion' (Fig. 1). Total error is 0.601 mm, determined as accuracy plus precision (taken as the highest average precision; 0.181 mm).

Table 1. Precision statistics of close-range structure from motion photogrammetry. All statistics presented in mm.

	Field test precision			Lab test precision		
AOI #	1	2	3	1	2	3
n	425	819	504	682	513	396
Mean	−0.139	−0.150	−0.033	−0.025	0.001	−0.181
Median	−0.124	−0.206	−0.038	0.009	0.020	−0.104
q95	0.650	0.938	0.710	0.959	0.856	0.725
Max	2.628	4.430	2.506	3.372	3.323	2.803
q5	−1.095	−1.211	−0.774	−1.040	−0.961	−1.250
Min	−2.823	−3.103	−1.533	−2.765	−1.956	−3.533
RMSE	0.619	0.826	0.490	0.660	0.601	0.758

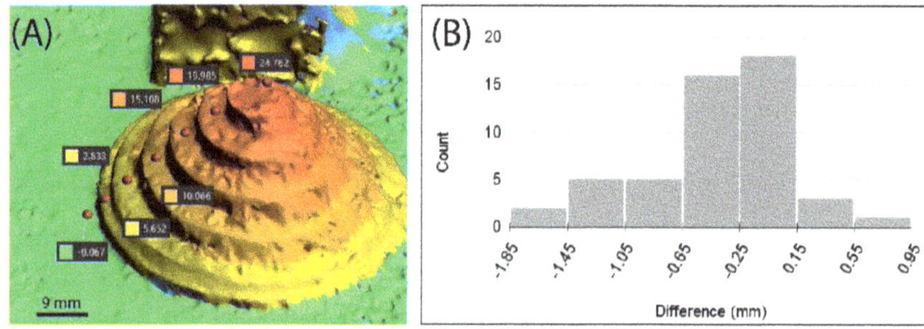

Fig. 1. (A) Snippet from the lab test heatmap, showing one of the objects (yellow-orange) that was used for accuracy testing and associated measurements derived in VX Elements. (B) Distribution of differences (actual − measured growth) in each measurement (n = 50) used to generate accuracy of photogrammetry technique.

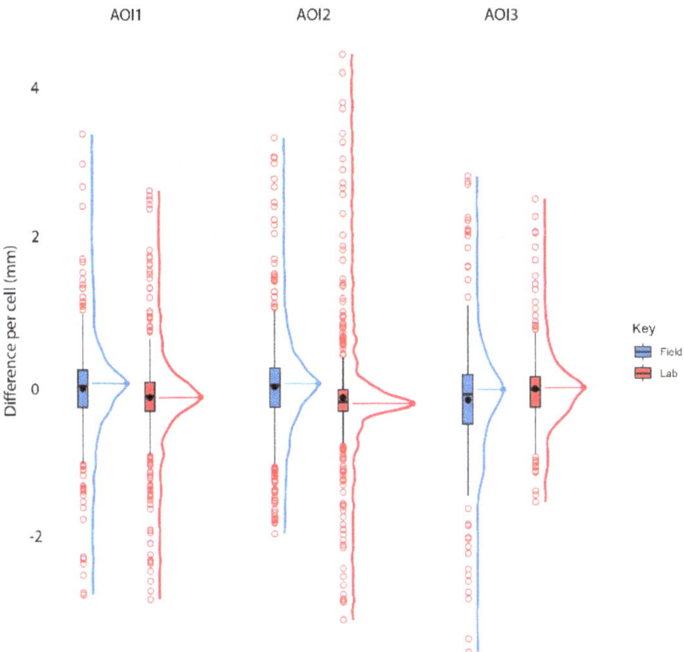

Fig. 2. Measured difference (mm) between two repeat Digital Surface Models generated in the field and the lab. Coloured histogram lines show relative abundance of measured cells. Outliers are shown by red circles.

3.2   Reef Vertical Accretion

We use site RE3 to present an example of quantifying one year of reef growth on an outer reef flat at Nuku'alofa. The comparison of the DSMs for 2023 and 2024 resulted in an overall average reef surface change of 8.42 ± 0.601 mm (i.e. averaged across the entire quadrat area, ± total error). The rate of accretion varied across the different substrate types. While ongoing studies are exploring variations in rates across coral genera, coralline algae and reef framework this paper demonstrates that digitate corals (of the genus *Acropora*) grew faster than massive corals and faster growth was observed around the perimeter of digitate coral colonies compared with the centre (Fig. 3). In general, reef framework was relatively stable across the study period. These results are comparable with studies that have measured reef accretion *in-situ*, such as an average accretion rate of 6.6 ± 12.5 mm/yr measured at a reef crest in the Maldives [2]. Minimal erosion was observed at RE3 in Nuku'alofa across the study period (Fig. 3). No tropical cyclones impacted this reef during the study period. Such events would likely cause coral breakage at the reef crest and impact on long-term reef accretion rates.

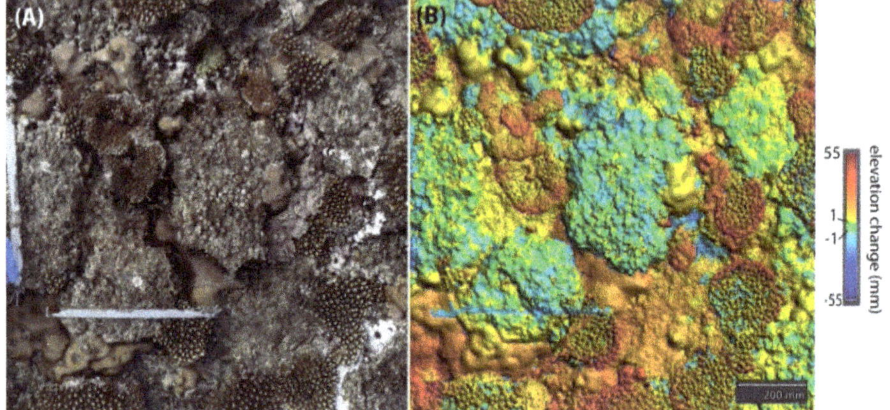

Fig. 3. (A) Digital Surface Model of site RE3 in July 2023. (B) Corresponding change in surface of reef at site RE3 between July 2023 and July 2024.

This study demonstrates that close-range photogrammetry is suitable for monitoring reef flat accretion rates with sub-mm error. We measured accretion of the reef flat at Nuku'alofa cross one year, but these findings cannot be scaled up spatially without collection of data from many more quadrats across the reef crest. Future work will focus on this, along with scaling up this methodology using drone-based photogrammetry. Furthermore, extending the temporal length of such datasets is required to capture meaningful long-term accretion rates and understand whether reefs are keeping pace with sea-level rise.

References

1. Cornwall CE et al (2021) Global declines in coral reef calcium carbonate production under ocean acidification and warming. Nature 5:2021
2. Kench PS, Beetham EP, Turner T, Morgan KM, Owen SD, McLean RF (2022) Sustained coral reef growth in the critical wave dissipation zone of a Maldivian atoll. Commun Earth Environ 3(1):1–12
3. Perry CT, Smithers SG, Gulliver P, Browne NK (2012) Evidence of very rapid reef accretion and reef growth under high turbidity and terrigenous sedimentation. Geology 40(8):719–722
4. Perry CT, Murphy GN, Graham NAJ, Wilson SK, Januchowski-Hartley FA, East HK (2015) Remote coral reefs can sustain high growth potential and may match future sea-level trends. Sci Rep 5:18289
5. Perry CT et al (2018) Loss of coral reef growth capacity to track future increases in sea level. Nature 558:396–400
6. Gordon S, et al (2023) Reef restoration and adaption program (EcoRRAP): standard operational procedure number 1: overview and in-field workflow, 14(14):62
7. Jackson-Bué T, Williams GJ, Walker-Springett G, Rowlands SJ, Davies AJ (2021) Three-dimensional mapping reveals scale-dependent dynamics in biogenic reef habitat structure. Remote Sens Ecol Conserv 7(4):621–637
8. Bythell J, Pan P, Lee J (2001) Three-dimensional morphometric measurements of reef corals using underwater photogrammetry techniques. Coral Reefs 20(3):193–199

9. Aston EA, Duce S, Hoey AS, Ferrari R (2022) A protocol for extracting structural metrics from 3d reconstructions of corals. Front Mar Sci 9(April):1–14
10. Burns JHR, Delparte D, Kapono L, Belt M, Gates RD, Takabayashi M (2016) Assessing the impact of acute disturbances on the structure and composition of a coral community using innovative 3D reconstruction techniques. Methods Oceanography 15–16:49–59
11. Lange ID, Perry CT (2020) A quick, easy and non-invasive method to quantify coral growth rates using photogrammetry and 3D model comparisons. Methods Ecol Evol 11(6):714–726
12. Joyce KE, Duce S, Leahy SM, Leon J, Maier SW (2018) Principles and practice of acquiring drone-based image data in marine environments. Mar Freshw Res 70(7):952–963
13. Stone A, Hickey S, Radford B, Wakeford M (2025) Mapping emergent coral reefs: a comparison of pixel-and object-based methods. Remote Sens Ecol Cons 11(1):20–39
14. Storlazzi CD, Dartnell P, Hatcher GA, Gibbs AE (2016) End of the chain? Rugosity and fine-scale bathymetry from existing underwater digital imagery using structure-from-motion (SfM) technology. Coral Reefs 35(3):889–894

Tidal Modulation of Topographic Eddies Drives Marine Megafauna Aggregations

Harvey Cairns[1(✉)], Daniel Conley[1], Phil Hosegood[1], and Christopher Stokes[2]

[1] University of Plymouth, Plymouth, UK
`harvey.cairns@plymouth.ac.uk`
[2] Met Office, Exeter, UK

Abstract. Hanifaru Bay, in Baa Atoll, Maldives, is globally recognised for hosting the largest known feeding aggregations of reef manta rays (*Mobula alfredi*). These aggregations are spatially and temporally discrete, often associated with flood tides, and are thought to result from hydrodynamic processes that elevate zooplankton concentrations. This study investigates the role of flow-topography interactions, specifically around the small headland-like feature in the bay's inlet (the 'nodule'), in creating conditions conducive to these feeding events.

A Delft3D hydrodynamic model of Baa Atoll was developed to simulate tidal flows within the bay, with high-resolution mesh focusing on Hanifaru Bay. Calibration and validation against *in situ* observations ensured the model reliably captured tidal dynamics. Results indicate that a retentive eddy forms behind the nodule during the flood tide, consistent with the timing of manta ray feeding events. This closed-core eddy retains zooplankton, allowing for their accumulation within the bay. By contrast, during the ebb tide, flow-through conditions prevent zooplankton aggregation.

Keywords: Tidal Hydrodynamics · Topographic Eddies · Zooplankton Retention · Hydrodynamic Modelling

1 Introduction

Hanifaru Bay, in Baa Atoll, Maldives, is unique as the location of the largest known feeding aggregations of reef manta rays (*Mobula alfredi*). These aggregations are spatially and temporally discrete, confined to the small bay (approx. 700×400 m) which is surrounded by shallow reef. They are often characterised by a circular or 'cyclonic' feeding pattern, and last only a few hours. It is thought that these feeding aggregations are associated with conditions that result in locally elevated primary productivity [1].

Previous biologically focused studies have identified the tide as a key predictor of manta feeding presence [2, 3], with feeding aggregations occurring during the flood tide, and approximately 2 h into the ebb. We hypothesise that flow-topography interactions are responsible for these aggregations, with the tidal currents interacting with the unique geometry of the bay – in particular the small 'headland' type feature in the inlet referred to as the 'nodule'.

© The Author(s) 2026
C. Coelho et al. (Eds.): CD 2025, CRL 41, pp. 584–589, 2026.
https://doi.org/10.1007/978-3-032-15473-6_89

Topographic eddies, such as headland eddies, form in the wake of coastal features like headlands due to interactions with sharp topography. When driven by tidal flows, their timing can also be predictable, often recurring across tidal cycles. While such eddies have been documented in similar settings (e.g. [4–6]), they have not been previously measured or modelled in Hanifaru Bay. These recurring features are known to have biological significance [7, 8].

Despite this, the specific hydrodynamic processes which arise as a result of these conditions, promoting aggregations, remain poorly understood. This study aims to investigate these processes in Hanifaru Bay to better understand the conditions leading to manta ray feeding aggregations, contributing to the broader understanding of marine animal aggregations and their reliance on specific hydrodynamic conditions.

2 Methods

We employ a Delft-3D D-Flow FM model to simulate hydrodynamic processes within Baa Atoll, covering approximately 62×62 km Fig. 1. The model uses a flexible mesh grid, with resolution varying from 1100 m at the open ocean boundaries to 9 m in Hanifaru Bay. This mesh was built using OceanMesh2D, a MATLAB based set of scripts for triangular mesh generation [9], removing much of the manual labour involved in mesh creation using the RGFGRID tool in Delft3D FM. This approach ensures high resolution in the area of interest while maintaining computational efficiency. The bathymetry used combines a Maldives-wide remote-sensed dataset [10] with high-resolution in situ multibeam and UAV-derived bathymetry for Hanifaru Bay.

Boundary conditions include tidal elevations and current velocities from TPXO9 [11], simulating tidal propagation from the open ocean through the complex atoll. Harmonic analysis of *in situ* data showed that four tidal constituents—M2, K1, O1, and S2—accounted for 97.85% of the total tidal energy. As a result, the model was simplified to use only these constituents for forcing. The forcing was imposed through water level and velocity conditions, using an advection boundary. Advection was enforced at inflow, while a Neumann condition guided by currents within the domain governed the outflow.

The model was run for 30 days (01/09/2022–30/09/2022) to align with the operation of all in situ ADCPs for maximum data comparison. Calibration involved adjusting bed friction and eddy viscosity to optimise agreement between observed and modelled velocities and water levels. A zero-phase elliptic filter was applied to the *in situ* data to remove signals above 4 cycles per day, attenuating the influence of wind and waves, on the *in situ* data.

Model performance was evaluated using four metrics: root mean square error (RMSE, Eq. 1), normalised root mean square error (nRMSE, Eq. 2), Pearson's correlation coefficient (r, Eq. 3), and bias (Eq. 4), defined as follows:

$$RMSE = \sqrt{\frac{1}{n} \sum_{i=1}^{n} (obs_i - model_i)^2} \tag{1}$$

$$nRMSE = \frac{RMSE}{Range\ of\ observed\ values} \tag{2}$$

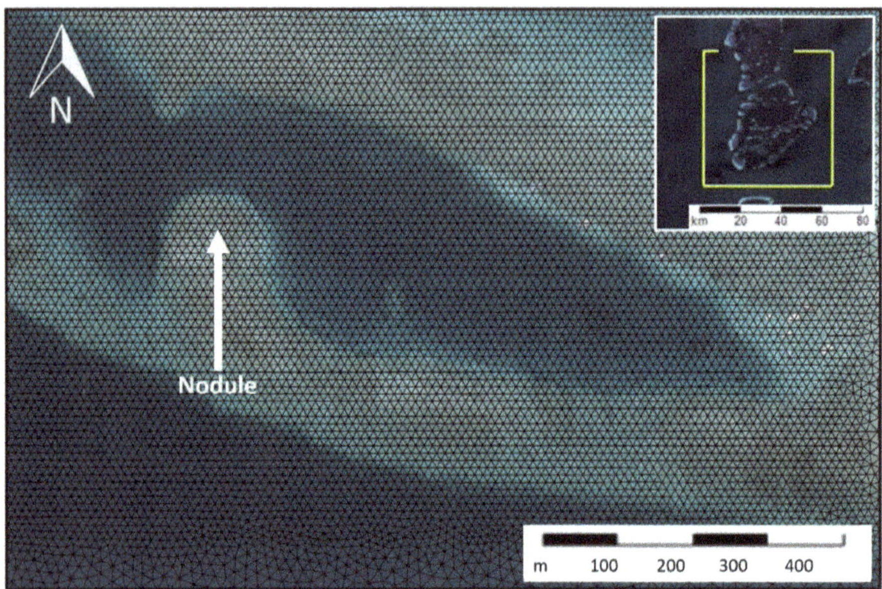

Fig. 1. Aerial view of Hanifaru Bay, with black grid indicating the computational mesh resolution. The feature known as the 'Nodule' is labelled. The yellow box in the inset map depicts the extent of the model domain.

$$r = \frac{\sum_{i=1}^{n}(model_i - \overline{model})(obs_i - \overline{obs})}{\sqrt{\sum_{i=1}^{n}(model_i - \overline{model})^2 \sum_{i=1}^{n}(obs_i - \overline{obs})^2}} \tag{3}$$

$$Bias = \frac{1}{n}\sum_{i=1}^{n}(obs_i - model_i) \tag{4}$$

3   Results

Model simulations were repeated using Manning friction coefficients from 0.023 to 0.15 and evaluated with validation metrics such as nRMSE, r, and bias. Comparisons with in-situ observations from three different locations adjacent to, and inside the bay, showed that a friction coefficient of 0.075 provided the best balance of high r and low bias, despite a slightly higher nRMSE than the 0.1 run. Thus, the 0.075 friction model was selected as the final configuration. Mean validation statistics across the locations are shown in table 1, indicating that observed velocities (water levels) were captured to within ±16% (±7%) with low bias. However, some of the velocity phasing from the complex tidal signal was missed (indicated by the low r).

Preliminary model runs suggest that an eddy forms behind the nodule on the flood tide (Fig. 2b). The streamlines suggest that this eddy is closed core (shown by the closed ring in Fig. 2b), meaning that particles may become retained within the eddy.

Table 1. Mean validation statistics averaged across all available locations.

Parameter	nRMSE (%)	r	Bias
u velocity	16	0.49	−0.05 m/s
v velocity	13	0.40	0.02 m/s
Water level	07	0.96	0.02 m

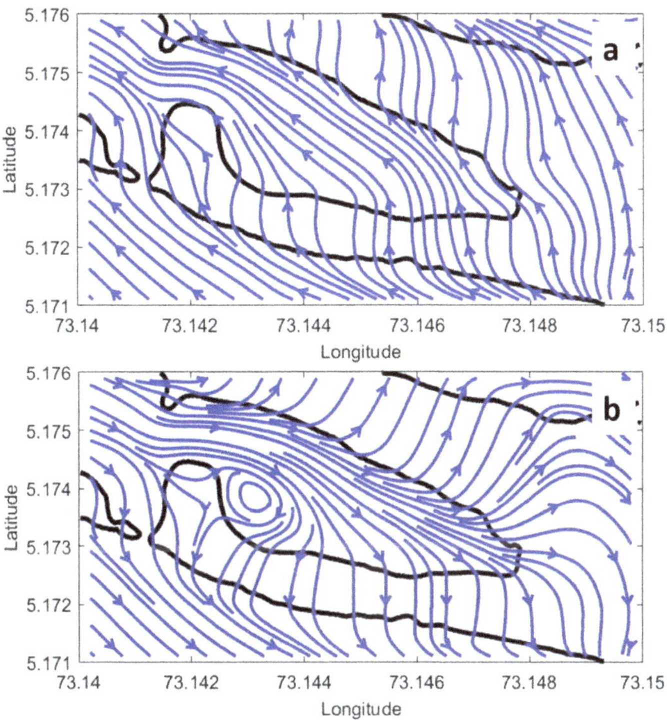

Fig. 2. a) Example circulation streamlines during the ebb tide. b) Example circulation streamlines during the flood tide. Arrows on the lines indicate direction. Black line depicts the 4m depth contour, showing the outline of the bay.

This same circulation pattern is not observed on the ebb tide (Fig. 2a). Instead, any flow that enters the bay continues to be advected through and out again, indicating that any zooplankton carried by the current would pass through the bay rather than being retained. This lack of aggregation and/or retention would likely result in a lower concentration of zooplankton, which would reduce feeding likelihood [2].

The temporal distribution of this eddy aligns with previous observations of flood-tide manta presence, as well as observations of zooplankton density. This finding supports our hypothesis that flow-topography interactions, particularly around the nodule, create conditions conducive to manta ray feeding. The closed-core structure of the eddy suggests that zooplankton entering the bay during the flood tide are retained, allowing for their accumulation within the bay. This retention mechanism likely explains the elevated zooplankton densities observed during feeding events, as retentive eddies can aggregate biological material [7, 12], and provide attractive feeding opportunities for zooplanktivorous organisms such as manta rays. In contrast, the absence of a retentive feature during the ebb tide supports the hypothesis that flow-through conditions inhibit zooplankton aggregation, thereby reducing feeding opportunities.

4 Conclusion

The identification of a retentive eddy forming behind the nodule during the flood tide provides a mechanistic explanation for the observed manta ray feeding aggregations in Hanifaru Bay. This study underscores the role of predictable, tidally driven hydrodynamic features in shaping marine ecological patterns. While explanations are provided for the high-frequency (tidal) mechanisms influencing manta presence, further investigation is needed into the mechanisms that control their seasonal presence and absence, particularly in relation to the monsoon.

References

1. Anderson RC, Adam MS, Goes JI (2011) From monsoons to mantas: seasonal distribution of Manta alfredi in the Maldives. Fish Oceanogr 20(2):104–113. https://doi.org/10.1111/J.1365-2419.2011.00571.X
2. Armstrong AO et al (2021) Reef manta rays forage on tidally driven, high density zooplankton patches in Hanifaru Bay, Maldives. PeerJ 9:e11992. https://doi.org/10.7717/PEERJ.11992/SUPP-1
3. Harris JL, Stevens GMW (2021) Environmental drivers of reef manta ray (Mobula alfredi) visitation patterns to key aggregation habitats in the Maldives. PLoS One 16(6):e0252470. https://doi.org/10.1371/journal.pone.0252470
4. Davies PA, Dakin JM, Falconer RA (1995) Eddy formation behind a coastal headland. J Coast Res 11(1):154–167
5. Pawlak G, MacCready P, Edwards KA, McCabe R (2003) Observations on the evolution of tidal vorticity at a stratified deep water headland. Geophys Res Lett 30(24):1–5. https://doi.org/10.1029/2003GL018092
6. Signell RP, Geyer WR (1991) Transient eddy formation around headlands. J Geophys Res 96(C2):2561–2575
7. Murdoch RC (1989) The effects of a headland eddy on surface macro-zooplankton assemblages North of Otago Peninsula, New Zealand. Estuar Coast Shelf Sci 29(4):361–383. https://doi.org/10.1016/0272-7714(89)90034-6
8. Russell P, Vennell R (2017) High-resolution observations of secondary circulation and tidally synchronized upwelling around a coastal headland. J Geophys Res Oceans 122(2):890–913. https://doi.org/10.1002/2016JC012117
9. Roberts KJ, Pringle WJ, Westerink JJ (2019) OceanMesh2D 1.0: MATLAB-based software for two-dimensional unstructured mesh generation in coastal ocean modeling. Geosci Model Dev 12(5):1847–1868. https://doi.org/10.5194/gmd-12-1847-2019

10. Rasheed S, Warder SC, Plancherel Y, Piggott MD (2020) An improved gridded bathymetric data set and tidal model for the maldives archipelago. Earth Space Sci 8(5):e2020EA001207. https://doi.org/10.1029/2020EA001207
11. Egbert GD, Erofeeva SY (2002) Efficient inverse modeling of barotropic ocean tides. J Atmos Ocean Technol 19(2):183–204. https://doi.org/10.1175/1520-0426(2002)019%3c0183:EIMOBO%3e2.0.CO;2
12. Hudson K et al (2021) A recirculating eddy promotes subsurface particle retention in an antarctic biological hotspot. J Geophys Res Oceans 126(11):e2021JC017304. https://doi.org/10.1029/2021JC017304

Coral Forereef Spurs and Grooves Facilitate Lagoon Outflow in Platform Reef Environments

Lachlan Perris[1,2(✉)], Ana Paula da Silva[1,2], Thomas E. Fellowes[3], Jody M. Webster[1,2], Alisha M. Thompson[4], Tristan Salles[1,2], and Ana Vila-Concejo[1,2]

[1] School of Geosciences, The University of Sydney, Sydney, Australia
lachlan.perris@sydney.edu.au
[2] Marine Studies Institute, Faculty of Science, The University of Sydney, Sydney, Australia
[3] Water Research Laboratory, School of Civil and Environmental Engineering, UNSW Sydney, Sydney, Australia
[4] Wenona School, North Sydney, NSW 2060, Australia

Abstract. The hydrodynamics of coral reefs control the transport and availability of sediments, nutrients, coral larvae and consequently, all life across coral reef systems and across all spatial and temporal scales. In this study, we investigate the hydrodynamics of a distinctive feature of forereef morphology known as spurs and grooves (SaG). Despite their presence across coral forereefs, there is limited *in-situ* hydrodynamic data of SaG, and no current data reported from the Great Barrier Reef (GBR). Here, we analyse a hydrodynamic dataset obtained in prevailing weather conditions with low wave energy over SaG at One Tree Island in the southern GBR, a platform reef with a large lagoon isolated from the open ocean during low tides. Our data show that, for the conditions measured, flow velocities in forereef SaG are controlled by tidal water levels, with greater flow velocities observed during low tides. Throughout the full tidal cycle, offshore flow occurs at the upper forereef groove, indicative of lagoonal outflow. Further investigation into the influence of these hydrodynamic conditions can assist in assessments of forereef morphological evolution.

Keywords: Coral reefs · Eco-morphodynamics · Forereef morphology

1 Introduction

On platform reef systems, lagoon systems fill and drain based on the relative control of the forereef and reef flat morphology over the open ocean hydrodynamics. Among these morphological controls, spurs and grooves (SaG) stand out as a common yet insufficiently studied feature of forereefs. Spurs are elongated, finger-like protrusions of carbonate reef structure that extend seaward from the reef flat, while grooves are channels that periodically separate the spurs [1, 2]. SaG formations are prevalent across all forms of coral reefs, such as platform reefs, fringing reefs, barrier reefs, and atolls, and their morphology varies by reef type [3, 4]. These features have been observed in coral-rich regions worldwide, including the Red Sea [5], the Atlantic Ocean [6], the

© The Author(s) 2026
C. Coelho et al. (Eds.): CD 2025, CRL 41, pp. 590–594, 2026.
https://doi.org/10.1007/978-3-032-15473-6_90

Indian Ocean [7], the Caribbean Sea [8], and the Pacific Ocean [2]. While their presence is well-documented, *in-situ* studies of hydrodynamic processes within SaG systems on platform reefs remain limited, likely due to the difficulty of accessing these shallow and highly turbulent environments.

Previous research on SaG hydrodynamics has largely focused on shore-normal waves and their role in wave energy dissipation and transformation [16, 18]. However, on platform reefs, waves often approach from multiple directions. Additionally, most studies have concentrated on SaG in wave-dominated environments, overlooking their occurrence in tidally influenced reef systems [16, 18]. This has led to limited understanding of the interactions between their morphology and hydrodynamic processes. In this study, we explore current flows obtained from an intensive field study on a forereef under oblique low wave energy conditions. By analysing current flows in context with lagoon processes we establish the controlling mechanisms of the current flows observed in SaG. Our research provides new insights into the dynamics driving SaG development in diverse reef settings.

2 Methods

2.1 Study Site

One Tree Reef (OTR) is located 84 km offshore in the southern GBR. It has a relatively large lagoon (12.7 km^2) with 65% infilled with sediments [19] that is isolated from the open ocean for 6 h of the tidal cycle [9]. The reef flat is continuous and unbroken by large tidal channels.

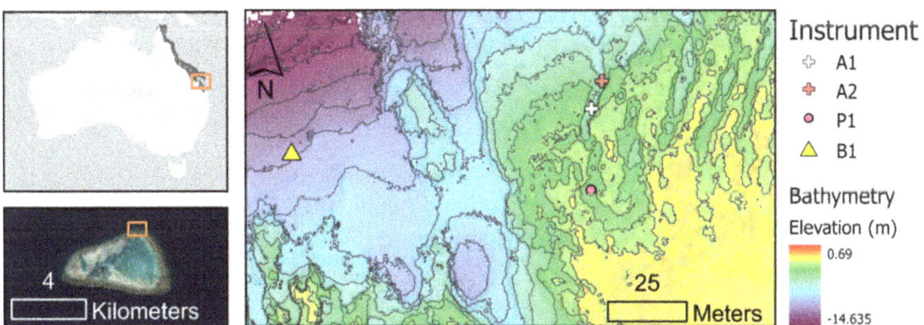

Fig. 1. Location and LiDAR derived bathymetry of the study site and instrument deployments on the northern forereef of OTI. A1 and A2 are Nortek Vector Velocimeter, P1 is a RBR Virtuoso 3 Pressure Transducer and B1 is a SOFAR Spotter wave buoy with smart mooring (RBR Conda3) (LiDAR [11]).

2.2 Data Collection

Hydrodynamic field data were collected from 6 to 8 October 2022 along a cross-shore transect within a forereef groove (Fig. 1, Table 1). Three dimensional current and pressure measurements were obtained using two Nortek Vectors (A1 and A2) and pressure with an RBR Virtuoso3 pressure transducer (P1), all set to continuously sample at a frequency of 8 Hz. Current meters, A1 and A2, were deployed with positive X-axis aligning with the groove (Fig. 1). In addition to the primary dataset, a supplementary survey was carried out to examine the hydrodynamic connectivity between the lagoon and the open ocean and to characterise representative wave conditions at the site. This additional deployment utilised a Sofar Spotter Buoy and a Smart Mooring system, equipped with a near-bed-mounted RBR Conda3, which recorded both temperature and pressure (B1).

Table 1. Instrument deployment details

Name	Deployment period	Instrument	Mean deployment depth (m)	Distance from crest (m)
A1	06/10/2022 – 08/10/2022	Nortek Vector	3.6	100.2
A2	06/10/2022 – 08/10/2022	Nortek Vector	3.3	92.4
P1	06/10/2022 – 08/10/2022	RBR Virtuoso	2.4	66.6
B1	31/10/2023 – 27/03/2024	Sofar Spotter Buoy with Smart Mooring (RBR Conda3)	10.5	97

2.3 Data Processing

Field data were processed using the Doppler Oceanography Library for Python (DOL-fYN) [12] and despiked to remove Doppler noise [13]. Wave data from pressure sensors on instruments A1, A2 and P1 were processed with a zero-crossing method [14] and wave power calculated from linear wave theory for intermediate waves.

3 Preliminary Results and Discussion

Under low wave-energy conditions (mean significant wave height ($H_{s\ mean}$) = 0.25 m), flow regimes in the groove were controlled by tidal water level, with flow velocity at low tide on average 75% higher (Fig. 2b). At high tides, the flow is less constrained to the groove axis, and currents are likely driven by surf-zone processes of oblique waves. Consequently, there is no net cross-shore flows observable in the groove at instrument A2 (furthest offshore from the reef crest), however slight net offshore flow is still observable at instrument A1 (closest to the reef crest). At low tides, with the flow constrained within the groove, the mean cross-shore current velocity was directed offshore (Fig. 2a) with a maximum velocity of 2 m/s, which is relatively high considering the low energy conditions during the study period. Notably, net offshore currents have not previously been observed in forereef SaG, and numerical models or field experiments have consistently defined onshore currents in upper forereef SaG [15, 16]. Our findings suggest that high mean offshore current velocities in the groove are linked to outflow from the lagoon (Fig. 2a).

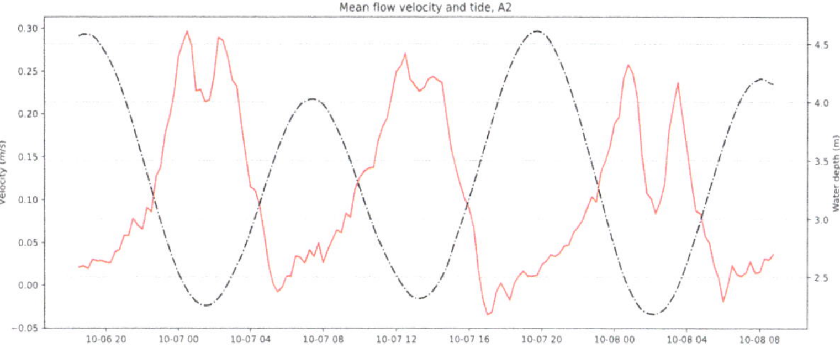

Fig. 2. Field measurements of measured water levels at instruments A1 (black dashed line) and mean offshore flow velocity (red line).

Our findings demonstrate the interconnectivity between forereef morphology and the reef platform, particularly the influence of lagoon ponding on forereef hydrodynamics. Future studies should consider the influence this exchange has on forereef platform evolution [17]. Long-term deployments that capture both fair-weather and storm conditions may assist in understanding the frequency and structure of offshore and onshore flow events. Understanding these exchanges may be key to understanding the morphological evolution of reef platforms.

References

1. Duce S, Vila-Concejo A, Hamylton SM, Webster JM, Bruce E, Beaman RJ (2016) A morphometric assessment and classification of coral reef spur and groove morphology. Geomorphology 265:68–83
2. Munk WH, Sargent MC (1948) Adjustment of Bikini Atoll to ocean waves. Trans Am Geophys Union 29(6):855
3. Cloud Jr, PE (1959) Geology of Saipan, Mariana Islands, Part 4: Submarine topography and shoal-water ecology. U.S. Geological Survey Professional Paper 280-K, 361–445
4. Schlager W, Purkis S (2015) Reticulate reef patterns - antecedent karst versus self-organization. Sedimentology 62(2):501–515
5. Sneh A, Friedman GM (1980) Spur and groove patterns on the reefs of the northern gulfs of the Red Sea. J Sediment Petrol 50(3):981–986
6. Shinn E (1963) Spur and groove formation on the Florida Reef Tract. SEPM J. Sediment. Res. 33(2):291–303
7. Sheppard C (1981) The groove and spur structures of Chagos atolls and their coral zonation. Estuar Coast Shelf Sci 12(5):549–560
8. Blanchon P, Jones B (1997) Hurricane control on shelf-edge reef architecture around Grand Cayman. Sedimentology 44(3):479–506
9. Silverman J et al (2012) Carbon turnover rates in the one tree island reef: a 40-year perspective. J Geophys. Res. Biogeosciences 117(3):1–16
10. Frith CR, Mason LB (1986) Modelling wind-driven circulation, one tree reef, southern great Barrier Reef. Coral Reefs 4(4):201–211

11. Harris DL, Webster JM, Vila-Concejo A, Duce S, Leon JX (2023) Geomorphology defining multi-scale surface roughness of a coral reef using a high-resolution LiDAR digital elevation model. Geomorphology 439:108852
12. Klise GT, et al (2020) MHKiT (Marine and Hydrokinetic Toolkit) - Python
13. Goring DG, Nikora VI (2002) Despiking acoustic Doppler velocimeter data. J Hydraul Eng 128(1):117–126
14. Karimpour A, Chen Q (2017) Wind wave analysis in depth-limited water using OCEANLYZ, a MATLAB toolbox. Comput Geosci 106:181–189
15. Roberts HH, Murray SP, Suhayda J (1977) Physical processes on fore-reef shelf environment. In: Proceedings of Third International Coral Reef Symposium, pp 507–516
16. da Silva RF, Storlazzi CD, Rogers JS, Reyns J, McCall R (2020) Modelling three-dimensional flow over spur-and-groove morphology. Coral Reefs
17. Perris LA, et al (In prep.) Field measurements of current flow in coral forereef spurs and grooves on the Southern Great Barrier Reef
18. Perris LA, Salles T, Fellowes TE, Duce S, Webster JM, Vila-Concejo A (2024) The influence of coral reef spur and groove morphology on wave energy dissipation in contrasting reef environments. J Geophys Res Earth Surf 129(8):1–17
19. Vila-Concejo H, Webster DF (2022) Lagoon infilling by coral reef sand aprons as a proxy for carbonate sediment productivity. Geology 50(12):1427–1431

Shallow Reef-Lagoon Exchange Flow on a Coral Atoll

Mathilde Lindhart[(⊠)], Gerd Masselink, Ashley Holsclaw, Tim Scott, Peter Ganderton, Timothy Poate, and Edward Robinson

School of Biological and Marine Sciences, University of Plymouth, Plymouth, UK
`mathilde.lindhart@plymouth.ac.uk`

Abstract. Coral atoll lagoons are typically connected to the bounding ocean by both deep channels and shallow reef flats. Reef flats are ubiquitous on atoll rims, interspersed between atoll islands, and flows here are primarily unidirectional; setup from wave breaking on reef flats create a lagoon-ward water level gradient that drives a flow into the lagoon. While this flow has a clear tidal variability, the magnitude is determined by the incoming wave energy and modified by the tide through depth-limited wave breaking. Results from a four-month deployment in Huvadhoo atoll in the Maldives, Indian Ocean, explore the relation between reef flat velocities and external forcing from offshore swell waves and tidal depth. Results show that tidal phase is an additional, important parameter, creating a lag between lagoon and ocean tides that opposes wave-generated setup on rising tides, and amplifies it on falling tides.

Keywords: Atoll island · Coral reef · Wave · Current · Tide

1 Introduction

Atolls are biodiversity hotspots, teeming with life and providing habitats for a myriad of species. Sitting atop these shallow atolls are reef flats, over which wave- and tidal forces work continuously to shift and move sand, sediment, and other matter. Flows across reef flats also connect shallow, warm lagoons to the cooler ocean, and create pathways for larvae and nutrients to move between exposed fore reefs and sheltered lagoon ecosystems. Thus, quantifying and understanding shallow reef flow is important at both small scales in predicting island morphodynamics, and at greater scales, estimating exchange flows between large atoll ecosystems and the surrounding ocean basins.

Studies using both theory and observations suggest that cross-reef velocities are driven by a combination of wave setup due to breaking and tidal depth [e.g., 1–3]. Based on a reef-flat (inshore of surf zone) momentum balance, the local reef velocity is given by

$$\frac{\tau_b}{\rho} = -gh\frac{\partial\overline{\eta}}{\partial x},\tag{1}$$

$$\tau_b = C_D u|u|,\tag{2}$$

© The Author(s) 2026

C. Coelho et al. (Eds.): CD 2025, CRL 41, pp. 595–600, 2026.

https://doi.org/10.1007/978-3-032-15473-6_91

where τ_b is the bed shear stress often modeled as quadratic in the mean velocity u, ρ is the density of water, g the gravitational acceleration, h the total water depth, and $\partial\bar{\eta}/\partial x$ the local water level gradient. Often, $\partial\bar{\eta}/\partial x \approx \Delta\eta_w/\Delta x$, where $\Delta\eta_w$ is assumed to be the setup due to wave breaking [1–5]. Thus, much previous work has focused on parameterizing wave setup on reefs [e.g., 5–7], although this has proven to be nontrivial, depending on, among other parameters, wave angle and period, roughness, and reef slope. Nevertheless, assuming Eq. (1) is satisfied, and the water level gradient is equivalent to wave setup, studies predict that the cross-reef velocity reaches a maximum magnitude at $h_b/h \approx 2$, where the depth at breaking is $h_b = H/\gamma$, H is the offshore wave height, γ the breaking parameter, and h the depth on the reef flat [1, 4].

2 Methods

Between April and August, 2024, an array consisting of two Nortek Acoustic Doppler Velocimeters (ADVs, C_ADV1 closest to the ocean and C_ADV2 closer to the lagoon, see Fig. 1) an array of RBR Solo pressure transducers (PTs, C1-C6), and one offshore RDI ADCP (C_ADCP) in approximately 20 m depth quantified the strength, direction, and variability of flow over a shallow reef flat and offshore wave conditions, see Table 1 and Fig. 2.

Fig. 1. Map of the shallow atoll reef flat and location near Fares-Mathooda, southern Huvadhoo (see inserted map in lower, right corner).

Table 1. Instrument deployments.

Instrument	Sampling	Name
Nortek ADVs	5 min @ 1 Hz every half hour	C_ADV1, C_ADV2
RBR Solo PTs	1 Hz continuous	C1, C2, C3, C4, C5, C6
RDI ADCP 600 kHz	10 min average currents and wave directional spectra every hour	C_ADCP

Depth on the reef flat (Fig. 2a) varies between 0–1 m, representative of an approximately 1 m tidal range in the southern Maldives. Observed current velocities in the along-channel direction were up to 0.65 m/s with a tidal dependency (Fig. 2b and Fig. 2c). Cross-channel flows were significantly weaker (up to 0.15 m/s). Velocities measured at the oceanward ADV were larger than those closer to the lagoon, however, the former of the two (C_ADV1) was mounted approximately 25 cm above the bed, compared to 15 cm at the lagoon-ward ADV (C_ADV2). Thus, part of the difference in magnitude may be attributed to C_ADV1 sampling farther from the bed, higher in the boundary layer. Offshore root-mean-square wave heights (H_{rms}) were approximately 0.5–1.5 m most of the deployment period, with some periods of up to 2.9 m (Fig. 2d). Along-channel velocity maxima coincide with maximum offshore wave heights.

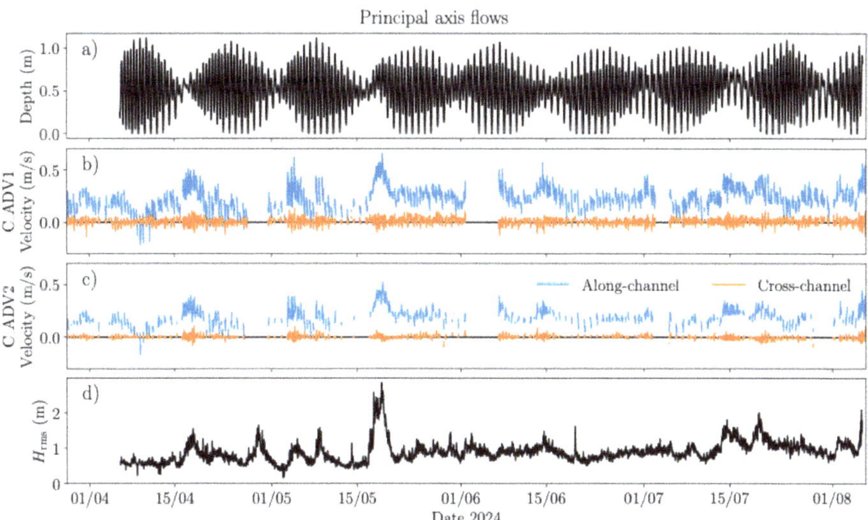

Fig. 2. Time series of observations. a) depth on the reef flat; b) and c) along-channel and across-channel velocities at C_ADV1 (b, closer to the ocean) and C_ADV2 (c, closer to the lagoon); d) offshore root-mean-square wave heights at C_ADCP.

3 Results

Preliminary results show that along-channel velocity, is related to offshore wave conditions and tidal height and phase, such that a larger offshore wave height and lower tidal water level increases wave breaking and wave setup, which increases $\Delta\eta$. Figure 3 shows this dependence of the along-channel velocity u as a function of the ratio between the depth at breaking ($h_b = H_{rms}/\gamma$, where $\gamma = 0.5$ from observations) and the depth on the reef flat (h).

Maximum velocities are predicted to occur at a ratio of $\frac{h_b}{h} \approx 2$ [1, 4]. Due to large scatter in the data, it is not clear from the results in Fig. 3 whether this is the case. Note, however, periods of negative velocities (directed from the lagoon towards the ocean) that seem to coincide with $\Delta h_L/\Delta t < 0$ (h_L is the depth in the lagoon), i.e., a falling tide and periods of small wave heights. This indicates that the phase of the water level offshore relative to the lagoon is important. Likely, this phase lag contributes a pressure gradient force that opposes wave setup on the falling tide and amplifies it on the rising tide. Thus, during periods of calm offshore wave conditions and falling tides, shallow reef flats that are often considered unidirectional become bidirectional, allowing for exchange flow along the reef flats as well as the deep channels.

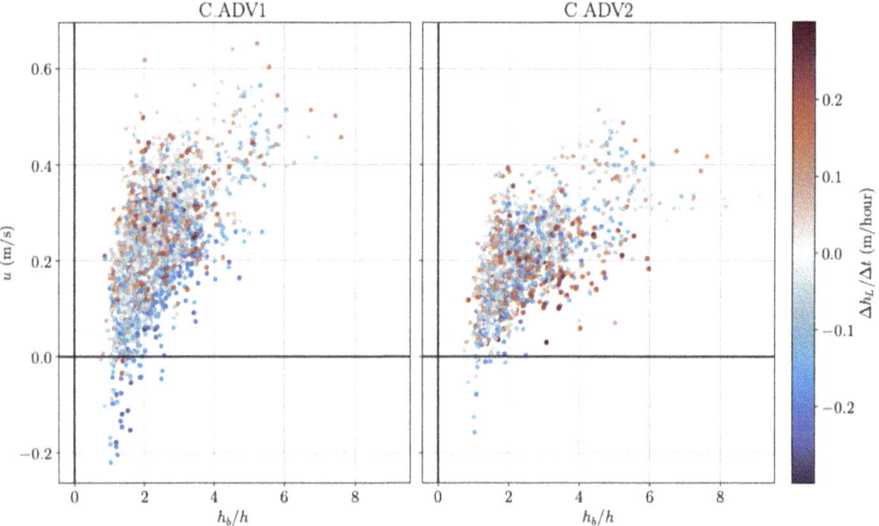

Fig. 3. Along-channel velocity as a function of the ratio of depth at breaking to reef flat depth (h_b/h). Colored according to tidal phase in the lagoon ($\Delta\eta/\Delta t$), where positive values indicate a rising tide and negative values indicate a falling tide.

To parameterize reef flow, first the local momentum balance is estimated at each ADV to 1) evaluate how the validity of applying Eq. (1), and (2) to estimate the drag coefficient C_D in Eq. (2), see Fig. 4. Here, the velocity u and depth h are measured at each ADV, and $\Delta\eta/\Delta x$ is the gradient across pressure sensors C2-C6 estimated by a least-squares, linear fit. Results show a reasonable agreement between observed velocities and local water level gradients (r^2 of 0.76 and 0.73) and drag coefficients of $C_D = 0.012$ and $C_D = 0.026$ for C_ADV1 and C_ADV2, respectively.

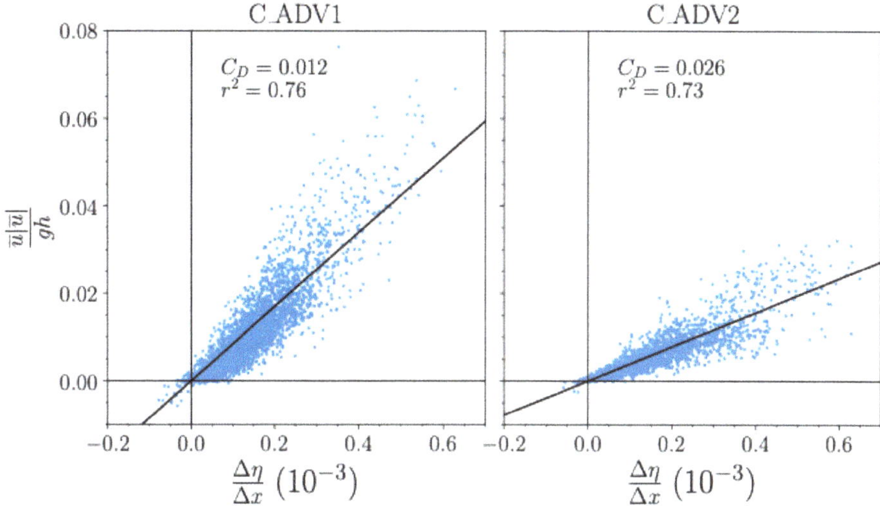

Fig. 4. Evaluation of the pressure gradient-friction momentum balance at each ADV. The slope is equivalent to the drag coefficient, and the coefficient of determination (r^2) is estimated for each deployment.

That is, locally, the velocities at each ADV can be reasonably estimated given the water surface gradient and a representative drag coefficient, C_D.

4 Discussion and Conclusion

Atoll islands accrete, erode, and transform on shallow reef flats, driven and shaped by waves and shallow flows. Understanding the relation between external forcing through tides and waves on reef flat flows enables the understanding of which conditions may allow for these island building processes to occur. It also allows insight into nutrient exchange and lagoon habitat variability through salinity and temperature changes.

This work explores the dependence on reef flat flow in southern Huvadhoo atoll in the Maldives. In particular, the importance of tidal phase lag between the lagoon and ocean is considered, showing that ebb tides and low wave conditions facilitate windows of a reversal of reef flat flow from the lagoon to the surrounding ocean. Further work will seek to parameterize channel flow through external forcing (tides and waves) and relate these in particular to: 1) periods of maximum velocity, and 2) periods of flow reversal.

Acknowledgements. This work was conducted under the UKRI-funded ARISE project 'Natural Adaptation of Atoll Islands to Sea-Level Rise Offering Opportunities for Ongoing Human Occupation' (EP/X029506/1).

References

1. Symonds G, Black KP, Young IR (1995) Wave-driven flow over shallow reefs. J Geophys Res 100:2639–2648
2. Gourlay MR (1996) Wave set-up on coral reefs. 1. Set-up and wave-generated flow on an idealised two dimensional horizontal reef. Coast Eng 27:161–193
3. Callaghan DP, Nielsen P, Cartwright N, Gourlay MR, Baldock TE (2006) Atoll lagoon flushing forced by waves. Coast Eng 53:691–704
4. Lindhart M, Rogers JS, Maticka SA, Woodson CB, Monismith SG (2021) Wave modulation of flows on open and closed reefs. J Geophys Res Oceans 126
5. Tait RJ (1972) Wave set-up on coral reefs. J Geophys Res 77:2207–2211
6. Vetter O, et al (2010) Wave setup over a Pacific Island fringing reef. J Geophys Res 115:2010JC006455
7. Becker JM, Merrifield MA, Ford M (2014) Water level effects on breaking wave setup for Pacific Island fringing reefs: water level effects on wave setup. J Geophys Res Oceans 119:914–932
8. Atkinson M, Smith SV, Stroup ED (1981) Circulation in enewetak atoll lagoon. Limnol Oceanogr 26:1074–1083

Large Scale Experiments on the Effect of Artificial Reef Restoration on Wave-Driven Flooding

Marion Tissier[1]([✉]), Vincent Takens[1], Kjell Albers[1,3], Gerd Masselink[2], Ad Reniers[1], Floortje Roelvink[2,3], Robert T. McCall[3], Samuel T. Rose[4], Damien Sous[5], Nadia Fani[6], and Samantha Haage[7]

[1] Delft University of Technology, Delft, Netherlands
m.f.s.tissier@tudelft.nl
[2] Plymouth University, Plymouth, UK
[3] Deltares, Delft, Netherlands
[4] University of Bath, Bath, UK
[5] Université de Pau et des Pays de l'Adour, Anglet, France
[6] Coastruction, Rotterdam, Netherlands
[7] Boskalis, Papendrecht, Netherlands

Abstract. This paper presents a new set of experiments conducted at near-prototype scale to investigate the impact of artificial reef flat restoration on wave transformation and runup. The experiments were carried out in the 300m-long Delta Flume, which allowed the representation of the entire fringing reef profile and sandy island at a 1: 3 scale. The reef restoration consisted of rows of complex-shaped roughness elements arranged in different layouts. In total, three configurations were tested under five irregular wave conditions and two water levels. All tests were also repeated on a bare (unrestored) reef flat to assess the impact of the restoration. Preliminary data analysis indicates that the artificial reef restoration reduced significant wave heights by up to 13% near the shore for the high-density restoration configuration and the lowest water level. This reduction in wave height was similar for both the sea-swell and infragravity wave frequency bands. Next steps are to examine how the spectral distribution of the (incident) wave energy is impacted by the reef restoration for the different configurations, relate this to flow observations within the artificial canopy and assess the impact on runup characteristics.

Keywords: Fringing reef · Reef restoration · Wave transformation · Physical modelling

1 Introduction

Atoll islands are often characterized by maximum elevations of only a few meters above mean sea-level making them extremely vulnerable to coastal flooding. Coral degradation decreases the reef roughness and elevation, and thus its ability to dissipate wave energy

© The Author(s) 2026
C. Coelho et al. (Eds.): CD 2025, CRL 41, pp. 601–606, 2026.
https://doi.org/10.1007/978-3-032-15473-6_92

and protect the coast. In this context, coral reef restoration measures, traditionally motivated by ecological reasons, are now brought forward to increase the resilience of coral islands to wave-driven flooding.

Because of costs and operational constraints, reef restorations are typically of limited spatial extent (a few 10s of meters at most in the cross-shore direction). A few studies have investigated how reef restoration design could be optimized for coastal protection [1, 2]. However, these studies often relied on numerical modelling in which the reef restoration was highly schematized [1]. For instance, wave-driven oscillatory flows within and around the (artificial) reef structures were not always represented. This could lead to different responses in terms of wave dissipation but also wave reflection, and thus to modifications of both the total amount of wave energy reaching the coast, the frequency distribution of this energy [3] and the mean water level response [4] which are all key parameters to wave-driven flooding. How the combination of these commonly neglected effects ultimately influences water levels at the reef-fronted beach is still largely unknown, and there is still limited experimental data allowing a quantification of these processes.

Here, we present large-scale experiments recently carried out in a 300 m-long wave flume (Delta Flume, Deltares) to characterize the impact of artificial reef restoration on wave transformation and wave runup. The experiments were conducted at near-prototype scale (geometric scaling factor of 1 on 3), allowing the modeling of the 3D flow around complex-shaped reef structures with minimal scale effects.

2 Experimental Set-Up

2.1 Reef Restoration Test Series

The reef restoration tests were part of a larger experimental program described in [5]. For these experiments, a 60 m-wide sandy island was built behind a 45 m-wide horizontal reef flat (see bed profile in Fig. 2c). The reef restoration tests were part of the so-called hydrodynamic test series, where the sandy island was elevated to prevent overwash and covered by a geotextile to measure runup on a permeable island while limiting morphodynamic changes.

The same wave and water level conditions were repeated with and without artificial reef structures on the reef flat. Five irregular wave conditions ($H_s = 0.5$–1.5 m and $T_p = 5.7$–9.5 s) were generated for two different water levels, corresponding to still water depths of 0.33 m and 0.67 m above the reef flat. Additionally, bichromatic wave conditions with fixed wave heights but increasing group periods were generated for the largest water level to study resonance behavior with and without artificial reef structures. The bichromatic cases are not discussed further in this contribution.

2.2 Artificial Reef Configurations

The hydrodynamic conditions were repeated for three configurations of the artificial reef restoration (called S1, S2 and S3 in the following). For S1 and S3, the restored area spread over 10m in cross-flume direction (between x = 136 and x = 146 m, with x = 0 at the wavemaker, see also Fig. 2c), and consisted exclusively of the 3D printed structures shown in Fig. 1a and 1b. These were 25 cm high concrete structures, with a base diameter of about 35 cm. The only difference between S1 and S3 was the spacing of the reef structures. For the 'low-density' configuration S1, the distance between the centers of the structures in both x- and y-direction was ~70 cm, leading to a total of 91 3D printed structures to cover the 10 m * 5 m area (Fig. 1a). For the 'high-density' configuration S3, the distance between the structures was reduced to ~50 cm from center to center, leading to the installation of 153 structures over the same area (Fig. 1b). The last configuration (S2) is an extension of the low-density configuration S1, with 11 larger structures installed in three additional rows shoreward of the main restoration area (see Fig. 1c).

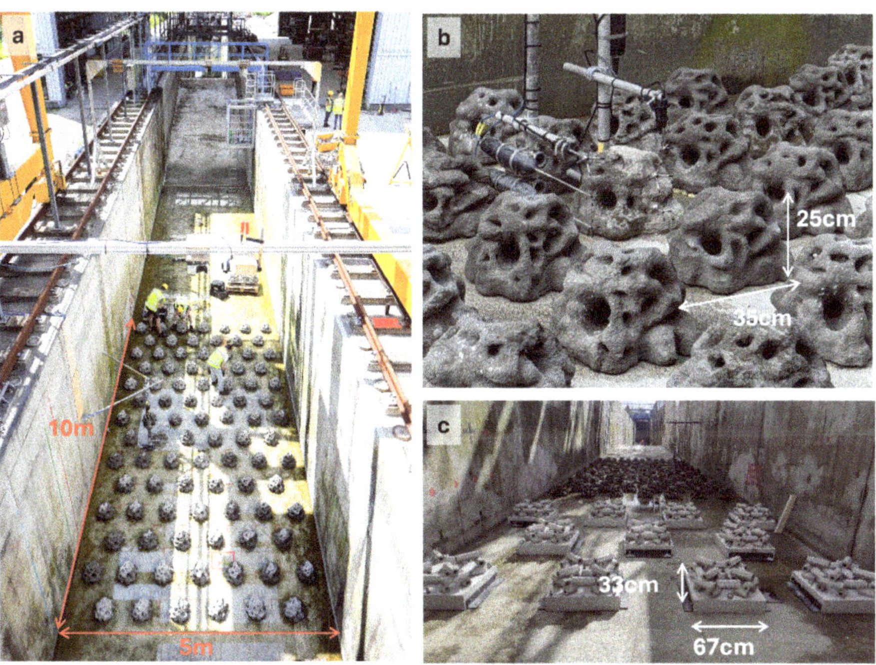

Fig. 1. (a) Top view of the reef restoration S1 (low-density configuration of the 3D printed structures designed and produced by Coastruction); (b) Zoom on the 3D printed reef structures, now deployed in the high-density configuration S3; This picture also shows the 6 acoustic doppler velocimeters deployed to measure the flow between the structures; (c) Configuration S2, where 11 additional, larger, reef structures designed by ReefSystems were added shoreward of the 3D printed structures.

2.3 Instrumentation

Wave transformation and runup was measured by a series of pressure sensors and electro-magnetic current meters as well as cameras and line-scanning lidars (see [5] for a detailed overview of the instrument setup). At 10 locations across the reef profile (including locations right before and right after the restoration), pressure sensors and electromagnetic current meters were collocated and can be used to separate incoming and reflected wave components. Two vertical arrays of 3 Acoustic Doppler Velocimeters were additionally installed during the reef restoration tests to measure the 3D flow between and right above the structures at elevations of 10, 20 and 30 cm (Fig. 1b). These velocimeters were installed around x $\approx$ 141 m, at the same y-position but about 30 cm apart in x-direction.

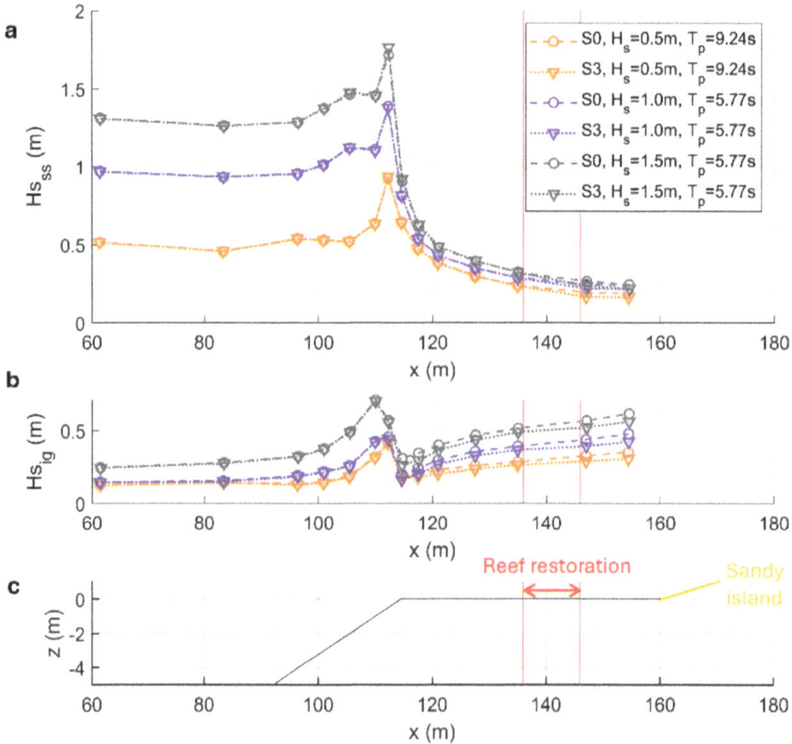

Fig. 2. Cross-shore evolution of the sea-swell significant wave height (a) and infragravity wave height (b) for 3 irregular wave conditions (one color per wave condition) for a still water depth of 33 cm above the reef flat. Both results without reef structures (circles) and with reef structures in the high-density configuration S3 (triangles) are plotted; (c) bathymetric profile with the 10m-long restoration area indicated in red.

3   First Results

Figure 2 presents a first quantification of the effect of reef restoration for the high-density configuration S3. The figure compares the cross-shore evolution of the significant wave height obtained from the pressure sensors for 3 irregular wave conditions, with and without artificial reef structures. Note that, as a first step in the analysis, the total wave height is examined, i.e. including both the incident and reflected components. The cases shown in Fig. 2 correspond to the lowest water level considered in the reef restoration experiment. The still water depth above the reef flat was then 33 cm, leading to mean water depths larger than 45 cm above the restoration for all wave conditions due to setup, such that the structures (and the ADVs) were always fully submerged.

Overall, the figure shows a relatively small but consistent effect of the reef restoration on the significant wave height evolution, for both the sea-swell frequency band (defined as $f > f_p/2$, where f_p is the offshore peak frequency) and the infragravity wave band (defined as $f < f_p/2$). For the sea-swell wave height (Fig. 2a), differences between the bare (S0) and restored (S3) reef appear for $x > 136$ m, i.e. at the sensors located shoreward of the restoration area. For the infragravity waves, the significant wave heights start differing on the outer reef already ($x > 120$ m). This is likely a result of the significant long wave reflection at the beach but needs to be investigated after decomposition of the signals between incident and reflected wave components. The reduction rate for the sea-swell significant wave height at $x = 154.5$ m, calculated as the relative difference between the measured wave height with and without restoration, ranges between 10 and 13% for the different wave conditions considered. A similar reduction rate is observed for the infragravity wave heights at the same location (9–13%). Increasing the still water depth above the reef flat to 67 cm leads to a decrease by roughly two of the wave height reduction rates (not shown).

4 Conclusion

In this paper, we presented a unique dataset allowing a systematic quantification of the effect of a localized reef flat restoration on wave transformation and runup at the back-reef beach. The experiments were performed in large-scale laboratory facilities, in which the reef profile was schematized at near-prototype scale (1: 3), significantly limiting scale effects in the representation of the interaction between (wave-induced) flow and the complex-shaped reef structures. The newly collected data can be used to improve our fundamental understanding of the interaction between waves and roughness elements, and guide further numerical model development.

Preliminary analysis of the data revealed that the significant wave height was reduced by up to 13% near the beach for the high-density configuration of the reef restoration, and that this reduction in wave height was of similar magnitude for both the sea-swell and infragravity waves. The next steps are to extend the analysis to the other reef restoration configurations and to examine how the spectral distribution of the (incident and reflected) energy is impacted by the reef restoration in these different cases. The observed spectral dissipation will furthermore be linked to flow characteristics within the artificial reef canopy. Finally, the impact on run-up will be quantified and the consequences for island evolution and coastal protection designs investigated.

Acknowledgements. We thank Chantal Willems, Arie van der Vliet, Pieter van der Gaag and Arno Doorn from the TU Delft Hydraulic Engineering Lab for their invaluable support during the design and execution phase of the experiments. We furthermore acknowledge the support of the Deltares Delta Flume team, and in particular the coordinating role of Madelief Doeleman and Suzanna Zwanenburg during the experiments. This research is supported by the TKI Delta Technology-funded CREST project (TU12) and by the UKRI-funded ARISE project (EP/X029506/1).

References

1. Roelvink F, Storlazzi C, Van Dongeren A, Pearson S (2021) Coral reef restorations can be optimized to reduce coastal flooding hazards. Front Marine Sci 8
2. Norris BK, Storlazzi CD, Pomeroy AWM, Reguero BG (2024) Optimizing infragravity wave attenuation to improve coral reef restoration design for coastal defense. J Mar Sci Eng 12:768
3. Lowe R, Falter J, Koseff J, Monismith S, Atkinson M (2007) Spectral wave flow attenuation within submerged canopies: implications for wave energy dissipation. J Geophys Res: Oceans 112
4. Buckley M, Lowe R, Hansen J, van Dongeren A, Storlazzi C (2018) Mechanisms of wave-driven water level variability on reef-fringed coastlines. JGR: Oceans 123
5. Masselink G, Roelvink F, Rose S, Tissier M, Zwanenburg S, Doeleman M (2025) Large-scale modelling of hydro- and morphodynamics associated with reef platform and island systems. In: Proceedings of coastal dynamics

Frictional Dissipation of Incident Waves Over a Spatially-Varying Rough Barrier Reef in Mayotte, Indian Ocean

Mila Geindre[1]([⊠]), Damien Sous[2], France Floc'h[1], Héloise Michaud[3], Kévin Martin[4], Marc Pezerat[3], Xavier Bertin[4], Matthieu Jeanson[5], and Aline Aubry[5]

[1] UMR 6523 Geo-Ocean, IUEM/UBO/Ifremer/IRD Plouzané, Brest, France
mila.geindre@univ-brest.fr
[2] Université de Pau Et Des Pays de L'Adour, E2S-UPPA, SIAME, Anglet, France
[3] Shom, Brest, France
[4] UMR 7266 LIENSs, CNRS-La Rochelle Université, La Rochelle, France
[5] Univ. Mayotte, Dembéni, Mayotte, France; Espace-Dev, Univ Montpellier, IRD, Univ Guyane, Univ La Réunion, Univ Antilles, Montpellier, France

Abstract. Coral reefs serve as a highly effective natural barrier against incoming ocean waves. However, climate change and human-induced degradations are threatening reefs, reducing their capacity to protect coastal populations by diminishing the frictional processes that dissipate wave energy. The most common approach to represent wave energy dissipation through bottom friction relies on the estimation of a representative roughness length, whose definition on coral environments remains to be elucidated. As a consequence, a generic parametrization for friction processes on coral reef systems is still lacking. In this study, high-resolution hydrodynamical and topographical data collected at the South-West barrier reef of Mayotte, Indian Ocean, were used to perform an estimation of wave friction. The hydraulic roughness estimated across the reef varies spatially, and attempts are made to connect this information with the diverse bed morphologies observed in the field.

Keywords: Friction · Roughness definition · Reef hydrodynamics

1 Introduction

Low-lying tropical islands are particularly exposed to coastal hazards such as tropical cyclones and storms which can lead to disastrous flooding. Those islands often exhibit coral reef systems that act as a natural barrier against ocean swell by dissipating their energy through breaking and friction processes (Lowe et al., 2005; Sous et al., 2019). In the context of climate change, which is leading to a global decline in coral reef ecosystems (Walker et al., 2024), friction processes linked to the geometric complexity of reefs will be modified (Carlot et al., 2023), potentially resulting in greater vulnerability to submersion risks. In spectral wave models, friction processes are modeled through the wave friction coefficient f_w (Jonsson, 1966) that is often related to a single length metric that does not

© The Author(s) 2026
C. Coelho et al. (Eds.): CD 2025, CRL 41, pp. 607–612, 2026.
https://doi.org/10.1007/978-3-032-15473-6_93

include bio-ecological characteristics of corals (Lowe et al., 2005, Lentz et al., 2016). However, reefs display high variability of f_w (between 0.01 up to 5) due to both varying hydrodynamic condition and geometrical structure (Monismith et al., 2015), leading to a poor representation of friction in rough context. A clear understanding of the frictional processes appears to be essential; this work aims to improve the understanding of friction and extend the link between friction and coral structural complexity based on a new comprehensive field experiments carried out in Mayotte, Indian Ocean. The experimental barrier reef is exposed to a meso-tidal regime which offers a unique opportunity to examine frictional dissipation under conditions where wave breaking can be confidently excluded. This site avoids the constraints of studying only mild waves, enabling the investigation of a broader range of orbital excursion amplitudes than typically addressed in previous studies.

2 Material and Methods

2.1 Study Site

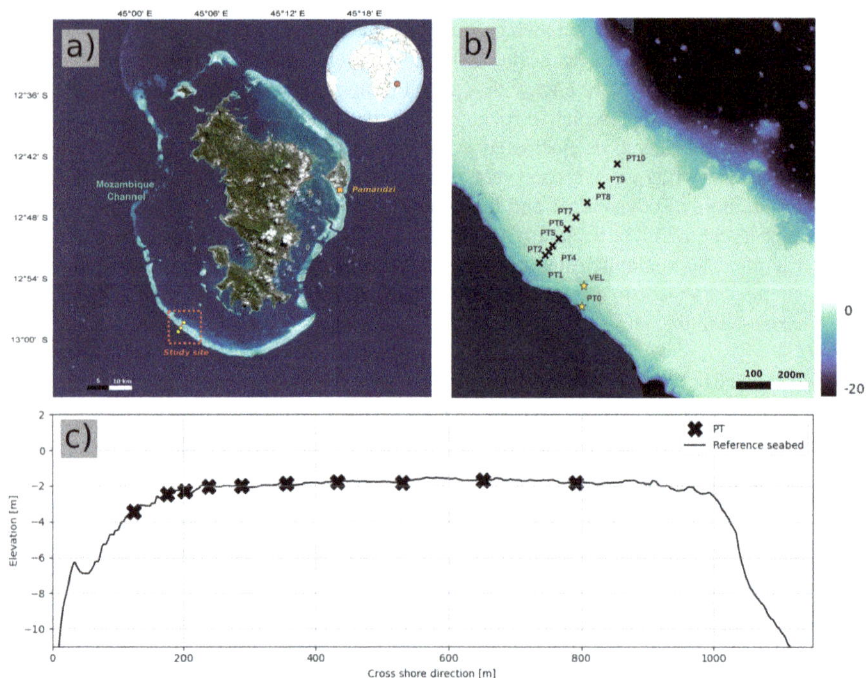

Fig. 1. Picture of Mayotte Island and barrier reef showing a) satellite view and location; b) instrument deployment and bathymetry with respect to mean sea level (MSL); c) cross section of the reference transect.

This study has been conducted on the southern barrier reef of Mayotte, France, located in the Mozambique Channel (Fig. 1a). This barrier is typical of reef systems with very rough elements or meso and macro-geomorphological formations like spur and groove. On the offshore side of the barrier reef, three distinct slopes can be identified: the lower part of the reef slope, characterized by a gradient steeper than 1:10, the upper part of the reef slope, with a slope of approximately 1:20, and the reef flat, which exhibits a slope of approximately 1:1000. The site is meso-tidal with tidal range reaching 4 m; the reef is exposed to trade winds, southern long swells and is regularly threatened by tropical cyclones.

An array of bottom-fixed pressure transducers, namely PT1 to PT10, was deployed along a SW-NE (39.4°, nautical convention) cross-barrier transect from 25-August to 1-October 2023. This transect is visible on Fig. 1.b) with corresponding cross shore elevation profile shown on Fig. 1.c). The sensors were configured to continuously monitor waves and local water depth across the reef flat (RF) at a sampling frequency of 2 Hz. An Acoustic Doppler Current Profiler (hereafter VEL) was deployed on the forereef in 5.6 m depth and configured in an hourly burst mode, recording 2,048 samples at 2 Hz. The pressure measured by the ADCP instrument was used to compare the performance of different reconstruction methods, using the signal measured by the Acoustic Sensor Transducer (AST) colocated with the pressure sensor as a reference. Incident wave conditions were measured by an additional RBR Virtuoso (hereafter PT0) deployed at roughly 8 m depth with respect to MSL.

2.2 Data Processing

Pressure data were corrected from atmospheric pressure and local depth corrections were applied (Sous et al. 2023, Marques et al. 2024). Corrected pressure timeseries were reconstructed into free surface elevation (FSE) data using the non-hydrostatic non-linear method of Martins et al. (2021), which granted best reconstruction of the higher frequency harmonics and the asymmetric shape of individual waves. In order to quantify friction over the study site, a 1D frequency-integrated energy balance model was used over pairs of adjacent sensors, assuming no breaking (only bursts where the ratio between the wave heights and the depth H_s/D is smaller than 0.25 are kept) and with consideration of non-linear transfers between triad frequencies. The wave friction coefficient f_w was calculated for each hourly burst and pair by combining discrete wave energy balance and frictional energy dissipation rate. To facilitate comparison, the resulting temporal series of median f_w were averaged on log-bins of wave orbital excursion A_b. For each pair, the associated Nikuradse roughness k_s^{opt} was estimated using the least-square optimization method of Sous et al. (2023). Spatially-averaged global $\widetilde{k_s^{opt}}$ was also computed as the median of all the local k_s^{opt}. All the roughness lengths were then compared to roughness statistics extracted from high-resolution bed elevation signal, following the methodology of Sous et al. (2024).

3   Results and Discussion

A reduction of the short wave (SW, 0.05–0.35 Hz) heights up to 92% was observed at Mayotte southern reef which thus stands as a protection against incoming waves. Reduction of SW heights was proven to be modulated by both the tide and the power of the incident waves. The barrier serves as a low-pass filter as the SW are reduced while the infra-gravity waves (IG, 0.003–0.05 Hz) can reach the mainland shore, since they either do not dissipate sufficiently or even grow across the reef flat of the outer barrier (average growth of +30% at hide tide). Non-linear energy transfers are observed from the SW towards the high frequencies mostly, thus contributing to the modification of the energy of the SW frequency band.

Figure 2 shows the estimation of wave friction factor f_w computed from observations of energy flux gradients and non-linear transfers. The bin-averaged observations are in the range 0.06–1.3, which are typical values observed in field experiments (Gerritsen et al., 1980, Péquignet et al., 2011, Monismith et al., 2015, Sous et al., 2023). Friction estimates follow global trends with decreasing f_w associated with increasing $\frac{A_b}{k_s}$ ratio.

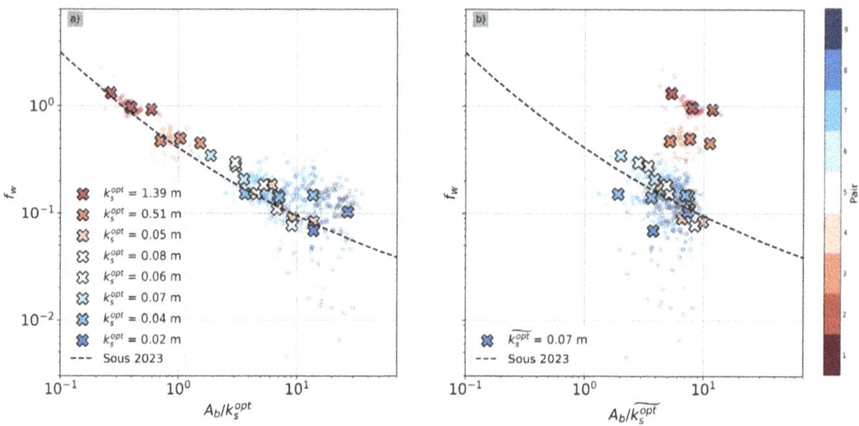

Fig. 2. Wave friction coefficient f_w computed for each pair against orbital excursion A_b and hydraulic roughness k_s^{opt} deriving from the least-square optimization of Sous et al. (2023) formulation with a) Pair-specific k_s^{opt} and b) Median $\widetilde{k_s^{opt}}$ for all pairs.

For each pair of sensors, the friction factors vary temporally due to variations in incident wave condition and tide; the range of f_w values also vary spatially due to wave attenuation and/or different roughness scales. The width of the barrier implies that spatially-integrated friction contributes from 25 up to 94% of the total dissipation (breaking included) of SW energy.

The roughness length k_s have been optimized to fit with existing parametrization of friction; the computed values range from 2 cm to 1.4 m with decreasing values from the reef crest to the lagoon. Usually, k_s is uniform for a considered site due to the lack of spatial resolution (Lentz et al., 2016). In this study, the spatial sensor coverage of the reef flat has allowed us to estimate k_s^{opt} at different location, as the geomorphology of a barrier reef exhibits spatially non-uniform coral characteristics. Optimized k_s^{opt} were greater (~ 1 m) at the beginning of the reef flat and the reef crest, coinciding with visually found coral corymbose; coral skeletons and sand were generally observed in the middle of the reef flat, leading to lower k_s^{opt} (~ 1 cm). When using global $\tilde{k}_s^{opt} = 0.07$ m, friction estimates show poor agreement to predictions, especially for pair 2 and 3 (Fig. 2b). Considering the same $\tilde{k}_s^{opt}$ for all pairs leads to diverging f_w values for similar $\frac{A_b}{k_s}$ conditions for all pairs. Those results highlight the need to consider pair-specific estimates of roughness k_s^{opt} to make observations fit empirical laws (Fig. 2a).

The optimized roughness lengths need to be compared with bed metrics computed over the high-resolution bathymetry. This ongoing work aims at verifying whether or not the law proposed by Lowe et al. (2005) linking the roughness scale to the standard deviation of the bed height applies at Mayotte site. This will offer the possibility to linkage the theoretical hydraulic roughness to other bed metrics like structural complexity (Carlot et al., 2023), spectral parameters (Sous et al., 2024) or roughness statistics (Dealbera et al., 2024).

Acknowledgments. This work was supported by FUTURISKS (ANR-22-POCE-0002), WEST (ANR-20-
CE01-0009 WEST) and Geo-Ocean lab (UMR6538). Acknowledgment to Nicolas.
Lachaussée, Yann Mercky, Shom GHOA and IES teams, DEALM Mayotte and Mayotte naval base.

References

Carlot J et al (2023) Coral reef structural complexity loss exposes coastlines to waves. Sci Rep 13(1):1683

Dealbera S, Sous D, Morichon D, Michaud H (2024) The role of roughness geometry in frictional wave dissipation. Coast Eng 189:104478

Jonsson IG (1967) Wave boundary layers and friction factors. In: Coastal engineering 1966, pp 127–148

Gerritsen F (1980) Wave attenuation and wave set-up on a coastal reef. In: Coastal engineering 1980, pp 444–461

Lentz SJ, Churchill JH, Davis KA, Farrar JT (2016) Surface gravity wave transformation across a platform coral reef in the Red Sea. J Geophys Res: Oceans 121(1):693–705

Lowe RJ, et al (2005) Spectral wave dissipation over a barrier reef. J Geophys Res: Oceans 110(C4)

Marques OB, Feddersen F, MacMahan J (2024) An effective water depth correction for pressure-based wave statistics on rough bathymetry. J Atmos Oceanic Tech 41(11):1047–1062

Martins K, Bonneton P, Lannes D, Michallet H (2021) Relation between orbital velocities, pressure, and surface elevation in nonlinear nearshore water waves. J Phys Oceanogr 51(11):3539–3556

Monismith SG, Rogers JS, Koweek D, Dunbar RB (2015) Frictional wave dissipation on a remarkably rough reef. Geophys Res Lett 42(10):4063–4071

Péquignet AC, Becker JM, Merrifield MA, Boc SJ (2011) The dissipation of wind wave energy across a fringing reef at Ipan, Guam. Coral Reefs 30:71–82

Sous D et al (2019) Wave transformation over a barrier reef. Cont Shelf Res 184:66–80

Sous D, Martins K, Tissier M, Bouchette F, Meulé S (2023) Spectral wave dissipation over a roughness-varying barrier reef. Geophys Res Lett 50(5):e2022GL102104
Sous D et al (2024) Quantifying the topographical structure of rocky and coral seabeds. PLoS ONE 19(6):e0303422
Walker AS, Kratochwill CA, van Woesik R (2024) Past disturbances and local conditions influence the recovery rates of coral reefs. Glob Change Biol 30(1):e17112

Field Observations of Island's Hydrodynamics: Insights from Lakshadweep Island, India

Rajith Kenoth[1]([envelope]), Subeesh Meethale Puthukkottu[1], Adithya Sajeevan[1], Smitha Bal Raj[2], and Jossia Joseph[3]

[1] Naval Physical and Oceanographic Laboratory, Cochin, India
rajith.npol@gov.in
[2] Centre for Marine Living Resources and Ecology, Cochin, India
[3] National Institute of Ocean Technology, Chennai, India

Abstract. The Lakshadweep Islands, situated in the South Eastern Arabian Sea, exhibit unique hydrodynamic characteristics influenced by monsoonal variations and regional oceanographic processes. This study investigates the variability in hydrography and circulation within a shallow region of the islands, focusing on Agatti Island during the transition period (March–May 2019). High-resolution observations were conducted using a moored Acoustic Doppler Current Profiler (ADCP) and buoy-deployed sensors, capturing 45 days of continuous data on wind, tide, current, and temperature profiles. Results reveal moderate wind speeds averaging below 4 m/s, predominantly northeasterly, with notable diurnal variability in air and sea surface temperatures (SST). SST rose from 28.5 °C to 30.5 °C during the observation period, correlating with increased atmospheric heating. Currents displayed moderate amplitudes below 0.5 m/s, with strong tidal currents peaking at 15 cm/s. The mixed semidiurnal tide dominated, marked by pronounced spring-neap cycles. Barotropic and baroclinic currents were identified, the latter showing vertical structures indicative of internal tides and inertial oscillations. Subsurface temperature analyses highlighted significant heat accumulation influenced by vertical mixing and baroclinic activity, posing implications for coral reef ecosystems in a warming climate. These findings highlight the intricate interplay of atmospheric and oceanographic forces shaping the region's hydrodynamics. Understanding these processes is crucial for assessing coral reef vulnerability and informing coastal management strategies in the context of global climate change.

Keywords: Acoustic Doppler Current Profiler · Barotropic and baroclinic currents · internal tides · vertical mixing

1 Introduction

The Lakshadweep Islands, located between latitudes 8°N–14°N and longitudes 71°E–74°E, form part of the northern section of the Chagos-Laccadive Ridge in the western Indian Ocean. The physical oceanography here is primarily influenced by seasonally reversing monsoon winds, which determine the circulation and hydrography of the

C. Coelho et al. (Eds.): CD 2025, CRL 41, pp. 613–621, 2026.
https://doi.org/10.1007/978-3-032-15473-6_94

region. The region experiences three distinct seasons: the northeast monsoon (December–February), a transition period (March–June), and the southwest monsoon (June–November). During the northeast monsoon, surface currents flow northward, and the upper water column is strongly influenced by low-salinity waters from the Bay of Bengal and equatorial regions. These inflows significantly affect the salinity, temperature, and nutrient dynamics in the region. As the monsoon transitions between March and June, the surface currents reverse direction, resulting in the advection of higher-salinity waters from higher latitudes.

Research on the physical oceanographic parameters of the Lakshadweep Islands is relatively limited. Jayaraman et al. (1960) conducted one of the earliest hydrographic studies of the Lakshadweep offshore waters using reversing thermometers and analytical methods. Their findings indicated the presence of isothermal water extending up to 50 m, with surface salinity reaching 35 PSU. They attributed this to the influence of Red Sea water in the Arabian Sea. Varkey et al. (1979) further explored the physical characteristics of the Lakshadweep waters. Their study involved temperature and salinity measurements along a cruise track from Kochi to Lakshadweep using bathythermographs and analytical methods. They identified a two-layer structure within the water column, with the mixed layer extending to a depth of 100 m and characterized by weak thermal gradients. Chandramohan et al. (1993) focused on the shoreline dynamics of the Lakshadweep Islands. They employed visual wave estimates from volunteer ships to assess longshore sediment transport. Their findings highlighted significant coastal erosion across various islands, primarily due to sediment removal for construction purposes.

Prakash et al. (2015) analyzed the hydrodynamics of the Lakshadweep Sea. Their study revealed that northwesterly winds predominate throughout the year, with higher speeds of 5 to 8 m/s observed during the southwest monsoon season from June to August. They reported that the wave climate is primarily influenced by the southwest monsoon, and inner-shelf currents in the region are generally weak, flowing from northeast to southwest. An observational study on the environmental variables of the coral reef ecosystems in the Lakshadweep Sea was conducted by Saravanan Kumaresan et al. (2018) between 2014 and 2015. Their investigation in Kavaratti and Agatti Islands revealed elevated levels of nutrients such as NO_2, NO_3, inorganic phosphates (IP), and reactive silicates (RS), which were attributed to anthropogenic activities in the region.

Dinakaran et al. (2022) investigated the extreme wave climate around Agatti Island using spectral wave modeling based on a 10-year hindcast dataset from 2011 to 2020. They utilized a third-generation spectral wave model to simulate the wave climate over the Arabian Sea and analyzed the spatial variability of wave power distribution around Agatti Island. Studies on coastal erosion in the Lakshadweep Islands remain sparse and are primarily based on modeling and satellite-derived imagery. Jinoj et al (2020). Examined nearshore sediment dynamics around Kavaratti Island using an integrated modeling system. Their study estimated that 83.43% of the island's coastline experiences coastal erosion. Recent research has increasingly focused on the relationship between coral bleaching and sea surface temperature (SST), as well as extreme climatic events such as El Niño-Southern Oscillation (ENSO). Arthur R. (2000) documented coral bleaching and mortality across three Indian reef regions during the 1997–1998 ENSO event, during which SSTs rose by more than 3 °C above the seasonal average, causing widespread coral bleaching. He found that bleaching-related mortality was notably high during these extreme climatic events. Idrees (2016) examined the status and changing trends of coral reefs in the Lakshadweep archipelago following the 1998 mass bleaching event. His study emphasized that SST is a critical factor influencing the health of coastal marine habitats, particularly coral reefs.

Despite the general knowledge of broad oceanographic features in the Lakshadweep Islands, our understanding of the oceanography in shallow coastal regions of Islands remains limited. This study utilized 45 days of high-resolution temperature and current data collected during the transition period (March–May) of 2019 from a shallow region of the Lakshadweep Islands. The observations are important for understanding the variability of temperature and currents in the region. Such insights are important to understand the processes driving circulation, heat accumulation to assess the vulnerability of coral reef ecosystems in the global warming scenario.

2 Area of Study and Methodology

Agatti Island, one of the 11 inhabited islands, was selected for addressing the hydrodynamic processes (Fig. 1a). Key parameters controlling these processes include wind speed and direction, temperature, current, tide, and waves. A time series measurement of current, tide, and bottom temperature was conducted using a Teledyne Marine 300 kHz Workhorse Sentinel ADCP. The ADCP was deployed upwardly using a specially designed mooring frame (Fig. 1b). The ADCP was set up to record the currents at 2 Hz at 10-min intervals over 45 days from March 15, 2019, to May 1, 2019, representing the transition period for the southwest coast of India. Additionally, air temperature, sea surface temperature, and seawater temperature at 3 m below the surface were recorded using a buoy deployed 1 km away from the ADCP site by the National Institute of Technology (NIOT), Chennai, India.

3 Results

3.1 Wind, Air Temperature, and Ocean Temperature

As a first step, the zonal (U) and meridional (V) components of wind velocity and wind stress were evaluated (Fig. 2). During the measurement period, wind speeds were moderate, ranging from 0 to 7 m/s, with an average speed of less than 4 m/s, The predominant direction was northeasterly. The maximum zonal wind component (U) reached 4.66 m/s on April 29, 2019, at 14:30, with a corresponding wind direction of 282°. The maximum meridional wind component (V) was recorded at 3.57 m/s on March 29, 2019, at 05:30, with a wind direction of 191°. The wind stress components exhibited significant variation. The zonal wind stress ranged from a maximum of 0.034 N/m^2 to a minimum of -0.064 N/m^2, with an average value of -0.01 N/m^2. The meridional wind stress had a maximum of 0.02 N/m^2 and a minimum of -0.062 N/m^2. The air temperature shows notable variability during the observation period. The maximum air temperature recorded was 30.70 °C, while the minimum was 26.88 °C. This range reflects the typical atmospheric conditions during the transition period (March–May) in the Lakshadweep

Islands. The relatively high mean air temperature highlights the potential for increased heat flux from the atmosphere to the ocean surface, contributing to elevated sea surface temperatures.

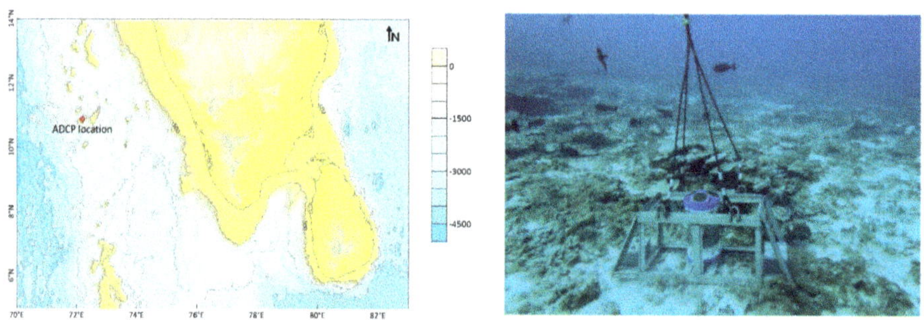

Fig. 1. a. Location of ADCP measurement, b. ADCP deployed at 30m water depth using specially designed mooring frame

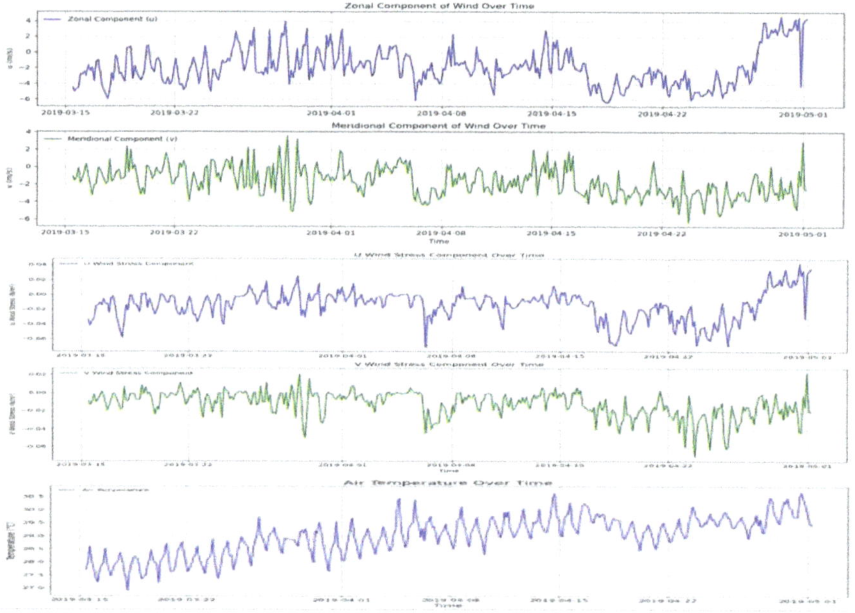

Fig. 2. The top panel displays the zonal and meridional components of wind and wind stress, while the bottom panel shows the air temperature.

The time series distribution of SST, sea water temperature at 3 m and at 30 m is analyzed. Observations indicated a continuous increase in air temperature as summer progressed, rising from 27.5 °C to 30.5 °C, over the 45 days. Similarly, the SST (skin)

increased from 28.5 °C to 30.5 °C, and the temperature at 3 m below the surface varied from 28.5 °C to 30.5 °C. Notably, the water temperature at 30 m below the surface was almost identical to the air temperature (Fig. 2). The analysis of ocean heat content at different depths reveals significant variations in thermal energy across the water column. At the surface, the ocean heat content ranged from a minimum of 1.17×10^8 J/m^2 to a maximum of 1.25×10^8 J/m^2, with an average value of 1.21×10^8 J/m^2. At a depth of 3 m, the ocean heat content increased substantially, with values ranging from a minimum of 3.54×10^8 J/m^2 to a maximum of 3.81×10^8 J/m^2, and a mean of 3.67×10^8 J/m^2. At this depth, the heat content is likely influenced by both surface interactions and the vertical mixing of water. At 30 m depth, the ocean heat content further increased, with a minimum value of 3.43×10^9 J/m^2, a maximum of 3.74×10^9 J/m^2, and a mean of 3.58×10^9 J/m^2. The increased heat content at deeper depths indicates the accumulation of thermal energy in the subsurface layers, likely influenced by long-term advection, vertical mixing, and the seasonal heat buildup from the surface. These observations are essential for understanding the distribution of heat throughout the water column and its potential impact on ocean dynamics and ecosystem responses (Figs. 3 and 4).

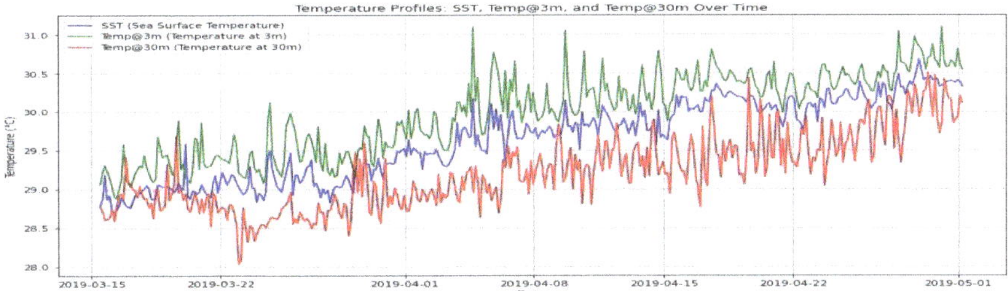

Fig. 3. Temperature evolution at surface (blue), 3 m (green) and 30 m (red) during observation period.

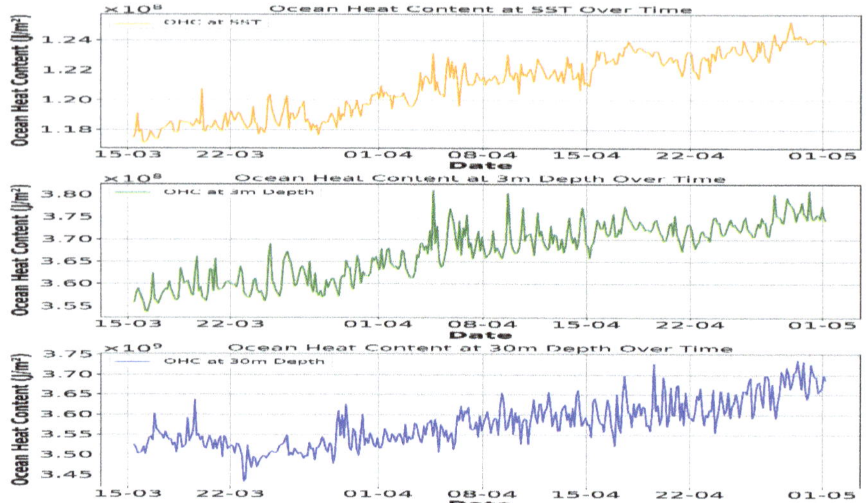

Fig. 4. Heat content at surface, 3 m and 30 m estimated from mooring temperature.

3.2 Tide

Time series measurement of current, tide, and bottom temperature was conducted using a Teledyne Marine 300 kHz Workhorse Sentinel ADCP. The ADCP was deployed upward using a specially designed mooring frame. The tide measurement was carried out at a 15 min interval. The water level fluctuation with respect to M.S.L is plotted as Fig. 5. Maximum tidal range during the measurement period was 1.4 m. The tide in the study area was a mixed semidiurnal type.

3.3 Currents

The ADCP was set up to record the currents at 2 Hz at 10-min intervals with 2m bin size. The depth at the station is 30m allowing for the mounting height of the ADCP and the blind areas on the surface and bottom, the layers that can be actually observed by the ADCP are limited between 2 and 28 m. During post processing, two other parameters, side lob and disturbances due to surface reflection was considered and first and last two bins were excluded. The currents were vertically averaged to obtain the barotropic currents. Baroclinic currents were derived by subtracting the barotropic currents from the total currents. Tidal components in both barotropic and baroclinic currents were separated using harmonic analysis. Figure 6 presents the residual currents at 4 m, with tides removed. The eastward current component is significantly stronger than the northward component, with the maximum eastward current reaching 30 cm/s near the surface. Figure 6b illustrates the U and V components of barotropic tidal currents over the observation period. A clear spring-neap cycle is evident in both the U and V components, with the maximum velocity observed in the U component, which reaches approximately 15 cm/s during spring tide. M2 is the dominant tide, followed by S2 and K1 (not shown).

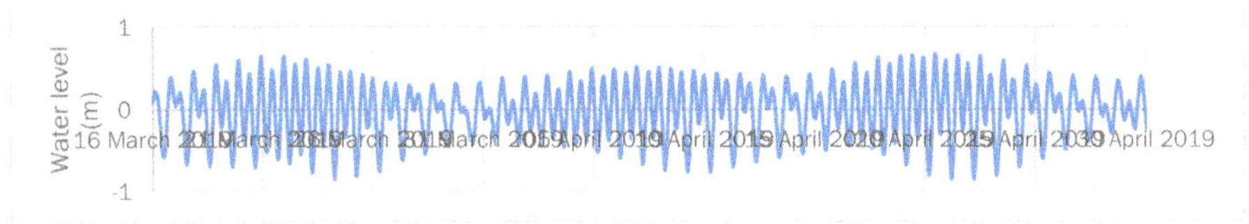

Fig. 5. Water level fluctuation with respect to M.S.L

The residual U and V currents exhibit similar vertical structures, with the daily-averaged currents being predominantly barotropic (Fig. 7). Figure 8 presents the baroclinic currents, which may include internal tides and other baroclinic flow features. The baroclinic currents are particularly strong in the U component, with velocities ranging from 6 to 18 cm/s, while the V component remains relatively weak. A clear intensification of baroclinic currents in the upper water column is observed. The daily variability is likely due to tidal oscillations.

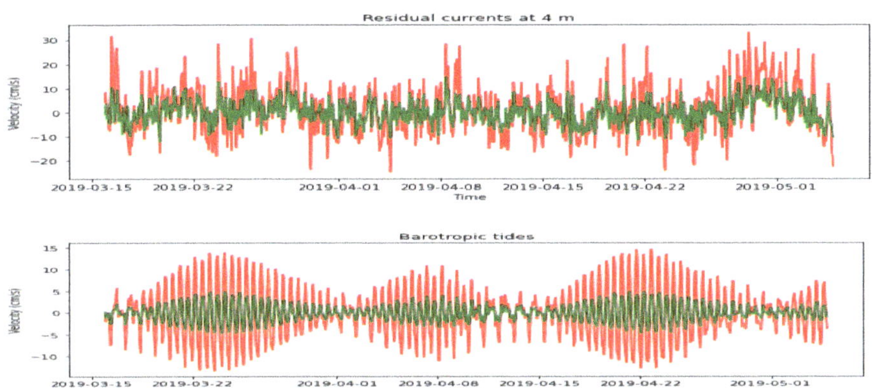

Fig. 6. Zonal (red) and meridional (green) components of (a) residual currents at 4 m (b) barotropic tidal currents.

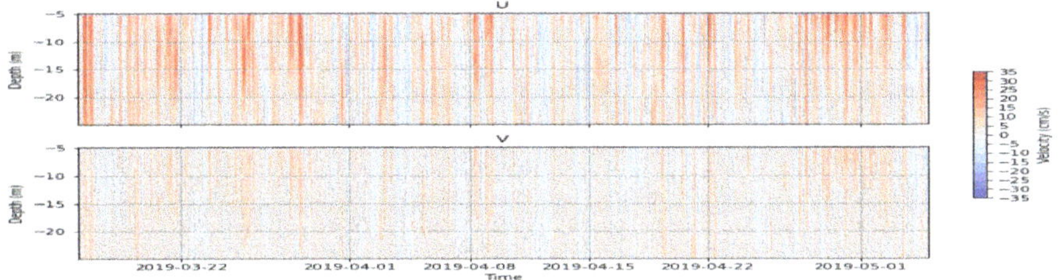

Fig. 7. Residual currents of zonal and meridional components of currents

4   Summary and Conclusion

We analyzed 45 days of moored observations of temperature and currents to investigate variability in hydrography and circulation in a shallow region of the Lakshadweep Islands. Atmospheric parameters, including wind speed and air temperature, were also utilized alongside oceanographic data. The basic characteristics of wind and currents during the observation period are presented.

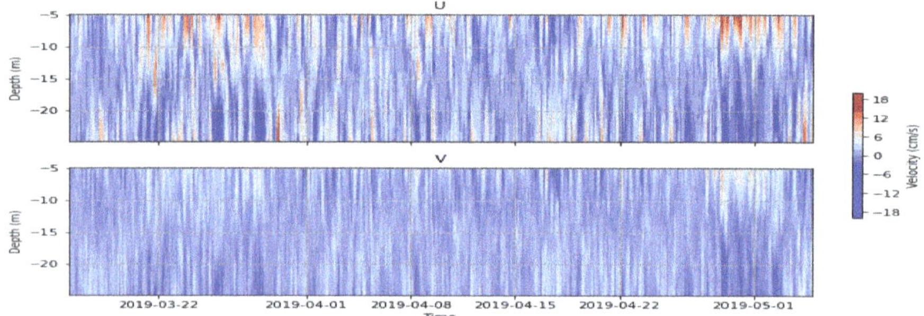

Fig. 8. (a-b) Baroclinic currents for U and V components.

The wind was predominantly northeasterly, aligning with the large-scale wind patterns typical of the transition period. Temperature observations revealed notable diurnal variability of approximately 1 °C, reflected in both air and sea temperatures. An upward trend in temperature was observed from March to May, coinciding with increasing air temperature. Notably, warm waters were present in the subsurface, likely due to the downward mixing of surface waters driven by strong baroclinic currents in the region.

Currents in the study area were moderate, with amplitudes below 0.5 m/s. Tidal currents were relatively strong, with amplitudes reaching up to 15 cm/s and significant spring-neap variability dominated by M2 tidal constituents. Baroclinic currents were also prominent, exhibiting varying vertical structures that suggest the presence of internal tides, inertial oscillations, and other internal wave processes. However, a detailed frequency-specific analysis is beyond the scope of this study.

Acknowledgement. Authors wish to express our deep sense of gratitude to Director, NPOL, Director CMLRE and Director, NIOT for their constant encouragement and for the facilities extended for the research work.

References

Arthur R (2000) Coral bleaching and mortality in three Indian reef regions during an El Niño southern oscillation event. Curr Sci 79(12):1723–1729

Chandramohan P, Anand NM, Nayak BV (1993) Shoreline dynamics of the Lakshadweep islands. Indian J Marine Sci 22:198–202

Dinakaran S, Alluri K, Joseph K, Murthy M, Venkatesan R (2022) Modelling and simulation of extreme wave heights around agatti island of lakshadweep, west coast of India. Front Built Environ 8:991768. https://doi.org/10.3389/fbuil.2022.991768

Idrees Babu KK, Kumar SS (2016) Status and changing trends of coral reefs in Lakshadweep archipelago after 1998 mass bleaching event–long term monitoring survey. Int J Appl Pure Sci Agric 163–175

Jayaraman R, Ramamirtham CP, Sundararamam KV, Nair CPA (1960) Hydrography of the Laccadives offshore waters. J Marine Biol Assoc India 2(1):24–34

Jinoj TP, Bonthu S, Robin RS, Babu KK, Purvaja R, Ramesh R (2020) Nearshore sediment dynamics of Kavaratti Island, Lakshadweep archipelago using integrated modelling system. Indian J Geo-Marine Sci 49:845–857

Prakash TN, Sheela Nair L, Shahul Hameed TS (2015) Hydrodynamics of lakshadweep sea. In: Geomorphology and physical oceanography of the lakshadweep coral islands in the Indian ocean. Springer briefs in earth sciences. Springer, Cham. https://doi.org/10.1007/978-3-319-12367-7_2

Kumaresan S, Shekhar S, Chakraborty S, Sundaramanickam A, Kuly N (2018) Environmental variables and nutrients in selected islands of Lakshadweep Sea. Address Coral Bleach Regional Stud Marine Sci 22:38–48

Varkey MJ, Kesava-Das V, Rama-Raju DV (1979) Physical characteristics of the Laccadive Sea (Lakshadweep). Indian J Marine Sci 8:203–210

Measurements of Wave Runup on an Atoll Island Using LiDAR

Samuel T. Rose[1]([✉]), Chris E. Blenkinsopp[1], Gerd Masselink[2], Ian L. Turner[3], Kévin Martin[4], and Curt D. Storlazzi[5]

[1] University of Bath, Bath, UK
str24@bath.ac.uk
[2] University of Plymouth, Plymouth, UK
[3] University of New South Wales, Sydney, Australia
[4] UMR 7266 LIENSs, CNRS - La Rochelle University, La Rochelle, France
[5] United States Geological Survey, Santa Cruz, USA

Abstract. Coral atoll islands are highly susceptible to flooding and overwash due to their low-lying nature and the impacts of climate change. This study presents the first long-term, high-resolution field dataset of wave runup on a coral atoll island, collected over 6 weeks using a shore-mounted LiDAR scanner. The LiDAR data captured swash dynamics on a steep coral rubble beach fronted by a conglomerate platform. Results demonstrate the limitations of depth-based swash extraction methods, particularly for thin swash events, where depth thresholds lead to significant underprediction. Further analyses show that low-frequency oscillations (infragravity and very-low-frequency bands) and high-frequency components (transitional bores) dominate the spectra on the reef platform and are reflected in the swash signal. The displayed swash spectra highlight the critical role of low-frequency oscillations in extreme runup events, emphasizing their contribution to island flooding and overwash.

Keywords: Atoll Islands · Wave-Runup · LiDAR · Swash

1 Introduction

Coral atoll islands are wave-formed accumulations of carbonate sediment on top of a reef platform that are low-lying, often less than 2 m above mean sea level at their highest point. Consequently, they are extremely susceptible to flooding and island overwash. Climate change will only exacerbate this problem, with predicted sea-level rise (SLR), coupled with the potential for an increase in the frequency and/or magnitude of extreme storm wave events. Additionally, coral degradation due to rising sea temperatures may decrease the energy dissipation potential of the reef platform and further escalate the magnitude, severity, and rate of island flooding. As such, atoll islands are considered one of the most vulnerable environments to climate change across the world [1].

The reefs fronting coral atolls dissipate large amounts of energy, with incident seaswell (SS, periods < 25 s) wave heights typically reduced by more than 90%, depending

© The Author(s) 2026
C. Coelho et al. (Eds.): CD 2025, CRL 41, pp. 622–628, 2026.
https://doi.org/10.1007/978-3-032-15473-6_95

on the tide level. However, substantial SS wave energy can shift to lower frequencies across the reef, with infragravity (IG, periods 25–250 s) and very-low frequencies (VLF, periods > 250 s) dominating the nearshore spectra on the reef flat. While previous studies have examined how wave forcing mechanisms affect wave runup on atoll islands, this work has generally relied on hydrodynamic models, with long-term field observations of wave runup in these environments currently lacking.

2 Methodology

This study presents a 6-week continuous runup dataset, collected between March and May 2024 using a shore-mounted LiDAR scanner sampling at 10 Hz on a coral atoll island. These data were obtained as part of a 6-month deployment on Dighelaabaadhoo, Maldives, alongside 25 other oceanographic instruments that were deployed to fully explore the complex hydrodynamic processes on reef platforms. The cross-shore transect explored in this paper included an array of 6 pressure transducers and a velocimeter located on the fore reef and reef flat (Fig. 1).

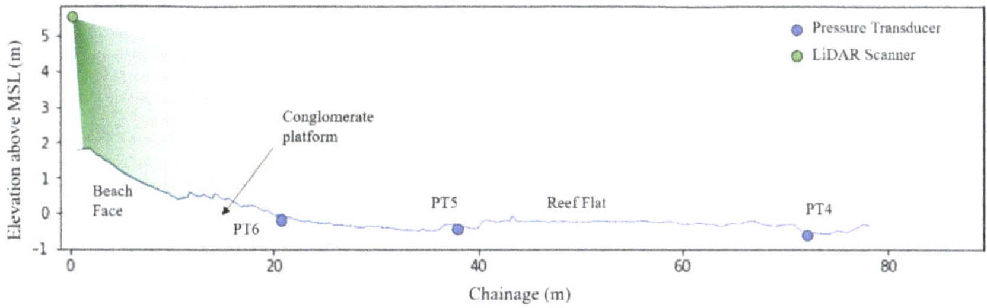

Fig. 1. (Left) Cross shore LiDAR transect showing the three most shoreward pressure transducers and Livox Avia LiDAR scanning a 70.4° angle continuously at 10 Hz.

The LiDAR transect captured a short, steep coral composite beach, fronted by a conglomerate platform, with an adjacent intertidal reef flat extending c.90 m to the reef crest. A Livox Avia LiDAR sensor was mounted atop a 4-m-tall tower, situated 1 m landward of the beach crest. The 3D data collected was mapped onto a 0.1 m regular grid using 4D interpolation for increased accuracy and to minimize any potential gaps in the data. Runup data was extracted using the intensity threshold method outlined by Hofland [2]. Due to the presence of the conglomerate platform (Fig. 1), the water depth across the inner reef flat must exceed 0.34 m above mean sea level (MSL) for swash to reach the beach face. A total of 169 datasets, each 34 min in duration, were analysed during periods when runup occurred on the beach face, corresponding to the pressure transducer sampling bursts.

3 Results

Due to the nature of the thin swash and coarse coral rubble beach, depth-based extraction methods were found to be unable to capture the full extent of the swash limit. The exceptionally low water turbidity creates a significant contrast in return intensity between sand and water. This makes the environment particularly well-suited for extracting the swash limit using the received signal strength indicator (RSSI) measured by the LiDAR. Here the swash limit is obtained using RSSI < 18. Figure 2 compares the results obtained using a signal intensity threshold with those extracted using the more common depth threshold approach for a range of depths between 0.0025 and 0.1 m.

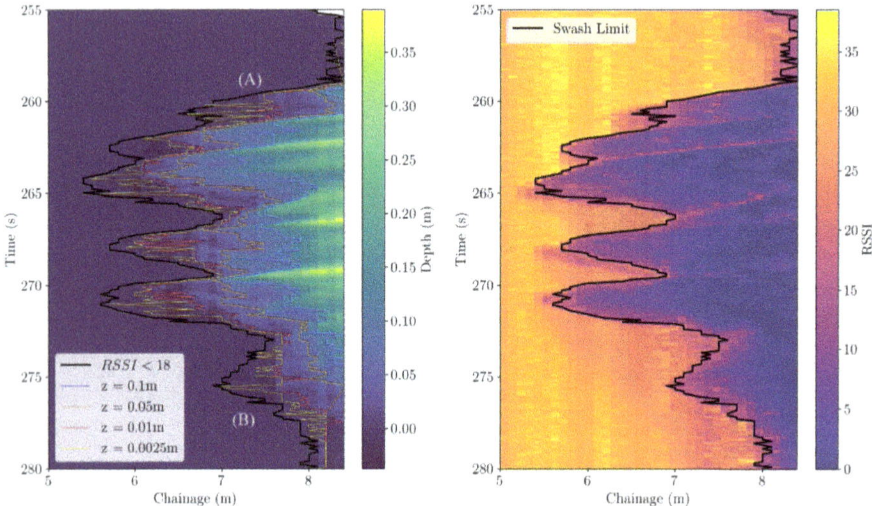

Fig. 2. (*Left*) Time stack of water depth with intensity derived swash limit (black) alongside the Water depth contours between 0.0025 to 0.1 m. The depth method significantly underpredicts the runup extents for all thresholds with smaller thresholds leading more accurate but increasingly noisy runup limit. (*A, B*) show swash events that have been significantly underpredicted. (*Right*) RSSI time stack of the same time series with intensity-derived swash limit.

Even with a small depth threshold of 1-2cm, the value commonly cited in other literature, the results show that there is a significant underprediction of the swash limit and irregularity in the swash tip trajectory. As this depth threshold tends to zero, the LiDAR intermittently penetrates the thin film of water measuring the beam returning from the bed as opposed to the water surface resulting in a swash limit with significant noise. As seen in Fig. 2 (*Right*), this thin film significantly reduces the RSSI leading to a defined limit of the swash extent. These effects are especially prominent on smaller runup events with the intensity method performing significantly better as the depth method underpredicts events (*A*), and (*B*) by as much as 50% (see Fig. 2 (*Left*)).

Due to the differing wave dynamics and presence of the conglomerate platform at the beach toe, the swash signal differs significantly from those observed on sandy beaches. Figure 3 displays a sample time series of swash depths over the beachface. The conglomerate platform traps much of the rundown resulting in significant pooling at the beach toe which then flows alongshore through a crack in the conglomerate. This pooling results in the swash limit remaining almost constant moving slowly seaward as the swash flow drains (Fig. 3 (*1*) and (*2*)). Another key feature of the swash signal is the regular pattern of high-frequency oscillations that follow large runup events. The process of bore development across the platform plays a key role in the structure of the runup signal. The reef platform at the site is c.90 m wide with a nearly constant elevation. As the energy dissipates, turbulent bores generated at the breakpoint transform into transitional bores with trailing wavelets. Due to nonlinear and dispersive effects, these wavelets spread out into a series of smaller, more regular oscillations. With each oscillation, both amplitude and wavelength decrease as the distance from the wavefront increases [3]. This presents in the runup signal as a series of higher frequency oscillations reducing in magnitude following a larger bore.

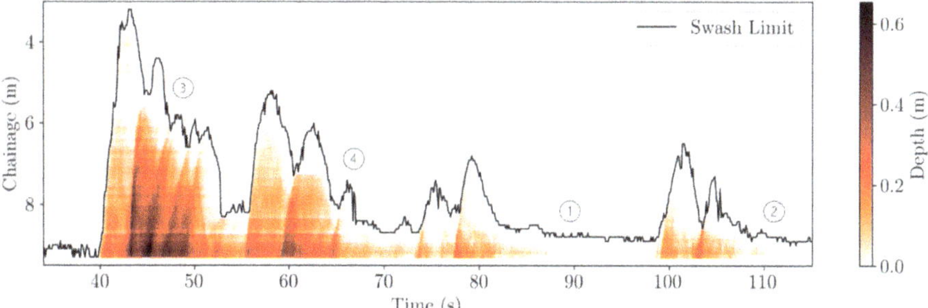

Fig. 3. Typical Runup signal extracted from 4th May 2024 with a H_{m0} = 1.78 m, T_p = 17.7 s. *(1–2)* Indicating water trapped by the conglomerate beach rock slowly draining. *(3–4)* Transitional bores are responsible for the consistent structures present in the swash signal.

Figure 4 *(top)* displays the spectra of all 169 offshore incident wave conditions, with the average spectrum shown in blue. The offshore spectrum, measured by a PT located 10.54 m below MSL, is characterized by short-period waves, with the majority of energy concentrated in the sea-swell band (SS, periods < 25 s) around a clear peak frequency. There is almost no infragravity (IG, periods 25–250 s) and very-low frequency (VLF, periods > 250 s) components. In comparison to Fig. 4 (bottom), measured at PT 5 in a depth of 0.67 m below MSL, a significant portion of the energy is dissipated during wave breaking. However, much of this energy is redistributed, with the spectrum dominated by IG and VLF bands as low-frequency waves are generated. It is also shown that significant energy is shifted to higher frequencies. The average spectrum has a defined peak in the SS band (f_{p1} : 0.048 Hz) with the presence of higher harmonics of this frequency seen at ($2f_{p1}$: 0.090 Hz, $3f_{p1}$, 0.139 Hz). The generation of the higher harmonics on the reef platform in Fig. 4 will be investigated further as part of ongoing research.

Swash spectra for all 169 datasets are shown in Fig. 5, with saturation occurring in the sea-swell band between 0.1 and 0.5 Hz. While some previous literature has suggested a f^{-3} energy decay, all spectra show a f^{-4} energy roll-off at higher frequencies supporting modelling by Peláez-Zapata on a reef environment [4] and measurements on a range of dissipative–reflective beaches [5]. The low-frequency bands exhibit the greatest variability highlighting that the IG and VLF components of the swash are unsaturated and highly dependent on incident conditions.

Figure 5 (Left) shows the spectra in relation to the wave forcing parameter, $W_0 = \sqrt{H_{m0}L_0}$. For smaller values of W_0 the spectra resemble that of a dissipative beach [5],

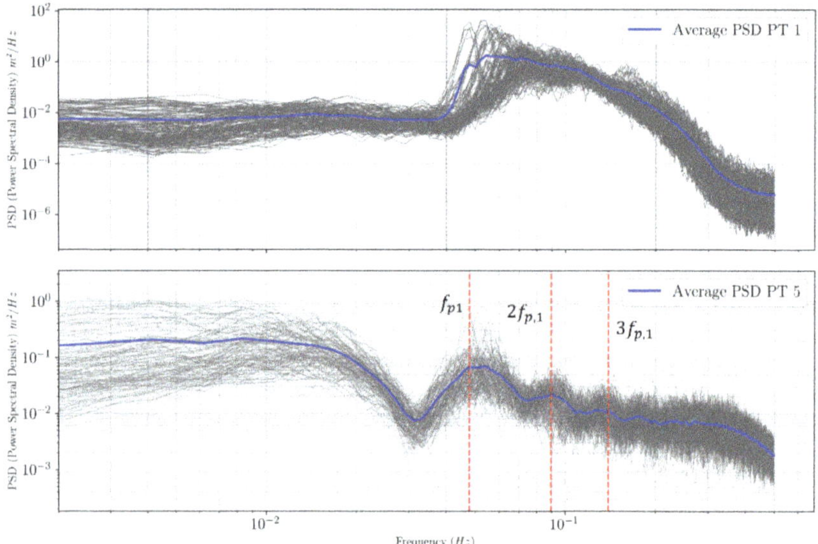

Fig. 4. (*Top*) PSD of the most offshore PT at a depth of 10.54 m below MSL and (*bottom*) the PT located on the reef platform PT5 at a depth of 0.67 m below MSL.

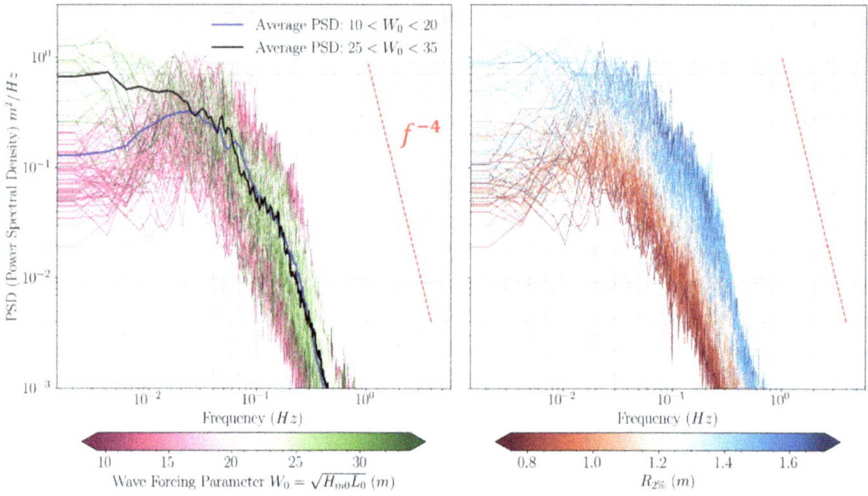

Fig. 5. (*Left*) Swash spectra compared to the wave forcing parameter (W_0), showing a shift from a typical dissipative beach spectrum to low-frequency dominance as wave energy increases. (*Right*) Swash spectra compared to $R_{2\%}$, highlighting that extreme events are dominated by low-frequency oscillations.

with a broad peak situated inside the low-frequency band. However, there is a small defined peak at $f_{p,s}$: 0.052 Hz in the SS band matching a peak frequency in both the offshore spectrum and reef flat (Fig. 4). In contrast, for higher values of W_0, the average

PSD displays no defined peak but a significant increase in low-frequency oscillations. These results are comparable to those obtained by the modelling study of Pelaéz-Zapata [4] and may be explained due to increased IG wave generation at the breakpoint.

These low-frequency dominated swash spectra are similar to those previously obtained for sandy beaches under high-energy, long-period incident wave conditions [6] and appear to contribute substantially to wave runup as seen in Fig. 5 (*Right*). As expected, swash spectra with higher energy across all bands resulted in a higher $R_{2\%}$, however, it can also be observed that the most extreme runup events exhibit a spectral shape dominated by low frequencies. While this relationship shows some variability, indicating that extreme runup is influenced by additional parameters, it emphasizes the importance of low-frequency oscillations in extreme swash events.

4 Conclusion

This paper presents the initial analysis of the first long-term high-resolution field runup dataset collected in a reef environment. The study highlights the significant differences between swash signals on sandy beaches and in this environment, examining the complex wave dynamics that lead to unique features such as bore decomposition and interactions with the conglomerate platform. The results emphasize the unique nature of low-frequency-dominated swash spectra, showing a correlation with extreme runup events. This supports the crucial role of low-frequency oscillations in island flooding and overwash.

References

1. Storlazzi C, Elias E, Berkowitz P (2015) Many atolls may be uninhabitable within decades due to climate change. Sci Rep 5:14546
2. Hofland B, Diamantidou E, van Steeg P, Meys P (2015) Wave runup and wave overtopping measurements using a laser scanner. Coast Eng 106:20–29
3. Gallagher B (1972) Some qualitative aspects of nonlinear wave radiation in a surf zone. Geophys Fluid Dyn 3(4):347–354
4. Pelaéz-Zapata DS, Montoya RD, Osorio AF (2018) Numerical study of run-up oscillations over fringing reefs. J Coastal Res 34(5):1065–1079
5. Hughes MG, Aagaard T, Baldock TE, Power HE (2014) Spectral signatures for swash on reflective, intermediate, and dissipative beaches. Mar Geol 355:88–97
6. Fiedler JW, Brodie KL, McNinch JE, Guza RT (2015) Observations of runup and energy flux on a low-slope beach with high-energy, long-period ocean swell. Geophys Res Lett 42:9933–9941

Morphodynamics and Mobility of Coral Rubble Tracts: Huvadhu Atoll, Maldives

Tim Scott[✉], Aitana Gea-Neuhaus, Gerd Masselink, Mathilde Lindhart, and Joseff Saunders

Coastal Processes Research Group, University of Plymouth, Plymouth, UK
timothy.scott@plymouth.ac.uk

Abstract. Transport of sand and coral-derived rubble (gravel/cobbles/boulders) across coral platforms could be a critical sediment supply for island maintenance under a rising sea level. While several studies have focused on cross-platform sand transport, less attention has been paid to understanding modes, mobility and rates of transport of rubble across different environments. Here we examine the distribution and morphodynamics of rubble tracts located on intertidal platforms associated with low lying atoll islands in the Huvadhu atoll, Maldives. Broad scale assessment of rubble tract distributions throughout the atoll (satellite images) identified the greatest abundance of rubble features in the high-energy SW of atoll. Local-scale morphodynamic assessment of a representative rubble tract was conducted by collecting 6 repeat RTK UAV flights over a 5-month period, combined with bed current measurement (ADV) at the tract and offshore wave observations (ADCP). Offshore significant wave heights and mean current conditions on the tract ranged from 0.48 to 3.91 m, and from -0.22 to 0.65 m/s, respectively, during the study period. DEMs of difference indicated that the studied rubble tract feature was geomorphologically stable (no significant change in total sediment budget) over a 5-month period. Although large portions (10–50%) of tract was mobile over weekly to 5-monthly timescales. These findings provide some evidence to suggest the rubble tract forms a contemporary sediment transport pathway under modal forcing conditions where material is being input from the ocean-side and exported at the lagoon-side while landform is being maintained. The key finding is that mesoscales are important for the understanding of sediment transport pathways and island maintenance processes in these environments.

Keywords: Atoll Islands · Coral Rubble · Sea-Level Rise · UAV Survey

1   Introduction

Understanding the contemporary sediment dynamics and transport pathways throughout low-lying atoll island environments is becoming increasingly important due to sea level rise and changes to the extreme wave climate. Transport of sand and coral-derived rubble (gravel/cobbles/boulders) across coral platforms could be a critical sediment supply for island maintenance under a rising sea level [1]. While several studies have focused on cross-platform sand transport [2, 3], less attention has been paid to understanding modes,

C. Coelho et al. (Eds.): CD 2025, CRL 41, pp. 629–634, 2026.
https://doi.org/10.1007/978-3-032-15473-6_96

mobility and rates of transport of rubble across different environments (e.g., One Tree Island, Great Barrier Reef; [4]).

Rubble deposits are globally ubiquitous throughout these aforementioned environments, both within the islands themselves, but also across the extensive coral platforms upon which the islands are often perched. Rubble is commonly derived from the forereef from where it is transported to the reef crest and platform under episodic high-energy conditions [5]. On the reef platform, mobilised rubble tends to occur in sheets or organize into raised tract features orthogonal to the shoreline ([6]; Fig. 1).

This study examines the morphodynamics and mobility of rubble strips observed within Huvadhu atoll, with a particular focus on a platform region between the islands of Keliahutta and Ekalonda on the exposed southern part of the atoll (Fig. 1). The research aims to establish a baseline understanding of the geomorphological behaviour and mobility of these important features across timescales of weeks to months.

2 Methodology

Local scale rubble tract dynamics are investigated through a time-series (weekly to monthly) of very high-resolution (0.5–2 cm) RTK-GNSS UAV derived ortho-rectified imagery and digital elevation models (Keliahutta platform). Six UAV datasets on an individual rubble tract were collected during field experiments between March and August 2024 (Fig. 1). Surveys were flown using a DJI Phantom 4 RTK UAV at an altitude of 25 m with image overlaps of 80% in along- and cross-track directions. Four GCPs (ground control points) were used for each flight for Structure-from-Motion solution optimization and as independent validation. GCPs were surveyed using a Trimble R10 RTK- GNSS. A local RTK reference station was established (NTRIP via 4G) with 2 km of the site. MSL was established through a 6-month tide gauge deployment in Fares-Maathodaa harbour. DEMs were computed from UAV images using SfM/MVS techniques using Agisoft MetaShape Pro, with final point cloud densities on the order of 0.01 m.

DEMs of difference (DoDs) were computed between each epoch and end members to examine geomorphological change and total mobility across the rubble tract. The rubble tracts under examination are intertidal in nature experiencing a maximum tidal range of ~ 1 m. The rubble strips are fully exposed on the lowest spring tides ($<$ -0.4 m relative to Mean Sea Level (MSL)); hence, all UAV flights were conducted during spring low tides. To avoid impact of errors due to refraction, in this analysis DEMs where thresholded above the observed water level of each flight and DoDs were computed from a combined common areal coverage for all epochs. To account for measurement uncertainty, a Limit of Detection (LoD) approach was taken to change detection. Observed vertical checkpoint errors from photogrammetry were 0.005 m, and RTK GNSS GCP observations were 0.025 m (from specification) therefore LoD was set at 0.03 m for this analysis.

These morphological observations were combined with current observations across the platform at the rubble strip location providing important insights into the environmental conditions. 3D current measurements were collected 0.25 m above the bed using a Nortek Acoustic Doppler Velocimeter (ADV) sampling at 1 Hz for 5 min every 30 min for 5 months (Fig. 1).

3 Results

Broad scale (atoll) rubble strip distribution within the context of wave climate was analysed through morpho-sedimentary classification of satellite imagery and extreme wave climate data [7]. Initial findings indicate that rubble tract presence/abundance appears linked to extreme wave exposure throughout the studied atoll with significant geomorphological change being episodic, likely linked to extreme wave/water level events. Qualitative classification of the rubble strip presence, scale and geometry within the Huvadhu atoll indicated that the highest abundance of (and largest scale) rubble features are associated with the SW-facing section of the atoll, exposed to the largest long-period swell wave events monsoon (Fig. 1). All atoll regions exposed to high energy SW to SE waves have abundant rubble features that form regular tract landforms. There was a significant reduction in abundance of rubble deposits in the most sheltered parts of the oceanward atoll rim (NE facing).

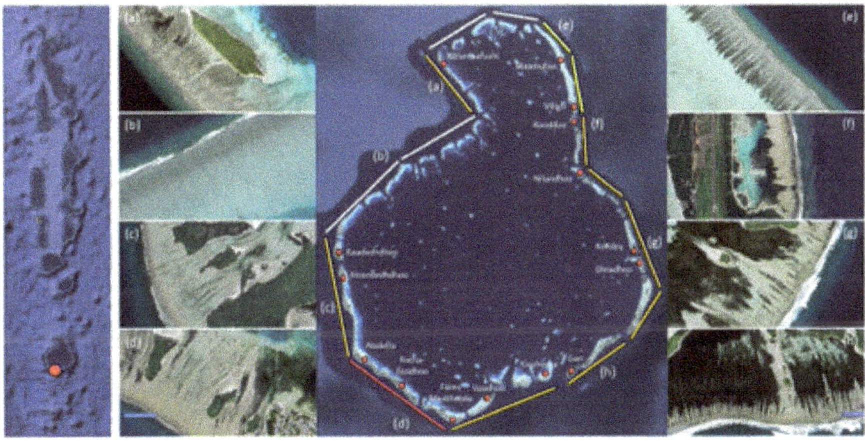

Fig. 1. (left) Location of Huvadhu atoll in the Maldives; (right) qualitative assessment (with examples) of rubble strip distribution and prominence around the atoll ocean facing island rim. Red (white) represents high (low) density of rubble strips.

On a local-scale, the mobility and morphodynamics of a single rubble strip was examined, located within the high energy SW-facing section of the atoll. The tract sits on a carbonate pavement with length, width and height dimensions on the order of 350 m, 40 m and 0.4 m, respectively (Fig. 2 & 3). Mean and max offshore significant wave heights were 1.34 m 3.91 m respectively. Mean and max lagoonward cross-shore current velocities (5-min averaged) on the platform (at rubble tract) were 0.24 m/s and 0.65 m/s, respectively. Morphological observations indicate widespread mobility of particles throughout the rubble tract under wave conditions that fall within typical annual return periods for the region [7], but with limited feature (tract) scale morphological change. Over a 5-month period there was an observed net morphological change of only 9.68 m^3 (0.0045 m/m^2), but tract-scale mobility (% surface area experiencing vertical changes > 0.03 m) was 49%. On a shorter timescale, weekly tract-scale mobility (28/3, 1/4, 7/4, 12/4) was between 2.34% and 11.78% (Fig. 3).

Mobility is largely driven by quasi-unidirectional flows across the platform during the upper half of the tide. Offshore waves heights are significantly correlated with observed bed flow velocity on the rubble tract. The epoch with the highest recorded max current velocity (0.65 m/s) was associated with the highest mobility (39%; 12/4 – 6/7; Fig. 3).

None of the epochs observed any total net change to the sediment budget above propagated measurement uncertainties, but significant elevation changes across the tract were observed with net accretion on the oceanward side to the rubble tract and net erosion at the lagoonward end. Current velocities indicate that the tract is dominated by onshore currents therefore morphological changes can be interpreted as an influx of sediment from the oceanward end and erosion via flattening or lagoonward transport of material beyond the region of analysis.

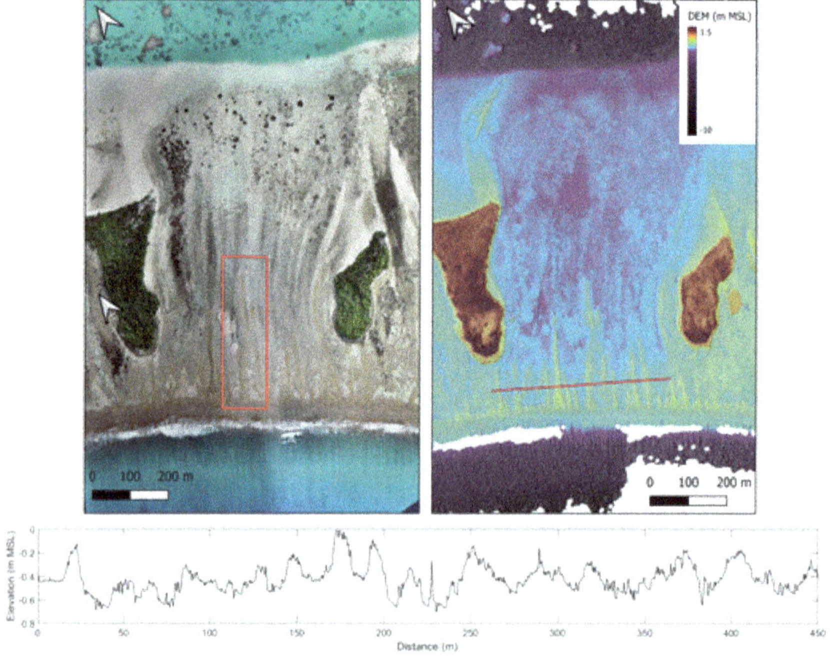

Fig. 2. (left) orthomosaic of local-scale rubble tract study region of Ekalonda (left island) Keli-ahutta (right island), red box is rubble tract being investigated; (right) DEM of same region identifying location of along-platform profile; and (bottom) example profile through a series of rubble tracts illustrating geometry of features.

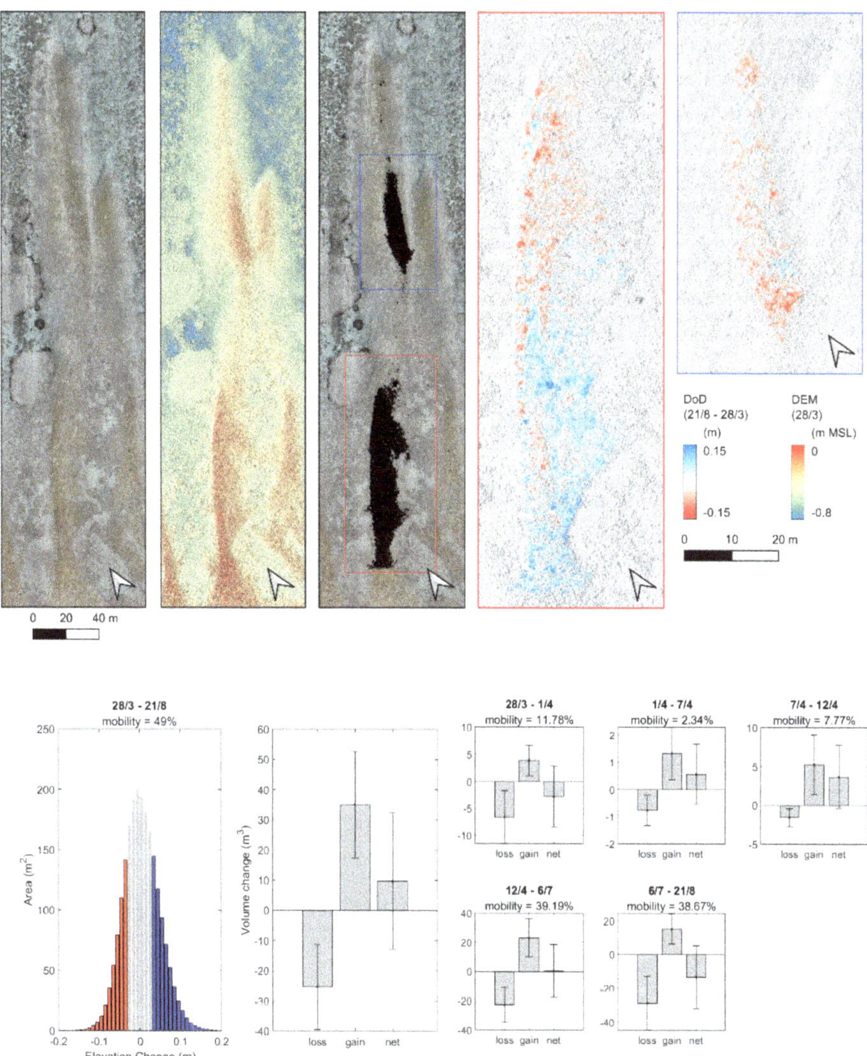

Fig. 3. (1) Orthomosaic and (2) DEM of rubble tract; (3) with black regions showing combined common areal coverage for all epochs that were used for analysis; (4–5) DoD for period 28/3 – 21/8; and (6) results of geomorphic change detection analysis for all epochs, where error bars are propagated measurement uncertainty for cells > LoD.

4   Discussion and Conclusions

This study provides some new baseline understanding into the morphodynamics and mobility of rubble tract features within low lying atoll islands. In the Huvadhu atoll these features are intertidal, exposed during spring low water and fully submerged during high water. Over the study period the site experienced a range of wave conditions typical of an annual return period. We found that over a 5-month period the analysed rubble tract feature was geomorphologically stable (no significant change in total sediment budget), but observed persistence of lagoonward quasi-unidirectional flows over the tract driven by offshore waves suggest material is being sourced from the ocean side and exported at the lagoon side, but do not explain why the landform is being maintained. Therefore, it is important for the understanding of sediment transport pathways and island building processes that large portions (10–50%) of these features are mobile over weekly to 6-monthly timescales.

Acknowledgements. This work was conducted under the UKRI-funded ARISE project 'Natural Adaptation of Atoll Islands to Sea-Level Rise Offering Opportunities for Ongoing Human Occupation' (EP/X029506/1).

References

1. Kench PS, Thompson D, Ford MR, Ogawa H, McLean RF (2015) Coral islands defy sea-level rise over the past century: records from a central pacific atoll. Geology 43:515
2. Kench PS (1998) Physical controls on development of lagoon sand deposits and lagoon infilling in an Indian Ocean atoll. J Coastal Res 14(3):1014–1024
3. Vila-Concejo A, Hamylton S, Webster J, Duce S, Fellowes T (2022) Lagoon infilling by coral reef sand aprons as a proxy for carbonate sediment productivity. Geology 50(12)
4. Talavera L et al (2021) Morphodynamic controls for growth and evolution of a rubble coral island. Remote Sens 13(8):1582
5. Rasser M, Riegl B (2002) Holocene coral reef rubble and its binding agents. Coral Reefs **21**
6. Thornborough KJ (2012) Rubble-dominated reef flat processes and development: evidence from one tree reef, southern great barrier reef. Ph.D. thesis. The University of Sydney
7. Amores A, et al (2021) Coastal flooding in the Maldives induced by mean sea-level rise and wind-waves: from global to local coastal modelling. Front Mar Sci 8

Dynamic Islands

Coral Rubble Mobility on an Intertidal Reef Flat Huvadhu Atoll, Maldives

Aitana Gea-Neuhaus[1]([✉]), Tim Scott[1], Gerd Masselink[1], Mathilde Lindhart[1], Floortje Roelvink[1], Ana Vila-Concejo[2], and Paul Kench[3]

[1] Coastal Processes Research Group, University of Plymouth, Plymouth, UK
aitana.geaneuhaus@plymouth.ac.uk
[2] Geocoastal Research Group, University of Sydney, Sydney, Australia
[3] Department of Geography, National University of Singapore, Singapore, Singapore

Abstract. Coral reef systems are highly vulnerable to anthropogenic climate change influencing sediment dynamics between islands and the surrounding reef. This study focuses on coral rubble mobilization of rubble sheet deposits on an intertidal platform in the Maldives. Rubble clasts display an overall migration and fining in a lagoonward direction, potentially contributing to island formation. To further understand mobilization patterns, a rubble tracer experiment was conducted using 180 rubble clasts of different sizes and shapes released at two locations *Ocean* and *Lagoon*. Their movements were tracked over 30 days under varying hydrodynamic conditions. Rubble clast size, shape and releasing location influenced displacement and mobility. The proportion of rubble mobilized during the experiment varied between 60–100% dependent on size and location. Small and medium rubble present higher median displacement and mobility rates than large clasts at both releasing locations with a significantly greater difference at the *Ocean* site ($p = 0.03$). Branching and massive clasts show distinctively higher displacement than tabular ones, especially for small clasts at both locations ($p = 0.03$ and $p = 0.01$) and medium at the *Lagoon* site ($p = 0.01$). These findings highlight the important interaction between rubble clast attributes and their role in mobilization.

Keywords: Coral Rubble Mobilization · Reef Flat Deposit · Tracer Experiment

1 Introduction

Coral reef systems are particularly vulnerable to the impacts of anthropogenic climate change, affecting ecological and physical processes governing sediment supply and distribution across reef zones [1, 2]. Important morphological features are reef flats, whether they are dominated by coral rubble or living coral [3]. Incident wave energy is substantially dissipated across the platform, establishing the mechanics for rubble entrainment and transport [4].

Here, we focus on the mobilization of rubble accumulations on an intertidal platform described as rubble strips or sheets [5]. These deposits are formed during high energetic

C. Coelho et al. (Eds.): CD 2025, CRL 41, pp. 637–643, 2026.
https://doi.org/10.1007/978-3-032-15473-6_97

events such as storms or tsunamis, when exceptional volumes of different sized coral rubble are transported from the outer reef slope and the algal rim to the reef flat [6]. The mobilized rubble clasts may be freshly generated by the physical breakdown of the reef framework or remobilized from previous offshore accumulations [6]. Once deposited, selective mobilization of the rubble is initiated, which is dependent on the size and, shape of each clast [7], and controlled by the hydrodynamic forces that act upon them [3]. Generally, this results in several concurring processes: a grain-size associated sorting, attrition of larger clasts and a large-scale lagoonward migration of the rubble deposit [3, 5]. These deposits may act as a sediment source, which can be incorporated into the island framework, therefore playing an important role in island formation and maintenance [8].

The limited research quantifying rubble clast mobilization in the field and flume has focused primarily on its role in physical disturbance in reef systems, particularly related to coral settlement and restoration in subtidal areas [9, 10], back reef [11] and reef flat algal communities [12]. From a morphodynamic perspective, mobility has been assessed for beach face transport and derived shoreline morphology [13], and formation and propagation of intertidal rubble strips at the Great Barrier Reef [14].

A predicted intensification in storminess paired with the ongoing ecological degradation of coral reefs will lead to an increase of coral rubble production [9]; thus, a better understanding of the parameters and morphodynamic settings governing mobilization thresholds is essential to understand future development of reef flats deposits and island maintenance. Here we examine the influence of: (1) distance to the oceanward algal rim; and (2) size and shape of rubble clasts on transport and mobilization during a 30-day field experiment.

2 Rubble Tracer Experiment

A rubble tracer mobilization experiment was conducted on an intertidal reef platform on the southernmost part of the Huvadhoo Atoll, Maldives (Fig. 2). A total of 180 coral rubble clasts of different sizes and shapes were collected at the study site using the random-walk method [15]. Once dried, the clasts were weighed, measured, painted, and tagged with a unique number for identification. The rubble tracers were released at two distinct locations of the intertidal rubble strip (*Ocean* and *Lagoon* site) and tracked with a RTK GNSS system between 28th of March and 29th of April 2024 (Fig. 2). During this period, the significant wave height (H_s) seaward of the reef crest ranged from 2.15 to 0.48 m (mean 0.97 m) with a period (T_p) of 4–18 s (mean 12 s), and a tidal range of ~ 1 m.

2.1 Rubble Clast Attributes

Dried rubble clasts were weighed, their axes measured (long *a*, intermediate *b* and short *c*) and the clasts were visually classified by size (*small*, *medium* and *large*) and shape (*tabular, branching, and massive*).

The size distribution of clasts ranged from small boulder to pebble [16] (Fig. 1).

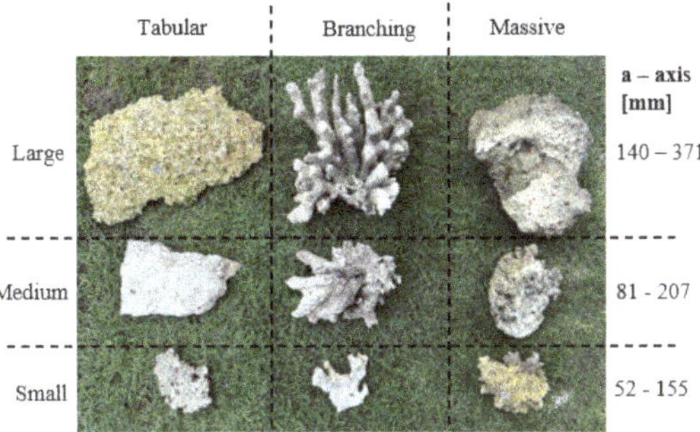

Fig. 1. Rubble Clast Attributes

2.2 Rubble Tracer Deployment and Tracking

The position of each rubble tracer was surveyed with a Trimble R10 RTK GNSS receiver (horizontal accuracy 1–2 cm) every 24h–48h at low tide (0.1–0.4 m relative to MSL) during 30 days, except for two neap periods when the reef flat was inaccessible (0.5–1.1 m relative to MSL). The survey pole was placed consistently at the outermost edge of the clast facing the lagoon and the identification number recorded to properly allocate each position.

Fig. 2. Research Location.

3 Results

3.1 Rubble Tracer Displacement and Mobility

The Role of Location and Size
Observations of rubble tracer movement (98% recovered) reveal a predominantly lagoon-ward transport direction across the reef flat at the *Ocean* and *Lagoon* releasing locations

(Fig. 3a, b). A statistically significant threshold of motion between measurements was determined to be 0.063 m (95% confidence level).

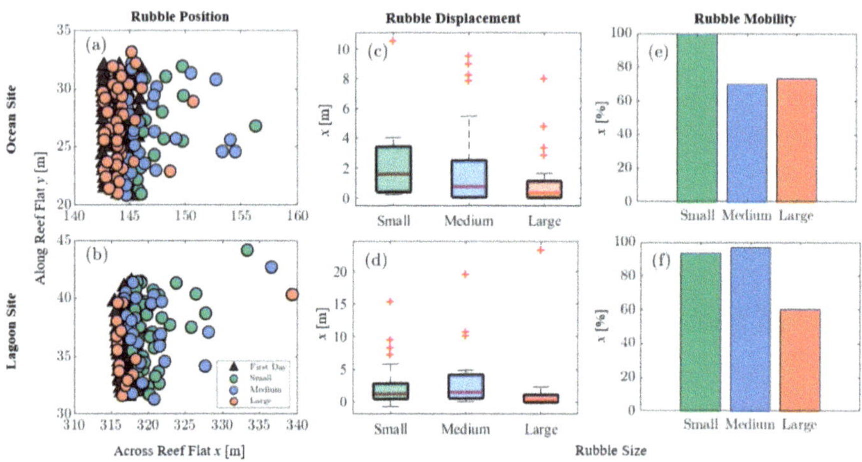

Fig. 3. Summary of rubble experiment: (a) and (b) initial (black) and last rubble position; (c) and (d) overall rubble displacement; and (e) and (f) overall rubble mobility. Green, blue and orange colors represent small, medium, and large size rubbles, respectively. Top and bottom panels represent Ocean and Lagoon site, respectively.

Throughout the experiment, rubble travelled varying distances depending on the releasing site and the size class as summarized in Table 1.

Higher transport distances and mobility were recorded at the *Lagoon* site for medium and large rubble clasts, whereas small clasts travelled further and were more mobile at the *Ocean* site (Fig. 3c–e).

One-way ANOVA analyses revealed a significant difference in rubble displacement between size classes at the *Ocean* site ($p = 0.03$), but not at the *Lagoon* site ($p = 0.55$). Comparisons of the same size class at the two locations reveal no significant difference regarding the overall cumulative displacement ($p \geq 0.05$).

Table 1. Summary of rubble displacement and mobility by location and size.

Location	Size	min. - max. Displ. [m]	median displ. [m]	Mobility [%]
Ocean site	Small	0.22–10.58	1.59	100
	Medium	0–9.60	0.77	70
	Large	0–8.05	0.38	73
Lagoon Site	Small	0–15.46	1.26	93
	Medium	0–19.63	1.51	97
	Large	0–23.38	0.45	60

The Role of Shape

Besides the role of location and size, the influence of clast shape on rubble displacement was analyzed. Observations show a higher cumulative displacement of branching and massive rubble tracer than tabular across all locations and size classes (Fig. 4a–f).

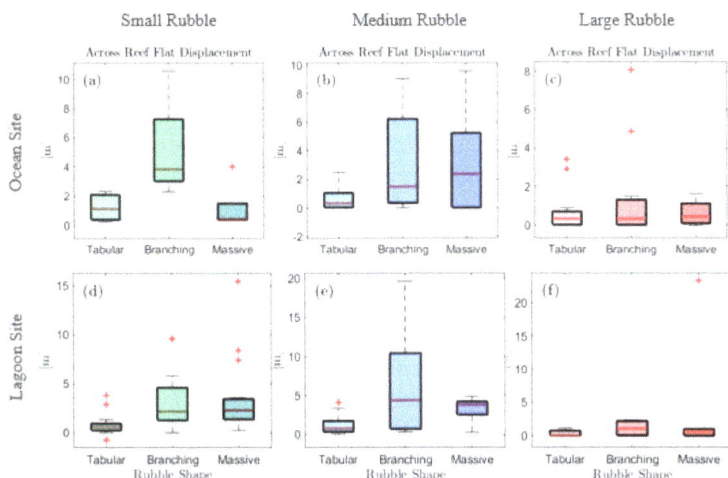

Fig. 4. Overall displacement by shape of rubble clast (tabular, branching, massive). (a) and (d) in green for small; (b) and (e) in blue for medium and (c) and (f) in red for large. Top and bottom panels represent Ocean and Lagoon site, respectively.

One-way ANOVA analyses, show significant difference between shapes for small rubble at the Ocean and Lagoon Site, $p = 0.03$ and $p = 0.01$ (Fig. 4a, d).

For medium rubble, the difference is observable, but only significant at the Lagoon Site ($p = 0.01$) (Fig. 4b, e). Generally, large rubble displays smaller displacement at both locations and the difference between shapes is not that remarkable with no significant variations ($p \geq 0.05$) (Fig. 4c, f).

4   Discussion

Reef flat deposits present a wide range of sizes and shapes of rubble clasts due to the heterogeneity of living coral species that act as primary source and the relative short transport distances between the outer reef slope and the depositional site [3]. During this rubble tracer experiment on an intertidal reef flat, we could successfully determine rubble displacement and mobility in relation to three important mobilization factors: distance to the algal rim, size, and shape.

Foremost, it is important to highlight that the majority of rubble tracer clearly mobilized during this monthly experiment under modal hydrodynamic conditions at both releasing locations (small: 93 – 100%; medium: 70–97% and large: 60–73%).

In terms of median displacement, medium and large rubble tracer display higher values at the *Lagoon* site (1.51 m and 0.45 m) than at the *Ocean* site (0.77 m and 0.38 m, respectively). Whereas small rubble shows an opposite trend with higher median displacement at the *Ocean* site (1.59 m) than at the distant *Lagoon* site (1.26 m). These findings suggest that the proximity to the algal rim and thus expected higher energy conditions may not result directly in greater overall displacement and mobility rates as proposed in previous research across different reef zones [9–14].

A more comprehensive analysis of the role of size, reveals that small and medium sized rubble tend to travel further distances and are more mobile than large clasts at both releasing locations, being significantly different at the *Ocean* site ($p = 0.03$). A trend of higher mobilization of smaller sized rubble has also been observed in previous studies [9, 10]. On the contrary, a rubble experiment conducted at One Tree Island on the southern Great Barrier Reef [14], concluded that boulder-sized rubble displays higher displacement than smaller clasts, which tend to accommodate more easily into the rubble matrix [14].

Accurate observations of the influence of shape determined a clear correlation, where branching and massive rubble clasts present higher median displacement than tabular across all locations and size classes. Significant differences regarding overall displacement could be established for small sizes at both sites ($p = 0.03$ and $p = 0.01$) and medium at the *Lagoon* site ($p = 0.01$). Comparison between shape classes to previous research is not feasible for different reasons: the role of shape is not directly considered [10, 12], only cylindrical or branching shapes are examined [9, 11] or the size and shape are intrinsically linked not allowing a clear differentiation between the effect of the parameters, i.e. small branching and tabular and massive boulder. [14].

Apart from these parameters, hydrodynamic forcing acting upon the rubble clasts and the underlying type of substrate are key drivers in their mobilization [9–14] and need to be addressed in more detail in further analysis of this experiment.

Acknowledgements. This work was conducted under the UKRI-funded ARISE project 'Natural Adaptation of Atoll Islands to Sea-Level Rise Offering Opportunities for Ongoing Human Occupation' (EP/X029506/1).

References

1. Perry CT, Kench PS, Smithers SG, Riegl B, Yamano H, O'Leary MJ (2011) Implications of reef ecosystem change for the stability and maintenance of coral reef is-lands. Glob Change Biol 17:3679–3696
2. Woodroffe CD (2008) Reef-island topography and the vulnerability of atolls to sea-level rise. Glob Planet Change 62:77–96
3. Thornborough KJ, Davies PJ (2011) Reef flats. In: Hopley D (ed) Encyclopedia of modern coral reefs, encyclopedia of earth sciences series. Springer, Dordrecht, pp 869–876
4. Brander RW, Kench PS, Hart D (2004) Spatial and temporal variations in wave characteristics across a reef platform, Warraber Island, Torres Strait, Australia. Mar Geol 207:169–184

5. Pirazzoli PA (2011) Boulder zone/ramparts. In: Hopley D (ed) Encyclopedia of modern coral reefs, encyclopedia of earth sciences series. Springer, Dordrecht, pp 167–168

6. Scoffin TP (1993) The geological effects of hurricanes on coral reefs and the interpretation of storm deposits. Coral Reefs 12:203–221

7. Rasser M, Riegl B (2002) Holocene coral reef rubble and its binding agents. Coral Reefs 21:57–72

8. Shannon A, Power H, Webster J, Vila-Concejo A (2012) Evolution of coral rubble de-posits on a reef platform as detected by remote sensing. Remote Sens 5:1–18

9. Kenyon TM, et al (2023) Mobilisation thresholds for coral rubble and consequences for windows of reef recovery. Biogeosciences 20:4339–4357

10. Viehman T, Hench J, Griffin S, Malhotra A, Egan K, Halpin P (2018) Understanding differential patterns in coral reef recovery: chronic hydrodynamic disturbance as a limiting mechanism for coral colonization. Mar Ecol Prog Ser 605:135–150

11. Cameron CM, Pausch RE, Miller MW (2016) Coral recruitment dynamics and substrate mobility in a rubble-dominated back reef habitat. Bull Mar Sci 92:123–136

12. Cheroske AG, Williams SL, Carpenter RC (2000) Effects of physical and biological disturbances on algal turfs in Kaneohe Bay, Hawaii. Exp Mar Biol Ecol 248:1–34

13. Kench PS, et al (2017) Nearshore hydrodynamics, beachface cobble transport and morphodynamics on a Pacific atoll motu. Marine Geol 389:17–31

14. Thornborough KJ (2012) Rubble-dominated reef flat processes and development: evidence from Tree Reef, Southern Great Barrier Reef. Ph.D. thesis. The University of Sydney, 121 p

15. Wohl EE, Anthony DJ, Madsen SW, Thompson DM (1996) A comparison of surface sampling methods for coarse fluvial sediments. Water Resour Res 32(10):3219–3226

16. Wentworth CK (1922) A scale of grade and class terms for clastic sediments. J Geol 30(5):377–392

Barrier Islands Facing Sea Level Rise in the Northern Adriatic Sea

Annelore Bezzi[(✉)], Giulia Casagrande, Saverio Fracaros, Simone Pillon, Davide Martinucci, Chiara Popesso, Sebastian Spadotto, Stefano Sponza, and Giorgio Fontolan

Department of Mathematics, Informatics and Geosciences, University of Trieste, Via Weiss 1, 34128 Trieste, Italy
bezzi@units.it

Abstract. Because of the presence of a unique freely-developing sandy barrier system, the Marano and Grado Lagoon provides as a useful case study in the debate on barrier island responses to relative sea level rise. To analyze the multidecadal evolution of these 10-km-long landforms, the coastline configuration in 16 different years was examined using historical cartography, aerial photos and topographic surveys. After subdividing the coast into 50 cells, the differences in the subaerial barrier area (AREA) and shoreline position (SHO) were computed for each temporal step and examined as morphodynamic indicators. The results show that barrier islands have been preserved throughout time as the AREA has increased, along with a general pattern of coastal retreat, indicating a dominant rollover process as a response to sea level rise. This process becomes significantly more complex when the behavior of indicators is examined at various time scales. On a four-year time scale, there are both typical rotational trends and periods of significant variability due to the coexistence of transgressive process and intense longshore constructive dynamics. The multidecadal history of Grado barrier islands demonstrates the complexity of barrier island response to sea level rise, often resulting in contemporary small-scale opposite dynamics due to either erosive or depositional phenomena.

Keywords: Barrier Island · sea-level rise · adaptive response · washover · rollover

1 Introduction

Barrier islands have unique environmental, socioeconomic, and cultural values that make challenging the understanding and predicting of their coastal response to relative sea level rise [1, 2]. The ability to adapt of barrier islands is not uniform and assumes significant local variability in response to the coastal forcing, sediment supply and human constraints [3–5]. Thanks to the equilibrium between tidal forcing and waves, the coasts of the northern Adriatic (Italy) feature notable lagoon and barrier island systems. Although the majority of these barriers are fully fixed due to engineering works, the easternmost sandy barrier islands of the Marano and Grado Lagoon evolve rather freely. A quantitative

© The Author(s) 2026
C. Coelho et al. (Eds.): CD 2025, CRL 41, pp. 644–651, 2026.
https://doi.org/10.1007/978-3-032-15473-6_98

multidecadal analysis of the sandbanks evolution makes this study case useful in the debate of barrier island responses to relative sea level rise.

2 Study Area

The Marano and Grado Lagoon (NE Italy), part of the Natura 2000 network (IT3320037) is protected in its easternmost sector by about 10 km of very dynamic, narrow (160 m), and low emerged barrier islands (Grado sandbanks, Fig. 1), only in some tracts fixed by littoral vegetation [6]. The back barrier area is represented by a small para-lagoonal basin limited landward by an ancient system of barrier islands, which was stabilized by engineering works in the early 1900s. The local wind climate is affected by the strong wind from the ENE (Bora), and the subordinate Sirocco wind from the SE, which, however, generates the highest waves, both because of the 10 times longer fetch and because of the geographical arrangement of the sandbanks. Tides are semi-diurnal with a mean range of 76 cm and mean spring-neap tide ranges of 105 and 22 cm. Because of their high degree of naturalness, the sandbanks provide breeding habitats for endangered coastal bird species.

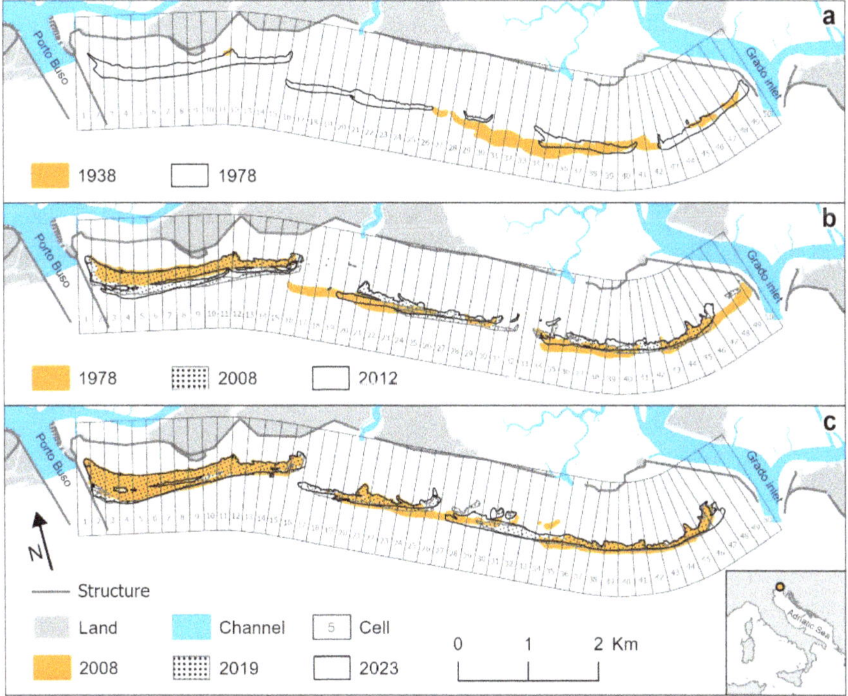

Fig. 1. The barrier islands located between the tidal inlets of Porto Buso and Grado: the maps report the evolution of the sandbanks during different time spans.

3   Methods

Starting from the work of [7], we revised and updated the shoreline multidecadal evolution of the Grado sandbanks. We derived the shoreline from historical cartography (1891/1938), aerial photos (from 1954 to 2014), and five "ad hoc" topographic surveys (from 2015 to 2023). After subdividing the coast into 50 cells, the variations of subaerial barrier area (AREA) and the shoreline position (SHO) were computed for each temporal step and then analysed as morphodynamic indicators. In addition, two surveys (1968 and 2015) along 11 bathymetric sections allowed for the analysis of shoreface volumetric changes down to a depth of 6 m.

4   Results and Discussion

A multidecadal qualitative and quantitative analysis of the landforms' changes described by AREA and SHO has allowed to reconstruct the evolution of the Grado sandbanks starting from old, imprecise geographic maps (IGM from 1891 and 1936) to the most recent GNSS survey in 2023 (Fig. 1).

The first phase of barrier island formation, from 1891/1938 to 1978, was strongly influenced by coastal defense works: a 1200-m-long seawall along the western side of the Grado inlet (1928–1934) and two jetties to fix the entrance of Porto Buso (1960–1964). In response to the interruption of the longshore sand drift, sandbanks emerged in a previously shallow water area, first in the eastern sector and later in the western one. In 1978, the barrier island system appeared to have fully formed over the entire coastal sector between the two tidal inlets, marking the end of this initial phase (Fig. 1a). In the second phase (from 1978 to 2023), the increased availability of data allows us to observe the banks' evolution at various time scales. Considering a medium-term interval (45 years, Fig. 1b, c and Fig. 2) a rise in AREA is the most prevalent pattern along the 50 bank cells, while SHO exhibits a relevant variability. The trap effect of the Porto Buso jetty with respect to the dominant longshore drift is attested to by the western sector (cells 1–15), which shows the highest progradation values close to the jetty and gradually decreasing towards the east.

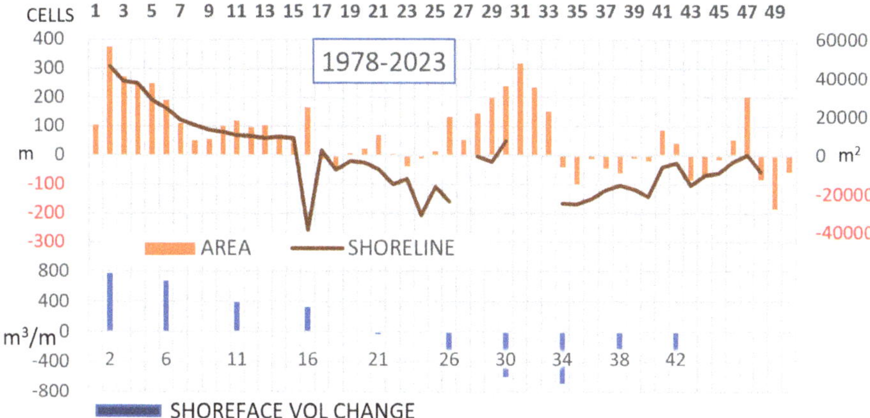

Fig. 2. Spatial distribution of the AREA and SHO indicators along the 50 cells in a medium-term interval (1978–2023).

The middle and western sectors (cell 16–50) exhibit a large degree of variability, significant shoreline retreat, and little concordance between the two morphodynamic indicators. The shoreface budget, which displays a typical rotational effect on the entire barrier island system, validates the spatial differentiation (Fig. 2). As examples on a shorter time scale, we provide two datasets with the same temporal span (4 years, Fig. 3). The cell data distribution in the 2008–2012 period clearly shows longshore variability resulting from a typical cannibalisation. The three distinct barrier islands exhibit a rotational pattern, as evidenced by the strong correlation between the two morphodynamic indicators (AREA and SHO), which show coastline progression and area gain in the downdrift sectors, as well as shoreline retreat and area loss in the updrift ones.

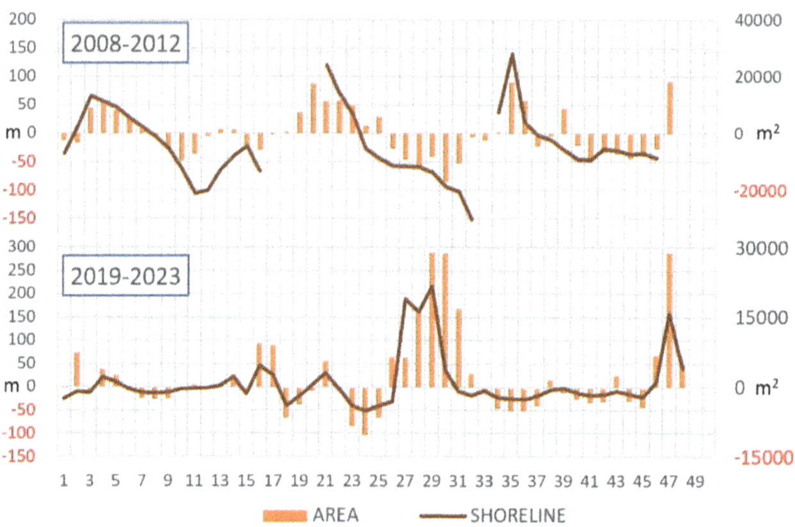

Fig. 3. Spatial distribution of the AREA and SHO indicators along the 50 cells in two short-term intervals (2008–2012 and 2019–2023).

A high spatial variability prevails in many of the time periods we have examined, and the 2019–2023 period is no exception. Different combinations of the two indicators are feasible within this variability, and they are associated with distinct geomorphic behaviours. Among the transgressive processes, the simplest is identifiable thanks to negative values of both indicators, i.e., the erosional response induces a shoreline retreat resulting in barrier thinning with surface loss (Fig. 4a). AREA and SHO are both negative also when overwash events lead to island breaching and landward spreading of sediments to form a subaqueous fan (Fig. 4b). When the washover deposit emerges, the indicators become opposite (SHO negative and AREA positive, Fig. 4c).

Progradational patterns coexist in other cells nowadays. Spit gemination and/or lengthening (Fig. 4d, e), as well as nearshore bar welding (Fig. 4f), are typical long-shore constructive processes that give positive values for both indicators. Depressions between newly formed emerged bars evolve into parallel swales or cat's-eye ponds (Fig. 4g). An unusual combination of the indicators or extremely high anomalous values for both can arise from more complicated circumstances brought on by distinct evolu-tionary processes taking place at a scale smaller than the one taken into consideration for the analysis.

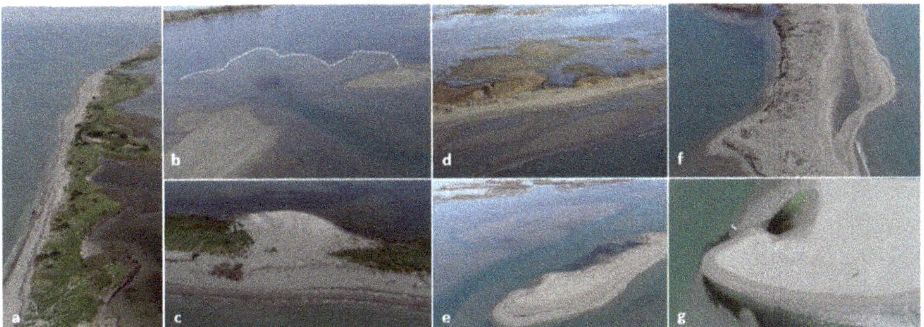

Fig. 4. Various morphodynamic scenarios in the sandbank evolution (see the text for details).

Finally, we can observe the overall signal (mean data of SHO and AREA) of the barrier island evolution and investigate if there is a relationship with the rate of sea level rise. The general trend, albeit through alternating phases, is towards an increase in the barrier area over time (Fig. 5a). This is explained by the fact that a washover tends to reduce the accommodation space through the rollover process. Despite the limitation of this type of analysis, the movement of the shoreline (Fig. 5b) appears to be related to the sea level fluctuation by alternating periods of stasis with times of severe retreat, according to a similar stepped variation of sea level, which significantly increased after 1989. The morphological adaptation seems to occur with a delay of around 4 years.

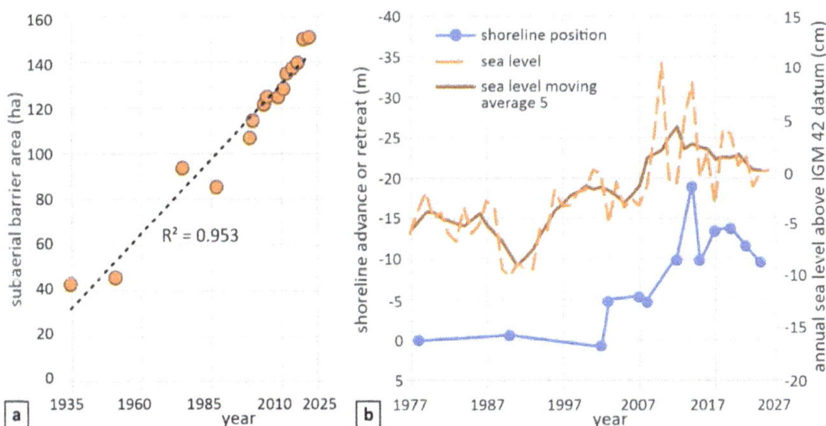

Fig. 5. Overall evolutive signal (mean value) for the entire sandbank system with regards to subaerial barrier area (a) and shoreline position (b) in relation to sea level rise. Sea level annual data from Trieste, Molo Sartorio tide gauge, available through SEANOE https://doi.org/10.17882/62758).

5 Conclusions

The multidecadal development of the Grado sandbanks demonstrates the potential of barrier islands to adapt to sea level rise, which may result in a dynamic coexistence of erosive and depositional processes. With a sea level rise rate of 1.4 mm/y (century-long) to 3.5 mm/y (after 1989), the barrier island system is self-sustaining via a rollover process and a compensatory longshore mechanism that benefits from sediment supplies deriving from the eastern riverine coastal tract. The degree of freedom and maintenance of sedimentary sources is inextricably linked to the preservation of these landforms and their role as ecosystem services. Given that the bulk of landforms, both submerged and subaerial, are destined to change swiftly and even suddenly, conservation programs must take on a unique meaning and application in this context. It should ideally be founded on the fundamentals of adaptive management and centered on the preservation of morphodynamic systems.

Acknowledgements. This research has been funded by the European Union granted by the NextGeneration EU, project "Interconnected Nord-Est Innovation Ecosystem iNEST" (CUP J43C22000320006) and by the Biodiversity Office, Regione Friuli Venezia Giulia.

References

1. Leatherman S (1983) Barrier dynamics and landward migration with Holocene sea-level rise. Nature 301:415–417
2. FitzGerald DM, Hein CJ, Hughes Z, Kulp M, Georgiou I, Miner M (2018) Runaway barrier island transgression concept: global case studies. In: Moore LJ, Murray AB (eds) Barrier dynamics and response to changing climate. Springer, Cham, pp 3–56
3. Morton RA (2008) Historical changes in the Mississippi-Alabama barrier-island chain and the roles of extreme storms, sea level, and human activities. J Coast Res 24:1587–1600
4. Kombiadou K, Matias A, Ferreira Ó, Carrasco AR, Costas S, Plomaritis T (2019) Impacts of human interventions on the evolution of the Ria Formosa barrier island system (S. Portugal). Geomorphology 343:129–144
5. Bezzi A et al (2021) From rapid coastal collapse to slow sedimentary recovery: the morphological ups and downs of the modern Po Delta. Estuar Coast Shelf Sci 260:107499
6. Fontolan G et al (2012) Human impact and the historical transformation of saltmarshes in the Marano and Grado Lagoon, northern Adriatic Sea. Estuar Coast Shelf Sci 113:41–56
7. Brambati A, De Muro S, Marocco R, Selivanov A (1998) Barrier island evolution in relation to sea level changes: the example of the Grado Lagoon (northern Adriatic Sea, Italy). Bollettino Geofisica Teorica Appl 39(2):145–161

Understanding the Drivers of Cay Morphodynamic Activity Within the Great Barrier Reef, Australia

Emily Lazarus[1]([✉]), Stephanie Duce[1], Stephen Lewis[2], and Scott Smithers[1]

[1] College of Science and Engineering, James Cook University, Townsville, QLD 4814, Australia
emily.lazarus@my.jcu.edu.au

[2] TropWATER, James Cook University, Townsville, QLD 4814, Australia

Abstract. Cays on the Great Barrier Reef (GBR) are dynamic low-lying carbonate sedimentary landforms that adopt various morphologies in response to the complex interplay between physical processes and sediments. This study presents a preliminary analysis of the drivers which most strongly influence cay morphodynamic activity. Focusing on two contrasting examples from the GBR, a small unvegetated cay (Taylor Cay) and a large, vegetated cay (Masthead Island), we assessed cay morphodynamic activity against geomorphic variables and variables associated with exposure to energy. We found significant ($p < 0.05$) negative correlation between cay activity and significant wave height and maximum annual days of cyclone exposure, and positive correlation with geomorphic predictors (average cay area, length, width and volume). The outcomes indicate that cays are most geomorphologically active when exposed to greater hydrodynamic energy over the tidal cycle. Geomorphic characteristics including larger size can buffer against shoreline perturbations driven by hydrodynamic energy exposure. Thus, more vulnerable cays can be identified and prioritized for management using the quantitative spectrum of drivers of cay morphodynamic activity.

Keywords: reef islands · geomorphic characteristics · hydrodynamics

1   Introduction

Cays on the Great Barrier Reef (GBR) are low-lying accumulations of biogenic carbonate sediments deposited on the host reef platform [1]. Cays are dynamic landforms, adopting various morphologies in response to the complex interplay between physical processes (waves, tides, currents, and sea level) and sediments (production, transport, and deposition) [2, 3]. Here we define cay morphodynamic activity as the magnitude of cay morphological adjustment in response to seasonal, annual and longer-term variations in environmental and physical drivers. It is necessary to understand and quantify the present-day drivers of cay morphodynamic activity to anticipate how sensitive [sensu 4] a cay's morphology is to changes in future conditions.

© The Author(s) 2026
C. Coelho et al. (Eds.): CD 2025, CRL 41, pp. 652–658, 2026.
https://doi.org/10.1007/978-3-032-15473-6_99

There are more than 300 cays within the GBR which support a range of high-value ecological and cultural services, including marine turtle and seabird nesting sites [5, 6]. Thus, it is important to understand and quantify the influence of drivers on the rate and magnitude of cay morphodynamic activity. Here we present a preliminary analysis of the drivers of cay morphodynamic activity, focusing on using two contrasting examples from a broader study conducted for 16 GBR cays [see 7].

2   Methods

Cay morphodynamic activity was assessed for 16 cays within the GBR. Here we present Taylor Cay, a small (0.84ha), unvegetated cay in the northern section of the GBR, and Masthead Island, a large (47ha), vegetated cay in the southern GBR (Fig. 1). Monthly cay toe of beach (TOB) shoreline position footprints were digitized from Sentinel-2 imagery (10m resolution) between August 2015 and February 2023 (91 months) by a single operator in ArcMap 10.8.2. The TOB was chosen as the shoreline proxy as it can be applied to both unvegetated and vegetated cays [8]. Images were removed where cloud cover obscured the TOB resulting in a dataset of 83 monthly footprints at Taylor Cay and 86 at Masthead Island (Fig. 1b, c).

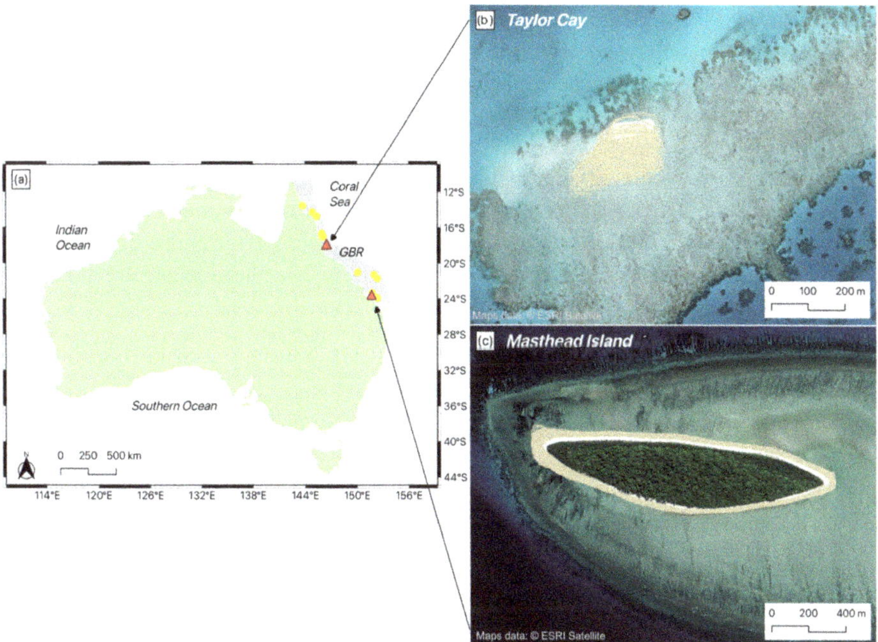

Fig. 1. Geographical context for all 16 study sites within (a) the Great Barrier Reef and the broader Australian context, highlighting (b) Taylor Cay with grey lines denoting the monthly shoreline footprints (n = 83) and (c) Masthead Island (n = 86).

Cay morphodynamic activity was measured using the largest temporal footprint overlap metric from the Reef Island Stability Assessment (RISA) approach [9]. This metric measures the area where there is the greatest duration of overlap for all digitized cay footprints and calculates that area as a proportion of the historical cay position envelope for the analysis period (aggregated area of all cay footprints). This proxy for cay morphodynamic activity captures event-based, seasonal, annual and longer-term perturbations to cay morphology. The largest temporal footprint overlap (response) was assessed against 41 environmental and physical variables (predictors) to determine the dominant drivers of cay morphodynamic activity [7]. A standardized approach was applied by normalizing these variables between 0 and 1[10]. Variables with significant ($p < 0.05$) moderate to strong ($<-0.5, >0.5$) Spearman's Rank linear correlations with morphodynamic activity were grouped into exposure and geomorphic predictors. The normalised scores for each variable in the two predictor groups were then added together to create a score reflecting their overall influence on cay morphodynamic activity.

3   Results

3.1   Cay Morphodynamic Activity

Geomorphological changes observed over the 91-month analysis period at Taylor Cay and Masthead Island demonstrate that morphological changes on GBR cays vary markedly both spatially and temporally (Fig. 2). At Taylor Cay the cay footprint is highly mobile, with no area where all cay footprints overlapped during the study period. The maximum number of overlapping cay footprints is 52 (of 83 monthly footprints) with the overlapping area representing just 1.94% of the historical cay positional envelope (Fig. 2a). The results reflect the history of high-frequency accretion, erosion, and active migration of Taylor Cay since mid-2015. In contrast, Masthead Island has been relatively stable, with 78.13% of the cay area being in place for the entire analysis period (i.e., all cay footprints overlapped). The mobile areas on Masthead Island are around the cay margins, particularly around the NW spit (Fig. 2b).

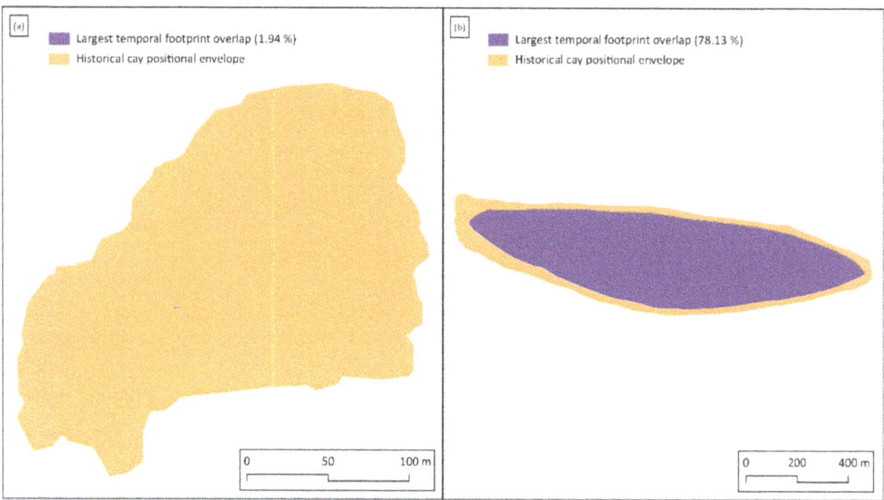

Fig. 2. The largest temporal footprint overlap area (proxy of cay morphodynamic activity) as a percentage relative to the historical cay footprint envelope at (a) Taylor Cay (1.94%) and (b) Masthead Island (78.13%).

3.2 Statistical Analyses

Based on a linear correlation matrix for all 41 predictor variables, the variables with significant ($p < 0.05$) negative linear correlations with cay morphodynamic activity are maximum number of annual days of cyclone exposure ($r = -0.73$), latitude ($r = -0.51$), HsMin ($r = -0.68$), HsMean ($r = -0.59$) and mean low water springs (MLWS) ($r = -0.48$). The variables with significant positive linear correlations are average cay area ($r = 0.91$), cay length ($r = 0.86$), cay width ($r = 0.87$), cay volume ($r = 0.91$), cay to reef area ratio ($r = 0.82$), existing cay classification ($r = 0.62$), and maximum cay elevation ($r = 0.64$). The normalized scores for the selected exposure predictors (negatively correlated) and geomorphic predictors (positively correlated) are compared for each cay (Fig. 3). Taylor Cay has a higher (3.95) combined exposure score when compared to Masthead Island (0.37). Conversely, Masthead Island demonstrates a higher combined geomorphic score (4.93) than Taylor Cay (0.15).

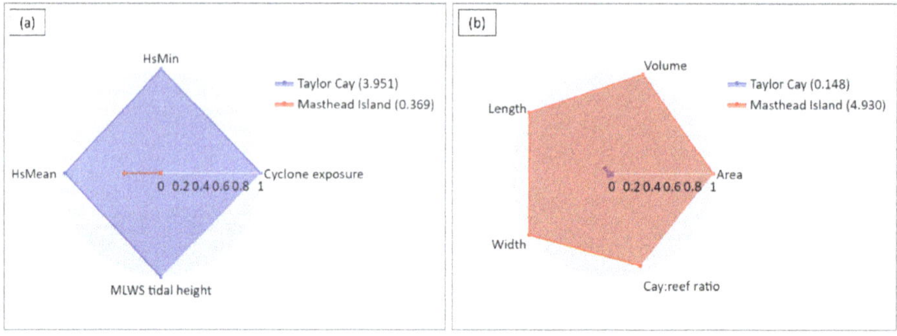

Fig. 3. The normalized scores of the (a) exposure predictors and (b) geomorphic predictors based on the variables with significant ($p < 0.05$) correlation to cay morphodynamic activity. The combined scores for each cay are given in brackets in the legend.

4 Discussion

The divergent morphodynamic behaviour of Taylor Cay and Masthead Island is not unprecedented or undocumented. However, within the GBR, there is a paucity of data to quantify such behaviour over seasonal and interannual temporal scales when compared to global atoll reef island settings [e.g., 11, 12]. Such dynamism of unvegetated cays has been documented over shorter timeframes; Taylor Cay was observed to migrate 90 m in six weeks during a period of strong winds [13] and Hopley [14] documented rapid migration of Wheeler Cay in response to Tropical Cyclone Althea. Similar patterns of shoreline accretion and contraction have been observed at several vegetated cays [e.g., 15, 16], where most of the change is typically associated with the movement of axial spits and relatively minor oscillations of the remaining shorelines.

4.1 Drivers of Cay Morphodynamic Activity

Exposure to hydrodynamic energy is well recognized as a primary driver of geomorphic change for cays within the GBR [1, 8]; however, the energy regimes have not been previously quantified as outlined here. Our findings confirm that cays exposed to greater hydrodynamic energy for longer periods over the tidal cycle are more likely to undergo geomorphic change on a more frequent basis. Geomorphic change potential is correlated with exposure to higher HsMin and HsMean, as well as exposure to a greater number of annual cyclone days (Fig. 3a). Taylor Cay is a very dynamic cay and exhibits a high degree of morphodynamic activity which is likely a consequence of its exposure to HsMin of 0.47 m and up to 20 days of annual cyclone exposure. Additionally, the MLWS tidal height is 0.82 m. These conditions result in exposure to hydrodynamic energy at the cay shoreline for a greater proportion of the tidal cycle. Cays less sensitive to morphodynamic change, such as Masthead Island, are located in settings where the exposure to hydrodynamic energy is lower (Fig. 3a). For example, HsMin around Masthead Island is less than 0.05 m. Additionally, Masthead Island is exposed to less than 10 days of cyclone conditions annually and the MLWS tidal height is 0.4 m.

It has been determined that large, elevated atoll motu with high sediment volume were generally more stable [17]. Our results demonstrate the same trends for GBR cays; Masthead Island and Taylor Cay represent opposing ends of the geomorphic predictors and morphodynamic activity spectrum. With a larger cay area, length, width, elevation, and volume, Masthead Island is able to buffer against shoreline perturbations [5]. In contrast, Taylor Cay has a small area, length, width, elevation, and volume which limits its capacity to buffer against shoreline perturbations. Overall, these geomorphic predictors reflect the scale of cay morphodynamic activity.

5 Conclusion

Cays are highly dynamic sedimentary landforms, undergoing change at monthly, seasonal, event-based and annual timescales. This study presents the first quantification of the exposure and geomorphic predictors over a spectrum of cay morphodynamic activity within the broader context of the Great Barrier Reef. Our findings indicate that exposure to hydrodynamic energy, including waves and storms, strongly influence cay morphodynamic activity. Additionally, larger cays by area, length, width and volume are more likely to demonstrate lower rates of morphodynamic activity. This detailed understanding of the drivers of cay morphodynamic activity will enable stakeholders to rapidly identify and prioritise vulnerable cays for management.

References

1. Gourlay MR (1988) Coral cays: products of wave action and geological processes in a biogenic environment. In: Proceedings of the 6th international coral reef symposium, Australia, vol 2, pp 491–496
2. Dawson JL, Smithers SG (2010) Shoreline and beach volume change between 1967 and 2007 at Raine Island, Great Barrier Reef, Australia. Glob Planetary Change 72(3):141–154. https://doi.org/10.1016/j.gloplacha.2010.01.026
3. Webb AP, Kench PS (2010) The dynamic response of reef islands to sea-level rise; evidence from multi-decadal analysis of island change in the central pacific. Glob Planet Change 72(3):234–246. https://doi.org/10.1016/j.gloplacha.2010.05.003
4. Tyler-Walters H, Tillin HM, d'Avack EAS, Perry F, Stamp T (2023) Marine evidence based sensitivity assessment (MarESA) – guidance manual. In: M. L. I. N. (MarLIN) (ed) Marine Biological Association of the UK, Plymouth, p 170
5. Hamylton S, Puotinen M (2015) A meta-analysis of reef island response to environmental change on the Great Barrier Reef. Earth Surf Proc Land 40(8):1006–1016. https://doi.org/10.1002/esp.3694
6. Hopley D, Smithers SG, Parnell KE (2007) The geomorphology of the great barrier reef: development, diversity, and change. Cambridge University Press, New York
7. Lazarus E, Duce S, Lewis S, Smithers S (in prep) Systematic classification of cay morphodynamic sensitivity within the Great Barrier Reef, Australia
8. Kench PS, Brander RW (2006) Wave processes on coral reef flats; implications for reef geomorphology using Australian case studies. J Coastal Res 22(1):209–223. https://doi.org/10.2112/05A-0016.1
9. Lazarus E, Duce S, Lewis S, Smithers S (in review) The reef island stability assessment: a new approach to quantify cay geomorphic change

10. Pradhan B (2011) Use of GIS-based fuzzy logic relations and its cross application to produce landslide susceptibility maps in three test areas in Malaysia. Environ Earth Sci 63(2):329–349. https://doi.org/10.1007/s12665-010-0705-1

11. Duvat VKE, Salvat B, Salmon C (2017) Drivers of shoreline change in atoll reef islands of the Tuamotu Archipelago, French Polynesia. Glob Planet Change 158:134–154. https://doi.org/10.1016/j.gloplacha.2017.09.016

12. Kench PS, Sengupta M, Ford MR, Owen SD (2024) Island change framework defines dominant modes of atoll island dynamics in response to environmental change. Commun Earth Environ 5(1):585. https://doi.org/10.1038/s43247-024-01757-1

13. Taylor T (1924) Movement of sand cays. Queensland Geogr J 39:38–39

14. Hopley D (1978) Geographical studies of the Townsville area (Monograph series, Occasional paper/James Cook University of North Queensland, Department of Geography; no 2). Department of Geography, James Cook University, Townsville

15. Flood P (1988) Shoreline changes on coral cays, Capricornia section, Great Barrier Reef marine park, Australia. In: Proceedings of the 6th international coral reef symposium, Australia, pp 219–224

16. Hopley D (1981) Sediment movement around a coral cay, Great Barrier Reef, Australia

17. Duvat VKE, Volto N, Costa S, Maquaire O, Pignon-Mussaud C, Davidson R (2021) Assessing atoll island physical robustness: application to Rangiroa Atoll, French Polynesia. Geomorphology 390:107871. https://doi.org/10.1016/j.geomorph.2021.107871

Large-Scale Modelling of Hydro- and Morphodynamics Associated with Reef Platform and Island Systems

Gerd Masselink[1]([✉]), Floortje Roelvink[1,2], Samuel T. Rose[3], Marion Tissier[4], Suzanna Zwanenburg[2], and Madelief Doeleman[2]

[1] University of Plymouth, Plymouth, UK
gmasselink@plymouth.ac.uk
[2] Deltares, Delft, The Netherlands
[3] University of Bath, Bath, UK
[4] Delft University of Technology, Delft, The Netherlands

Abstract. Hydro- and morphodynamic processes associated with reef platform and island systems are complex and challenging to measure under the energetic conditions that are commonly associated with significant island change. Numerical models of reef platform hydrodynamics are well developed and widely applied to predict the impact of sea-level rise (SLR) on future flood risk and island habitability. Morphodynamic modelling of atoll island response to SLR is much less well developed, partly due to a lack of suitable observational hydrodynamic and morphological data of island overwash required for developing a robust modelling capability. Hence, such modelling is largely based on uncalibrated models. Here, we describe a large-scale experiment in the Delta Flume where a 1:3 scale reef platform and island system was constructed out of concrete and sand, and subjected to a range of sea-level and wave conditions. It was found that the cross-platform changes in the hydrodynamics (wave energy dissipation, low-frequency wave energy, wave setup, wave runup) were represented very well by the 1D XBeach-NH model. During physical model simulations with SLR, the island was found to more or less keep pace with the rising sea level through overwash-induced deposition.

Keywords: Atoll Islands · Physical modelling · Overwash · Sea-Level Rise

1 Introduction

Atoll islands are considered amongst the world's most vulnerable environments to climate change as enhanced flooding due to sea-level rise (SLR) is expected to make most of them uninhabitable by 2050 [1]. However, island flooding can also result in overwash-induced deposition that increases island elevation and freeboard [2], and this may make the islands more resilient to SLR.

We presently do not have a robust modelling capability to investigate the future evolution of atoll islands due to SLR and this makes it difficult to design and evaluate

© The Author(s) 2026
C. Coelho et al. (Eds.): CD 2025, CRL 41, pp. 659–664, 2026.
https://doi.org/10.1007/978-3-032-15473-6_100

climate adaptation strategies for these islands. The general approach is to ignore any morphological consequences of island flooding and only conduct hydrodynamic modelling. The challenge of long-term modelling of atoll island evolution due to SLR is that there is a paucity of data on overwash hydro- and sediment-dynamic processes to develop such models. To address this gap, a 1:3 scale model of an atoll island and reef platform was constructed in the Delta Flume, Netherlands, and subjected to a range of wave and water level conditions.

2 Experimental Set-Up

The large-scale validation tests were performed in Deltares' Delta Flume, which has a length of 300 m, a depth of 9.5 m and a width of 5 m. A 110-m wide 'reef platform' with a steep 'fore reef' was constructed out of sand capped with a 10-cm layer of concrete (Fig. 1). The fore reef was located 90 m from the wave board and had a slope of 1:4.5. The elevation of the platform' was 5 m above the flume floor. A 60-m wide 'atoll island' was created on top of the platform. The island consisted of a 1-m layer of fine sand ($D_{50} = 0.25$ mm) and its maximum elevation was 6 m above the flume floor. The island was placed on top of the platform with 45 m and 10 m of exposed platform at the front and the back of the island, respectively. At the back of the island was a 'lagoon' that drained in a reservoir by means of an adjustable weir. The weir was adjusted to the height corresponding to the 'ocean' water level in front of the island. Water draining landward into the lagoon during the testing, e.g., through overwash or groundwater seepage, thus flowed into the reservoir where the data collected by a pressure sensor was used to derive the associated discharge.

The main instrumentation deployed during the testing included: 24 pressure transducers (PTs) for measuring water levels, 12 electromagnetic current meters (EMCM) for recording mean and wave-driven currents, including overwash flows, 7 LiDAR scanners for measuring complete water surface elevation profiles across the reef/island systems, as well as the island morphology, 3 CCTV cameras for recording wave runup and overwash characteristics and 3 Optical Backscatter Sensors (OBS) for measuring suspended sediment concentrations.

The maximum significant wave height that can be generated at the Delta Flume is 2.0 m, but due to the limited water depth during the testing (5–6 m), the maximum significant wave height was 1.5 m. In all tests, the wave field was forced with a JONSWAP spectrum. The wave board is equipped with an Active Reflection Compensation (ARC) to ensure waves reflected from the reef platform/island model are not re-reflected towards the model. Second-order wave generation was used in all tests, meaning that in addition to the random incident waves, a bound long wave associated with wave groupiness is also generated at the wave board.

3 Test Conditions

The experimental test series consisted of 3 months of testing, divided into three phases:

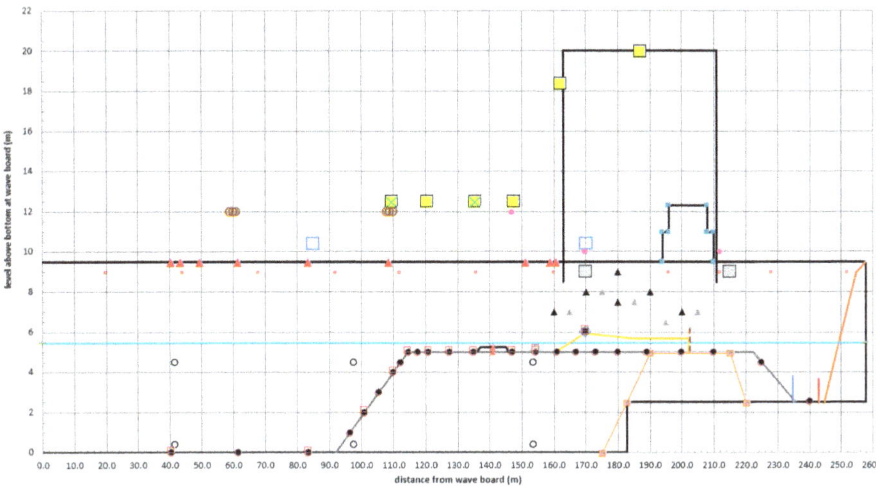

Fig. 1. Cross-section through the Delta Flume model: red triangles = wave gauges; yellow squares = LiDAR; grey squares = cameras; red squares = current meters; black dots = pressure sensors; black triangles = GCPs; blue horizontal line is default sea level; orange line = atoll island; black line = reef platform.

1. **TS1:** Hydrodynamic tests to measure wave transformation across the platform and runup on the island. Tests were conducted with and without artificial reef structures (Fig. 2) on the platform.
2. **TS2:** Morphodynamic calibration tests to identify different morphological response regimes (erosion, collision, overwash, inundation), also with and without artificial reef structures.
3. **TS3:** Sea-level rise morphodynamic tests to investigate the morphological response of the island to slowly increasing water levels

During **TS1** and **TS2**, the influence of artificial reef structures on wave transformation processes and island evolution was investigated. Two types of reef structures were deployed on the reef platform (Fig. 2 shows one type). During **TS1** an elevated berm (2.25 m above the platform) covered with geotextile cloth was added to the island to prevent overwash and morphological change, but during **TS2** and **TS3** the island is allowed to evolve and frequent reshaping will ensure that different forcing conditions start with the same island morphology.

Wave conditions used during the testing were based on an analysis of the modelled wave climates in the Maldives and the Pacific (and applying 1:3 Froude scaling). From the full wave climate at both sites, 11 representative wave conditions were extracted and these were used during the simulations. The most energetic wave condition modelled was $H_s = 1.5$ and $T_p = 5.77$ s.

During **TS1** and **TS2**, only a few sea level conditions were used, representing water depths across the reef platform of 0.17, 0.33, 0.50 and 0.67 m. During **TS3**, the sea level was increased from 0.33 m to 0.67 m in 1- or 2-cm steps per hour. These SLR simulations lasted a full week.

Fig. 2. Deployment of more than 100 artificial reef structures on the reef platform (see [3] for a description of the reef restoration runs).

4   Results

The measured cross-reef variation in mean sea level and significant wave height was compared with that modelled by a 1D XBeach-NH model during one of the **TS1** runs with a still water level of 0.33 m and offshore waves characterized by $H_s = 1$ m and $T_p = 7.51$ s (Fig. 3). The numerical model was run for two different bed roughness values ($C_f = 0.01$ and $C_f = 0.001$), both representative for smooth surfaces. The dissipation of incident wave energy across the reef platform, the wave setup profile and the increase in very-low frequency energy is reproduced very well by the numerical model in both cases, but the landward increase in infragravity energy is best modelled with the lower value of C_f.

Figure 4 shows the results of a SLR simulation conducted during **TS3**. During this run, sea level increased by 2 cm per hour and a total SLR of 0.34 m was achieved over 18×1-h runs. Whole-island accretion of c. 0.2 m occurred as a result of overwash. In addition, the front of the island retreated by c. 10 m and the lower part of the beachface became flatter. The crest of the island (the highest point) only accreted by a few cm. It could thus be inferred that despite the overwash-induced accretion, the island was losing freeboard and could not keep up with rising sea level. This is only partly confirmed by the time series of island overwash, where it can be seen that for the same wave forcing (same colored symbols in the right panel), overwash discharge only moderately increased during the SLR simulation.

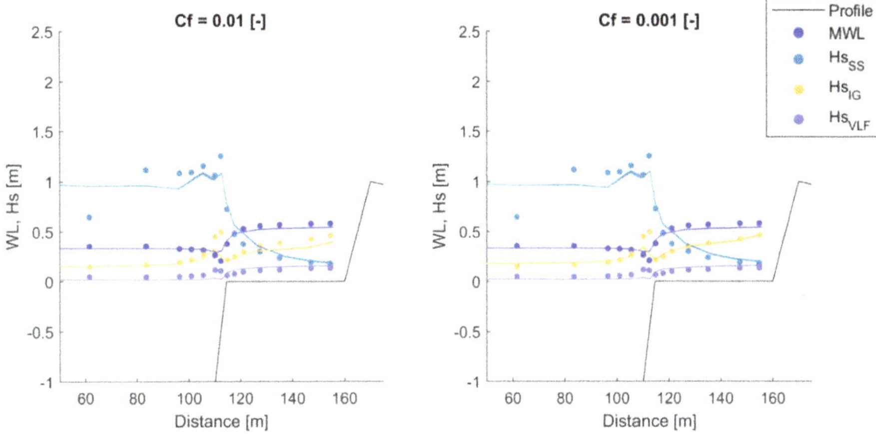

Fig. 3. Cross-reef variation in measured and XBeach-modelled mean sea level (WL) and significant wave height (H_s). Left and right panel show numerical model results for Colebrook-White roughness $C_f = 0.01$ and 0.001, respectively. Symbols and lines represent measurements and model results, respectively. Different colors represent (from top to bottom): grey blue = incident wave height; dark blue = mean water level; orange = infragravity wave height; and light blue = very-low- frequency wave height.

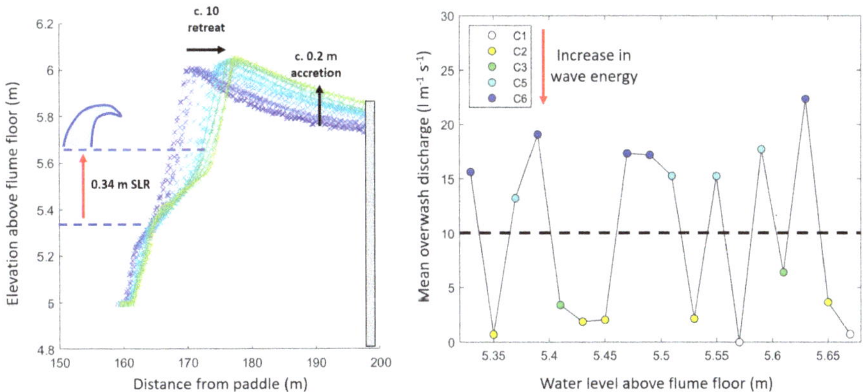

Fig. 4. Island development (left panel) and mean overwash discharge at the island crest (right panel) for one of the SLR simulations during **TS3** with the rate of sea-level rise set to 2 cm per hour. The colors in the left panel represent a progression in time from blue to green. In the right panel, the overwash discharge represents hourly averages and the different colored symbols represent different wave conditions, with wave energy increasing from C1 to C6.

5   Conclusions

A unique data set on reef platform wave transformation and atoll island dynamics was collected at the almost proto-type scale (1:3) in the Delta Flume. So far, the wave transformation across the reef platform is produced very well with a 1D XBeach-NH model using a smooth bed. The next steps are to extend the comparison to the measurements of runup and overwash discharge. The real challenge will be to reproduce the observed island development with a morphodynamic model. Such model can then be used to investigate long-term atoll island development in response to sea-level rise.

Acknowledgements. This work was conducted under the UKRI-funded ARISE project 'Natural Adaptation of Atoll Islands to Sea-Level Rise Offering Opportunities for Ongoing Human Occupation' (EP/X029506/1).

References

1. Storlazzi CD, et al (2018) Most atolls will be uninhabitable by the mid-21st century because of sea-level rise exacerbating wave-driven flooding. Sci Adv 4:eaap9741
2. Masselink G, Beetham E, Kench P (2020) Coral reef islands can accrete vertically in response to sea level rise. Sci Adv 6:eaay3656
3. Tissier M, et al (2025) Large-scale experiments on the effect of artificial reef restoration on wave-driven flooding. In: Proceeding coastal dynamics 2025

Attenuation of Geomorphic Capital Along an Ebb-Dominant Mixed-Energy Barrier Island

Patrick Barrineau[✉]

Coastal Science and Engineering, Inc., Columbia, SC 29205, USA
patrick@coastalscience.com

Abstract. Debidue Beach and North Inlet are a coupled mixed-energy inlet-barrier system along the South Atlantic Bight in South Carolina, USA. Long-term chronic erosion along much of Debidue throughout the 20th century is the result of a shoreline adjustment triggered by an avulsion in the main channel of North Inlet occurring between 1926 and 1934. We use a historical database of shoreline positions compiled by the United States Geological Survey (USGS) and the Analysis of Moving Boundaries Using R (AMBUR) package to quantify changes in the Debidue Beach shoreline over nearly 150 years from 1872 to 2011. This analysis documents relatively large-scale shoreline changes (and equivalent volumetric changes above local depth of closure 'DOC') following the shift in channel position, and a logarithmic decrease in erosion rates over the following decades. Mixed-energy ebb-dominant inlets have considerable effect on adjacent beaches due to their ability to retain and shed large quantities of sand relatively quickly. This study demonstrates that even systems exhibiting long-term stability – like North Inlet – are indeed migrational landforms and should be considered as such when formulating shoreline management plans along adjacent beaches.

Keywords: ebb-tidal delta · barrier evolution · shoreline adjustment

1 Introduction

Debidue Beach is an 8-kilometer-long mixed energy barrier located along the South Atlantic Bight in South Carolina, USA. It is bordered to the north by the relatively small Pawleys Inlet and to the south by the relatively large North Inlet. Net longshore sediment transport is from north to south. Because Pawleys Inlet is a relatively small migratory inlet, it continuously bypasses sediment towards Debidue Beach. North Inlet, however, is a larger system exhibiting relative stability over longer periods of time [1]. As a result, sand tends to accumulate on the updrift (Debidue) side of North Inlet between large bypass episodes during which material is lost to the south.

Measured changes in shoreline position along Debidue Beach from 1872 to 2011 suggest a large-scale bypass in the North Inlet system occurred between 1926 and 1934. Using measured horizontal shoreline changes and an arithmetic conversion to estimate volumetric changes, we generate reasonable estimates of large-scale beach volume changes from 1872 to 2011. Additionally, we compare these results to annual

C. Coelho et al. (Eds.): CD 2025, CRL 41, pp. 665–670, 2026.
https://doi.org/10.1007/978-3-032-15473-6_101

monitoring data collected via survey beyond the local depth of closure (DOC) by Coastal Science & Engineering, Inc. over the same period. This comparison shows the trends identified in the shoreline position analysis are a reasonable proxy for the equivalent volumetric changes out to DOC.

The findings presented herein demonstrate shoreline position changes are a reasonable – though imperfect – substitute for volumetric changes. Additionally, the results demonstrate that large shoreface deposits like relict ebb-tidal deltas can alter local sediment budgets for more than a century following abandonment. This second finding is particularly relevant for researchers exploring the volume of sediment available for shoreward transport through future periods of accelerated sea level rise (SLR). Modern barrier islands with a deficit of relict sediment available for post-storm recovery and filling of landward accommodation space as sea levels rise tend to exhibit rapid rates of transgression [2]. Determining the locations and volumes of such deposits of 'geomorphic capital' is therefore a useful exercise due to the nonlinearities inherent in a barrier system's reaction to future SLR [3].

2 Methods

Historical shorelines are determined using the position of mean high water (MHW) from a mixture of older publicly available and younger manually delineated sources. A database developed by the United States Geological Survey (USGS) contains shapefiles of shorelines for South Carolina from 1852 to 2000. The 1852 shoreline is not complete along Debidue Beach, but shorelines are available from 1872 to 2000. Additional USACE shorelines from 2006 and 2011 are included as well as newer shorelines (2014, 2017, 2020, 2024) digitized from orthophotography collected as part of ongoing beach monitoring efforts along the study area.

The shoreline positions and change rates are measured using the Analyzing Moving Boundaries Using R (AMBUR) package [4]. The horizontal change in shoreline position is converted into volumetric changes using the formula:

$$\partial V = H \times \partial S \tag{1}$$

where ∂V is volumetric change, H is the shoreface height, and ∂S is the horizontal change in shoreline position. This formulation is commonly used in engineering practice, particularly at longer time scales [5]. As a means of testing this formulation, horizontal changes are averaged across ~1- to 2-km reaches and compared against surveyed volumetric changes above local DOC.

3 Results

Accretion between 1872 and 1926 advanced MHW nearly 260 m seaward (4.8 m/yr) along Debidue Island without any longshore migration in the primary ebb channel of North Inlet. Between 1926 and 1934, the primary ebb channel moved nearly 2 km south. From 1934 to 1973, the rate of erosion along Debidue Beach decreased logarithmically from 6 m/yr to 2 m/yr with a net linear average of -2.7 m/yr and a total average landward retreat of 71 m over 39 years. From 1973 to 2011, erosion rates hovered around 2 m/yr with a net linear average of -1.6 m/yr and a total average landward retreat of 61 m over 38 years. These results are depicted in graphical (Fig. 1) and map (Fig. 2) formats below.

Trends in horizontal shoreline change across averaged by reach compared well (average $R^2 = 0.8$) to volumetric changes measured via survey profiles. Correlations between horizontal and volumetric changes calculated along individual profiles were weaker (commonly $R^2 < 0.5$), though inlet and bar-related processes appear to influence results along individual profiles. Assuming the correlation between horizontal and volumetric changes is acceptable for reach scale conversions, the 1872 to 1926 accretion represents nearly 13 million m^3 of sediment accumulation: approximately $+50$ m^3/m/yr. From 1934 to 1973, annual losses along the barrier island decreased logarithmically from -60 m^3/m/yr to approximately -10 m^3/m/yr where the long-term change rates have settled into a dynamic equilibrium from 1973 through 2024.

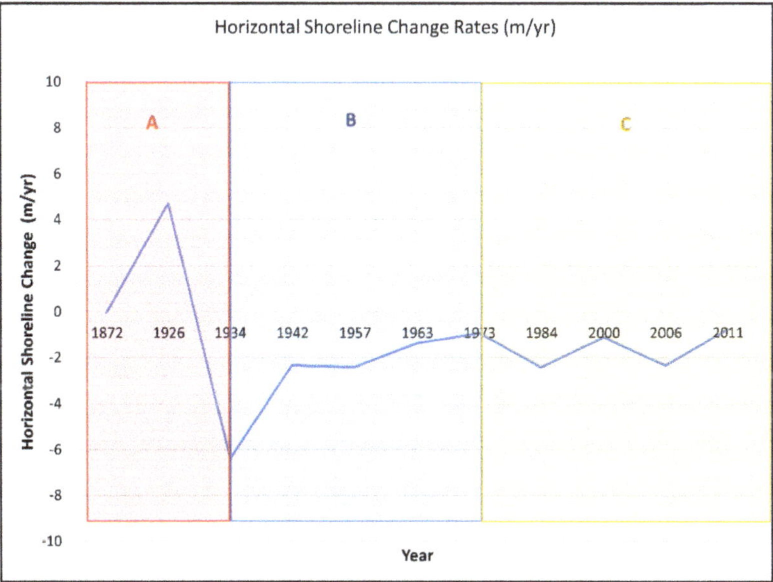

Fig. 1. Three distinct phases of system response are demonstrated in the results presented herein. These consist of an initial avulsion in volumes and inlet position ('A'), attenuation of the accumulated geomorphic capital of the abandoned ebb tide delta ('B'), and resumption of the natural 'background' long-term erosion rates ('C').

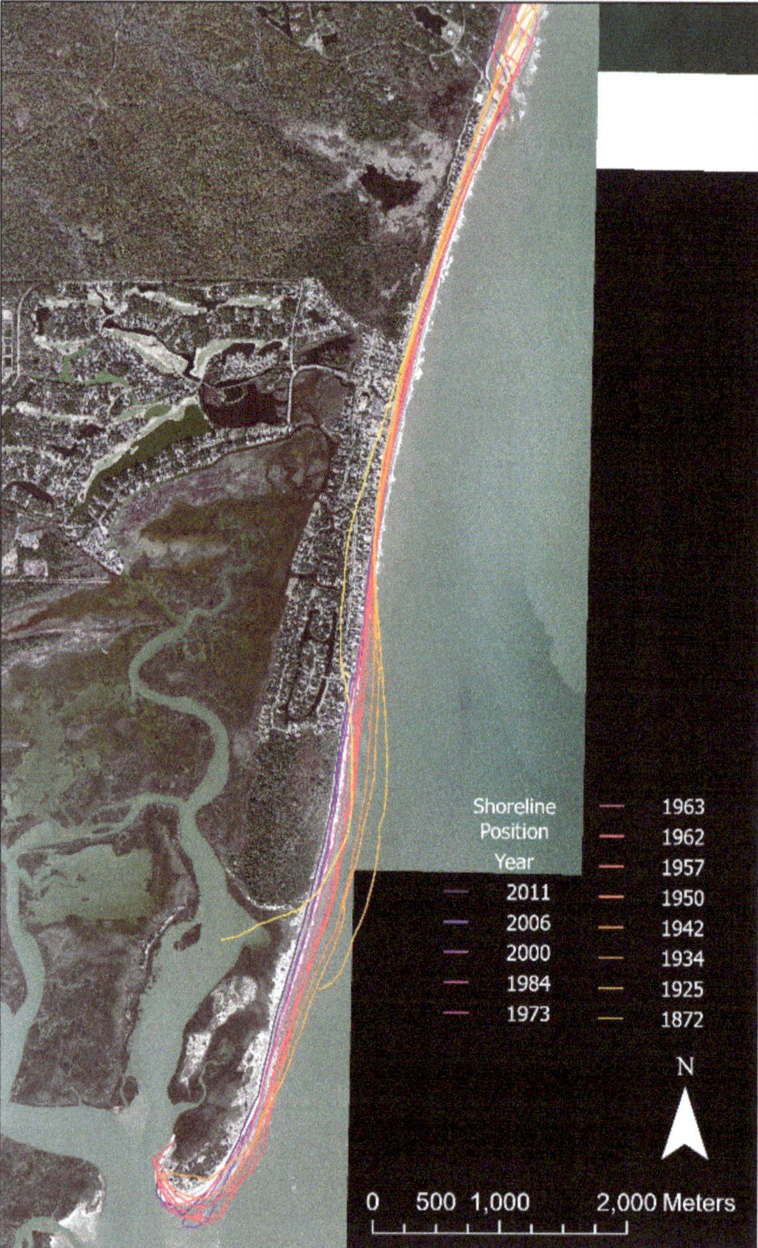

Fig. 2. Significant volume increases from 1872 to 1925 were followed by spit extension from 1925 to 1934, and subsequent drawdown in the geomorphic capital of the abandoned ebb tide delta.

4 Discussion

The measurement of horizontal change rates from 1872 to 2024, and likewise volumetric changes above local DOC, demonstrate natural bypassing at North Inlet can alter shoreface volumes by 10^7 m^3. For comparison, this volume change is greater than the inferred and measured total of all anthropogenic shoreface manipulations (e.g. hardened structures, nourishment) over the same period. Determining the exact mechanism for the large bypass is beyond the scope of this study, but it is unlikely that any anthropogenic activity in the area would have caused such an avulsion – especially given the similarities in forms and rates of change across similar mixed-energy settings.

The increase in volume along southern Debidue Beach from 1872 to 1926 represents the buildup in 'geomorphic capital' via ebb delta deposits at the former North Inlet site [2, 3, 6]. The logarithmic decrease in annualized loss rates from 1934 through 2024 reflects gradual attenuation of the stored capital and a resumption of the long-term background erosion rates due to prevailing wave conditions. The effects of sea level rise on nearshore processes at Debidue Beach have been outstripped for nearly a century by this accumulation and attenuation of stored sediment triggered by the migration of North Inlet a century ago.

Understanding the frequency and magnitude of background changes to shoreface volumes is important for long-term coastal planning along developed mixed-energy shorelines. Relict deposits generated from historical bypass events – or inherited from previous regimes of sediment transport and storage – can cause decadal-to-centennial-scale chronic erosion under the right conditions. In an era of widespread transgression and cascading impacts to barrier systems [7, 8], this study demonstrates two useful lessons for coastal managers in mixed-energy settings. First, historical understanding of a system is useful for determining scopes of work for coastal projects. Second, boundary conditions for risk assessments and models along mixed-energy settings should be examined closely.

5 Conclusions

The findings here demonstrate that even relatively large mixed-energy inlets exhibiting longer-term positional stability experience cyclical changes related to shoal bypass cycles, which in turn affect long-term erosion and accretion rates along adjacent shorelines. This conclusion suggests more mixed-energy inlets should be examined to identify the long-term envelope of shoreline change along adjacent beaches.

With increasing rates of sea level rise [9], tidal prism and in-channel hydraulics in these systems will change. These changes will impact the rates of bypassing in mixed-energy systems, which will trigger changes in shoreline position along adjacent beaches.

The availability (or unavailability) of geomorphic capital to be recycled and transported landward from the nearshore is likely to represent an important element in long-term planning for developed coastal areas like the South Atlantic Bight and other mixed-energy barrier systems.

More broadly, planning for regional sediment management along developed portions of mixed-energy barrier systems will necessitate identification and analysis of these deposits for use in future renourishment events.

References

1. Gaudiano DJ, Kana TW (2001) Shoal bypassing in South Carolina tidal inlets: geomorphic variables and empirical predictions for nine mesotidal inlets. J Coastal Res 17
2. Mariotti G, Hein CJ (2022) Lag in response of coastal barrier-island retreat to sea-level rise. Nat Geosci 15
3. Hein CJ, Kirwan ML (2023) Marine transgression in modern times. Annu Rev Marine Sci 16(1)
4. Jackson CW, Alexander CR, Bush DM (2012) Application of the AMBUR R package for spatio-temporal analysis of shoreline change: Jekyll Island, Georgia, USA. Comput Geosci 41
5. Farris AS, List JH (2007) Shoreline change as a proxy for subaerial beach volume change. J Coastal Res 23(3)
6. Luijendijk A, Hagenaars G, Ranasinghe R, Baart F, Donchyts G, Aarninkhof S (2018) The state of the world's beaches. Sci Rep 8
7. Anarde KA, Moore LJ, Murray AB, Reeves IRB (2024) The future of developed barrier systems: 2. Alongshore complexities and emergent climate change dynamics. Earth's Future 12
8. Barrineau P (2024) One hundred years of inlet processes and barrier adjustments at North Inlet, SC. Shore Beach 92(3)
9. Dangendorf S, Qiang S, Wahl T, Thompson P, Mitrovica JX, Hamlington B (2024) ESSD – ocean/physical oceanography. https://doi.org/10.5194/essd-2024-46

Using a Unique 150-Year Geomorphological Cycle in Modern Beach Nourishment Design

Scott L. Douglass[1]([✉]), Bret M. Webb[2], and Thomas "Beau" Buhring[1]

[1] South Coast Engineers, PO Box 72, Fairhope, AL 36533, USA
scott@southcoastengineers.com
[2] Department of Civil, Coastal and Environmental Engineering, University of South Alabama, Mobile, AL, USA

Abstract. A unique, repeating 150-year geomorphological cycle of migration of a small island onto the beaches of Dauphin Island is being considered in the design of modern beach nourishment projects. We are in the midst of the third such migration in recorded history (since 1699) but there is geomorphological evidence (extremely high sand dunes) that the same cycle has occurred for millennia. The modern beach nourishment projects (2016, 2024, and a third being planned for 2026–27) are being designed to work with this ongoing migration.

Keywords: barrier island · geomorphology · inlet dynamics · beach nourishment · Dauphin Island

1 Introduction

Dauphin Island, Alabama, USA, has a rich cultural history. Located just west of the mouth of Mobile Bay, it is the easternmost island in the Mississippi-Alabama barrier island chain that extends roughly 100 km to the west.

A repeating cyclical geomorphological phenomenon has been discovered that has impacted both the island's history and its beaches. The repeating phenomenon is now influencing the design of the island's adaptation strategies to protect infrastructure and mitigate against sea level rise. This paper describes the 150-year cyclical phenomenon, which is happening right now for the third time in recorded history, and how our understanding of geomorphology is informing the design of beach nourishment projects to protect both the island's natural habitats and built infrastructure.

2 The Unique 150-Year Geomorphological Cycle

We have long recognized that the shoreline of Dauphin Island is influenced by the location of several small islands and shoals, commonly called Pelican Island and Sand Island, immediately to the south that are, essentially, on the outer shield of the extremely large Mobile Pass ebb-tidal shoal. The locations of the emergent portions of these small,

© The Author(s) 2026
C. Coelho et al. (Eds.): CD 2025, CRL 41, pp. 671–676, 2026.
https://doi.org/10.1007/978-3-032-15473-6_102

ephemeral islands have moved around south of Dauphin Island for centuries. A lighthouse built on Sand Island in the 19th century is still standing but it has been surrounded by water for the past 8 or 9 decades (i.e., the island has eroded away from the lighthouse structure). These shoals usually feed sand to the beaches of Dauphin Island through stable-inlet processes including sand movement along a subaqueous sand bar system between the offshore shoals and the larger island [1].

Sanchez and Douglass [2], however, realized that there is a history of major migrations of Pelican Island onto Dauphin Island and predicted another migration which subsequently occurred about a decade later in 2008 (see Fig. 1). This is a natural mechanism for episodic sand bypassing at the tidal inlet. Most of the time, Mobile Pass fits into Fitzgerald's Stable Inlet Processes Model, where sand is moved onto the downdrift island by wave-driven transport underwater [3]. However, once every 150 years, the bypassing takes an extreme and modified form of Fitzgerald's ebb-tidal shoal breaching mechanism, where Pelican Island migrates or "welds" onto the downdrift island en masse. Many inlets have ebb-tidal breaching processes where the main ebb channel migrates downdrift and then realigns, releasing a large underwater sand bar to migrate to the downdrift island. At this location, however, it happens on a uniquely long time-cycle and with an emergent island [4, 5]. Historical charts dating from 1717 and 1864 depict similar migrations.

Fig. 1. The presently ongoing migration of Pelican Island (now called "Pelican Peninsula") onto Dauphin Island, Alabama. Similar migrations occurred in 1717 and 1864 (this image is circa 2010).

The 150-year duration between episodes of this documented cycle at Dauphin Island is truly unique. One other episodic coastal geomorphology cycle has been documented with recurrence every 40–80 years; a quasi-cyclic, repeating breach in Nauset Beach near Chatham, Massachusetts on Cape Cod [6].

Dauphin Island has sand dunes that are over 15 m high along the shoreline where Pelican Island is collapsing onto Dauphin Island today. These sand dunes protect a mature maritime forest as well as the old "Village" portion of Dauphin Island, where most of the houses are today and where the original European settlers started the Town in 1699. Direct descendants of those original settlers still live on the island today. The large dune field has always protected the village from major hurricanes. These massive sand dunes (some of the largest along the northern Gulf of Mexico coast) are likely here because of repeated migrations of Pelican Island onto these beaches for millennia.

The presently ongoing migration of Pelican Island onto Dauphin Island completely sealed the former Pelican Passage between the islands in 2008 and now the new "Pelican Peninsula" is gradually getting shorter as the sand moves north driven by waves (an excellent, online illustration of this recent migration is at [7]). This is essentially a very large, natural, very slowly developing beach nourishment for Dauphin Island. The recent migration left the former municipal fishing pier completely landlocked (about 400 m north of the shoreline by 2024!). The present-day shoreline shape (see Figs. 1 and 2) bears an uncanny resemblance to the shoreline in 1717. The 1717 migration sealed off the entrance to the port that was briefly the capital of the French Louisiana Territory. The 1864 closure/migration was noted on the surveys used by Admiral Farragut in the Battle of Mobile Bay with his famous command to "damn the torpedoes, full speed ahead".

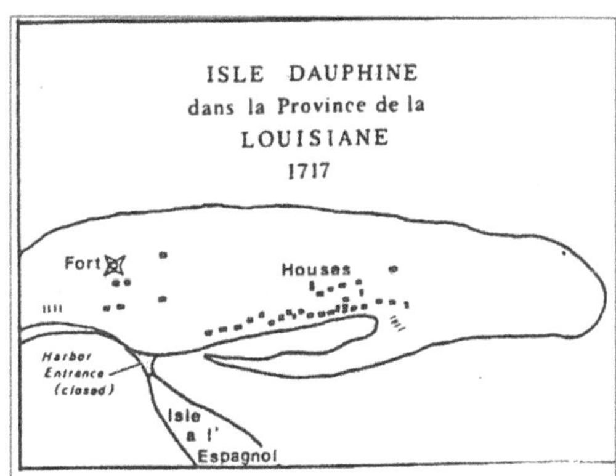

Fig. 2. 1717 Shoreline of the east end of Dauphin Island [8]

3 Modern Beach Nourishment Projects

The Town of Dauphin Island manages the beaches of the island today. The Town's management decisions are framed by the geomorphological dynamics of the barrier island with the recognition that sea level is rising. Two major beach nourishment projects have been constructed on the east end of the island, and one is being designed now for the west end of the island. These projects have been designed to take full advantage of the larger natural geomorphological phenomenon, and the designs are being informed with modern numerical modeling including applications of ADCIRC-SWAN, XBeach, and EDGR (e.g., [9]).

The first major, engineered beach nourishment project in island history was constructed in 2016 on the easternmost 1.5 km of south-facing beaches with 250,000 m^3 of sand mined from a portion of the Mobile Pass ebb-tidal shoals over 6 km south of Dauphin Island. The sand was moved to the beaches by hydraulic pipeline. That 2016 project included new coastal structures to retain some of the new sand at the far eastern end of the project where the public has the best access to the beach and where the shoreline recession had been the greatest due to wave-driven longshore sand transport. The new structures were built where an old groin field, originally constructed in 1909, was completely flanked by the shoreline recession and had been standing in the water for decades. The new structures are three shore-parallel breakwaters designed for two to function as shore-attached breakwaters and the westernmost one to function as an offshore breakwater. The 2016 project performance was excellent both in terms of habitat creation/protection and longevity. The primary reason for the perceived success of the project was the high quality of the nourishment sands, which matched the native beach sands almost perfectly.

The second major beach nourishment project in island history was recently completed along 2.5 km of the East End in 2024 with 920,000 m^3 of sand from the same offshore shoal area. This project renourished the 2016 project and extended it farther west. These projects have widened the beach by over 100 m. Over 600,000 plants, primarily sea oats grown from native seeds, were planted on the newly constructed sand dunes.

Figure 3 shows the new beach about six months after construction. In the foreground/left are the three coastal structures built in 2016. Both the middle and western breakwaters are shore-attached in the photograph, but the tombolo attached to the westernmost breakwater is expected to detach and become a salient in the next year or so. The new pocket beaches are extremely popular with residents and visitors. This beach is used for summer evening movie nights, sandcastle building contests, wading and fishing. The mature vegetation line at the right of Fig. 2 (with vegetated sand dunes) is about where the 2015 (pre-project) shoreline was located. This beach had receded over 200 m in the decades prior to the 2016 project. New sand fencing and vegetation is visible on the new 2024 beach project in the middle of the photograph. In the far background, a new sand dune in front of houses is visible. Those houses were in the surf prior to the 2024 project construction.

Fig. 3. The 2024 Dauphin Island East End Beach and Dune Restoration Project (photo looking west: October 2024).

The 2024 East End project was designed to work with the ongoing collapse and migration of Pelican Peninsula. As the peninsula gets shorter and shorter, more wave energy from southwest is expected to strike these East End beaches and drive some of that sand back east toward these structures.

A third major beach nourishment project is now being designed for the west end of Dauphin Island for construction in 2026–27. That project will be built to the west of Pelican Peninsula (to the left side of Fig. 1). It is being designed to take full advantage of the sand coming to Dauphin Island because of the ongoing migration of Pelican Island onto Dauphin Island. About 3,000,000 m^3 of sand is being fed to these beaches during this multi-decadal migration. Severely eroded beaches west of this area (not shown here) will be nourished with beach-quality sands from a different part of the Mobile Pass ebb-tidal shoal. The expectation is that future maintenance of Dauphin Island's beaches will be partly accomplished with improved artificial bypassing of sands dredged to maintain the Mobile Ship Channel that runs through the Mobile Pass ebb-tidal shoal.

Acknowledgements. The Town of Dauphin Island, Alabama has funded much of the work used in developing the opinions outlined in this presentation with grants from the Alabama Department of Conservation & Natural Resources, NOAA, the National Fish and Wildlife Foundation's Gulf

Environmental Benefit Fund, the Resources and Ecosystems Sustainability, Tourist Opportunities, and Revived Economies of the Gulf Coast States Act (RESTORE Act: some of the penalty monies from the 2010 Deepwater Horizon oil spill), and others. This long-term support is gratefully acknowledged. South Coast Engineers was the lead planning and design firm that developed the modern beach nourishment projects with significant assistance from Weeks Marine, Ardea Environmental Consultants, Goodwyn Mills Cawood, APTIM, Coastal Planning & Engineering, Coastal Technology Corporation, Moffatt & Nichol, and others.

References

1. Douglass SL (1994) Beach Erosion and Deposition on Dauphin Island, Alabama, U.S.A. J Coastal Res 10(2):306–328
2. Sanchez TA, Douglass SL (1996) Stop 3a: the Pelican/Sand Island Shoal Complex. In: Hummell, Haywick (eds.) Coastal deposition and ecosystems of Alabama, A guidebook for the 33rd annual field trip of the Alabama geological society
3. Fitzgerald DM (1988) Shoreline erosional-depositional processes associated with tidal inlets. In: Aubrey DG, Weishar L (eds) Hydrodynamics and sediment dynamics of tidal inlets. Springer, Berlin, pp 186–225
4. Douglass SL, Webb BM, Buhring B (2022) Coastal processes and geomorphology of the east end of Dauphin Island, South coast engineers report to the town of Dauphin Island, January 20, 2022 (rev. 1), 107 p
5. Douglass SL, Webb BM, Buhring B (2023) Coastal processes and geomorphology of the west end of Dauphin Island, South coast engineers report to the town of Dauphin Island, October 23, 2023, 119 p
6. Giese GS, Mague ST, Rogers SS (2009) A geomorphological analysis of nauset beach/pleasant bay/Chatham Harbor for the purpose of estimating future configurations and conditions, Report to the Pleasant Bay Resource Management Alliance, 32 p
7. Cassidy E (2024) Dynamic Dauphin Island. https://earthobservatory.nasa.gov/images/152444/dynamic-dauphin-island
8. Giraud M (1974) A history of French Louisiana. Translation copyright by Louisiana State University Press
9. Passeri DL, Bilskie MV, Hagen SC, Mickey RC, Dalyander SP, Gonzalez VM (2021) Assessing the effectiveness of nourishment in decadal barrier island morphological resilience. Water 13(7):944. https://doi.org/10.3390/w13070944

Emerging Technology for Surfzone Observations

Surf-Zone Rip Current Measurements Using Robotic Drifters

Adam M. Collins[1]([⊠]), A. Spicer Bak[1], Dylan J. Anderson[1], Ian R. Conery[1], Grant R. Fischer[1,2], and Katherine L. Brodie[1]

[1] U.S. Army Corps of Engineers, Duck, NC, USA
Adam.M.Collins@erdc.dren.mil
[2] University of North Carolina - Wilmington, Wilmington, NC, USA

Abstract. Rip current measurements in the surf zone are essential for understanding nearshore circulation, improving coastal hazard assessments, and refining wave-current interaction models. Traditional in-situ approaches—such as dye tracking, optical imagery, and GPS-enabled drifters—provide valuable but often limited data, as drifters can be rapidly ejected offshore and yield sparse measurements within the surf zone, decreasing efficiency. This study investigates a novel robotic drifter system designed to overcome these limitations by combining high-mobility surface operations with station-keeping and drift-timer modes. These autonomous drifters collect repeat velocity measurements in precisely defined areas, enabling the creation of statistically robust flow fields. Initial results from field experiments at Duck, NC, demonstrate the system's capacity to resolve fine-scale rip current features, and reveal time-averaged flow patterns. Overall, this robotic drifter system offers a platform for comprehensive, high-resolution mapping of surf-zone currents, facilitating more accurate assessments of rip current behavior and improving safety and operational decision-making in coastal environments.

Keywords: surf-zone · surface currents · robotics

1 Introduction

Surf-zone currents are a fundamental contributor to many areas of coastal interest, from altering bathymetry to inducing high velocity flows that can imperil equipment and lives. Historically rip currents have been measured through PIV methods and low cost GNSS based drifters [1–3] to enable the measurement of surface flows.

Traditional Lagrangian drifters are affected by flow changes on small spatio-temporal scales, such as transient eddy features, creating drift tracks that are an integration of instantaneous velocities not representative of a time-averaged current field (a more useful scientific quantity). This is mitigated by using multiple passes of drift tracks through a grid cell. Inevitably this leaves cell coverage and density to chance depending on short-term changes in turbulence, often coalescing around points of low-flow, which is less than ideal. Studying rip-currents, where flows are directed offshore can make traditional drifters an energy intensive exercise (and/or dangerous) with a lot of discarded

© The Author(s) 2026
C. Coelho et al. (Eds.): CD 2025, CRL 41, pp. 679–685, 2026.
https://doi.org/10.1007/978-3-032-15473-6_103

data. Additionally, drifter GNSS measurements experience noise from wave-by-wave interactions that can lift the drifter and propel it toward shore. Filtering the noisy *surfing* motions from the desired Lagrangian surface current measurements is difficult, complicating GNSS drifter derived velocity maps to electro-optical and infrared imagery derived estimates [2–5].

We propose the utilization of a novel robotic drifter system to address some of these limitations of the 'classical' drifter model for tracking potentially hazardous surf-zone currents. These systems, while very similar, have motors, and control systems that allow for both traditional long drifts, while also being able to focus short drifts in a prescribed area, allowing creation of dense drift paths and enhancing sampling in conditions or locations that would be difficult to sample with traditional methods. These drift paths can be discretized and averaged over varying length and space scales, to aid in analysis of a range of coastal features from time-averaged bulk-flow maps, investigations into small transient eddy features, and rip current identification. Additionally, the GNSS data can be used to filter unrealistic velocity and orientation changes that correspond to flow features of low interest (e.g. surfing). These data can then be used to validate, or train nearshore numerical or predictive current models.

2 Methodology

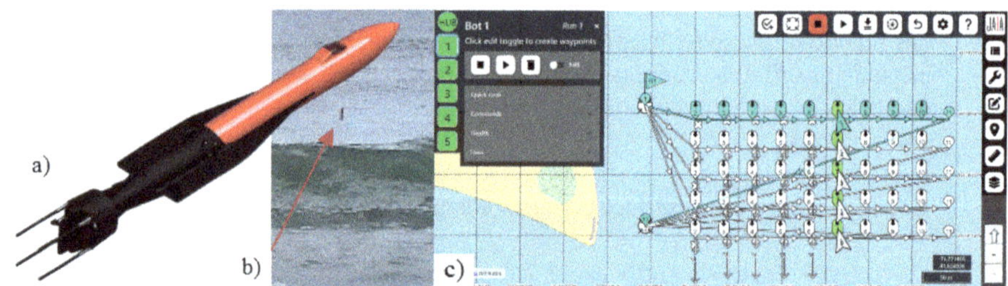

Fig. 1. (a) Image of the Jaia robot, (b) floating JaiaBots offshore of the breakpoint in Duck, NC (indicated with red arrow), (c) screenshot of the mission planning interface.

The robotic drifters (Fig. 1a), built by Jaia Robotics, are equipped with a variety of sensors, including GPS, IMU, conductivity and pressure sensors and utilize open-source software/data. They transit along the surface of the water at high speeds (up to 10 knots), while also having the ability to dive below the surface to collect water depth measurements by utilizing the pressure sensor. For collecting current measurements, the vehicle will turn its engine off and float (Fig. 1b). The web-based command and control interface is accessed through an Android tablet connected to a radio hub on the beach. (Fig. 1b) Communications with the bots are over 900 MHz band radio allowing for mission programming and on-the-fly adjustments while in the field.

2.1   Collection Modes

For collecting currents, we use two different modes of operation: 'Drift Timer' and 'Station Keep'. Drift timer mode turns the motor off for a user defined period of time (e.g. 60–600 s). After the end of this timer, the robot will then return to the initial position, repeating the drift timer. This cycle can be repeated in as many locations and times as desired. Preliminary results suggest this is useful for tracking ejection patterns in rip currents, and instantaneous currents over larger distances (25+ m). The 'Station Keep' mode will return to an inner radius once a bot leaves an outer radius (both of which are user prescribed, 5 and 15 m for this effort).

2.2   Processing Drifts to Grids

GNSS tracks are recorded over the whole deployment (sometimes hours) and broken up in to 1-h chunks across multiple bots (up to 4 at a time). This manuscript examines a single hour 11 October 2023 at 12:00 UTC. Raw GNSS tracks are filtered to 5 Hz to account for dunking before converted to a shore-perpendicular coordinate reference frame, and then the velocities are calculated in a cross/alongshore dimension approximated using a forward differencing scheme. Velocities moving shoreward > 1 m/s were removed as an initial cut to minimize *surfing* motions.

These flow fields can be quantified and averaged over varying time and space scales, to aid in analysis of a range of coastal features from time-averaged bulk flow maps, investigations into small transient eddy features, and identification of rip currents. Filtered velocities are binned to a 5m-by-5m grid and velocities are averaged. Grid size was determined by looking at static GNSS position over 15 min of collect, showing a noise floor of ± 0.05 m.

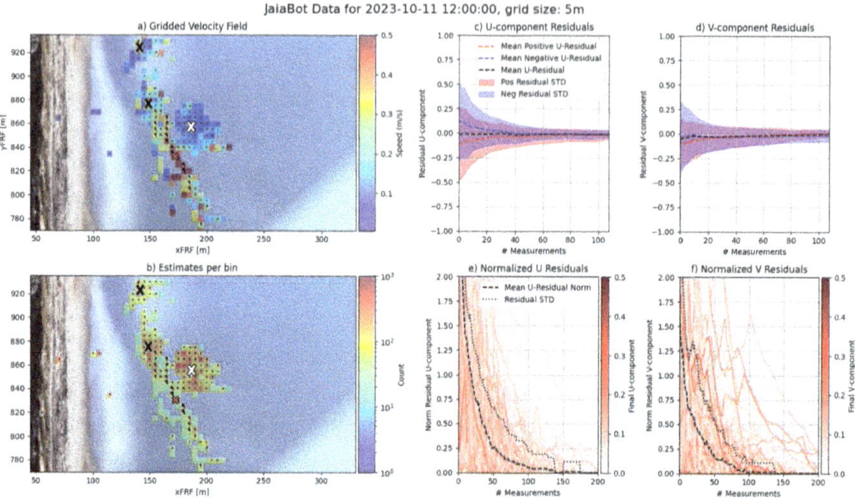

Fig. 2. Velocities for a collection on 2023-10-11 at 12:00 ET, where xFRF is the cross-shore coordinate and yFRF is the along-shore coordinate. (a) shows a nadir view on rectified tower based imagery of velocity grid cells. Color of cells indicate magnitude of the velocity in a particular cell, while black arrows also represent magnitude as well as direction. Cells with black arrows have 50 measurements, indicating confidence in the vector. Lower speeds have shorter arrows, where arrows at very low speeds (<0.1 m/s) look similar to black dots. Black x's show the target initial location for drift timer, and a white x shows the location of a station keep. (b) shows the number of measurements made inside each bin during the collection, note that cells with counts < 50 do not have a black arrow or dot. (c, d) shows the cumulative average residual of the u and v component of grid cells with a velocity measurement count above 50 within each cell. (e, f) shows the normalized residual value of each individual grid cells as well as the residuals mean (dashed black) and standard deviation of the mean (dotted black).

3   Results and Discussion

The gridded velocity product for a collection ($H_s = 0.85$ m, $T_p = 14$ s, $T_m = 8$ s) at the Field Research Facility in Duck, NC is shown in Fig. 2. For this collection we utilized both the 'Drift Timer' and 'Station Keep' collection modes with two robots deployed simultaneously. Figure 2a shows the velocities collected by the bots, with colors showing increasing magnitude of speed. The first bot was allowed to travel and be ejected from the surf from its initial position along the rip current starting at x, y = 145, 875 (black x). The flow patterns show a transverse rip moving to the southeast. Drift timer tracks were also initiated at x, y = 145, 925 and second bot was set to station keep at x, y = 195, 850, to show the northern boundary of the rip current, where there is little net flow. Naturally, higher counts occur in cells with lower velocities (Fig. 2b) demonstrating the challenge associated with drifting measurements, as higher velocity bins will naturally see less measurements as the bots are swept through more quickly.

To better understand the data count required to produce confident velocity estimates we examine the residual between instantaneous velocity measurement and the cell's final average in both cross-shore velocity (u-Fig. 2c), and along-shore velocity (v-Fig. 2d) as a function as the number of velocity measurements in each bin. The mean and standard deviation of the residual of the positive and negative velocity cells are more uniform for the cross-shore direction (Fig. 2d), while there is a distinct spread in the directionality of the residual based on whether it is on-shore or off-shore directed (Fig. 2c). As more measurements are collected, these residuals asymptote toward zero and the final average velocity. Figure 2e, f shows the normalized residual of the u, v, components decreasing over time. For this collection, after 50 measurements the mean error of 25%, which reduces to the GNSS noise floor at around 100 measurements.

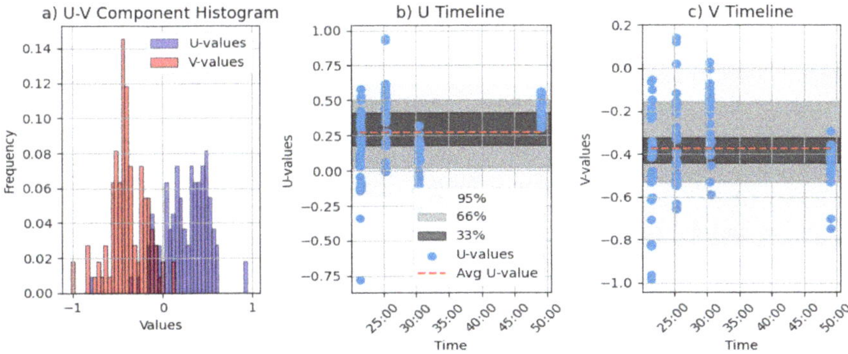

Fig. 3. Velocity measurements for 2023-10-11 at grid cell x, y = 175, 830 (marked by a red x in Fig. 2a, b). (a) shows a histogram of the u and v values measured in this grid cell during the collect. (b, c) shows timeline plots of the measurements taken within the grid cell, as well as the average value. The 95, 66, and 33% of instantaneous values measured during the grid cell are shaded in greys.

Figure 3 examines the contents of a single cell, in the neck of the rip current from this day at x, y = 175, 830 m (marked with a red x in Fig. 2a. b). Figure 3a is a histogram of the u and v components of the individual velocity measurements that fall within that grid cell. Figure 3c and d shows a timeline plot of the measurements for each component of the velocity. This grid cell was revisited four separate times, with the first three visits happening within 10 min (20–30 min after start), and then another visit at 50 min after start. The histogram (Fig. 3a) and timeline plots (Fig. 3b, c), highlights the noise in the measurements native to drifter based measurements (driven by pulsating and turbulent flows). While the spread in a single pass can be large, looking at the average and standard deviation of the velocity measurements (Fig. 3b, c – varying grey bars), we gain confidence in our estimates.

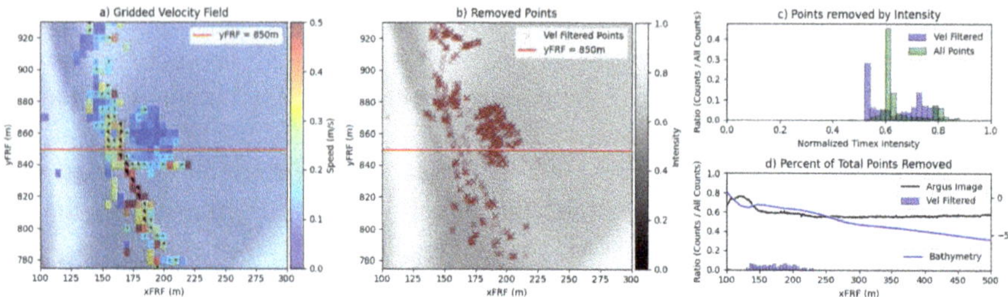

Fig. 4. Points removed by a shoreward velocity filter. (a) shows velocities for a collection on 2023-10-11 at 12:00 ET. (b) shows points removed from the collection before gridding, on a channel averaged image. (c) likelihood of all of the points, and of velocity filtered points to be in a particular intensity bin. (d) shows percent of removed points compared to the total points in the collect as a function of cross-shore position, as well as the Argus intensity (black) and bathymetry profile (blue).

Current filtering approaches, while still under development, use sampling frequency ($dt = 0.2$ s) and shoreward directed velocity filter (>1 m/s) to remove for dunking and 'surfing' effects from the bot being caught up and riding on breaking waves. Figure 4a shows the velocity grid, and 4b highlights where our filters are removing points spatially across the surfzone overlaid on a grayscale timex image, where lighter values indicate wave breaking. Figure 4c shows points removed relative to total points collected when looking at the timex intensity of the location of the removed point. For intensity bins where the all point probability is significant ($>.05$), we see a peak of points removed in the intensity range (0.74–0.8), these slightly elevated intensities when compared to background (~.60) likely indicate intermittent wave breaking lightening the image. This is also demonstrated when the cross-shore distributions of measurements and points removed (Fig. 4d) show an even distribution of removed measurements around the sandbar. On small wave days like the collection here, filter removal is fairly cross-shore uniform and strongly correlated with total points in the cross-shore. When wave heights are larger the filter becomes more important to remove shore-ward velocities from the surface current signal when breaking is less intermittent.

4   Conclusions

These initial field observations illustrate the promise of using robotic drifters' station-keeping to capture bulk flow patterns filtering out transient flows in the surf zone. By accumulating multiple velocity measurements in precisely defined areas—rather than relying solely on chance drift pathways—researchers or practitioners can build more robust datasets that characterize fine-scale nearshore circulation. This approach also significantly reduces the inefficiencies encountered when traditional drifters are ejected offshore, simultaneously providing a richer record of spatiotemporal variability. As demonstrated by these initial deployments, the robotic drifters' ability to perform targeted sampling can enhance our understanding of nearshore processes.

References

1. Schmidt WE, Guza RT, Slinn DN (2005) Surf zone currents over irregular bathymetry: drifter observations and numerical simulations. J Geophys Res. 110(C12):2004JC002421. https://doi.org/10.1029/2004JC002421
2. MacMahan J, Brown J, Thornton E (2009) Low-Cost handheld global positioning system for measuring surf-zone currents. J Coastal Res 253:744–754. https://doi.org/10.2112/08-1000.1
3. Spydell MS, Feddersen F (2012) A Lagrangian stochastic model of surf zone drifter dispersion. J Geophys Res. 117(C3):2011JC007701. https://doi.org/10.1029/2011JC007701
4. Anderson D, Bak AS, Brodie KL, Cohn N, Holman RA, Stanley J (2021) Quantifying optically derived two-dimensional wave-averaged currents in the surf zone. Remote Sens 13(4):690
5. Dooley C, Elgar S, Raubenheimer B, Gorrell, L (2025) Estimating surfzone currents with near-field optical remote sensing. J Atmos Oceanic Technol 42(1):33–46

Operational Modeling of Surfing Transport for Buoyant Objects in OpenDrift

E. J. Rainville[1(✉)], Jim Thomson[1], Melissa Moulton[1,2], Morteza Derakhti[1], Gaute Hope[3], Knut-Frode Dagestad[3], and Øyvind Breivik[3]

[1] University of Washington – Applied Physics Laboratory, Seattle, WA, USA
erainvil@uw.edu
[2] National Science Foundation – National Center for Atmospheric Research, Boulder, CO, USA
[3] Meteorologisk Institutt, Bergen, NO, USA

Abstract. Surfing transport occurs when a buoyant object is caught on the face of a breaking wave and travels with the breaking wave. Surfing transport can affect the transport of marine debris, plastics, search and rescue objects, among other objects of interest. Operational trajectory models, such as OpenDrift, allow users to quickly predict where an object will go or where it came from as it drifts in the ocean. These tools are useful for predicting where plastics or oil will go after a spill so effective cleaning efforts can be launched along with effective search patterns deployed for search and rescue objects. The OpenDrift model is a lightweight, highly modular, Python-based modeling framework used to predict the trajectory of a wide variety of objects in the ocean. The model is object-oriented, where the Python objects represent a model class that can be configured based on a desired set of transport physics and object characteristics. A surfing transport function is added to the set of physics methods based on observations of surfing transport made using small drifters in the nearshore. Including surfing transport has been shown to increase the accuracy of the trajectory modeling for small buoyant plastics in the nearshore. The additional functionality is also expanded for use in deep water but has not yet been validated against similar observations.

Keywords: Surfing Transport · Trajectory Modeling · Plastics Fate and Transport

1 Introduction

Marine debris, plastics, and other buoyant objects often wash up on beaches, causing a hazard to beachgoers and coastal ecosystems. It is crucial to understand how macroplastics, a source of microplastics as they break down in the surf zone, and other debris move near the coastlines to launch effective cleanup and mitigation efforts [1]. Buoyant objects in the nearshore region move with the wind, waves, and currents [2]. Similar to prior studies in deep water [3–5], observations from drifting buoys in the surf zone indicate that breaking waves cause enhanced dispersal of buoyant objects [6]. This transport mechanism is known as surfing transport and has not previously been included in particle tracking models. As surfing is inherently a wave-resolved process, it must be included in wave-averaged operational models as a stochastic process.

© The Author(s) 2026
C. Coelho et al. (Eds.): CD 2025, CRL 41, pp. 686–690, 2026.
https://doi.org/10.1007/978-3-032-15473-6_104

Surfing events manifest as "jumps" in the position of drifting objects, where an object moves very quickly for a short duration. Small, free-drifting buoys called microSWIFTs were deployed in the nearshore as part of the During Nearshore Events Experiment (DUNEX) in October 2021 in Duck, NC [6]. Jumps in the position of the microSWIFTs are identified and characterized from the drift trajectories of the buoys, and a surfing parameterization is developed [7]. The parameterization is then implemented in the operational trajectory modeling framework, OpenDrift, which has been applied to model the drift of plastic, oil, biota, search and rescue targets, and other objects [8].

2   Surfing Algorithm Description and Implementation

The OpenDrift trajectory modeling framework is Python-based and is highly adaptable to use in many applications [8]. The basic structure of the OpenDrift framework is that an object-oriented model reads in forcing data (using Reader functions), and then a drifter object (application-specific) is advected based on a set of methods depending on the object's properties. All drift modules inherit functionality from the BaseModel class that contains basic functions such as element seeding, managing and referencing Readers, and functions to update positions. The application-specific modules are then set apart from the base model by the sets of physics methods applied to the object. Examples of these physics methods include wind-driven transport, Stokes drift, and current-driven transport. The goal of this project is to add an additional physics method to simulate the process of surfing transport for buoyant objects.

A module is developed and described in Fig. 1 to simulate buoyant drifters transport in the nearshore. The module has the characteristics of only surface-trapped buoyant objects (no vertical mixing), and the physics affecting the drifter are wind drift (with a 3% wind sensitivity factor), drift from surface currents, Stokes drift, and surfing transport. The forcing conditions are output from a coupled Regional Ocean Modeling System (ROMS) and Simulating Waves Nearshore (SWAN) model at the field site. The ocean and wave models are two-way coupled through the COAWST modeling framework [9].

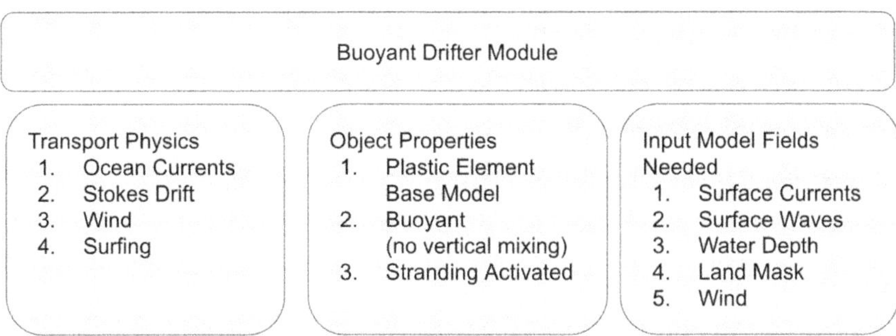

Fig. 1. Description of the buoyant drifter module within the OpenDrift framework describing the associated transport physics, properties assigned to the drifters, and the input model details.

The surfing parameterization that is added to this module is described in Fig. 2. The surfing function acts in conjunction with the typical wind-, current-, and Stokes-drift-driven motion. As the object drifts, a boolean value "catch_wave" is updated by using a random number generator and comparing that value to the fraction of breaking waves at the current location of the drifter. The fraction of breaking waves is estimated within the OpenDrift model and is a method associated with the BaseModel. If the random number generator produces a value less than the fraction of breaking, the boolean "catch_wave" switch is activated, and the drifters jump in the dominant wave direction. The duration of the jump is set to $0.69T_m$ at a speed of $0.72c$ where T_m is the energy weight mean wave period, and c is the linear phase speed of the wave. Once the jump is complete, the drifter returns to drifting with the wind, currents, and Stokes drift for a full wave period in which the "catch_wave" value is checked again.

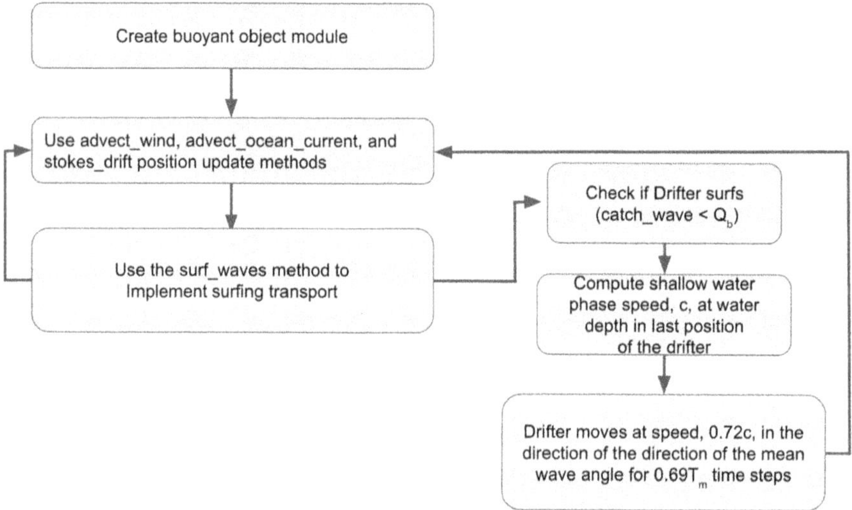

Fig. 2. Flow chart describing the parameterization used to represent the surfing transport mechanism for the OpenDrift implementation. The parameterization is informed by the results for jump sizes from (Figure is modified from [7]).

3   Results and Discussion

The surfing function is implemented in the OpenDrift model and one of the deployed microSWIFT drifters is simulated using different sets of physics. The sets of physics tested are a wind-only case, currents only, wind and currents, and wind, currents, and surfing. The results of the simulated trajectory and the true observed trajectory of the microSWIFT are shown in Fig. 3 overlaid on the observed bathymetry. This simulation shows that including surfing transport qualitatively improves the accuracy of the trajectory modeling compared to the other models. The results of [7] show that including surfing transport in nearshore trajectory models improves the accuracy of the trajectory model with regard to beaching location and time to beaching. This functionality is now added to the OpenDrift framework and can be used for any nearshore drifting applications.

Further work is needed to understand how different objects interact with surfing transport and how surfing transport affects objects in deep water. For now, this parameterization defaults to affect all buoyant objects the same as in shallow water and as if they were microSWIFT drifters. The default values can be updated in the future as the transport mechanism is better understood for different objects and conditions but the implementiation in OpenDrift is flexible for now.

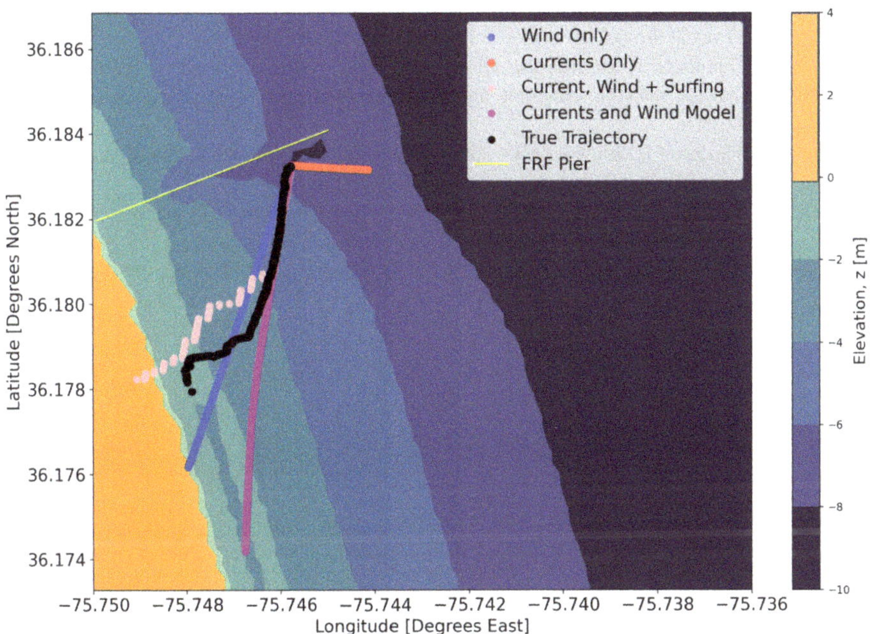

Fig. 3. Example of a true trajectory of a microSWIFT drifter (black line) being deployed near the 6-m isobath and drifting towards the beach compared to modeled trajectories from the OpenDrift framework with varying sets of physics applied. All trajectories are overlaid on the measured bathymetry for the field site.

4   Conclusions

This study has taken a parameterization for surfing transport developed for small free-drifting buoys and successfully implemented it in an operational trajectory model. Through this implementation, the physics of surfing transport are now available to aid in other applications for studying drift in the nearshore and determining where objects may be stranded. This parameterization is flexible and can be modified by the user. The surfing transport function has been validated for shallow water and nearshore surfing of buoyant objects, but can be expanded to deep-water surfing and other objects as the surfing transport mechanism is further studied.

References

1. Van Sebille E, Aliani S, Law K, Maximenko N (2020) The physical oceanography of the transport of floating marine debris. Environ Res Lett 15(2):023003
2. Wagner T, Eisenman I, Ceroli A, Constantinou N (2022) How winds and ocean currents influence the drift of floating objects. J Phys Oceanogr 52(5):907–916
3. Deike L, Pizzo N, Melville W (2017) Lagrangian transport by breaking surface waves. J Fluid Mech 829:364–391
4. Eeltink D, Calvert R, Swagemakers J, Xiao Q, Van Den Bremer T (2023) Stochastic particle transport by deep-water irregular breaking waves. J Fluid Mech 971:A38
5. Pizzo N, Melville W, Deike L (2019) Lagrangian transport by nonbreaking and breaking deep-water waves at the ocean surface. J Phys Oceanogr 49(4):983–992
6. Rainville E, Thomson J, Moulton M, Derakhti M (2023) Measurements of nearshore ocean surface kinematics through coherent arrays of free-drifting buoys. Earth Syst Sci Data 15(11):5135–5151
7. Rainville E, Thomson J, Moulton M, Derakhti M: Surfing transport of buoyant objects observed in the nearshore. J Geophys Res Oceans (Submitted 2025)
8. Dagestad K-F, Rohrs J, Breivik O, Adlandsvik B (2018) OpenDrift v1.0: a generic framework for trajectory modeling. Geosci Model Dev 11(4):1405–1420
9. Warner JC, Armstrong B, He R, Zambon JB (2010) Development of a coupled ocean– atmosphere–wave–sediment transport (COAWST) modeling system. Ocean Model 35(3):230–244

Spaceborne Observations of the Coastal Zone

Erwin W. J. Bergsma[(⊠)], Stéphanie Artigues, Florian Wery, and Jean-Marc Delvit

EB/SA: The Earth Observation Lab, FW: Altimetry-Radar Department and JMD: Strategy
Department, The French Space Agency (CNES), Avenue Edouard Belin, 31400 Toulouse,
France
erwin.bergsma@cnes.fr

Abstract. The dynamic nature of coastal zones necessitates regular and precise
monitoring to understand their ever-changing states and address critical societal and environmental challenges. Advances in spaceborne technologies, including optical imagery, radar, and altimetry, have significantly enhanced our ability to observe and analyse coastal and littoral dynamics. Current approaches,
such as Satellite-Derived Shorelines (SDS), provide indirect measurements of
coastal changes, though more complex products like bathymetry and topography are gaining accuracy and traction. Emerging missions like CO3D and SWOT
introduce transformative capabilities. CO3D employs instantaneous stereo mode
to achieve high-resolution, instantaneous 3D reconstructions of wave dynamics,
ocean currents, and coastal topography, while SWOT's swath altimetry delivers
direct measurements of the free-surface, limitedly affected by typical challenges
of optical sensors like cloud cover. These innovations enable detailed insights into
wave attenuation, sandbar interactions, and large-scale coastal processes. This
paper reviews the current and future potential of spaceborne sensors to monitor
coastal zones, emphasizing their role in providing robust, large-scale observations to address key scientific and societal questions. Integration of these technologies offers a comprehensive framework for advancing coastal research and
management strategies, supporting sustainable development and risk assessment
in vulnerable coastal areas.

Keywords: Remote Sensing · Coastal zone observations · bathymetry ·
topography · hydro and morphodynamics from space

1 Introduction

The highly dynamic and often energetic nature of the coastal zone means that these
areas need to be measured regularly to understand their dynamic and instantaneous
state, while at the same time limiting their accessibility. Depending on the question at
hand, monitoring strategies or the type of missions can be adapted. For example, for
hydro and morphodynamics, or storm impact, inter-storm recovery and risk assessment,
we can use all available satellite data, whereas for climate studies weekly measurements
may be sufficient. Here we present the current and future capabilities of optical imagery,
radar and altimeter in the coastal and littoral zone. This includes current CNES precursor

© The Author(s) 2026
C. Coelho et al. (Eds.): CD 2025, CRL 41, pp. 691–695, 2026.
https://doi.org/10.1007/978-3-032-15473-6_105

missions such as optical CO3D and swath altimetry missions such as SWOT [1], but also a path towards the integration of complete coastal and littoral measurements from space to answer pressing societal questions.

2   Current State of Coastal Observations from Space

Spaceborne measurements are emerging in the coastal dynamics community. Most efforts focus on Satellite Derived Shorelines (SDS) e.g. [2, 3] – as this is the most mature and the most direct measurement from an spaceborne image. A variety of approaches and indices exists with a similar goal; to separate everything that is water and land. More sophisticated inverse or indirectly derived products such as littoral bathymetry [4, 5] and coastal topography [6] as shown in Fig. 1, are gaining in accuracy and traction. However, these techniques are not yet operational – with the exemption of 3D topography that will be operational for the CO3D programme [7]. Nonetheless, the most advanced measurements (the shoreline) are rather a proxy of beach dynamics since it is not a direct morphological measurement and, if not interpreted correctly, only tells part of the story [8]. Instead, we focus here on direct measurements of the wave conditions and elevations within the littoral and coastal zones acquired during a single overpass of optical and radar satellites.

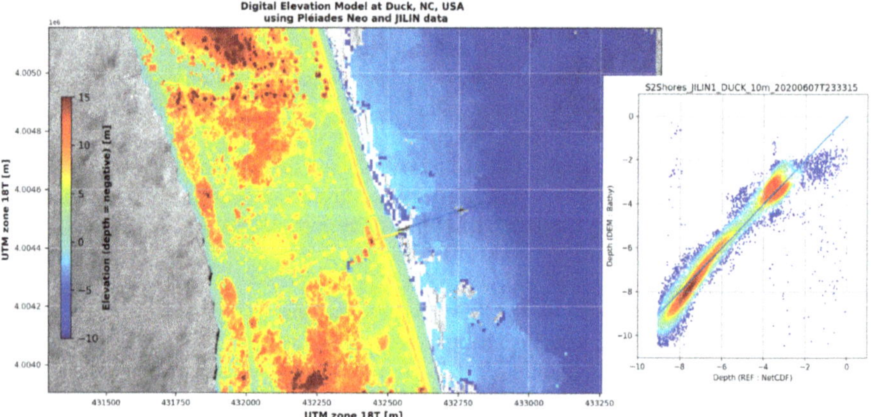

Fig. 1. Current state of on demand coastal observations of the morphological setting, single pass topography (CARS) and bathymetry (S2Shores). On the right, unfiltered direct bathymetry estimations are compared to in-situ measurements (y-axis is satellite derived vs x-axis measured).

In terms of spaceborne sensors, a large part of existing work is based on the use of optical imagers. This is often related to the fact that these data resemble what we, humans observe by eye, and are thus perceived to work with compared to radar or altimeter data sources. The passive optical sensors can be divided into publicly available high resolution data (15 to 5 m resolution) and commercial very high resolution data (2 m to 30 cm resolution). Public data, such as Sentinel-2 and/or Landsat, are acquired in nadir mode, using a push-broom sensor and aim to cover the entire globe, while commercial data, being higher resolution, inherently cover a smaller area, not necessarily nadir as most of these satellites are agile. It is this agility that enables coastal zone monitoring using stereo photogrammetry [9] and, more importantly, videos from space [10].

3   Novel and Future Approaches

Single satellites, although potentially agile, are limited to observing stationary objects to derive topography, whereas for bathymetry it is these dynamic wave features that we track to obtain local depth information. For the bathymetry, however, it is in a flat horizontal plane; where the real information of the free surface is missing. The future CO3D mission includes two pairs (4 satellites in total) flying in formation to obtain an instantaneous image of the same scene from different angles as illustrated in Fig. 2, allowing instantaneous 3D reconstruction. Including a free-surface in sequential stereo-mode at 50 cm resolution. This sequential 3D stereo mode allows for the acquisition of realistic wave energy spectra, wave deformation and attenuation in the near shore, as well as ocean currents.

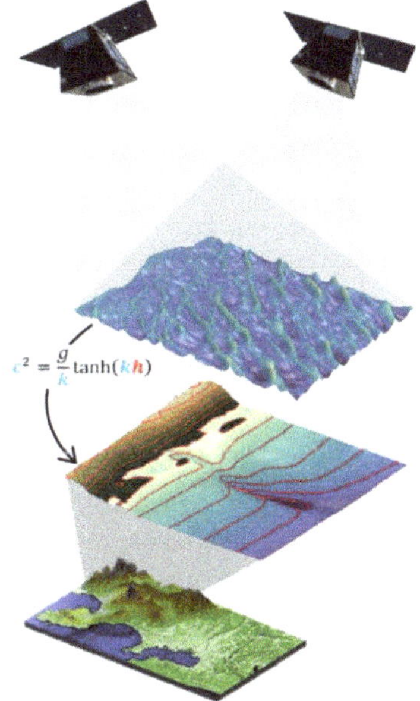

Fig. 2. The principle of CO3D for free surface estimation. Two out-of-phase satellites pairs observe the Earth simultaneously, mimicking human vision and enabling 3D stereo photogrammetry. Once the free-surface is estimated, depth can be derived using the dispersive wave properties.

Optical data alone is limited by its incapacity to sense through clouds, making it a major challenge to obtain reliable measurements over large areas, particularly around the equator. SWOT, a swath altimeter launched in December 2022, is much less affected by cloud cover and provides an image of the coastal zone every 21 days (without taking into account overlapping swaths). The swath altimeter in high resolution mode (Fig. 3) shows a very clear image of the complex wave fields near the coast. In particular, we can see the complex wave attenuation in the nearshore as the waves interact with the sandbars, refract, shorten and finally break. In addition, the SWOT measurements provide a direct measurement of the topography with centimetre accuracy. These large scale measurements will allow understanding of large scale morphodynamics and provide feedback to regular optical measurements to improve depth and topography estimates of the nearshore and coastal zone.

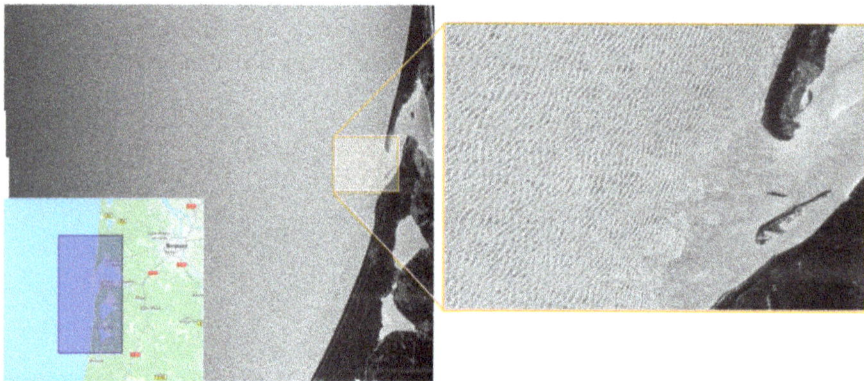

Fig. 3. Example SWOT data near the French Aquitaine coast. SWOT, being a swath altimetry is capable of resolving the instantaneous free surface

4   Conclusion

Coastal zones are highly dynamic environments that require precise and regular monitoring to address a variety of scientific and societal challenges. Spaceborne technologies have become critical in advancing our understanding of coastal and littoral dynamics. Current optical, radar, and altimetry sensors are essential tools for observing coastal changes, though challenges like technological limitations persist. Innovations in space borne technology, such as the upcoming CO3D mission and the recently launched SWOT satellite, offer significant advancements. CO3D's sequential stereo mode enables high-resolution, instantaneous 3D reconstructions of coastal areas, while SWOT's swath altimetry provides precise topographic measurements unaffected by cloud cover. Together, these technologies promise more comprehensive and accurate assessments of wave dynamics, morphodynamics, and large-scale coastal processes. These developments highlight the critical role of satellite missions in supporting effective monitoring and addressing pressing environmental and societal questions in the coastal zone.

Acknowledgements. CO3D is a satellite co-designed by the French Space Agency CNES and Airbus Defence and Space via a partnership. SWOT is a satellite altimeter jointly developed and operated by NASA and CNES, the French space agency, in partnership with the Canadian Space Agency and UK Space Agency.

References

1. Fu L.-L et al (2024) The surface water and ocean topography mission: a breakthrough in radar remote sensing of the ocean and land surface water. Geophys Res Lett 51
2. Vos K, Harley MD, Splinter KD, Simmons JA, Turner IL (2019) Sub-annual to multi-decadal shoreline variability from publicly available satellite imagery. Coast Eng 150:160–174
3. Bergsma EWJ, et al (2024) Shoreliner: A sub-pixel coastal waterline extraction pipeline for multi-spectral satellite optical imagery. Accepted for publication in Remote Sensing
4. Stumpf RP, Holderied K, Sinclair M (2003) Determination of water depth with high resolution satellite imagery over variable bottom types. Limnol Oceanogr 48(1part2):547–556
5. Almar R, et al (2024) Satellite-derived bathymetry from correlation of Sentinel-2 spectral bands to derive wave kinematics: qualification of Sentinel-2 S2Shores estimates with hydrographic standards. Coastal Engineering. 189
6. Bergsma EWJ, Almar R, Rolland A, Binet R, Brodie KL, Bak AS (2021) Coastal morphology from space: a showcase of monitoring the topography-bathymetry continuum. Remote Sens Environ 261:112469
7. CO3D: A constellation to map the world in 3D. https://cnes.fr/en/projects/co3d. Accessed 8 Jan 2025
8. Graffin M, et al Contemporary patterns of shoreline change along the North American West Coast
9. Melet O et al (2020) Co3d mission digital surface model production pipeline. Int Arch Photogram Remote Sens Spatial Inf Sci 43:143–148
10. Klotz AN et al (2024) Nearshore satellite-derived bathymetry from a single-pass satellite video: Improvements from adaptive correlation window size and modulation transfer function. Remote Sens Environ 315:114411

Measuring Storm Waves and Water Levels from a Fixed Structure with a Rapidly Deployable Oceanographic Radar

Jenna A. Brown[1](✉), Bryce McClenney[2], and Pat Dickhudt[3]

[1] U.S. Geological Survey, St. Petersburg Coastal and Marine Science Center, 600 4 th St. S., St. Petersburg, FL 33701, USA
jennabrown@usgs.gov
[2] U.S. Geological Survey, South Atlantic Water Science Center, 3916 Sunset Ridge Rd., Raleigh, NC 27607, USA
[3] U.S. Army Corps of Engineers, Engineer Research and Development Center, Field Research, 1261 Duck Rd., Kitty Hawk, NC 27949, USA

Abstract. A new oceanographic radar instrument package was developed by the U.S. Geological Survey (USGS) to measure storm waves and water levels in the nearshore, capable of being deployed rapidly and transmitting data in near real-time. To test the performance and accuracy of the sensor, multiple years of data were collected over various hydrodynamic conditions and compared to long-term monitoring data collected at the U.S. Army Corps of Engineers (USACE) Field Research Facility in Duck, North Carolina, USA. The oceanographic radars were highly reliable, with less than 1% of the record being erroneous spikes or missing data points. At the end of the pier, the radar was highly accurate, with nearly perfect agreement in water level ($r^2 = 0.997$) compared to a nearby National Oceanic and Atmospheric Administration (NOAA) tide gauge, and good agreement in significant wave height ($r^2 = 0.98$) and peak wave period ($r^2 = 0.65$) compared to a nearby USACE sensor. This work demonstrates the potential of the USGS radar for rapid response storm deployments and collecting reliable and accurate hydrodynamic measurements in the nearshore for validating coastal impact models.

Keywords: Oceanographic Radar · Waves · Water Levels · Real-Time Data

1 Introduction

Extreme storms can generate elevated water levels and dangerous wave conditions that can cause extensive flooding, significant landscape changes, and destruction of property at the coast. Documenting the height, extent, and timing of storm surge and waves during hurricanes is critical for validating and improving coastal impact models and forecasts, used by decision makers to mitigate risks to humans from severe storms. However, deploying sensors in the dynamic open ocean environment is notoriously challenging, especially during rapid response conditions when time is limited. One approach is to

© The Author(s) 2026
C. Coelho et al. (Eds.): CD 2025, CRL 41, pp. 696–702, 2026.
https://doi.org/10.1007/978-3-032-15473-6_106

attach a pressure sensor to beachfront infrastructure (e.g., beach walkways, pier pilings) and measure water levels as the sensor is submerged in time. These sensors tend to be deployed over dry land and often stay dry until extreme water levels occur and do not stay submerged for the entire tidal cycle, which limits the observations to peak values, makes computing wave statistics difficult, and does not provide information on the hydrodynamic conditions as the storm approaches [1]. Installing in-situ sensors seaward of the shoreline requires timing of ideal wave and tide conditions as well as the ability to secure the sensor platform to the bottom (usually done by divers), which is complicated even under calm conditions. Shore-based remote sensing techniques, such as light detection and ranging (lidar) and camera imagery, can also be used to estimate coastal water levels and waves; however, they have limitations during poor weather conditions and require time-consuming calibration at installation [2].

In this work, we use existing pier infrastructure as a platform to attach instruments for making remotely sensed water level and wave measurements seaward of the shoreline. An instrument package was developed by the U.S. Geological Survey (USGS) that: 1) can sample quickly and measure the full spectrum of storm waves, 2) can transmit data in near real-time, 3) is self-powered by solar and/or battery, and 4) is capable of being rapidly deployed with temporary and non-intrusive hardware. Here we show the performance and accuracy of the instrument package with data collected over multiple years and various hydrodynamic conditions, compared to nearby long-term monitoring data sites. This technology allows for collecting measurements to document the storm-induced transformation and dissipation of waves and surge in the nearshore.

2 Field Data

A non-contact oceanographic radar sensor (Geolux LX-80-15 10 Hz, https://www.geo lux-radars.com/lx-80-ocean) was employed, which directly measures the elevation of the sea surface using a frequency-modulated electromagnetic (radio) wave to detect the distance from the sensor to the water surface and relates it to the surveyed elevation of the radar sensor. To make oceanographic measurements, the radar samples continuously at 10 Hertz (Hz) then computes the mean water level elevation and wave parameters over a user-defined period on-board and transmits the data in near real-time. To preserve the raw sea surface elevation measurements for post-processing, a data logger (Campbell Scientific CR1000X) is implemented to save down-sampled 5 Hz timeseries data at the station; these data are also transmitted in near real-time over a cellular network connection in the event the equipment is non-recoverable. The complete "USGS radar" instrument package consists of the radar sensor, data logger, cellular modem, and an external battery housed inside a water-tight box, with a cellular antenna and optional solar panel affixed on the outside of the lid. For collecting measurements in the nearshore, the box is attached to pier railings using specially fabricated stainless-steel L-shaped mounting arms that can be installed quickly and do not require drilling into the structure. The USGS radar is deployed such that the sensor extends over the water surface with no obstructions below (e.g., pier deck, piling, rocks) and with the box secured and as level as possible. The elevation of the sensor is then surveyed to a known datum.

The USGS radar was deployed on August 18, 2022, at the U.S. Army Corps of Engineers (USACE) Field Research Facility (FRF) in Duck, North Carolina, USA, to compare observations of water levels and waves in the nearshore with other well-established, co-located instruments (Fig. 1) for evaluating instrument performance and accuracy. Instrument locations are given in a local FRF coordinate system in meters (m), where cross-shore (x) increases offshore and alongshore (y) increases to the north, with the pier at y = 524 m and the USACE cross-shore instrument array at y = 940 m. The USGS radar was mounted on the south-facing railing near the offshore end of the FRF research pier at x = 600 m, in approximately 6-m water depth. Mean water level measurements are compared to a co-located National Oceanic and Atmospheric Administration (NOAA) tide gauge (https://tidesandcurrents.noaa.gov/waterlevels.html?id=8651370), which uses a microwave radar to measure the distance to the water surface. Mean wave parameter measurements are compared to a USACE bottom-mounted, upward-looking, acoustic Doppler current profiler (ADCP, https://chldata.erdc.dren.mil/thredds/catalog/frf/oceanography/waves/sig940-600/catalog.html) deployed approximately 400 m north of the pier as part of a cross-shore array, also in nominal 6-m water depth.

Over two years of data were collected by the USGS radar at the end of the pier during varying hydrodynamics conditions for instrument validation [3]. Additionally, from January to April 2024, two USGS radars were installed closer to shore to examine measurements over shallower water and during adverse conditions, such as breaking waves and sea spray. The radars were also mounted on the south-facing railing, at x = 200 m and x = 300 m, in similar cross-shore locations as instruments in the USACE cross-shore array in 2-m (https://chldata.erdc.dren.mil/thredds/catalog/frf/oceanography/waves/par os940-250/catalog.html) and 3-m (https://chldata.erdc.dren.mil/thredds/catalog/frf/oce anography/waves/sig940-300/catalog.html) water depth.

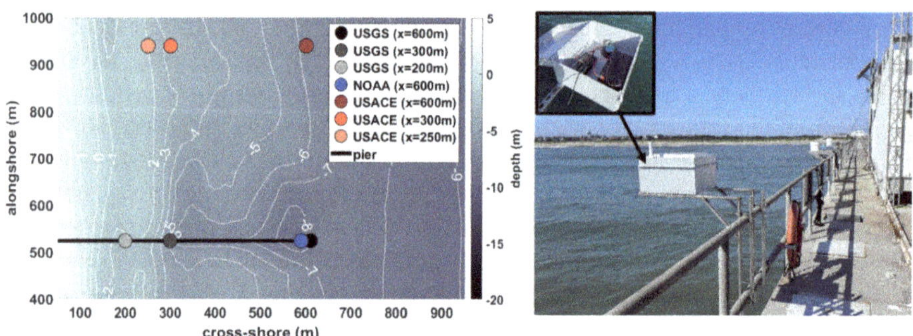

Fig. 1. (Left) Map of instrument locations and bathymetry contours (Jan 2024, https://chldata. erdc.dren.mil/thredds/catalog/frf/geomorphology/DEMs/surveyDEM/data/catalog.html) at the U.S. Army Corps of Engineers (USACE) Field Research Facility (FRF). (Right) Photo of the U.S. Geological Survey (USGS) radar near the National Oceanic and Atmospheric Administration (NOAA) tide gauge at the end of the FRF pier.

3   Methods

The elevation of the water level referenced to a given datum is measured as the elevation of the sensor minus the distance measured by the sensor to the instantaneous water surface; both the timeseries of measured distances and computed water levels are recorded by the

USGS radar. Due to the fast sampling-frequency and constant changing motion of the water surface, some erroneous spikes appear in the raw timeseries data. These outliers have default values of water level = 7999 and distance = 15 and are identified during post-processing and set to not-a-number (NaN). Gaps in the radar timeseries data less than 1 min long were linearly interpolated, and longer gaps were set to NaN. Then, the mean water level was computed for 6-min windows, and significant wave height, H_s, and peak wave period, T_p, were computed for one-hour windows using timeseries and wave spectra analysis (Fig. 2).

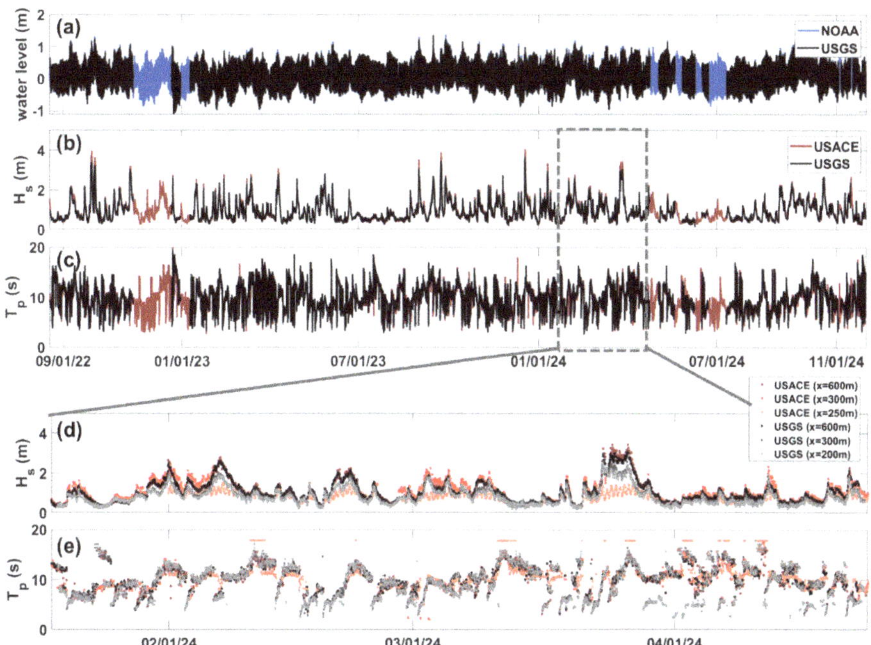

Fig. 2. Timeseries of (a) 6-min mean water levels, relative to NAVD88, and hourly-mean (b) significant wave height, H_s, and (c) peak wave period, T_p, measured by the U.S. Geological Survey (USGS) (black), National Oceanic and Atmospheric Administration (NOAA) (blue) and U.S. Army Corps of Engineers (USACE) (red) sensors at x = 600 m. The gray dashed box indicates when mid-pier radars were deployed, with the timeseries shown of hourly-mean (d) significant wave height, H_s, and (e) peak wave period, T_p, measured at different cross-shore locations by USGS radars on the pier at y = 524 m and USACE sensors in the cross-shore array at y = 940 m.

4   Results

The performance of the USGS radar was evaluated by examining the number of outlier and missing data points in the 5 Hz timeseries data. Less than 1% of the entire record was an erroneous spike or missing data point, indicating a high reliability of the data and sufficiently dependable data to make accurate sea surface elevation measurements for computing mean water levels and waves. Although there were outlier points measured consistently throughout the record, the number increased during large wave events. Additionally, there were more outliers in the data from the mid-pier radar in shallower water than the end-of-pier radar, likely due to more active hydrodynamics near the surf zone, but they were still less than 1% of the record.

The accuracy of the end-of-pier USGS radar was evaluated by performing linear regressions of the mean observations collected from each instrument (Fig. 3). The 6-min mean water level was nearly identical to the NOAA tide gauge measurements ($r^2 = 0.997$). Hourly-mean H_s and T_p were compared to the USACE ADCP, and there was good agreement in H_s measurements ($r^2 = 0.98$), with increased discrepancy for larger waves, while there was less agreement in T_p measurements ($r^2 = 0.65$).

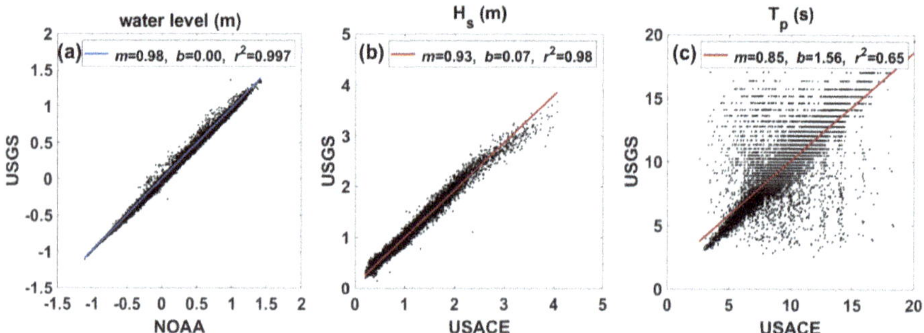

Fig. 3. Linear regressions of (a) the 6-min mean water level, relative to NAVD88, measured by the end-of-pier U.S. Geological Survey (USGS) radar and the National Oceanic and Atmospheric Administration (NOAA) tide gauge, and the hourly-mean (b) significant wave height, H_s, and (c) peak wave period, T_p, measured by the end-of-pier USGS radar and the U.S. Army Corps of Engineers (USACE) acoustic Doppler current profiler. Colored lines indicate the line of best fit, with the slope (m), y-intercept (b) and r^2 values given.

The mean wave parameters measured by the USGS radar at the mid-pier in shallower water depths was compared to corresponding instruments in the USACE cross-shore array at similar distances offshore (Fig. 2). Noticeable differences are apparent in the timeseries data, with much larger waves measured at x = 300 m and smaller waves measured at x = 250 m by the USACE ADCPs in the cross-shore array. This is most likely due to differences in the bathymetric elevation profiles at the pier and the cross-shore array (Fig. 1), where scouring around the pier pilings results in the water depths being much deeper offshore of about x = 270 m; bathymetry contours are alongshore uniform seaward of the pier. Therefore, the waves at the end of the pier are representative of the offshore conditions, but at the shoreward sites, the waves are smaller due to less shoaling and more wave energy refracting away from the pier.

5 Discussion and Summary

An instrument package utilizing a commercially available oceanographic radar sensor and data logger was developed that can be quickly attached to fixed infrastructure (e.g., pier railing) and send 5 Hz timeseries data over cellular connection to collect hydrodynamic measurements during storm events. It was tested at the USACE FRF over various hydrodynamic conditions from August 2022 through November 2024. Less than 1% of the timeseries data points were erroneous or missing, and the mean measurements of water levels and waves compared well to nearby long-term monitoring instruments. The primary limitation is the requirement of existing infrastructure to attach the instrument package. The infrastructure could affect the local hydrodynamic and morphologic conditions, so we note that water level and wave conditions measured by a USGS radar may not be entirely representative of the conditions away from the pier. Additionally, the radar sensor measures water levels relative to the sensor and/or datum, and not the water depth above the seabed. However, this is not an issue for numerical model validations since model inputs and predictions must also be relative to a known elevation (datum), unless the bathymetry used for the model input is not accurate. The radar sensor performed consistently well throughout the deployment, suggesting precipitation and wind did not adversely impact the measurements. Overall, the USGS radar instrument package provides reliable and accurate observations in near real-time and is therefore well-suited for rapid response storm deployments from fixed structures.

Acknowledgements. Any use of trade, firm, or product names is for descriptive purposes only and does not imply endorsement by the U.S. Government. The authors thank Geolux, the U.S. Geological Survey field team, and the U.S. Army Corps of Engineers Field Research Facility for their support. Funding provided by the National Oceanographic Partnership Program Hurricane Coastal Impacts Project (Office of Naval Research Grant N0001421IP00064).

References

1. Birchler JJ et al (2024) Skill assessment of a total water level and coastal change forecast during the landfall of a hurricane. Coast Eng 193:104590
2. Holman R, Haller M (2013) Remote sensing of the nearshore. Annu Rev Mar Sci 5:95–113
3. Brown JA, McClenney B, Suttles S (2025) Oceanographic Radar Time Series Measurements Collected at Duck, North Carolina. U.S. Geological Survey Data Release. https://doi.org/10.5066/P1NLPWRP

Assessment of Surfzone Bathymetry Measurements from Yellowfin ASV in a Range of Conditions and Beaches

Kara M. Koetje[1]([✉]), A. Spicer Bak[1], Peter Traykovski[2], Nick Desimone[1], and Ian R. Conery[1]

[1] U.S. Army Engineer Research and Development Center, Duck, NC 27949, USA
kara.m.koetje@usace.army.mil
[2] Woods Hole Oceanographic Institution, Falmouth, MA 02543, USA

Abstract. This work assesses the performance of the Yellowfin Autonomous Surface Vessel (ASV) in measuring surfzone bathymetry at two beaches under a range of conditions. The Yellowfin ASV, equipped with a precision GNSS and a Cerulean S500 echosounder, conducted 8 surveys at sites in Duck, NC, and Martha's Vineyard, MA. Surveys comprised of repeated lines were conducted in order to evaluate for uncertainty in the measurement. Statistical analyses indicate good agreement between repeated transects and are evaluated for total vertical uncertainty. Results demonstrate the vehicle's ability to navigate surf zones and produce accurate bathymetric data, with the highest uncertainties primarily observed in very shallow areas.

Keywords: Uncrewed Systems · Surfzone Bathymetry · Measurement Uncertainty

1 Background

Bathymetry changes rapidly along wave-dominated sandy coastlines, necessitating accurate state assessments to understand sediment transport and morphodynamics and subsequently to predict nearshore hydrodynamics and wave-induced flooding. Typical methods producing highly precise bathymetry estimates require human-operated vessels to work in dangerous breaking wave environments, making the process labor intensive and high-risk.

Birkemeier (1984) highlighted the Coastal Research Amphibious Buggy (CRAB) which used a known geometry between survey method and wheels contacting the seafloor to accurately map the surf zone. Common state-of-the-art methods utilize single-beam acoustic sonar operated on maneuverable vessels such as the Lighter Amphibious Resupply Cargo (LARC) or a Personal Water Craft (Forte, 2017; Dugan, 2001). While more capable in nearshore environments, an onboard crew is still required, and the presence of turbidity and bubbles in the surf zone pose significant challenges. Recent work (Francis, 2021; Bak, 2022; Bak, 2023) has begun exploring the use of uncrewed systems for collecting measurements in the surf zone. Francis and Traykovski (2021) documented the

C. Coelho et al. (Eds.): CD 2025, CRL 41, pp. 703–709, 2026.
https://doi.org/10.1007/978-3-032-15473-6_107

development of an Uncrewed Surface Vehicle (USV) for surfzone operation, a modified version of which will be considered here. Bak (2023) utilized an Amphibious Uncrewed Ground Vehicle (AUGV) with an RTK GNSS producing RMSD values of 8.7 cm when compared to LARC/CRAB measurements. Also explored was a method for evaluating variability in measured bathymetry by surveying transects repeatedly within a single mission, here referred to as a repeatability assessments. The resultant statistical comparison of variance across the repeated lines are used to quantify uncertainty in the measurements. This research aims to build on that prior work.

2 Methods

2.1 Platform Description

The platform utilized in this work is the Yellowfin ASV developed by Integrated Coastal Solutions, an update to the autonomous survey vehicle prototype by Francis and Traykovski (2021). This platform weighs approximately 14 kg and is 1.5 m long, 40 cm wide and 40 cm tall. The innegra-basalt laminated hull is stiffened with a series of bulkheads and stringers and has a fiberglass lid for radio frequency transparency. The hull shape has been modified from Francis and Traykovski (2021) to a semi-displacement hull with a sharp bow to enhance vehicle wave piercing. Additionally, the foam super-structure is replaced with a hardened bulbous shell above the main deck, maintaining the center of buoyancy significantly above the center of gravity, allowing for self-righting capability in the event of an overturn or broach. The vehicle still uses a jet drive to provide the ample propulsion maneuverability required for safe and successful surfzone operation.

Fig. 1. Deployment of Yellowfin ASV at the ERDC Field Research Facility (left) and demonstration of wave piercing capability (right).

The Yellowfin is powered by lithium-ion batteries with an operating range of 32 km (or 6 h). Waypoint based autonomy is provided by the Pixhawk Cube, with remote operation capable at any time. The vessel had been equipped with a Cerulean S500 (500 kHz) single-beam echosounder and an Emlid M2 precision GPS/GNSS module in order to create highly detailed shallow water bathymetric maps. The Yellowfin can be hand-deployed from the beach (Fig. 1) and has been found to successfully navigate the surfzone in wave heights exceeding 1 m in some cases.

2.2 Field Sites and Survey Method

This effort includes surveys conducted at two different locations. Five repeatability tests were conducted at the U.S. Army Engineer Research and Development Center (ERDC) Field Research Facility (FRF) in Duck, NC. Three additional tests were conducted on the south shore of Martha's Vineyard, MA near Long Point Beach. Both are open coast, sandy beaches with 180-degree exposure to waves, from the east in Duck and from the south at Martha's Vineyard. Deployments were targeted to span a range of conditions with an intent to understand uncertainty in data quality as a function of wave conditions.

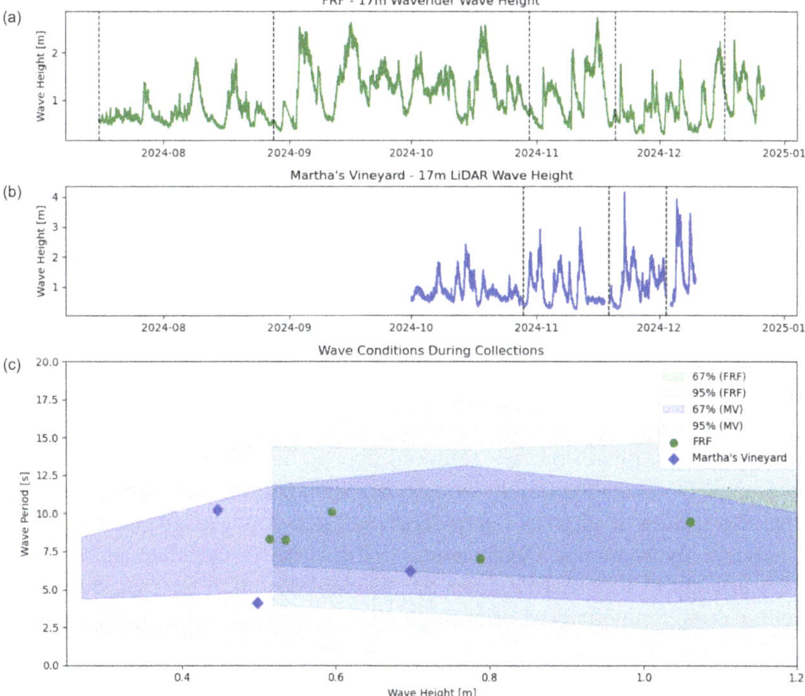

Fig. 2. Offshore wave conditions, wave height and peak period, during times of repeatability test at FRF (green) and at Martha's Vineyard (blue). Timing of repeatability tests are represented by vertical dashed lines in panels (a) and (b).

Repeatability tests were conducted using pre-programmed missions where the same pattern (out and back) was run multiple times (target 5 times). The pattern was comprised of approximately a 200m cross-shore and alongshore component for both sites.

Wave condition data, including significant wave height (H_s) and peak period (T_p), is available through long-term instruments at both sites; the 17-m Waverider buoy at the FRF and the 17-m LiDAR wave gage that is part of the Martha's Vineyard Coastal Observatory, each located approximately 3 km offshore. While offshore wave measurements do not precisely reflect the conditions experienced by the vehicle while operating in the nearshore (for a number of reasons to include shoaling, non-linear interactions, direction of propagation, etc.), these data provide useful context for the bulk conditions at the time of the survey. Additionally, monthly LARC surveys were available at the FRF as ground-truth verification for FRF tests.

2.3 Processing Methods

By pairing the sonar readings with data from a precision post-processed kinematic (PPK) GNSS, the ASV may be used to make detailed bathymetric maps of the surfzone. First, recorded RINEX files are PPK processed using a local NOAA Continuously Operating Reference Station (CORS) at the Field Research Facility and Javad Triumph-1M base station set at a marked reference location on Martha's Vineyard. The Cerulean S500 automated bed detection algorithm reports at 10 Hz, though acoustic backscatter is logged. The unfiltered reported bottom is cross-correlated to the PPK GNSS position to ensure proper time synchronization. Poor GNSS (float state) and poor sonar quality (quality $< 60\%$) data are removed before combining to an elevation product in post-processing.

Once a raw corrected product is produced, sonar depth is de-spiked using two rounds of rolling-window median filtering to remove outliers with simple linear interpolation. In the case of datasets where a second surface is present inside the very nearshore region, as seen in Fig. 3(a), datapoints are removed using a manual data selection tool. Finally, repeated transect lines are selected, separating cross-shore and alongshore segments.

Repeatability assessments were done by making a statistical comparison across the 10 repeated transects and computing variance in those measurements in order to assess the uncertainty. In this effort, we followed approaches outlined by the International Hydrographic Organization (IHO), who defines standards in the nearshore environment for characterization of error as depth-dependent and depth-independent uncertainties (International Hydrographic Organization 2020). Here we calculate survey Total Vertical Uncertainty (TVU) as 1.96 times the standard deviation for each transect at 1-m resolution, considering transect orientation relative to the shoreline as well as direction of travel.

3 Results

A test conducted on July 16, 2024 at the FRF is presented to illustrate our method and representative results, where offshore wave conditions were observed to have 0.5 m wave height and a peak period of 8 s. In this example, repeatability results are presented for both pre- and post-median filtering and data cleaning, referred to as *cleaned* and *uncleaned* data. A reflected surface, likely due to the influence of bubbles in the surfzone, can be seen in the first 50m of the transect. Once filtered and cleaned, each transect is considered based on the vehicle direction of travel, where offshore directed travel is presented in cool colors and onshore directed is presented in warm colors. The reduced length of the onshore directed transects is a result of vehicle turning radius. Transects can be seen to have broad agreement in the depth profile (as validated by comparison to available LARC survey) and maintain $\pm$ less than 1 m of path variance in the y-direction.

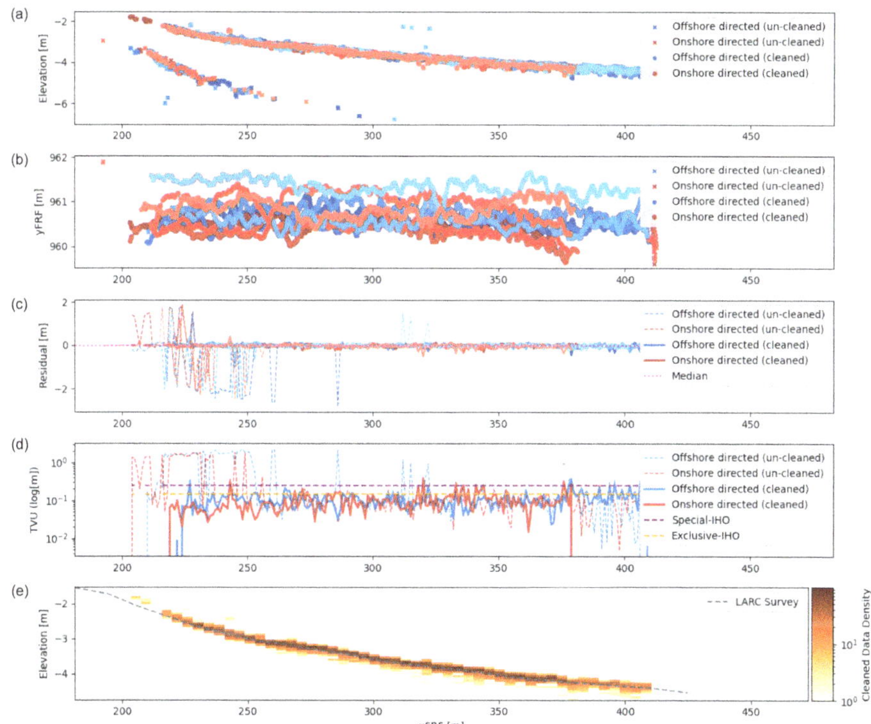

Fig. 3. (a) Yellowfin repeatability study conducted on July 16, 2024 with elevations presented for cleaned and *un-cleaned* data (cool colors for offshore directed path of travel; warm colors for onshore directed); (b) plan view of comparison with individual transect lines colored as in (a); (c) elevation residuals between individual transect elevation and median elevation of the population at each cross-shore location, colored as in (a) with median elevation shown in magenta; (d) uncertainty (at 95% confidence) in elevation at each cross-shore location, divided by offshore (warm) and offshore (cool) populations; and (e) cleaned repeatability study elevation profile with elevation data densities shaded orange and the corresponding profile from the nearest LARC survey.

Residuals are computed relative to the median value for each 1-m bin along the transect. High residual in the very nearshore portion of the *un-cleaned* data can be seen, but *cleaned* depth profiles show good agreement and low residuals in both the onshore and offshore directed transects. Onshore directed transects of *cleaned* data are shown to have a slightly lower median TVU (0.082 m) as compared to offshore (0.096 m), with increased uncertainly in the most offshore 30 m of the transect. This increased uncertainty corresponds to a decreased data density due to the shortened onshore-directed transects and is not present in other datasets with greater data density in this region. Lower data density is also observed in the most very shallow nearshore due to the removal of reflected surface depth returns. Transects in both directions of travel fall within IHO uncertainty standards with few exceptions.

4 Conclusions

Data presented in this work indicate the Yellowfin ASV to be a capable platform able to provide high-accuracy bathymetric measurements with good agreement across repeated transects as shallow as 2m water depth. The highest uncertainty is observed in the very shallow nearshore, likely due to bubble-induced returns caused by breaking waves. Repeatability tests also show good agreement with other trusted survey datasets when available. Alongshore profile lines were not presented here but have been evaluated and speak to the influence of vehicle roll on the variability of measured depth. Further aggregation of bulk results and comparison to conditions, vehicle motion, and position relative to wave breaking are forthcoming.

Limitations due to the quality of manufacturer provided bed-detection and bottom-tracking algorithms provide an avenue for development and could improve uncertainty results. Surfzone bubbles and turbidity are a known source of noise and present a challenge for surfzone mapping broadly. Addressing that challenge is not within the scope of this work, however, straightforward human-guided QA/QC tools could be automated to improve the quality of results.

References

Bak AS, Durkin P, Bruder B, Saenz MJ, Forte MF, Brodie KL (2023) Amphibious uncrewed ground vehicle for coastal surfzone survey. J. Surv. Eng. ASCE

Bak AS, Durkin P, Saenz MJ, Brodie KL, Bruder BL, Forte MF (2022) Evaluating autonomous system performance in the surfzone. In: OCEANS 2022, Hampton Roads, pp 1–8, IEEE

Birkemeier WA, Mason C (1984) The CRAB: a unique nearshore surveying vehicle. J Surv Eng 110(1):1–7

Dugan JP, Morris WD, Vierra KC, Piotrowski CC, Farruggia GJ, Campion DC (2001) Jetski-based nearshore bathymetric and current survey system. J. Coast Res. 900–908

Francis H, Traykovski P (2021) Development of a highly portable unmanned surface vehicle for surf zone bathymetric surveying. J Coastal Res 37(5):933–945

Forte MF, Birkemeier WA, Mitchell JR (2017) Nearshore survey system evaluation. U.S. Army Engineer Research and Development Center, ERDC/CHL TR-17-19

IHO (International Hydrographic Organization). 2020. International hydrographic organization standards for hydrographic survey. Number 44 in 6.0.0. Monaco: IHO

Satellite-Derived Bathymetry for Shallow Waters, in Regions of High Optical Complexity, Using Empirical Methods and Convolutional Neural Network

Mario Luiz Mascagni[1,2(✉)], Antonio Henrique da Fontoura Klein[1,2],
Anita Maria da Rocha Fernandes[3], Andrigo Borba dos Santos[3], Dennis Kerr Coelho[3],
and Lais Pool[1,2]

[1] Federal University of Rio Grande Do Sul (UFRGS), Porto Alegre, RS 91540000, Brazil
`mario.mascagni@ufrgs.br`
[2] Federal University of Santa Catarina (UFSC), Florianópolis, SC 88040900, Brazil
[3] University of Vale do Itajaí (UNIVALI), São José, SC 88032005, Brazil

Abstract. Recent advancements in remote sensing instrumentation and technics have positioned satellite-derived bathymetry as one of the most promising methods for monitoring coastal environments. This study compares two approaches for estimating shallow water depths from Sentinel-2 satellite imagery: (1) an empirical method based on linear models correlating water depth with the interaction of different bands of the electromagnetic spectrum within the water column, and (2) a machine learning approach employing convolutional neural networks to capture nonlinear relationships to address the complex influence of suspended particulate matter and colored dissolved organic matter in optically complex waters. The CNN-based model demonstrated significant accuracy in turbid coastal environments, achieving a mean absolute error of approximately 0.1 m in depths up to 30 m. These findings highlight the potential of combining empirical and machine learning methods to enhance bathymetric assessments in regions where traditional surveys are costly and challenging.

Keywords: Remote Sensing · Machine Learning · Bathymetric Inversion · Turbid Waters · Sentinel-2

1 Introduction

Monitoring shallow water depths is crucial for understanding the morphodynamics of coastal systems. Such knowledge is the basis for developing sustainable strategies for the management of waterways and coastal space use, as well as for implementing mitigation and/or restoration measures for coasts affected by erosion, flooding or degradation of coastal habitats [1].

However, hydrographic surveys using traditional bathymetric mapping techniques are generally very expensive and require specialized vessels, equipment and personnel

C. Coelho et al. (Eds.): CD 2025, CRL 41, pp. 710–716, 2026.
https://doi.org/10.1007/978-3-032-15473-6_108

for these activities; in very shallow environments close to the coast, the challenge is even greater due to the incidence of waves and tidal variations that increase the risk of accidents and often lead to inconsistencies in the data collected. Therefore, bathymetric data from shallow waters are rare and most of the time restricted to areas of port interest. On the other hand, remote sensing techniques using public satellite image databases are a promising source of geospatial information that can contribute to estimating depth values from bathymetric inversion techniques, with low cost and without operational risk [2].

The present study investigates two different satellite-derived bathymetry (SDB) techniques from the Sentinel-2 image collection, available on the Google Earth Engine (GEE) platform. The bathymetric inversion techniques used in this study involve the application of both: (1) empirical models based on the physics of the interaction of electromagnetic spectrum bands with the water column in shallow waters up to 30 m deep, and (2) machine learning (ML) models based on Convolutional Neural Network (CNN).

2 Methods

2.1 Study Area

The selected study area for this research was Babitonga Bay, located along the Santa Catarina coast in southern Brazil. Babitonga Bay is a RIA-type estuary characterized by drowned valleys and situated in a region with high average rainfall [3]. The bay exhibits significant optical complexity due to the interaction of Suspended Particulate Matter (SPM) and Colored Dissolved Organic Matter (CDOM), which result from the mixing of marine and continental waters contributed by rivers flowing into the estuary [4]. The estuary spans approximately 2 km in width at its mouth and extends about 25 km from the upper to the lower estuary sectors. Figure 1 illustrate the study area, highlighting the section where the bathymetric inversion tests were conducted.

The data used for training and validating the bathymetric inversion models were collected by traditional hydrographic surveys conducted in May 2018, using single-beam echo sounders. The bathymetric measurements were interpolated using the Inverse Distance Weighting (IDW) method with an anisotropy factor of 1.2x to produce a regular 10-m grid for the construction of the CNN kernels' scanning grids (Fig. 2).

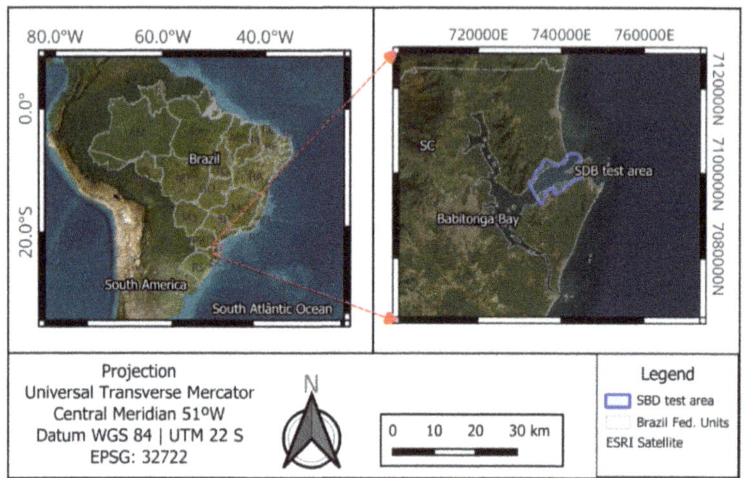

Fig. 1. Study site for SDB – Babitonga Bay, Santa Catarina coast in southern Brazil.

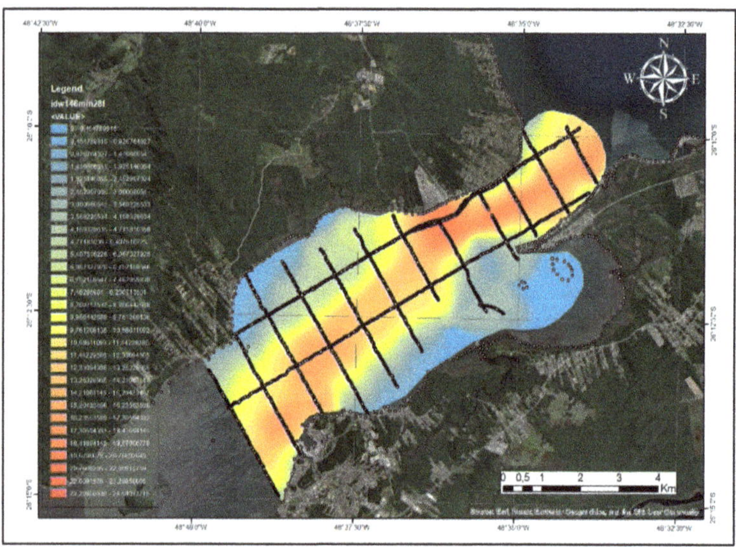

Fig. 2. Interpolation of bathymetric field survey data using IDW with anisotropy, for 10m resolution.

2.2 Satellite-Derived Bathymetry Models

Satellite images from the Sentinel-2 mission, level L1 - Top of Atmosphere (TOA), were preprocessed using JavaScript cloud-based routines in GEE platform. The preprocessing steps included: (1) masking terrestrial pixels using the Normalized Difference Water Index (NDWI); (2) performing atmospheric corrections using the SIAC-2 functions [5]; (3) masking pixels affected by cloud cover and cloud shadows; (4) resampling bands B5, B6, B7, B8A, B11, and B12 from 20 m to 10 m per pixel for enhanced spatial resolution;

and (5) generating pixel-by-pixel geometric median composite images [6] to mitigate the influence of SPM and CDOM in the analyzed scenes. These steps were designed to ensure optimal image quality and improve the accuracy of the bathymetric inversion models.

The empirical model employed in this study [7, 8] and [9] assumes that water properties differentially absorb and scatter the different bands of the electromagnetic spectrum, thus the ratio between different bands in the satellite images, when transformed logarithmically (Eqs. (1) and (2)), can be correlated through a linear regression to estimate depths in shallow waters. In this framework, red bands are predominantly absorbed and scattered near the ocean surface, while green and blue bands penetrate further into the water column and are absorbed at greater depths, up to a limit of 30 m in clear waters.

$$SDB = m_1 pSDB - m_0 \tag{1}$$

where,

- **SDB** is the final depth derived from the satellite (meters);
- **m1** and **m0** are constants for the adjustments of the linear model;
- **pSDB** is the relative depth of the satellite (dimensionless);

$$pSDB = \frac{ln(ncRrs(\lambda_i)}{ln(ncRrs(\lambda_j)} \tag{2}$$

where,

- **Rrs** is the Reflectance of the satellite image bands (B2, B3, B4, B5 and B8);
- **c** is a correction constant for the reflectance dependent on the satellite sensor;
- **n** $= 1000$ is a constant to ensure that both logarithms will be always positive.

The empirical model was implemented alongside the GEE pre-processing routines using band ratios from the near-infrared (B8), red edge (B5), red (B4), green (B3), and blue (B2) Sentinel-2 bands.

In parallel, a Convolutional Neural Network (CNN) model was constructed using all available "B" bands from Sentinel-2 TOA satellite images. Three versions of input layers were configured, comprising 12, 36, and 72 input channels, respectively. In the 12-channel configuration, each channel corresponded to the reflectance values of individual Sentinel-2 image "B" bands, in the closest date of the bathymetric survey of May 20, 2018. The 36- and 72-channel configurations incorporated multiple Sentinel-2 TOA images acquired respectively within a one-month and a three-month period, before and after the survey date. For these versions, geometric median composites of the pixels were generated to construct the CNN input channels.

The hidden layers of the CNN model were configured with two 2D convolutional layer with 128 and 256 filters and kernel size of 5×5, employing a ReLU activation function, one average pooling (2×2) layer and three dense layers of size 144, 72 and 12 respectively. The output layer contained approximately 600,000 depth points derived from the IDW interpolation (N sample), and from this dataset, 70% was used for training, while the remaining 30% was allocated for testing and validation. The loss function was set to Mean Absolute Error (MAE), and the ADAM optimizer was applied with a learning rate of 0.0001. Training was conducted over 100 epochs using GPU-based processing to enhance computational efficiency.

3   Results

The best empirical model results produced a Root Mean Square Error (RMSE) of 2.99 m and a MAE of approximately 2.30 m for bathymetric data samples ranging from 0 to 16 m in depth across the study area. The model demonstrated superior performance when using the Sentinel-2 single image captured closest to the date of the field survey.

In contrast, the CNN model showed improved performance for the 72-channel configuration, which used geometric medians of Sentinel-2 images acquired within a three-month period before and after the bathymetric survey. This configuration achieved a Spearman correlation coefficient of 1.0, an RMSE of 0.2 m, and an MAE of 0.09 m for the validation dataset.

Figure 4 presents the bathymetric map generated through the ML model, displayed in the larger frame on the left. For comparison, the scatterplot graphs illustrating the performance of the (a) empirical model with single image (upper right frame) and (b) CNN model with median composite images (lower right frame).

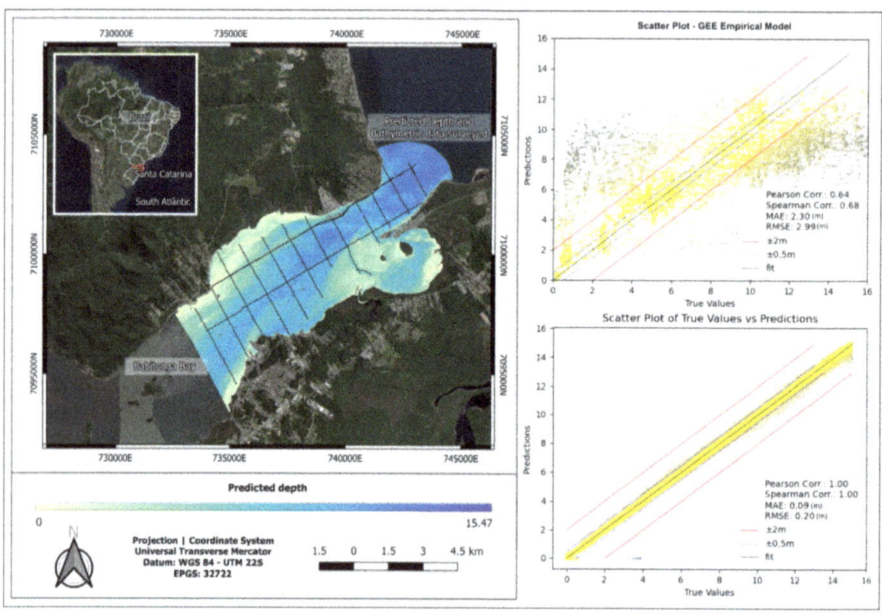

Fig. 4. Bathymetric data generate through ML applied to SDB in the test area (left frame) and the scatterplots with the statistics, RMSE and MAE errors, and linear correlations calculated for the empirical model (upper right) and the CNN (lower right).

4 Conclusions

The results achieved by the empirical model were similar to those achieved in turbid waters by [10]. The overall results indicate that the CNN model, particularly in its 72-layer configuration, provides a more accurate estimate of bathymetric depths, benefiting from nonlinear inferences in the convolution process and temporal aggregation of data to reduce pixel noise associated with interferences in the water column, caused mainly by SPM and CDOM.

The integration of empirical and ML models plays a crucial role in enhancing the accuracy of SDB in optically complex regions. Empirical models are well known to contribute to a more robust understanding and description of light penetration processes in clear waters by leveraging well-established physical principles. Complementarily, ML models, particularly the CNNs, excel in adjusting the nonlinear influences of SPM and CDOM on light propagation in the water column of optically complex regions.

This complementary relationship underscores the potential for significant advancements through the combined use of these approaches to produce reliable data for improving the understanding of morphological evolution in shallow water environments.

References

1. Bosboom J, Stive MJF (2022) Coastal Dynamics. TU Delft Open. Delft University of Technology, Delft, The Netherlands. Version 1.1, p. 579
2. Cheng J et al (2023) A comprehensive evaluation of machine learning and classical approaches for spaceborne active-passive fusion bathymetry of coral reefs. ISPRS Int J Geo-Inf 12:381
3. Noernberg MA, Rodrigo PA, Luersen DM (2020) Seasonal and fortnightly variability of the hydrodynamic regime at Babitonga Bay, Southern of Brazil. Reg Stud Mar Sci 40:101518
4. Bonetti J (2002) Diagnóstico Ambiental do sistema ecológico costeiro de São Francisco do Sul (Santa Catarina). In Relatório Final Petrobras/UFSC – subprojeto: Caracterização ambiental do sistema ecológico costeiro de São Francisco do Sul. Brasil
5. Yin F, Philip EL, Gomez-Dans JL, Wu Q (2019) Bayesian atmospheric correction over land: sentinel-2/MSI and landsat 8/OLI. Geosci Model Dev 15:7933–7976
6. Roberts D, Norman M, Mcintyre A (2017) High-dimensional pixel composites from earth observation time series. IEEE Trans Geosci Remote Sens 55(11):6254–64 (2017)
7. Lyzenga DR (1978) Passive remote sensing techniques for mapping water depth and bottom features. Appl Opt 17(3):379
8. Stumpf RP, Holderied K, Sinclair M (2003) Determination of water depth with high-resolution satellite imagery over variable bottom types. Limnol Oceanogr 48:547–556
9. Caballero I, Stumpf RP (2020a) Towards routine mapping of shallow bathymetry in environments with variable turbidity: contribution of sentinel-2A/B satellites mission. Rem Sens 12
10. Traganos D, Poursanidis D, Aggarwal B, Chrysoulakis N, Reinartz P (2018) Estimating satellite-derived bathymetry (SDB) with the google earth engine and sentinel-2. Remote Sens 10:859

A Smartphone Rip-Detection Tool to Improve Rip Current Awareness in Australia

Mitchell D. Harley[1]([envelope]), Shenyang Qian[1], Yang Song[1], Imran Razzak[1], Toby Walsh[1], Rob Brander[1], Amy Peden[1], Rachel Irvine[1], Jasmin Lawes[2], and Shane Daw[2]

[1] UNSW Sydney, Kensington, NSW 2052, Australia
m.harley@unsw.edu.au
[2] Surf Life Saving Australia, Bondi, Sydney, NSW 2026, Australia

Abstract. Rip currents claim more lives annually in Australia than floods, bushfires, cyclones, or shark attacks, with fatalities disproportionately affecting young males and occurring at unpatrolled beaches. Despite ongoing investments in surf lifesaving, safety infrastructure, and educational campaigns, rip current-related deaths are rising, highlighting the need for innovative solutions. Addressing this challenge, the "RipEye" project leverages computer vision technology to detect rip currents via smartphone cameras. This collaborative initiative, involving Surf Life Saving Australia (SLSA) and multidisciplinary university researchers, focuses on three objectives: (1) compiling a rip current detection dataset specific to Australian beaches, (2) optimizing image and video analysis techniques for accurate rip detection, and (3) evaluating the framework's feasibility to enhance public awareness and safety practices. By targeting unpatrolled beaches and high-risk demographics, the RipEye project seeks to supplement traditional lifesaving methods, providing accessible tools for rip current identification. This paper outlines the methodology behind RipEye, presents preliminary findings, and discusses its potential impact in improving beach safety.

Keywords: rip currents · drowning prevention · deep learning · computer vision

1 Introduction

Rip currents claim more Australian lives on average each year than floods, cyclones, bushfires and shark attacks combined. Despite numerous efforts to reduce rip current risk through investments in surf lifesaving, safety infrastructure and education campaigns, recent statistics indicate that rip current-related deaths in Australia are still on the rise. Of these, 15–39 year-old males (45%) and unpatrolled locations (62%) make up a disproportionate number of deaths [1]. Clearly, a new approach is needed to reach these high-risk demographics, recognizing that not all of Australia's 11,200 beaches can be patrolled by lifeguards or surf lifesavers at all times.

This paper presents a new initiative that harnesses computer vision algorithms to detect rip currents from smartphones. The project – known as "RipEye" – is a partnership between Surf Life Saving Australia (the primary organization responsible for surf life

© The Author(s) 2026
C. Coelho et al. (Eds.): CD 2025, CRL 41, pp. 717–721, 2026.
https://doi.org/10.1007/978-3-032-15473-6_109

saving in Australia)) and multi-disciplinary university researchers. It aims to address the following: (1) to develop an extensive rip current detection dataset for algorithm training purposes, with a particular focus on Australian beach systems; (2) to determine the optimum image/video analysis methods for reliable and robust rip current detection from smartphone cameras; and (3) to assess the feasibility and acceptability of this framework increase rip current awareness and identification skills among key stakeholder groups. A further description of the project methodology and preliminary results are presented below.

2   Methodology

2.1   A New Dataset of Rip Current Imagery for Algorithm Testing

Recent efforts to develop rip-current identification from video have focused primarily on detection from aerial viewpoints such as those taken from UAV (e.g. [2]) or plan-view coastal imaging (e.g. [3]). Given the need for ground-based rip current detection to aid beach goers, the first phase of the project comprises the development of a new database of annotated rip-currents from a range of viewpoints. As shown in Fig. 1, this database classifies rip current imagery into 6 different classes: 3 classes related to elevation (low-view, high-view and top view) and 3 classes related to viewpoint (front-on, diagonal and side-on) to the rip channel. It is anticipated that each class has unique challenges in rip current detection as has been identified from community surveys of rip current identification skills [4].

Rip current imagery for this new dataset is sourced from a number of techniques, including the existing network of images obtained from the CoastSnap community beach monitoring program [5], the Argus coastal imaging [6] and new collection efforts undertaken by the team and community volunteers. To significantly enhance the dataset, a new GAN-based data augmentation method ("RipGAN") is also applied that produces synthetic rip current imagery based on a Fast Fourier Fusion Module.

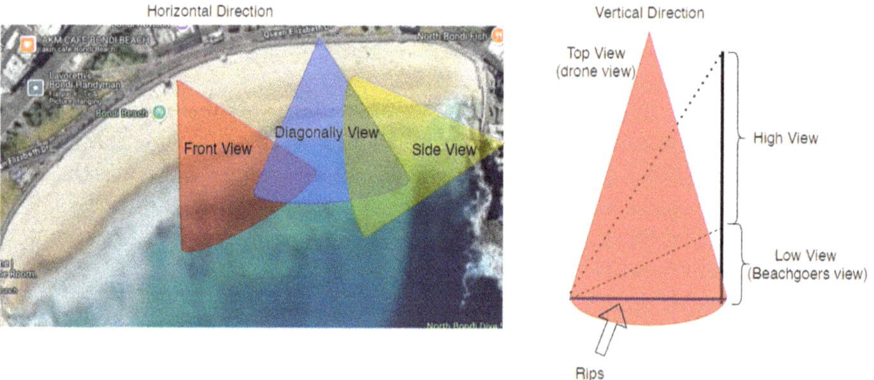

Fig. 1. Classification of rip current imagery into six different classes based on viewpoint (left) and elevation (right).

Fig. 2. Example of rip current detection using YOLO at a CoastSnap station in Cape Byron, Southeast Australia

All imagery is subsequently annotated (bounding box + polygon) using the online annotation software CVAT.

2.2 Algorithm Testing

A range of common algorithms to computer vision algorithms are first tested to benchmarch detection performance on the new dataset. This includes the popular YOLO series, RCNN series, SSD series and DETR series of object detectors. An example of rip current detection applied using YOLO for a CoastSnap image at Cape Byron (Tallow Beach) in SE Australis is shown in Fig. 2.

Following [2], performance is evaluated using AP50 and AP50:95, which are the most common metrics for object detection. AP50 refers to the mean average precision when Intersection over Union (IoU) between the detected and annotated rip current is 50%. AP50:95 meanwhile is the average precision with an IoU threshold between 50% and 95% (interval = 5%). Recall and precision are also used in the performance evaluation.

2.3 Stakeholder Engagement and Co-Design

While the initial phase of the project is focused on algorithm development, the overarching aim is to improve rip current awareness among Australian beachgoers – particularly high-risk groups. To this end, a series of stakeholder engagement consultations are to be conducted throughout 2025 that will identify key challenges related to rip current detection among various stakeholder groups. These stakeholder engagements will aid co-design of a rip current education tool and online *"Challenge the RipEye"* game for use in lifesaver training and on-beach and external rip current education applications.

Fig. 3. Rip current annotation results using the annotation software CVAT

3 Preliminary Results and Conclusions

To date, a total of 9,670 rip current images have been annotated using the CVAT software for algorithm testing (refer Fig. 3 for an example). Of this, 62% are from a high view according to Fig. 1, 20% from a low view and 18% from a top view. This distribution represents a bias towards elevated viewpoints characteristic of CoastSnap stations, but significantly more ground-based angles relative to previous datasets in the literature. Initial results based on application of common computer vision algorithms YOLO, RCNN, DINO and DETR indicate detection performance on the test set of P50 > 0.9, providing encouraging results for further development of the RipEye tool.

By leveraging cutting-edge computer vision technology and establishing a robust dataset tailored to Australian beach systems, this project lays the foundation for smarter, more accessible rip current detection tools. The preliminary results underscore the potential of smartphone-based solutions not only to enhance real-time identification of hazardous rip currents but also to empower key at-risk demographics with greater awareness and safety skills. Moving forward, the integration of this technology with educational outreach and stakeholder collaboration promises a scalable framework to improve beach safety across Australia's vast (and mostly unpatrolled) coastline.

References

1. Cooper B, Ledger J, Daw S, Lawes S (2021) Coastal Safety Brief: Rip currents. Surf Life Saving Australia
2. Dumitriu A, Tatui F, Miron F, Ionescu RT, Timofte R (2023) Rip current segmentation: a novel benchmark and YOLOv8 baseline results. In: IEEE/CVF Conference on Computer Vision and Pattern Recognition (CVPR) Workshops
3. Maryan C, Tamijidul Hoque M, Michael C, Ioup E, Abdelguerfi M (2019) Machine learning applications in detecting rip channels from images. Appl Soft Comput 78

4. Uebelhoer L, Koon W, Harley M, Lawes J, Brander R.: Characteristics and beach safety knowledge of beachgoers on unpatrolled surf beaches in Australia. Nat Hazards Earth Syst Sci 22(3) (2022)
5. Harley M, Kinsela M (2022) CoastSnap: a global citizen science program to monitor changing coastlines. Cont Shelf Res 234:104796
6. Turner I, Arrninkhof S, Holman R (2006) Coastal imaging applications and research in Australia. J Coastal Res 22

Development of Unmanned Surface Vessels for Surf-Zone Bathymetric Survey

Peter Traykovski[1](✉) and Inna Shapovalenko[2]

[1] Woods Hole Oceanographic Institution, Woods Hole, MA 02540, USA
`ptraykovski@whoi.edu`
[2] Franklin and Marshall University, Lancaster, PA 17604, USA

Abstract. A series of unmanned surface vessels for surf-zone bathymetric surveys have been developed. Improvements to the original design, based on an off-the-shelf scaled offshore racing hull, include optimization for efficiency and control in various environments, including those with floating debris. Upgraded data acquisition and transmission systems allow real-time transmission of survey data to assess quality and change survey plans. Algorithms have been developed to evaluate the accuracy of gridded surfaces based on the track-line spacing and surface fitting parameters. Future directions include refined simulation procedures to enhance vehicle control in energetic wave conditions.

Keywords: Nearshore Bathymetry · Unmanned Surface Vessel · Echosounders

1 Introduction

Modeling and predictions of coastal morphodynamic response to storms and other nearshore processes are often limited by bathymetry measurements before, during, and after storms. While beach topography is simple to measure with aerial vehicles or direct in-situ survey techniques, nearshore bathymetry is much more difficult. Aerial techniques, such as water penetrating LIDAR, require clear water to penetrate more than several meters and thus are often unsuitable during or immediately after periods of high energy. Manned surveys by small boats or personal watercraft (Jet skis) are the industry standard but are inefficient due to the use of human operators and logistical constraints [1]. Unmanned surface vessels offer an attractive alternative to crewed craft for nearshore surveys. Small surf-zone capable USVs have been in development at Woods Hole Oceanographic Institution and Integrated Coastal Solutions since 2018 and have been deployed in various conditions.

2 Surf-Zone Vehicle Design Overview

2.1 1st Generation - RoboTurtle

The first system was developed in 2018 by modifying a commercially available remote-control boat with a self-righting high-buoyancy deck structure and a jet drive for safe launch and recovery in very shallow water [2]. This vehicle was named RoboTurtle due

C. Coelho et al. (Eds.): CD 2025, CRL 41, pp. 722–728, 2026.
https://doi.org/10.1007/978-3-032-15473-6_110

to the shape of the self-righting deck. The vessel is controlled by an Ardupilot cube controller, which allows a seamless transition from remote-controlled modes to autonomous GNSS waypoint following modes. The vessel payload consisted of an Echologger model ECS 24 200/450 kHz single beam echosounder and a precision dual-band (L1/L2) GPS capable of 2 to 5 cm accuracy PPK processing. The vessel was a scale model of an offshore racing boat designed for high-speed planning operation with a deep-v hull for reduced slamming in steep short-period waves. This design led to a bow-up attitude at survey speeds (Fig. 1a). While this vehicle performed well in the surf zone and produced high-quality bathymetric data, it was limited by its short endurance of 1.5 to 2 h before requiring a battery swap. Survey speeds of a compact USV are typically around 1.5 to 2 m/s. Hull speed of a 2 m length (L) vessel ($V_{Hull} = \sqrt{gL/2\pi}$) is 1.5 m/s; thus, at speeds above this the vehicle is planing. Unless the water surface is very flat while planning, the echosounder data quality decreases due to either bubble entrained as the vessel bounces over waves or as the vehicle temporarily bounces out of the water. The bow-up attitude of RoboTurtle at survey speed, combined with a relatively small battery pack (580 Wh), led to insufficient endurance for practical use in nearshore surveying.

Fig. 1. a. First generation RoboTurtle surf-zone USV developed in 2018, b. Yellowfin USV c. Vectored jet drive propulsion and d. Dual thruster propulsion system.

2.2   2nd Generation - Yellowfin

To improve the endurance of the RoboTurtle, a 2nd generation vehicle, Yellowfin, was designed with a semi-displacement hull shape and a narrower beam than RoboTurtle. The hull was generally rounded, similar to a full displacement hull, with a small flat section in the center for mounting sensors flush to the hull. The center of gravity was located further forward than the C.G. of the RoboTurtle design, and Yellowfin has a wave-piercing bow design that does not produce as much lift in the forward section. Both features are aimed at making a level trim at survey speeds and delay the onset of planing to speeds well above typical survey speed (Fig. 1b). The efficiency of these vehicles was compared with the same jet drive system (Fig. 1c), motor and motor controller by running the vehicles through a series of speeds and measuring amps drawn from the battery. At a typical survey speed of 2 m/s, the average power consumption of Yellowfin is 208 Watts compared to the 352 Watts of RoboTurtle (Fig. 2a). Due to its increased efficiency, combined with a larger battery pack 1180 Wh, the endurance of Yellowfin is increased to 5.7 h compared to the 1.7 h of Robot Turtle (Fig. 2a). This is sufficient for most nearshore surveys that are ideally performed at high tide. The Ardupilot navigation control system on Yellowfin is the same as RoboTurtle, while the single beam echosounder was swapped for a Cerulean s500 echosounder. A significant improvement was a data acquisition system that allows real-time transmission of bathymetry data to the base station, allowing resurvey of areas with poor data quality or features of interest that were unknown in the original survey plan.

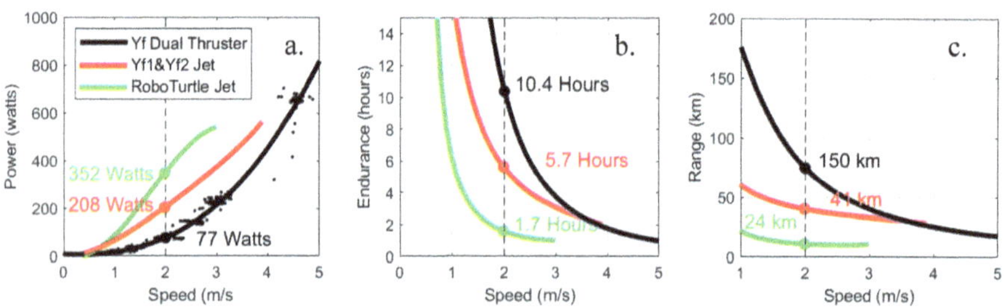

Fig. 2. a. Power consumption vs. speed for each system, b. Endurance vs. speed, c. Ranges vs. speed.

2.3 3rd Generation - Yellowfin Dual Thruster

One significant drawback of the Jet Drive propulsion system (Fig. 1c) is that it is susceptible to clogging in environments with floating seaweed or other debris. To rectify this, the standard Yellowfin Hull was fitted with two BlueRoboticsTM M200 thrusters (Fig. 1d). The swept-back blades of these thrusters were evaluated in various environments with floating debris. They were found to consistently shed debris in the forward direction. The dual thruster configuration also allows reverse, and small-radius pivot turns. The disadvantage of the dual thruster configuration is that it is more susceptible to damage if the vessel runs aground in the swash zone, and the pivot turns lead to periods when the vehicle does not have forward momentum. In our experience, it is advantageous always to maintain forward momentum in regions with breaking waves. The dual thruster configuration is also more efficient, with a power consumption of 77 Watts at 2 m/s, leading to an endurance of 10.4 h with an 800 Wh battery pack. Thus, in our experience, the jet drive system is preferred in energetic surf conditions. The dual thruster configuration can have advantages in confined areas, particularly with floating debris, such as seaweed.

3 Gridded Survey DEM Uncertainty

To assess the uncertainty of the surveys produced by this system, the along-track accuracy of the combined single beam echosounder and PPK-GNSS heave compensation system was previously evaluated against direct GNSS measurements of the seafloor position with a 5 m long survey pole in calm conditions resulting in 5 to 7 cm r.m.s error between the two measurements [2].

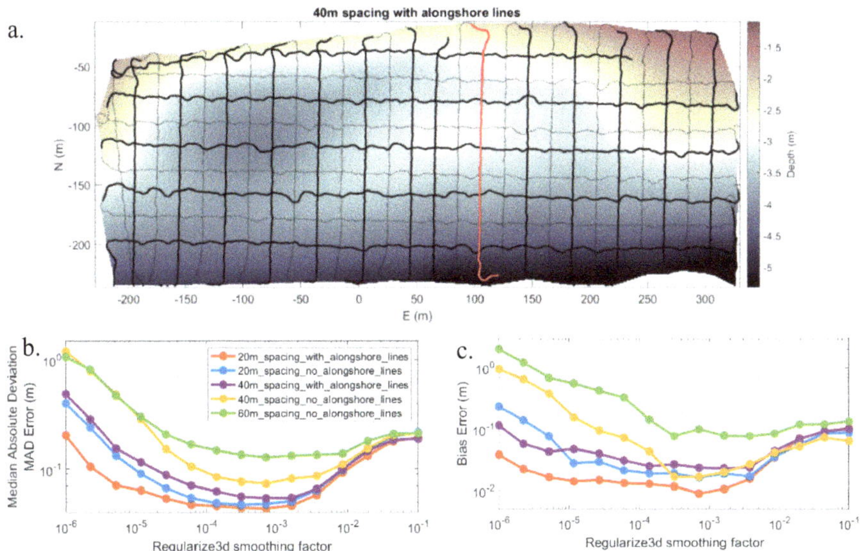

Fig 3. a. DEM surface based on 40 m spaced along and across-shore lines (solid black). The original 20-m spacing is shown as dashed black. b. *MAD* error vs smoothing factor and c. *Be* vs. smoothing factor for a variety of line spacing options.

In slightly more energetic conditions with wave heights of 50 cm, the along-track measurements from the USV were compared to measurements from the well-established USACE ERDC Duck FRF LARC System [3] with similar r.m.s. error of 5 cm. Often, the goal of a nearshore survey is not simply along-track data but a gridded digital elevation model (DEM) based on the combined beach topography and nearshore bathymetry. A surface fitting procedure is required to produce a DEM from the single beam echosounder along-track data. We have been using the RegularizeData3D [4] algorithm available in the Matlab Community file exchange. This highly efficient routine can fit a surface to a typical 500 m by 300 m survey grid with $O(1e5)$ along-track points in less than 10 s. This surface fitting algorithm uses a smoothing factor to determine the closeness of the surface to data points in the least squares sense. A smoothing factor that is too low fits all the noise in the measured along-track data and produces a very rough surface that does not capture the true bathymetry. Conversely, a smoothing factor that is too high smooths out actual features present in the data, such as sand bars. To determine the optimum smoothing factor, we use a procedure that divides the survey into a series of along and across-shore lines (Fig. 3). The surface fit with one line (data points at locations x_i) not included in the data input to the fit (e.g., the red line in Fig. 3a). The statistics of error between the predicted data from the surface ($S(x_i)$) and the measurements from the withheld line ($D(x_i)$), $E(x_i) = D(x_i) - S(x_i)$, is calculated with two metrics. The Median Absolute Deviation (*MAD*) captures the variability of the error term and *Be* captures the bias.

$$MAD = med\left(|(E(\boldsymbol{x_i}) - med\left(E(\boldsymbol{x_i})\right)|\right) \tag{1}$$

$$Be = med\left(E(\boldsymbol{x_i})\right) \tag{2}$$

Once these error statistics are calculated with one line withheld, the algorithm repeats the procedure for all the lines in the survey by removing subsequent individual lines and using the others to fit a surface. It averages the resulting *MAD* and *Be* statistics over all realizations. This procedure identifies the optimum smoothing factor and the *MAD* and *Be* resulting from that smoothing factor. The procedure was carried out with 1) using all 20 m spaced along and across-shore lines (Fig. 3.), 2) using only the 20 m spaced across-shore lines, 3) dropping every other line for 40-m spacing and retaining both along and across-shore lines, 4) using every other across-shore line resulting in 40 m spacing, and 5) using every third across-shore line resulting in 60 m spacing. The optimum smooth factor remains the same at 10^{-4} to 10^{-3} for all the different line spacings, but the quality of the surface fit as quantified by *MAD* and *Be* varies significantly. With the most densely spaced lines, the *MAD* values are in the range of 5 to 7 cm, similar to the along-track accuracy, and the bias is less than 3 cm. With 40 m spacing and retaining along and across-shore lines, the MAD increases to 8 cm, which is larger than the along-track error. Once line spacing is increased to 60 m with no along-shore lines, the error statics become large enough to make conducting change analysis over a storm cycle difficult.

4   Survey Results

A typical morphodynamic change based on three surveys over a month with several wave events ranging from 2.5 to 3.5 m in height occurring in between surveys is shown in Fig. 4. The surveys combine the bathymetry data from the USV with a backpack-based PPK-GNSS surveys on the beach to make a combined topobathy map. The difference map shows most change occurs near the 0 (re NAVD88) contour level, and there is more change on the east side of the survey region, where a shoreface attached shelf transitions to an offshore sand bar.

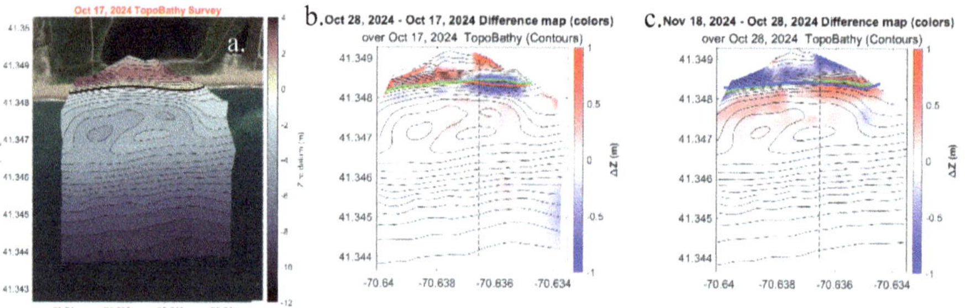

Fig. 4. a. Initial combined topobathy survey on Oct. 17, 2024, b. difference map from Oct. 28, 2024 – Oct 17, 2024, and c. difference map from Nov. 18, 2024 – Oct 28, 2024. Red colors are deposition, and blue represents erosion.

5 Future Directions

To date, we have conducted approximately 35 surveys with the Yellowfin systems with good results. We have mainly targeted calm conditions before and after storms to map bathymetric change. However, some surveys after storms have been conducted with up to 1.5 m high long period swells as the storm moves offshore. The vehicle performs well in these conditions and successfully transits sand bars with intermittent wave breaking. Several passes are needed to ensure high-quality echosounder data in wave-breaking regions. Navigation and echosounder performance deteriorate as the breaking becomes more consistent on shallower sand bars. In larger waves, the inner waypoints must be set further offshore to prevent the vehicle from being washed ashore in the swash zone. This has happened several times during surveys, and the vehicle can be quickly relaunched. Remote control navigation modes can be used to enter the swash zone between sets of waves.

To improve performance in more energetic conditions, we have recently upgraded the Ardupilot simulation environment to include horizontal orbital velocities from waves. The default motorboat simulator in the Ardupilot software only includes vertical motions and pitch and roll effects. The simulations show that when orbital velocities are less than the set survey speed, the vehicle path is perturbed slightly, but when the orbital velocities are equal to or larger than the set survey speed, significant perturbations occur (Fig. 5). This new simulation capability provides a path forward for improving control algorithms in energetic wave conditions.

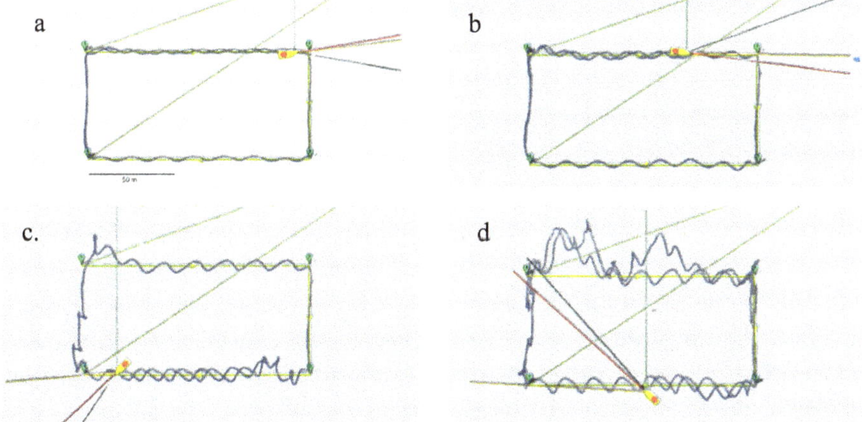

Fig. 5. Simulations including horizontal wave orbital velocities. Survey speed was set to 2 m/s, and wave orbital speed amplitude was set to a) 50 cm/s, b) 100 cm/s, c) 200 cm/s, d) 300 cm/s while the vehicle attempted to navigate through the same waypoints.

References

1. Dugan JP, Morris WD, Vierra KC, Piotrowski CC, Farruggia GJ, Campion DC (2001) Jetski-based nearshore bathymetric and current survey system. J Coast Res 900–908

2. Francis H, Traykovski P (2021) Development of a highly portable unmanned surface vehicle for surf zone bathymetric surveying. J Coastal Res 37(5):933–945
3. Forte MF, Birkemeier WA, Mitchell JR (2017) Nearshore survey system evaluation. U.S. Army Engineer Research and Development Center, ERDC/CHL TR-17–19
4. Jamal, 2025. RegularizeData3D (https://www.mathworks.com/matlabcentral/fileexchange/46223-regularizedata3d), MATLAB Central File Exchange. Accessed 14 Jan 2025

X-Band and Doppler-Radar Observations of Waves, Currents and Morphologies on a Sandy Beach at the Research Pier Hors

Satoshi Takewaka[1]($\boxtimes$), Yugo Ohtsuka[1], and Masayuki Banno[2]

[1] University of Tsukuba, Tsukuba 305-8573, Japan
takewaka@kz.tsukuba.ac.jp
[2] Port and Airport Research Institute, Yokosuka 239-0826, Japan

Abstract. We use X-Band radar and Doppler radar velocimeter to observe waves, currents, and morphologies on a sandy beach at the research pier HORS facing the Pacific Ocean in Japan. The area is an almost straight coast with fine sands, and a 400 m-long research pier is installed on the beach. X-Band radar is an imaging radar that can capture wave propagations and run-ups. A Doppler radar velocimeter is installed at the pier deck to estimate the current variation in the longshore. We detected the longshore migration of coastal morphological features, like shoals, troughs, and beach cusps, in the surf zone from the X-Band radar observation and compared it with the variation of the longshore current. Results for twelve months show that coastal features migrate to the downflows depending on the longshore current direction and speed. We expect this to serve as a proxy signal of longshore sediment motion.

Keywords: X-Band Radar · Doppler Velocimeter · Research Pier HORS · Longshore Migrations of Coastal Features

1 Introduction and Methods

We combine two remote sensing techniques to observe and analyze the longshore migration of coastal features in the surf zone along a sandy beach at a research pier. One method is X-Band radar observation that can capture wave propagations and run-ups, and the other is continuous Doppler radar velocimeter measurement of longshore current from the pier deck throughout the year. This work highlights the longshore movements of wavy patterns detected in the radar images, representing shoals, troughs, and beach cusps, and relates them to the longshore current speeds and directions.

1.1 Research Pier HORS

Hasaki Oceanographical Research Station (HORS) is a coastal research facility facing the Pacific Ocean, operated by the Port and Airport Research Institute (PARI, www. pari.go.jp). It has a 400 m-long research pier for coastal science and engineering studies

C. Coelho et al. (Eds.): CD 2025, CRL 41, pp. 729–734, 2026.
https://doi.org/10.1007/978-3-032-15473-6_111

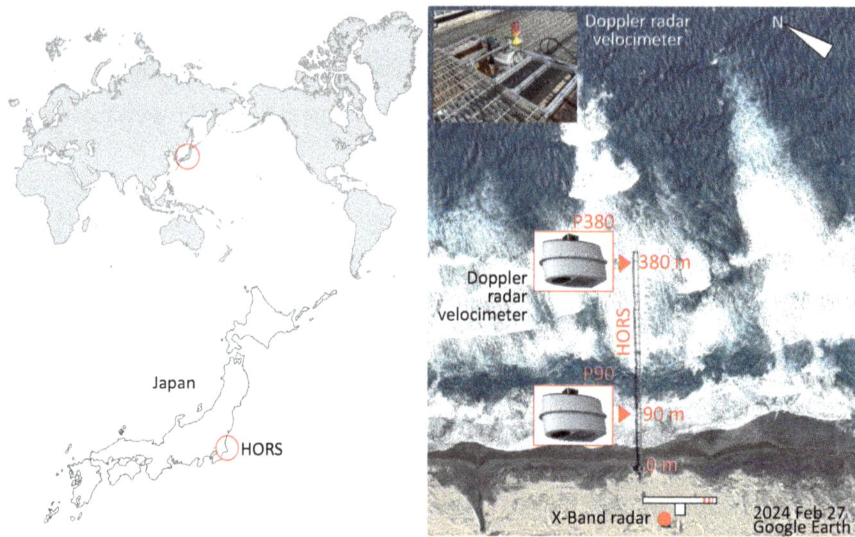

Fig. 1. Research pier HORS

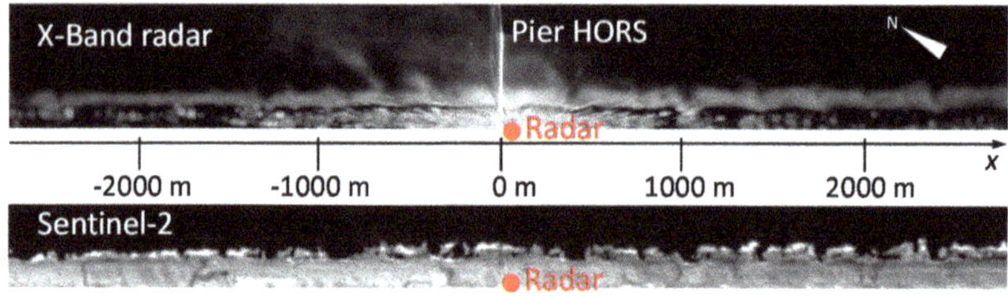

Fig. 2. (Upper panel) Time-averaged radar image: Pixel resolution = 5.6 m. (Lower panel) Satellite image: Sentinel-2, near-infrared, pixel resolution = 10 m. 2022 December 8, 10h (JST).

(Fig. 1). The pier is on an almost straight beach that extends from north to south with nearly uniform sand with a median grain size of 0.18 mm. The research team of the HORS regularly surveys the bathymetry of the area [1].

The mean tidal range at spring tides is approximately 1.5 m. Incident waves are mainly from the southern in the summer and the northern in the winter. The beach is stable and is not in an eroding or accumulating trend. The coast typically has a single bar and a mean beach slope of approximately 1/50 in the foreshore and 1/20 in the deeper region.

1.2 X-Band Radar

X-Band marine radar is an imaging radar capable of tracking the movements of wave crests over an area spanning several kilometers [2]. The most attractive feature of using an X-band radar is its ability to collect data on coastal processes continuously and remotely, even in bad weather conditions that typically accompany erosive high-wave conditions.

The X-Band radar is installed on the roof of the research building, which is 17 m above the mean sea level and approximately 100 m backward from the mean shoreline position. Figure 1 shows the layout of the pier, research building, and radar location.

The radar captures raw echo images every two seconds. In this image, we see the wave crests and can understand the refraction sequence of the incident waves. We get a time-averaged image or so-called time exposure if we average the raw images over a few minutes. In the time-averaged image, individual waves disappear, and an edge extending in the longshore direction becomes visible. Further, we can interpret several features, such as the breaker zone, shoreline position, and bar crest locations.

Figure 2 compares a time-averaged radar image and a satellite scene captured during the same period. In the center of the radar image, we see the pier extending vertically. We observe similar shoreline undulations in both, especially in the near range close to the pier. Bright portions have higher backscatters, mainly from breaking waves. Since the radar beam emission is at a grazing angle, the far view becomes blurred and darker compared to the near.

1.3 Doppler Radar Velocimeter

Two Doppler radar velocimeters are installed at the pier deck, one close to the shoreline (P90) and the other at the offshore tip (P380), as shown in Fig. 1. The instrument is developed originally to measure bi-directional surface current speeds remotely in wastewater treatment plants (SQ-R flow meter, SOMMER, https://www.sommer.at/en/products/sewage-wastewater/sq-flow-meters). It emits K-band radio waves and detects the Doppler shifts of the backscatters from the water surface to estimate the surface flow speed. We apply this instrument to estimate the longshore current variation.

We verified the estimated current direction and speed using the observational result and wave stat and confirmed that it could monitor the variation of the longshore current. Figure 3 compares longshore current velocities estimated by the radar and measured manually by float release observations [3]. They are in a linear trend. Since the radar detects the water-surface motion, it gives a higher result than the float observation, representing the motion in the middle of the water column.

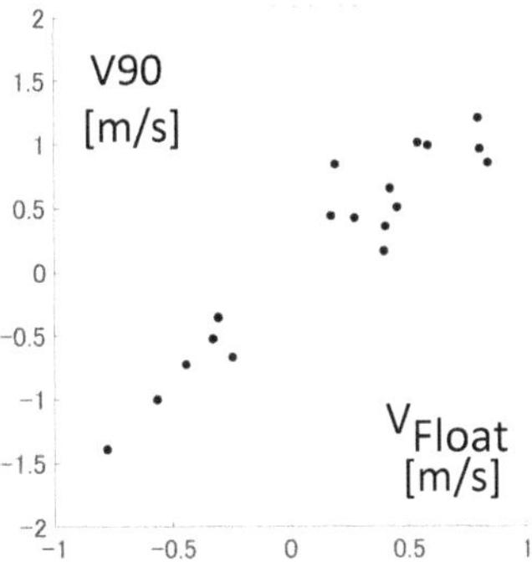

Fig. 3. Longshore current at P90 estimated by the Doppler radar velocimeter (V90) and observed from the float release observation (V_FLOAT).

2 Results

Beach cusps are frequently observed in the area, usually with wavelengths of several hundred meters [4]. The seawards of the cusps have shallow and deep portions under the water. These features are visible in the time-averaged radar images, as shown in Fig. 2. At the shallows, waves break and appear as bright portions in the radar image, and dark portions represent deeper areas where less wave breaking occurs.

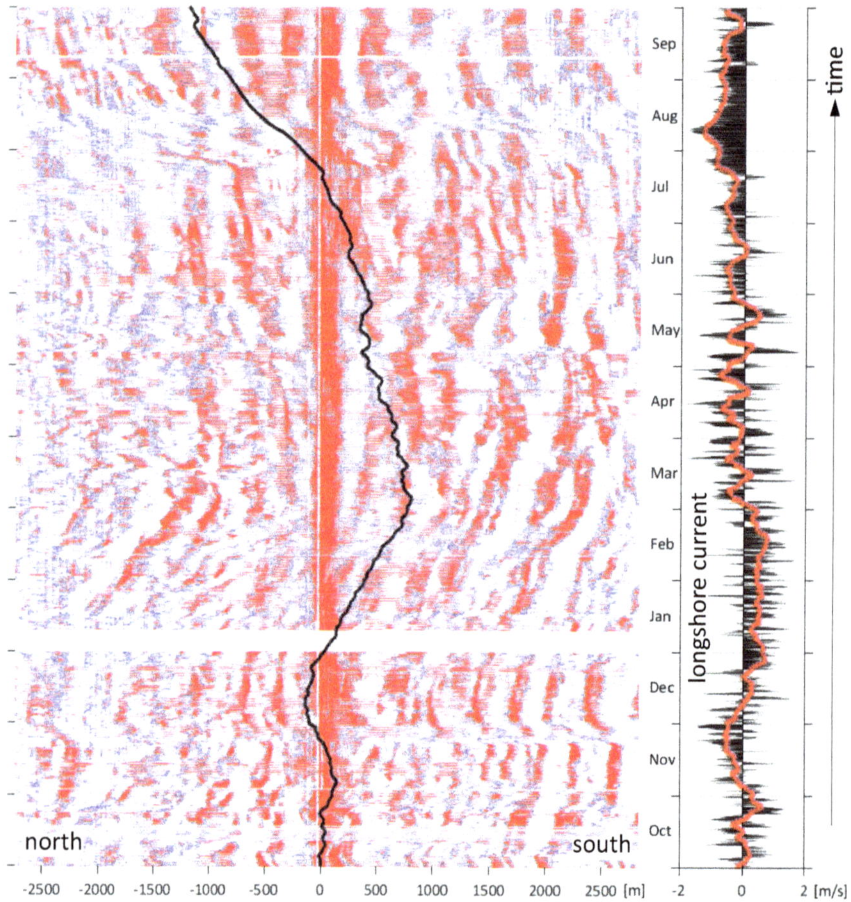

Fig. 4. Longshore migration of coastal features: 2022 October – 2023 September. (Main panel, left) Lateral axis = longshore extent, vertical axes = time. Hot color represents a higher backscatter from breakers on shoals, and cool color represents a weaker backscatter in deeper areas. The solid black line is the integral of the longshore current measured at P90. (Sub panel, right) Variation of longshore current measured at P90. Positive value = longshore current heading to the south. Black lines = raw hourly data, and the red thick line = moving average over one week.

Following the time-averaged images, we observe these features move along the shore [2]. We processed a time stack image for twelve months to visualize this motion, as shown in the main panel in Fig. 4. We processed this image in the following manner: First, we extracted pixel brightness from time-averaged images along a longshore transect fixed close to the mean shoreline position and stacked in the vertical. The lateral axis is the longshore extent, and the vertical is the elapsed time. Then, we adjusted the brightness to ensure the far-field visibility. Bright and dark portions exist along the shore at several hundred meters intervals, constantly migrating throughout the year.

Longshore current may be a factor driving these migrations. To examine this, we show the longshore current variation measured at the pier in the sub-panel of Fig. 4. The current is mostly heading north due to southern wave incidence in the summer months and vice versa. The overall migration pattern seems to follow the longshore current: Migrating to the north in summer and to the south in winter.

We integrated the longshore current velocity in time and displayed it with a solid black line in the main panel, which is scaled with an arbitrary factor. Surprisingly, it seems that the integral can follow the oblique pattern features, which change the direction and angle of the extension. Although the sedimentary processes of longshore migration of beach cusps and accompanying shallow and deep portions are poorly understood, we regard the longshore current as playing a role in longshore motion.

3 Concluding Remarks

We showed observation results with the X-Band radar and Doppler radar velocimeter at the research pier HORS. From the X-Band radar observation, we found shoreline undulations and accompanying shallow and deep portions are migrating to the down-waves throughout the year. This migration was also consistent with the record of longshore current measured by the Doppler radar velocimeter.

To the author's knowledge, no model (analytical or numerical) can explain the migrative motions observed in this study. If we can establish a model to reproduce this, we expect to evaluate the longshore sediment by observing the migrations of the coastal features.

References

1. Banno M, Nakamura S, Kosako T, Nakagawa Y, Yanagishima S, Kuriyama Y (2020) Long-term observations of beach variability at Hasaki, Japan. J Mar Sci Eng 8(11):871
2. Galal EM, Takewaka S (2008) Longshore migration of shoreline mega-cusps observed with X-Band radar. Coast Eng J 50:247–276
3. Kuriyama Y, Ito Y, Yanagishima S (2008) Cross-shore variation of long-term average longshore current velocity in the nearshore zone. Cont Shelf Res 28:491–502
4. Tabasi M, Suzuki T, Cox DT, Okayasu A (2024) Artificial neural network based model to predict mega cusp characteristics in Hasaki, Japan. Geomorphology 464:109377

Estuarine and Coastal Lagoons' Processes

Numerical Analysis of Sand Bar Flushing and River Mouth Terrace Formation During Floods Due to Differences in Sand Bar Shapes

Akira Sakaue[(✉)], Daichi Matsuo, and Tatsuhiko Uchida

Hiroshima University, Kagamiyama 1-4-1, Higashihiroshima, Japan
`m240509@hiroshima-u.ac.jp`

Abstract. Coastal erosion is caused by decreased sediment supply from rivers and the volume reduction in river mouth terraces, which serve as sources of fine sediment. Predicting morphological variations in coastal areas, including river mouth sandbars and terraces, requires comprehending sediment transport dynamics caused by flood flows and waves. As the first step in analyzing fine sediment transport in coastal areas, flood flow and riverbed variation analyses were conducted, including the downstream river sections with the surrounding coastal area. The initial shape of the river mouth sandbar is crucial when it comes to how it affects the morphological changes in the river mouth. However, it is difficult to give the initial topography accurately due to its considerable temporal variation and was a focal point of this analysis. Analyses of the several initial shapes of the river mouth sandbar were compared with direct observational after the flood. The processes of river mouth sandbar flushing and the terrace formation were investigated. The results show that the temporal change in water and river mouth terrace formation vary significantly depending on the initial shape of the river mouth sandbar and three-dimensional flows. This highlights the need to understand the shape of the river mouth sandbar before the flood.

Keywords: river mouth sandbar · river mouth terraces · GBVC3 method · flood flow · sediment transport

1 Introduction

Coastal erosion is often caused by the reduction in the sediment supply from rivers to the sea. During short-term flooding events at an estuary, river mouth bars are flushed away, and fine-grained sediment is discharged into the sea, forming river mouth terraces [1, 2]. The river mouth terraces play a crucial role as sources of sand, and wave action subsequently restores river mouth bars and supplies sand to the adjacent coastal areas [3].

However, comprehensive studies covering short-term topographic changes during floods and medium- to long-term wave-action topographic changes have not been conducted extensively. To serve as a foundation, this study aims to investigate the effects of sand bar topography and three-dimensional flows on the dynamics of fine sediment in a river mouth, using quasi-three-dimensional flood flow and sediment transport analysis.

© The Author(s) 2026
C. Coelho et al. (Eds.): CD 2025, CRL 41, pp. 737–743, 2026.
https://doi.org/10.1007/978-3-032-15473-6_112

2 Target of Analysis

The target river for this study is the Tenjin River in Tottori Prefecture, Japan, focusing on the flood event on 7–9 July 2021. A single-beam bathymetric survey of the Tenjin River was conducted before (19 October 2020) and after (26 July 2021) the July 2021 flood [4]. The survey revealed that the flood breached the river mouth bar at the center and scoured the bed, resulting in the formation of a tongue-shaped river mouth terrace in front of the river mouth (see Fig. 3 for 26 July 2021). Figure 1 shows the observation results and satellite photographs [5]. The upstream discharge and tide level for the analysis period are shown in Fig. 2.

Fig. 1. Pre- and post-flood observations and satellite photographs

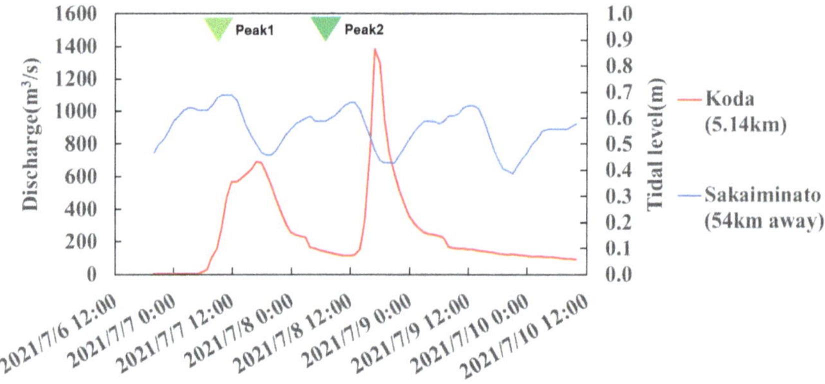

Fig. 2. The boundary conditions

3 Methodology

To analyze riverbed deformation during floods in estuarine and marine areas, the simulation utilized a mesh encompassing both river channels and the sea which continuously tracked the topographic changes. Comparative analyses examined the impact of three different initial sandbar shapes on sediment dynamics and explored discrepancies between flood flow analysis methods.

3.1 Flood Flow Analysis Method

To analyze topographic changes including sandbar flushing and river mouth terrace formation, evaluating three-dimensional flow and riverbed shear stress is essential. This study employed the General Bottom Velocity Calculation (GBVC3) method, which solves depth averaged horizontal vorticity and surface motion equations without assuming shallow water flow [6, 7], capturing near-bottom flow with non-equilibrium velocity and non-hydrostatic pressure distributions and sediment transport characteristics. The salinity and wave effects were excluded in this research, which are future research to be addressed.

3.2 Riverbed Deformation Analysis Method

The temporal variation in riverbed height was calculated using the continuity equation for sediment transport. The critical bed-load shear stress was determined using Iwagaki's formula [8], while the bed-load sediment transport rate was calculated using the Ashida-Michiue formula [8]. This study examined the flushing of river mouth sandbars and intense movement of fine sediment particles, which are transported to the sea and contribute to estuarine terrace formation.

To simplify the treatment of fine sediment dynamics, without considering the variation in the vertical profile, the depth-averaged suspended sediment concentration was computed by solving the advection-diffusion Eq. (1).

$$\frac{\partial Ch}{\partial t} + \frac{\partial CU_i h}{\partial x_i} = \frac{\partial}{\partial x_i}\left(Dh\frac{\partial C}{\partial x_i}\right) + Q_{su} - Q_{sw} \tag{1}$$

where, C: depth-averaged suspended sediment concentration, h: water depth, U_i: depth-averaged flow velocity in the i-direction, D: suspended sediment diffusion coefficient, Q_{su}: upward flux of suspended sediment, Q_{sw}: downward flux of suspended sediment. The downward flux of suspended sediment was calculated using the bottom concentration c_b, which is derived from the depth-averaged suspended sediment concentration, and the sediment settling velocity w_0, as expressed in Eq. (2).

$$Q_{sw} = c_b * w_0 \tag{2}$$

$$c_b = C * \frac{\Gamma}{1.0 - \exp(-\Gamma)}, \Gamma = \frac{hw_0}{\varepsilon}, \varepsilon = 1.2\alpha u_* h$$

where, the settling velocity was evaluated using Rubey's formula [8]. The upward flux of suspended sediment was expressed using the equilibrium bottom concentration c_{be}, as given in Eq. (3).

$$Q_{su} = c_{be} * w_0 \tag{3}$$

The calculation of the bottom concentration c_{be} used Eq. (4) from the two-phase flow model for flow and sediment, as provided by Uchida and Fukuoka [9].

$$c_{be} = \frac{C_B}{1 + (w_0/\alpha u_*)}\frac{1 - a(w_0/\alpha u_*)}{1 + (1 - a)(w_0/\alpha u_*)} \tag{4}$$

$$C_B = 0.6, a - 0.08, \alpha - \kappa/6, \kappa - 0.4$$

3.3 Analysis Conditions

Calculations Run 0, 1 and 2 were performed with three different initial sandbar shapes (a), (b), and (c), respectively, as illustrated in Fig. 3 to investigate the impact of sandbar shape on the flushing of sandbars and the formation process of river mouth terraces during the flood. (a) used the initial topography based on the pre-flood survey on 19 October 2020. Regarding the sandbar height, which was difficult to estimate from satellite images, (b) set the height approximately 1 m lower than (a) to ensure that sandbar flushing would occur. In (c), based on the riverbed shape of (b), the initial sandbar height was adjusted gradually to reproduce the observed water level at the river mouth (Nagase, 0.0 km) using the GBVC3 method. As a result, the sandbar height in (c) was approximately 0.3 m lower than in (a). In addition, the particle size was 0.7 mm, and the amount of equilibrium suspended sediment in the upstream flow was supplied.

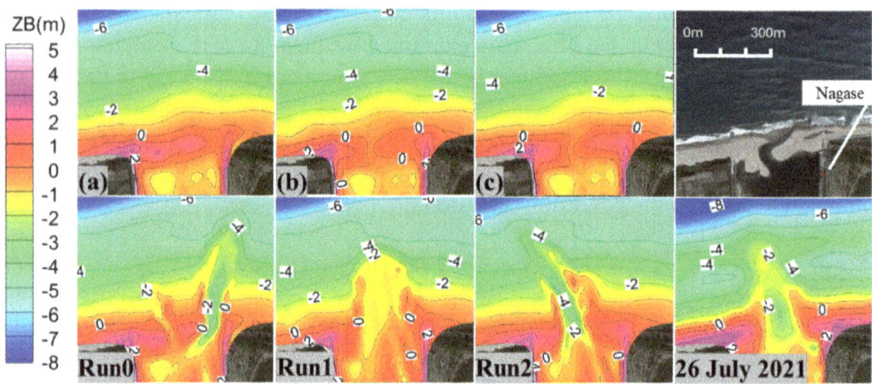

Fig. 3. Topography contour of the estuary before (upper figures) and after (lower figures) the flood

4 Results and Discussions

The topographic height contours of the river mouth before and after the analyses are shown in Fig. 3. Run0 with the survey results on 19 October 2020 failed to reproduce the central location of the sandbar opening and the tongue-like terrace shape observed in the post-flood survey results. On the other hand, Run1 and 2 partially reproduced the post-flood estuary topography. This indicated that the reliability of the sediment dynamics analysis during flooding was improved when incorporating accurate initial topography.

4.1 Comparison of Water Levels

The simulated water levels near the Nagase Station (0.0km) at the river mouth indicate that due to the flushing of the sandbar, the water level during Peak 2—despite a higher discharge—was comparable to that at Peak 1 in all analyses. Figure 4 shows the results of each analysis as a difference from the observed water level, and the accuracy of Run2 is the best at peak 2. By adjusting the shape of the estuarine sandbar, the sandbar flushing and terrace formation process during flooding were reproduced. To better improve accuracy, the shape of the initial sandbar should match the water levels observed under both normal flow and flood conditions.

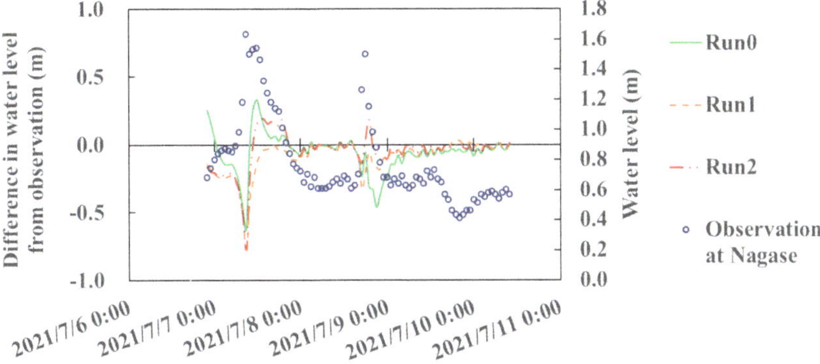

Fig. 4. Comparison of water levels at Nagase (0.0 km)

4.2 Effect of Quasi-Three-Dimensional Flow Considerations

Figure 5(A) and (B) shows riverbed deformation of the Run2 and the difference between the final riverbed deformation from the analysis result by the 2D method, respectively.

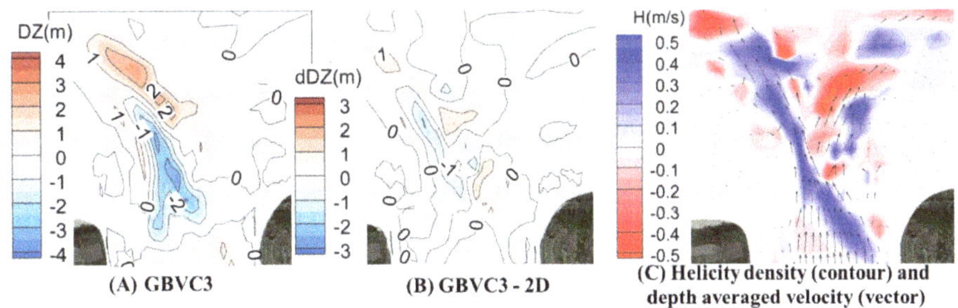

Fig. 5. Riverbed modifications and secondary flow intensity

Figure 5(C) shows the secondary flow intensity and a depth average flow velocity vector. Here, H is Helicity density defined with the depth integrated vorticity $\Omega_{x,y}$ in the x, y-direction.

$$H = \frac{(U_x * \Omega_x + U_y * \Omega_y) * h}{\sqrt{(U_x^2 + U_y^2)}} \tag{5}$$

The cooler color of Fig. 5(C) represent stronger clockwise secondary flows, while warmer colors indicate stronger counterclockwise flows. At flow concentration points, the development of clockwise secondary flows, combined with counterclockwise flows on the right side, enhanced the bottom flow velocity and sediment transport vector deflection. It was found that the amount of riverbed deformation in the GBVC method is more pronounced than in the 2D method, especially where H is large, and the 2D method does not take into account these 3D flow effects, so the transport of sediment is underestimated.

5 Conclusions

This study developed an enhanced depth integrated model for fine sediment dynamics during floods and the formation of estuarine terraces in a river's downstream region. The analysis using pre-flood survey data could not reproduced the observed central sandbar opening or tongue-shaped terrace due to sandbar deformation from wave action and discharge changes before the flood.

By modifying the sandbar shape using aerial photographs and water level data just before the flood, post-flood topography was more accurately reproduced, improving sediment dynamics analysis reliability. However, discrepancies in scour depth and sandbar flushing on the right bank persisted, likely due to backfilling effects and model simplifications, such as ignoring wave influences, saline mixing, and mixed grain sizes. The findings highlight the importance of accurate initial sandbar surveys and analytical methods considering three-dimensional flow effects. Additionally, monitoring water level fluctuations near sandbars is also critical for better understanding of flood-induced morphological changes.

References

1. Tateyama M, Uchida T, Fukuoka S, Tabe N (2013) River mouth sandbar and bed variation analysis during large flood: July 2011 Flood in the Agano River. Ann J Hydraulic Eng JSCE 69(4):1009–1014
2. Richard JM, Deirdre EH, Thomas AC, Hicks DM (2020) Processes controlling river-mouth lagoon dynamics on high-energy mixed sand and gravel coasts. Mar Geol 420:106082
3. Katayama T, Kuroiwa M, Yamamoto S, Kajikawa Y, Tajima D (2020) A river mouth bar morphodynamics due to waves and river floods at Tenjin River in Tottori. J JSCE 76(2):613–618
4. Katayama T, Ishii K, Kojima T, Kuroiwa M, Kajikawa Y (2023) Field investigation and numerical model on formation process of river_mouth bar. J JSCE 79(17):23–17084
5. Google Earth, Image © 2024 Maxar Technologies, © 2024 Google LLC
6. Uchida T, Fukuoka S (2014) Numerical calculation for bed variation in compound-meandering channel using depth integrated model without assumption of shallow water flow. Adv Water Resour 72:45–56
7. Lugina FP, Uchida T, Kawahara Y (2024) Numerical calculations for curved open channel flows using advanced depth-integrated models KSCE. J Civ Eng 28(3):1026–1040
8. Hydraulic Formulas 2018 Edition, JSCE (2019)
9. Uchida T, Fukuoka S (2019) Quasi-3D two-phase model for dam-break flow over movable bed based on a non-hydrostatic depth-integrated model with a dynamic rough wall law. Adv Water Resour 129:311–327

Insights into Monitoring Turbidity and Nutrient Dynamics in a Large Mesotidal Estuary

Anne Fleur Mineke van Leeuwen[1]([✉]), Emma Ryan[1], Mark Dickson[1], Alice Della Penna[1], and Joanne O'Callaghan[2]

[1] The University of Auckland, 23 Symonds Street, Auckland 1142, New Zealand
avan519@aucklanduni.ac.nz
[2] Oceanly Science, 161 Owen Street, Wellington 6021, New Zealand

Abstract. With the increase of climate change and local stressors on estuarine ecosystem health, there is an increased interest and demand for better understanding of estuarine physical and biogeochemical processes. New Zealand estuaries experience a high sediment input and are often undersampled. The large fine resuspendable sediment pool is expected to interact with nutrients within the water column and therefore the Estuary Turbidity Maximum (ETM) is expected to play a large role in estuarine biogeochemical processes. Here we present our experience in setting up a measurement scheme in a large, shallow, mesotidal and undersampled estuary. We monitor turbidity, salinity, and nutrients both via transects and deployments. This provides a spatial and temporal snapshot of general trends and relevant processes within an estuary tributary. In addition, this method allowed the localization of the ETM and subdivision of the tributary into zones to assist in future sample site selection and post-processing.

Keywords: ETM · Methodology Development · Kaipara Harbour

1 Introduction

Estuaries, at the interface between land, freshwater, and ocean, provide key ecosystem services. Not only do estuaries support an abundance of habitats and species, but they protect the connected marine ecosystem by filtering and buffering terrestrial nutrient runoff [1, 2]. However, increasing terrestrial nutrient and sediment runoff is threatening estuarine ecosystem functioning [3, 4]. Without a functioning estuary system, estuarine filtering and buffering are reduced. This puts additional pressures on both estuarine and marine ecosystems on top of those imposed by humans and climate change [5, 6].

In New Zealand, estuaries have experienced high sedimentation rates in the past and continue to do so in the present [7]. Terrestrial sediment input is higher for catchments with fine sediments and poor drainage, especially after and during increasingly frequent episodic flood events [8–10]. New Zealand's local governments, iwi (Māori tribes) and non-profit organizations have put in effort to reduce sediment and nutrient run-off, predominantly via riparian planting and fencing [11]. The effectiveness of these efforts on estuarine water quality remains difficult to assess due to 1) large quantities

© The Author(s) 2026
C. Coelho et al. (Eds.): CD 2025, CRL 41, pp. 744–750, 2026.
https://doi.org/10.1007/978-3-032-15473-6_113

of pre-existing sediment within the estuaries, 2) the complex internal biogeochemical processes that govern estuarine filtering, and 3) the spatial variation of those variables. Consequently, predicting how existing sediment and future sediment load reductions will influence water clarity and nutrient concentrations is challenging, highlighting the uncertainty in the outcomes of runoff mitigation practices.

A key location of biogeochemical processes within estuaries is the Estuary Turbidity Maximum (ETM), where river and ocean water mix [12]. Here, the continuous mixing of sediment and different waters enables a large range of biogeochemical reactions between (re)suspended sediment, bacteria, and the water column. These biogeochemical processes alter water quality and thus increase or decrease the risk of estuarine eutrophication and hypoxia [13]. Understanding the net effect of biogeochemical processes within the ETM on nutrient concentrations and water quality, however, remains difficult: the formation of the ETM and the internal biogeochemical processes are governed by various interacting mechanisms with different temporal and spatial scales: e.g. tides, season, and catchment characteristics [14, 15]. Currently the major estuaries in NZ are monitored monthly, lacking the resolution needed to capture these mechanisms and understand the interaction between nutrient dynamics and the ETM [7].

This paper reflects on the setup of a sampling scheme in an undersampled estuary to capture the physical and biogeochemical processes of the ETM and its role in estuarine filtering and nutrient dynamics. Thus, providing valuable insights setting up research in similar systems and conditions. The Kaipara Harbour (Fig. 1), located in the North Island of Aotearoa New Zealand, is an example of a large, complex, and undersampled estuary that has been the focus of growing concern from the local community due to its degrading ecosystem health. Actionable guidelines are required for site selection, mapping processes at play, and the expected results of said processes on water quality. Here we provide a critical reflection on the initial sampling design and suggest improvements.

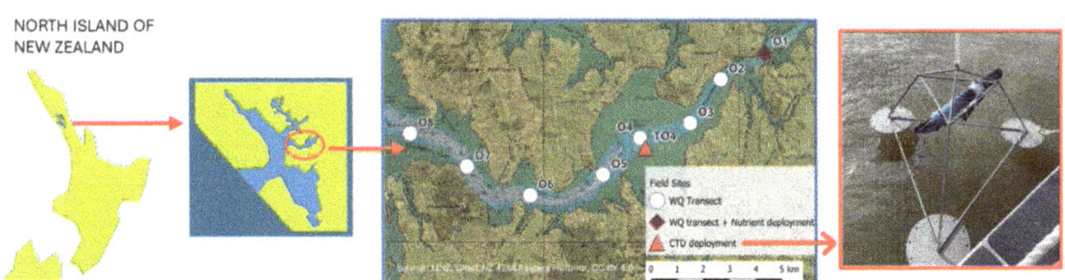

Fig. 1. Location of the study area, sample sites and the CTD sensor deployment structure (note: a different CTD was used in this study). *CTD = conductivity, temperature and depth profiler. WQ = water quality sampling of turbidity, salinity and nutrients.*

2   Methods

2.1   Study Site

The Kaipara Harbour is one of the largest estuaries in the southern hemisphere (over 90000 ha). It is classified as a shallow, tide-dominated and well-mixed drowned river valley. The Kaipara Harbour receives runoff from surrounding agricultural catchments totalling to 5836 km^2 [16] and, as a result, it has one of the highest (fine) sediment inputs in New Zealand, with rates of approximately 2–5 mm/year [7, 17]. Through the ETM, 95% of these fine sediments are reworked and recycled through the estuary until they accumulate in low energy environments [18]. Half of the tributaries in the Kaipara have a similar elongated geomorphology and have a low river discharge compared to the total estuary volume. Therefore, the ETM formation in the upper reaches of these estuary tributaries should be similar. After consultation with local stakeholders, we selected one tributary, the Ōruawharo (Fig. 2), as a case study. The Ōruawharo river has a relatively low discharge (~ 2.25 m^3/s) and its water quality is relatively good compared to other tributaries.[16, 19].

2.2 Data Collection

To get relevant spatial and temporal coverage, we combined transect sampling along a brackish- to saltwater gradient and long-term deployment of stationary equipment. These complementary methods included the collection of turbidity, salinity and nutrient data. The water samples were analyzed for NOx, PO4 and NH4, which are common in agricultural runoff [20].

For the transect data, eight sites were selected within the estuary arm, at roughly equal distances along the transect. The endpoint of the transect is where marine dominance is expected and therefore fluvial influence is negligible. In the Kaipara, that is where the estuary arm lines up with deeper well flushed channels that are significantly less turbid. Using the nautical charts of the estuary and satellite images, this endpoint was selected. At each site an RBR Conductivity Temperature Depth (CTD) profiler equipped with an RBR turbidity sensor was used to profile the water column (8 Hz; Concerto, RBR, Canada). In addition, 40 ml of water for nutrient samples were collected from the top, midway and bottom of the water column using a 2 litre Niskin bottle. The samples were filtered through a 0.43-micron nylon filter, frozen and analyzed with a Flow Injection Analyzer.

For the deployments, the same model CTDs with turbidity sensors in burst mode (8 Hz, every 5 min) were deployed for 2.5 weeks at site TO4 (Fig. 1). That is long enough to encompass half a spring-neap cycle, but short enough to avoid excessive biofouling. The deployment structures had to be solid enough to endure strong tidal currents, yet not sink into the mud. A triangle tripod structure was found to be most appropriate (Fig. 1). The CTD-deployment sites were selected along the brackish-salt water line and varying degrees of wind/wave exposure. Nutrients were sampled using Teledyne Isco auto-sampler at site O1 (Fig. 1). Samples were collected every half-hour at a fixed depth above the substrate for the duration of one tidal cycle.

3 Results

3.1 Transect Data: CTD Casts & Nutrient Concentrations

Salinity increases with distance from the wharf (Site O1). Turbidity decreases as salinity increases, except for O6 and O7. Site O6 demonstrates a steep halocline at 5 m depth, which corresponds with a ~ 10 NTU increase in turbidity. O2 depicts a similar pattern as O6, albeit with weaker gradients. The increase in turbidity, yet consistent salinity at the bottom of some casts indicate the presence of fluid mud (Fig. 2).

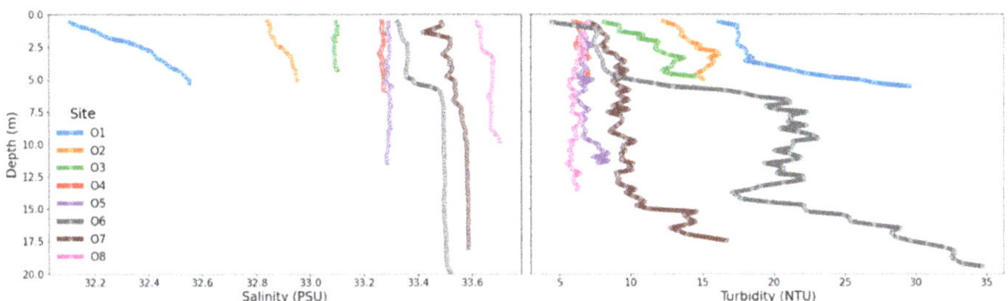

Fig. 2. CTD depth profiles of salinity (left) and turbidity (right) taken in a transect.

Nutrient concentrations vary both with depth and between sites (Fig. 3). Both NOx and PO4, especially at the surface, decrease with distance from the wharf after peaking at O2. NOx bottom and midway concentrations increase at O6, overtaking NOx-surface concentrations. NH4 measurements are highly variable, surface concentrations decrease with distance from the start of the transect, showing a ~ 55% decrease at O2, whereas bottom and midway samples fluctuate strongly, peaking at O1 and O2.

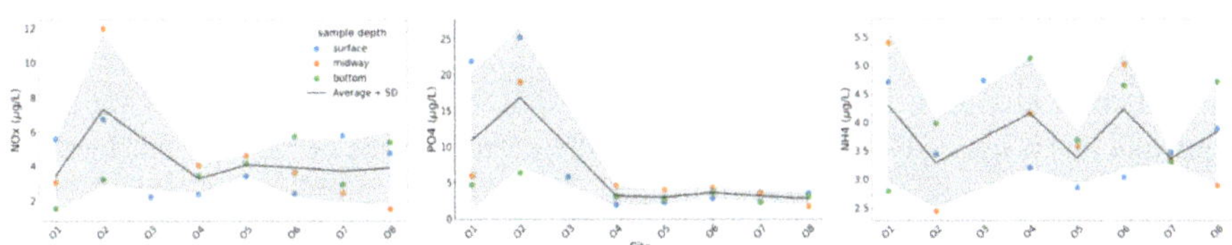

Fig. 3. NOx, PO$_4$, and NH$_4$ concentrations at three depths across eight sites. Black lines indicate site-specific depth-averaged concentrations; grey shading shows ±1 SD.

3.2 Deployment Data: CTD Data & Nutrient Concentrations

Turbidity data at site TO4 indicates turbidity is highest during the low (22 NTU) & rising tide (28 NTU) but drops by 68% at high and falling tide. Salinity is lowest at low tide around 33.05 psu, indicating more fluvial influence. It increases up to 33.35 psu during high tide. NOx peaks significantly during the rising tide, reaching 7 µg/L, while remaining relatively consistent at ± 4 µg/L during the other tidal stages. No significant change was observed in PO4 and NH4 concentrations.

4 Discussion and Conclusions

Within the CTD transects, two profiles stood out: O2 and most importantly O6. These two profiles coincide with what can be expected in an ETM: less turbid and less saline water at the surface followed by a strong transition to a more turbid & saline environment due to the continuous mixing. It is unusual to encounter two ETM within one estuary due to the different processes that can form an ETM [12]. These observations along with the general in/decrease in salinity/turbidity with distance permits a subdivision of the estuary arm into four zones: a fluvial zone (O1-O3), a non-mixing zone (O4-O5), marine ETM zone (O6–07) and a marine zone (O8). These zones align with the observed changes in nutrients, both between sites and depths. The separation of the estuary in such zones provides better guidelines for other sampling techniques (e.g., CTD deployment) and the effect of different processes on the collected data.

This zonation of the estuary arm should be treated with caution. Since turbidity and salinity concentrations fluctuate with the tide, the zones would shift either up or downstream depending on the tidal stage. This highlights the importance of CTD-cast timing. Due to logistical challenges, the CTD-cast data presented here was during the high tide (O1-O3), falling tide (O4-O7) and low tide (O8). It is possible to estimate the direction of the shift in the zones, however the formation and strength of the ETM may be linked more to tidal currents, than differences in water mass densities, especially in estuaries with low fluvial discharge. In which case, the halocline observed at O6 would be weaker during slack tides. Ideally, CTD-cast transects would be repeated during the four tidal stages. Long term deployments of a CTD can mitigate this issue as it captures tidal variation. Due to the CTD deployment set-up (fixed depth in a shallow area), only relative comparisons between the two datasets can be made. Suggested improvements include adding: (1) CTD deployment in the other zones to capture the spatial heterogeneity and (2) a bottom mounted ADCP to determine the effect of tidal currents on resuspension.

Having both nutrient data from the entire river arm (transect) and tidal cycle (deployment) is essential to disentangle the effects of dilution and biogeochemical processes on nutrient concentrations. The transect data does show nutrient concentrations vary with depth and this is not consistent between sites, therefore it would be better if nutrient data is collected at a specific water depth instead of a fixed depth above the seabed. Now, further analysis is needed to determine if the NOx fluctuations seen in the deployment data (3.2) are due to water depth variations or turbidity differences.

The combination of transect data and deployment data of nutrients, turbidity and salinity encapsulates a range of different processes and timescales that influence ETM formation and estuarine filtering. In addition, the combination of methods allows us to double check how turbidity and nutrient chemistry interact. Lastly, this methodology allows for a subdivision of the estuary arm into zones which could be applied to determine optimal deployment locations and increased understanding of spatial heterogeneity within the estuary arm. However, the zonation is dependent on the tidal regime, highlighting the importance of CTD casts and nutrient sampling timing.

References

1. Thrush SF, Townsend M, Hewitt JE, Davies K, Lohrer AM, Lundquist C, Cartner K (2013) The many uses and values of estuarine ecosystems. In: Dymond J (ed) Ecosystem services in New Zealand: conditions and trends. Manaaki Whenua Press, Lincoln, pp 226–237
2. Mackenzie F, De Carlo E, Lerman A (2011) 5.10-coupled C, N, P, and O biogeochemical cycling at the land–ocean interface. Treatise Estuar Coast Sci 317–342

3. Lotze H, Lenihan H, Bourque B, Bradbury R, Cooke R, Kay M, Kidwell S, Kirby M, Peterson C, Jackson J (2006) Depletion, degradation, and recovery potential of estuaries and coastal seas. Science 312:1806–1809

4. Sun J, Liu L, Lin J, Lin B, Zhao H (2020) Vertical water renewal in a large estuary and implications for water quality. Sci Total Environ 710:135593

5. Nicholls RJ, Wong PP, Burkett VR, Codignotto JO, Hay JE, McLean RF, Ragoonaden S, Woodroffe CD (2007) Coastal systems and low-lying areas. In: Parry ML, Canziani OF, Palutikof JP, van der Linden PJ, Hanson CE (eds) Climate Change 2007: Impacts, adaptation and vulnerability. Contribution of Working Group II to the Fourth Assessment Report of the Intergovernmental Panel on Climate Change. Cambridge University Press, Cambridge, UK, pp 315–356

6. Townsend M, Davies K, Hanley N, Hewitt JE, Lundquist CJ, Lohrer AM (2018) The challenge of implementing the marine ecosystem service concept. Frontiers in Marine Science 5:359

7. Pinkerton MH, Gall M, O'Callaghan J, Thoral F, Swales A (2022) Monitoring coastal suspended sediment in Aotearoa New Zealand: utility of satellite remote sensing. NIWA Client Report No. 2022106WN_rev1, NIWA Project DOC21305, Wellington, New Zealand

8. Haggitt T, Mead ST, Bellingham M (2008) Review of Environmental Information on the Kaipara Harbour Marine Environment

9. Liu Y, Zhang K, Li Z, Liu Z, Wang J, Huang P (2020) A hybrid runoff generation modelling framework based on spatial combination of three runoff generation schemes for semi-humid and semi-arid watersheds. Journal of Hydrology 590:125440

10. Ministry for the Environment (2020) National Climate Change Risk Assessment for Aotearoa New Zealand: Main Report – Arotakenga Tūraru Mō Te Huringa Āhuarangi o Āotearoa: Pūrongo Whakatōpū 133

11. Matheson F (2024) Riparian margin establishment/protection. In: Lohrer D, et al (eds) Information stocktakes of fifty-five environmental attributes across air, soil, terrestrial, freshwater, estuaries and coastal waters domains. Prepared by NIWA, Manaaki Whenua Landcare Research, Cawthron Institute, and Environet Limited for the Ministry for the Environment. NIWA Report No. 2024216HN (Project MFE24203, June 2024)

12. Burchard H, Schuttelaars HM, Ralston DK (2018) Sediment trapping in estuaries. Ann Rev Mar Sci 10:371–395

13. Howarth RW, Chan FK-M, Conley DJ, Garnier J, Doney SC, Marino R, Billen G (2011) Coupled biogeochemical cycles: eutrophication and hypoxia in temperate estuaries and coastal marine ecosystems. Frontiers in Ecology and the Environment 9:18–26

14. Heathwaite AL (2010) Multiple stressors on water availability at global to catchment scales: understanding human impact on nutrient cycles to protect water quality and water availability in the long term. Freshw Biol 55:241–257

15. Khalil B, Ouarda TBMJ, St-Hilaire A (2011) Estimation of water quality characteristics at ungauged sites using artificial neural networks and canonical correlation analysis. J Hydrol 405:277–287

16. Swales A, Gibbs M, Ovenden R, Costley K, Hermanspahn N, Budd R, Rendle D, Hart C, Wadhwa S (2011) Patterns and rates of recent sedimentation and intertidal vegetation changes in the Kaipara Harbour. NIWA Client Report No. ARC10224, NIWA Project HAM2011-040, Wellington, New Zealand

17. Swales A, Hume TM, McGlone MS, Pilvio R, Ovenden R, Zviguina N, Hatton S, Nicholls P, Budd R, Hewitt J, Pickmere S, Costley K (2002) Evidence for the physical effects of catchment sediment runoff preserved in estuarine sediments: phase II (field study). NIWA Client Report No. HAM2002-067, NIWA Project ARC01272

18. Allis M, Bosserelle C, Edhouse S (2019) Assessment of Coastal Sediment 120
19. Clay K, Easton S (2019) Ara Tūhono Project, Warkworth to Wellsford Section; Water Technical Report 3: Catchment Sediment Modelling Technical Report
20. Monaghan RM, Hedley MJ, Di HJ, McDowell RW, Cameron KC, Ledgard SF (2007) Nutrient management in New Zealand pastures—recent developments and future issues. New Zealand Journal of Agricultural Research 50:181–201

Recent Dynamics of the Mouth Bar in Ebro River (NW Mediterranean Sea)

Benjamí Calvillo[⊠], Manel Grifoll, and Vicente Gracia

Laboratori d'Enginyeria Marítima, Universitat Politècnica de Catalunya (UPC-Barcelona Tech), Barcelona, Spain
`benjami.calvillo@upc.edu`

Abstract. Deltas are highly dynamic and play a crucial role in socio-economic activities and environment, yet they are very vulnerable to climate changes and human pressures. This contribution focuses on the recent dynamics of the mouth bar in the Ebro River (NW Mediterranean Sea) due to changes in hydrodynamics conditions. The low river discharge period occurred from July 2022 to November 2023 and the increase of south and southwest waves have significantly influenced the hydro-morphodynamics of the river mouth and mouth bar. These conditions caused littoral transport to dominate the morphology of the beach, generating a salient on the left side of the mouth and causing the mouth bar to emerge and deflect towards the left. Using Sentinel 2 images processed with Google Earth Engine, orthophotos, wave and river discharge data along with data from lagrangian buoys launched in field campaigns indicates a directional shift of the main river channel discharge from northward to eastward, driven by morphological changes on the coast and seabed due to alterations on the alongshore sediment transport.

Keywords: Shoreline · Morphology · Mouth bar · Hydro-morphodynamics · Deltas

1 Introduction

Coastal areas are a very dynamic environment and are one of the most important zones of human activities. These areas concentrate several activities like tourism, maritime transport, fishing, and important economic activities. The deltas result from river and ocean interaction resulting in a complex morphology. According to Galloway (1975), the morphology of deltas is mainly controlled (and classified) by the sedimentary contribution of the rivers, the mobilization of sediments by the tides and the incidence of waves. Deltas are one of the most vulnerable areas to climate change and are also very impacted by human pressure [2, 3]. Climate change predicts tangible impacts due to sea level rise, increase of coastal erosion, and eventual flooding of coastal areas [4, 5]. On the other hand, the construction along the coast, including ports, hardens the coast and alters the sediment transport. As well as the construction of dams along the river courses, which retain sediment, together with the high-water demand has generated artificial flow regimes, changing the pattern of water and sediment discharge that fed the

© The Author(s) 2026
C. Coelho et al. (Eds.): CD 2025, CRL 41, pp. 751–757, 2026.
https://doi.org/10.1007/978-3-032-15473-6_114

coastal systems. Due to the complexity of hydro-morphodynamics processes of deltas, we still do not understand their development and evolution. The formation and dynamics of the mouth bar remain a complex issue with many questions. Waves play a significant role in mouth bar formation and evolution [6] and are crucial in the delta morphology leading the formation of offshore spits and alongshore sand waves [7] also, the dynamics of river jet can control the dynamics of sediment deposition [8]. For these reasons, this contribution focuses on how the recent changes in wave climate and river discharge regime impacts on the dynamics of the Ebro mouth bar. The analysis is carried out on the Ebro Delta mouth (Fig. 1a,b) from July 2022 to November 2023, when the river discharge remained very low, giving the minimum flow of 79 m^3/s ever recorded. During this period, the wave climate has varied: waves from the South (S) and Southwest (SW) have predominated over those from the East (E) and Northeast (NE), which corresponds the typical wave climate observed in the region (Fig. 1c,d).

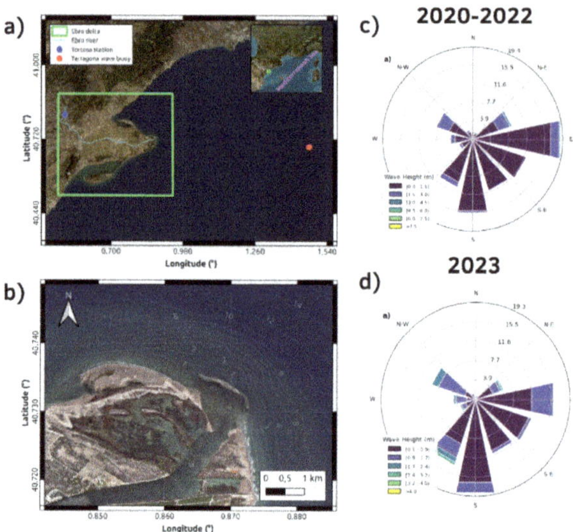

Fig. 1. a) Location of the Ebro delta, blue mark indicates Tortosa freshwater flow station and red mark the Tarragona wave buoy. b) Ebro mouth including bathymetry. c) Wave rose: significant wave height (m) for the period 2020 to 2022. d) Wave rose: significant wave height (m) for the period 2023.

2   Study Area

The Ebro Delta is located in the NW Mediterranean Sea (Fig. 1a,b) and covers an emerging surface area of 325 km^2 and the mean flow discharge from 1913 to 2010 is 425 m^3/s annual average. During the first half of the 20th century, due to the construction of dams along the river course, the sedimentary contribution of the Ebro River has been reduced to 99% of the sedimentary load compared to before constructions of the dams [9]. This lack of sedimentary contributions shifted the Ebro Delta from a river-wave dominated delta to a wave-dominated delta [10]. The Ebro delta is a microtidal (0.03 m) coast and fetch limited wave-dominated ($H_s = 0.3$ m, $T_p = 4 - 10$ s). The wave climates are seasonal and principal waves come from the E and NW, as well as from the S and SW to a lesser extent. The river mouth is a very highly dynamics area, Aranda et al., (2022), point out that since the second half of the 20th century, the mouth of the Ebro River has shown a strong retreat on the right side of the main river channel, but a significant sediment accumulation on the left side.

3   Data and Methods

Wave data was extracted from the Tarragona wave buoy (Fig. 1a), maintained by *Puertos del Estado*. In this study, hourly wave height (Hs), period (Tp), and direction ($\varnothing$) was used. The daily discharge average of the Ebro River was provided by *Estado General de la Cuenca del Ebro* from Tortosa station (Fig. 1a). The study period of wave climate and the Ebro River discharge is between 2020 to 2024. The information on the direction of the main river flow discharge and its water velocity was obtained from two field campaigns carried out at the mouth of the Ebro River between November 2023 and March 2024, where lagrangian buoys were launched (Fig. 3b,c). We also use the aerial orthophotos provided by *Institut Cartogràfic i Geològic de Catalunya* (ICGC) to analyze the shoreline during the period 2020 to 2024 and satellite images from Sentinel 2 extracted and processed with Google Earth Engine (GEE).

The changes over the Ebro River discharge and the incident waves of the delta will influence the hydro-morphodynamic processes of the mouth. The fluvial dominance ratio R [12] has been calculated (Eq. 1) to determine if the delta is dominated by waves; this occurs when the maximum littoral transport is greater than the fluvial sand flux.

$$R = \frac{Q_r}{Q_{s,max}} \tag{1}$$

Maximum littoral transport $Q_{s,\,max}$ (kg/s) was calculated for the right and left shores of the river mouth for the last 4 years. Longshore sediment transport (Q_s) was calculated using the modified CERC formulation for deep waters proposed by Ashton and Giosan (2011). The maximum littoral transport along the coast on the river mouth is the sum of the maximum flows to-wards the left and right flank of the delta in function of $E(\varnothing)$, being the wave energy probability density distribution (see Eq. 2). Fluvial sand flux Qr (kg/s) was calculated with Meyer-Peter and Muller (1948) formulation (Eq. 3).

$$Q_{s,max} = max\left[E(\varnothing) * Q_{s,right}(\varnothing - \theta)\right] - min\left[E(\varnothing) * Q_{s,right}(\varnothing - \theta)\right] \tag{2}$$

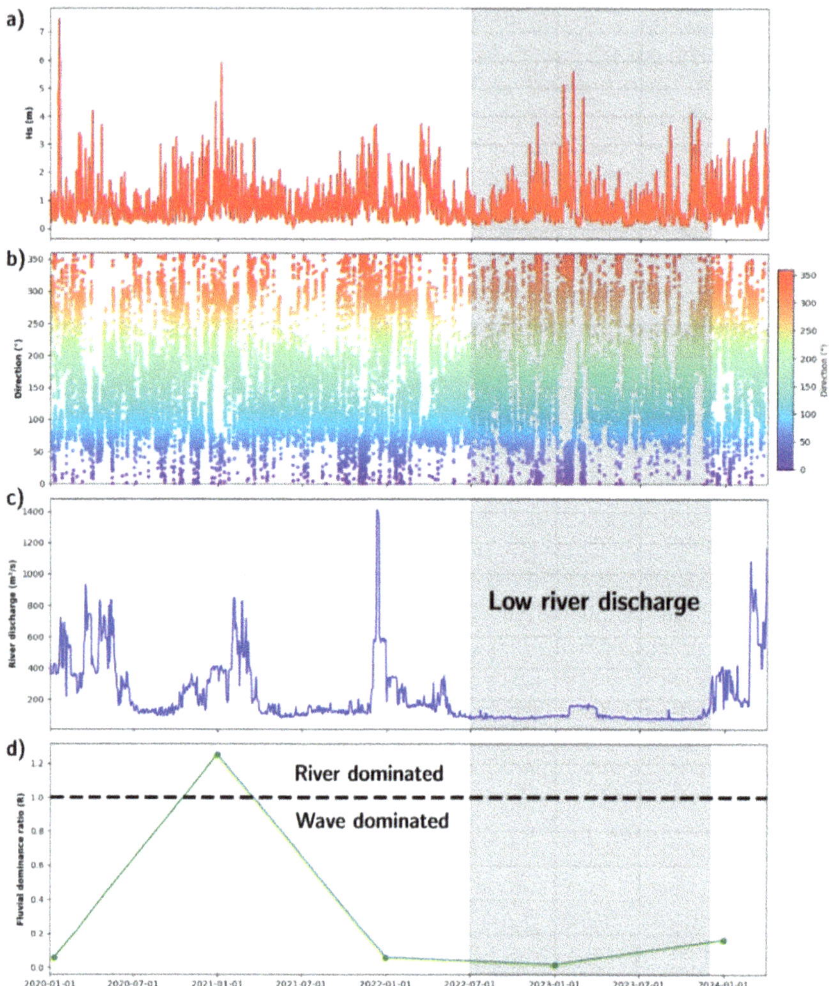

Fig. 2. a) Significant wave height (m). b) Wave direction (°), color bar indicates the direction. c) River discharge (m³/s). d) Fluvial dominance ratio (R). Grey shadow over the plots covers the period of low river discharges.

$$Q_r = I p_s B \sqrt{RgD} D \alpha_t \left\{ \left(\frac{Q^2 k_c^{1/3}}{\alpha_r^{1/3} g B^2} \right)^{3/10} \frac{S^{7/10}}{RD} - \tau_C \right\}^{2.5} \tag{3}$$

To examine the changes produced in the coastline of the river mouth between 2020 to 2024, a JavaScript code in GEE is implemented to obtain the Normalized Difference Water Index (NDWI) to differentiate between water and sand (https://code.ear thengine.google.com/c65246b1c78761976905f90ee21aa5be). We use the Otsu's algorithm who determines the NDWI, maximizing the variance between the water and sand distributions. We extract the shorelines from orthophotos of ICGC with 25 cm pixel resolution.

4 Results and Discussion

Results illustrate that during the periods from July 2022 to November 2023, low river flow conditions occurred (i.e. 80 m^3/s), along an increase in waves from the S and SW, and a decrease in waves from the E and NE during 2023 (Fig. 1c,d and Fig. 2b), which caused significant wave dominance processes controlling the morphology of the Ebro Delta during 2023. The results from the lagrangian buoys heading N indicate lower surface velocity (∼0.3 m/s) than those heading E (∼0.4–1.2 m/s). These results show that the direction of the main channel discharge shifts from N to E. The shorelines visible in Fig. 3a show how a salient has been generated on the left side of the mouth. This salient began to form in October 2022, this was visualized by reviewing the processed images from Sentinel 2 using GEE. In front of the mouth, a mouth bar was formed, deflected to the left, eventually merging with the salient, resulting in a tombolo in July 2024 (Fig. 3a).

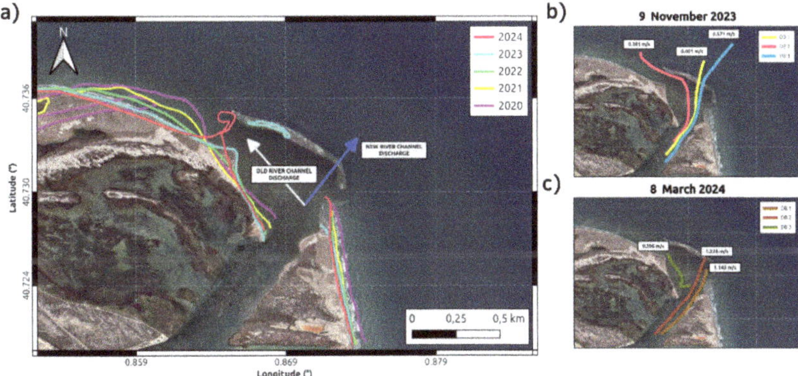

Fig. 3. a) Shorelines from 2020 to 2024 extracted from ICGC orthophotos, the arrows indicate the old and new river channels. b) Direction and velocity of drift buoys launched in November 2023. c) Direction and velocity of drift buoys launched in March 2024.

The lowest part of the delta and the river mouth suffered natural-induced shifting in the last century according to Aranda et al., (2022), with a final configuration in which there is a single river mouth oriented towards the northern lobe. In the last 2 years, the extremely low river discharges, in the order of 80 m^3/s, have led to the emerge of the mouth bar which subsequently suffered a deflection to the left of the Ebro mouth and the formation of a salient. This forced the main channel to be diverted to the South. The low discharge levels of the Ebro from July 2022 to November 2023 and the increase of the south waves and the decrease of the east waves, generated that the longitudinal transport was the main forcing factor in the configuration of the morphology of the mouth, being the Fluvial Dominance Rate R = 0.02 (Fig. 2d). In October 2022 with low flows and longitudinal transport being dominant, salient formation begins, and the river mouth emerges in December 2022. The river mouth deflects to the left and closes in July 2024 forming a tombolo. The difference in velocities of the lagrangian buoys suggest the formation of a new river channel discharge. The formation of this new river discharge channel is explained by morphological changes occurring in the river mouth and its bathymetry. The development of the salient and the emerged mouth bar, driven by the longitudinal transport of sediment to the left of the delta mouth and the observed low flows, weakened the river's ability to maintain the old discharge channel, leading to the formation of a new channel.

Acknowledgments. This paper is funded by EU research project DANUBIUS-IP (grant number 101079778). We would like to thank project PITACORA (TED2021-129776B-C21) and as a group, we would like to thank the Departament de Recerca i Universitats de la Generalitat de Catalunya (2021GR00600).

References

1. Galloway, W. E.: Process framework for describing the morphologic and stratigraphic evolution of deltaic depositional systems. Journal Name 87–98 (1975)
2. Tessler ZD et al (2015) Profiling risk and sustainability in coastal deltas of the world. Science 349(6248):638–643
3. Nienhuis JH et al (2020) Global-scale human impact on delta morphology has led to net land area gain. Nature 577(7791):514–518
4. Sánchez-Arcilla A et al (2011) Climatic drivers of potential hazards in Mediterranean coasts. Reg Environ Change 11:617–636
5. Sánchez-Artús X, Gracia V, Espino M, Sierra JP, Pinyol J, Sánchez-Arcilla A (2023) Present and future flooding and erosion along the NW Spanish Mediterranean Coast. Front Mar Sci 10:1125138
6. Nardin, W., Fagherazzi, S.: The effect of wind waves on the development of river mouth bars. Geophysical Research Letters 39(12) (2012)
7. Ashton, A. D., Giosan, L.: Wave-angle control of delta evolution. Geophysical Research Letters 38(13) (2011)
8. Canestrelli A, Nardin W, Edmonds D, Fagherazzi S, Slingerland R (2014) Importance of frictional effects and jet instability on the morphodynamics of river mouth bars and levees. Journal of Geophysical Research: Oceans 119(1):509–522
9. Guillén J, Palanques A (1992) Sediment dynamics and hydrodynamics in the lower course of a river highly regulated by dams: the Ebro River. Sedimentology 39(4):567–579
10. Jiménez JA, Sánchez-Arcilla A (1993) Medium-term coastal response at the Ebro delta. Spain. Marine Geology 114(1–2):105–118
11. Aranda García, M., Gracia Prieto, F. J., Rodríguez-Santalla, I.: Historical morphological changes (1956–2017) and future trends at the mouth of the Ebro River delta (NE Spain). (2022)
12. Nienhuis JH, Ashton AD, Giosan L (2015) What makes a delta wave-dominated? Geology 43(6):511–514
13. Meyer-Peter, E., Muller, R.: Formulas for Bed Load Transport. In: Proceedings of the 2nd Meeting of the International Association for Hydraulic Structures Research, pp. 39–64, IAHR, Delft (1948)

Unravelling the Response of the Guadiana Ebb-Tidal Delta to Extreme Wave Events

Carlos Rubio[1]([envelope]), Erwan Garel[2], Carmen Zarzuelo[1], and Alejandro López-Ruiz[1]

[1] Departamento de Ingeniería Aeroespacial y Mecánica de Fluidos, Universidad de Sevilla, Seville, Spain
`crubio1@us.es`
[2] Centre for Marine and Environmental Research - CIMA, Universidade do Algarve, Faro, Portugal

Abstract. This study investigates the morphodynamic response of the Guadiana ebb-tidal delta to extreme wave storms using a process-based numerical model. Field data including wave climate, wind and bathymetry were integrated into the Delft3D model, calibrated and validated with hydrodynamic data collected at the lower estuary in 2013, 2015 and 2023 and bathymetric maps of the ebb delta from 2016 and 2017. The analysis focuses on the impact of Storm Emma (February-March 2018), characterised maximum wave height up to 7.1 m. Results show that Emma induced a significant shift in sediment dynamics, causing retreat and erosion of the updrift shoal complex (with a net sediment volume loss of 2.40%), sediment transfer to the downdrift complex (that gained sediment volume of 0.32%), and channel rotation.. These changes contrast with the seaward growth of the shoal under mild conditions observed during non-extreme conditions, as between 2016 and 2017. The results highlight the importance that large extreme wave events can have on the morphodynamic evolution of ebb tidal deltas.

Keywords: numerical model · ebb tidal delta · Emma storm · morphology

1   Introduction

Ebb-tidal deltas are among the most dynamic coastal areas in the world. Understanding their morphodynamics under various forcing conditions, such as wind waves, tidal flows or littoral drift, has become increasingly important for effective management. This need arises from growing pressures exerted by climate change and human interventions, such as development of navigation channels, which have significantly threatened their environmental quality in recent decades. This study aims to characterise the morphodynamic response of the Guadiana ebb-tidal delta (GETD) to extreme wave storms using a process-based numerical model. The results are crucial for understanding how changes in the frequency and intensity of these extreme events may affect its morphodynamic evolution in the coming decades.

The GETD is a mixed-energy, tide-dominated delta on the southern border between Spain and Portugal. In the region, tides are semidiurnal with a mean tidal range of 2 m.

C. Coelho et al. (Eds.): CD 2025, CRL 41, pp. 758–764, 2026.
https://doi.org/10.1007/978-3-032-15473-6_115

The offshore wave climate is dominated by W–SW waves (71% of occurrences) and SE sea waves (23% of occurrences) of moderate energy, with yearly averaged significant wave height of 1 m and peak period of 8.2 s [1]. The littoral sand transport in the area is eastward, feeding the delta by an average of 100 Mm^3/year, with little contribution from the Guadiana River.

The delta (Fig. 1) typically consists of a main inlet channel and outer ebb shoal featuring subparallel lobed swash bars, flanked by shallower areas affected by waves. The updrift area consists of a narrow and straight lateral bar, while the downdrift area consists of a broad complex, which is the remnant of the historic delta (before the jetty was built), which widespread erosion nourishes adjacent beaches [2].

Storm Emma made landfall in the southwest of the Iberian Peninsula on 28 February 2018 and lasted about 4 days during a spring tide period. Propagating in a WNW direction, the storm reached a maximum wave height of 7.1 m, with a mean wave height of 4.1 m, causing beach erosion and damage to several coastal infrastructures along the coastline of the Gulf of Cádiz.

a)   b)

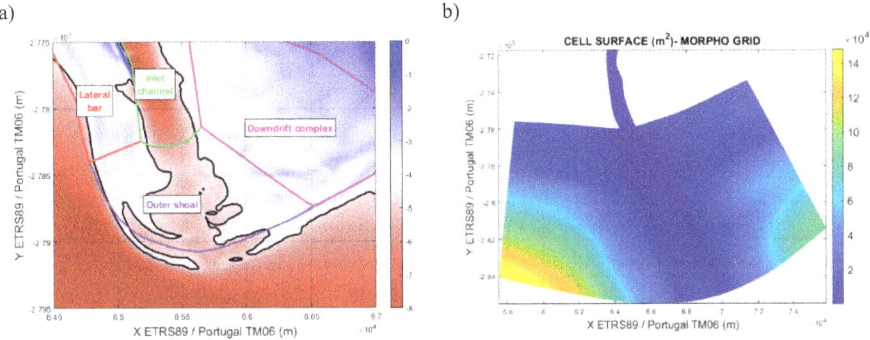

Fig. 1. (a) Initial bathymetric grid (2016). The figure shows areas of motion transfer according to [3]: updrift lateral bar, outer shoal, inlet channel and downdrift complex. (b) Cell surface (m^2) of the morphological grid.

2 Methods

2.1 Field Data

Wave observations recorded every three hours were retrieved from the Faro buoy (IH - Instituto Hidrográfico de Portugal), located at 50 km west of the study site. Gaps in the hourly time-series were replaced by hindcasts extracted nearby from the SIMAR (point 1049047) numerical model (Spanish National Port Authority). Significant wave heights from the model (Hs*) were corrected using the correlation Eq. (1) obtained through comparison with available buoy observations (Hs) for the period 1958–2024:

$$H_s^* = 0.987 \cdot H_s + 0.468 \tag{1}$$

Hourly wind data were obtained from the SIMAR model. The daily average Guadiana River discharge was obtained from a hydrographic station (Pulo do Lobo) near the estuary head [4]. Both data for 2016–2018.

Hydrodynamic observations included water level and depth averaged velocities collected at the lower estuary (i.e., 4 km upstream from the mouth) in 2013, 2015 and 2023 at 15 min intervals with an Acoustic Doppler Current Profiler (ADCP). Bathymetric data were collected (in RTK mode) in June 2016 and 2017 using a single beam echo sounder along parallel transect of 50 m intervals. The water depths were tide-corrected and referenced to the mean sea level (MSL) in ETRS89, Portugal TM06 coordinate system (Fig. 1a).

2.2 Numerical Model

The Delft3D model was implemented at the study site to simulate the morphodynamics of the GETD, using both the hydrodynamic (Flow) and wave propagation (Wave) modules. The simulations were carried out in two dimensions, solving the depth-averaged flow. The van Rijn [1993] equations were used to calculate the suspended and bedload sediment transport due to waves and currents.

The grid of the model covered the ebb-tidal delta and lower estuary stretch lying from the mouth to 7 km upstream (Fig. 1b). A non-uniform grid with minimum grid size of 20 × 20 m was used [6] over the ebb delta. The boundary conditions were: (1) the fourteen main tidal constituents for the offshore boundary, (2) Neumann conditions for the lateral cross-shore boundaries, and (3) depth averaged velocities at the upstream boundary. These velocities were obtained from prior hydrodynamic simulations carried out on a different grid covering the complete estuary (from the head at Mértola to 10 km offshore) that included the Guadiana River discharge.

The numerical model was calibrated and validated for hydrodynamics and calibrated morphodynamics.

For the former, water level and depth-averaged velocity data from 2015 [5] and 2023 were used (Fig. 2). The validation was performed using data from a period of high river discharge in 2013, which peaked at 2,331 m³/s.

The results of the hydrodynamic validation for the depth averaged velocity along the estuary channel are excellent. The results were observed for the period 01 March 2013 to 01 May 2013 with a root mean square error of 0.28 m/s, a correlation coefficient of 0.93 and a skill score of 0.91 for the northward velocity and 0.07 m/s, 0.93 and 0.89 for the eastward velocity.

For the morphodynamic calibration, bathymetric maps of the GETD from 2016 and 2017 were used, simulating the period between the two field surveys and comparing predicted and observed bed levels. This period was characterized by a wave climate with mean wave energy consistent with the historical data and no major storms occurred. The BSS (Brier Skill Score) index [6] was used to quantify the agreement between observations and model results, complemented by the adjustment of the mean-squared error skill score [7].

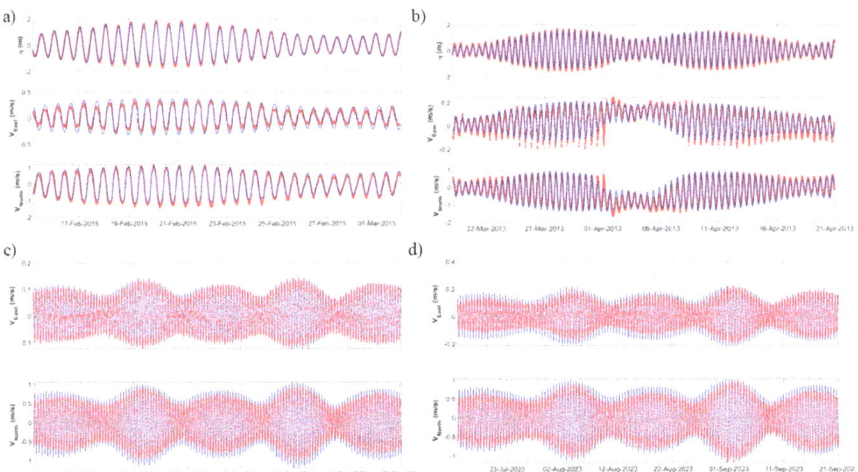

Fig. 2. Hydrodynamic model calibration and validation results. ADCP field data (blue) and DELFT3D model data (red) are compared with water level η (m), north velocity, and east velocity. (a) Model calibration (Feb 15, 2015 - Mar 02, 2015). (b) Strong river discharge event (Mar 01, 2013 - May 01, 2013). (c, d) Model Jul 14, 2023 - Sep 15, 2023.

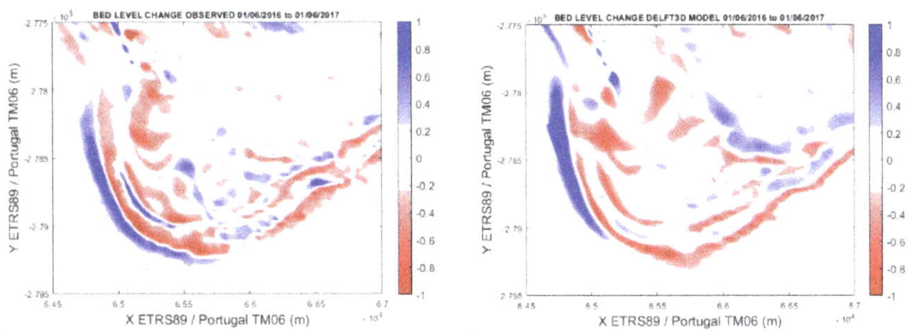

Fig. 3. Difference in bottom depth (in meters) between 2017 and 2016 from observations (left) and simulations (right). The color indicates changes in bed elevation > 0.2 m Negative values indicate erosion (red), while positive results indicate sedimentation (blue).

For the morphodynamic calibration, a BSS of 0.59 (good) was obtained for the GETD (Fig. 3). If only grid cells with bed level changes > 0.5 m are considered, the BSS result is 0.84 (excellent, according to the Sutherland scale). Furthermore, the erosion and deposition patterns observed during 2016–2017 period are qualitatively well captured (Fig. 3), although the erosion processes are slightly underestimated.

3   Results

The morphodynamic behaviour observed between 2016 and 2017 corresponds to a typical year with non-extreme energy conditions [3]. Broadly, it is characterized by a seaward growth of the outer shoal and landward retreat of the downdrift complex. In contrast, run between June 2017 and June 2018 show that the Emma storm produced a markedly different pattern of erosion and sedimentation (Fig. 4). During the storm, the updrift flank of the delta retreated, and sediment was transferred to the downdrift complex that extended offshore (Fig. 4). There was also a noticeable shift (up to 30 m) and clockwise rotation of the channel of about 6° (dashed lines Fig. 4).

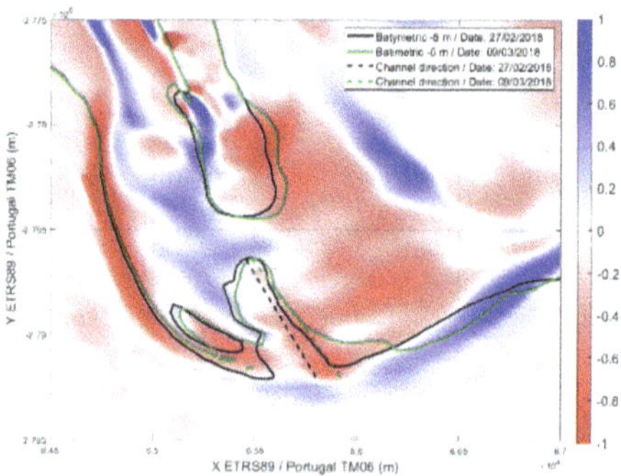

Fig. 4. Result of the Delft3D model. Bed level change between June 2017 and June 2018, showing the influence of Emma Storm. Erosion in red and sedimentation in blue. The continuous lines represent the -5m depth contours. The black lines correspond pre-Emma (27/02/2018) while the green lines represent, after the Emma Storm (09/03/2018). The dashed lines indicate the navigation channel direction for both years.

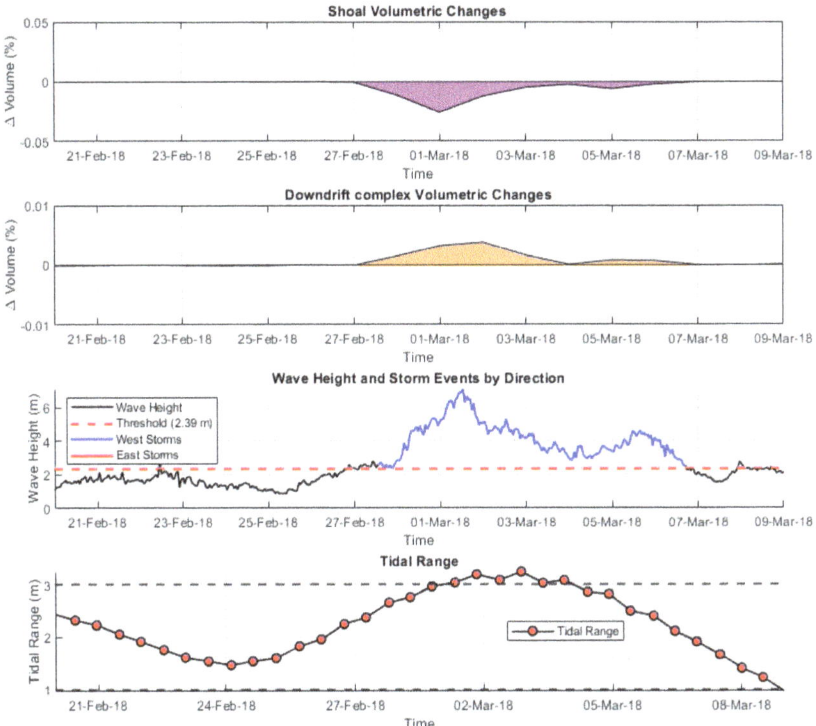

Fig. 5. From top to bottom: (1) Volumetric change in the shoal complex, (2) Volumetric change in the downdrift complex, (3) Significant wave height, (4) The tidal range.

Average daily volume changes in the GETD analysis areas (shoal and downdrift complex) were calculated, showing that the largest volume changes (increase/decrease) occurred during the Emma storm (Fig. 5). The simulation revealed a total net sediment volume loss of 2.40% in the updrift area (Fig. 5a) during Storm Emma (February-March 2018), offset by an increase of 0.32% in the downdrift complex (orange area, Fig. 5). These quantitative results are consistent with the qualitative trends observed in Fig. 4. In addition, the data show that sediment transport is significantly correlated with storm events. Figure 5 clearly shows that volume changes coincide with both storm events and periods when tidal conditions deviate from the average range (more than 3m above the average maximum range).

4 Conclusions

This study analyses the impact of the Emma storm on the mophodynamics of the GETD. Numerical model results show that Emma caused large areas of erosion and deposition in the delta, with patterns that contrast with those observed under mild wave energy conditions. Overall, the storm induced a clockwise rotation of the entire delta (with retreat of the west flank and accretion of the eastern flank), including the navigation channel, associated to significant changes in sediment volume across the main features of the delta.

These findings have critical implications for the effective management of the delta, its navigational channel and adjacent beaches, particularly in the context of increasing intensity and frequency of extreme wave events driven by climate change. Furthermore, this study highlights the value of numerical modelling as a robust tool for understanding and predicting the complex dynamics of the GETD.

Acknowledgements. This work was supported by the Spanish Ministry of Economy and Competitiveness, Project PID2021-125895O/AEI/https://doi.org/10.13039/501100011033/FEDER, UE (RESILIENCE).

References

1. Costas S et al (2018) Surficial sediment texture database for SW Iberia Atlantic Margin. Earth Syst Sci 10(2):1185–1195
2. Garel E, Sousa C, Ferreira Ó, Morales JA (2014) Decadal morphological response of an ebb-tidal delta and down-drift beach to artificial breaching and inlet stabilisation. Geomorphology 216:13–25
3. Garel E, López-Ruiz A, Ferreira Ó (2019) A method to estimate the longshore sediment transport at ebb-tidal deltas based on their volumetric growth: application to the Guadiana (Spain–Portugal border). Earth Surface Process Landforms **44**(13)
4. López-Ruiz A, Garel E, Ferreira Ó (2020) The effects of high river discharges on the morphodynamics of the guadiana ebb-tidal delta. J Coastal Res 95(sp1):558–562
5. Garel E, Sousa C, Ferreira Ó (2015) Sand bypass and updrift beach evolution after jetty construction at an ebb-tidal delta. Estuarine Coast Shelf Sci **167**(A):4–13
6. Sutherland J, Peet A, Soulsby R (2004) Evaluating the performance of morphological models. Coast Eng 51:917–939
7. Bosbomm J, Reniers AJHM, Luijendijk A (2014) On the perception of morphodynamic model skill. Coast Eng 94:112–125

The Influence of Water Levels and Salinity in Coastal Wetlands on the Conservation of Threatened Amphibians in New South Wales, Australia

Kate Tunstill[1]([✉]), Hannah Power[1], Alex Callen[1], Gabriel C. Rau[1], and Matt Hayward[1,2]

[1] The University of Newcastle, Newcastle, Australia
kate.tunstill@uon.edu.au
[2] Centre for African Conservation Ecology, Nelson Mandela University, Port Elizabeth, South Africa

Abstract. Coastal environments provide critical ecosystem services to support biodiversity for a variety of organisms through transitions between marine, freshwater and land habitats. However, these systems are threatened by climate change and human development, especially regarding densely populated regions such as Australia's eastern coastline of New South Wales (NSW). This area contains 184 estuaries, including Closed and Open Lakes and Lagoons (ICOLLS) which provide critical habitats for these species. Amphibians are heavily understudied in the context of coastal environment despite the threats from habitat loss, pollution and chytrid fungus. This study investigates the relationship between salinity, water level and frog call intensity of the endangered green and golden bell frog *(Litoria aurea)* in brackish coastal wetlands, using Avoca Lagoon (NSW) as a case example. Data from collecting audio recordings and water parameter measurements determined significant negative correlations between frog call intensity, salinity, and water level, with a notable interaction effect. It was found that optimal breeding conditions occurred within a range for salinities (5.34–6.96 ppt) and water levels (1.24–1.33 mAHD), highlighting the importance of interdisciplinary monitoring techniques. This study provides insight into managing highly dynamic coastal systems to enhance habitat suitability for species such as the green and golden bell frog, ensuring their resilience in the face of future challenges.

Keywords: Coastal Wetlands · Salinity · Estuaries · Amphibians

1 Introduction

Coastal ecosystems are globally recognized for their invaluable ecosystem services (Costanza et al., 2014), supporting vulnerable terrestrial and aquatic species through complex transitions between freshwater and marine environments (Dias et al., 2006). Despite their biodiversity richness and essential services, these dynamics make coastal systems challenging to predict and manage amid threats from climate change and development (Hughes et al., 2022; McSweeney et al., 2017).

© The Author(s) 2026
C. Coelho et al. (Eds.): CD 2025, CRL 41, pp. 765–770, 2026.
https://doi.org/10.1007/978-3-032-15473-6_116

In Australia, there are 184 estuaries along the New South Wales (NSW) coastline, mostly Intermittently Closed and Open Lakes and Lagoons (ICOLLs) with extensive coastal wetlands (Roper et al., 2011). With 80% of NSW's population living within 50 km of the coastline (Head et al., 2014), human-induced habitat degradation poses risks to species reliant on these habitats, including fish, insects, and birds (Balakrishnan et al., 2014). Amphibian species, however, are understudied in this context (Felipe Bairos Moreira et al., 2015).

Globally, amphibians have suffered severe declines due to factors like chytrid fungus, invasive species, habitat loss, and pollution (Houlahan et al., 2000). In Australian coastal settings, estuaries with brackish coastal wetlands provide refuge for many frog species by reducing chytrid fungus infection rates. For instance, 36% of the 184 estuaries house records of the endangered green and golden bell frog (*L. aurea*), which has lost 90% of its historical range and now exists in fragmented populations along the coastline (Goldingay, 1996).

This multidisciplinary study aims to understand how inundation regimes and tidal dynamics of NSW estuaries and associated wetlands impact the conservation of the green and golden bell frog by monitoring water quality parameters and wetland inundation.

1.1 Study Site

Twelve study sites, including interactions with nine estuary systems along the NSW coastline, were selected. These sites comprise of five ICOLLs, two wave-dominated estuaries, one freshwater estuary, and one tide-dominated, based on the classification by Roy et al. (2001). All sites have historical records of bell frogs according to the Atlas of Living Australia (ALA), although data on bell frogs is often outdated due to limited conservation resources and funding.

Avoca Lagoon was selected as a case study, this site is located in Avoca, NSW, Australia and is classified as an ICOLL. The target species is found within a wetland located next to the entrance of the estuary; this area is named Bareena Wetland (Fig. 1).

2 Methodology

2.1 Fieldwork

All selected sites had audio devices installed to monitor frog calls during the breeding season in 2023, 2024 and 2025. These recorded from 8pm – 12am for 5min on the hour from September to April.

Water level and salinity data were collected using pressure transducers and water quality meters installed in selected wetlands. This data aimed to track hydrological changes over a two-to-four-month period coinciding with the bell frog breeding season (September-April). Audio recordings, which corresponded with the installation of water monitoring equipment, were analyzed to determine the relationship between frog call intensity and water parameter readings.

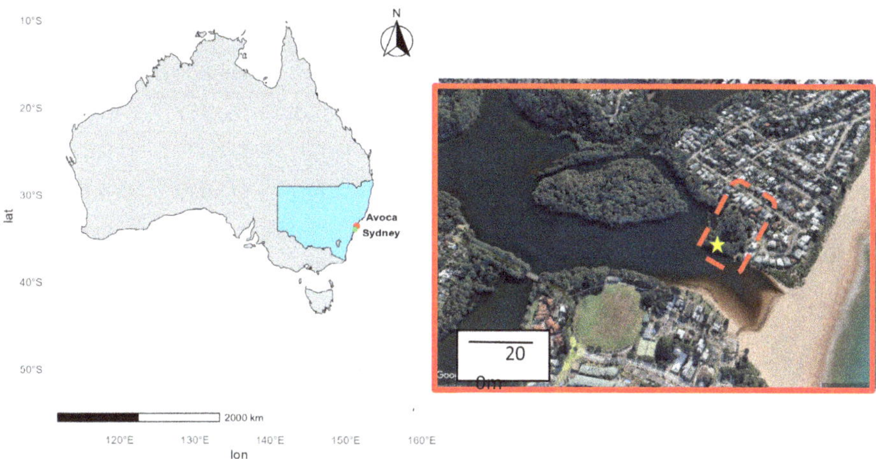

Fig. 1. Study site map, Avoca Lagoon, Avoca, NSW, Australia. Satellite image (Google Earth) on right shows Avoca lagoon and Bareena wetland in red dotted area. Yellow pin represents location of audio device, water parameter meter and pressure transducer.

2.2 Data Processing

Audio data at Avoca was processed manually by the author and a volunteer university student, Yasmine Webb. Approx. 120 h of recordings were processed via Excel and Kaleidoscope software with bell frogs categorized into calling intensity of 0 – absent, 1 - single call, 2 – pair call, 3 – trio calling, 4 – chorus (4 + frogs).

Water data processing followed protocol by Palmer et al. (2019), where pressure readings were reduced to water depth and vertically corrected to water level. Conductivity was converted to salinity (ppt) using the UNESCO 1983 conversion. Data points where pressure transducers or water parameter meters were above the water level were excluded. All data was processed using code established in R (Version 4.4.1).

3   Results and Discussion

Preliminary findings from the Negative Binomial (NB) model show a significant negative relationship between salinity and frog call intensity (−35.182 estimate) and between water level and frog call intensity (−262.53 estimate) both of which are statistically significant (P-value < 0.001) (Table 1). Further to this, there is a significant interaction effect between salinity and water level (−89.479), which suggests the negative effects of salinity on frog calling becomes stronger as water level increases (Table 1). The correlation between salinity, water level and frog call intensity suggest that these environmental stressors limit calling behavior, which is crucial for breeding success.

Table 1. NB analysis results

Predictor	Estimate	Std. Error	Z value	P-value	Conf. Lower	Conf. Upper
Intercept	−104.25	28.62	12.374	< 0.001	−160.455	−48.049
Salinity	−35.182	0.091	−1.745	<0.001	−53.687	−16.677
Water level (mAHD)	−262.53	0.493	−8.261	<0.001	−411.020	−114.040
Salinity x Water level	−89.479	0.05	0.711	<0.001	−138.785	−40.173

A range was determined by the max and min parameters for water level (1.24–1.33 mAHD) and salinity (5.34 - 6.96 ppt) in which frog calling occurred (Fig. 2). The study found significant differences in salinity and water level between the non-calling and calling groups of *Litoria aurea*. Salinity was noted to be significantly lower when frogs were heard calling and when the water level was within a specific threshold range of 1.25–1.3 mAHD.

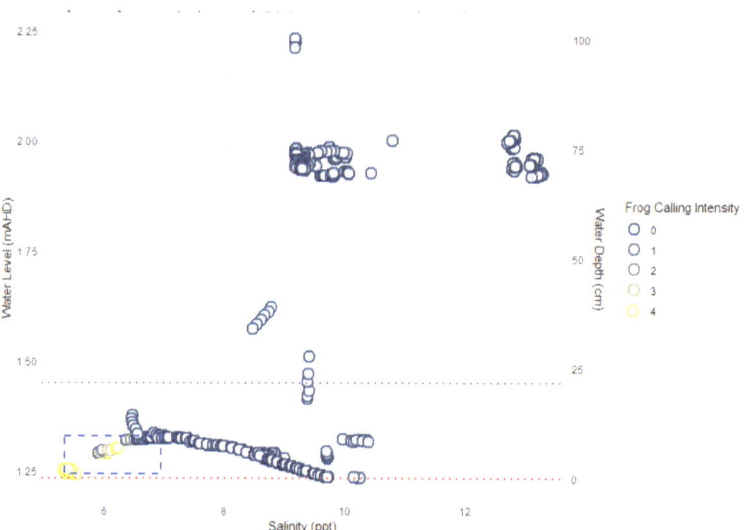

Fig. 2. Frog call intensity (0–4) represented in coloured circles, compared with water level (mAHD), water depth (cm) and salinity (ppt). The blue box highlights the threshold area where frog calling occurs. Purple dotted line represents water level in which wetland hydrologically connects with the lagoon. The red dotted line is the dry level of the wetland.

The results of the NB model suggest that salinity and water level can independently reduce frog calling intensity (significant negative relationship), while the combined effect amplifies this reduction, leading to the sharp decline of calling intensity.

The findings in this study indicate that the wetland becomes hydrologically connected with the estuary at approximately 1.45 mAHD, leading to an influx of salinity above

8 ppt within the breeding area of the bell frog. However, it is important to note that salinity levels can be altered by the amount of freshwater run off into the system as well as deposits from groundwater. This information is critically important for understanding the ideal breeding conditions for *Litoria aurea*, which in this case, is outside the ideal range for successful breeding without adverse effects on the tadpoles (5–6 ppt) and can lead to total clutch failure (Beranek et al., 2020, Callen et al., 2023). These findings have significant implications for the overall management of the estuary and associated wetland, as well as ongoing efforts to conserve the species. Considering that Avoca lagoon is manually opened to mitigate localized flooding, potential conclusions can be drawn regarding the frequency and timing of this practice to avoid disrupting potential breeding events.

4 Conclusion

Specific ranges for water level and salinity were identified as essential for potential *Litoria aurea* breeding success, considering saltwater intrusion. This study highlights the importance of understanding hydrological dynamics for amphibian conservation in coastal wetlands. It offers a multi-disciplinary approach by integrating hydrological and ecological monitoring tools to manage these dynamic systems. Future research should examine different estuary types and apply a similar method to other aquatic and terrestrial species in these habitats.

References

Balakrishnan S, Srinivasan M, Mohanraj J (2014) Diversity of some insect fauna in different coastal habitats of Tamil Nadu, southeast coast of India. J Asia-Pacific Biodivers 7(4):408–414

Beranek CT, Clulow J, Mahony M (2020) A simple design feature to increase hydro-period in constructed ephemeral wetlands to avoid tadpole desiccation-induced mortality. Ecol Manag Restor 21(3):250–253

Callen A, Pizzatto L, Stockwell MP, Clulow S, Clulow J, Mahony MJ (2023) The effect of salt dosing for chytrid mitigation on tadpoles of a threatened frog, Litoria aurea. J Comp Physiol 193(2):239–247

Costanza R et al (2014) Changes in the global value of ecosystem services. Glob Environ Chang 26(1):152–158

Dias MP, Granadeiro JP, Lecoq M, Santos CD, Palmeirim JM (2006) Distance to high-tide roosts constrains the use of foraging areas by dunlins: Implications for the management of estuarine wetlands. Biol Cons 131(3):446–452

Moreira LFB, Knauth DS, Maltchik L (2015) Intermittently closed estuaries and tadpole communities: influence of artificial breaching. Estuaries Coasts 38(3):979–987

Goldingay RL (1996) The Green and Golden Bell Frog Litoria aurea - From riches to ruins: Conservation of a formerly common species. Aust Zool 30(2):248–256

Head L, Adams M, Mcgregor HV, Toole S (2014) Climate change and Australia. Wiley Interdisc Rev: Clim Change 5(2):175–197

Houlahan JE, Fidlay CS, Schmidt BR, Meyer AH, Kuzmin SL (2000) Quantitative evidence for global amphibian population declines. Nature 404(6779):752–755

Hughes MG, Glasby TM, Hanslow DJ, West GJ, Wen L (2022) Random forest classification method for predicting intertidal wetland migration under sea level rise. Front Environ Sci 10

McSweeney SL, Kennedy DM, Rutherfurd ID, Stout JC (2017) Intermittently Closed/Open Lakes and Lagoons: their global distribution and boundary conditions. Geomorphology 292:142–152

Palmer K (2019) Tasmania tide monitoring project. https://doi.org/10.25959/GEFX-FV62

Roper T et al (2011) Assessing the condition of estuaries and coastal lake ecosystems in NSW, Estuaries and coastal lakes, Technical report series; Monitoring, evaluation and reporting program. State of NSW and Office of Environment and Heritage, Sydney

Time-Averaged Flow and Turbulence Structure Over Low-Angle Asymmetric Estuarine Dunes Under Reversing Steady Flows

Kevin Bobiles[1,2](), Bernhard Kondziella[3], Christina Carstensen[3], Elda Miramontes[1,2], and Alice Lefebvre[1]

[1] MARUM – Center for Marine Environmental Sciences, University of Bremen, Bremen, Germany
kbobiles@marum.de
[2] Faculty of Geosciences, University of Bremen, Bremen, Germany
[3] Federal Waterways Engineering and Research Institute (BAW), Hamburg, Germany

Abstract. Large-scale (1:15), high resolution flume experiments were conducted to characterise in detail the time-averaged flow and turbulence structures over a 2D fixed low-angle estuarine dune field under steady currents from two flow directions. Over a lee side with a mean slope of $12°$ including a steep section of $22°$, a smaller permanent flow separation is detected compared to the typically well-developed flow separation over angle-of-repose dunes. The associated turbulent wake is weak and located close to the bed. When flow is reversed towards the gentle slope of $4°$, neither permanent nor intermittent flow separation and no definite turbulent wake structure were detected. Turbulent events characteristics are assessed through quadrant analysis and indicate the presence of macroturbulence such as coherent structures. This study provides further insights into the flow and turbulence properties over low-angle dunes and demonstrates the importance of large-scale dune models and near-bottom flow measurements to detect the details of flow and turbulence processes over a low- to intermediate-angle dunes.

Keywords: Estuarine dunes · flow dynamics · turbulence structure · bedform morphology · high-resolution flume experiments

1 Introduction

Flow properties over angle-of-repose $(30°)$ triangular dunes, commonly found in laboratory flumes and small rivers, have been already extensively studied and well-documented [1, 2]. Over such dunes, a permanent flow separation and intense turbulent wake is usually observed. However, an intermittent or even no flow separation and a weak turbulent wake is believed to occur over low- to intermediate-angle bedforms [3, 4] generally observed in large rivers, estuaries and shallow seas. The shape of river and estuarine dunes differ: river dunes have a rounded crest and steep lee side portion close to the trough [5], whereas estuarine dunes have their steepest slope near the crest and a gentle lee side close to the trough [6]. Finally, the complex interaction between the reversing tidal flows

© The Author(s) 2026
C. Coelho et al. (Eds.): CD 2025, CRL 41, pp. 771–776, 2026.
https://doi.org/10.1007/978-3-032-15473-6_117

and the natural morphology typically present in an estuarine environment (compared to unidirectional flow in a river) has not been investigated extensively. Therefore, flow properties over low-angle estuarine dunes are currently not well known.

The aim of this study is to characterise flow properties over representative two-dimensional fixed estuarine dunes. To accomplish this, large-scale flume experiments are conducted and the time-averaged and turbulent flow are measured over low- to intermediate-angle sharp-crested estuarine bedforms representing dunes (1:15 scale) observed in the Weser Estuary, Germany [6]. Such a large-scale flume experiment to study flow properties over low-angle estuarine dunes has been demonstrated to be feasible [7].

2 Experimental Methodology

2.1 Experimental Facility, Setup and Modelled Dune

Laboratory experiments were conducted in the large recirculating flume of the Federal Waterways Engineering and Research Institute (BAW), Hamburg. The recirculating flume has an overall length of 220 m, consisting of two straight sections connected by a semi-circular segment at both ends. The straight test section has a length of 70 m, a width of 1.5 m and a maximum water depth of 1.3 m. Figure 1 shows the modeled dunes for the flume experiments. Ten fixed concrete dunes (length and height of 3 m and 0.15 m) were deployed in the flume. Their surface was fine-sandblasted to provide natural grain roughness. A water depth of 1.0 m and a flow velocity 0.3 m/s are imposed. Measurements were conducted in two opposite flow directions (i.e., reversed steady flow, F1 & F2 flow setups) in order to mimic the interaction of dunes with a steady tidal flow.

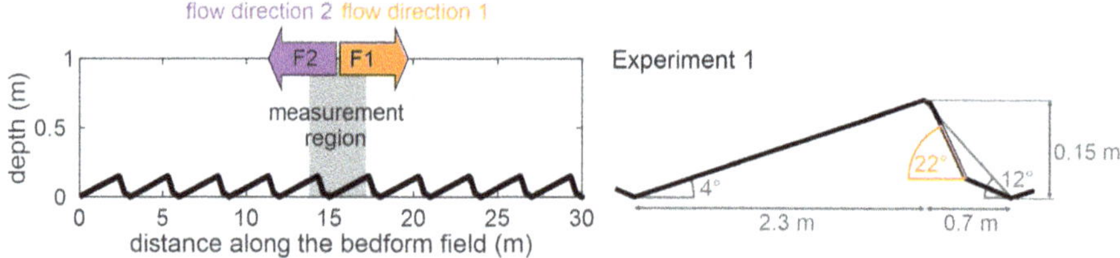

Fig. 1. Modelled estuarine dune and experimental flume setup.

2.2 High-Resolution Flow Measurements

High-resolution instantaneous flow measurements with a sampling rate of 100 Hz and a duration of 120 s and ranging 1300–1500 points per flow setup are conducted using an Acoustic Doppler Velocimeter (ADV) Vectrino by Nortek at the lateral centerline of the flume over the 5th and 6th dunes (see Fig. 1) where the flow is already fully adapted to the bedform field. Detailed characteristics of the time-averaged and turbulent flow over the intermediate-angle dunes are investigated for the two flow setups. More specifically, the presence (i.e. permanent, intermittent or inexistent) and size of the flow separation zone together with the shear layer are identified. The intermittency factor (IF) which indicates the frequency of flow reversal at each location is also provided. Furthermore, the size and intensity of the turbulent wake is identified based on the turbulent kinetic energy distribution.

3 Experimental Results and Discussion

3.1 Time-Averaged Flow and Turbulence Structures

Figure 2 shows the time-averaged flow field over the 2D intermediate-angle estuarine dune for both representative ebb (F1) and flood flow (F2) setups.

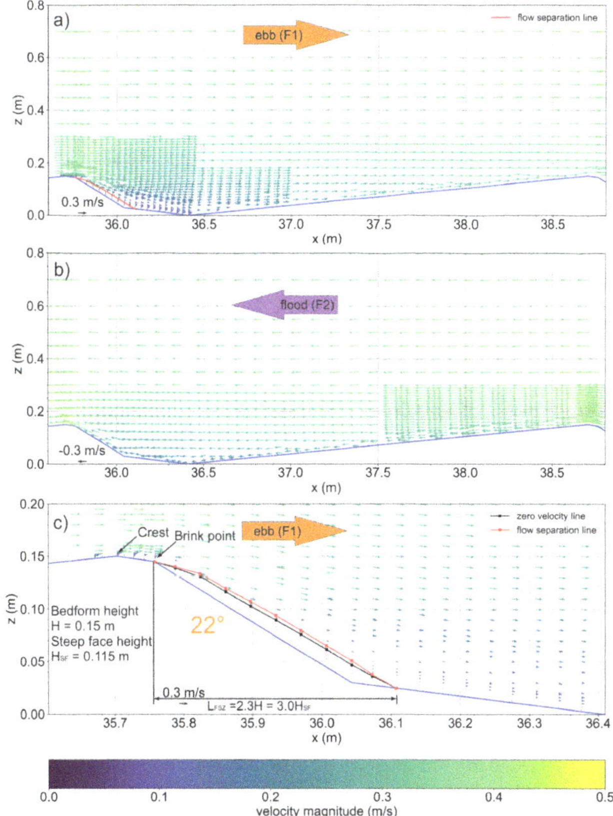

Fig. 2. Time-averaged flow field. a) ebb flow (F1 setup) b) flood flow (F2 setup) c) lee side segment of F1 setup showing the flow separation zone.

For F1 setup (see Figs. 2 a & c), a permanent flow separation zone is detected only over the steep portion of the lee side. The length of the flow separation (L_{FSZ}) is compared to the height of the bedform (H) or the height of the steep face (H_{SF}). In our experiment, over the steep face of 22°, the flow separation is estimated to be 2.3 H or 3 H_{SF}. This flow separation is shorter than that of estuarine dune with a steep slope of 25° starting directly at the crest which is 4.3 H or 6.5 H_{SF} [7], or that of angle-of-repose (30°) dune which is 4 − 6 H [8].

Over the very gentle lee side under F2 setup (see Fig. 2b), no flow separation zone is observed and the flow field is similar to that of low-angle dune found in large rivers [4].

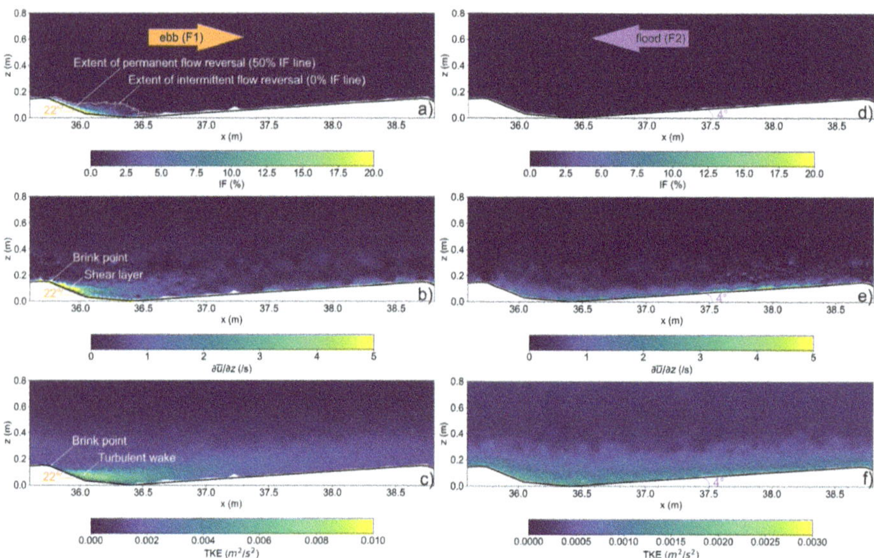

Fig. 3. Time-averaged flow and turbulence structures. a) Intermittency factor, IF (%), b) Velocity gradient, $\partial\overline{u}/\partial z$ (/s), c) Turbulent kinetic energy, TKE (m²/s²) for ebb flow (F1 setup). d) Intermittency factor, IF (%), e) Velocity gradient, $\partial\overline{u}/\partial z$ (/s) and f) Turbulent kinetic energy, TKE (m²/s²) for flood flow (F2 setup). Note different scale used for TKE.

Under F1 setup, the intermittency factor in the lee side is over 50% IF (> 50%) where a permanent reversed flow is detected (Fig. 3a). Above it, a region with a decreasing amount of intermittent flow is observed. This region is also characterized by a high horizontal velocity gradient, $\partial\overline{u}/\partial z$, , or shear layer, which develops past the brink point, expands downward to the trough before dissipating over the stoss side of the next dune (Fig. 3b). A well-define turbulent wake is also detected developing over the steep lee side and expanding downward to the trough before dissipating further upstream (Fig. 3c). The intensity of the wake is weaker in comparison to observations above steeper dunes [4, 7, 8].

When the current direction is reversed and flows towards the gentle lee side of 4° (F2 flow setup), no intermittent flow separation is observed. A thin shear layer can be identified close to the bed but no define turbulent wake is observed (Figs. 3d, e & f).

3.2 Characteristics of Turbulent Events (Quadrant Analysis)

To investigate the associated macroturbulence such as eddies and coherent flow structures that might be generated over the tested dune design, a quadrant analysis is performed to identify the significant turbulent events. Figure 4 shows the spatial distribution of the percentage of observation at each significant quadrant event applying a hole size (HS) of 2.0 [3].

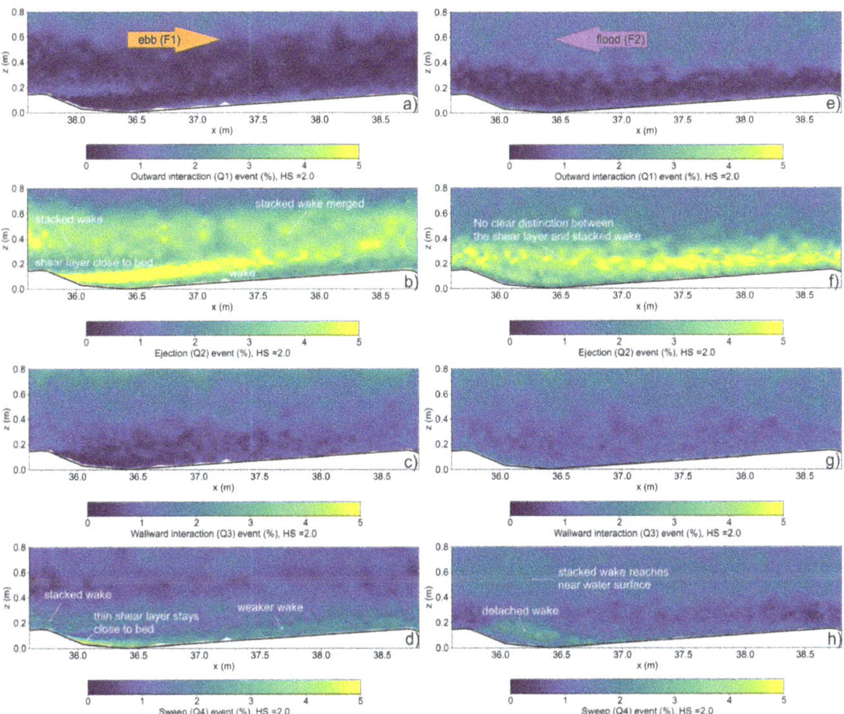

Fig. 4. Percentage of observation at each quadrant event. a) Outward interaction event, Q1 (%), b) Ejection event, Q2 (%), c) Wallward interaction event, Q3 (%), d) Sweep event, Q4 (%) for ebb flow (F1 setup). e) Outward interaction event, Q1 (%), f) Ejection event, Q2 (%), g) Wallward interaction event, Q3 (%), h) Sweep event, Q4 (%) for flood flow (F2 setup).

Ejection (Q2) and sweep (Q4) events which are contributors to turbulence via Reynolds stress are seen to be dominant turbulent events for both flow setups (Figs. 4b, d, f & h). A turbulent wake that merges with stacked wakes from adjacent dunes and later rises up is observed, especially identifiable from ejections events (Fig. 4b). However, sweep turbulent events are seen to be dominant only within close proximity to the bed (Figs. 4d & h).

4 Concluding Remarks

High-resolution flow measurements were carried out to characterise in detail the time-averaged flow and turbulence structures over a 2D large-scale (1:15) intermediate-angle asymmetric estuarine dune under steady flows in one direction, then in the opposite direction. When the current is flowing towards the steeper (maximum slope of 22°) lee side (F1 setup), a short (2.3H or $3H_{SF}$) permanent flow separation zone is observed close to the bed over the steepest portion of the lee side with a region of an intermittent flow separation above it. A distinct turbulent wake that is almost attached to the bed and extends downward before dissipating further upstream is also present. However, neither permanent nor intermittent flow separation and no definite turbulent wake structure are detected when the current direction is reversed and flowing towards the gentle (slope of 4°) lee side (F2 setup). Furthermore, an investigation of the characteristics of turbulent events through quadrant analysis suggests the presence of coherent structures, such as especially ejection (Q2) and sweep (Q4) turbulent events.

References

1. Best J (2005) The fluid dynamics of river dunes: a review and some future research directions. J Geophys Res 110(F04S02):21
2. Venditti JG (2013) Bedforms in sand-bedded rivers. In: Shroder J, Wohl E (eds) Treatise on geomorphology. Academic Press, pp 137–162
3. Best J, Kostaschuk R (2002) An experimental study of turbulent flow over a low-angle dune. J Geophys Res 107(C9):3135
4. Kwoll E, Venditti JG, Bradley RW, Winter C (2016) Flow structure and resistance over subaqueous high- and low-angle dunes. J Geophys Res Earth Surf 121:545–564
5. Cisneros J, et al (2020) Dunes in the world's big rivers are characterized by low-angle lee-side slopes and a complex shape. Nat Geosci 13(2):156-+
6. Lefebvre A, Herrling G, Becker M, Zorndt A, Krämer K, Winter C (2021) Morphology of estuarine bedforms, Weser Estuary, Germany, Earth Surf. Processes Landforms
7. Carstensen C, Holzwarth I (2023) Flow and Turbulence over an Estuarine Dune - large-scale flume experiments. Die Küste 93:17
8. Lefebvre A, Paarlberg AJ, Winter C (2014) Flow separation and shear stress over angle of repose bedforms: a numerical investigation. Water Resour Res 50(2):986–1005

Infragravity Wave Propagation and Transport
in a Small Estuary

Maricarmen Guerra[1(✉)], Raúl Flores[2,4], Rodrigo Cienfuegos[3,4], Patricio A. Catalán[2,4], Felipe Lucero[5], Paola Díaz[3], Camila Bastías[1], and Rodrigo Zuñiga[1]

[1] Dpto. Ingeniería Civil, Universidad de Concepción, Concepción, Chile
marguerra@udec.cl
[2] Dpto. Obras Civiles, Universidad Técnica Federico Santa María, Valparaiso, Chile
[3] Dpto. Ingeniería Hidráulica y Ambiental, Pontificia Universidad Católica de Chile, Santiago, Chile
[4] Centro de Investigación Para La Gestión Integrada del Riesgo de Desastres (CIGIDEN), Macul, Chile
[5] Marine Energy and Innovation Center (MERIC), Valdivia, Chile

Abstract. This study investigates infragravity wave propagation in the Nilahue River estuary, a small Intermittently Open/Closed Estuary (IOCE) on Chile's Pacific coast (34.48° S; 72.00°W). The estuary, critical for ancient sea-salt production, experiences artificial breaching due to reduced river discharges and energetic swells that enhance sediment accumulation at the river mouth. Field measurements conducted in August 2023, during low river discharge, spring tides, and high swells, revealed that infragravity waves propagate at least 3 km upstream the estuary during flood tides. Data from acoustic Doppler velocimeters confirmed infragravity wave-driven scatterer fluctuations potentially affecting sediment transport dynamics. A second experiment conducted in July 2024 expands on these findings by incorporating turbulence, echosounder, salinity and turbidity measurements to improve our understanding of IG waves' role in sediment transport and mixing in IOCEs.

Keywords: IOCE · Infragravity waves · acoustic measurements · turbidity

1 Introduction

Small estuaries characterized by seasonally variable freshwater outflows are commonly found in energetic wave-dominated microtidal mediterranean coasts. These small bar-built estuaries typically follow a natural periodic mouth opening/closing cycle and are known as Intermittently Open Close Estuaries (IOCE) [1]. Recently, due to climate change and long dry seasons often leading to persistent lower river discharges, artificial openings are performed to reconnect these estuaries to the ocean to improve water quality within the estuary lagoon and to reduce flood hazard [2].

Infragravity waves (IG) are water surface oscillations with periods from 30 s to 300 s or longer [3]. IG waves have been observed to propagate upstream in estuaries as the

© The Author(s) 2026
C. Coelho et al. (Eds.): CD 2025, CRL 41, pp. 777–782, 2026.
https://doi.org/10.1007/978-3-032-15473-6_118

river mouth acts a natural low-pass filter to incident short surface gravity waves [4]. Their upstream propagation occurs during flood tides, as wave blocking occurs during strong ebb tidal currents [5]. Large velocity fluctuations in the IG wave frequency band have been observed as IG waves propagate upstream tidal inlets and estuaries being linked to sediment transport, river inlet morphology, and mixing [6, 7].

In this investigation we present field measurements using acoustic Doppler velocimeter and current profilers are used to identify and characterize infragravity wave propagation upstream an IOCE during flood tide and its relation to sediment transport. The study site, field measurements and initial analysis and results are shown in the following sections.

2 Methods

2.1 Study Site

The Nilahue river estuary is an IOCE located in the Pacific coast of central Chile (Fig. 1). The Nilahue river drains a 1611 km^2 coastal watershed with a mean annual discharge of 8 m^3/s and 2-year return period peak discharge of 198 m^3/s. The site experiences mixed semi-diurnal tides of 2 m maximum tidal range. The predominant wave climate is from the SW-SSW with a 2.5 m mean annual significant wave height and a 13 s peak period. The lower portion of the Nilahue river, just upstream the river mouth is known as Laguna Cahuil. The Nilahue river is culturally and socially important in this region. About 4 km upstream the river inlet there are 500 years old salt production marshes, which rely on saltwater availability upstream the river. The Nilahue river mouth is artificially opened when needed to guarantee salinity levels at the salt production marshes and to reduce flood risk prior to winter storms.

In this investigation we present recent field measurements along the Nilahue river estuary focusing on infragravity wave propagation upstream the river inlet and their implications for sediment transport. The observations are part of a long-term monitoring program of this small IOCE to establish protocols that guarantee the success of artificial inlet openings in terms of water quality, tide propagation, and flood hazard reduction.

2.2 Data Collection and Processing

The field experiment occurred in the Nilahue river estuary between August 1 and August 3rd, 2023. Observations occurred in a period of low river discharge (~1 m^3/s), spring tides (mixed semi diurnal, 2 m range) and high swells (3 m significant wave height and 15 s peak period from the SW). The inlet had been artificial breached weeks prior to the field experiment and remained opened during the extent of the measurements.

Water pressure, velocities, and acoustic backscatter were sampled at 1 Hz using three Nortek Vector acoustic Doppler velocimeters (ADVs) distributed along this small estuary as shown in Fig. 1. All data was quality controlled prior to analysis. Data from the rising portion of the higher flood tide that occurred on August 2nd, 2023, are used in the analysis. In addition, velocity and pressure data were collected using an RDI Teledyne 1200 kHz acoustic current Doppler profiler located between ADV 1 and ADV 2 during the field measurements. These ADCP data are included here to provide context on the hydrodynamic conditions within the lagoon during the experiment.

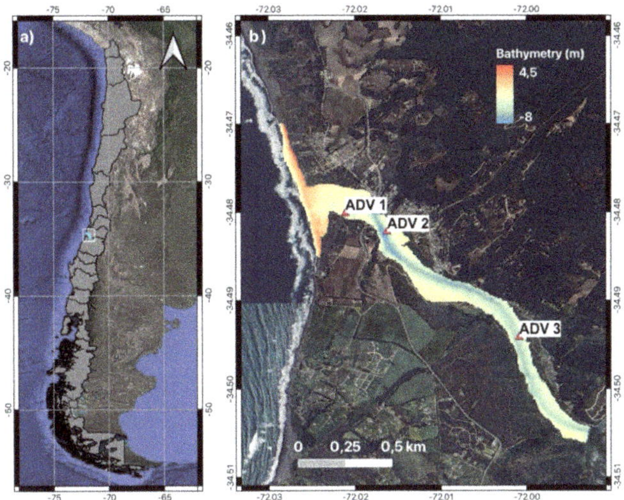

Fig. 1. a) Map of Chile and the location of the study area. b) Study area overview, instrument location and Nilahue river bathymetry.

3   Results

Figure 2 shows a time series of water depth and along-channel velocity (positive landward) collected with the RDI ADCP. Water levels and flow velocity in the estuary are strongly influenced by the tide. A typical two-layer estuarine circulation structure with positive (landward) velocities near the bottom and negative (seaward) in the upper part of the water column develops during flood tide. During ebb tide, the flow is seaward throughout the water column as the lagoon empties. We note that despite the small flood tide barely affecting water levels in the Laguna, the two-layer flow structure still develops, albeit weaker than during the large flood.

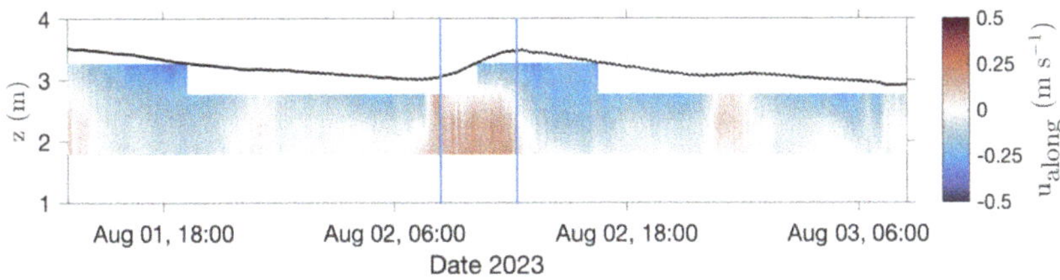

Fig. 2. Hydrodynamic conditions within the Nilahue estuarine lagoon collected with a bottom-mounted ADCP located between ADV 1 and ADV 2. Black line corresponds to water depth, and colors corresponds to along-channel velocity, positive landward. Vertical blue lines denote the timing of the data shown used in the IG waves analysis.

Figure 3 shows raw time series of pressure (Fig. 3a), along-channel velocity (Fig. 3b), and acoustic backscatter (Fig. 3c) from ADV 1 during the rising portion of the larger flood tide rise of August 2nd, 2023 (Fig. 2). Long-wave surface and velocity motions are observed in all three instruments during this flood tide, albeit of decreasing amplitude as distance from the river mouth increases. No long-wave fluctuations were observed during ebb tide as supercritical flow was observed at the river mouth, preventing the propagation of waves upstream the estuary. The observed long-wave motions are accompanied by similar fluctuations in the acoustic backscatter at all instruments, indicative of variations of scatterer concentrations in the water column.

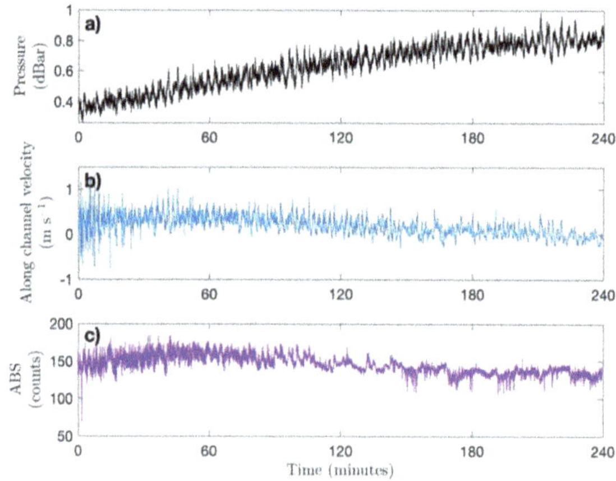

Fig. 3. Raw time series of a) pressure, b) along-channel velocity and c) acoustic backscatter from ADV 1.

Figure 4a to c shows pressure, velocity, and acoustic backscatter spectra obtained from the de-tided time series at all three measurement locations. Spectra from all quantities show peaks in the IG frequency band (below 1/30 Hz) confirming the propagation of infragravity waves upstream the estuary. The peak magnitude decreases with distance from the river mouth, but it is still noticeable 3 km upstream the estuary a ADV 3 location.

Figure 4d shows wavelet coherence between along-channel velocity and acoustic backscatter at ADV 1 location. The peaks in the wavelet coherence between along-channel velocity and acoustic backscatter observed at the IG wave frequency band suggest that IG waves are driving scatterers fluctuations at this measurement location thus potentially influencing sediment transport within the estuary during flood tide.

Future work includes the analysis of synchronous free-surface, turbulent velocities, and acoustic backscatter observations from a Nortek Signature 1000 Hz acoustic current Doppler profiler together with turbidity measurements collected approximately 500 m upstream the river mouth between July 31st and August 2nd 2024. Long-wave oscillations were observe during the rising portion of the large flood tides of the experiments. Data collection and analysis from both field experiments will be include in the talk.

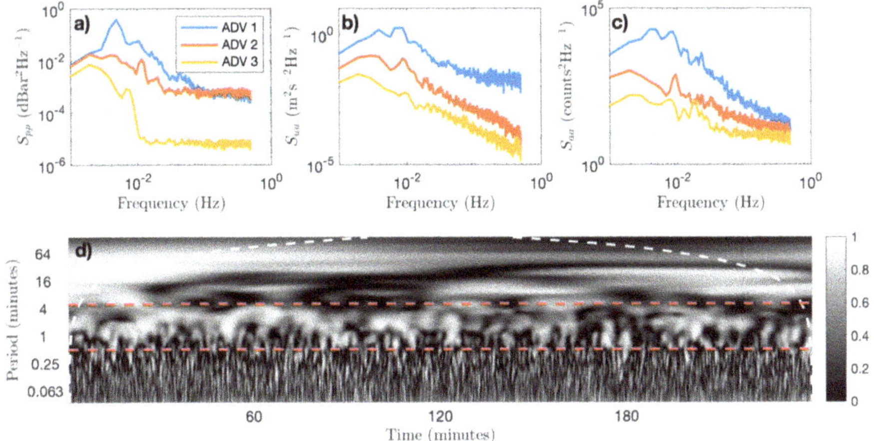

Fig. 4. a) Pressure, b) along-channel velocity, c) acoustic backscatter spectra, and d) along-channel velocity and acoustic backscatter wavelet coherence from ADV 1.

References

1. McSweeney SL, Kennedy DM, Rutherfurd ID, Stout JC (2017) Intermittently Closed/Open Lakes and Lagoons: their global distribution and boundary conditions. Geomorphology 292:142–152
2. McSweeney SL, Stout JC, Kennedy DM (2020) Variability in infragravity wave processes during estuary artificial entrance openings. Earth Surf Proc Land 45(13):3414–3428
3. Guza RT, Thornton EB (1982) Swash oscillations on a natural beach. J Geophys Res: Oceans 87(C1):483–491
4. Melito L et al (2018) Wave–current interactions and infragravity wave propagation at a microtidal inlet. In: Multidisciplinary digital publishing institute proceedings, vol 2, no 11, p 628
5. Bertin X, Olabarrieta M (2016) Relevance of infragravity waves in a wave-dominated inlet. J Geophys Res: Oceans 121(8):5418–5435
6. Williams ME, Stacey MT (2016) Tidally discontinuous ocean forcing in bar-built estuaries: the interaction of tides, infragravity motions, and frictional control. J Geophys Res: Oceans 121(1):571–585
7. Bertin X, Mendes D, Martins K, Fortunato AB, Lavaud L (2019) The closure of a shallow tidal inlet promoted by infragravity waves. Geophys Res Lett 46(12):6804–6810

Alternatives for Cost-Effective Shoaling Reduction at Navigation Channels Near Inlets

Michael B. Kabiling[✉]

Taylor Engineering, Inc., Jacksonville, FL 32256, USA
mkabiling@taylorengineering.com

Abstract. This study developed and applied dynamically-integrated MIKE21 Flexible Mesh hydrodynamic, wave, sediment transport and morphology, and particle tracking models to analyze the effect of sediment impoundment basins and Intracoastal Waterway (IWW) channel rerouting near Bakers Haulover Inlet in Miami-Dade County, Florida, USA. The study aims to find alternatives that reduce shoaling and IWW channel maintenance dredging cost. For computational efficiency and to complete the alternative evaluations within schedule, the study used a month-long representative period from a representative year to estimate shoaling rates and then prorated the computed shoaling rates for an average year. The prorated shoaling rate offers a good approximation of the shoaling rates of channel shoaling reduction alternatives relative to the shoaling rate for baseline (existing) conditions. Model results show the best performing shoaling reduction alternative requires maintenance dredging once every eight years and provides a substantial annualized cost savings of $249,000 when compared with the baseline dredging frequency of once every four years. The cost of this study (including data collection, integrated modeling, and related analysis work) is a very small percentage (1%–2%) of the total cost savings with the implementation of any of the viable shoaling reduction alternatives.

Keywords: Shoaling Reduction · Bakers Haulover Inlet · Intracoastal Waterway · Cut-DA9 · Channel Rerouting

1 Introduction

The Intracoastal Waterway (IWW) near Bakers Haulover Inlet in Miami, Florida passes 0.64 km west of the inlet. Wave- and tide-generated currents entrain and transport some of the littoral sediments through the inlet and the rest to downdrift nearshore areas. These currents can move sediments inshore until flow velocities decrease below a critical value, at which point sediments fall out of suspension and can form shoals in the IWW and surrounding areas (Fig. 1). A Florida Department of Environmental Protection study [1] estimates a total net transport into the inlet of approximately 46,410 m^3/yr for 2007–2016.

A review of U.S. Army Corps of Engineers (USACE) data of IWW maintenance dredging events near Bakers Haulover Inlet from 1969 to 2017 shows shoaling volumes total 314,754 m^3. Of this total volume, at least 71% was dredged between 1969 and

C. Coelho et al. (Eds.): CD 2025, CRL 41, pp. 783–789, 2026.
https://doi.org/10.1007/978-3-032-15473-6_119

1998 from within the 1.77-km portion of Cut-DA9 adjacent to Haulover Sandbar (North Flood Shoal in Fig. 1) and South Flood Shoal. Thus, this portion of the IWW Cut-DA9 is vulnerable to large shoaling and historically required dredging on average once every 4–5 years to maintain the authorized IWW navigation depth of 3.05 m below mean lower low water.

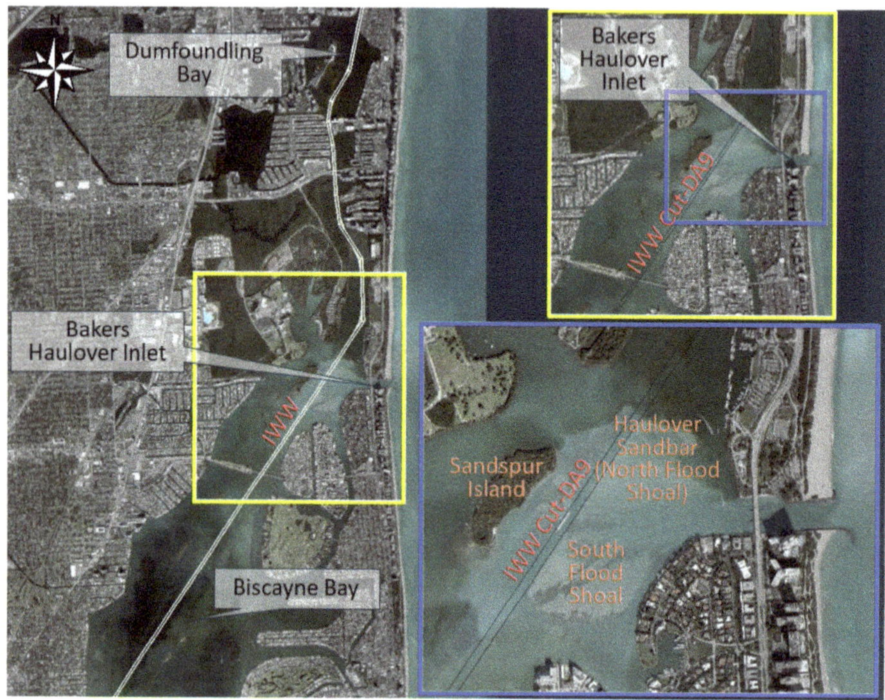

Fig. 1. Bakers Haulover Inlet and location of the area of interest

2   Shoaling Reduction Alternatives

At the IWW near Bakers Haulover Inlet, construction of deposition basins and/or removal of nearby shoal sediment sources (e.g., Haulover Sandbar) could, with periodic dredging, reduce shoaling in the IWW. Alternatively, rerouting of the IWW Cut-DA9 to west of Sandspur Island (Fig. 2) could also reduce maintenance dredging requirements and improve navigation safety as rerouting would move the navigation channel farther away from the inlet and Haulover Sandbar to an area with low flow velocity and sediment transport. The prospect of these cost-saving outcomes encouraged the Florida Inland Navigation District (FIND) to request Taylor Engineering to perform a sediment deposition basin and IWW Cut-DA9 rerouting study at Bakers Haulover Inlet and adjacent waterways.

Fig. 2. Model bed elevations, model mesh at the area of interest, calibration stations, and analyzed alternatives including Alternative G reroute (shown by red dots)

3 Numerical Modeling and Methodology

To analyze the effect of sediment impoundment basins and IWW channel rerouting, this study developed and applied dynamically-integrated MIKE21 Flexible Mesh (FM) hydrodynamic (HD), wave (SW), sediment transport and morphology (ST), and particle tracking (PT) models. The PT model provided the sediment transport pathways to help locate the sediment deposition basin.

Mesh development for this study takes advantage of existing Federal Emergency Management Agency (FEMA) South Florida ADCIRC model Version 11 mesh. Using available shoreline and IWW delineation data, this study re-generated the MIKE21 FM HD model mesh in the IWW and in areas between the IWW and the Stranahan River, Dumfoundling Bay and Biscayne Bay shorelines; and if needed, either maintained the ADCIRC model mesh or generated new mesh elements outside of these areas. To construct the existing condition model mesh, multiple sources provided the topographic and bathymetric data—(a) the FEMA 2016 South Florida ADCIRC model Version 11 mesh; (b) USACE 2019 Cut-DA9, Bakers Haulover Inlet, and potential IWW rerouting area bathymetric survey data; (c) 2018 FDEP survey of Broward and Miami-Dade Counties beaches; and (d) Olin Hydrographic Solutions, Inc. 2018 bathymetric survey of the flood and ebb shoals at Bakers Haulover Inlet.

As this study measured inshore validation data in August–September 2020, this study applied the existing condition model mesh for the hydrodynamic model validation in the inshore areas. As a previous study [2] collected model validation data just east of the inlet's mouth in 2018, this study applied the Olin Hydrographic Solutions, Inc. 2018 bathymetry survey data to replace the Bakers Haulover Inlet, flood shoal, and ebb shoal bathymetries for the model validation in the nearshore area.

3.1 Hydrodynamic Model

This study performed the HD model calibration and verification in two parts—the first part is the calibration/verification for the inshore areas and the second part is the calibration/verification for the nearshore areas—because the model calibration and model verification data come from two data sets measured at different times. Calibration/verification for the model's inshore area applied the measured inshore data in August–September 2020 and calibration/verification for the model's nearshore area applied a previous study's [2] May–June 2018 measured nearshore data.

As a Manning's n value of 0.035 s/m$^{0.33}$ provides the best overall water level comparison with water level measurements during model inshore calibration, this study selected a model Manning's n value of 0.035 for bed resistance calculation for Bakers Haulover Inlet, Stranahan River, Biscayne Bay, and smaller inshore waterways. For the second part of the calibration, a finer nearshore calibration changed the Manning's n near the inlet area from 0.035 s/m$^{0.33}$ to 0.045 s/m$^{0.33}$ and provided better comparisons with the previous study's [2] measured data in the nearshore area. Station TB3 modeled water level comparisons with measured water level provided a mean error of -0.012 m, root mean square error of 0.020 m, and a correlation coefficient of 0.998. The small mean error and root mean square error comprise less than 2% and 3% of the mean tidal range, respectively. Figure 3 shows the modeled flow velocity at Station VB3 (see Fig. 2) compared well with measurements in speed and direction.

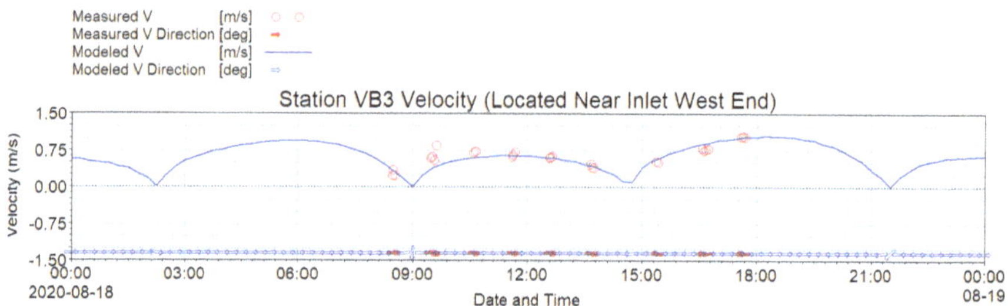

Fig. 3. Comparison of modeled and measured velocity at Station VB3 for HD model calibration period

3.2 Wave Model

The wave model applied Wavewatch III wave output offshore. Among the wave model parameter options available for including wave breaking in the model, this study specified the default gamma option—equal to a constant 0.8—consistent with the breaking criterion associated with the ratio of wave height and water depth (0.78) at the wave breaking point. When compared with measurements from a previous study [2], the modeled significant wave height shows a small mean error of 0.048 m, a small root mean square error of 0.157 m, and a good positive correlation of 0.778. The modeled peak wave period follows the general trend in measured peak wave period. The modeled mean wave direction compared well with the measurements and mostly lies within the range of measured mean wave directions—an acceptable comparison given the substantial noise in the measured mean wave direction.

3.3 Sediment Transport Model

Surface grab samples in the IWW, Bakers Haulover Inlet, and nearby waterways provided the sediment characteristics applied in the model with averaged median grain size diameter of 0.23 mm and grading coefficient of 1.80 for consistent sedimentation patterns at the area of interest and for the evaluation of shoaling at the area of interest. This study applied combined currents from tides and waves as the driving mechanism for sediment transport. The model calculated sediment transport rates by interpolating from a sediment transport table. The Generation of Q3D Sediment Transport Tables tool in MIKEZero created the sediment transport table.

4 Representative Period

Littoral transport magnitude and direction can vary by season, day, or hour because waves can vary with the same time scale. Thus, the estimation of the shoaling rate in a sediment impoundment basin and channel reroute requires simulation of sediment transport over, at least, a full year. However, such long-term simulations require lengthy computational time that goes beyond the schedule of this study. Alternatively, this study simulated a month-long representative period from a representative year to estimate shoaling rates and then prorated the computed shoaling rates for an average year (i.e., a year with average gross littoral transport). Thus, using the Littoral Processes (LP) model, this study (a) estimated the annual gross littoral transport for each year in the available 1980–2014 data set of the USACE Wave Information Study; (b) calculated the average annual gross littoral transport for the period 1980–2014 at transects that originate from selected beach transects; and (c) selected 2006 as the year with available Florida Current data that provides an annual gross littoral transport nearest to the calculated annual average at these transects. This study then used the LP model to calculate the monthly gross littoral transport for January–December 2006. At the selected transects, February provides the monthly gross littoral transport that is nearest to the calculated average monthly gross littoral transport in 2006. Therefore, this study applied the predicted tides and hindcasted waves in February 2006 as offshore wave boundary conditions for model calculation of baseline and design alternatives tides, waves, sediment transport, and shoaling rates for the representative month.

Notably, the model calculated shoaling rates associated with the representative month will likely not reflect the long-term basin and in-channel shoaling rates—an acceptable approximation given that future hydraulic conditions (water levels and waves) are unknown. Further, the basin sediment impoundment efficiency diminishes as sediment deposits in the basin. However, the prorated shoaling rate offers a good approximation of the shoaling rates of alternatives to reduce in-channel shoaling relative to the shoaling rate for baseline conditions (i.e., no basins and no IWW rerouting). Moreover, the prorated shoaling rate provides an effective way of determining which alternative performs best to reduce in-channel shoaling. Comparison of the shoaling rates for baseline and shoaling reduction alternatives provided the means to estimate shoaling rates at Cut-DA9 with sediment impoundment basins construction, Haulover Sandbar removal, and Cut-DA9 westward rerouting.

5 Results of Analysis

Model results show the best performing deposition basin alternative (Alternative D in Fig. 2) would likely change the frequency of dredging to approximately once every nine years. Each dredging operation would remove approximately 57,840 m^3 at an equivalent uniform annual cost of $507,000. Compared to the baseline equivalent uniform annual cost (i.e., the existing maintenance dredging volume and frequency at the time of this study), the estimated annual savings of $177,000 is substantial and the second largest among the analyzed alternatives. However, Alternative D is very vulnerable to large shoaling from storms.

Model results show the best performing IWW Cut-DA9 reroute alternative (Alternative G in Fig. 2) requires an additional initial dredging of 66,200 m^3 for the new channel and each succeeding maintenance dredging operation would remove approximately 29,460 m^3 once every eight years or more at an equivalent uniform annual cost of $435,000. Compared to the baseline equivalent uniform annual cost, the estimated annual savings of $249,000 is the largest among the analyzed alternatives.

6 Conclusions

This study recommends IWW rerouting Alternative G for final channel/basin modifications, geotechnical/environmental analysis, and a boat navigation analysis of the route. Evaluate the long-term shoaling rate for better estimation of the variation of the sediment impoundment performance and to better account for more potential long-term variations of waves and water levels. The cost of the study (including data collection, integrated modeling, and related analysis work) is a very small percentage (e.g., 1%–2%) of the total cost savings with the implementation of any of the viable shoaling reduction alternatives.

References

1. (2021) Florida Department of Environmental Protection: Bakers Haulover Inlet Management Plan 08-2021, p 35
2. (2019) Moffat & Nichol, Inc.: Bakers Haulover Inlet Feasibility Study, Miami-Dade County, Florida, p 218

Development of Hybrid Approaches for Predicting and Estimating Salt Wedge Intrusion in Estuaries

Mirian Jiménez[1]([✉]), Laura Cagigal[1], Alba Ricondo[2], Beatriz Pérez-Díaz[1], Javier Sopelana[3], Cielo Fernández[3], Sonia Castanedo[1], and Fernando Méndez[1]

[1] Geomatics and Ocean Engineering Group, University of Cantabria, Santander, Spain
mirian.jimenez@unican.es
[2] Oregon State University, Corvallis, OR State, USA
[3] Aquática Ingeniería, Vigo, Spain

Abstract. Estuarine systems are particularly vulnerable to climate change, as sea-level rise and altered river discharge drive inland saltwater intrusion. This intrusion disrupts biogeochemical cycles, ecosystems, and human activities, such as agriculture and industry, by increasing salinity levels. Addressing these challenges, we present a novel hybrid methodology combining advanced numerical modeling with statistical techniques to predict salinity behavior in estuaries. This methodology enables fast and robust predictions compared to other techniques currently in use. The approach begins with identifying seven key parameters influencing estuarine salinity, including tidal range, river discharge, and storm surges. Using Latin Hypercube Sampling (LHS), synthetic scenarios are generated, subsequently, representative cases are selected through the Maximum Dissimilarity Algorithm. Those selected cases are modeled using the Delft3d numerical model, calibrated and validated using field data collected from an estuary in northern Spain. Principal Component Analysis (PCA) is applied to reduce data dimensionality, identifying dominant salinity modes, followed by interpolation through Radial Basis Functions (RBFs) to create a interpolation surface. Results demonstrate that the hybrid methodology captures salinity concentration dynamics under diverse conditions, providing rapid and reliable predictions. This will enable the establishment of more efficient management systems to address different uses for industrial and agricultural activities in estuarine environments.

Keywords: Hybrid Methodology · Salt-wedge intrusion · Estuaries

1 Introduction

Over the past decade, extensive research has highlighted the impacts of climate change, particularly sea level rise (SLR) and altered river discharge, on saline intrusion in estuarine environments. Saltwater intrusion is the leading edge of sea-level rise, preceding tidal inundation, but leaving its salty signature far inland. With climate change, saltwater is shifting landward into regions that previously had not experienced or adapted to salinity, leading to novel transitions in biogeochemistry, ecology, industrial and human uses.

© The Author(s) 2026
C. Coelho et al. (Eds.): CD 2025, CRL 41, pp. 790–795, 2026.
https://doi.org/10.1007/978-3-032-15473-6_120

These changes in salt-wedge behavior threaten various ecosystems and human activities. This includes industrial and agricultural operations, as they contribute to increased ionic strength, alkalinization, and sulfidation [1]. Previous studies, such as those by [2], have identified river flow, sea level, and estuarine geometry as the primary factors influencing salt wedge intrusion. For salinity intrusion, several numerical and empirical models have been developed to predict the saltwater intrusion length within an estuary as a function of measurable parameters, such as estuarine geometry, river inflow, and tidal parameters [3]. The complexity of this variable makes its prediction computationally very expensive. Studies such as [4] and [5] explore the use of machine learning tools in simplified geometries to minimize these computational costs. That is why this work takes a step further, presenting a hybrid predictive model to simulate saline behavior in a real estuary located in northern Spain.

2 Methodology

The methodology proposed in this study is a versatile tool that allows for the prediction of salinity concentration at any point within an estuary (see Fig. 120.1). The described methodology is applied to an estuary located in northern Spain, which has a length of 15 km. It is an intertidal valley estuary with marine dominance and is regulated upstream by a reservoir. The methodology begins with an identification of the main parameters influencing estuarine salinity behavior, including factors such as river discharge and the influence of tidal dynamics. This step ensures that all significant variables affecting salinity behavior within the estuary are accounted. A total of seven key parameters are identified as relevant factors to consider, including: tidal range (m), tidal period (h), storm surge (m), river discharge (m^3/s), reservoir peak discharge (m^3/s), duration of reservoir peak discharge (h), and time lag (h) (the time elapsed between high tide and the reservoir peak discharge).

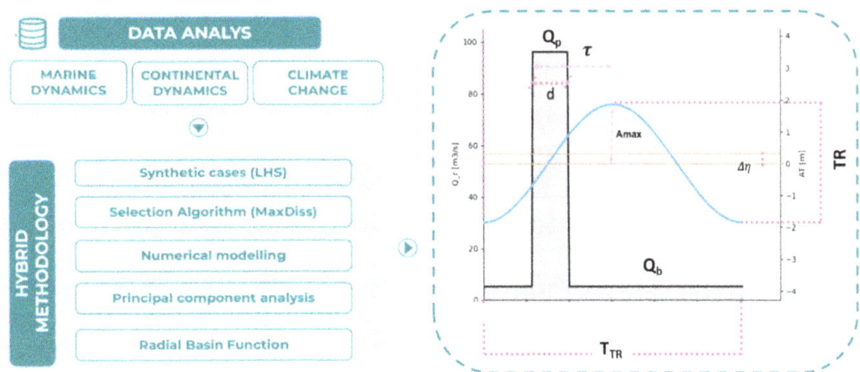

Fig. 120.1. Methodological workflow diagram.

Subsequently, a large set of synthetic cases (N = 100,000) is generated using the Latin Hypercube Sampling. This sampling method ensures that the entire range of each parameter is adequately explored, providing a comprehensive set of scenarios for analysis. From these synthetic cases, 1,500 representative cases are selected through the Maximum Dissimilarity Algorithm [6]. This algorithm prioritizes the most dissimilar cases, ensuring a diverse and representative sample that captures the full spectrum of potential behaviors within the estuary. Next, these selected cases are dynamically downscaled using three-dimensional numerical simulation using the Delft3D model [7]. The numerical model is run on a grid of 89*859 cells in the horizontal discretization and 20 layers in vertical discretization. This model has been previously calibrated and validated under various meteorological conditions in an estuary in northern Spain. Field campaigns were conducted during two distinct periods, lasting three months. During these campaigns, multiple CTD sensors were deployed at different points along the estuary to measure pressure, temperature, and conductivity along its main axis. The upper panel of Fig. 120.2 shows the locations of the sensors placed at various points in the estuary, which were used for calibrating the salinity concentration in the numerical model. The lower panel displays scatter plots comparing the modeled salinity concentration with the measured salinity concentration collected over five days during the field campaigns.

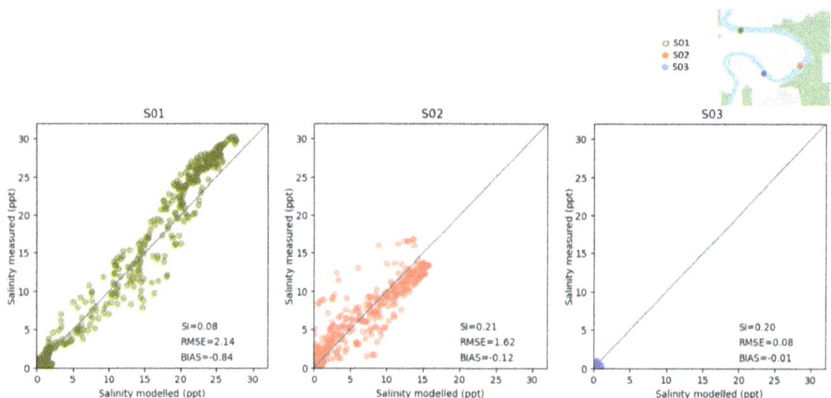

Fig. 120.2. Scatter plot of salinity concentration at different locations in the estuary.

This calibration and validation process ensures the numerical model's accuracy in capturing salinity behavior in the estuary. Following the simulations, the dimensionality of the results is reduced using Principal Component Analysis (PCA). PCA has been widely used in many fields, such as oceanography to identify dominant variability patterns ([8]; [9]). By projecting the original data onto a new space, PCA captures the maximum variance in the sample data. This analysis helps to streamline the complex dataset, emphasizing the most relevant patterns and trends in salinity behavior. Finally, an interpolation surface for salinity concentration is created using Radial Basis Functions (RBFs), that is a very common method in data mining, these functions are used for data interpolation in multidimensional spaces, capturing the relationship between one or more independent variables and a response variable. To obtain this, different Gaussian functions are centered on points whose response function is known and by adjusting the width of these bells, the final relationship function is obtained. This mathematical technique allows the generation of a smooth and continuous surface that accurately represents the salinity distribution at any location within the estuary.

3 Results

Once the interpolation surface for salinity concentration is obtained using Radial Basis Functions (RBFs), the response variable, salinity, is automatically calculated for any given set of values for the seven parameters described in the first step of the methodology. Figure 120.3 illustrates the evolution of salinity concentration at a specific point in the estuary during a tidal cycle. Results from the numerical model are shown in dark purple, while those from the methodology described in Sect. 2 are represented in light purple. The figure includes three different panels, each depicting a different scenario based on varying values of the seven parameters that influence salinity dynamics in the estuary. As observed, the hybrid methodology effectively captures both the shape and magnitude of salinity behavior throughout an entire tidal cycle and is capable of producing a wide range of salinity values.

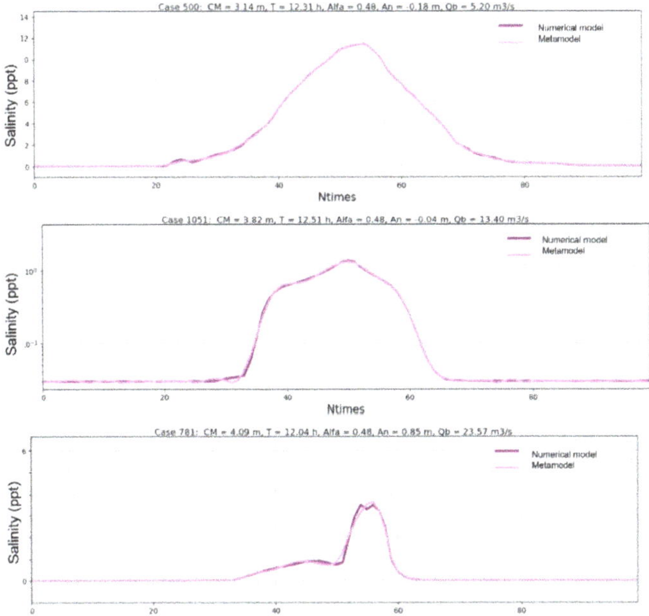

Fig. 120.3. Comparative between numerical model results (dark purple) and hybrid methodology results (light purple) of salt concentration during a tidal cycle.

4 Conclusions

Given the complexity of behavior in estuarine systems, resulting from the interactions between coastal dynamics, riverine dynamics, and the anthropogenic pressures to which they are subjected, providing reliable and rapid responses can often become highly complex and computationally expensive. Hybrid methodologies that combine advanced numerical models with statistical techniques, such as the one proposed in this study, prove to be robust and effective tools for predicting variables influenced by multiple factors and governed by nonlinear interactions, as is often the case in estuarine environments. Specifically, the integration of techniques such as Latin Hypercube Sampling (LHS), Principal Component Analysis (PCA), and Radial Basis Functions (RBFs) significantly reduces data dimensionality and consequently decreases response times in predictive systems. This hybrid approach has provided excellent results in reconstructing sea level and wave heights ([10; 11; 12]), but it had not been as widely developed to address the study of the evolution of the salt wedge and the regime of water levels. It is worth highlighting that this study incorporates both instrumental and numerical validations, ensuring the robustness and applicability of these approaches in managing sensitive ecosystems and planning adaptive measures for economic, industrial and agricultural uses.

Acknowledgments. The development of this work is framed within: Methodologies for risk assessment of the industrial sector and adaptation to climate change in the estuarine environment (MyEst) that has the support of the Biodiversity Foundation of the Ministry for Ecological Transition and Demographic Challenge, through the Call for grants for the implementation of projects that contribute to implement the National Plan for Adaptation to Climate Change (2021–2030).

References

1. Tully K, Gedan K, Epanchin-Niell R, Strong A, Bernhardt ES, BenDor T (2019) The invisible flood: the chemistry, ecology, and social implications of coastal saltwater intrusion. Bioscience 69:368–437
2. Tian R (2019) Factors controlling saltwater intrusion across multi-time scales in estuaries, Chester River, Chesapeake Bay. Estuarine Coastal Shelf Sci 223:61–73
3. Prandle D (2004) Saline intrusion in partially mixed estuaries. Estuar Coast Shelf Sci 59:385–397
4. Hendrickx GG, Antolínez JAA, Herman PMJ (2023) Predicting the response of complex systems for coastal management. Coast Eng 182:104327
5. Maglietta R et al (2025) Advancing estuarine box modeling: a novel hybrid machine learning and physics-based approach. Environ Model Softw 183:106223
6. Camus P, Méndez FJ, Medina R (2011) A hybrid efficient method to downscale wave climate to coastal areas. Coast Eng 58(9):851–862
7. (2021) Deltares Delft3D-FLOW User Manual: Simulation of Multi-Dimensional Hydrodynamic Flows and Transport Phenomena, Including Sediments. Version 3.15. Deltares, Delft, The Netherlands
8. Eckert-Gallup AC, Sallaberry CJ, Dallman AR, Neary VS (2016) Application of principal component analysis (PCA) and improved joint probability distributions to the inverse first-order reliability method (I-FORM) for predicting extreme sea states. Ocean Eng 112:307–319
9. Camus P, Mendez FJ, Medina R, Tomas A, Izaguirre C (2013) High resolution downscaled ocean waves (DOW) reanalysis in coastal areas. Coast Eng 72:56–68. https://doi.org/10.1016/j.coastaleng.2012.09.002
10. Anderson DL et al (2021) Projecting climate dependent coastal flood risk with a hybrid statistical dynamical model. Earth's Fut 9(12)

11. Ricondo A, Cagigal L, Rueda A, Hoeke R, Storlazzi CD, Méndez FJ (2023) HyWaves: hybrid downscaling of multimodal wave spectra to nearshore areas. Ocean Model 184
12. Sopelana J, Cea L, Ruano S (2018) A continuous simulation approach for the estimation of extreme flood inundation in coastal river reaches affected by meso-and macrotides. Nat Hazards 93:1337–1358

Initiation of Motion for Sand-Mud Bed Types

P. S. Miranda[1,2](✉), J. J. van der Werf[1,2], and S. J. M. H. Hulscher[1]

[1] University of Twente, 7522 NB Enschede, The Netherlands
`p.s.miranda@utwente.nl`
[2] Deltares, Boussinesqweg 1, 2629 HV Delft, The Netherlands

Abstract. The initiation of motion of sand-mud bed mixtures remains inadequately understood and not accurately predicted. This study sought to compare and analyze the methodology and results of different sand-mud erosion experiments. Data from different erosion experiments were combined to create a dataset of varying sand-mud combinations with an emphasis on varying clay-silt ratios. The framework proposed by Van Ledden [1] was used to separate the data into six different bed types. These bed types are characterized by their cohesive property and network structure. The cohesive property is based on the mass fraction of clay and the network structure is based on the dominant sediment fraction. The different bed types demonstrated different ranges of critical bed shear stress for various ranges of median grain size or mass fraction of mud. This indicates that distinct initiation of motion mechanisms are dominant in different bed types.

Keywords: sand-mud · initiation of motion · sediment transport

1 Introduction

Sand-mud is a ubiquitous type of sediment in estuaries and deltas. Estuaries and deltas function as locations for global ports around the world and as vital habitat for marine ecology. Analyzing the properties of sand-mud mixtures, particularly its erosion potential, facilitates in the planning and management of these systems.

Sand-mud erosion occurs in different modes (e.g. particle, surface, and mass erosion) as discussed by van Rijn [2]. However, identifying the initiation of motion remains without quantifiable standards. As a result, recent sand-mud erosion experiments continue to use different methods.

The proposed mathematical models for sand-mud initiation of motion, donated by the critical bed shear stress parameter or τ_{crit}, have been developed based on available sand-mud data from researchers developing those models. These equations describe the initiation of motion of sand-mud beds through bulk parameters, such as median grain size diameter, d_{50}, or mass fraction of mud, P_{mud}. However, mud is composed of silt and clay. Clay particles are naturally cohesive while silt particles exhibit apparent cohesion due to their reduced permeability. At present, initiation of motion equations for sand-mud mixtures do not account for different quantities in silt and clay.

© The Author(s) 2026
C. Coelho et al. (Eds.): CD 2025, CRL 41, pp. 796–802, 2026.
https://doi.org/10.1007/978-3-032-15473-6_121

To date, an erosion experiment using wide variations in sand-silt-clay combinations has not been undertaken. As a result, the analysis of the initiation of motion of sand-mud beds with different mud fractions using large variations in silt and clay quantities has not been done. However, recent erosion experiments have been performed with limited but different ranges of clay-silt combinations.

This study was conducted to combine the results of sand-mud erosion experiments that used various clay-silt ratios through an analysis of methodologies. This study also examined a framework that highlights the contribution of each sediment fraction in the initiation of motion process. This paper is organized as follows. Section 2 explains the methodology used in the selection of data and framework. Section 3 lists the erosion experiments whose results are included in the combined dataset and the application of the chosen framework to the combined dataset. Section 4 gives conclusions on the created dataset and implications of applying the selected framework.

2 Data and Framework Selection

The methodology for dataset selection needed criteria to assess the discrepancies in methods between different erosion experiments. The methodology for framework selection needed to isolate the different sediment fractions based on physical characteristics.

2.1 Data Selection

The selection process defined criteria to analyze methodologies and results across erosion experiments. Datasets from erosion experiments using sediment less than 2 mm were used. The definition of each sediment fraction was based on their cohesive properties (sand: non-cohesive; silt: apparently cohesive; clay: cohesive) and grain size diameter. Specifically, the grain size diameter definition has the diameter lower limit of sand at 63 μm and the diameter upper limit of clay at 2 μm. Furthermore, datasets must have reported measured bulk geotechnical parameters such as water content, and bulk density. Next, a criterion for experimental setups limited the selection to those that used unidirectional flow since this type of forcing replicates the tidal forcing in estuaries. The method of obtaining derived parameters was also reviewed to ensure that parameters such as bed shear stress were similarly calculated. The derivation of bed shear stress from erosion experiments lacks a standardized methodology, therefore, multiple methods were accepted. Such methods included experiments that derived bed shear stress from law of the wall velocity profiles or experiments that estimated from the measured near-bed turbulence. A more subjective derived parameter is the τ_{crit} or the bed shear stress at initiation of motion. Similar to bed shear stress, there is no standard definition for determining τ_{crit}. Consequently, multiple methods were also deemed accepted. These methods included: visual observation, extrapolation from the relationship between erosion rate and bed shear stress, a threshold erosion rate, and a threshold suspended sediment concentration value. These criteria did not yield an absolute like-for-like comparison across sand-mud erosion experiments. However, the criteria provided a dataset with a large variation in sand-mud composition, bulk geotechnical parameters, and initiation of motion information.

2.2 Framework

The work by Van Ledden [1] proposed a framework delineating the erosion behavior of sand-mud that accounts for each sediment fraction. The framework categorizes according to 2 characteristics: cohesion and network structure. Cohesion is based on the empirically observed threshold for mass fraction of clay or P_{clay}. Sand-mud is considered cohesive with $P_{clay} \geq 5\text{--}10\%$. In subsequent analyses, this threshold is set at 7%. Network structure refers to the sediment fraction that may or may not form a network of particles within the sand-mud bed. From fluidization experiments, sand particles form a network when the volume fraction of sand, ϕ_{sand}, exceeds 40%. Similarly, silt particles form a network when the volume fraction of silt, ϕ_{silt}, , exceeds 40% of the pore volume around sand particles. In contrast and due to the cohesive nature, a clay-water matrix is formed instead of a clay network. This matrix is present when sand-mud bed is neither sand- or silt-dominated, and cohesive. Finally, a mixed network is formed when neither sand nor silt form a network and the bed is non-cohesive. A mixed network is characterized as having contributions from all sediment fractions. The proposed framework results in six bed types: (I) sand-dominated non-cohesive, (II) sand-dominated cohesive, (III) mixed network non-cohesive, (IV) clay-water matrix cohesive, (V) silt-dominated non-cohesive, and (VI) silt-dominated cohesive. The framework is adopted to describe the initiation of motion based on cohesion and dominant network structure of a sand-mud bed type.

3 Results

3.1 Selected Datasets

After applying the proposed criteria, the following datasets provided a wide range of sand-mud compositions with sufficient experimental and geotechnical information.

The work by Jacobs [3, 4] investigated sand-mud erosion using artificially created sand-mud based on clay-silt ratios between 0.04–0.44 and sand-silt ratios between 0.31 and 18.6.

The work at the US Army Corp of Engineers [5] investigated the erosion thresholds of sand-mud using both artificial and natural sand-mud. The sediment had mud content from 0 to 100% with different types of clay. Experiments were conducted using purely Kaolinite. Other experiments were conducted using a combination of Kaolinite and Bentonite. While another set of experiments used mud taken from the Mississippi river.

The previous two datasets provided data containing mud with significant clay content. It was necessary to find datasets that conducted erosion experiments for mud with low clay content ($P_{clay} < 7\%$). Part of the PhD work by te Slaa [6] was useful in filling this gap in data since the PhD work focused on the erosion of silt-rich environments. In this dataset, the clay content was 0 while the silt content was between 90–100%.

Another erosion experiment that focused on the behavior of silt-dominated systems is the work by Yao [7]. The recalculated P_{silt} of the sediment samples range from 20–82% while P_{clay} was approximately 0%. This dataset provided additional samples with significant silt content and insignificant amounts of clay content.

Finally, the dataset from the MuSa project in Deltares [8] collected and tested sand-mud samples from within the Netherlands and from sites abroad. The range of sand-silt-clay composition had mud content between 12–95%, and composed of different clay-silt ratios, 0.19–0.76.

The geotechnical information provided in the experiments were not consistently reported. To complete the information, known geotechnical relationships were used to calculate for any gaps. In the end, it was imperative that median grain size diameter, (d_{50}, , m), water content (W%), volume fraction of water (ϕ_{water}), dry bulk density (kg/m^3), and bulk density (kg/m^3) were known or calculated for each data point.

In total, the dataset contains 107 data points. This dataset has different combinations of sand and mud as well as sufficient data containing different types of mud.

3.2 Bed Type Classification

The framework proposed by Van Ledden is applied to the collected datasets. The results are shown in Fig. 121.1: 54 datapoints in bed type I, 3 datapoints in bed type II, 13 datapoints in bed type III, 25 datapoints in bed type IV, 12 datapoints in bed type V, and 0 datapoints in bed type VI. Despite the absence of data in bed type VI, the result of applying the framework provides an opportunity to investigate parameters and processes in bed types I to V.

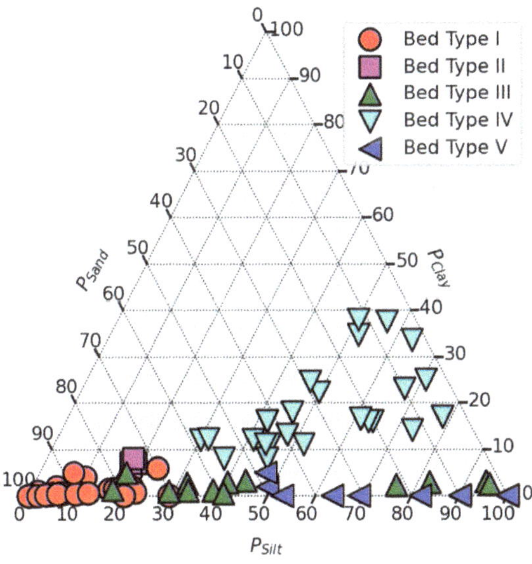

Fig. 121.1. Ternary diagram showing the spread of data from the datasets that fulfilled the selection criteria. Each data point is marked and colored according to its Van Ledden bed type [1].

Classification of sand-mud into different bed types shows its effect on the range of τ_{crit}. Figure 121.2 shows a plot of each bed type in terms of its τ_{crit} against d_{50} and volume fraction of mud, P_{mud}. Bed type I has the largest d_{50} and lowest P_{mud} with a wide range of τ_{crit} between 0.1–2.4 Pa. Despite limited datapoints, bed type II has smaller d_{50} but larger P_{mud} than bed type I with a narrower range of τ_{crit} between 0.9–1.0 Pa. Bed types III and IV have a large spreads of d_{50}'s and of critical bed shear stresses despite containing the same fraction of mud. Bed type III shows that an increase in P_{mud} may result in an increase in τ_{crit}. Conversely, bed type IV appears to decrease in τ_{crit} with increasing P_{mud}. Finally, bed type V does have a large range of d_{50} but its τ_{crit} values are, on average, comparable to the values of pure sand. The bed types with different cohesive and network structure properties are shown to affect the magnitude and spread of τ_{crit} across d_{50} and P_{mud}.

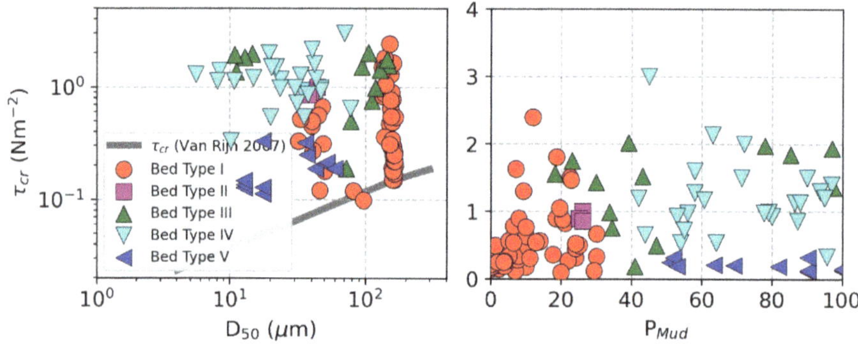

Fig. 121.2. Left plot presents d_{50} and τ_{crit}. Data points are colored according to bed type [1]. The τ_{crit} of pure sand from van Rijn [9] is also plotted for reference. Right plot presents the τ_{crit} as P_{mud} varies. The data points are similarly colored [1].

This classification helps identify which mechanisms may be relevant during the complex initiation of motion process. The complexity is evident in the different forms of proposed critical shear stress equations [9, 10, 11]. Using a framework helps reduce complexity by classifying sand-mud mixtures on a physical basis. In this classification, a better understanding of the controlling factors for different sand-mud combinations can be done. Furthermore, 1st order equations can be estimated based on the subdivision. These can serve as guidelines for future erosion experiments or analyses.

4   Conclusions

The objective of this study was to build a dataset with a large spectrum of sand-silt-clay combinations from various erosion experiments in an attempt to obtain like-for-like comparisons between experimental setups and results. This objective was sought in lieu of performing erosion experiments with different mud content using a large variation in clay-silt ratios. Since there is no standardized methodology for measuring parameters, an absolute fair comparison was not achieved. This is especially true for derived parameters such as bed shear stress and critical bed shear stress. Nevertheless, the selection of methods and reported sediment information are reasonable criteria to build a coherent dataset of various sand-silt-clay combinations from different erosion experiments.

The framework proposed by Van Ledden classified this dataset into six bed types based on cohesion and network structure characteristics. This approach avoids hidden or compound effects that is possible in mathematical models that describe initiation of motion based on the variations of a single parameter, such as d_{50} or P_{mud}. This framework reduces the complexity in sand-mud bed initiation of motion using physical characteristics by classifying sand-silt-clay combinations into bed types. The framework also allows subsequent research to focus on parameters dominant in describing processes in initiation of motion per bed type.

Acknowledgements. This study is financially support by the European Union's (EU) Horizon Europe Framework Programme (HORIZON) via SEDIMARE (Grant Agreement No. 101072443), an MSCA Doctoral Network (HORIZON-MSCA-2021-DN-01).

The researchers would like to acknowledge dr. Bas van Maren and Roy van Weerdenburg for their input in sand-mud, and specifically, silt erosion research. The researchers would also like to acknowledge David Perkey, PhD for his generosity in sharing the dataset from the US Army Corp of Engineering report [5].

References

1. Van Ledden M et al (2004) A conceptual framework for the erosion behaviour of sand-mud mixtures. Cont Shelf Res 24(1):1–11
2. van Rijn LC (2020) Erodibility of mud–sand bed mixtures. J Hydraul Eng 146(1)
3. Jacobs W et al (2011) Erosion threshold of sand–mud mixtures. Cont Shelf Res 31(10):S14–S25
4. Jacobs W (2011) Sand-Mud Erosion from a Soil Mechanical Perspective. [Doctoral thesis, Delft University of Technology]
5. Perera C et al (2020) Erosion rate of sand and mud mixtures. Intl J Sediment Res 35(6):563–575
6. te Slaa S (2020). Deposition and erosion of silt-rich sediment-water mixtures [Doctoral thesis, Delft University of Technology]
7. Yao P et al (2022) Erosion behavior of sand-silt mixtures: revisiting the erosion threshold. Water Resour Res 58(9)
8. Albernaz M B et al (2025) Results of laboratory experiments: Results Phase 1A & B. https://publicwiki.deltares.nl/display/TKIP/DEL112+-+MUSA. Accessed 15 Jan 2025
9. van Rijn LC (2007) Unified view of sediment transport by currents and waves. i: initiation of motion, bed roughness, and bed-load transport. J Hydraul Eng 133(6):649–667
10. Wu W et al (2018) Critical shear stress for erosion of sand and mud mixtures. J Hydraul Res 56(1):96–110
11. Chen D et al (2018) Unified formula for critical shear stress for erosion of sand, mud, and sand–mud mixtures. J Hydraul Eng 144(8)

Extreme Flood Dynamics: A Modelling Approach of the 2021 Record-Breaking Flood of the Amazon Estuary

Paul Coulet[1,2]([⊠]), Fabien Durand[1,2], Laurent Testut[3], Alice Fassoni-Andrade[1,2], Jamal Khan[1], Florence Toublanc[1], Leandro Guedes Santos[4], and Daniel Medeiros Moreira[5,6]

[1] LEGOS, Université Toulouse, IRD, CNRS, CNES, UPS, Toulouse, France
`paul.coulet@univ-tlse3.fr`
[2] Institute of Geosciences, University of Brasília (UnB), Brasília, Brazil
[3] LIENSs UMR 7266, CNRS- La Rochelle University, 17000 La Rochelle, France
[4] CPRM, Serviço Geológico do Brasil, Avenida Doutor Freitas, Marco, Belém 3645, Brazil
[5] CPRM, Serviço Geológico do Brasil, Avenida Pasteur 404, Urca, Rio de Janeiro, Brazil
[6] GET, CNRS/CNES/IRD/UPS, 31400 Toulouse, France

Abstract. The Amazon estuary is the eastern terminal connection of this hydrosystem, linking the Amazon watershed to the open ocean. At Óbidos, the upstream limit of the estuary, the 2021 May-June flood was recorded as the highest flood in 12 years, with a peak discharge of 257,661 $m^3.s^{-1}$. The impact of this record flood has not been yet quantified along the estuary. This study aims to quantify its signature on estuarine hydrodynamics, using a cross-scale hydrodynamic model of the Amazonian estuarine continuum based on SCHISM. It turns out that the 2021 discharge anomaly (10% above the seasonal climatology) has a prominent influence on the Amazon River from 800 km to 380 km inland, inducing water level maxima typically 0.3 m higher than during a normal flood year, and about 1 m higher than during a weak flood year. Analysis of the various hydrodynamic factors conducive to the water level maxima along the estuary (i.e. discharge, oceanic tide and atmospheric forcing) shows that this is largely due to the discharge contributing twice as much in 2021 as in a normal year. In contrast, from 380 km inland to the oceanic mouth, the 2021 flood has no significant impact on the water level maxima dynamics, as the variability is dominated by the oceanic tide.

Keywords: Amazon · estuary · flood · tide

1 Introduction

Situated in the Amazon lowlands, the Amazon estuary connects the vast upstream watershed to the Atlantic Ocean. It conveys the world's largest freshwater discharge (around 20%, 200′000 m3.s-1 on average) to the Atlantic Ocean [1].

The Amazon River discharge shows significant seasonal and interannual variability (Fig. 1c) due to rainfall variability over the Amazon basin. In recent years, the Amazon

C. Coelho et al. (Eds.): CD 2025, CRL 41, pp. 803–809, 2026.
https://doi.org/10.1007/978-3-032-15473-6_122

River has been facing a significant change in its hydrological regime, with an intensification of the extreme flood occurrences [2]. Compared to the recent extreme floods of the last decade, a record flood occurred in 2021 [2].

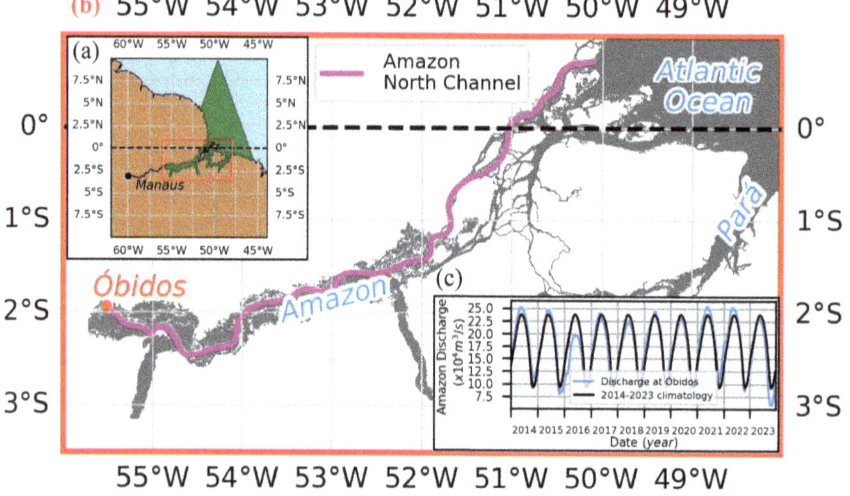

Fig. 122.1. (a) The inset map shows the location of the Amazon estuary on the Northern Coast of Brazil. The domain of the model is shown in green. The black dot locates Manaus hydrological station (b) The Amazon estuary, the location of the main water bodies and of the gauging station of Óbidos. The Amazon North channel is displayed in pink. (c) Amazon Discharge observed at Óbidos, the upper limit of the Amazon-Pará estuary (blue line). The black line represents the seasonal climatology of the discharge over the past decade (2014–2023). Data provided by ANA (Agência Nacional de Águas, Brazilian water agency).

The Amazon estuary stretches from the open ocean to 800 km upstream around Óbidos (Kosuth et al., 2009), where the tidal signal vanishes. On the 16th of June 2021, 600 km upstream of Óbidos, the water level measured at Manaus hydrological station was recorded as the highest water level ever recorded during the 1903–2021 period (Espinoza et al., 2022). At Óbidos hydrological station, the river discharge reached 258,000 m3.s-1 on the 30th of May 2021 (www.snirh.gov.br/hidrotelemetria/Estacoes. aspx). To date, no comprehensive analysis of the consequences of the 2021 flood on the Amazon-Pará estuary hydrodynamics has been achieved. To tackle this, we used a hydrodynamical model of the estuary to study the 2021 flood influence on estuarine hydrodynamics, encompassing the whole region from Óbidos to the open ocean. The model used in this study to assess the hydrodynamic signature of the 2021 flood over the estuary is based on SCHISM code. It is the same as the one previously used to infer the seasonal dynamics of the Amazon estuary [5].

The objective of the present study is to address the three following questions: what is the dynamics of the Amazon estuary water level during the 2021 flood? how far did this

2021 extreme flood extend over the estuary towards downstream? How different was it from a flood occurring during a normal year?

2 Data and Methods

2.1 2DH Hydrodynamic Model of the Amazon Estuary

This section focuses on the SCHISM [4] cross-scale hydrodynamic model of the Amazon estuary. The present work is based on the modeling framework that we developed in a previous study [5]. The modeling domain encompasses the entire hydraulic continuum from the upstream limit of the estuary, in Óbidos, to the deep Atlantic Ocean. It covers the Amazon River and its adjoining floodplains as well as the Pará River, as is depicted by Fig. 1a. We use SCHISM v5.11 [4] in depth-integrated mode. Our model domain is discretized on an unstructured triangular mesh with 1,039,864 nodes with a node spacing varying from 40 m to 5 km. The model is forced at its open boundaries by the XTRACK altimetric tidal constituents in the open ocean (https://www.aviso.altimetry.fr/fr/donnees/produits/produits-hauteur-de-mer/reg ional/x-track-sla/x-track-l2p-sla-version-2022.html), and by observed discharge at the upstream open boundaries inland. The atmospheric forcing is prescribed over the model domain with atmospheric pressure and wind from a 6-hourly CFSR reanalysis at 0.5° resolution (https://rda.ucar.edu/datasets/d094001/). This model has been extensively calibrated and validated against in situ tide gauge data and spaceborne altimetry [5].

To study the water level dynamics over the past decade and keeping in mind our specific interest in the 2021 flood, we simulated the period 01/01/2014 to 01/01/2024. A reference simulation (*REF*) was conducted with all the water level forcing factors considered in our model (river discharge, tide, atmospheric pressure and surface wind). We also conducted a sensitivity experiment without atmospheric forcing (*NOATM*), deactivating the effects of the atmospheric pressure and surface wind.

2.2 Assessment of the Water Level Dynamics

To study the signature of the 2021 flood on the estuarine hydrodynamics, we will focus on the yearly water level maxima reached during this year, along the mainstems of the estuarine network (namely the North channel of the Amazon, and the Pará river). Yearly maxima are defined as the water levels above the 99.5 percentile of the yearly water level distribution. This approach has already been applied in our previous study about the dynamics of yearly maxima across the Amazon estuary [5].

We will also quantify the various forcing factors at play in generating extreme events. Our previous study [5] showed that in the Amazon estuary, the discharge, the tide and the atmosphere (essentially the wind) altogether play a significant role in generating water level yearly maxima. Building on this methodology, we isolate the various forcing factors of the water level maxima, as follows. The tide contribution $TIDE_{contribution}$ is the modeled tide during the maxima (computed from a harmonic analysis of *REF* simulation, considering the 5 dominant tidal constituents which are M2, S2, M4, MSf and Mm).

The atmospheric contribution $ATM_{contribution}$ is the water level difference between the REF and the $NOATM$ simulations:

$$ATM_{contribution} = REF - NOATM \qquad (122.1)$$

Finally, the discharge contribution $DISCH_{contribution}$ will be assessed in the following way:

$$DISCH_{contribution} = \left(NOATM - climato_{2014\_2023}(NOATM) - TIDE_{contribution}\right) \qquad (122.2)$$

where $climato_{2014\_2023}(NOATM)$ is the seasonal climatology of the $NOATM$ simulation water level computed over 2014–2034 period after filtering the effect of the tide by applying a 28-day running mean. We term as $ANOMALY$ the water level difference between REF and $climato_{2014\_2023}(NOATM)$:

$$ANOMALY = REF - climato_{2014\_2023}(NOATM) \qquad (122.3)$$

3   Results: Consequences of 2021 Flood on the Estuarine Hydrodynamics

Every year, the flood season takes place in May in the Amazon. The May 2021 peak discharge observed in Óbidos is 10% above the seasonal climatology computed over the past decade. Firstly, we investigated the impact of the discharge excess of the 2021 flood on the total water level along the estuary. For this, we compared the yearly maxima simulated by our model (REF simulation) during three contrasting years in terms of Amazon discharge (Fig. 1c): May 2021 (year of the record-breaking flood), May 2019 (a near-normal year for the past decade in terms of Amazon discharge, with a peak discharge hardly 3% above climatology), and May 2016 (a year of minor flood, the weakest flood over the past decade, 15% below climatology). Figure 2a shows the mean value of yearly water level maxima along the estuary (solid lines) and the difference between the maximum and the minimum values reached during yearly maxima (seen as the shaded envelope of each curve), separately for 2016, 2019 and 2021 in the Amazon North Channel.

In 2021, water level maxima range from 11 m in Óbidos at the upstream limit of the estuary, to 2.9 m in the downstream part close to the ocean mouth. Throughout the upper 300 km of the estuary, the 2021 maxima lie 30 to 40 cm higher than those of 2019. Further downstream, around 380km from the ocean, the 2021 and 2019 maxima values become similar, within their overlapping envelopes. They remain similar throughout the downstream half of the estuary, down to the ocean. 2021 maximum exceeds 2016 maximum by 1.78 m in Óbidos. This excess remains larger than 0.4 m until 370 km upstream of the ocean. In contrast, over the 150 km downstream-most reach of the estuary, the maxima of 2021 are not significantly higher than those of 2016. Therefore, in the Amazon North Channel, the 2021 flood has a sensible signature on the maximum water level (of order 10% larger than the extreme reached during a normal year) that is basically restricted to the upper half of the estuary, upstream of 380 km.

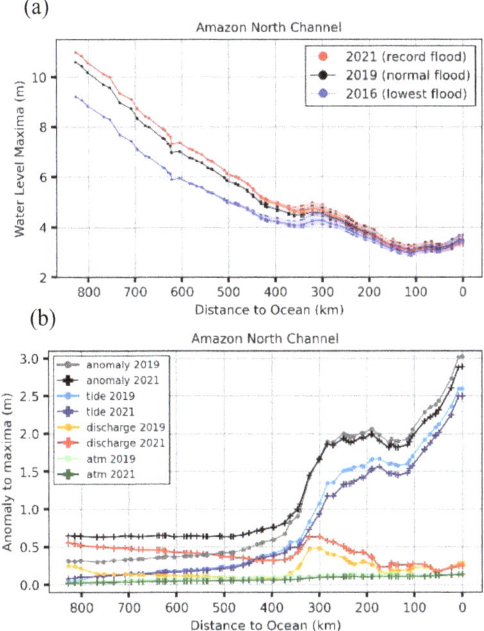

Fig. 122.2. (a) Mean value of yearly maxima distribution in Amazon North Channel for 2016, 2019 and 2021. Min-max deviation is also displayed. Yearly maxima is calculated with respect to Mean Sea Level. (b) Mean anomaly of total water level (Eq. 122.3) during yearly maxima with respect to 2014–2023 climatology (black and grey lines) in Amazon North for both years. The contribution of the forcing factors are shown in colors, namely the tide (light and dark blue), the discharge (red and orange) and the atmosphere (light and dark green).

Secondly, to gain a comprehensive view of the imprint of the 2021 flood on the whole estuary, we quantified the role of the various forcing factors on the 2021 and 2019 water level maxima from upstream to downstream, considering the various sensitivity experiments (see Eqs. 122.1–122.3). Along the Amazon North Channel, the model suggests three successive regimes, from the upstream estuary to the downstream region (Fig. 2b). Over the upstream part (viz. The upper 400 km), the anomaly of water level during the occurrences of the yearly maxima is around 60 cm higher in 2021 than in 2019 (black and grey curves in Fig. 2b). This is mostly explained by the discharge contribution (red curve), that is twice higher in 2021 vs. 2019 (80% vs 40% for the most part (not shown)). The tide contribution is nearly the same for both years, accounting for around 5cm upstream to 50cm downstream. As for the atmospheric contribution, it is higher in 2019, ranging from 7 to 10 cm, vs. 4 to 7 cm in 2021, translating into a much higher relative contribution in maxima generation for 2019 (around 15% (not shown)) compared to 2021 (around 5% (not shown)).

In the mid-estuary, between 350 km and 280km, the anomaly is roughly the same between 2019 and 2021. There, the absolute contribution of the tide increases more vividly as the discharge contribution starts to diminish.

Further downstream, over the terminal part of the estuary, from 280km to the open ocean, the anomaly from 2019 gradually surpasses 2021 from 2 cm to 14 cm, caused by the drastic reduction of the discharge contribution for both years, along with an important increase in the tidal contribution. The 2019 tidal contribution is typically 5 to 10 cm larger than the 2021 contribution.

4 Conclusion

At the upstream limit of Amazon estuary in Óbidos, the 30th of May 2021 a peak flood was recorded as the highest in 12 years, and 2nd highest in 56 years. Thanks to a state-of-the-art cross-scale modeling of the whole estuarine continuum of the Amazon estuary, encompassing the various branches of its terminal delta, the present study shows that this record-breaking flood had a contrasted spatial influence in the water level dynamics along the estuary. In the Amazon North Channel, the imprint of this anomaly in terms of extreme water level extends from Óbidos (800 km upstream of the oceanic mouths) towards downstream, until 380 km upstream of the oceanic mouths. The magnitude of the water level maxima, around 60 cm higher than the seasonal climatology, is very significant. The underlying mechanism of water level maxima generation largely results from the effect of the anomalous Amazon discharge. Further downstream, the water level maxima of 2021 do not significantly differ from the regime of a normal year (2019). This can be explained by the growing influence of the tide in the overall variability of the water level, over the lower half of the estuary.

References

1. Callède J, Cochonneau G, Alves FV, Guyot J-L, Guimarães VS, De Oliveira E (2010) Les apports en eau de l'Amazone à l'Océan Atlantique. J Water Sci 23:247–273
2. Espinoza J-C, Marengo JA, Schongart J, Jimenez JC (2022) The new historical flood of 2021 in the Amazon River compared to major floods of the 21st century: Atmospheric features in the context of the intensification of floods. Weather Climate Extremes 35:100406
3. Fassoni-Andrade AC et al (2023) Seasonal to interannual variability of the tide in the Amazon estuary. ISSN: 0278–4343
4. Zhang YJ, Ye F, Stanev EV, Grashorn S (2016) Seamless cross-scale modeling with SCHISM. Ocean Model 102:64–81
5. Coulet P et al (2025) Dynamics of yearly maximum water levels in the Amazon estuary. Estuaries and Coasts. in press

Continuous Interannual Monitoring of Artificial and Natural Breachings of an Intermittently Open/Closed Estuary in Central Chile

Rodrigo Cienfuegos[1,2(✉)], Patricio A. Catalán[2,3], Felipe Lucero[4], Paola Díaz[1], Raúl Flores[2,3], Maricarmen Guerra[5], Fernanda Venegas[3], Álvaro Ossandón[3], and Luis Araya[6]

[1] Depto. Ingeniería Hidráulica y Ambiental, Pontificia Universidad Católica de Chile, Santiago, Chile
rcienfue@uc.cl
[2] Centro de Investigación Para La Gestión Integrada del Riesgo de Desastres (CIGIDEN), Macul, Chile
[3] Depto. Obras Civiles, Universidad Técnica Federico Santa María, Santa María, Chile
[4] Marine Energy and Innovation Center (MERIC), Valparaíso, Chile
[5] Facultad de Ingeniería, Universidad de Concepción, Concepción, Chile
[6] Ministerio de Medioambiente, Programa GEF Humedales Costeros, Santiago, Chile

Abstract. We present the analysis of continuous remotely sensed data (hourly collections of high-resolution daytime videos) of the Nilahue Intermittently Open/Closed Estuary (IOCE) on the central Chilean Pacific seaboard (34.48°S, 72.00°W). Videos are automatically processed on-the-fly using state-of-the-art Artificial Intelligence algorithms to produce time series of river mouth metrics, which, in combination with measured and modeled hydrological (rainfall, river discharge, lagoon water level), oceanographic (tides and waves), and environmental (salinity) variables, provide unprecedented insights for understanding, diagnosing, and forecasting multi-scale processes in these fragile and rapidly evolving ecosystems. The initial analyses cover a time frame from April 2023 to November 2024, spanning two complete winter seasons and one summer. In particular, during 2023, significant rainfall associated with the El Niño/Southern Oscillation (ENSO) led to extreme flooding and a complete river mouth opening and reset, allowing us to track the subsequent morphodynamical processes and gradual closure of the mouth. In this talk, we also report on periods of artificial openings during spring-summer-autumn (December to April), summarizing the different conditions and outcomes. Based on the captured data and analysis, we discuss the possibilities of building physics-based and data-driven forecasting capabilities to understand lagoon functioning and, from there, provide guidance for IOCE management to ensure their ecological integrity amidst uncertain environmental changes.

Keywords: Intermittently Open/Closed Estuaries · river inlet dynamics · remote sensing · coastal monitoring

© The Author(s) 2026
C. Coelho et al. (Eds.): CD 2025, CRL 41, pp. 810–816, 2026.
https://doi.org/10.1007/978-3-032-15473-6_123

1 Introduction

Transitions from fluvial to marine environments, shaped by hydrologic and coastal processes, create diverse geomorphological and ecosystemic configurations at river mouths [1]. Intermittently Closed/Open Lakes and Lagoons (ICOLLs) also termed Intermittently Open/Closed Estuaries (IOCEs) are notable examples in wave-dominated and microtidal mid-latitude settings, where river mouth connections to the ocean occur intermittently following dry and wet seasons [2]. These systems have gained increasing scientific attention due to their environmental significance, flood mitigation capabilities, and cultural relevance [3, 4]. However, climate change and urban expansion have disrupted natural opening and closing cycles, necessitating artificial interventions to manage floods and water quality [5]. Unfortunately, these interventions often lack pre-established objectives and protocols and are hampered by inadequate continuous monitoring of essential physical and environmental variables.

The critical monitoring gaps regarding these systems arise from increasing anthropogenic pressures and the impact of climate change, all of which require high-resolution spatiotemporal data for effective understanding and designing adaptation measures. As such, continuous monitoring is essential, especially to scale-up short-scale disruptions from extreme natural events or anthropogenic interventions. In recent decades, the use of remote sensing systems has opened new opportunities for understanding these processes [6, 7].

In this talk, we present initial analyses of the dynamics of natural and artificial river mouth openings and closings, utilizing continuous video data alongside measured and modeled hydrological, oceanographic, and environmental variables. The study site, data sources and methods, and initial analysis and findings are summarized in the following sections.

2 Study Site, Data Sources, and Continuous Automated Capturing-Processing Video System

2.1 The Nilahue River Catchment and Data Sources for Observed Variables

The study site is located on the central Chilean Pacific seaboard (34.48°S, 72.00°W; Fig. 1-a) and has a Mediterranean climate. The Nilahue River catchment is a relatively small watershed draining the coastal mountain range (Fig. 1-b), where a shallow bar-built lagoon system forms an IOCE at its river mouth (Laguna Cáhuil, Fig. 1-c). It is exposed to energetic swells predominantly coming from the SW-SSW and experiences microtidal conditions. The average hydrologic and oceanographic characteristics are summarized in Table 1. Over the last few decades, central Chile has suffered prolonged droughts [8], resulting in reduced mean river flow discharges, as evidenced in Fig. 1-d for this particular river catchment.

The different measured and modeled oceanographic, hydrologic, and environmental variables are summarized in Table 2. Wave climates are hindcasted using the WW3 model [9] using nesting and model configurations previously validated over Chilean oceanographic conditions [10, 11]. Tides, rainfall intensities, river discharge, levels and

Table 1. Average hydrologic and oceanographic characteristics at the Nilahue River mouth.

Watershed area and Max. Altitude	Tidal Range	Average Annual River Discharge	2-year Return Period of Flood Discharge	Mean Annual Significant Wave Height and Peak Period
1,700 km2 900 masl	1–2 m	8.0 m^3/s	148 m^3/s	2.5 m 13 s

salinity within the lagoon/estuary are measured using in-situ instruments. River mouth dynamics is captured using video images as described in the next section.

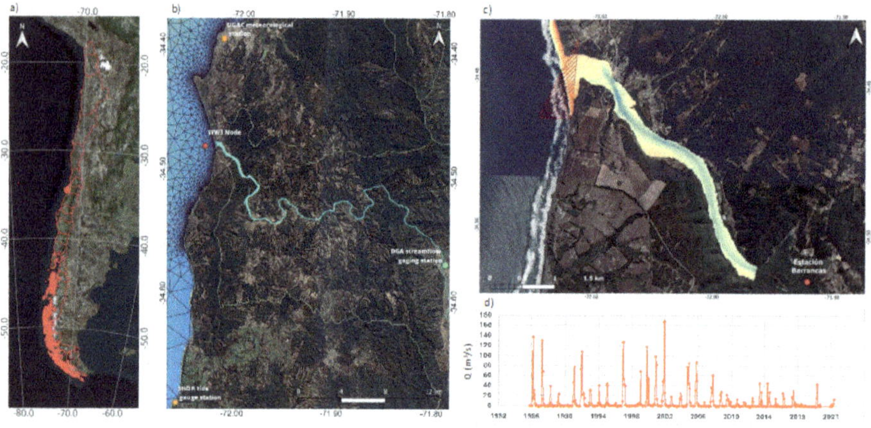

Fig. 1. Study site at the Laguna Cáhuil. a) Location in central Chile shown in a red dot. b) Overview of the location of the gauging stations for rainfall and wind (DGAC), tides (SHOA), river discharge (DGA), and the WW3 numerical grid and node (WW3). c) Laguna Cáhuil and the coverage of the video camera (hatched red lines) at the river mouth. c) Average monthly river discharges measured at the DGA station.

Table 2. Observed variables, temporal acquisition rates, and locations.

Observed variable	In-situ instrument/Model	Temporal acquisition rate	Location
Wave heights, peak periods, mean direction	**Model** WW3 [9]	1 h	WWIII node in front of the Laguna Cáhuil at a depth of 18 m (-34.480755°, -72.035415°)
Tides	**In-situ** SHOA-IOC	1 min	Boyeruca station (SHOA) (-34.6873055°, -72.05786°)

(continued)

Table 2. (*continued*)

Observed variable	In-situ instrument/Model	Temporal acquisition rate	Location
River discharge	**In-situ** DGA	30 min	Nilahue en Santa Teresa station (DGA) (-34.573625°,-71.793869°)
Rainfall intensity	**In-situ** DGAC	6 h	Aeródromo de Pichilemu (DGAC) (-34.39417°, -72.01861°)
Water level and salinity	**In-situ** GEF Humedales Costeros	10 min	Barrancas Estero Nilahue (-34.506128°, -71.984223°)
River mouth evolution	**In-situ** video footages	Hourly collections of high-resolution daytime videos	River mouth at Laguna Cáhuil (-34.479791°, -72.024905°)

2.2 Remote Sensing of the River Mouth Evolution

The dynamics of the river mouth are captured by the high-resolution video monitoring station, part of the "SIStema de MONitoreo y Anticipación de la Resiliencia Costera - SIMONA Costa" (www.simonacosta.cl). The acquisition system comprises a microcontroller (Raspberry Pi 4) and a 12.3-megapixel camera (Raspberry Pi HQ), which utilizes a Sony IMX477 sensor in conjunction with a 16 mm focal length Arducam lens. Camera calibration and ortho-rectification are conducted following [12], implemented using the OpenCV library [13], with 58 ground control points obtained using a Trimble R3 GPS system, all referenced to the local datum. The system captures 20-min videos every hour between 08:00 am and 6:00 pm at a rate of 7 fps. Shoreline and river mouth evolution are computed from Timex images (one image each hour, Fig. 2) using the SAM segmentation model [14].

Fig. 2. Example of Timex image of the river mouth obtained on August 12th 2023 at 8:00 am.

3   Preliminary Analysis and Findings

In Fig. 3, we present the time series of both in situ and remotely sensed variables for the year 2023, a year marked by intense rainfall events associated with the El Niño/Southern Oscillation (ENSO). Indeed, an extreme flood event in late August destroyed all instruments installed within the river channel and lagoon. Despite this, remotely sensed data and in situ measurements enabled us to monitor the river mouth evolution, lagoon levels, and salinity, covering periods of artificial openings during summer-autumn (January to April) and natural sandbar breaching (from swells overtopping and river floods) during autumn-winter. After the first intense winter rainfall in June, the river remained fully connected to the ocean and began functioning as an estuary, as the salinity time series accounts for (Fig. 3).

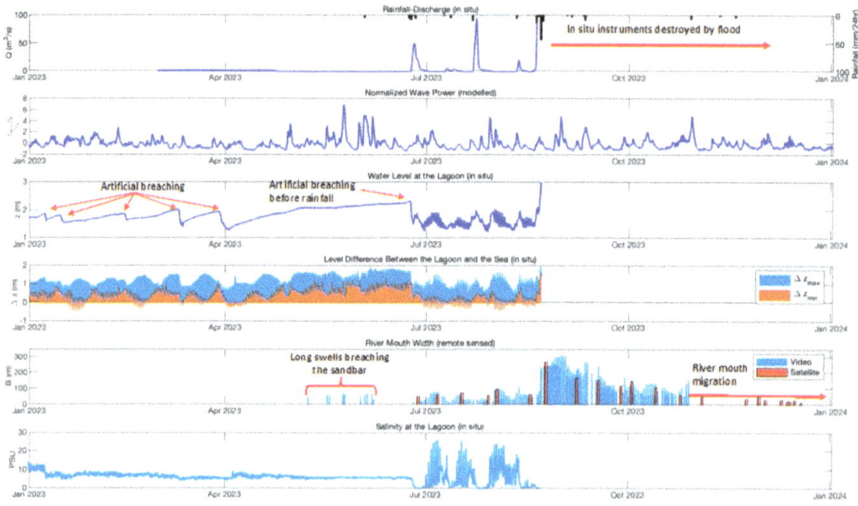

Fig. 3. Time series of the different observed variables over year 2023. Artificial breaching interventions, and the destruction of in-situ instruments by the large flood of late August 2023 are highlighted. Planet images (https://www.planet.com) are used for satellite derived river mouth-width estimation.

The river mouth width then oscillated between 50 m and 100 m, following the tidal cycles and some moderate floods (maximum around 100 m³/s) until a larger storm occurred in mid-August. The resulting flood destroyed the river stream gauge station, whose last record exceeded 600 m³/s (not shown). The river mouth experienced a complete opening and reset, with and extremely fast initial mouth width expansion (~150 m/day) before the increasing tidal range cycle took over to modulate the opening rate at a much slower pace during the remainder of August to reach a maximum of nearly 350 m (Fig. 4). The closing process began shortly after and continued uninterrupted at two distinct rates: first ~ 8 m/day during the remainder of September, slowing down to ~ 1 m/day from October until the complete closing of the river mouth in late December. This process occurred with substantial changes in the river mouth orientation and location, migrating fast into the north (not shown). Indeed, the river mouth migration implied that the camera vision was unable to follow the final stage of this process and satellite images were used to complement the picture (Fig. 3–4). Once the river mouth was closed, significant sediment accretion on the beach was observed leading to a rapid onshore shoreline migration (not shown).

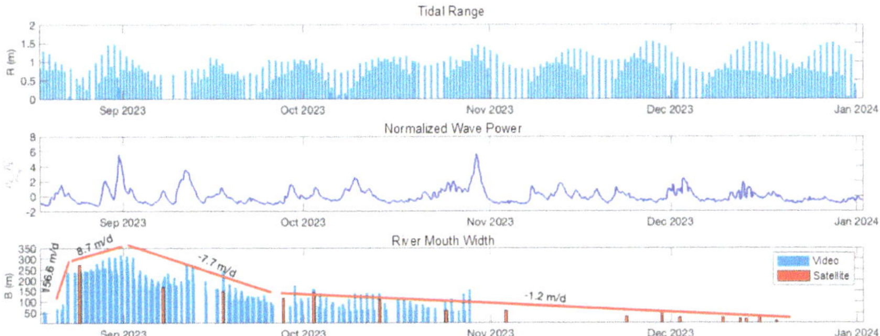

Fig. 4. Time series of the different observed variables over year 2023. Artificial breaching interventions, and the destruction of in-situ instruments by the large flood of late August 2023 are highlighted. Planet images (https://www.planet.com) are used for satellite derived river mouth-width estimation.

In the talk, we will further describe and analyze the morphodynamical changes observed during the winter seasons of 2023 and 2024 following the natural openings that occurred in these periods. In addition, we have prepared a catalogue of the artificial openings conducted during the dry seasons of 2023 and 2024, which provides a means to perform a quantitative assessment of their observed outcomes while analyzing hydrological, oceanographic, and environmental variables together. Based on the captured data and analysis, we discuss the possibilities of building physics-based and data-driven forecasting capabilities to provide guidance for IOCE management to ensure their ecological integrity amidst uncertain environmental changes.

References

1. Boyd R, Dalrymple R, Zaitlin BA (1992) Classification of clastic coastal depositional environments. Sed Geol 80(3–4):139–150
2. McSweeney SL, Kennedy DM, Rutherfurd ID, Stout JC (2017) Intermittently Closed/Open Lakes and Lagoons: their global distribution and boundary conditions. Geomorphology 292:142–152
3. Thorne KM, Buffington KJ, Jones SF, Largier JL (2021) Wetlands in intermittently closed estuaries can build elevations to keep pace with sea-level rise. Estuar Coast Shelf Sci 257:107386
4. Mayjor M, Reichelt-Brushett AJ, Malcolm HA, Page A (2023) Water quality fluctuations in small intermittently closed and open lakes and lagoons (ICOLLs) after natural and artificial openings. Estuar Coast Shelf Sci 281:108208

5. McSweeney SL, Stout JC, Kennedy DM (2020) Variability in infragravity wave processes during estuary artificial entrance openings. Earth Surf Proc Land 45(13):3414–3428
6. Vos K, Harley MD, Splinter KD, Simmons JA, Turner IL (2019) Sub-annual to multi-decadal shoreline variability from publicly available satellite imagery. Coast Eng 150:160–174
7. Nichols CR et al (2019) Collaborative science to enhance coastal resilience and adaptation. Front Mar Sci 6:404
8. CR2: La Megasequía 2010–2019: Una lección para el futuro (2020) Informe a las naciones, Centro del Clima y la Resiliencia (CR2)
9. Tolman, H. L.: User manual and system documentation of WAVEWATCH III TM version 3.14. Technical note, MMAB contribution, 276(220) (2009)
10. Lucero F, Catalán PA, Ossandón A, Beyá J, Puelma A, Zamorano L (2017) Wave energy assessment in the central-south coast of Chile. Renew Energy 114:120–131
11. Mazzaretto OM, Lucero F, Besio G, Cienfuegos R (2020) Perspectives for harnessing the energetic persistent high swells reaching the coast of Chile. Renew Energy 159:494–505
12. Holland KT, Holman RA, Lippmann TC, Stanley J, Plant N (1997) Practical use of video imagery in nearshore oceanographic field studies. IEEE J Oceanic Eng 22:81–92
13. Bradski G (2000) The OpenCV Library. Dr. Dobb's J Softw Tools 120:122–125 R
14. Kirillov A et al (2023) Segment anything. Meta AI Res

Emergent Sediment-Sharing Cells in a Barrier Island-Lagoon System

Stuart G. Pearson[1]([⊠]), Roy van Weerdenburg[1,2], Hassan Shafiei[1], Johan Reyns[1,3], Edwin Elias[2], Zheng Bing Wang[1,2], Quirijn Lodder[4], and Bram van Prooijen[1]

[1] Department of Hydraulic Engineering, TU Delft, Delft, The Netherlands
`s.g.pearson@tudelft.nl`
[2] Deltares, Delft, The Netherlands
[3] IHE Delft, Delft, The Netherlands
[4] Rijkswaterstaat, Lelystad, The Netherlands

Abstract. Coastal sediment budgets are a foundational source of information for coastal management decision-making. To quantify these budgets, coastal systems are often divided into "cells" based on jurisdictional boundaries or topography. However, such divisions do not account for the pathways that water and sediment particles actually take. In this study we quantify cell boundaries that emerge from numerical simulations of sand and water pathways in a barrier island-lagoon system in the Netherlands (the Western Wadden Sea). By quantifying Lagrangian particle pathways as a network, we can derive internally well-connected but externally disconnected modules. Here we show that large (O(10 km)) coherent modules develop from flow patterns at tidal timescales (12.5 h), and are persistent through varying tide and weather conditions. Conversely, modules derived from 100 μm sand pathways are less coherent and highly spatially fragmented. The difference in patterns likely relates to the longer timescales associated with sediment transport. These emergent patterns could be used to better inform coastal and estuarine management by providing physics-based sediment cell boundaries.

Keywords: sediment connectivity · network analysis · modularity · barrier island · lagoon

1 Introduction

Coastal sediment budgets are a foundational source of information for coastal management decision-making. These budgets are often estimated based on bathymetric changes, and their internal boundaries or "cells" delimited based on jurisdictional boundaries or topography. However, such approaches neglect a key consideration: do these cell boundaries actually correspond to the pathways that water and sediment particles take? What are the internal boundaries that naturally emerge in coastal systems? To answer these questions, we apply a Lagrangian sediment transport model to the Western Wadden Sea, a barrier island-lagoon system in the Netherlands, and derive emergent sediment-sharing cells using the connectivity approach of Pearson et al. [1].

© The Author(s) 2026
C. Coelho et al. (Eds.): CD 2025, CRL 41, pp. 817–822, 2026.
https://doi.org/10.1007/978-3-032-15473-6_124

2 Methodology

2.1 Hydrodynamic and Sediment Transport Modelling

We used a Delft3D-FM model of the Western Wadden Sea [2] to hindcast the hydrodynamic conditions at 30 min intervals from August 29th to October 9th, 2017. The model accounts for tidal and wind-driven forcing but not the effects of waves, focusing for now on larger-scale flow patterns. To estimate particle pathways, we use the Sed-TRAILS model [3]. We released particles on a uniform 667x667 m grid (9194 particles total) every 12.5 h (80 simulations total), and tracked their movement for 12.5 h (~1 tidal cycle). We first consider passive tracers (moving with depth-averaged velocity) as a baseline (Fig. 1a), and then 100 μm sand. The sediment velocity field was defined from bed shear stress and flow velocity using the method of Soulsby et al. [4]. Particle burial and mixing within the seabed are neglected to focus on sediment pathways of maximum potential lengths.

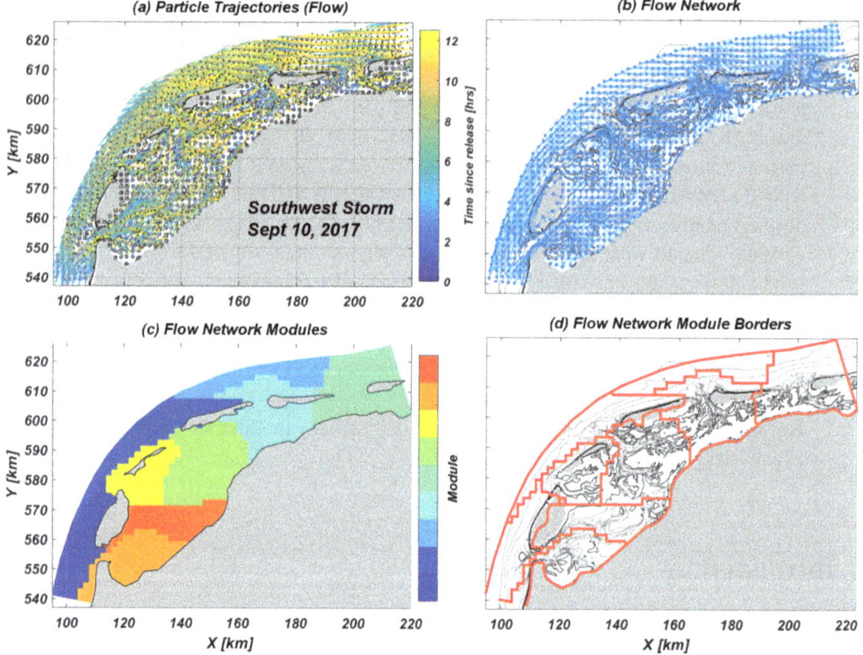

Fig. 1. Example workflow to derive network module borders, based on a single tidal cycle on September 10, 2017, during a storm with winds primarily from the southwest. (a) Passive particle trajectories advected by depth-averaged flow and coloured by time since release. (b) Network diagram indicating connections between each node. (c) Network modules, defined by high connectivity within modules and limited connectivity between modules. (d) Borders of network modules in panel (c). The identical procedure was repeated for 80 tidal cycles.

2.2 Network Analysis

We then compiled the particle pathways from each SedTRAILS run into a connectivity network (Fig. 1b) as per [1]. Network nodes are spaced out on a uniform 2000x2000 m grid (963 total, containing ~ 9.5 particle sources each), and the connections between them are estimated based on the number of particles from a given source node that reach a given sink node. Using the Leicht & Newman algorithm for directed networks [5], we define groups of nodes ("modules") that have high within-group connectivity and low between-group connectivity (Fig. 1c). Following Thiemann et al. [6], we can define the borders of these modules (Fig. 1d). By linearly superimposing the borders corresponding to each tidal cycle, we can visualize which borders are more persistent through time.

3 Results

Here we first consider how the global network properties vary in time for both flow and 100μm sand networks. The flow network is better connected during windier storm events and spring tide, as quantified by network density (Fig. 2a,c). Overall system connectivity is persistently lower for sand, although it still varies with storms and spring-neap cycles.

This behaviour is further reflected in the networks' tendency to form separate modules, as quantified by modularity (Fig. 2c). Disconnected modules tend to join up into larger, more coherent modules during storms, leading to reduced modularity. Given the shorter transport pathways and reduced connectivity of the sand network, its nodes tend to cluster in smaller modules.

By considering an ensemble of module borders over the study period, we can better understand the persistent water- and sediment-sharing cells that emerge in the Western Wadden Sea (Fig. 3). Flows in the Western Wadden Sea exhibit a tendency to form large modules (O(10 km)) at tidal timescales (~12.5 h) that are persistent over the simulated 40-day period (Fig. 3a). Although geographic coordinates of each node in the network are not included in the modularity calculation, spatially coherent cells nevertheless emerge, indicating strong local connectivity and mixing. The persistent modules have a spatial scale closer to the mean excursion length ΣX travelled by particles (14.7 km) than to their net displacement ΔX (3.8 km) (Fig. 2d). Over a single tidal cycle, modules spanning the entire domain (O(100 km)) do not develop, even under the most energetic conditions.

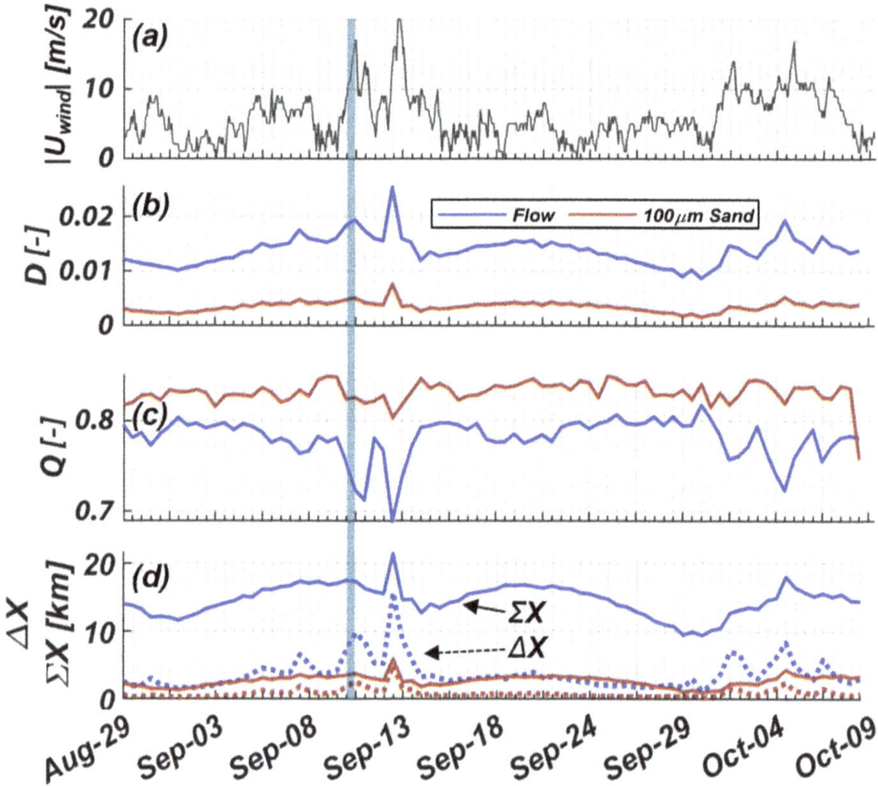

Fig. 2. (a) Windspeed magnitude measured at Terschelling (Hoorn). (b) Network density (D), which represents the fraction of actual connections out of all possible connections. (c) Network modularity (Q), where high values indicate a greater tendency to form separate modules. (d) Net particle displacement ΔX and particle excursion length ΣX, averaged across all particles in each simulation. Light blue vertical line corresponds to example in Fig. 1 from September 10, 2017.

Although many of these borders coincide with topographically-defined tidal watersheds between adjacent inlets (the most persistent borders occur around the Eierlandse Gat sub-basin and Terschelling watershed), they are quite variable depending on weather conditions (Fig. 3a). Furthermore, the Marsdiep and Vlie basins feature several persistent sub-modules. Equally relevant as the barriers are the areas without any barriers (e.g., Ameland basin), suggesting areas of strong internal mixing.

Conversely, the 100 μm sand network shows a much weaker spatial coherence, forming many smaller (O(1 km)) and more spatially-fragmented nodes (Fig. 3b). As with flows, persistent sand modules have a spatial scale similar to the mean sand excursion length (2.7 km). The "fuzzy" module borders of the sand network also indicate that module connections are not as temporally consistent as in the flow network. This is supported by the weak trend in sand modularity through time in Fig. 2c. The landward side of Ameland basin in Fig. 3b shows a sharp border at the edge of quiescent intertidal areas where sand is less frequently mobilized.

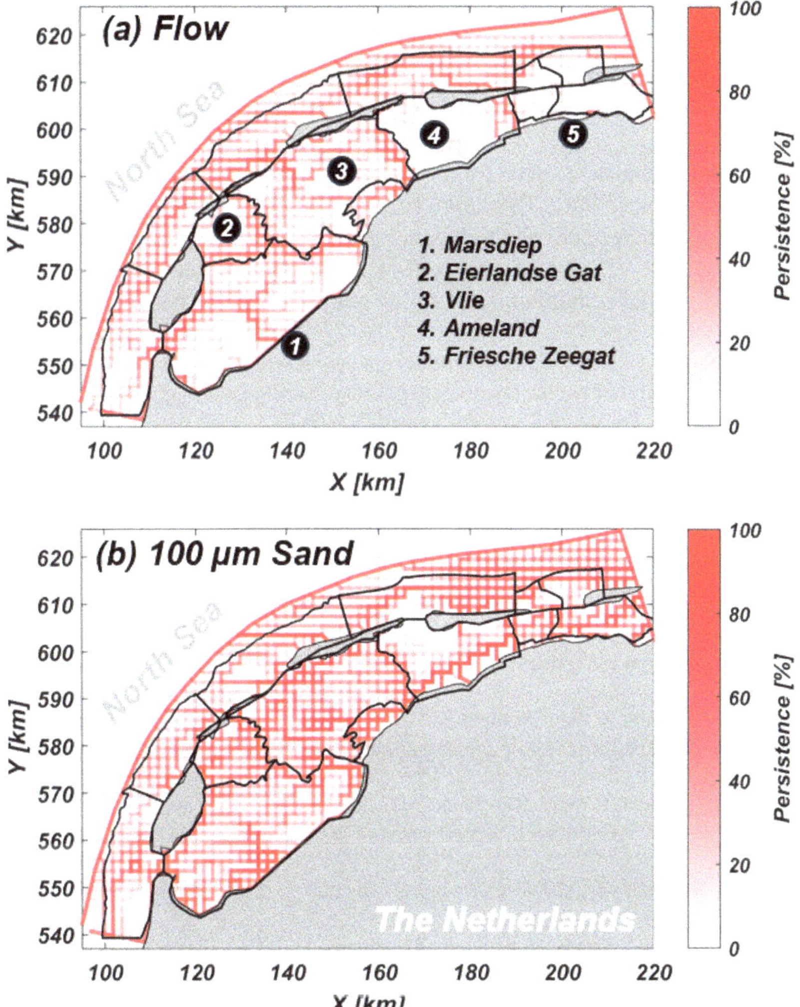

Fig. 3. Module borders for **(a)** flow and **(b)** 100µm sand networks. Borders from each 12.5 h period are linearly superimposed on each other, with darker red lines representing more persistent borders. Black lines indicate management boundaries based largely on topography, used by Elias & Wang [7] to derive sediment budgets.

4 Discussion

Coherent modules emerge from flow patterns in the Western Wadden Sea, while sediment transport produces much more fragmented modules at similar timescales. Unlike passive water particles, sand particles are limited by their critical threshold of motion, and respond nonlinearly to flows. This explains why some areas (e.g., tidal inlets where velocities are strong) are well-connected with more coherent modules, while other areas (e.g. within the Wadden Sea) show limited motion, low connectivity, and fragmented modules (Fig. 3b). While 12.5 h is apparently a sufficiently long time to generate coherent modules in the flow network, we hypothesize that longer simulation times may be needed to produce similar patterns in sediment networks. The relationship between module size and particle path length suggests that module properties also vary as a function of grain size.

Our findings point to the need for a deeper examination of the spatiotemporal evolution of marine connectivity. This approach could be used to better delineate sediment cells for coastal management and sediment budgets, bridging between small and large spatiotemporal scales. Future research will focus on the influence of longer simulation periods, higher spatial resolution, and varying grain sizes (including fine sediment).

Acknowledgments. This research stems from techniques developed as part of the TRAILS project (with project number 17600) in the research programme 'Living Labs in the Dutch Delta', which is (partly) financed by the Dutch Research Council (NWO).

References

1. Pearson SG, Prooijen BC, Elias EPL, Vitousek S, Wang ZB (2020) Sediment connectivity: a framework for analyzing coastal sediment transport pathways. J Geophys Res: Earth Surf 125(10)
2. van Weerdenburg R et al (2021) Field measurements and numerical modelling of wind-driven exchange flows in a tidal inlet system in the Dutch Wadden Sea. Ocean Coast Manag 215:105941
3. Pearson SG, Reniers A, Van Prooijen BC (2023) Revealing the hidden structure of coastal sediment transport pathways using Lagrangian coherent structures. Proc Coast Sed 2023:1212–1221
4. Soulsby RL, Mead CT, Wild BR, Wood MJ (2011) Lagrangian model for simulating the dispersal of sand-sized particles in coastal waters. J Waterw Port Coast Ocean Eng 137(3):123–131
5. Leicht EA, Newman ME (2008) Community structure in directed networks. Phys Rev Lett 100(11):118703
6. Thiemann C, Theis F, Grady D, Brune R, Brockmann D (2010) The structure of borders in a small world. PLoS ONE 5(11):e15422
7. Elias EPL, Wang ZB (2020) Sedimentbalans Waddenzee; Synthese Ten Behoeve Van Technisch Advies Kustgenese 2.0. Report 1220339–007-ZKS-0010,. Deltares, Delft

Author Index

© The Editor(s) (if applicable) and The Author(s) 2026
C. Coelho et al. (Eds.): CD 2025, CRL 41, pp. 823–828, 2026.
https://doi.org/10.1007/978-3-032-15473-6

GPSR Compliance
The European Union's (EU) General Product Safety Regulation (GPSR) is a set
of rules that requires consumer products to be safe and our obligations to
ensure this.

If you have any concerns about our products, you can contact us on

ProductSafety@springernature.com

In case Publisher is established outside the EU, the EU authorized
representative is:

Springer Nature Customer Service Center GmbH
Europaplatz 3
69115 Heidelberg, Germany